Springer Theses

Recognizing Outstanding Ph.D. Research

Aims and Scope

The series "Springer Theses" brings together a selection of the very best Ph.D. theses from around the world and across the physical sciences. Nominated and endorsed by two recognized specialists, each published volume has been selected for its scientific excellence and the high impact of its contents for the pertinent field of research. For greater accessibility to non-specialists, the published versions include an extended introduction, as well as a foreword by the student's supervisor explaining the special relevance of the work for the field. As a whole, the series will provide a valuable resource both for newcomers to the research fields described, and for other scientists seeking detailed background information on special questions. Finally, it provides an accredited documentation of the valuable contributions made by today's younger generation of scientists.

Theses may be nominated for publication in this series by heads of department at internationally leading universities or institutes and should fulfill all of the following criteria

- They must be written in good English.
- The topic should fall within the confines of Chemistry, Physics, Earth Sciences, Engineering and related interdisciplinary fields such as Materials, Nanoscience, Chemical Engineering, Complex Systems and Biophysics.
- The work reported in the thesis must represent a significant scientific advance.
- If the thesis includes previously published material, permission to reproduce this must be gained from the respective copyright holder (a maximum 30% of the thesis should be a verbatim reproduction from the author's previous publications).
- They must have been examined and passed during the 12 months prior to nomination.
- Each thesis should include a foreword by the supervisor outlining the significance of its content.
- The theses should have a clearly defined structure including an introduction accessible to new PhD students and scientists not expert in the relevant field.

Indexed by zbMATH.

David Gregory Weissbrodt

Engineering Granular Microbiomes

Bacterial Resource Management for Nutrient Removal in Aerobic Granular Sludge Wastewater Treatment Systems

Doctoral Thesis accepted by
École Polytechnique Fédérale de Lausanne (EPFL),
Lausanne, Switzerland

Springer

Author
Prof. Dr. David Gregory Weissbrodt
Laboratory for Environmental
Biotechnology, School of Architecture
Civil and Environmental Engineering
Institute of Environmental Engineering
Ecole Polytechnique Fédérale de Lausanne
Lausanne, Switzerland

Faculty of Natural Sciences, Department
of Biotechnology and Food Science
Norwegian University of Science
and Technology
Trondheim, Norway

Supervisor
Prof. Dr. Christof Holliger
Laboratory for Environmental
Biotechnology, School of Architecture
Civil and Environmental Engineering
Institute of Environmental Engineering
Ecole Polytechnique Fédérale de Lausanne
Lausanne, Switzerland

ISSN 2190-5053 ISSN 2190-5061 (electronic)
Springer Theses
ISBN 978-3-031-41008-6 ISBN 978-3-031-41009-3 (eBook)
https://doi.org/10.1007/978-3-031-41009-3

This Springer imprint is published by the registered company Springer Nature Switzerland AG
The registered company address is: Gewerbestrasse 11, 6330 Cham, Switzerland

Paper in this product is recyclable.

Supervisor's Foreword

The Ph.D. thesis of David Weissbrodt is a very comprehensive contribution to our understanding of a novel wastewater treatment technology based on aerobic granular sludge biofilms for high-rate removal of carbon, nitrogen, and phosphorus. After the recent 100th anniversary of the most widespread activated sludge wastewater treatment technology, this area of public services will go through major changes for different reasons. On the one hand, one tries to move away from the concept of removing pollution to minimize the environmental impact of wastewater and move towards recovery of resources from these effluents of our society, hence transforming wastewater treatment plants into so-called biorefineries. On the other hand, micropollutants have clearly been identified as an important threat to our aquatic ecosystems and possibly human health which lead in Switzerland to new regulations obliging large wastewater treatment plants to implement an additional treatment step and to upgrade those that do not yet nitrify the ammonium loads to plants that do so. In countries where a big part of the population has already been connected to wastewater treatment plants, footprint constraints will lead to the construction of new and more compact installations and aerobic granular sludge technology is one of the most promising solutions to this problem.

The work presented in the thesis covers many different topics of wastewater treatment by aerobic granular sludge technology, going from a general introduction including an economical assessment of this technology compared with traditional wastewater treatment with activated sludge to the assessment of the parameters that govern the competition between phosphate- and glycogen-accumulating organisms using an experimental planning design. A very interesting part is the state-of-the-art chapter that is a very comprehensive literature review that has not been published elsewhere but that provides an excellent overview of topics involved in the work of the thesis, e.g. biofilm and its application in wastewater treatment, self-aggregation of microorganisms, aerobic granular sludge, microbial ecology of wastewater treatment systems, and mathematical modelling of wastewater treatment systems. A most integrated approach was applied involving economical assessment, environmental bioprocess design, multifactorial experiments, molecular biology, microscopy, biochemistry, multivariate analyses, and mathematical modelling. A conceptual model of the aerobic granule ecosystem has been developed based on the dynamics of the microbial populations and the corresponding treatment performances and operation conditions that proposes aerobic granules being a community of producers and consumers of exopolymeric substances. The work resulted in the formulation of a methodology for an efficient management of the bacterial resource in this new-generation wastewater treatment system in order to select for a stable, active, and cooperative bacterial community for an efficient removal of nutrients. This methodology is an excellent tool to guide scientists and engineers in their research on aerobic granular sludge as well as practitioners operating wastewater treatment plants with this novel technology.

Lausanne, Switzerland
January 2016

Prof. Christof Holliger

Concepts developed in this thesis contributed to the following publications:

Critical Review

Winkler MKH, Meunier C, Henriet O, Mahillon J, Suárez-Ojeda ME, Del Moro G, De Sanctis M, Di Iaconi C, Weissbrodt DG (2018) **An integrative review of granular sludge for the biological removal of nutrients and of recalcitrant organic matter from wastewater**. Chem Eng J 336: 489–502. 10.1016/j.cej.2017.12.026

Original Research Articles

Weissbrodt DG, Holliger C, Morgenroth E (2017) **Modelling hydraulic transport and anaerobic uptake by PAOs and GAOs during wastewater feeding in EBPR granular sludge reactors**. Biotechnol Bioeng 114(8):1688–1702. 10.1002/bit.26295

Weissbrodt DG, Maillard J, Brovelli A, Chabrelie A, May J, Holliger C (2014) **Multilevel correlations in the biological phosphorus removal process: from bacterial enrichment to conductivity-based metabolic batch tests and polyphosphatase assays**. Biotechnol Bioeng 111(12):2421–2435. 10.1002/bit.25320

Weissbrodt DG, Shani N, Holliger C (2014) **Linking bacterial population dynamics and nutrient removal in the granular sludge biofilm ecosystem engineered for wastewater treatment**. FEMS Microbiol Ecol 88(3):579–595. 10.1111/1574-6941.12326

Weissbrodt DG, Schneiter GS, Fuerbringer JM, Holliger C (2013) **Identification of trigger factors selecting for polyphosphate- and glycogen-accumulating organisms in aerobic granular sludge sequencing batch reactors**. Water Res 47(19):7006–7018. 10.1016/j.watres.2013.08.043

Weissbrodt DG, Neu TR, Kuhlicke U, Rappaz Y, Holliger C (2013) **Assessment of bacterial and structural dynamics in aerobic granular biofilms**. Front Microbiol 4:175. 10.3389/fmicb.2013.00175

Weissbrodt DG, Lochmatter S, Ebrahimi S, Rossi P, Maillard J, Holliger C (2012) **Bacterial selection during the formation of early-stage aerobic granules in wastewater treatment systems operated under wash-out dynamics**. Front Microbiol 3:332. 10.3389/fmicb.2012.00332

Weissbrodt DG*, Shani N*, Sinclair L, Lefebvre G, Rossi P, Maillard J, Rougemont J, Holliger C (2012) **PyroTRF-ID: a novel bioinformatics methodology for the affiliation of terminal-restriction fragments using 16S rRNA gene pyrosequencing data**. BMC Microbiol 12:306. 10.1186/1471-2180-12-306 (*equal contribution)

Technical Articles

Weissbrodt DG (2018) **StaRRE—Stations de Récupération des Ressources de l'Eau: Intensification, Bioraffinage et Valorisation**. Aqua & Gas 1: 20–24. https://www.aquaetgas.ch/fr/eau/eaux-us%C3%A9es/ag2018-01_fa_weissbrodt/

Weissbrodt DG (2017) **Moi, je travaille pour les StaRRE!—Intensification et bioprospection à haute valeur ajoutée en Stations de Récupération des Ressources de l'Eau**. Bulletin de l'ARPEA—Journal Romand de l'Environnement 272:40–45.

Weissbrodt DG, Holliger C (2013) **Intensification du traitement biologique des eaux usées : Technologie à boues granulaires et Gestion des ressources bactériennes**. Bulletin de l'ARPEA—Journal Romand de l'Environnement 256:12–22.

Peer-Reviewed Conference Proceedings

Keynote Presentations and Invited Talks

Guimarães LB*,°, Wagner J, Akaboci TRV, Daudt GC, Nielsen PH, van Loosdrecht MCM, Weissbrodt DG*,**,°, da Costa RHR** (2017) **From failures to successful granular sludge process: Hints for real wastewater treatment under coastal warm climate**. In: Alvarez *et al.* (eds.) Proceedings of 14th IWA Leading Edge Conference on Water and Wastewater Technologies—Innovative Technology Solutions to Address Challenges at the Water-Energy-Food Interface, Florianopolis, Brazil. (*equal contribution; **co-senior authors; °fusion keynote)

Guimarães LB*,°, Gubser NR, Lin Y, Pronk M, Welles L, Albertsen M, Daudt GC, Geleijnse MA, da Costa RH, Nielsen PH, van Loosdrecht MC, Weissbrodt DG*,° (2016) **Exopolysaccharides biorefining from used water: an enterprise in the microbiome of granular sludge**. In: van Loosdrecht *et al.* (eds.) Proceedings of 13th IWA Leading Edge Conference on Water and Wastewater Technologies—Evaluating Impacts of Innovation, Jerez de la Frontera, Spain. (*equal contribution; °fusion keynote)

Weissbrodt DG°, Holliger C (2014) **Towards management of the bacterial resource for nutrient removal in granular sludge biofilm systems**. In: Fatone *et al.* (eds.), Proceedings of 2nd IWA Specialized International Conference on Ecotechnologies for Wastewater Treatment (EcoSTP), University of Verona, Verona, Italy. (°keynote)

Weissbrodt DG°, Derlon N, Holliger C, Morgenroth E (2014) **A consolidated approach of flocculent and granular sludge systems under the perspective of bacterial resource management**. In: WEF (ed.), Proceedings of Water Environment Federation's Annual Technical Exhibition and Conference (WEFTEC), Knowledge Development Forum 'Flocs versus Granules—The Ultimate Match', New Orleans, USA. (°invited talk)

Oral Presentations

Weissbrodt DG°, Neu TR, Derlon N, Szivák I, Holliger C, Morgenroth E (2014) **Fluorescence lectin-binding analysis reveals the complexity of extracellular glycoconjugate matrices in aerobic granular sludge biofilms**. In: Flemming *et al.* (eds.), Proceedings of IWA Biocluster Conference: The Perfect Slime—Nature, Properties, Regulation and Dynamics of EPS, Conference in the honour of Prof. Dr. Hans-Curt Flemming, University of Duisburg-Essen, Essen, Germany. (°oral presentation)

Weissbrodt DG°, Shani N, Holliger C (2013) **Linking microbial population dynamics and nutrient removal during wastewater treatment**. In: Love *et al.* (eds.), Proceedings of 5th International Conference on Microbial Ecology and Water Engineering (MEWE), University of Michigan, Ann Arbor, USA. (°oral presentation)

Weissbrodt DG°, Lochmatter S, Holliger C (2013) **Bacterial selection for nutrient removal in aerobic granular sludge wastewater treatment systems**. In: van Loosdrecht *et al.* (eds.), Proceedings of 2nd Water Research Conference, Singapore Expo, Singapore. (°oral presentation)

Weissbrodt DG°, Gabus S, Lochmatter S, Rohrbach E, Rossi P, Ebrahimi S, Holliger C (2009) **The choice of inoculum is key for aerobic granular sludge process development**. In: Wuertz *et al.* (eds.), Proceedings of IWA Biofilm Specialist Conference: Fundamentals to Applications, University of California Davis, USA. (°oral presentation)

Poster Presentations

Weissbrodt DG°, Holliger C, Morgenroth E (2014) **Modelling bacterial selection during the plug-flow feeding phase of aerobic granular sludge biofilm reactors**. In: Nopens *et al.* (eds.), Proceedings of 4th IWA/WEF Wastewater Treatment Modelling Seminar (WWTMod), Ghent University, Spa, Belgium. (°poster presentation)

Weissbrodt DG, Shani N, Holliger C° (2014) **Linking microbial population dynamics and nutrient removal during wastewater treatment by aerobic granular sludge**. In: Kim *et al.* (eds.), Proceedings of 15th International Symposium on Microbial Ecology (ISME15), Seoul, Korea. (°poster presentation)

Weissbrodt DG°, Neu TR, Holliger C (2014) **The biofilm granulation mechanisms depend on the predominant bacterial populations involved**. In: Battin *et al.* (eds.), Proceedings of Biofilm 6 International Conference (Biofilm6), University of Vienna, Austria. (°poster presentation)

Weissbrodt DG°, Holliger C (2013) **Bacterial resource management in aerobic granular sludge for nutrient removal**. In: Love *et al.* (eds.), Proceedings of 5th International Conference on Microbial Ecology and Water Engineering (MEWE), University of Michigan, Ann Arbor, USA. (°poster presentation)

Weissbrodt DG, Shani N, Sinclair L, Lefebvre G, Rougemont J, Rossi P, Maillard J, Holliger C° (2012) **PyroTRF-ID: a novel bioinformatics approach for the identification of terminal-restriction fragments using microbiome pyrosequencing data**. In: Sorensen *et al.* (eds.), Proceedings of 14th International Symposium on Microbial Ecology (ISME14): The Power of Small, Copenhagen, Denmark. (°poster presentation)

Weissbrodt DG°, Lochmatter S, Neu TR, Holliger C (2011) **Significance of *Rhodocyclaceae* for the formation of aerobic granular sludge biofilms and nutrient removal from wastewater**. In: Qi *et al.* (eds.), Proceedings of IWA Biofilm Specialist Conference: Processes in Biofilms, Tongji University, Shanghai, China. (°poster presentation)

Peer-Reviewed Conference Abstracts and Talks

Oral Presentations

Weissbrodt DG°, Neu TR, Holliger C (2013) **The biofilm granulation mechanisms depend on the predominant bacterial populations involved**. 71st Annual Congress of Swiss Society for Microbiology (SSM-SGM), Interlaken, Switzerland. (°oral presentation)

Weissbrodt DG*,°, Shani N*, Sinclair L, Lefebvre G, Rossi P, Maillard J, Rougemont J, Holliger C (2013) **PyroTRF-ID: a novel bioinformatics methodology for the affiliation of terminal-restriction fragments using 16S rRNA gene pyrosequencing data**. 5th Swiss Microbial Ecology Meeting of Swiss Society for Microbiology (SME), Murten, Switzerland. (*equal contribution) (°oral presentation)

Weissbrodt DG*,°, Winkler MKH*,°, Barr JJ*,° (2011) **Microbial diversity of aerobic granular sludge, Is it different from activated sludge?** IWA Biofilm Specialist Conference: Processes in Biofilms, Aerobic Granular Sludge Workshop, Tongji University, Shanghai /CN

Weissbrodt DG°, Lochmatter S, Shani N, Holliger C (2011) **Favoring polyphosphate-accumulating Rhodocyclaceae in aerobic granular sludge biofilms for nutrient removal from wastewater**. 4th Swiss Microbial Ecology Meeting of Swiss Society for Microbiology (SGM-SSM), Engelberg, Switzerland. (°oral presentation)

Weissbrodt DG°, Gabus S, Lochmatter S, Rohrbach E, Rossi P, Ebrahimi S, Holliger C (2010) **Predominance of *Zoogloea* spp. during aerobic granular sludge biofilms development for wastewater treatment**. 69th Annual Congress of Swiss Society for Microbiology (SGM-SSM), ETH Zürich, Switzerland. (°oral presentation)

Poster Presentations

Weissbrodt DG*, Shani N*,°, Sinclair L, Lefebvre G, Rossi P, Maillard J, Rougemont J, Holliger C (2012) **PyroTRF-ID: A novel bioinformatics approach for the identification of terminal-restriction fragments using microbiome pyrosequencing data**. Joint Annual Meeting of Swiss Societies for Infectious Diseases, Hospital Hygiene, Microbiology, Tropical Medicine and Parasitology, St. Gallen, Switzerland. (*equal contribution, °poster presentation)

Weissbrodt DG°, Gabus S, Lochmatter S, Rohrbach E, Rossi P, Ebrahimi S, Holliger C (2009) **The choice of inoculum is key for the formation of dense fast-settling aerobic granular biofilms**. 68th Annual Congress of Swiss Society for Microbiology (SGM-SSM), Université de Lausanne, Switzerland. (°poster presentation)

Acknowledgments

This book was driven by my doctoral research carried out at the Laboratory for Environmental Biotechnology at Ecole Polytechnique Fédérale de Lausanne (EPFL), situated in the Swiss alpine scenery overlooking Lake Geneva. EPFL is part of the ETH domain of the Swiss Federal Institutes of Technology and Research. The high academic standards and dynamic environment of the institution stimulated my personal development and fostered (inter)national collaborations.

This work was made possible with the unconditional support of my family and friends.

I express my gratitude to Christof Holliger for his support and mentorship, as well as for our enriching collaboration on the integration of microbial ecology and environmental biotechnology in the ecological engineering of granular sludge and wastewater treatment processes. This engaging project significantly contributed to my academic development.

I extend my acknowledgements to Rizlan Bernier-Latmani (EPFL, president of the jury), Eberhard Morgenroth (ETH Zürich and Eawag, Switzerland), Mark van Loosdrecht (Delft University of Technology, The Netherlands), and Per Halkjær Nielsen (Aalborg University, Denmark). Their high-quality evaluations and contributions as opponents were invaluable, and I am grateful for the excellent collaborations that followed.

The research was enriched by interactions at EPFL, involving Samuel Lochmatter and Graciela Gonzalez-Gil on granular sludge, Jean-Marie Fürbringer on design of experiments, Julien Maillard and Alessandro Brovelli on the biochemistry and modelling of PAO metabolisms, and Noam Shani and Pierre Rossi on the microbial ecology of environmental systems. The development of bioinformatics strategies was enhanced through the contributions of Lucas Sinclair at the Bioinformatics and Biostatistics Core Facility of the EPFL School of Life Sciences, Grégory Lefebvre at the Nestlé Institute of Health Sciences, as well as Ioannis Xénarios and Jacques Rougemont at the Vital-IT cluster of the Swiss Institute for Bioinformatics. At Eawag, reflections with Eberhard Morgenroth and Nicolas Derlon resulted in translation of microbial ecology principles into process engineering and modelling.

My research stay by Thomas Neu and Ute Kuhlicke at Helmholtz-UFZ in Germany was an important step toward integrating CLSM in granular sludge research.

I developed a cross-disciplinary skillset encompassing molecular biology, analytical chemistry, bioprocess design, and system automation through collaborative work with Emmanuelle Rohrbach, Elena Rossel, Jean-Pierre Kradolfer, and Marc Deront. My scientific concepts were consolidated during my doctoral studies at ETHZ and EPFL. At ETHZ, I gained important insights from Willy Gujer on Biological Wastewater Treatment and Systems Analysis for Water Technology. Jean-Marie Fürbringer at EPFL guided me in the principles of Design of Experiments, and Philippe Zaza from BASF contributed to my understanding of Process Development. My scientific curiosity was nurtured through participation in the Advanced Biofilm Course 2011, organized by Thomas Neu (UFZ, Germany), Harald Horn (KIT, Germany), Cristian Picioreanu (TU Delft, The Netherlands), and Michael Kühl (University of Copenhagen, Denmark).

My academic motivation was cultivated through teaching contributions to foundation, undergraduate, and graduate courses at EPFL. I worked alongside Chantal Seignez and Emmanuelle Rohrbach (EPFL) on Microbiology for the Engineer, Jacques Besse and Alain-François Grogg (HES-SO Valais/Wallis) on Process Engineering, Christof Holliger (EPFL) on Water and Wastewater Treatment, Marc Deront (EPFL) on Environmental Bioprocess Design, and Hansruedi Siegrist (Eawag and ETH Zürich) on Wastewater Treatment Plant Simulation and Instrumentation. Guidance from Jean-Louis Ricci (EPFL) in didactics and workshops organized by the Consulting, Education, and Evaluation Network of Western Switzerland (RCFE) played a pivotal role in supporting my teaching aspirations.

I had the privilege of supervising the vocational education and training in biology of Corinne Weis, Audrey Ducrey, Yoan Rappaz, and Guillaume Schneiter (EPFL). Additionally, I had the pleasure of mentoring the semester projects of Jonathan Habermacher and Simon Taillard (EPFL) and the master projects of Jonathan May (AgroSup Dijon), Christelle Petit (Polytech' Clermont-Ferrand), and Alexandre Chabrelie (Polytech Grenoble) from France.

I am thankful to the EPFL ENAC School for the inaugural Ph.D. mobility award 2011 and funding, which facilitated the research collaboration at Helmholtz-UFZ. I am also appreciative of the International Water Association (IWA) for the best poster presentation award received at Tongji University in China, and to the Luce Grivat Foundation for the doctoral prize recognizing a high level and innovative share in the field of environmental science and engineering. The selection of this book for publication in the Springer Theses Series "The Best of the Best"—Recognizing Outstanding Ph.D. Research is equally fulfilling.

This research was financed by the Swiss National Science Foundation, under grants no. 205321-120536 and 200020-138148, supporting the project "Understanding and tailoring aerobic granular sludge wastewater treatment systems".

I extend the reach of this book to all colleagues, students, practitioners, and policymakers with whom we have been developing outstanding links on the field of microbial ecology and water engineering, toward better environmental biotechnologies. Engineering microbiomes relies on the integration of knowledge and expertise, further enhanced through collaborative efforts. This fundamental aspect is crucial for the implementation of sustainable solutions aimed at safeguarding health and the environment, as well as preserving water and resources for a circular economy within an ecologically balanced society.

David Weissbrodt

Contents

Symbols and Abbreviations

Base

The notation framework of this book is based on the symbolizations developed for activated sludge models ASM1 to ASM3 (Henze et al. 2000), used for good practice in biofilms and biofilm reactor modelling (Morgenroth 2008), used in microbial ecology and metabolic modelling (Oehmen et al. 2010), and standardized for integration in wastewater treatment modelling (Corominas et al. 2010).

Acronyms

1-, 2-, 3-D	One-, two-, three-dimensional
AGS	Aerobic granular sludge
AND	Alternated nitrification and dentrification
ANOVA	Analysis of variance
AOP	Advanced oxidation processes
ARISA	Automated ribosomal intergenic spacer analysis
ASM	Activated sludge model
AUSB	Aerobic up-flow sludge blanket
BC	Bubble column
BNR	Biological nutrient removal
CHF	Swiss franc
CLSM	Confocal laser scanning microscopy
CSTR	Continuous-flow stirred-tank reactor
DGGE	Denaturing gradient gel electrophoresis
DLVO	Derjaguin–Landau–Verwey–Overbeek adhesion approach
DO	Dissolved oxygen
EBPR	Enhanced biological phosphorus removal
EDTA	Ethylenediaminetetraacetic acid

EGSB	Expanded granular sludge bed
EPS	Exopolymeric substances
FISH	Fluorescent in situ hybridization
FLBA	Fluorescence lectin-binding analysis
IC	Internal circulation
IWA	International Water Association
LB	Luria–Bertani broth medium
LOD	Limit of detection
LOQ	Limit of quantification
MBBR	Moving bed bioreactor
MBR	Membrane bioreactor
NMR	Nuclear magnetic resonance
OTU	Operational taxonomic unit
PCR	Polymerase chain reaction
PVC	Polyvinyl chloride
qFISH	Quantitative FISH
qPCR	Quantitative real-time PCR
RAPD-PCR	Random amplification of polymorphic DNA
RT-PCR	Reverse transcription PCR
RT-qPCR	Reverse transcription qPCR
SBBR	Sequencing batch biofilm reactor
SBR	Sequencing batch reactor
SCADA	Supervisory control and data acquisition
SFASS	Sterile filtered activated sludge supernatant
SND	Simultaneous nitrification and denitrification
SNDPR	Simultaneous nitrification, denitrification, and phosphorus removal
STR	Stirred-tank reactor
TR	Time relays
T-RFLP	Terminal restriction fragment length polymorphism
UASB	Up-flow anaerobic sludge blanket
UCT	University of Cape Town process
UNEP	United Nations Environment Programme
USB	Up-flow sludge blanket
WWC	World Water Council
WWTP	Wastewater treatment plant
XDLVO	Extended DLVO approach

Lumped Variables

B	Biodegradable substrates
Bio	Organisms (biomass)
BOD5	Biological oxygen demand, 5 days
COD	Chemical oxygen demand

Ig	Inorganic compound
ISS	Inorganic suspended solids
Org	Organic compound
Stor	Cell-internal storage compound
TKN	Total Kjeldahl nitrogen
TN	Total nitrogen
TP	Total phosphorus
Tot	Total
TSS	Total suspended solids
U	Undegradable substrates
VSS	Volatile suspended solids

Organisms

AMO	Anaerobic ammonium-oxidizing organism
ANO	Autotrophic nitrifying organism
AOO	Aerobic ammonium-oxidizing organism
DGAO	Denitrifying GAO
DHO	Denitrifying heterotrophic organism
DPAO	Denitrifying PAO
GAO	Glycogen-accumulating organism
GAODEF	GAO of the genus *Defluviicoccus*
GAODDEF	Denitrifying GAO of the genus *Defluviicoccus*
GAOGB	GAO of the genus "*Candidatus* Competibacter"
GAODGB	Denitrifying GAO of the genus "*Candidatus* Competibacter"
GO	G-Bacteria-related organism
NOO	Nitrite-oxidizing organism
OHO	Ordinary heterotrophic organism
PAO	Polyphosphate-accumulating organism

Chemical Compounds (Used as Abbreviations or Indices)

Ac	Acetate
Alk	Alkalinity
ATP	Adenosine triphosphate
ATU	Allylthiourea
Bu	Butyrate
C	Carbon
Ca	Calcium
cDNA	Complementary DNA
CHO	Carbohydrates

CO_2	Carbon dioxide
DNA	Deoxyribonucleic acid
eDNA	Extracellular DNA
Gly	Glycogen
H_2	Hydrogen
H_2O	Water
HAc	Acetic acid
HCO_3^-	Bicarbonate
HNO_2	Nitrous acid
HPr	Propionic acid
K	Potassium
Mg	Magnesium
N	Nitrogen
N_2	Dinitrogen
NH_4^+	Ammonium
NO_x^-	$NO_2^- + NO_3^-$
NO_2^-	Nitrite
NO_3^-	Nitrate
N_2O	Nitrous oxide
O	Oxygen
O_2	Dioxygen
P	Phosphorus
PHA	β-polyhydroxyalkanoate
PHB	β-polyhydroxybutyrate
PH_2MV	β-polyhydroxy-2-methylvalerate
PHV	β-polyhydroxyvalerate
Pi	Orthophosphate residues
PO_4^{3-}	Orthophosphate
PP	Polyphosphate
Pr	Propionate
RNA	Ribonucleic acid
rRNA	Ribosomal RNA
SO_4^{2-}	Sulphate
VFA	Volatile fatty acids

Mathematical Symbols and Abbreviations, with Units

a	Specific surface area ($m^2\ m^{-3}$)
$a_{PHB/PHA}$	Fraction of PHB in PHA (C-mol $C\text{-}mol^{-1}$)
A	Area (m^2)
A_{530}, A_{630}	Absorbances at 530 and 630 nm (-)
b_{dec}	Decay rate ($kg_{CODx}\ d^{-1}\ kg_{CODx}^{-1}$ or $C\text{-}mol_X\ d^{-1}\ C\text{-}mol_X^{-1}$)
$b_{PHV/PHA}$	Fraction of PHV in PHA (C-mol $C\text{-}mol^{-1}$)

B_A	Surface loading rate (kg s^{-1} m_F^{-2})
c	Concentration (kg m^{-3} or mol m^{-3})
$c_{PH2MV/PHA}$	Fraction of PH2MV in PHA (C-mol C-mol^{-1})
COD/N	COD-to-nitrogen ratio (kg_{COD} kg_N^{-1} or mol_{O2} mol_N^{-1})
COD/P	COD-to-phosphorus ratio (kg_{COD} kg_P^{-1} or mol_{O2} mol_P^{-1})
d	Diameter (m)
D	Diffusion coefficient (m^2 s^{-1})
f	Fraction (-)
F/M	Food-to-microorganism ratio (kg_{CODs} kg_{CODx}^{-1} or C-mol_S C-mol_X^{-1})
GFO	General factory overhead (CHF)
H/D	Height-to-diameter ratio (m m^{-1})
HRT	Hydraulic retention time (h)
i	Composition coefficient (kg kg^{-1} or mol mol^{-1})
J	Flux (kg s^{-1} m_F^{-2})
k_La	Volumetric oxygen mass transfer coefficient (s^{-1} or d^{-1})
K	Saturation coefficient (kg m^{-3} or mol m^{-3})
K_1	ATP for biomass synthesis from Acetyl-CoA* (mol_{ATP} C-mol^{-1})
K_2	ATP for biomass synthesis from Propionyl-CoA* (mol_{ATP} C-mol^{-1})
m	mass (kg)
m	Biomass-specific maintenance rate (kg d^{-1} kg_{CODx}^{-1} or mol d^{-1} C-mol_X^{-1})
MW	Molecular weight (g mol^{-1})
NLR_v	Volumetric ammonium nitrogen loading rate ($kg_{N\text{-}NH4}$ d^{-1} m_R^{-3})
N/M	Nitrogen-to-microorganism ratio (kg_N kg_{CODx}^{-1})
NSP	Number of spherical particles or granules (-)
OLR_v	Volumetric organic loading rate (kg_{CODs} d^{-1} m_R^{-3})
ORP	Oxidation reduction potential (mV)
PE	Person-equivalents (cap)
pH	Potential hydrogen (-)
P/M	Phosphorus-to-microorganism ratio (kg_P kg_{CODx}^{-1})
P/O	Phosphate–oxygen ratio oxidative phosphorylation (mol_{ADP} mol_O^{-1})
PPC	Periodical production cost (CHF)
q	Biomass-specific transformation rate (kg d^{-1} kg_{CODx}^{-1} or mol d^{-1} C-mol_X^{-1})
q*	Apparent biomass-specific transformation rate (kg d^{-1} kg_{CODx}^{-1} or mol d^{-1} C-mol_X^{-1})
Q	Flowrate (m^3 s^{-1})
r	Volumetric transformation rate (kg d^{-1} m_R^{-3} or mol d^{-1} m_R^{-3})
Re	Reynold's number (-)
S	Soluble component (kg m^{-3} or mol m^{-3})
SA	Sludge age or SRT (d)
SLV	Superficial liquid velocity (m s^{-1})
SGV	Superficial gas velocity (m s^{-1})
S/N	Signal-to-noise ratio (-)
SRT	Sludge retention time or SA (d)

SVI Sludge volume index (mL g_X^{-1})
ThOD Theoretical oxygen demand (kg_{O2} m^{-3} or mol_{O2} m^{-3})
TPC Total production cost (CHF)
V Volume (m^{-3})
VPC Variable production cost (CHF)
VSP Volume of spherical particles or granules (m_F^3)
X Particulate component (kg m^{-3} or mol m^{-3})
XC Particulate and colloidal components (kg m^{-3} or mol m^{-3})
Y Yield (kg kg^{-1} or mol mol^{-1})
z Ion charge (mol of positive charge)

Greek Mathematical Symbols

α ATP for transport VFA across cell membrane (mol_{ATP} C-mol_{VFA}^{-1})
β Fraction of Propionyl-CoA* in PHA (C-mol C-mol^{-1})
γ Degree of reduction (e-mol mol^{-1}, or e-mol C-mol^{-1})
δ Yield of ATP produced per NADH oxidized, or P/O ratio, or Y_{NADH_ATP} (mol_{ATP} mol_{NADH}^{-1})
ε Phosphate transport coefficient ($mol_{P\text{-}PO4}$ mol_{NADH}^{-1})
ε Porosity (-)
η Dynamic viscosity (Pa s, or N m^{-2} s, or kg m^{-1} s^{-1})
η_{Ax} Factor of reduction of process rates under anoxic conditions (-)
θ Temperature correction factor (-)
λ Fraction of Acetyl-CoA* in PHA (C-mol C-mol^{-1})
Λ Molar electrical conductivity (m^2 S mol^{-1})
μ Biomass-specific growth rate (kg_{CODx} d^{-1} kg_{CODx}^{-1} or C-mol_X d^{-1} C-mol_X^{-1} or d^{-1})
ν Kinematic viscosity (m^2 s^{-1})
ρ Density (kg m^{-3})
ρ Biomass-specific process rate (kg s^{-1} kg_{CODx}^{-1} or mol s^{-1} C-mol_X^{-1})

Processes

ab Acid–base reaction
dec Decay
endo Endogenous respiration
gro Growth
hyd Hydrolysis
lys Lysis
stor Storage of cell-internal compounds

Environmental Conditions

A2/O	Anaerobic–anoxic–oxic (or aerobic)
A/O	Anaerobic–oxic (or aerobic)
A/O/A	Anaerobic–oxic (or aerobic)–anoxic
An	Anaerobic
Ax	Anoxic (nitrite and nitrate present)
Ax2	Anoxilic (nitrite present)
Ax3	Anoxalic (nitrate present)
Ox	Oxic, or aerobic

Compartments

F	Inner biofilm
G	Gas
L	Liquid
LF	Liquid–biofilm interface, or biofilm surface

Other Indices

B	Bulk liquid phase
Eff	Effluent
F	Film or biofilm
In	Entering
Inf	Influent
L	Liquid
LF	Liquid–film interface
Max	Maximum
Min	Minimum
Out	Leaving
P	Particle
R	Reactor
S	Substrate
V	Volumetric
X	Biomass

References

Corominas L, Rieger L, Takacs I, Ekama G, Hauduc H, Vanrolleghem PA, Oehmen A, Gernaey KV, van Loosdrecht MCM, Comeau Y (2010) New framework for standardized notation in wastewater treatment modelling. Water Sci Technol 61:841–857

Henze M, Gujer W, Mino T, van Loosdrecht MCM (2000) Activated sludge models ASM1, ASM2, ASM2d and ASM3. IWA Publishing, London

Morgenroth E (2008) Modelling biofilms. In: Henze M, van Loosdrecht MCM, Ekama GA, Brdjanovic D (eds) Biological wastewater treatment: principles, modelling and design. IWA Publishing, London, pp 457–492

Oehmen A, Lopez-Vazquez CM, Carvalho G, Reis MAM, van Loosdrecht MCM (2010) Modelling the population dynamics and metabolic diversity of organisms relevant in anaerobic/anoxic/aerobic enhanced biological phosphorus removal processes. Water Res 44(15):4473–4486

Chapter 1
General Introduction and Economic Analysis

Apart from being better, an innovation needs to be cheaper as well.
(van der Roest and van Loosdrecht 2012)

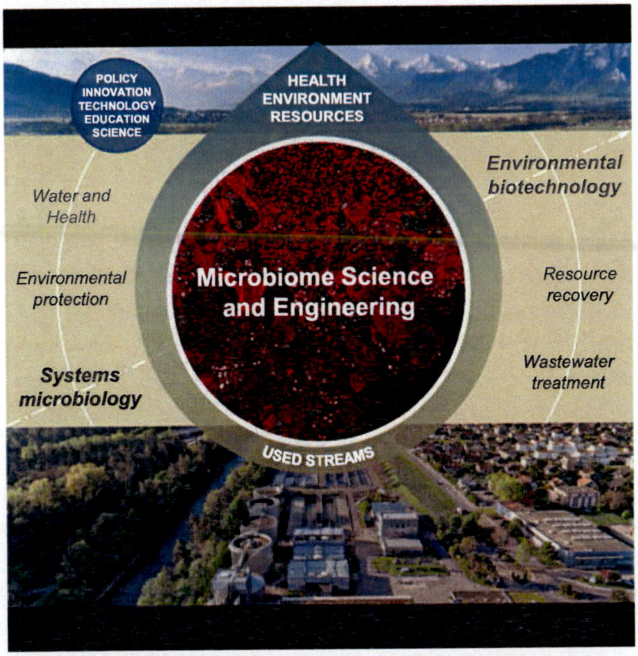

Engineering wastewater microbiomes[1]

[1] Background photo from WWTP Thunersee at the foothills of Eiger, Mönch, and Jungfrau mountains in the Swiss Alps. This WWTP was closely linked to research investigations of this book. It is one of the few WWTPs in Switzerland that removes all nutrients biologically. It served as the source of seed sludge for the research chapters of this work.

D. G. Weissbrodt, *Engineering Granular Microbiomes*, Springer Theses,
https://doi.org/10.1007/978-3-031-41009-3_1

1.1 Introduction

Water is essential for life, the environment, and health. Urbanization and industrialization of the society over the last century resulted in six-fold increase in water use, mainly for food production and industrial activities. Rapid population growth and water demand are challenging the water sustainability and security on the twenty-first century. Action must be taken at different levels for a safe water access and a sustainable water chain (Zehnder et al. 2003). Protection of water resources for a sustained water supply is a strategic target of the Sustainable Development Goals of the United Nations, and the 2030 Agenda for Sustainable Development.[2]

The quality of surface water and groundwater is one major concern of industrialized countries, impacted by the increased water use and wastewater discharge. Sanitation and sewage treatment are central for protecting aquatic ecosystems and human health, such as stressed by the World Water Council on '*Supporting political action to improve water and sanitation services and water management*' (WWC 2010). Wastewater treatment is shifting towards holistic principles for achieving sustainability besides environmental protection (Kehrein et al. 2020a; Zhou et al. 2011). Enhancing conversion capacities, resource recovery, energetic efficiency, and land use is required along reduced investments and operating costs.

Environmental science and engineering established as key educational programs and professions (Alha et al. 2000; Morgenroth et al. 2004). Drinking water adduction and treatment, urban water management, wastewater treatment, and environmental biotechnology are important disciplines at the water nexus. A systems approach for sustainability connects principles of physics, chemistry, and the life sciences with environmental, energy, and economy endpoints, as well as sociology, psychology and policy for public acceptance and implementation (Jeffrey et al. 2004). Paradigms evolve for managing urban water and residuals (Daigger 2009) and reclaiming drinking water from wastewater by (in)direct reuse (Chfadi et al. 2021; Hurlimann and Dolnicar 2010; Peterson et al. 2011; Salgot 2008; Wang et al. 2017). While centralized and decentralized technologies are available and affordable (Guo et al. 2014; Opher and Friedler 2016; Tang et al. 2018), concepts for direct potable water reuse request citizen support and policy incentives (Dishman et al. 1989). This is a grand challenge in the design of sustainable cities (Binz et al. 2016; Ma et al. 2015; Poortvliet et al. 2018).

Better sanitation and water treatment increased life expectancy increased by about 50% over the last century. Wastewater treatment plants (WWTPs) protect water resources. By driving the largest biotechnological processes, they clean the loads of nutrients that eutrophicate aquatic ecosystems. Located downstream of sewers at the end of catchment areas, they are challenged by the cocktail of chemical and biological pollutants discharged by anthropogenic activities (Eggen et al.

[2] https://sdgs.un.org/publications/transforming-our-world-2030-agenda-sustainable-development-17981.

2014; Gillings et al. 2018). Analytical and (eco)toxicological advances highlight the ever-increasing number of contaminants to suppress. Novel physical, chemical, and biological methods are integrated in (de)centralized installations and intensified for treating wastewater, removing xenobiotic and xenogenetic contaminants, and recovering resources to reinject in the circular economy. WWTPs transition toward water resource recovery facilities (WRRFs) (Daigger 2011, 2017; Guest et al. 2009), water resource factories (WRFs) (Kehrein et al. 2020b), or sewage treatment and resource recovery (STARRE) installations (fr., *stations de récupération des ressources de l'eau*; StaRRE) (Lalumière 2016; Vanrolleghem 2015; Weissbrodt 2017, 2018). Responsible science, engineering and education is needed to support these incentives and bridge the academic and engineering sectors (Weissbrodt et al. 2020b).

1.2 New Challenges for Wastewater Treatment Industries

Conventional WWTPs using flocculent activated sludge have been operated during the last 30 years in industrialized countries as standards for the treatment of carbon (C), nitrogen (N) and phosphorus (P) loads transported by sewage. The chronology of pollution emitted with urban wastewater, their effects on aquatic environments, and the technical measures taken at plant level are given in Table 1.1. Ammonium (NH_4^+) and nitrite (NO_2^-) are toxic for aquatic ecosystems, while phosphorus (e.g., total phosphorus and orthophosphates, PO_4^{3-}) and nitrate (NO_3^-) lead to their eutrophication. Algae and aquatic plants over-proliferate under high nutrient load. The decay of this biomass in excess results in the depletion of dissolved consumption in the water system at the expense of indigenous fauna and flora. Strict regulations have been adopted to control eutrophication. The implementation of quality control criteria on phosphorus removal since the 1970s has resulted in the restauration of the health state of receiving water bodies, like the three sub-alpine Lakes Geneva, Annecy, and Lugano (Span et al. 1994). Mitigation measures against nitrate are required for emissions in the catchment basins of sensitive areas like the North Sea (Fux et al. 2003; Howarth et al. 1996; Lancelot 1995; Skjaerseth 2000, 2006; van der Voet et al. 1996).

Sewers and WWTPs are confronted to aging of infrastructures, and investments are required for their rehabilitation (Abraham 2003; Qasem et al. 2010). The Swiss wastewater sector shifts from construction to maintenance, and extension phases (Maurer 2009; Stamm et al. 2015). New strategies and technologies are required to decrease public expenditures while solving treatment challenges. More stringent effluent standards require the extension of WWTPs with biological and physical-chemical processes to remove all C–N–P macropollutants (Barnard and Steichen 2006; Neethling et al. 2010) and xenobiotic organic micropollutants (Eggen et al. 2014; Eriksson et al. 2011; Janssens et al. 1997; Stamm et al. 2015; Ternes 2007). Decentralized treatment help minimize emissions at the source (Abegglen et al. 2008; Bell 2015; Daigger et al. 2005; Grotehusmann et al. 1994; Guest et al. 2010;

Table 1.1 Chronology of pollution emitted from urban wastewater in industrialized countries, associated environmental phenomena, and technical measures taken; adapted, and complemented from Maurer (2007)

Detection	Phenomenon	Undesired pollutant	Technical measure	Application
1920	Accumulation of mud in rivers	Total suspended solids (TSS)	Mechanical treatment	1920
1950	Anoxia in rivers	Organic matter as biological oxygen demand (BOD_5) equivalents	Biological carbon removal	1950
1965	Eutrophication of lakes	Total phosphorus	Chemical dephosphatation Enhanced biological phosphorus removal (EBPR)	1965 2000
1975	Fish toxicity	Ammonium	Biological nitrification	1975
1980	Pollutants in agriculture	Heavy metals	Interdiction of sewage sludge farming	2000
1990	Eutrophication of the North Sea	Nitrate	Biological denitrification	1995
2000	Toxicity and hormonal disturbances	Micropollutants	Advanced oxidation (ozonation, bio-oxidation) Sorption on active carbon	2018
2010	Accumulation in biological systems	Nanoparticles	Membrane filtration	Foreseen 2030?
2020	Accumulation in water, biological, and food systems	Microplastics	Filtrations, sorption, chemical and biological treatments	Foreseen 2030?
Continuous	Antibiotic resistance	Antibiotics, antibiotic resistant bacteria, antibiotic resistance genes	Membrane filtration, advanced oxidations, sorption	Foreseen 2035?

Applications notably relate the context of Switzerland

Iribarnegaray et al. 2018; Kovalova et al. 2012; Langergraber and Muellegger 2005; Machado et al. 2017; Nansubuga et al. 2016; Orner and Mihelcic 2018; Orth 2007; Otterpohl et al. 1997; Parkinson and Tayler 2003; Tchobanoglous et al. 2004; van Lier and Lettinga 1999; Wilderer 2004). Regional techno-economic analyses are needed to address the pros and cons of (de)centralized treatments (Capodaglio et al. 2017; Eggimann et al. 2015; Jung et al. 2018; Massoud et al. 2009; Maurer et al. 2005; Roefs et al. 2017; Singh et al. 2015).

The fate, effects, and removal of nanoparticles (Ahmed et al. 2021; Arnaout and Gunsch 2012; Colman et al. 2014; Enfrin et al. 2019; Gottschalk and Nowack 2011; Kaegi et al. 2013; Mohammad et al. 2015; Nowack 2010; Yang et al. 2013) and microplastics (Browne et al. 2011; Carr et al. 2016; Leslie et al. 2017; Mason et al. 2016; Mintenig et al. 2017; Murphy et al. 2016; Padervand et al. 2020; Rhein et al.

2022; Sun et al. 2019; Vuori and Ollikainen 2022; Zhang et al. 2022; Ziajahromi et al. 2017) from wastewater are importantly addressed.

Wastewater and WWTPs are further pointed out as hotspots for the dissemination of antibiotic resistance and multi-resistant pathogens (Czekalski et al. 2014; Garner et al. 2018; Ginn et al. 2021; Graham et al. 2019; Michael et al. 2013; Miłobedzka et al. 2021; Rizzo et al. 2013). Antibiotic resistant bacteria and antibiotic resistance genes are importantly surveilled at wastewater lines (Baquero et al. 2008; Berendonk et al. 2015; Bürgmann et al. 2018; Ginn et al. 2021; Gothwal and Shashidhar 2015; Lu and Guo 2021; Pallares-Vega et al. 2019, 2021; Pruden et al. 2018; Russell and Yost 2021; Schwartz et al. 2003; Yang et al. 2014; Zhu et al. 2021). This entails the tracking of antimicrobial resistance determinants, mobile genetic elements, and genetically modified materials present on both intracellular DNA and extracellular free DNA pools (Calderón-Franco et al. 2020, 2021, 2022; Gillings et al. 2018; Ikuma and Rehmann 2020; Slipko et al. 2019; Woegerbauer et al. 2020). These problematic biological pollutants, collectively referred to as "xenogenetic elements", can replicate and transfer between microorganisms, which can re-grow after treatment. This poses a serious challenge for the management of wastewater and water resources. In the risk assessment framework, addressing the exposure pathways is important for predicting effects on environmental and human health (Fatta-Kassinos et al. 2011; Lenart-Boroń et al. 2020; Manaia 2017; Meng et al. 2021; Miłobedzka et al. 2021).

Paradigms are changing for an integrated management of the wastewater resource (Daigger 2008; Lazarova 1999). WWTPs are reconceived as factories for the recovery of valuable products, in addition to their main function for the treatment of sewage and for the protection of environmental and public health (Hultman and Plaza 2010). Additional processes are required to close the resource loop (Sutton et al. 2011; Verstraete et al. 2009; Verstraete and Vlaeminck 2011; Wilsenach et al. 2003), to produce energy (Kleerebezem and van Loosdrecht 2007; Rabaey and Verstraete 2005; Zeeman et al. 2008), and to reclaim and reuse water (Daigger 2009; Grommen and Verstraete 2002; Lazarova et al. 2012; Reungoat et al. 2012; Suzuki and Minami 1991; UNEP 2005; Wang et al. 2006), such as exemplified by Fig. 1.1.

On top, WWTPs are facing a footprint challenge for the implementation of additional processes with reduced land area availability resulting from the growth of cities (O'Connor et al. 2010). Intensified processes like biofilm and granular sludge systems are required for a high-rate biological nutrient removal (BNR) from wastewater.

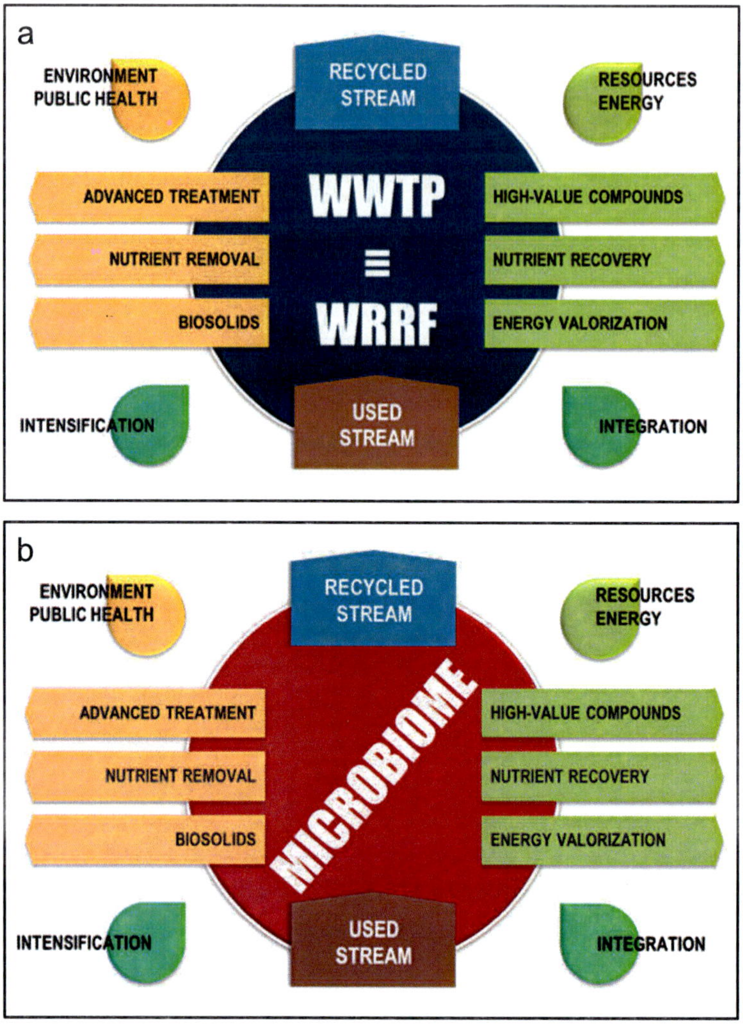

Fig. 1.1 Changing paradigms in the wastewater treatment sector. **a** Wastewater treatment plants (WWTPs) embrace key challenges to enhance the protection of the environment, water resources, and public health. Besides removing pathogens and nutrients like organic matter, nitrogen and phosphorus from sewage, new incentives aim at evolving WWTPs as water resource recovery facilities (WRRFs) to capture nutrients, treat emerging chemical, physical and biological pollutants, and short-cycle resources by recovering energy, chemicals and biomaterials from used streams. Achieving these different targets on the site of a WWTP requires process intensification and integration. Granular sludge offers key opportunities to this end. Modified from Weissbrodt (2018), inspired from the "Turning wastewater into safe water" concept (Monterey Regional Water Pollution Control Agency, California, USA, 2012). **b** The biological processes of WWTPs and WRRFs rely on the metabolic efficiency of microorganisms present in their biomasses (e.g., activated sludge, biofilms, granular sludge). These biomasses comprise complex microbiomes to elucidate and manage using ecological engineering principles. Systems microbiology and environmental biotechnology are key disciplines to integrate in wastewater engineering

1.3 Peak Phosphorus and Biological Methods of Urban Mining

The evolution of concepts related to used water resources relates to the metabolism of the society (Beck and Cummings 1996). The challenge of "peak phosphorus" (Beardsley 2011) is an excellent illustration. Together with water and nitrogen, phosphorus is essential is for life and biological systems. About 80% of the phosphate rock mined is manufactured as mineral fertilizer for world food production (Heffer et al. 2006). The phosphorus resource is not renewable because of its non-gaseous environmental cycle. Scarcity in phosphorus reserves in concentrated form is predicted. Peak phosphorus is scarier than peak oil since this essential element cannot be substituted. The vision of urban mining is of upmost importance in national environmental programs toward the recycling and reuse of phosphorus from used streams as a sustainability basis to minimize losses and conserve existing resources (Cohen et al. 2011). Phosphorus recovery from wastewater has therefore become a necessity. However, no economic incentives have so far been set for technological implementations. The selling price of rock phosphate indeed still remains lower than the cost of phosphorus recovered from sewage (Molinos-Senante et al. 2011). Accounting for environmental externalities remains a target. Phosphorus recovery is a sustainable measure that is economically viable. Switzerland will make phosphorus recycling obligatory (Jedelhauser et al. 2018).

Technologies for an enhanced biological phosphorus removal (EBPR) (Lopez-Vazquez et al. 2020; Wentzel et al. 2008) and full BNR (Barnard and Abraham 2006) from wastewater are progressively gaining more importance in this context. In contrast to chemical precipitation of orthophosphate, these biological treatment flows allow to remove it from the liquid phase and to potentially couple them with recovery units (Foley et al. 2010; Hao and van Loosdrecht 2003; Hirota et al. 2009; Keller 2008; Ludwig 2009; Marti et al. 2008; Muryanto and Bayuseno 2012; Nielsen et al. 2011; Paul et al. 2001). According to life-cycle assessment (LCA), EBPR is more environmentally efficient than chemical dephosphatation: achieving an effluent quality of 0.1–0.5 $mg_{P\text{-}PO4}$ L^{-1} by EBPR induces an up to 5–13% lower impact on global warming (Coats et al. 2011). Still, chemical precipitation of phosphate in the form of vivianite can foster new avenues for phosphorus recovery from wastewater (Frossard et al. 1997; Wilfert et al. 2018).

Implementation of full BNR in conventional WWTPs using flocculent activated sludge however requires almost fifteen times higher working volume and land area than for the treatment of organic matter alone. Novel intensive and high-rate technologies are therefore required for the biological treatment and recovery of phosphorus loads with lower reactor volumes and footprint.

1.4 Transitioning to a More Sustainable Management of Wastewater

Over the last 100 years of activated sludge systems (Jenkins and Wanner 2014; Sheik et al. 2014), industrialized countries have developed wastewater treatment technologies to overcome of water pollutions (Table 1.1). Some countries are fine-tuning the control of micropollutants and environmental impacts on sensitive areas (Eggen et al. 2014; Gavrilescu et al. 2015; Petrie et al. 2015; von Sperling et al. 2008).

In contrast, transition countries such as China (Qiu et al. 2009; Wang and Jin 2006; Wu et al. 1999, 2012), India (Ginn et al. 2021; Rajendra Prasad 2009; Sarkar et al. 2007; Schellenberg et al. 2020; Vij et al. 2021), Brazil (von Sperling et al. 2008), Russia (Likhacheva 2011), or South Africa (Buckley et al. 2011; Howard et al. 1997; van Ginkel 2011) face major environmental challenges in parallel to (fast) growth, modernization, and urbanization. These countries are under a constant pressure resulting from the international trends for stricter environmental standards and from the difficulty to reverse environmental degradation.

Directly transposing the experience of developed countries can sound as unrealistic because of specific economic and social circumstances (Gao et al. 2008), as well as issues related to scalability, implementation, and expertise under local constraints. Transition countries still use this knowledge by anticipating future regulation issues and technology needs in the wastewater treatment sector. They benefit from this opportunity for developing and testing novel technologies for wastewater treatment from bench to pilot and full scale. South Africa makes a leading contribution to international research on wastewater treatment technologies since several decades. The University of Cape Town (UCT) has been active on EBPR development (Dold et al. 1980; Hu et al. 2007; Wentzel et al. 2008). The aerobic granular sludge technology was demonstrated on a first pilot in South Africa (van der Roest et al. 2011).

Novel, intensive, and adaptable technologies for wastewater treatment are globally required to face the environmental, resource, and sustainability challenges across economical regions. Technical solutions result from a strong connection between fundamental science, applied research, techno-economics, and incentives by water authorities.

1.5 Aerobic Granular Sludge Technology

The aerobic granular sludge (AGS) technology has been highlighted as an attractive, intensive alternative to conventional activated sludge plants for full BNR and secondary clarification in single sequencing batch reactors (SBR) (Beun et al. 2000; Morgenroth et al. 1997). This technology is based on mobile, dense, and fast-settling

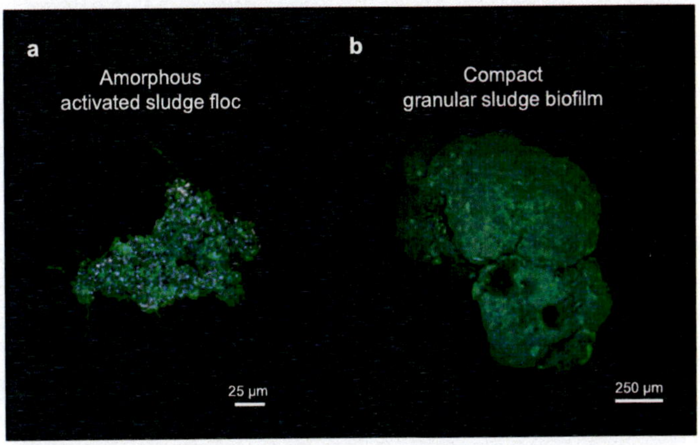

Fig. 1.2 Comparison of the morphologies and metrics of activated sludge flocs and granular sludge biofilms from BNR wastewater treatment systems. Activated sludge flocs display an amorphous structure and loose aggregation of microorganisms along filamentous backbones (**a**). Granular sludge biofilms exhibit a particulate morphology with dense microbial aggregation (**b**). Granules and flocs display significantly distinct metrics, with granules being at least tenfold bigger than flocs. The digital images were obtained after biomass staining with Rhodamine 6G (*green color allocation*) and examination with a confocal laser scanning microscope (CLSM) (Weissbrodt et al. 2013, Chap. 8). The reflection signals (*grey*) display the overall structures

particulate bioaggregates, called "granules", of different metrics on the μm–mm range (Fig. 1.2). Granules are special cases of biofilms that form by spontaneous self-immobilization of microorganisms originating from the wastewater environment within a matrix of extracellular polymeric substances (EPS) in a somewhat spherical shape without the need of a carrier material (de Kreuk et al. 2005b; Grotenhuis et al. 1991).

The use of granular sludge has successfully been experienced since the eighties for high-rate anaerobic digestion of high-loaded wastewaters in up-flow anaerobic sludge blanket (UASB), expanded granular sludge bed (EGSB), or internal circulation (IC) reactors (Lettinga 1995; van Lier et al. 2020). Aerobic granules are the counterpart to the strict anaerobic ones cultivated in UASB systems and are a much more recent development. Anoxic granular sludge processes have been developed for anaerobic ammonium oxidation (anammox) (van der Star et al. 2007). Granules are also used in phototrophic mixed-culture processes for (in)organic carbon and nutrient capture (Abouhend et al. 2020; Blansaer et al. 2022; Brockmann et al. 2021; Cerruti et al. 2020; Stegman et al. 2021).

While the terms "aerobic" (for BNR) and "anoxic" (for anammox) and "anaerobic" (for anaerobic digestion) have been associated to different types of granular sludge, one should keep in mind that granules experience various redox conditions across their biofilm depth and across reactor operations (Weissbrodt et al. 2020a).

The integration of knowledge on granular sludge from disciplines across the redox continuum was recently promoted.[3]

The term granular sludge should be used in a generic way like activated sludge (Winkler et al. 2018). Depending on process configurations, additional hints on redox conditions can be provided, like A/O (anaerobic/aerobic), A2/O (anaerobic-anoxic-aerobic), EBPR, and BNR processes (Fig. 1.3). In practice, "aerobic granules" are typically used in SBRs operated under temporal alternation of anaerobic and aerobic conditions to promote BNR. Simultaneous nitrification and denitrification (SND) (de Kreuk et al. 2005a) is enabled because of dissolved oxygen gradients in their biofilm architecture comprising aerobic (dissolved oxygen as terminal electron acceptor), anoxic (nitrogen oxides) and anaerobic (no electron acceptor) biovolumes. SND can get limited (Layer et al. 2020b). Alternating nitrification and denitrification (AND) is also promoted by on/off aeration patterns imposed on the SBR (Lochmatter et al. 2013).

Therefore, AGS systems and granules experience a multitude of redox conditions inside and outside the aggregates, that are not entirely "aerobic". AGS can be referred in a generic term as "BNR granular sludge" (Winkler et al. 2018). In literature, readers can find other denominations like aerated granular sludge (Cydzik-Kwiatkowska et al. 2014), anaerobic/aerobic granular sludge (De Vleeschauwer et al. 2021), or EBPR granular sludge (Weissbrodt et al. 2017), among others, matching with operational conditions and treatment objectives. Acknowledging this, aerobic granular sludge (AGS) is used as main term across the book and refers to granular sludge used for full BNR.

1.5.1 Advantages of Aerobic Granular Sludge for a High-Rate BNR

The use of aerobic granular sludge offers definite advantages for biological wastewater treatment and BNR (Pronk et al. 2020), such as described hereafter.

The dense biofilm architecture of granules confer them with fast settling velocities (Morgenroth et al. 1997). Granules have been defined in the first Aerobic Granular Sludge Workshop 2004 of the International Water Association (IWA) as '*aggregates of microbial origin, which do not coagulate under reduced hydrodynamic shear, and which subsequently settle significantly faster than activated sludge flocs*' (de Kreuk et al. 2005b, 2007). These bio-particles exhibit the following additional attributes (Adav et al. 2008): (i) regular, smooth, and nearly round shape, (ii) excellent settleability, (iii) dense and strong microbial structure, (iv) high biomass retention, (v) ability to withstand high organic loading, and (vi) tolerance to toxic compounds. Mature granules typically settle in the order of 1 m min^{-1} in contrast to slow-settling

[3] de Kreuk and Weissbrodt (conference chairs), International Water Association, Biofilms Specialist Group, "IWA Biofilms: Granular Sludge Conference 2018", Delft, The Netherlands. www.granularsludgeconference.org.

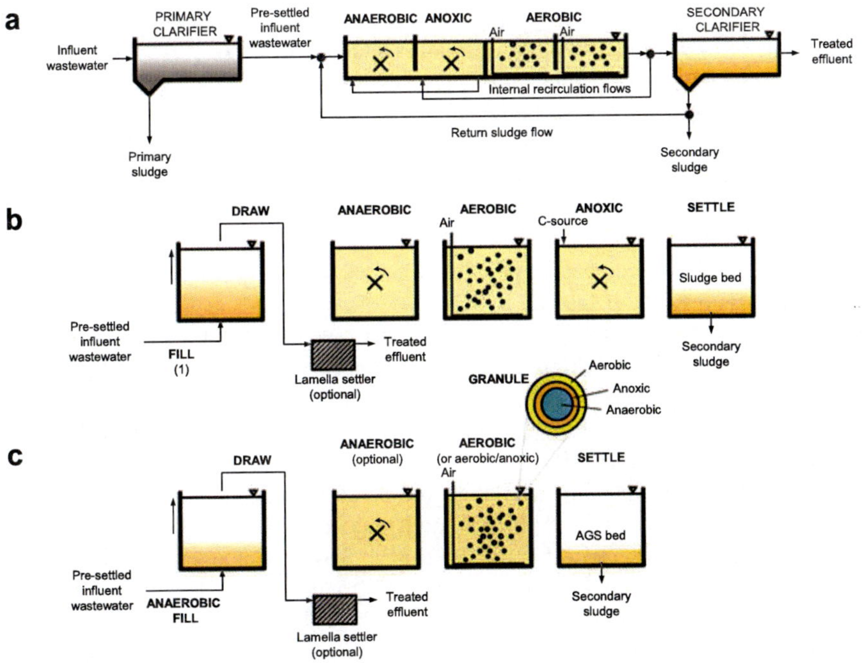

Fig. 1.3 Comparison of configuration of wastewater treatment processes using flocculent activated sludge and AGS for full BNR. **a** Flow-through configuration of a conventional A2/O system using flocculent activated sludge with spatial separation of anaerobic, anoxic, and aerobic redox phases. **b** Configuration of a SBR operated for full BNR using flocculent activated sludge with temporal alternance of redox phases. **c** Configuration of an anaerobic-aerobic AGS-SBR for full BNR. The anaerobic phase consists mainly of anaerobic feeding. This phase can be complemented by an optional anaerobic batch phase. During aeration, BNR is achieved by simultaneous nitrification and denitrification (SND) thanks to the spatial separation of microbial niches within redox layers within granules in function of the penetration depth of terminal electron acceptors (O_2, NO_x^-). It can also be achieved by alternating nitrification and denitrification (AND) by on/off aeration patterns

amorphous activated sludge flocs (1 m h^{-1}) that require a much longer hydraulic retention time (HRT) to sediment and get removed from the treated wastewater in secondary clarifiers of conventional flow-through WWTPs (Etterer and Wilderer 2001; Liu et al. 2005). The sludge volume index (SVI) of settled AGS is typically below 50 mL g^{-1}, which is significantly lower than the SVI of $\geq$ 120 mL g^{-1} of flocculent activated sludge (Toh et al. 2003). In AGS systems, the solid-liquid separation is achieved in short settling phases at lab scale (Beun et al. 1999), and by a good management of up-flow feeding regimes and particle settling velocities in fill-and-draw phases implemented at full scale (Pronk et al. 2015b; van Dijk et al. 2020).

Biofilm processes with high biomass densities can operate at high volumetric conversion rates and cope with high loading rates (Nicolella et al. 2000; Ramadori et al. 2006; van Loosdrecht 2011). Biomass concentrations of 15–30 kg_{VSS} m_r^{-3} are obtained in particle-based biofilm processes, i.e., 5–10 times higher than in activated sludge plants (3–4 $kg_{VSS}\ m^{-3}$). Long biomass retention times are generated in biofilm reactors even with low hydraulic loadings. This sustains the growth of slow-growing organisms and maintains metabolic activities at high level (Wilderer and McSwain 2004). Operation of AGS reactors with a prolonged sludge retention times (SRT, or sludge age) above 20 days can favorably result in a decreased sludge production when compared to activated sludge processes (typical SRT of 5–12 days) (Di Iaconi et al. 2009).

Distinct microbial niches responsible for the removal of the different nutrients can establish inside the biofilm architecture of granules depending on the penetration depth of substrates and on the internal redox gradient (Amann and Kuhl 1998; Sørensen and Morgenroth 2020; Williamson and McCarty 1976). Simultaneous BNR is obtained in single aerated batch cycles with these biofilm particles. In contrast, anaerobic (i.e., absence of any terminal electron acceptor), anoxic (i.e., presence of nitrate and nitrite), and aerobic (i.e. presence of dissolved oxygen) redox conditions (Weissbrodt et al. 2020a) should be established in the successive tanks of conventional WWTPs for full BNR (de Kreuk et al. 2005a). Optimized granular sludge systems now operate for full BNR with SBR cycles alternating anaerobic, anoxic, and aerobic conditions in time (Coma et al. 2012). BNR plants using AGS are successfully performing at full-scale (Pronk et al. 2015b). New AGS methods are currently developed in flow-through systems for the upgrade and intensification of existing WWTPs (Downing et al. 2022; Regmi et al. 2022; Strubbe et al. 2022; Wei et al. 2020).

1.5.2 The Flexibility of the SBR Technology

BNR granular sludge processes are optimally operated in SBR mode. The flexibility of the SBR technology is advantageous for implementing biofilm and AGS processes (Morgenroth et al. 1997; Morgenroth and Wilderer 1998; Wilderer and McSwain 2004). SBRs operate with time-controlled sequences of influent feeding, biological reaction, sludge settling, effluent decanting, and idle phase, whereas flow-through systems are designed with tanks in series (Fig. 1.3). Separation of sludge from treated wastewater is achieved inside the working volume of a SBR, while flow-through processes require an external secondary clarifier. The SBR flexibility enables the manual or automated adaptation of phases and lengths according to influent/environmental variations, treatment performances, and discharge criteria. It can allow to withstand fluctuations in wastewater compositions and hydraulic loadings, such as during rain events (Pronk et al. 2015b). Influent feeding and effluent decanting are achieved in full-scale AGS-SBR by simultaneous fill-and-draw (van Dijk et al. 2020).

The SBR technology has been developed a century ago, as the original wastewater treatment process, before the development of flow-through systems (Ardern 1927;

Jenkins and Wanner 2014). It is a widely accepted, technical, and cost-efficient alternative to conventional flow-through systems (Janczukowicz et al. 2001; Morgenroth and Wilderer 1998; Pace and Harlow 2000; Wilderer et al. 2001). The SBR operation mode allows for an efficient control of unsteady conditions. The constant repetition of process variations over SBR cycles selects for robust microbial communities with optimal substrate removal and settling properties. Substrate gradients generated over SBR cycles are favorable for granule formation (de Graaff et al. 2020; Pronk et al. 2015a; van Dijk et al. 2022). Spontaneous granulation has been observed in several SBR systems (Beun et al. 2000; Dangcong et al. 1999; Mosquera-Corral et al. 2011; Wilderer and McSwain 2004), as well in some flow-through WWTPs (Downing et al. 2022).

For instance, WWTP Thunersee is one of the few installations in Switzerland operated for full BNR in A2/O flow-through activated sludge process. This WWTP was used as source of inoculation sludge for the works conducted in this doctoral research. The formation of a dense granulating activated sludge (with a SVI as low as 35 mL g^{-1}) was observed in this installation, inducing problems to detect the sludge level (*Schlammspiegel*) in the tanks, and to manage the sludge age and the return sludge flow (Bangerter 2017). In this period, ammonium and nitrite surpassed of about 5 times the effluent quality criteria of 2 $mg_{N\text{–}NH4^+}$ L^{-1} and 0.3 $mg_{NO2\text{–}N}$ L^{-1}, respectively. In a sludge transplant approach (Dottorini et al. 2023), replacing the full biomass of the WWTP by a fresh sludge only led to a transient improvement of operation, but the same sludge densification problem arose one year later at the same period. Causes of the sludge densification have not been fully resolved, but one hypothesis can relate to the joint effects of the plug flow regime (i.e., strong substrate gradients (Pronk et al. 2015a)) and anaerobic selector (i.e., selection for PAOs which form dense microcolonies (Larsen et al. 2006; Weissbrodt et al. 2013)). A balance between biological (60%) and chemical (40%) phosphorus elimination is currently used.

Thus, granulation can be seen both as (dis)advantage depending on process targets. A good management of microbial selection, bioaggregation phenomena and BNR is important to manage the performance of the biomass in flow-through and SBR systems on the continuum from activated sludge to granular sludge.

1.5.3 Granular Sludge for BNR Process Intensification

The successful cultivation of aerobic granules (Morgenroth et al. 1997) has triggered the development of AGS-SBRs as an intensified and flexible processes for BNR from wastewater at high volumetric rates and low footprint (de Kreuk 2006), eventually getting installed at full scale (Pronk et al. 2015b).

The high volumetric rates and enhanced settling abilities of granular sludge result in significant savings of 75% on land area for the biological secondary treatment (de Bruin et al. 2004). Savings in land area and tank volumes can sustain the extension of WWTPs with additional processes for the treatment of organic micropollutants such as required in the up-coming application of the revised Swiss Federal Water Protection Ordinance (Eggen et al. 2014; OEaux 1998). Several installations also require the implementation of a full nitrification prior to ozonation.

The operation AGS reactors also results in lower pumping (no recycle and return sludge flows) and an at least 30% improved energy balance when compared to conventional activated sludge processes (de Bruin et al. 2004; van Haandel and van der Lubbe 2012; Watts et al. 2012).

AGS can be applied for the treatment of urban, food-industry, and industrial wastewater. This intensified process can be considered for both centralized and decentralized treatments of wastewater (Li et al. 2006). The AGS technology has further been shown by LCA to be environmentally more efficient than conventional activated sludge for the polishing of effluents from UASB reactors treating brewery wastewater, with significant gains on power supply and waste sludge production (de Bruin 2011; Keller and Giesen 2010; van Haandel and van der Lubbe 2012; Watts et al. 2012). A comprehensive LCA is still required for outlining the sustainability picture of granular sludge and activated sludge technologies. Such assessment should also integrate nitric and nitrous oxides (NO, N_2O) emissions (Gruber et al. 2021; van Dijk et al. 2021; Wunderlin et al. 2013) as global warming indicators.

World-wide interest has rapidly been attributed to the AGS technology because of its definite advantages. Different patents have been filed since the late nineties on methods for promoting self-aggregation of activated sludge into aerobic granules and on systems for purifying wastewater using AGS (Heijnen and van Loosdrecht 1998; Kim et al. 2003; Tay et al. 2004; van Loosdrecht and de Kreuk 2004; Cote 2006; de Bruin et al. 2007; Cote and Behmann 2008). The potential of granular mixed cultures is further interesting for bioaugmentation purposes, e.g. by integrating microbial populations selected for the biodegradation of xenobiotics from industrial process effluents (Duque et al. 2015; Glancer-Soljan et al. 2001; Jemaat et al. 2013).

The traditional AGS-SBR technology designed for full BNR has initially been commercialized under the name Nereda®[4] by referring to the oceanic deities of the Greek mythology, the Nereids sea nymphs (de Kreuk 2006; van der Roest et al. 2011). The first full-scale AGS plants have been implemented in The Netherlands and begins to populate worldwide. Intensive processes using AGS can notably establish in regions and cities where land area is scarce. A National Nereda Research Program has been set up by the Dutch water boards to develop it as an important export product of The Netherlands (Giesen et al. 2012) and to actively implement this technology at full scale (Inocencio et al. 2013; Pronk et al. 2015b). AGS installations are operating world-wide across regional contexts of wastewater (Giesen et al. 2015).

[4] Link to electronic website: www.dhv.com/nereda.

The AGS technology can become a new standard for biological wastewater treatment (Di Iaconi et al. 2007; van der Roest and van Loosdrecht 2012; Watts et al. 2012), among other technologies like biofilm processes and integrated fixed-film activated sludge processes for advanced nitrogen removal by partial nitritation and anammox (Lackner et al. 2014; Laureni et al. 2016; Veuillet et al. 2014), membrane aerated biofilm reactors (Bunse et al. 2020; He et al. 2021; Nerenberg 2016; Syron and Casey 2008), anaerobic membrane bioreactors (Hu et al. 2020; Maaz et al. 2019; Ozgun et al. 2013; Smith et al. 2012; Wu et al. 2021).

Besides BNR intensification, AGS propels an innovation for the recovery of (alginate-like) exopolymers that can be extracted out of the matrix of granules (Felz et al. 2016; Lin et al. 2010; Seviour et al. 2019), commercialized under the name Kaumera (Dutch Water Sector 2019; van der Roest et al. 2015). These biomaterials that can be used as components of high-tech materials and nanocomposites in the industry for civil engineering, coating, flame retardant, and textile applications among others (Lin et al. 2015; Zlopasa et al. 2015).

1.6 Economic Assessment of the Aerobic Granular Sludge Technology for a Swiss WWTP Operated for Full BNR

A simplified economic analysis was performed here to assess the potential of the AGS technology for the treatment of Swiss wastewaters. The total production cost (TPC) of 1 m^3 of treated wastewater was compared over the secondary biological treatment between the conventional activated sludge technology (Fig. 1.4a) and the AGS-SBR technology (Fig. 1.4b) under identical loading conditions. The analysis followed the traditional approach applied by chemical engineers for the economic assessment of industrial production plants (Vogel 2000). All costs are given in Swiss franc (1 CHF $\approx$ 1 EUR $\approx$ 1.05 USD).

1.6.1 Reference WWTP

The functional unit for the economic analysis was defined as a medium Swiss WWTP of 200,000 person-equivalents (PE) treating all nutrients biologically, namely the aforementioned WWTP Thunersee (Uetendorf, Switzerland). This plant occupies a total land area of 3.3 ha. The biological secondary treatment has been designed as an anaerobic-anoxic-aerobic process (A2/O) and has a footprint of 2.6 ha. This corresponds to a field investment cost of 0.910 mioCHF by considering 35 CHF m^{-2} in non-residential areas. Input information for the calculation of the TPC under conventional conditions (Table 1.2) was gained from the 2008 Annual Report of WWTP Thunersee (Boss 2008). It was compared to the representative national data given by the Report on State, Costs and Investments in Sewage Disposal from the

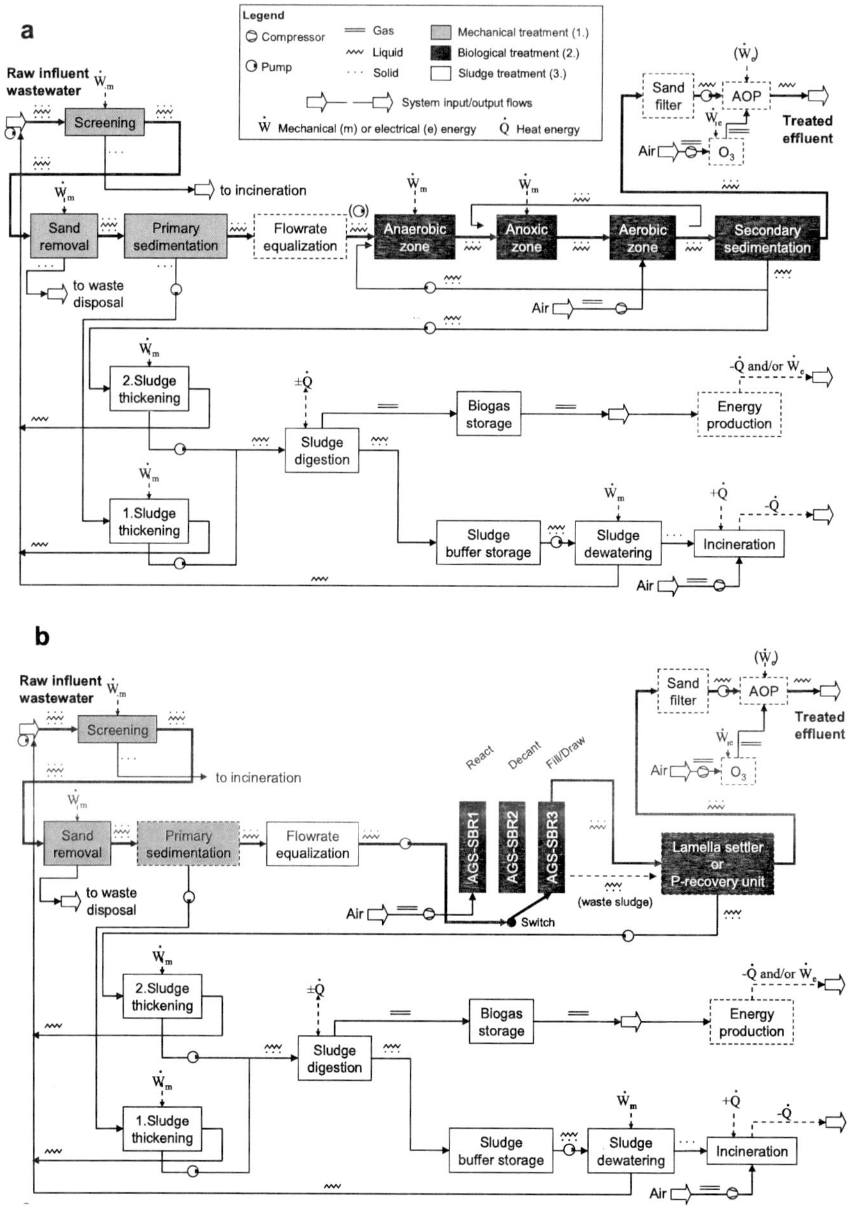

Fig. 1.4 Comparison of process flow diagrams of conventional activated sludge and AGS systems. **a** Scheme of a conventional flow-through plant using activated sludge in a secondary biological treatment designed as an anaerobic/anoxic/aerobic (A2/O) process. **b** Scheme of an AGS-SBR plant proposed with optional primary and secondary clarifiers. Legend: *AOP* advanced oxidation processes

Table 1.2 Input information gained from the annual report of WWTP Thunersee, Switzerland (Boss 2008) (1 CHF ≈ 0.93 EUR ≈ 1.02 USD)

Label	Value	Units
Person equivalents (PE)	200,000	cap
Amount of biologically treated wastewater	14,050,000	m^3
Amount of excess sludge produced	3573	tons of TS
Installation value (wastewater treatment)	136.800	mioCHF[a]
Installation value (connection to sewer system, pumping station)	27.100	mioCHF
Annual added replacement value	4.600	mioCHF
Gross investments for maintenance and projects	5.450	mioCHF
Earnings capital budgeting	1.120	mioCHF
Capital budgeting (county and partner part)	4.330	mioCHF
Cost (incl. wastewater fund)	6.520	mioCHF[b]
Net operating costs (county and partner part)	4.580	mioCHF[c]
Earnings	1.940	mioCHF
Working staff	20	cap
Occupation rate	1695	%

[a] An installation value of 120 mioCHF relates to 200 kPE plants (Maurer and Herlyn 2006)
[b] A cost value of 6 mioCHF was estimated for 200 kPE plants (Liebi 2007)
[c] A net operating cost of 5 mioCHF was assessed for 200 kPE plants (Maurer and Herlyn 2006)

Swiss Federal Office for the Environment (Maurer and Herlyn 2006) and by the Benchmarking Report for Swiss WWTPs (Liebi 2007).

1.6.2 Assumptions

The economic analysis included initial assumptions based on previous reports from AGS systems, described as follows.

1.6.2.1 No Impact of the Granular Sludge matrix

No impact of the granular matrix was accounted for the downstream processing of the waste granular sludge. No difference was assumed in the treatment efficiency of flocculent activated sludge and granular sludge. A similar anaerobic biodegradability (50%) of activated sludge and AGS has been reported (Val del Rio et al. 2011). A dependency has been highlighted between the anaerobic digestion potential and chemical composition of sludge in terms of dissolved organic carbon (DOC), ratio of chemical oxygen demand (COD) to total organic carbon (TOC), as well as carbohydrates, proteins and lipids fractions (Mottet et al. 2010).

A thermal pre-treatment at 190 °C can help solubilize organic compounds to improve the digestibility of the granular sludge, while heat energy requirement is balanced by the increased biogas production in the anaerobic digester (Val del Rio et al. 2011). The present economic analysis can be adapted with further scenarios accounting for waste AGS composition. Additional knowledge on waste sludge treatment unit operations such as thickening, dehydrating or anaerobic digestion can enhance the picture.

1.6.2.2 Savings of 75% on Land Area, 20–40% on Construction, 20–30% on Energy

Economical savings in land area (75%), construction costs (20–40%) and operation and energy costs (20–30%) have been predicted for the AGS technology (de Bruin et al. 2004). Two process scheme variants of (i) an AGS-SBR line with primary treatment and no post-treatment and (ii) an AGS-SBR line with post-treatment only with a lamella settler resulted in identical savings in plant footprint.

Relatively high concentrations of total suspended solids (TSS) between 20 and 100 mg_{TSS} L^{-1} have been measured in the effluent of bubble-column AGS-SBRs operated at lab or pilot scales with short settling times and effluent withdrawal from the middle of the column (Lemaire et al. 2008; Li et al. 2008; Ni et al. 2009; Schwarzenbeck et al. 2005), surpassing the Swiss quality criteria (15–20 mg_{TSS} L^{-1}) (OEaux 1998). At pilot and full scales, a post treatment with a lamella settler or a sand filter can help reduce the discharge of suspended solids. A concentration of 20–30 mg_{TSS} L^{-1} in the effluent (vs. 236 mg_{TSS} L^{-1} in the influent) has been reported from a full scale Nereda plant (Pronk et al. 2015b). Managing the flow hydraulics during the fill/draw phase in function of the settleability of the sludge will allow for the control of suspended solids in the effluent (van Dijk et al. 2020).

Raw wastewater from short and steep sewer systems, like in Switzerland, contains a relatively high amount of suspended solids (Huisman et al. 2003; Nielsen et al. 1992). Swiss wastewater is therefore composed of a substantial fraction of particulate organic matter (40–60% of total COD), whose effect on the AGS process should be addressed (de Kreuk et al. 2010; Derlon et al. 2016; Layer et al. 2020a; Wagner et al. 2015). A pre-fermentation step or the extension of the anaerobic contact time in the SBR can be required to allow for the hydrolysis of particulate organic matter into soluble substrates and their fermentation into VFAs prior to uptake by PAOs and GAOs (Weissbrodt et al. 2017). Impacts on process design, operation, and economics should be addressed.

1.6.3 *Economic Analysis*

The value of the biological secondary treatment installation of the reference conventional WWTP Thunersee amounts to 136.8 mioCHF (Boss 2008). It was used as basis for the calculation of the equipment, direct and indirect costs, administration costs, and contingency. Conversion factors were adapted from the chemical engineering practice (Vogel 2000). The costs of the AGS installation were calculated after integrating the saving factors (de Bruin et al. 2004). An AGS plant value of 95.1 mioCHF was obtained together with a global saving of 41.7 mioCHF when compared to the conventional plant value (Table 1.3).

The TPC of 1 m^3 of wastewater treated biologically was then calculated for both scenario by summing up the variable production costs (VPC), periodical production costs (PPC), and general factory overhead (GFO) (Table 1.4). The resulting annual cost for the treatment of 14.0 mio of m^3 of wastewater with the conventional activated sludge amounted to 18.0 mioCHF y^{-1} with a TPC of 1.28 CHF m^{-3}. With the AGS technology, 11.7 mioCHF would be spent per annum with a TPC of 0.83 CHF m^{-3} resulting in a potential saving of 6.3 mioCHF y^{-1} or 0.45 CHF m^{-3}. These values need validation at pilot and full scales.

1.7 Conclusion

Biological wastewater treatment processes are central elements of the water chain. Industrialized and transition countries are facing new challenges for an improved management of the wastewater resource. A shift in paradigm has been promoted toward the achievement of sustainability on top of a safe protection of environmental and public health. Urban growth leads to new challenges for WWTPs in terms of decreased land availability and emergence of new xenobiotic and xenogenetic pollutants to eliminate in addition to a full removal of C–N–P nutrients.

AGS is promising from the engineering and economic point of view for intensifying BNR processes. The AGS-SBR flexibility allows for high-rate BNR and secondary clarification in intensified systems. The use of the AGS technology for a Swiss reference WWTP of 200 kPE was linked to economical savings of 41.7 mioCHF on initial investment and 6.3 mioCHF on annual operation costs. This is of particular interest for municipalities since WWTPs are partly running on public fund. The AGS technology is attractive for the upgrade of wastewater treatment processes, e.g., from the land-use and flexibility perspective for the integration of new treatments for the elimination of organic micropollutants.

The AGS technology is tested with Swiss wastewater, with a close look at the effect of particulate organic matter that is abundant in the outlet of short, steep, and aerated sewers.

Besides the technological innovation, granular sludge drives the scientific curiosity on microbial processes assembled in granules. Harnessing microbial selection, granulation phenomena, and BNR conversions is essential for enhancing and intensifying biological wastewater treatment across regions. Microbial ecology and process engineering principles can be combined to sustain the ecological engineering of granular sludge processes. This is the motivation of this book.

Table 1.3 Comparison of the total value of the biological secondary treatment installation of the reference WWTP Thunersee (Switzerland) as conventional activated sludge (CAS) and aerobic granular sludge (AGS) schemes (1 CHF ≈ 0.93 EUR ≈ 1.02 USD)

Label	Formula[a]	CAS plant value[b] (mioCHF)	%[c]	AGS plant value[b] (mioCHF)	%[c]
(1) Field investment		*0.910*	*0.7*		
(1.1) Field investment CAS		0.910	0.7		
(1.2) Savings in land area AGS plant	75%[d] of (1.1)	(0.683)			
(1.3) Field investment AGS plant	(1.1)–(1.2)			0.228	0.2
(2) Plant value		*136.800*	*100.0*	*95.088*	*100.0*
(3) Equipment cost		*40.670*	*29.7*	*28.474*	*30.0*
(3.1) Equipment cost CAS plant		40.670	29.7		
(3.2) Equipment cost AGS plant	1/0.93 of (4.3)			28.474	30.0
(4) Direct costs		*52.880*	*38.7*	*37.017*	*38.9*
(4.1) Installation cost CAS plant	93% of (3.1)	37.830	27.7		
(4.2) Savings in construction costs AGS plant	30%[d] of (4.1)	(11.349)			
(4.3) Installation costs AGS plant	(4.1)–(4.2)			26.481	27.9
(4.4) Labor cost	37% of (3)	15.050	11.0	10.536	11.1
(5) Indirect costs		*28.580*	*20.9*	*20.015*	*21.1*
(5.1) Insurances, taxes, transport	8% of (3) and (4.1)	6.280	4.6	4.396	4.6
(5.2) General costs	70% of (4.2)	10.530	7.7	7.375	7.8
(5.3) Engineering costs	15% of (3) and (4.1)	11.770	8.6	8.243	8.7
(6) Administration costs and contingency		*14.664*	*10.7*	*10.266*	*10.8*
(6.1) Contingency fund	15% of (4) and (5)	12.220	8.9	8.555	9.0
(6.2) Administration costs and contingency	3% of (4) and (5)	2.444	1.8	1.711	1.8
Sum of costs	*(3) + (4) + (5) + (6)*	*136.794*	*100.0*	*95.771*	*100.0*

[a] Typical conversion factors used in the chemical engineering practice (Vogel 2000)
[b] The economical balance resolved with Mathcad™
[c] Percentage of the plant value
[d] Savings for the AGS plant obtained from de Bruin et al. (2004)

Table 1.4 Comparison of the total production cost (TPC) of 1 m^3 of wastewater treated in the biological secondary treatment installation of the reference WWTP Thunersee (Switzerland), with conventional activated sludge (CAS) and aerobic granular sludge (AGS) (1 CHF ≈ 0.93 EUR ≈ 1.02 USD)

Label	Formula	Total cost CAS treatment (kCHF)	Total cost AGS treatment[a] (kCHF)
(1) Variable production costs (VPC)		**(8020)**	**(8020)**
(1.1) Raw materials (raw wastewater)[b]	14.050 mio $m^3\ y^{-1}$ at − 0.60 CHF m^{-3}	(8430)	
(1.2) Products (chemicals)		410	
(2) Periodical production costs (PPC)		**24,804**	**18,762**
(2.1) Salaries	16.95 cap at 100 kCHF cap^{-1}	1678	
(2.2) Energies[c]			
Net electricity	1.29 mio kWh at 0.10 CHF kWh^{-1}; 20% savings AGS	129	103
Heating fuel		14	
Water		8	
Fuel and lubricants		18	
Aeration (accounted in electricity)	(0.10–0.60 CHF kg O_2^{-1})	–	
(2.3) Equipments[c]			
Consumables		34	
Acquisitions		67	
Spare parts		195	
Maintenance	5%[d] plant value (Table 1.3)	6840	4754
Storage	3%[d] of (1.1) and (1.2)	(241)	
Ecology (incineration, transport, disposal)		2035	
Depreciation	10%[d] plant value (Table 1.3)	13,680	9509
(2.4) Indirect costs			
Social costs		294	
Property insurance		53	
(etc., e.g. effluent quality control)		(...)	

(continued)

Table 1.4 (continued)

Label	Formula	Total cost CAS treatment (kCHF)	Total cost AGS treatment[a] (kCHF)
(3) General factory overhead (GFO)	**5%[d] of (2) PPC**	**1240**	**938**
(4) Total production cost (TPC)	**(1) VPC + (2) PPC + (3) GFO**	**18,024**	**11,680**
	Volumetric TPC in CHF m^{-3}	*1.28*	*0.83*

[a] Only changes for the AGS plant scenario are given in this column. These changes will have an impact on the economic analysis. Other costs are shared by the two technologies
[b] The raw wastewater is related to a gain for the WWTP (i.e. negative cost). The mean volumetric Swiss tax sewage of 2 CHF m^{-3} was allocated to 70% to the sewer system for the transport of the wastewater and to 30% (0.60 CHF m^{-3}) for its treatment (Maurer 2007)
[c] Energetic and equipment costs were taken from WWTP Thunersee's annual report (Boss 2008)
[d] Typical conversion factors used in the chemical engineering practice (Vogel 2000)

Granular sludge

Acknowledgements Philippe Zaza (BASF Suisse SA) and Thierry Meyer (EPFL, Institute of Chemical Sciences and Engineering) for their advice on the economic assessment of the granular sludge technology.

References

Abegglen C, Ospelt M, Siegrist H (2008) Biological nutrient removal in a small-scale MBR treating household wastewater. Water Res 42(1–2):338–346

Abouhend AS, Milferstedt K, Hamelin J, Ansari AA, Butler C, Carbajal-González BI, Park C (2020) Growth progression of oxygenic photogranules and its impact on bioactivity for aeration-free wastewater treatment. Environ Sci Technol 54(1):486–496

Abraham DM (2003) Life cycle cost integration for the rehabilitation of wastewater infrastructure. In: Molenaar KR, Chinowsky PS (eds) Contruction research congress winds of change: integration and innovation in construction. ASCE, Honolulu, pp 627–635

Adav SS, Lee DJ, Show KY, Tay JH (2008) Aerobic granular sludge: recent advances. Biotechnol Adv 26(5):411–423

Ahmed SF, Mofijur M, Nuzhat S, Chowdhury AT, Rafa N, Uddin MA, Inayat A, Mahlia TMI, Ong HC, Chia WY, Show PL (2021) Recent developments in physical, biological, chemical, and hybrid treatment techniques for removing emerging contaminants from wastewater. J Hazard Mater 416:125912

Alha K, Holliger C, Larsen BS, Purcell P, Rauch W (2000) Environmental engineering education—summary report of the 1st European Seminar. Water Sci Technol 41:1–7

Amann R, Kuhl M (1998) In situ methods for assessment of microorganisms and their activities. Curr Opin Microbiol 1(3):352–358

Ardern E (1927) The activated sludge process of sewage purification. J Soc Chem Ind 36:822–830

Arnaout CL, Gunsch CK (2012) Impacts of silver nanoparticle coating on the nitrification potential of *Nitrosomonas europaea*. Environ Sci Technol 46(10):5387–5395

Bangerter B (2017) Abwasserkennzahlen ARA Thunersee. ARA Thunersee, Uetendorf, p 27

Baquero F, Martínez JL, Cantón R (2008) Antibiotics and antibiotic resistance in water environments. Curr Opin Biotechnol 19(3):260–265

Barnard JL, Abraham K (2006) Key features of successful BNR operation. Water Sci Technol 53(12):1–9

Barnard JL, Steichen MT (2006) Where is biological nutrient removal going now? Water Sci Technol 53(3):155–164

Beardsley TM (2011) Peak phosphorus. Bioscience 61(2):91

Beck MB, Cummings RG (1996) Wastewater infrastructure: challenges for the sustainable city in the new millennium. Habitat Int 20(3):405–420

Bell S (2015) Renegotiating urban water. Prog Plan 96:1–28

Berendonk TU, Manaia CM, Merlin C, Fatta-Kassinos D, Cytryn E, Walsh F, Bürgmann H, Sørum H, Norström M, Pons MN, Kreuzinger N, Huovinen P, Stefani S, Schwartz T, Kisand V, Baquero F, Martinez JL (2015) Tackling antibiotic resistance: the environmental framework. Nat Rev Microbiol 13(5):310–317

Beun JJ, Hendriks A, van Loosdrecht MCM, Morgenroth E, Wilderer PA, Heijnen JJ (1999) Aerobic granulation in a sequencing batch reactor. Water Res 33(10):2283–2290

Beun JJ, van Loosdrecht MCM, Heijnen JJ (2000) Aerobic granulation. Water Sci Technol 41(4–5):41–48

Binz C, Harris-Lovett S, Kiparsky M, Sedlak DL, Truffer B (2016) The thorny road to technology legitimation—institutional work for potable water reuse in California. Technol Forecast Soc Change 103:249–263

Blansaer N, Alloul A, Verstraete W, Vlaeminck SE, Smets BF (2022) Aggregation of purple bacteria in an upflow photobioreactor to facilitate solid/liquid separation: impact of organic loading rate, hydraulic retention time and water composition. Bioresour Technol 348:126806

Boss H (2008) ARA Thunersee Jahresbericht 2008. ARA Thunersee, Uetendorf, p 15

Brockmann D, Gérand Y, Park C, Milferstedt K, Hélias A, Hamelin J (2021) Wastewater treatment using oxygenic photogranule-based process has lower environmental impact than conventional activated sludge process. Bioresour Technol 319:124204

Browne MA, Crump P, Niven SJ, Teuten E, Tonkin A, Galloway T, Thompson R (2011) Accumulation of microplastic on shorelines worldwide: sources and sinks. Environ Sci Technol 45(21):9175–9179

de Bruin LMM, Kraan MW, de Kreuk MK (2007) Process and apparatus for the purification of waste water. NL Patent WO 2007/089141, 20.01.2006

Buckley C, Friedrich E, von Blottnitz H (2011) Life-cycle assessments in the South African water sector: a review and future challenges. Water SA 37:719–726

Bunse P, Orschler L, Agrawal S, Lackner S (2020) Membrane aerated biofilm reactors for mainstream partial nitritation/anammox: experiences using real municipal wastewater. Water Res X 9:100066

Bürgmann H, Frigon D, Gaze WH, Manaia CM, Pruden A, Singer AC, Smets BF, Zhang T (2018) Water and sanitation: an essential battlefront in the war on antimicrobial resistance. FEMS Microbiol Ecol 94(9):fiy101

Calderón-Franco D, Lin Q, van Loosdrecht MCM, Abbas B, Weissbrodt DG (2020) Anticipating xenogenic pollution at the source: impact of sterilizations on DNA release from microbial cultures. Front Bioeng Biotechnol 8:171

Calderón-Franco D, van Loosdrecht MCM, Abeel T, Weissbrodt DG (2021) Free-floating extracellular DNA: systematic profiling of mobile genetic elements and antibiotic resistance from wastewater. Water Res 189:116592

Calderón-Franco D, Sarelse R, Christou S, Pronk M, van Loosdrecht MCM, Abeel T, Weissbrodt DG (2022) Metagenomic profiling and transfer dynamics of antibiotic resistance determinants in a full-scale granular sludge wastewater treatment plant. Water Res 219:118571

Capodaglio AG, Callegari A, Cecconet D, Molognoni D (2017) Sustainability of decentralized wastewater treatment technologies. Water Pract Technol 12(2):463–477

Carr SA, Liu J, Tesoro AG (2016) Transport and fate of microplastic particles in wastewater treatment plants. Water Res 91:174–182

Cerruti M, Stevens B, Ebrahimi S, Alloul A, Vlaeminck SE, Weissbrodt DG (2020) Enrichment and aggregation of purple non-sulfur bacteria in a mixed-culture sequencing-batch photobioreactor for biological nutrient removal from wastewater. Front Bioeng Biotechnol 8:557234

Chfadi T, Gheblawi M, Thaha R (2021) Public acceptance of wastewater reuse: new evidence from factor and regression analyses. Water 13(10):1391

Coats ER, Watkins DL, Kranenburg D (2011) A comparative environmental life-cycle analysis for removing phosphorus from wastewater: biological versus physical/chemical processes. Water Environ Res 83(8):750–760

Cohen Y, Kirchmann H, Enfält P (2011) Management of phosphorus resources—historical perspective, principal problems and sustainable solutions. In: Kumar S (ed) Integrated waste management, vol II. InTech, pp 247–268

Colman BP, Espinasse B, Richardson CJ, Matson CW, Lowry GV, Hunt DE, Wiesner MR, Bernhardt ES (2014) Emerging contaminant or an old toxin in disguise? Silver nanoparticle impacts on ecosystems. Environ Sci Technol 48(9):5229–5236

Coma M, Verawaty M, Pijuan M, Yuan Z, Bond PL (2012) Enhancing aerobic granulation for biological nutrient removal from domestic wastewater. Bioresour Technol 103(1):101–108

Cote PL (2006) Wastewater treatment with aerobic granules. US Patent WO 2006/642513, 21.12.2006

Cote PL, Behmann H (2008) Flow-through aerobic granulator. US Patent WO 2004/024638, 02.12.2008

Cydzik-Kwiatkowska A, Rusanowska P, Zielińska M, Bernat K, Wojnowska-Baryła I (2014) Structure of nitrogen-converting communities induced by hydraulic retention time and COD/N ratio in constantly aerated granular sludge reactors treating digester supernatant. Bioresour Technol 154:162–170.

Czekalski N, Gascón Díez E, Bürgmann H (2014) Wastewater as a point source of antibiotic-resistance genes in the sediment of a freshwater lake. ISME J 8(7):1381–1390

Daigger GT (2008) New approaches and technologies for wastewater management. Bridge Link Eng Soc Fall 38–45

Daigger GT (2009) Evolving urban water and residuals management paradigms: water reclamation and reuse, decentralization, and resource recovery. Water Environ Res 81(8):809–823

Daigger GT (2011) Changing paradigms: from wastewater treatment to resource recovery. Proc Water Environ Fed 2011(6):942–957

Daigger GT (2017) Flexibility and adaptability: essential elements of the WRRF of the future. Water Pract Technol 12(1):156–165

Daigger GT, Rittmann BE, Adham S, Andreottola G (2005) Are membrane bioreactors ready for widespread application? Environ Sci Technol 39(19):399A–406A

Dangcong P, Bernet N, Delgenes JP, Moletta R (1999) Aerobic granular sludge—a case report. Water Res 33:890–893

de Bruin LMM (2011) Scaling-up aerobic granular sludge technology—role of different players in the process. In: Zhou Q (ed) IWA biofilm conference processes in biofilms, Shanghai

de Bruin LMM, de Kreuk MK, van der Roest HFR, Uijterlinde C, van Loosdrecht MCM (2004) Aerobic granular sludge technology: an alternative to activated sludge? Water Sci Technol 49(11–12):1–7

de Graaff DR, van Loosdrecht MCM, Pronk M (2020) Stable granulation of seawater-adapted aerobic granular sludge with filamentous *Thiothrix* bacteria. Water Res 175

de Kreuk MK (2006) Aerobic granular sludge, scaling up a new technology. PhD thesis, Delft University of Technology

de Kreuk MK, Heijnen JJ, van Loosdrecht MCM (2005a) Simultaneous COD, nitrogen, and phosphate removal by aerobic granular sludge. Biotechnol Bioeng 90(6):761–769

de Kreuk MK, McSwain BS, Bathe S, Tay STL, Schwarzenbeck N, Wilderer PA (2005b) Discussion outcomes. In: Bathe S, de Kreuk MK, McSwain BS, Schwarzenbeck N (eds) Aerobic granular sludge. IWA, London, pp 153–169

de Kreuk MK, Kishida N, van Loosdrecht MCM (2007) Aerobic granular sludge—state of the art. Water Sci Technol 55:75–81

de Kreuk MK, Kishida N, Tsuneda S, van Loosdrecht MCM (2010) Behavior of polymeric substrates in an aerobic granular sludge system. Water Res 44(20):5929–5938

De Vleeschauwer F, Caluwé M, Dobbeleers T, Stes H, Dockx L, Kiekens F, Copot C, Dries J (2021) A dynamically controlled anaerobic/aerobic granular sludge reactor efficiently treats brewery/bottling wastewater. Water Sci Technol 84(12):3515–3527

Derlon N, Wagner J, da Costa RHR, Morgenroth E (2016) Formation of aerobic granules for the treatment of real and low-strength municipal wastewater using a sequencing batch reactor operated at constant volume. Water Res 105:341–350

Di Iaconi C, Ramadori R, Lopez A, Passino R (2007) Aerobic granular sludge systems: The new generation of wastewater treatment technologies. Ind Eng Chem Res 46 (21):6661–6665

Di Iaconi C, del Moro G, Ramadori R, Lopez A, Colombino M, Moletta R (2009) Influence of hydraulic residence time on the performances of an aerobic granular biomass based system for treating municipal wastewater at demonstrative scale. Desalin Water Treat 4(1–3):206–211

Dishman CM, Sherrard JH, Rebhun M (1989) Gaining support for direct potable water reuse. J Prof Issues Eng Educ Pract 115(2):154–161

Dold PL, Ekama GA, Van Marais GR (1980) A general model for the activated sludge process. Prog Water Technol 12(6):47–77

Dottorini G, Wágner DS, Stokholm-Bjerregaard M, Kucheryavskiy S, Michaelsen TY, Nierychlo M, Peces M, Williams R, Nielsen PH, Andersen KS, Nielsen PH (2023) Full-scale activated sludge transplantation reveals a highly resilient community structure. Water Res 229:119454

Downing L, Redmond E, Avila I (2022) When density is desirable. Water Online

Duque AF, Bessa VS, Castro PML (2015) Characterization of the bacterial communities of aerobic granules in a 2-fluorophenol degrading process. Biotechnol Rep 5:98–104

Dutch Water Sector (2019) World's first waste water treatment plant to produce biopolymer. Kaumera Water & Technology, Netherlands

Eggen RI, Hollender J, Joss A, Schärer M, Stamm C (2014) Reducing the discharge of micropollutants in the aquatic environment: the benefits of upgrading wastewater treatment plants. Environ Sci Technol 48(14):7683–7689

Eggimann S, Truffer B, Maurer M (2015) To connect or not to connect? Modelling the optimal degree of centralisation for wastewater infrastructures. Water Res 84:218–231

Enfrin M, Dumée LF, Lee J (2019) Nano/microplastics in water and wastewater treatment processes—origin, impact and potential solutions. Water Res 161:621–638

Eriksson E, Revitt DM, Ledin A, Lundy L, Holten Lutzhoft HC, Wickman T, Mikkelsen PS (2011) Water management in cities of the future using emission control strategies for priority hazardous substances. Water Sci Technol 64(10):2109–2118

Etterer T, Wilderer PA (2001) Generation and properties of aerobic granular sludge. Water Sci Technol 43:19–26

Fatta-Kassinos D, Kalavrouziotis IK, Koukoulakis PH, Vasquez MI (2011) The risks associated with wastewater reuse and xenobiotics in the agroecological environment. Sci Total Environ 409(19):3555–3563

Felz S, Al-Zuhairy S, Aarstad OA, van Loosdrecht MCM, Lin YM (2016) Extraction of structural extracellular polymeric substances from aerobic granular sludge. J Vis Exp 115

Foley J, de Haas D, Hartley K, Lant P (2010) Comprehensive life cycle inventories of alternative wastewater treatment systems. Water Res 44(5):1654–1666

Frossard E, Bauer JP, Lothe F (1997) Evidence of vivianite in $FeSO_4$ -flocculated sludges. Water Res 31(10):2449–2454

Fux C, Egli K, van der Meer JR, Siegrist H (2003) The anammox process for nitrogen removal from wastewater: the fruitful collaboration between microbiologists and process engineers. In: Eawag News, vol 56. Eawag, Duebendorf

Gao T, Chen H, Xia S, Zhou Z (2008) Review of water pollution control in China. Front Environ Sci Eng China 2(2):142–149

Garner E, McLain J, Bowers J, Engelthaler DM, Edwards MA, Pruden A (2018) Microbial ecology and water chemistry impact regrowth of opportunistic pathogens in full-scale reclaimed water distribution systems. Environ Sci Technol 52(16):9056–9068

Gavrilescu M, Demnerová K, Aamand J, Agathos S, Fava F (2015) Emerging pollutants in the environment: present and future challenges in biomonitoring, ecological risks and bioremediation. New Biotechnol 32(1):147–156

Giesen A, Niermans R, van Loosdrecht MCM (2012) Aerobic granular biomass: the new standard for domestic and industrial wastewater treatment? Water 21 4:28–30

Giesen A, van Loosdrecht M, Robertson S, de Bruin B (2015) Aerobic granular biomass technology: further innovation, system development and design optimisation. Proc Water Environ Fed 2015(16):1897–1917

Gillings MR, Westoby M, Ghaly TM (2018) Pollutants that replicate: xenogenetic DNAs. Trends Microbiol 26(12):975–977

Ginn O, Berendes D, Wood A, Bivins A, Rocha-Melogno L, Deshusses MA, Tripathi SN, Bergin MH, Brown J (2021) Open waste canals as potential sources of antimicrobial resistance genes in aerosols in Urban Kanpur, India. Am J Trop Med Hyg 104(5):1761–1767

Glancer-Soljan M, Ban S, Landeka Dragicevic T, Soljan V, Matic V (2001) Granulated mixed microbial culture suggesting successful employment of bioaugmentation in the treatment of process wastewaters. Chem Biochem Eng Q 15(3):87–94

Gothwal R, Shashidhar T (2015) Antibiotic pollution in the environment: a review. Clean Soil Air Water 43(4):479–489

Gottschalk F, Nowack B (2011) The release of engineered nanomaterials to the environment. J Environ Monit 13(5):1145–1155

Graham DW, Bergeron G, Bourassa MW, Dickson J, Gomes F, Howe A, Kahn LH, Morley PS, Scott HM, Simjee S, Singer RS, Smith TC, Storrs C, Wittum TE (2019) Complexities in understanding antimicrobial resistance across domesticated animal, human, and environmental systems. Ann N Y Acad Sci 1441:17–30

Grommen R, Verstraete W (2002) Environmental biotechnology: the ongoing quest. J Biotechnol 98(1):113–123

Grotehusmann D, Kheli A, Sieker F, Uhl M (1994) Alternative urban drainage concept and design. Water Sci Technol 29(1–2):277–282

Grotenhuis JTC, Smit M, Lammeren AAM, Stams AJM, Zehnder AJB (1991) Localization and quantification of extracellular polymers in methanogenic granular sludge. Appl Microbiol Biotechnol 36(1):115–119

Gruber W, von Känel L, Vogt L, Luck M, Biolley L, Feller K, Moosmann A, Krähenbühl N, Kipf M, Loosli R, Vogel M, Morgenroth E, Braun D, Joss A (2021) Estimation of countrywide N_2O emissions from wastewater treatment in Switzerland using long-term monitoring data. Water Res X 13:100122

Guest JS, Skerlos SJ, Barnard JL, Beck MB, Daigger GT, Hilger H, Jackson SJ, Karvazy K, Kelly L, Macpherson L, Mihelcic JR, Pramanik A, Raskin L, Van Loosdrecht MCM, Yeh D, Love NG (2009) A new planning and design paradigm to achieve sustainable resource recovery from wastewater. Environ Sci Technol 43(16):6126–6130

Guest JS, Skerlos SJ, Daigger GT, Corbett JRE, Love NG (2010) The use of qualitative system dynamics to identify sustainability characteristics of decentralized wastewater management alternatives. Water Sci Technol 61(6):1637–1644

Guo T, Englehardt J, Wu T (2014) Review of cost versus scale: water and wastewater treatment and reuse processes. Water Sci Technol 69(2):223–234

Hao XD, van Loosdrecht MCM (2003) A proposed sustainable BNR plant with the emphasis on recovery of COD and phosphate. Water Sci Technol 48:77–85

He H, Wagner BM, Carlson AL, Yang C, Daigger GT (2021) Recent progress using membrane aerated biofilm reactors for wastewater treatment. Water Sci Technol 84(9):2131–2157

Heffer P, Prud'homme M, Muirhead B, Isherwood KF (2006) Phosphorus fertilization: issues and outlook. In: IFS (ed) International fertiliser society conference, vol 586. The International Fertilizer Society, Cambridge, p 586

Heijnen JJ, van Loosdrecht MCM (1998) Method for acquiring grain-shaped growth of a microorganism in a reactor. NL Patent WO 98/37027, 27.08.1998

Hirota R, Kuroda A, Kato J, Ohtake H (2009) Bacterial phosphate metabolism and its application to phosphorus recovery and industrial bioprocesses. J Biosci Bioeng 109(5):423–432

Howard JR, Hodgson KG, Simpson DE (1997) Water quality monitoring and reporting compliance—meeting the challenges of effective water resources management in a developing country. Water Supply 15(4):65–74

Howarth RW, Billen G, Swaney D, Townsend A, Jaworski N, Lajtha K, Downing JA, Elmgren R, Caraco N, Jordan T, Berendse F, Freney J, Kudeyarov V, Murdoch P, Zhu ZL (1996) Regional nitrogen budgets and riverine N & P fluxes for the drainages to the North Atlantic Ocean: natural and human influences. Biogeochemistry 35(1):75–139

Hu ZR, Wentzel MC, Ekama GA (2007) A general kinetic model for biological nutrient removal activated sludge systems: model development. Biotechnol Bioeng 98(6):1242–1258

Hu Y, Cheng H, Ji J, Li Y-Y (2020) A review of anaerobic membrane bioreactors for municipal wastewater treatment with a focus on multicomponent biogas and membrane fouling control. Environ Sci Water Res Technol 6(10):2641–2663

Huisman JL, Krebs P, Gujer W (2003) Integral and unified model for the sewer and wastewater treatment plant focusing on transformations. Water Sci Technol 47(12):65–71

Hultman B, Plaza E (2010) Wastewater treatment—new challenges. In: Plaza E, Levlin E (eds) Research and application of new technologies in wastewater treatment and municipal solid waste disposal in Ukraine, Sweden and Poland, vol 3026. TRITA-LWR. Report, Stockholm, pp 1–11

Hurlimann A, Dolnicar S (2010) When public opposition defeats alternative water projects—the case of Toowoomba Australia. Water Res 44(1):287–297

Ikuma K, Rehmann CR (2020) Importance of extracellular DNA in the fate and transport of antibiotic resistance genes downstream of a wastewater treatment plant. Environ Eng Sci 37(2):164–168

Inocencio P, Coehlo F, van Loosdrecht MCM, Giesen A (2013) The future of sewage treatment: Nereda technology exceeds high expectations. Water 21(4):28–29
Iribarnegaray MA, Rodriguez-Alvarez MS, Moraña LB, Tejerina WA, Seghezzo L (2018) Management challenges for a more decentralized treatment and reuse of domestic wastewater in metropolitan areas. J Water Sanit Hyg Dev 8(1):113–122
Janczukowicz W, Szewczyk M, Krzemieniewski M, Pesta J (2001) Settling properties of activated sludge from a sequencing batch reactor (SBR). Pol J Environ Stud 10(1):15–20
Janssens I, Tanghe T, Verstraete W (1997) Micropollutants: a bottleneck in sustainable wastewater treatment. Water Sci Technol 35(10):13–26
Jedelhauser M, Mehr J, Binder CR (2018) Transition of the Swiss phosphorus system towards a circular economy—part 2: socio-technical scenarios. Sustainability 10(6):1980
Jeffrey P, Stephenson T, Temple C (2004) Ever deeper and wider: incorporating sustainability into a practitioner oriented engineering curriculum. Water Sci Technol 49(8):43–48
Jemaat Z, Suarez-Ojeda ME, Perez J, Carrera J (2013) Simultaneous nitritation and p-nitrophenol removal using aerobic granular biomass in a continuous airlift reactor. Bioresour Technol 150:307–313
Jenkins D, Wanner J (eds) (2014) Activated sludge—100 years and counting. London
Jung YT, Narayanan NC, Cheng YL (2018) Cost comparison of centralized and decentralized wastewater management systems using optimization model. J Environ Manage 213:90–97
Kaegi R, Voegelin A, Ort C, Sinnet B, Thalmann B, Krismer J, Hagendorfer H, Elumelu M, Mueller E (2013) Fate and transformation of silver nanoparticles in urban wastewater systems. Water Res 47(12):3866–3877
Kehrein P, van Loosdrecht M, Osseweijer P, Garfí M, Dewulf J, Posada J (2020a) A critical review of resource recovery from municipal wastewater treatment plants—market supply potentials, technologies and bottlenecks. Environ Sci Water Res Technol 6(4):877–910
Kehrein P, van Loosdrecht M, Osseweijer P, Posada J, Dewulf J (2020b) The SPPD-WRF framework: a novel and holistic methodology for strategical planning and process design of water resource factories. Sustainability 12(10):4168
Keller J (2008) Reduce, recover, and…? Chem Eng (804):24–25
Keller J, Giesen A (2010) Advancements in aerobic granular biomass processes. Paper presented at the Neptune and Innowatech end user conference, Gent, Belgium, 27 Jan 2010
Kim KS, Gee CS, Lee HJ, Kim CW, Seo BW, Ahn KH, Cho HH, Byun YS (2003) Sewage treatment apparatus using selfgranulated activated sludge and sewage treatment method thereof. Korea Patent WO 2003/10734342, 11.12.2003
Kleerebezem R, van Loosdrecht MC (2007) Mixed culture biotechnology for bioenergy production. Curr Opin Biotechnol 18(3):207–212
Kovalova L, Siegrist H, Singer H, Wittmer A, McArdell CS (2012) Hospital wastewater treatment by membrane bioreactor: performance and efficiency for organic micropollutant elimination. Environ Sci Technol 46(3):1536–1545
Lackner S, Gilbert EM, Vlaeminck SE, Joss A, Horn H, van Loosdrecht MCM (2014) Full-scale partial nitritation/anammox experiences—an application survey. Water Res 55:292–303
Lalumière A (2016) Le PEX StaRRE: un nouveau programme destiné aux ouvrages d'assainissement des eaux usées. Vecteur Environ Mars 58–59
Lancelot C (1995) The mucilage phenomenon in the continental coastal waters of the North Sea. Sci Total Environ 165:83–102
Langergraber G, Muellegger E (2005) Ecological sanitation—a way to solve global sanitation problems? Environ Int 31(3):433–444
Larsen P, Eriksen PS, Lou MA, Thomsen TR, Kong YH, Nielsen JL, Nielsen PH (2006) Floc-forming properties of polyphosphate accumulating organisms in activated sludge. Water Sci Technol 54(1):257–265
Laureni M, Falås P, Robin O, Wick A, Weissbrodt DG, Nielsen JL, Ternes T, Morgenroth E, Joss A (2016) Mainstream partial nitritation and anammox: long-term process stability and effluent quality at low temperatures. Water Res 101:628–639

Layer M, Bock K, Ranzinger F, Horn H, Morgenroth E, Derlon N (2020a) Particulate substrate retention in plug-flow and fully-mixed conditions during operation of aerobic granular sludge systems. Water Res X 9:100075

Layer M, Villodres MG, Hernandez A, Reynaert E, Morgenroth E, Derlon N (2020b) Limited simultaneous nitrification-denitrification (SND) in aerobic granular sludge systems treating municipal wastewater: mechanisms and practical implications. Water Res X 7:100048

Lazarova V (1999) Role of wastewater reuse for the integrated resource management: costs, benefits and technological challenges [Role de la reutilisation des eaux usees pour la gestion integree des ressources: couts, benefices et defis technologiques]. L'eau l'Ind Nuis (227):47–57

Lazarova V, Choo K-H, Cornel P (2012) Meeting the challenges of the water-energy nexus: the role of reuse and wastewater treatment. Water21 14:2–17

Lemaire R, Webb RI, Yuan Z (2008) Micro-scale observations of the structure of aerobic microbial granules used for the treatment of nutrient-rich industrial wastewater. ISME J 2(5):528–541

Lenart-Boroń A, Prajsnar J, Guzik M, Boroń P, Chmiel M (2020) How much of antibiotics can enter surface water with treated wastewater and how it affects the resistance of waterborne bacteria: a case study of the Białka river sewage treatment plant. Environ Res 191:110037

Leslie HA, Brandsma SH, van Velzen MJM, Vethaak AD (2017) Microplastics en route: field measurements in the Dutch river delta and Amsterdam canals, wastewater treatment plants, North Sea sediments and biota. Environ Int 101:133–142

Lettinga G (1995) Anaerobic digestion and wastewater treatment systems. Anton Leeuw Int J G 67(1):3–28

Li ZH, Kuba T, Kusuda T (2006) Aerobic granular sludge: a promising technology for decentralised wastewater treatment. Water Sci Technol 53(9):79–85

Li ZH, Kuba T, Kusuda T, Wang XC (2008) A comparative study on aerobic granular sludge and effluent suspended solids in a sequence batch reactor. Environ Eng Sci 25(4):577–584

Liebi C (2007) Kläranlagen-Benchmarking 2005: Zusammenfassung des Schlussberichtes für die ARA Kloten/Opfikon. Kläranlageverband Kloten/Opfikon, Kappeler Umwelt Consulting AG, Glattbrugg, CH, p 8

Likhacheva A (2011) Water industry in Russia: challenges and political priorities. In: Finger M, Kunneacke R, Christodoulou A, Scholten D (eds) 4th annual conference on competition and regulation in network industries. CNRI, Brussels, pp 1–17

Lin Y, de Kreuk M, van Loosdrecht MCM, Adin A (2010) Characterization of alginate-like exopolysaccharides isolated from aerobic granular sludge in pilot-plant. Water Res 44(11):3355–3364

Lin YM, Nierop KGJ, Girbal-Neuhauser E, Adriaanse M, van Loosdrecht MCM (2015) Sustainable polysaccharide-based biomaterial recovered from waste aerobic granular sludge as a surface coating material. Sustain Mater Technol 4:24–29

Liu Y, Wang Z-W, Liu Y-Q, Qin L, Tay J-H (2005) A generalized model for settling velocity of aerobic granular sludge. Biotechnol Prog 21(2):621–626

Lochmatter S, Gonzalez-Gil G, Holliger C (2013) Optimized aeration strategies for nitrogen and phosphorus removal with aerobic granular sludge. Water Res 47(16):6187–6197

Lopez-Vazquez CM, Wentzel MC, Comeau Y, Ekama GA, van Loosdrecht MCM, Brdjanovic D, Oehmen A (2020) Enhanced biological phosphorus removal. In: Chen G, van Loosdrecht MCM, Ekama GA, Brdjanovic D (eds) Biological wastewater treatment: principles, modelling and design, 2nd edn. IWA Publishing, London, pp 239–326

Lu J, Guo J (2021) Disinfection spreads antimicrobial resistance. Science 371(6528):474

Ludwig H (2009) Rückgewinnung von Phosphor aus der Abwasserreinigung. Eine Bestandesaufnahme. Bundes Amt für Umwelt, Bern, Schweiz

Ma X, Xue X, González-Mejía A, Garland J, Cashdollar J (2015) Sustainable water systems for the city of tomorrow—a conceptual framework. Sustainability (Switzerland) 7(9):12071–12105

Maaz M, Yasin M, Aslam M, Kumar G, Atabani AE, Idrees M, Anjum F, Jamil F, Ahmad R, Khan AL, Lesage G, Heran M, Kim J (2019) Anaerobic membrane bioreactors for wastewater

treatment: novel configurations, fouling control and energy considerations. Bioresour Technol 283:358–372
Machado AI, Beretta M, Fragoso R, Duarte E (2017) Overview of the state of the art of constructed wetlands for decentralized wastewater management in Brazil. J Environ Manage 187:560–570
Manaia CM (2017) Assessing the risk of antibiotic resistance transmission from the environment to humans: non-direct proportionality between abundance and risk. Trends Microbiol 25(3):173–181
Marti N, Ferrer J, Seco A, Bouzas A (2008) Optimisation of sludge line management to enhance phosphorus recovery in WWTP. Water Res 42(18):4609–4618
Mason SA, Garneau D, Sutton R, Chu Y, Ehmann K, Barnes J, Fink P, Papazissimos D, Rogers DL (2016) Microplastic pollution is widely detected in US municipal wastewater treatment plant effluent. Environ Pollut 218:1045–1054
Massoud MA, Tarhini A, Nasr JA (2009) Decentralized approaches to wastewater treatment and management: applicability in developing countries. J Environ Manage 90(1):652–659
Maurer M (2007) Infrastructure systems in urban water management—lecture notes. ETH Zurich, Zurich
Maurer M (2009) Infrastructure systems in urban water management—lecture notes. ETH Zurich, Zurich
Maurer M, Herlyn A (2006) Zustand, Kosten und Investitionsbedarf der schweizerischen Abwasserentsorgung. Eawag-BAFU, Duebendorf, CH, p 63
Maurer M, Rothenberger O, Larsen TA (2005) Decentralised wastewater treatment technologies from a national perspective: at what cost are they competitive? Water Sci Technol Water Supply 5(6):145–154
Meng Y, Liu W, Fiedler H, Zhang J, Wei X, Liu X, Peng M, Zhang T (2021) Fate and risk assessment of emerging contaminants in reclaimed water production processes. Front Environ Sci Eng 15(5):104
Michael I, Rizzo L, McArdell CS, Manaia CM, Merlin C, Schwartz T, Dagot C, Fatta-Kassinos D (2013) Urban wastewater treatment plants as hotspots for the release of antibiotics in the environment: a review. Water Res 47(3):957–995
Miłobedzka A, Ferreira C, Vaz-Moreira I, Calderón-Franco D, Gorecki A, Purkrtova S, Jan B, Dziewit L, Singleton CM, Nielsen PH, Weissbrodt DG, Manaia CM (2021) Monitoring antibiotic resistance genes in wastewater environments: the challenges of filling a gap in the One-Health cycle. J Hazard Mater 424:127407
Mintenig SM, Int-Veen I, Löder MGJ, Primpke S, Gerdts G (2017) Identification of microplastic in effluents of waste water treatment plants using focal plane array-based micro-Fourier-transform infrared imaging. Water Res 108:365–372
Mohammad AW, Teow YH, Ang WL, Chung YT, Oatley-Radcliffe DL, Hilal N (2015) Nanofiltration membranes review: recent advances and future prospects. Desalination 356:226–254
Molinos-Senante M, Hernandez-Sancho F, Sala-Garrido R, Garrido-Baserba M (2011) Economic feasibility study for phosphorus recovery processes. Ambio 40(4):408–416
Morgenroth E, Wilderer PA (1998) Sequencing batch reactor technology: concepts, design and experiences. J Chart Inst Water Environ Manag 12(5):314–321
Morgenroth E, Sherden T, van Loosdrecht MCM, Heijnen JJ, Wilderer PA (1997) Aerobic granular sludge in a sequencing batch reactor. Water Res 31(12):3191–3194
Morgenroth E, Daigger GT, Ledin A, Keller J (2004) International evaluation of current and future requirements for environmental engineering education. Water Sci Technol 49(8):11–18
Mosquera-Corral A, Arrojo B, Figueroa M, Campos JL, Mendez R (2011) Aerobic granulation in a mechanical stirred SBR: treatment of low organic loads. Water Sci Technol 64(1):155–161
Mottet A, Francois E, Latrille E, Steyer JP, Deleris S, Vedrenne F, Carrere H (2010) Estimating anaerobic biodegradability indicators for waste activated sludge. Chem Eng J 160(2):488–496
Murphy F, Ewins C, Carbonnier F, Quinn B (2016) Wastewater treatment works (WwTW) as a source of microplastics in the aquatic environment. Environ Sci Technol 50(11):5800–5808

Muryanto S, Bayuseno AP (2012) Wastewater treatment for a sustainable future: overview of phosphorus recovery. Appl Mech Mater 110–116:2043–2048

Nansubuga I, Banadda N, Verstraete W, Rabaey K (2016) A review of sustainable sanitation systems in Africa. Rev Environ Sci Biotechnol 15(3):465–478

Neethling JB, Clark D, Pramanik A, Stensel HD, Sandino J, Tsuchihashi R (2010) WERF nutrient challenge investigates limits of nutrient removal technologies. Water Sci Technol 61(4):945–953

Nerenberg R (2016) The membrane-biofilm reactor (MBfR) as a counter-diffusional biofilm process. Curr Opin Biotechnol 38:131–136

Ni BJ, Xie WM, Liu SG, Yu HQ, Wang YZ, Wang G, Dai XL (2009) Granulation of activated sludge in a pilot-scale sequencing batch reactor for the treatment of low-strength municipal wastewater. Water Res 43(3):751–761

Nicolella C, van Loosdrecht MCM, Heijnen SJ (2000) Particle-based biofilm reactor technology. Trends Biotechnol 18(7):312–320

Nielsen PH, Raunkjaer K, Norsker NH, Jensen NA, Hvitved-Jacobsen T (1992) Transformation of wastewater in sewer systems—a review. Water Sci Technol 25(6):17–31

Nielsen PH, Saunders AM, Hansen AA, Larsen P, Nielsen JL (2011) Microbial communities involved in enhanced biological phosphorus removal from wastewater—a model system in environmental biotechnology. Curr Opin Biotechnol 23(3):452–459

Nowack B (2010) Nanosilver revisited downstream. Science 330(6007):1054–1055

O'Connor TP, Rodrigo D, Cannan A (2010) Total water management: the new paradigm for urban water resources planning. In: World environmental and water resources congress 2010: challenges of change. Providence, pp 3251–3260

OEaux (1998) Swiss federal water protection ordinance. In: Environment FOft (ed). Bern, CH, p 60

Opher T, Friedler E (2016) Comparative LCA of decentralized wastewater treatment alternatives for non-potable urban reuse. J Environ Manage 182:464–476

Orner KD, Mihelcic JR (2018) A review of sanitation technologies to achieve multiple sustainable development goals that promote resource recovery. Environ Sci Water Res Technol 4(1):16–32

Orth H (2007) Centralised versus decentralised wastewater systems? Water Sci Technol 56:259–266

Otterpohl R, Grottker M, Lange J (1997) Sustainable water and waste management in urban areas. Water Sci Technol 35(9):121–133

Ozgun H, Dereli RK, Ersahin ME, Kinaci C, Spanjers H, van Lier JB (2013) A review of anaerobic membrane bioreactors for municipal wastewater treatment: integration options, limitations and expectations. Sep Purif Technol 118:89–104

Pace CB, Harlow R (2000) SBR vs. continuous flow: a cost comparison of waste treatment technologies. In: Proceedings of construction congress VI: building together for a better tomorrow in an increasingly complex world, vol 278, pp 948–957

Padervand M, Lichtfouse E, Robert D, Wang C (2020) Removal of microplastics from the environment. A review. Environ Chem Lett 18(3):807–828

Pallares-Vega R, Blaak H, van der Plaats R, de Roda Husman AM, Hernandez Leal L, van Loosdrecht MCM, Weissbrodt DG, Schmitt H (2019) Determinants of presence and removal of antibiotic resistance genes during WWTP treatment: a cross-sectional study. Water Res 161:319–328

Pallares-Vega R, Hernandez Leal L, Fletcher BN, Vias-Torres E, van Loosdrecht MCM, Weissbrodt DG, Schmitt H (2021) Annual dynamics of antimicrobials and resistance determinants in flocculent and aerobic granular sludge treatment systems. Water Res 190:116752

Parkinson J, Tayler K (2003) Decentralized wastewater management in peri-urban areas in low-income countries. Environ Urban 15(1):75–90

Paul E, Laval ML, Sperandio M (2001) Excess sludge production and costs due to phosphorus removal. Environ Technol 22(11):1363–1371

Peterson JD, Murphy RR, Jin Y, Wang L, Nessl MB, Ikehata K (2011) Health effects associated with wastewater treatment, reuse, and disposal. Water Environ Res 83(10):1853–1875

Petrie B, Barden R, Kasprzyk-Hordern B (2015) A review on emerging contaminants in wastewaters and the environment: current knowledge, understudied areas and recommendations for future monitoring. Water Res 72:3–27

Poortvliet PM, Sanders L, Weijma J, De Vries JR (2018) Acceptance of new sanitation: the role of end-users' pro-environmental personal norms and risk and benefit perceptions. Water Res 131:90–99

Pronk M, Abbas B, Al-Zuhairy SH, Kraan R, Kleerebezem R, van Loosdrecht MC (2015a) Effect and behaviour of different substrates in relation to the formation of aerobic granular sludge. Appl Microbiol Biotechnol 99(12):5257–5268

Pronk M, de Kreuk MK, de Bruin B, Kamminga P, Kleerebezem R, van Loosdrecht MCM (2015b) Full scale performance of the aerobic granular sludge process for sewage treatment. Water Res 84:207–217

Pronk M, van Dijk EJH, van Loosdrecht MCM (2020) Aerobic granular sludge. In: Chen GH, van Loosdrecht MCM, Ekama GA, Brdjanovic D (eds) Biological wastewater treatment: principles, modelling and design, 2nd edn. IWA Publishing, London

Pruden A, Alcalde RE, Alvarez PJJ, Ashbolt N, Bischel H, Capiro NL, Crossette E, Frigon D, Grimes K, Haas CN, Ikuma K, Kappell A, LaPara T, Kimbell L, Li M, Li X, McNamara P, Seo Y, Sobsey MD, Sozzi E, Navab-Daneshmand T, Raskin L, Riquelme MV, Vikesland P, Wigginton K, Zhou Z (2018) An environmental science and engineering framework for combating antimicrobial resistance. Environ Eng Sci 35(10):1005–1011

Qasem A, Zayed T, Chen Z (2010) A condition rating system for wastewater treatment plants infrastructures. Int J Civ Environ Eng 2(3):450–454

Qiu G, Xiang L, Song Y, Peng J, Zeng P, Yuan P (2009) Comparison and modeling of two biofilm processes applied to decentralized wastewater treatment. Front Environ Sci Eng China 3(4):412–420

Rabaey K, Verstraete W (2005) Microbial fuel cells: novel biotechnology for energy generation. Trends Biotechnol 23(6):291–298

Rajendra Prasad S (2009) Status of municipal wastewater treatment in some cities of India. In: Jiang Ying H, Praveen M (eds) International conference on environmental science and information application technology, vol 2. Wuhan, pp 346–350

Ramadori R, Di Laconi C, Lopez A, Passino R (2006) An innovative technology based on aerobic granular biomass for treating municipal and/or industrial wastewater with low environmental impact. Water Sci Technol 53(12):321–329

Regmi P, Sturm B, Hiripitiyage D, Keller N, Murthy S, Jimenez J (2022) Combining continuous flow aerobic granulation using an external selector and carbon-efficient nutrient removal with AvN control in a full-scale simultaneous nitrification-denitrification process. Water Res 210:117991

Reungoat J, Escher BI, Macova M, Argaud FX, Gernjak W, Keller J (2012) Ozonation and biological activated carbon filtration of wastewater treatment plant effluents. Water Res 46(3):863–872

Rhein F, Nirschl H, Kaegi R (2022) Separation of microplastic particles from sewage sludge extracts using magnetic seeded filtration. Water Res X 17:100155

Rizzo L, Manaia C, Merlin C, Schwartz T, Dagot C, Ploy MC, Michael I, Fatta-Kassinos D (2013) Urban wastewater treatment plants as hotspots for antibiotic resistant bacteria and genes spread into the environment: a review. Sci Total Environ 447:345–360

Roefs I, Meulman B, Vreeburg JHG, Spiller M (2017) Centralised, decentralised or hybrid sanitation systems? Economic evaluation under urban development uncertainty and phased expansion. Water Res 109:274–286

Russell JN, Yost CK (2021) Alternative, environmentally conscious approaches for removing antibiotics from wastewater treatment systems. Chemosphere 263:128177

Salgot M (2008) Water reclamation, recycling and reuse: implementation issues. Desalination 218(1–3):190–197

Sarkar SK, Saha M, Takada H, Bhattacharya A, Mishra P, Bhattacharya B (2007) Water quality management in the lower stretch of the river Ganges, east coast of India: an approach through environmental education. J Clean Prod 15(16):1559–1567

Schellenberg T, Subramanian V, Ganeshan G, Tompkins D, Pradeep R (2020) Wastewater discharge standards in the evolving context of urban sustainability—the case of India. Front Environ Sci 8:30

Schwartz T, Kohnen W, Jansen B, Obst U (2003) Detection of antibiotic-resistant bacteria and their resistance genes in wastewater, surface water, and drinking water biofilms. FEMS Microbiol Ecol 43(3):325–335

Schwarzenbeck N, Borges JM, Wilderer PA (2005) Treatment of dairy effluents in an aerobic granular sludge sequencing batch reactor. Appl Microbiol Biotechnol 66(6):711–718

Seviour T, Derlon N, Dueholm MS, Flemming H-C, Girbal-Neuhauser E, Horn H, Kjelleberg S, van Loosdrecht MCM, Lotti T, Malpei MF, Nerenberg R, Neu TR, Paul E, Yu H, Lin Y (2019) Extracellular polymeric substances of biofilms: suffering from an identity crisis. Water Res 151:1–7

Sheik AR, Muller EEL, Wilmes P (2014) A hundred years of activated sludge: time for a rethink. Front Microbiol 5:47

Singh NK, Kazmi AA, Starkl M (2015) A review on full-scale decentralized wastewater treatment systems: techno-economical approach. Water Sci Technol 71(4):468–478

Skjaerseth JB (2000) North Sea cooperation: linking international and domestic pollution control. Manchester University Press, Manchester

Skjaerseth JB (2006) Protecting the North-East Atlantic: enhancing synergies by institutional interplay. Mar Policy 30(2):157–166

Slipko K, Reif D, Wögerbauer M, Hufnagl P, Krampe J, Kreuzinger N (2019) Removal of extracellular free DNA and antibiotic resistance genes from water and wastewater by membranes ranging from microfiltration to reverse osmosis. Water Res 164:114916

Smith AL, Stadler LB, Love NG, Skerlos SJ, Raskin L (2012) Perspectives on anaerobic membrane bioreactor treatment of domestic wastewater: a critical review. Bioresour Technol 122:149–159

Sørensen K, Morgenroth E (2020) Biofilm reactors. In: Chen G, van Loosdrecht MCM, Ekama GA, Brdjanovic D (eds) Biological wastewater treatment: principles, modelling and design, 2nd edn. IWA Publishing, London, pp 813–839

Span D, Dominik J, Loizeau JL, Arpagaus P, Vernet JP (1994) Phosphorus evolution in three subalpine lakes: Annecy, Geneva and Lugano: influence of lake restoration managements. Eclogae Geol Helv 87(2):369–383

Stamm C, Eggen RIL, Hering JG, Hollender J, Joss A, Schärer M (2015) Micropollutant removal from wastewater: facts and decision-making despite uncertainty. Environ Sci Technol 49(11):6374–6375

Stegman S, Batstone DJ, Rozendal R, Jensen PD, Hülsen T (2021) Purple phototrophic bacteria granules under high and low upflow velocities. Water Res 190:116760

Strubbe L, Pennewaerde M, Baeten JE, Volcke EIP (2022) Continuous aerobic granular sludge plants: better settling versus diffusion limitation. Chem Eng J 428:131427

Sun J, Dai X, Wang Q, van Loosdrecht MCM, Ni BJ (2019) Microplastics in wastewater treatment plants: detection, occurrence and removal. Water Res 152:21–37

Sutton PM, Rittmann BE, Schraa OJ, Banaszak JE, Togna AP (2011) Wastewater as a resource: a unique approach to achieving energy sustainability. Water Sci Technol 63(9):2004–2009

Suzuki Y, Minami T (1991) Technological development of a wastewater reclamation process for recreational reuse: an approach to advanced wastewater treatment featuring reverse osmosis membrane. Water Sci Technol 23(7–9):1629–1638

Syron E, Casey E (2008) Membrane-aerated biofilms for high rate biotreatment: performance appraisal, engineering principles, scale-up, and development requirements. Environ Sci Technol 42(6):1833–1844

Tang CY, Yang Z, Guo H, Wen JJ, Nghiem LD, Cornelissen E (2018) Potable water reuse through advanced membrane technology. Environ Sci Technol 52(18):10215–10223

Tay JH, Tay STL, Show KY, Liu Y, Ivanov V (2004) Aerobic biomass granules for waste water treatment. Singapore Patent WO 2004/6793822, 21.09.2004

Tchobanoglous G, Ruppe L, Leverenz H, Darby J (2004) Decentralized wastewater management: challenges and opportunities for the twenty-first century. Water Sci Technol Water Supply 4(1):95–102

Ternes T (2007) The occurrence of micropollutants in the aquatic environment: a new challenge for water management. Water Sci Technol 55(12):327–332

Toh SK, Tay JH, Moy BYP, Ivanov V, Tay STL (2003) Size-effect on the physical characteristics of the aerobic granule in a SBR. Appl Microbiol Biotechnol 60(6):687–695

UNEP (2005) Water and wastewater reuse—an environmentally sound approach for sustainable urban water management. United Nations Environment Programme and Global Environment Centre Foundation, Osaka, Japan, p 48

Val del Rio A, Morales N, Isanta E, Mosquera-Corral A, Campos JL, Steyer JP, Carrere H (2011) Thermal pre-treatment of aerobic granular sludge: impact on anaerobic biodegradability. Water Res 45(18):6011–6020

van der Roest HF, van Loosdrecht MCM (2012) Water purification, the new standard: purely based on character. Delft outlook. Magazine of Delft University of Technology. TUDelft, Delft, pp 6–11

van der Roest HF, de Bruin LMM, Gademan G, Coelho F (2011) Towards sustainable waste water treatment with Dutch Nereda® technology. Water Pract Technol 6(3):59

van der Roest H, van Loosdrecht M, Langkamp EJ, Uijterlinde C (2015) Recovery and reuse of alginate from granular Nereda sludge. Water21 48

van der Star WRL, Abma WR, Blommers D, Mulder J-W, Tokutomi T, Strous M, Picioreanu C, van Loosdrecht MCM (2007) Startup of reactors for anoxic ammonium oxidation: experiences from the first full-scale anammox reactor in Rotterdam. Water Res 41(18):4149–4163

van der Voet E, Kleijn R, Udo De Haes HA (1996) Nitrogen pollution in the European Union—origins and proposed solutions. Environ Conserv 23(2):120–132

van Dijk EJH, Pronk M, van Loosdrecht MCM (2020) A settling model for full-scale aerobic granular sludge. Water Res 186:116135

van Dijk EJH, van Loosdrecht MCM, Pronk M (2021) Nitrous oxide emission from full-scale municipal aerobic granular sludge. Water Res 198:117159

van Dijk EJH, Haaksman VA, van Loosdrecht MCM, Pronk M (2022) On the mechanisms for aerobic granulation—model based evaluation. Water Res 216:118365

van Ginkel CE (2011) Eutrophication: present reality and future challenges for South Africa. Water SA 37(5):693–702

van Haandel AC, van der Lubbe JGM (2012) Handbook of biological wastewater treatment, design and optimisation of activated sludge systems, 2nd edn. IWA Publishing, London

van Lier JB, Lettinga G (1999) Appropriate technologies for effective management of industrial and domestic waste waters: the decentralised approach. Water Sci Technol 40(7):171–183

van Lier JB, Mahmoud N, Zeeman G (2020) Anaerobic wastewater treatment. In: Chen G, van Loosdrecht MCM, Ekama GA, Brdjanovic D (eds) Biological wastewater treatment: principles, modelling and design, 2nd edn. IWA Publishing, London, pp 701–756

van Loosdrecht MCM (2011) Biofilm reactors. In: Zhou Q (ed) IWA biofilm conference: processes in biofilms. IWA, Shanghai

van Loosdrecht MCM, de Kreuk MK (2004) Method for the treatment of waste water with sludge granules. NL Patent WO 2004/024638, 13.10.2003

Vanrolleghem P (2015) Les StaRRE de type conventionnel: une bonne option pour éliminer beaucoup d'azote. Vecteur Environ Janvier 56

Verstraete W, Vlaeminck SE (2011) ZeroWasteWater: short-cycling of wastewater resources for sustainable cities of the future. Int J Sustain Dev World 18(3):253–264

Verstraete W, Van de Caveye P, Diamantis V (2009) Maximum use of resources present in domestic "used water." Bioresour Technol 100(23):5537–5545

Veuillet F, Lacroix S, Bausseron A, Gonidec E, Ochoa J, Christensson M, Lemaire R (2014) Integrated fixed-film activated sludge ANITA™ Mox process—a new perspective for advanced nitrogen removal. Water Sci Technol 69(5):915–922

Vij S, Moors E, Kujawa-Roeleveld K, Lindeboom REF, Singh T, de Kreuk MK (2021) From pea soup to water factories: wastewater paradigms in India and the Netherlands. Environ Sci Policy 115:16–25

Vogel GH (2000) Process development, 2. Evaluation. In: Ullmann's encyclopedia of industrial chemistry. Wiley-VCH Verlag GmbH & Co. KGaA

von Sperling M, Schmidt M, Glasson J, Emmelin L, Helbron H (2008) Standards for wastewater treatment in Brazil. In: Schmidt M, Knopp L (eds) Standards and thresholds for impact assessment. Environmental protection in the European Union, vol 3. Springer, Berlin Heidelberg, pp 125–132

Vuori L, Ollikainen M (2022) How to remove microplastics in wastewater? A cost-effectiveness analysis. Ecol Econ 192:107246

Wagner J, Weissbrodt DG, Manguin V, Ribeiro da Costa RH, Morgenroth E, Derlon N (2015) Effect of particulate organic substrate on aerobic granulation and operating conditions of sequencing batch reactors. Water Res 85:158–166

Wang XC, Jin PK (2006) Water shortage and needs for wastewater re-use in the north China. Water Sci Technol 53(9):35–44

Wang XC, Qiu FG, Jin PK (2006) Safety of treated water for re-use purposes—comparison of filtration and disinfection processes. Water Sci Technol 53(9):213–220

Wang Z, Shao D, Westerhoff P (2017) Wastewater discharge impact on drinking water sources along the Yangtze River (China). Sci Total Environ 599–600:1399–1407

Watts S, de Kreuk M, Pijuan M, di Iaconi C, Ried A, Rossetti S, del Moro G, Mancini A, De Sanctis M, Giesen A, Pronk M, van Loosdrecht MCM, Keller J (2012) Aerobic granular biomass processes. In: Lopez A, di Iaconi C, Mascolo G, Pollice A (eds) Innovative and integrated technologies for the treatment of industrial wastewater. IWA Publishing, London, pp 1–86

Wei SP, Stensel HD, Nguyen Quoc B, Stahl DA, Huang X, Lee PH, Winkler MKH (2020) Flocs in disguise? High granule abundance found in continuous-flow activated sludge treatment plants. Water Res 179:115865

Weissbrodt DG (2017) Moi je travaille pour les StaRRE!: Intensification et bioprospection à haute valeur ajoutée en stations de récupération des ressources de l'eau. Bull l'ARPEA J Romand L'Environ 272:40–45

Weissbrodt DG (2018) StaRRE—stations de récupération des ressources de l'eau. Aqua Gas 1:20–24

Weissbrodt DG, Neu TR, Kuhlicke U, Rappaz Y, Holliger C (2013) Assessment of bacterial and structural dynamics in aerobic granular biofilms. Front Microbiol 4:175

Weissbrodt DG, Holliger C, Morgenroth E (2017) Modeling hydraulic transport and anaerobic uptake by PAOs and GAOs during wastewater feeding in EBPR granular sludge reactors. Biotechnol Bioeng 114(8):1688–1702

Weissbrodt DG, Laureni M, van Loosdrecht MCM, Comeau Y (2020a) Basic microbiology and metabolism. In: Chen GH, van Loosdrecht MCM, Ekama GA, Brdjanovic D (eds) Biological wastewater treatment: principles, modelling and design, 2nd edn. IWA Publishing, London

Weissbrodt DG, Winkler MKH, Wells GF (2020b) Responsible science, engineering and education for water resource recovery and circularity. Environ Sci Water Res Technol 6(8):1952–1966

Wentzel MC, Comeau Y, Ekama GA, van Loosdrecht MCM, Brdjanovic D (2008) Enhanced biological phosphorus removal. In: Henze M, van Loosdrecht MCM, Ekama GA, Brdjanovic D (eds) Biological wastewater treatment: principles, modelling and design. IWA Publishing, London, pp 155–220

Wilderer PA (2004) Applying sustainable water management concepts in rural and urban areas: some thoughts about reasons, means and needs. Water Sci Technol 49(7):8–16

Wilderer PA, McSwain BS (2004) The SBR and its biofilm application potentials. Water Sci Technol 50(10):1–10

Wilderer PA, Irvine RL, Goronszy MC (2001) Sequencing batch reactor technology. IWA Publishing, London

Wilfert P, Dugulan AI, Goubitz K, Korving L, Witkamp GJ, Van Loosdrecht MCM (2018) Vivianite as the main phosphate mineral in digested sewage sludge and its role for phosphate recovery. Water Res 144:312–321
Williamson K, McCarty PL (1976) A model of substrate utilization by bacterial films. J Water Pollut Control Fed 48(1):9–24
Wilsenach JA, Maurer M, Larsen TA, van Loosdrecht MCM (2003) From waste treatment to integrated resource management. Water Sci Technol 48(1):1–9
Winkler MKH, Meunier C, Henriet O, Mahillon J, Suarez-Ojeda ME, Del Moro G, De Sanctis M, Di Iaconi C, Weissbrodt DG (2018) An integrative review of granular sludge for the biological removal of nutrients and of recalcitrant organic matter from wastewater. Chem Eng J 336:489–502
Woegerbauer M, Bellanger X, Merlin C (2020) Cell-free DNA: an underestimated source of antibiotic resistance gene dissemination at the interface between human activities and downstream environments in the context of wastewater reuse. Front Microbiol 11:671
Wu C, Maurer C, Wang Y, Xue S, Davis DL (1999) Water pollution and human health in China. Environ Health Perspect 107(4):251–256
Wu H, Yuan Z, Zhang L, Bi J (2012) Eutrophication mitigation strategies: perspectives from the quantification of phosphorus flows in socioeconomic system of Feixi, Central China. J Clean Prod 23(1):122–137
Wu J, Kong Z, Luo Z, Qin Y, Rong C, Wang T, Hanaoka T, Sakemi S, Ito M, Kobayashi S, Kobayashi M, Xu K-Q, Kobayashi T, Kubota K, Li Y-Y (2021) A successful start-up of an anaerobic membrane bioreactor (AnMBR) coupled mainstream partial nitritation-anammox (PN/A) system: a pilot-scale study on in-situ NOB elimination, AnAOB growth kinetics, and mainstream treatment performance. Water Res 207:117783
Wunderlin P, Siegrist H, Joss A (2013) Online N_2O measurement: the next standard for controlling biological ammonia oxidation? Environ Sci Technol 47(17):9567–9568
WWC (2010) A new water politics—world water council 2010–2012 strategy. World Water Council, Marseille, p 24
Yang Y, Wang J, Xiu Z, Alvarez PJJ (2013) Impacts of silver nanoparticles on cellular and transcriptional activity of nitrogen-cycling bacteria. Environ Toxicol Chem 32(7):1488–1494
Yang Y, Li B, Zou S, Fang HHP, Zhang T (2014) Fate of antibiotic resistance genes in sewage treatment plant revealed by metagenomic approach. Water Res 62:97–106
Zeeman G, Kujawa K, de Mes T, Hernandez L, de Graaff M, Abu-Ghunmi L, Mels A, Meulman B, Temmink H, Buisman C, van Lier J, Lettinga G (2008) Anaerobic treatment as a core technology for energy, nutrients and water recovery from source-separated domestic waste(water). Water Sci Technol 57(8):1207–1212
Zehnder AJ, Yang H, Schertenleib R (2003) Water issues: the need for action at different levels. Aquat Sci 65:1–20
Zhang Y, Li Y, Su F, Peng L, Liu D (2022) The life cycle of micro-nano plastics in domestic sewage. Sci Total Environ 802:149658
Zhou J, McCreanor PT, Montalto F, Erdal ZK (2011) Sustainability. Water Environ Res 83(10):1414–1438
Zhu TT, Su ZX, Lai WX, Zhang YB, Liu YW (2021) Insights into the fate and removal of antibiotics and antibiotic resistance genes using biological wastewater treatment technology. Sci Total Environ 776:7234–7264
Ziajahromi S, Neale PA, Rintoul L, Leusch FDL (2017) Wastewater treatment plants as a pathway for microplastics: development of a new approach to sample wastewater-based microplastics. Water Res 112:93–99
Zlopasa J, Norder B, Koenders EAB, Picken SJ (2015) Origin of highly ordered sodium alginate/montmorillonite bionanocomposites. Macromolecules 48(4):1204–1209

Chapter 2
Granular Sludge—State of the Art

Looking for Interactions at Different Scales

Granular gains. But do bacteria have perceptions?
Adapted from Glancer et al. (2003) and Torley (2007)

Granules

D. G. Weissbrodt, *Engineering Granular Microbiomes*, Springer Theses,
https://doi.org/10.1007/978-3-031-41009-3_2

2.1 Introduction

The development of the aerobic granular sludge (AGS) technology for biological nutrient removal (BNR) combines fundamental science for harnessing microbial phenomena and engineering practice for technology scale-up and operation.

Over the last 20 years, AGS made breakthrough by moving from bench to pilot and full scales, becoming an important technology among other environmental engineering processes like activated sludge and biofilm systems.

Besides technological implementation, granular sludge offers several avenues for scientific research across environmental biotechnology, microbial ecology, biofilm engineering, and material sciences among others.

Understanding microbial, chemical, and physical phenomena in granular sludge is an essential step toward the development of engineering concepts. An important point to manage granular sludge and environmental biotechnology systems in general is to keep an integrated overview of the continuum of bioaggregates across flocs, granules, and biofilms. The same principles often apply.

This state-of-the-art review covers the fundamentals of nutrient removal, microbial selection, and bioaggregation from activated sludge to biofilms and granular sludge processes. It opens the black box of these biomasses by further highlighting the importance of microbial ecology and mathematical modelling to better harness these environmental biotechnologies. It highlights knowledge gaps toward an improved management of the microbial resource of AGS systems, by an ecological engineering of granular microbiomes.

2.2 Biological Nutrient Removal from Wastewater

The removal of carbon (C), nitrogen (N), and phosphorus (P) nutrients from wastewater is required for protecting the aquatic ecosystems. Compared to physicochemical methods, biological wastewater treatment is economically attractive. An extensive description of BNR processes was provided by different authors and books (Barnard and Abraham 2006; Barnard and Steichen 2006; Chen et al. 2020; Henze et al. 2008). BNR is based on the activity of environmental microorganisms enriched in activated sludge tanks and that grow on the C–N–P nutrients caried by the wastewater. The one-sludge A2/O plant configuration presented in Fig. 1.1 in the General Introduction is optimal for selecting for the dedicated functional groups for full BNR in activated sludge. BNR microbial guilds are further assembled in various biofilm and granular sludge processes to intensify wastewater treatment (Pronk et al. 2020; Sørensen and Morgenroth 2020).

2.2.1 *Microorganisms for Biological Nutrient Removal*

The success of BNR relates to the design of processes and operations that select for microorganisms and metabolisms of interest and sustain them on the long run. The following guilds of microorganisms are important for BNR, whose conversions are integrated in activated sludge models (Corominas et al. 2010; Henze et al. 2000; Weissbrodt et al. 2020a).

Aerobic ordinary heterotrophic organisms (OHOs, "heterotrophs") catalyze the oxidation of organic matter into carbon dioxide.

Aerobic autotrophic nitrifying organisms (ANOs, "nitrifiers") are composed of ammonium-oxidizing organisms (AOOs) which oxidize ammonium into nitrite and nitrite-oxidizing organisms (NOOs) which oxidize nitrite into nitrate. Recently, the nitrification network was revisited, with some *Nitrospira*-like NOOs described as complete ammonium-oxidizing organisms (CMOs, "comammox") (Daims et al. 2015; Fowler et al. 2018; Palomo et al. 2018; van Kessel et al. 2015).

Nitrogen removal is achieved by denitrifying heterotrophic organisms (DHOs, "denitrifiers") which reduce nitrogen oxides from nitrate (NO_3^-) to nitrite (NO_2^-) to nitric and nitrous oxides (NO, N_2O) to dinitrogen (N_2), using organic matter as electron donor. The nitrogen cycle is notably completed anaerobic ammonium oxidizing organisms (AMOs, "anammox") which achieve a chemolithoautotrophic nitrogen removal, leading to considerable savings in resource and energy (Jetten et al. 1998; Kartal et al. 2010).

Phosphorus is removed by polyphosphate-accumulating organisms (PAOs, "bio-P bacteria"), which can take up orthophosphate from the microenvironment to constitute intracellular energetic stocks in the form of inorganic polyphosphates (Comeau et al. 1986; Lopez-Vazquez et al. 2020). They compete with glycogen-accumulating organisms (GAOs, "G-bacteria") for the carbon source that they store as poly-β-hydroxylkanoates (PHAs) under anaerobic conditions (Crocetti et al. 2002; Zeng et al. 2003c). Both PAOs and GAOs grow on PHAs under aerobic conditions. Some clades of PAOs and GAOs can denitrify (DPAOs, DGAOs) under anoxic conditions. PAOs are not only important for an enhanced biological phosphorus removal (EBPR), but also for the cohesion of aerobic granules (de Kreuk and van Loosdrecht 2004; Guimarães et al. 2018; Weissbrodt et al. 2013a).

2.2.2 *The Cycling Metabolism of Polyphosphate-Accumulating Organisms*

PAOs and their cycling metabolism deserve special attention (Fig. 2.1) (Lopez-Vazquez et al. 2020). While PAOs can take up phosphorus by growing aerobically on soluble substrates (Pijuan et al. 2005), these organisms are preferentially selected over OHOs when the supplies of organic electron donors (e.g., volatile fatty acids, VFAs) and terminal electron acceptors (e.g., dissolved oxygen, nitrate or nitrite). This is achieved by alternating periods of anaerobic feast and aerobic or anoxic starvation in organic matter (Comeau et al. 1986; Kuba et al. 1993; Lemos et al. 1998; Mino et al. 1998; Satoh et al. 1996; van Loosdrecht et al. 1997a; Zhou et al. 2010). PAOs like the model specie "*Candidatus* Accumulibacter phosphatis" (Crocetti et al. 2000; Hesselmann et al. 1999) preferentially take up VFAs such as acetate and propionate. They have recently be reported to also metabolize glucose (Ziliani et al. 2023).

During the anaerobic feast phase, PAOs take up VFAs and condensed them into poly-β-hydroxyalkanoate (PHA) polymers. Cellular energy in the form of adenosine triphosphate (ATP) is required for the active transport of VFAs across the cell membrane. ATP is provided by the (i) hydrolysis of energetic phosphoanhydride bonds of inorganic polyphosphate into orthophosphate monomers which are excreted into the bulk liquid phase; and partly by the (ii) hydrolysis of intracellular

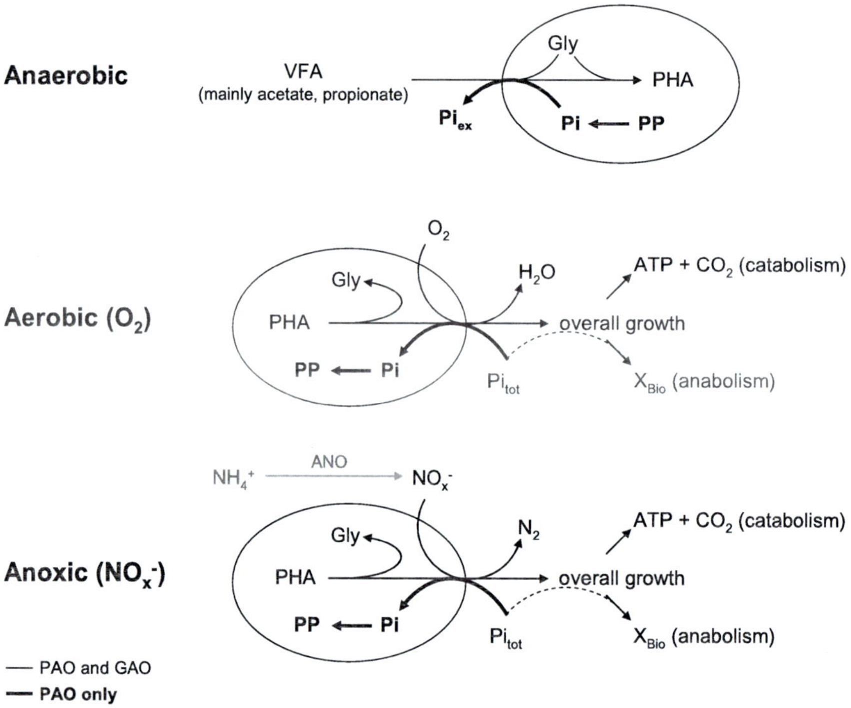

Fig. 2.1 Simplified representations of the metabolisms of PAOs and GAOs under anaerobic, aerobic (presence of dissolved oxygen as terminal electron acceptor), and anoxic (presence of nitrite or nitrate as terminal electron acceptors) conditions. In contrast to GAOs, PAOs have the additional metabolic property to accumulate luxurious amounts on inorganic polyphosphate by removing and polymerizing orthophosphate residues present in the wastewater. Adapted from Mino et al. (1998) and Oehmen et al. (2010b). *ANO* autotrophic nitrifying organisms, *Gly* glycogen, NO_x^- nitrification products (nitrite and nitrate), *PHA* poly-β-hydroxyalkanoates, *Pi* orthophosphate residues, Pi_{ex} excreted Pi, Pi_{tot} sum of Pi present in the influent wastewater and excreted from PAO cells, *PP* polyphosphate, *VFA* volatile fatty acids, X_{Bio} biomass

glycogen molecules. Nicotinamide adenine dinucleotide (NADH) formed during this metabolism is used as reducing power for the synthesis of PHAs.

During the subsequent starvation period under aerobic or anoxic conditions, PAOs use intracellular PHAs as electron-donor and carbon source for growth (catabolism and anabolism). While growing, they replenish their intracellular stocks of glycogen and polyphosphate. Since PAOs grow, the net uptake of orthophosphate from the medium exceeds the amount released during the anaerobic phase.

Phosphorus leaves the biological treatment system with the waste sludge. The thickened waste sludge can then be used for biogas production in anaerobic digesters or incinerated and disposed in landfills. Prior to anaerobic digestion, incineration or disposal, phosphorus can be recovered bacterial cells in a side stream system from by releasing it anaerobically in the presence of a carbon source prior to precipitation with calcium (Morse et al. 1998).

Operation with full uptake of the chemical oxygen demand (COD) equivalents under anaerobic conditions is required for selecting for PAOs over OHOs. Depending on the influent composition, the anaerobic contact time between the wastewater and the biomass can be extended to enable the hydrolysis of particulate substrates (X_S) into soluble substrates (S_S) and their fermentation of into VFAs prior to full uptake by PAOs.

Even if the leakage of organics into aeration tanks/phases is prevented, other organisms such as GAOs like "*Ca.* Competibacter phosphatis" (Crocetti et al. 2002) can compete with PAOs for the anaerobic uptake of VFAs. GAOs exhibit similar PHA-cycling ang glycogen-cycling metabolisms as PAOs, but do not involve polyphosphate cycling (Fig. 2.1). Since GAOs cannot remove phosphorus from wastewater, these organisms are not desired in EBPR systems. However, like PAOs, (D)GAOs can be used to stabilize the granule structures and to remove nitrogen, if EBPR is not a treatment objective.

More details on the microbial ecology of BNR activated sludge systems, on the competition of PAOs and GAOs, and on the diversity of these guilds are given further below (Sect. 2.7). Remarkably, PAOs are more diverse than the model "*Ca.* Accumulibacter" with other organisms like *Tetrasphaera* (or the closely related "*Ca.* Phosphoribacter") and *Dechloromonas* among others (Stokholm-Bjerregaard et al. 2017). *Dechloromonas* grows under the same conditions as "*Ca.* Accumulibacter", whereas *Tetrasphaera* ferments glucose and amino acids (Adler and Holliger 2020; Close et al. 2021; Kristiansen et al. 2013; Petriglieri et al. 2020; Singleton et al. 2022). PAO populations are tracked since the 1980s, and the saga continues, based on the bioanalytical advances made to unravel microorganisms and their metabolisms in microbiomes using culture-independent methods. "*Ca.* Accumulibacter" and most of PAOs relevant for wastewater treatment have not yet been successfully cultivated in pure cultures. While PAOs are importantly studied in environmental engineering systems, their ecological niches in nature remain to get unraveled (Akbari et al. 2021; Peterson et al. 2008; Saia et al. 2017, 2021).

Based on their fascinating resource-recycling metabolism, PAOs can be considered as key organisms for the urban circular economy (Weissbrodt 2022). They play an important role in granulation, BNR, and phosphorus recovery.

2.3 Biofilms

Biofilms are sophisticated living entities with a range of functionalities spanning from beneficial environmental services to detrimental processes for infrastructure and health. Biofilm technologies help intensify wastewater treatment processes (Morgenroth 2020; Regmi et al. 2017; Sørensen and Morgenroth 2020). Understanding their main characteristics is essential for managing not only biofilms, but also bioaggregates in general from flocs to marine snow and granules. Granules can be considered as mobile biofilms. A continuum from flocs to granules and biofilms should be considered when addressing bioaggregation phenomena (Aqeel et al. 2019; Weissbrodt et al. 2014a).

In natural and engineered environments, microorganisms can adopt different lifestyles like a free-living planktonic state, bioaggregates (flocs, granules, marine snow), or surface-attached biofilm communities (Alldredge and Silver 1988; Bruckner and Mosch 2012; Costerton 1999a; Hall-Stoodley et al. 2004; O'Toole 2004; Simon et al. 2002). Bioaggregates and biofilms comprise the normal environment for most microbial cells in natural and artificial environments (Sutherland 2001). On the Earth surface, biofilms dominate in most habitats (except the oceans) accounting for around 80% of bacterial and archaeal cells (Flemming and Wuertz 2019). Planktonic microorganisms constitute less than 0.1% of the total microbial life (Costerton et al. 1995). Direct observation of natural and engineered systems has revealed that bacteria preferentially grow in sessile biofilms composed of multiple species enclosed in a matrix (Battin et al. 2003; Characklis 1973; Costerton 2007; Costerton et al. 1978; Geesey et al. 1977; Henrici 1933; Jenkinson and Lappin-Scott 2001). Surface association protects microorganisms from washout and enables them to subsist in a nutritionally advantageous environment (Dunne 2002; Wanner et al. 2006; Watnick and Kolter 2000).

Different definitions of biofilms were proposed, e.g., arrays of microcolonies entrapped in a matrix of EPS (Costerton 2004), communities of microorganisms attached to a surface (O'Toole et al. 2000), populations of microorganisms concentrated at a (usually) solid-liquid interface and surrounded by an EPS matrix (Hall-Stoodley et al. 2004), or layers of prokaryotic an eukaryotic cells anchored to a substratum surface and embedded in an organic matrix of biological origin (Bos et al. 1999; Wilderer and Characklis 1989).

Depending on research questions, biofilms can be considered as simple matrices or approached by accounting for their intrinsic complexities (Wanner et al. 2006). Within biofilms, microorganisms form different types of habitats. They differentiate their niches along the diffusion of chemicals across the biofilm depth. Depending on substrate gradients and growth rates, they form homogeneous biofilms or heterogenous structures (Picioreanu et al. 2000a). Inside biofilms, fast-growing microorganisms like OHOs form homogeneous matrices, while slower-growing organisms like PAOs, ANOs or AMOs form compact microcolonies (Alpkvist et al. 2006; Kreft et al. 2001; Picioreanu et al. 2016; Weissbrodt et al. 2013a). Such level of microbial differentiation in biofilms fosters BNR in single process tanks.

2.3.1 Interactions at Interfaces

Biofilms are formed when microorganisms attach to surfaces and grow, multiply, and produce extracellular components while being attached (Christensen 1989). Biofilms are initiated by the interaction of planktonic cells with a surface leading to a phenotypic change in the colonizing bacteria which switch from planktonic to sessile growth (Sauer and Camper 2001). The planktonic-biofilm transition is a complex and highly regulated process (O'Toole et al. 2000). Surfaces in the natural environment are the predominant sites of microbial activity: initial bacterial adhesion to surfaces is governed by physicochemical interactions between the cell surface and the surface of the colonized environment (van Loosdrecht et al. 1990). Bacterial surface characteristics are influenced by growth conditions. The cell surface tends to become more hydrophobic at high growth rates. This increases their tendency to adhere to a surface.

Adhesion can impact the microbial cell physiology (Marshall and Goodman 1994). Genes are switched on or off when planktonic bacteria adhere to surfaces. Genetic regulation phenomena related to biofilm formation were linked to the physicochemical conditions at the solid-liquid interface and in the bulk liquid. In most of cases, a solid phase only has an indirect effect on microbial activity by influencing the surrounding bacterial medium (van Loosdrecht et al. 1990).

The initial formation of multispecies biofilms in a sequence of six steps (Bos et al. 1999): (i) adsorption of conditioning film components (organic molecules), (ii) microbial transport and co-aggregation, (iii) (reversible) adhesion of single organisms and of microbial aggregates, (iv) co-adhesion between microbial pairs, (v) anchoring and establishment of irreversible adhesion through exopolymer production, and (vi) growth.

Microbial adhesion results from physicochemical interactions prior to biofilm formation (Dunne Jr 2002; van Loosdrecht et al. 1989). Different theories were proposed to describe them. The thermodynamic approach considers the interfacial free energies between interacting surfaces (i.e., free energy of adhesion). The Derjaguin–Landau–Verwey–Overbeek (DLVO) theory focuses on the interaction energies between interacting surfaces, based on the balance between Lifshitz–van der Waals and repulsive/attractive electrostatic forces, and their decay with separation distance (Derjaguin et al. 1987). The extended DLVO theory (XDLVO) includes short-range Lewis acid-base interactions, and accounts for hydrophobic-attractive and hydrophilic-repulsive forces (i.e., polar or acid-base interfacial free energy balance) (Poortinga et al. 2002). The acid-base interactions are based on electron-donating and electron-accepting interactions between polar moieties in aqueous solution.

Biofilms are assemblies of microorganisms that interact with surfaces to establish their niches for capturing nutrients from their environment and sustaining their growth. Biofilms are sophisticated forms of microbial life. Biofilm assemblies result from not only biological but also chemical and physical phenomena.

2.3.2 *Functional Sophistication of Biofilms: The Role of EPS*

Biofilms exhibit a high level of structural and functional sophistication (Costerton 2004). They are sketched as 'cities of microbes' or 'microbial metropolises', hence exhibiting complex, highly differentiated, and multi-species communities (Watnick and Kolter 2000; Wimpenny and Poole 2009). Biofilms are not only composed of microorganisms but also of extracellular polymeric substances (EPS). Biofilms are hydrogels predominantly composed of water (70–95% wet weight) and held by embedding highly hydrated EPS (70–95% dry weight) that are secreted in situ by the biofilm microorganisms (Flemming 1993; Flemming et al. 2016). The EPS components form gel networks by means of chemical bonding (Wang et al. 2006c). The EPS matrix represent the 'house of biofilm cells' in which bacteria can organize their life (Flemming 2011).

The inherent characteristics of EPS such as porosity, density, water content, charge, sorption properties, hydrophobicity and mechanical stability determine the immediate conditions of microbial life in this microenvironment (Flemming et al. 2007). In natural environments, bacterial cells synthesize and elaborate at their wall a surrounding exopolysaccharide glycocalyx (Costerton et al. 1981). This highly hydrated surface matrix of fibrillar polyanionic polymers enables cell adhesion and development of adherent biofilms. It also functions as water-entrapping agent, protective barrier, and ion exchange resin for capturing nutrients from the surrounding environment (Costerton 1999b).

EPS compositions are more complex than only polysaccharides (Seviour et al. 2019; Weissbrodt et al. 2013a). EPS matrices are composed of proteins, humic compounds, carbohydrates, uronic acids, aminosugars, glycoproteins, glycolipids, exoenzymes, fibrils, adhesins, cellulose, inorganic residues and extracellular DNA (Allison 2003; Basuvaraj et al. 2015; Decho and Gutierrez 2017; Flemming et al. 2007; Frolund et al. 1995, 1996; Okshevsky and Meyer 2015; Rossi and De Philippis 2015; Sutherland 2001). The EPS slime is a highly sophisticated, dynamic and active system which endows the biofilm mode of life, beyond providing the biofilm glue (Flemming et al. 2007). EPS are therefore important components of biofilms that are targeted both for biofilm formation and disruption, e.g., in the case of biofouling phenomena on water filtration membranes (Bucs et al. 2018; Derlon et al. 2014; Desmond et al. 2018; Jafari et al. 2020).

Confocal laser scanning microscopy (CLSM) allows for an enhanced examination of the internal heterogeneous architecture of biofilms, that consist of microbial cells and aggregates of high genetic diversity, biogenic extracellular material, void space, and interspersing channels (Lawrence et al. 1996; Okabe et al. 1997; Staudt et al. 2004; Weissbrodt et al. 2013a). Water channels constitute primitive circulatory systems across the biofilms (Costerton 2004; Stoodley et al. 1994). The internal biofilm organization can regulate the flux of dissolved oxygen (DO) and nutrients leading to the formation of in situ chemical microenvironments. Microbial niches establish along redox, pH, and chemical gradients (Lawrence et al. 1996). The presence of microdomains inside biofilms can be visualized by CLSM combined with fluorescent lectin-binding analyses (FLBA) for labelling glycoconjugates and several other stainings targeting specific components present at the outer layer of the embedded microcolonies and cells (Lawrence et al. 2007; Neu and Lawrence 1999; Weissbrodt et al. 2013a).

Since EPS play a key role for microbial adhesion (Nielsen et al. 1997) in flocs, biofilms and granules, studies aiming at elucidating the microorganisms that metabolize EPS under both the production and consumption aspects (Albertsen et al. 2011b; Dueholm et al. 2023; Guimarães et al. 2016; Tomás-Martínez et al. 2022a, c).

2.3.3 *Mass Transfer Limited Ecosystems*

Diffusion phenomena and mass transfer limitations are a main characteristic of biofilms and granules, that differentiate them from suspended cultures (Perez et al. 2005; van den Berg et al. 2020, 2021). Diffusion in biofilms can be summarized in four points (Stewart 2003): (i) diffusion is the predominant solute transport process within cell clusters; (ii) the time scale for diffusive equilibration of a non-reacting solute ranges from a fraction of a second to tens of minutes; (iii) diffusion limitation leads to concentration gradients of reacting solutes and hence to gradients in physiology; (iv) water channels can carry solutes into or out of the depths of a biofilm, but do not guarantee access to the interior of cell clusters.

Diffusion limitations in biofilms therefore lead to local variations in the concentrations of substrates, microbial species, and metabolic products. Concentration profiles in biofilms can be measured with microelectrodes for specific compounds (Amann and Kuhl 1998; de Beer and Schramm 1999; de Beer et al. 1997; Gonzalez et al. 2011b; Kuhl and Jorgensen 1992; Lee et al. 2011; Lens et al. 1995; Yu et al. 2004). These analyses can be coupled to CLSM and other microscopy and molecular methods for the temporal and spatial mapping of microbial communities and activities in biofilms (Gieseke et al. 2001; Horn 1994; Kofoed et al. 2012; Lawrence et al. 1994; Okabe et al. 1999; Santegoeds et al. 1999; Schramm et al. 1998). Mathematical models can efficiently describe chemical gradients in biofilms (Morgenroth 2020).

2.3.4 *Ecologically and Metabolically Diverse Habitats*

Biofilm communities are composed of a large spectrum of fast- and slow-growing organisms (Costerton 2004). Each biofilm cell exists in its own microenvironmental habitat and communicates with its neighbors by quorum sensing (Allison and Gilbert 1995; Davies et al. 1998; Singh et al. 2000; West et al. 2006). Quorum sensing can be suppressed by quorum quenching (Grandclément et al. 2015). Synergistic and antagonistic interactions of sessile microorganisms impact the development and shape of the biofilm community (Elias and Banin 2012; James et al. 1995; Kolter and Losick 1998). Compared to planktonic cultures, selection pressures are increased in biofilms because of diffusional and mass transfer limitations. Genetic diversification is rapidly obtained in biofilms (Boles et al. 2004). The confined biofilm environment stimulates horizontal gene transfer between microbial populations, that not only develops multi-antibiotic-resistant pathogens but also to metabolic evolution of microorganisms (Aminov 2011; Balcázar et al. 2015; De Oliveira et al. 2020; Flemming and Wuertz 2019; Ghigo 2001; Hall and Mah 2017; Hausner and Wuertz 1999; Madsen et al. 2012; Molin and Tolker-Nielsen 2003; Sorensen et al. 2005).

Biofilms are 'hotbeds' of functional diversity that can promote microbial adaptation to environmental stresses and safeguard the microbial community (i.e., ecological insurance principle) (Boles et al. 2004). For instance, a minimal increase in genetic diversity can enhance resistance towards predation (Koh et al. 2012). The stability of an ecosystem like biofilms relates to the ability of communities to contain species or functional groups capable of differential responses (McCann 2000).

Due to their highly organized architectures and high degree of cell differentiation, microbial colonies and biofilms are considered as multicellular organisms (Aguilar et al. 2007; Shapiro 1988). In a conceptual model of multicellular behavior, microbial consortia are coordinated biochemically and thermodynamically (Shapiro 1998). They are spatially organized with inter-species metabolic cooperation and cellular division of labor. In biofilms, microbial proximity enhances metabolic interactions.

Microbial activities in biofilms result in microscale chemical and physical heterogeneities of the interstitial fluid (Stewart and Franklin 2008). Overlapping chemical gradients of solutes, nutrient, electron donors, electron acceptors, metabolic waste products, and signaling compounds result in unique microenvironmental niches. This drives microbial niche differentiation and spatial distribution (Eigentler et al. 2022). These contribute to the structural, genetic, and physiological heterogeneity of biofilms. Biofilms are therefore complex communities which are resolved spatially and temporally. They respond dynamically in function of microscale chemical gradients. Microenvironments in a biofilm can range from conditions allowing metabolic activity and growth to starvation and decline. A model for predicting substrate utilization rates in biofilms has early been described (Williamson and McCarty 1976). Mathematical models have then been developed in 0–1–2–3 dimensions to describe biological, chemical and physical phenomena in biofilms and engineer solutions to manage them (Morgenroth 2020; Nagarajan et al. 2022; Noguera et al. 1999a; Rittmann and McCarty 1980; Vallina et al. 2019; Van Loosdrecht et al. 2002; Wanner and Reichert 1996).

Heterogeneous physiological states can already occur within a biofilm composed of a single microbial specie. This has interestingly been shown for a facultative anaerobic bacterium that grow aerobically in the presence of oxygen and ferment in the absence of oxygen (Stewart and Franklin 2008). In a mature biofilm, such population can harbor different physiological states. Cells located near the biofilm-bulk liquid interface can access substrates and dissolved oxygen that are available in non-limiting concentrations and develop aerobically. After the rapid depletion of oxygen in the outer biofilm layer, a second zone develops below the oxic region, where bacteria ferment the substrate under anaerobic conditions. After successive depletion of both oxygen and substrate, a third layer exists at the biofilm base where cells are starved from substrates, become inactive, and lyse. This can lead to the conceptualization of biofilms and granules as 'layered onion models'. Measurements of the spatial distributions of bacterial species CLSM coupled with fluorescence in situ hybridization (FISH) however highlighted that individual populations are not compartmentalized in strict strata, but rather intermingled or present in clonal pockets throughout the biofilm. Bacterial scattering in biofilms can result in a spectrum of different physiological states within single populations, besides microbial communities.

2.4 High-Rate Biofilm Process Engineering

Because of their high cell density and functional assemblies, biofilms are important components for process intensification in environmental engineering. Numerous environmental biotechnologies use biofilms for a high-rate treatment of water, wastewater, and waste-air (Bishop 2007; Boltz et al. 2017; Heijnen et al. 1991; Regmi et al. 2017; Rittmann 2007; Sørensen and Morgenroth 2020; van Loosdrecht and Heijnen 1993).

Biofilms foster the anaerobic treatment of wastewater highly loaded by organics (Fuentes et al. 2009b; Henze and Harremoes 1983), the removal of nutrients (Arvin and Harremoes 1990; Boelee et al. 2011; Ebrahimi et al. 2005; Falkentoft et al. 2000; Lacamp et al. 1993; Morgenroth and Wilderer 1999; Odegaard et al. 1994; Rittmann 2006a; Van Hulle et al. 2010), and the elimination of xenobiotics from wastewater (Kaballo et al. 1995; Nicolella et al. 2005; Ohandja and Stuckey 2006). They are further used in biologically active filters for removing contaminants from drinking water (Zhu et al. 2010) and to recycle water in recirculating aquaculture systems (Navada et al. 2020; Schreier et al. 2010), as well as for producing electricity and products in microbial fuel cells and bioelectrochemical systems (Cabau-Peinado et al. 2021; Gajda et al. 2015; Rabaey and Verstraete 2005). Phototrophic biofilms are used to produce microalgal feedstocks for manufacturing biofuels and chemicals from residuals (Christenson and Sims 2011; Medipally et al. 2015; Roeselers et al. 2008).

Biofilm processes sustain the design of compact WWTPs, the upgrade of existing WWTPs towards additional environmental quality standards, as well as the implementation of compact and adaptable processes in coastal and mountain tourist areas with strong variations in person equivalents across the year (Andreottola et al. 2004; Desbos et al. 1990; Iwai et al. 1990; Odegaard et al. 1994; Odegaard and Storhaug 1990; Orhon et al. 2002; Rogalla et al. 1992; Veuillet et al. 2014).

2.4.1 Biofilm Process Configurations

Biofilms were implemented in various process configurations for wastewater treatment, such as non-submerged systems (biotrickling filters, rotating biological contactors), submerged fixed-bed biofilm reactors, fluidized-bed biofilm reactors (three-phase, bubble-column, air-lift, moving bed bioreactor—MBBR), and membrane biofilm reactors (Christensson et al. 2013; Nerenberg 2016; Odegaard 2006; Sørensen and Morgenroth 2020; van Loosdrecht and Heijnen 1993; Wanner et al. 2006). The main differences between the biofilm reactor types are the available specific surface area depending on the support media used (from 50 to 200 $m^2\ m^{-3}$ for biotrickling filters to 2000–3000 $m^2\ m^{-3}$ for granular sludge reactors up to 3000–4000 $m^2\ m^{-3}$ for fluidized bed reactors), mixing conditions, gas transfer, and mechanisms for removing excess biomass.

Collectively, biofilm reactors meet the following principles. Retention of microorganisms is obtained by attachment to a substratum, or by self-microbial aggregation in the case of granules. Nutrients, redox compounds (e.g., oxygen), or alkalinity can be amended to the systems when needed. Effective mass transfer of dissolved compounds from the bulk liquid to the biofilm surface is determined by local mixing conditions and turbulence. Transfer of chemicals inside the biofilm is governed by diffusion. Biofilm growth is balanced with detachment to avoid clogging and retain an active biomass in stable biofilms.

The shape of biofilm structures obtained in a biofilm reactor is determined by substrate concentration gradients at the biofilm-liquid interface and by the detachment forces (van Loosdrecht et al. 1997b; Wimpenny and Colasanti 1997a, b). The microbial ecology, biomass yield, and production of EPS in biofilms are additional factors that can affect their shape. For microorganisms with higher biomass and EPS production yields, the faster production rate of new biofilm material can counterbalance the shear and detachment rate, and results in a more porous and heterogeneous biofilm. For microorganisms with lower yield, less new biofilm material is produced, detachment forces have higher impact of the biofilm structures, and more homogenous structures are obtained.

Interactions between scales of a biofilm system from microbial to biofilm and process boundaries are therefore shaping its overall performance (Morgenroth and Milferstedt 2009; Young et al. 2016).

2.4.2 Features of Biofilm Reactor Systems

A generalized approach of biofilm reactors was developed in comparison with conventional activated sludge systems (Morgenroth 2020; Sørensen and Morgenroth 2020). Important features are given hereafter.

Similar treatment objectives can be achieved in activated sludge and biofilm reactor by selecting for the same BNR microorganisms exposed to the same microenvironmental conditions (i.e., electron donor, electron acceptor, C source, N source, pH, temperature). Wash-out of suspended biomass is key for initiating the development of biofilms. Biofilm growth of microorganisms is obtained when the wash-out rate of suspended cells is larger than the maximum growth rate (μ_{max}) of this population (Beun et al. 2000; Heijnen et al. 1992; Legner et al. 2019; Weissbrodt et al. 2020a, 2023). The wash-out rate relates to the dilution rate (D = Q/V) in a chemostat or to the inverse of the sludge retention time (SRT) in a system that uncouples it from hydraulic retention time (HRT). In other terms, biofilms from when $D > \mu_{max}$. Microorganisms that can stick to surfaces remain in the washed-out system. In addition, bacteria tend to become more hydrophobic at high growth rate (during exponential phases in batch cultures and at dilution rates close to maximum growth rates in chemostats) and tend to aggregate (van Loosdrecht et al. 1987a). Therefore, microorganisms frequently stick to surfaces of continuous culture vessels when high dilution rates are applied.

Inside biofilms, the mass transfer of substrates and redox compounds occurs by molecular diffusion. The rate of molecular diffusion is usually slower than substrate removal kinetics, resulting in substrate gradients within biofilms. Conversion processes are consequently limited in biofilm systems by mass transfer, and different ecological niches develop along gradient directions. In addition to the availability of substrates in the bulk liquid phase, the location of bacterial clusters inside the biofilm impacts microbial interactions and competition. Bacterial populations have a more direct access to substrates when present in the outer layers and are better protected from detachment when present in the deeper layers.

Mass transfer limitations and microbial ecology gradients inside biofilms impact on the design and operation of biofilm reactors. The specific biofilm area, substrate flux from the liquid phase into the biofilm, surface and volumetric substrate loading rates, and hydraulic loading are important parameters for design. In contrast to activated sludge systems, the substrate removal performance of a biofilm reactor is not determined by the total amount of biomass but by the available biofilm surface area and by the substrate flux from the liquid phase in the biofilm.

2.5 Self-aggregation of Microorganisms

2.5.1 Microbial Aggregation and Flocculation

Microbial aggregation in well-settling flocs is essential for activated sludge systems. These processes uncouple the HRT (several hours) and SRT (several days) by retaining sludge using secondary clarifiers and recirculating a sludge fraction in the activated sludge tank to maintain a certain biomass concentration and sludge age.

Under unfavorable conditions, filamentous or viscous bulking phenomena deteriorate the settleability of flocs, leading to their wash-out with the effluent (Eikelboom 1975, 2000; Jenkins and Richard 1985; Martins et al. 2004a; Palm et al. 1980; Sam et al. 2022; van Loosdrecht et al. 2020b). Aggregation of activated sludge should be monitored and managed by process operation and control, in function of hydraulic and nutrient loadings and seasonal variations.

Bioaggregation levels relate to gradients of substrates between the bulk liquid phase and the biomass solid phase (Martins et al. 2004c; Pronk et al. 2015a; Tijhuis et al. 1994). Gradients can span in different directions from outside to inside and from inside to outside the flocs. Soluble compounds like nutrients supplied by the wastewater are present in high concentrations in the liquid phase and diffuse inside the flocs. The organic matter can be composed of more than 50% of colloidal or particulate compounds, especially in regions equipped with short and steep sewers like Switzerland (Huisman et al. 2003; Nielsen et al. 1992). These macromolecular materials adsorb onto flocs and microorganisms grow around by hydrolyze them with extracellular enzymes. The concentration of hydrolyzed products increases inside flocs, resulting into a gradient between the floc and liquid phase (Bossier and Verstraete 1996; Martins et al. 2011). Filamentous bacteria can deploy themselves from inside to outside the flocs to reach low residual concentrations of substrates and nutrients in the bulk. Selectors can be designed to suppress the filamentous growth (van Loosdrecht et al. 2020b).

Floc formation depends on internal and external forces generated by macromolecular interactions similar to those involved in cellular adhesion and biofilm formation (Wang et al. 2006c). The SRT impacts the microbial physiological age, EPS composition, and physicochemical properties (hydrophobicity and surface charge) of activated sludge (Liao et al. 2001; Liss et al. 2002). Surfaces of activated sludge flocs are less negatively charged and more hydrophobic at high SRTs (16–20 days) than at lower SRTs (4–9 days). Increasing SRTs are accompanied by an increase in the proteins-to-carbohydrates ratio of the EPS (from 1.3 to 5.0). The evolution of this ratio relates to changes in both the growth rate and the microbial community composition. The surface properties, hydrophobicity, surface charge, and EPS composition govern flocculation, rather than the quantity of EPS.

Activated sludge flocs and biofilms are different manifestations of the same phenomenon governed by diffusion limitations (Martins et al. 2004c). According to individual-based 3-D model simulations, flocs expand uniformly in all directions when substrate is not limiting (i.e., high substrate concentration and/or low specific growth rate). This leads to a smooth and compact aggregation like biofilm morphologies developed on spherical carrier material. Like biofilm formation, flocculation can help bacteria survive in oligotrophic environments (Liss et al. 1996). EPS can trap the nutrients from the surrounding, whereas cell lysis inside aggregates forms a source of nutrients. Stable bioaggregates are generally obtained when fully penetrated by the substrates and rely on the activity of microorganisms inside them (Pronk et al. 2015a; Wilen et al. 2000).

Some bacterial maintained under prolonged starvation change their cell surface roughness and hydrophobicity (Kjelleberg and Hermansson 1984). In addition to increased cell hydrophobicity induced by high growth rate conditions, starvation also triggers EPS production and biofilm formation. Alternating feast-famine regimes in SBRs or in plug-flow systems with intermittent feeding leads to better aggregation and settleability of flocs and granules (Beun et al. 2000; Chiesa et al. 1985; de Graaff et al. 2020).

Adverse environmental conditions can further stimulate aggregation. Fluctuations in substrate concentrations can trigger the expression of genes involved in flocculation and aggregation (Bossier and Verstraete 1996). The alginate biosynthetic pathway is switched on under environmental stress. Predation by protozoans ca lead to bacterial aggregation. Lysis products and incompletely digested cell products can favor aggregation. The local turbulent water flow generated by the movement of ciliates can also stimulate aggregation (Bossier and Verstraete 1996; Taherzadeh et al. 2010; Zima 2008).

2.5.2 Microorganisms and Adhesins in Floc Formation

From the microbiology side of flocculation, some microorganisms were described as model floc formers (Lau et al. 1984). *Zoogloea* spp. were early considered as exopolysaccharide producers and floc-forming organisms (Finstein 1967; Friedman and Dugan 1968; Lee et al. 1996; Murgel et al. 1991; Rossello-Mora et al. 1995; Unz and Farrah 1976; van Niekerk et al. 1987). Their overgrowth in activated sludge flocs however results in zoogloeal bulking by entrapment of water in the excessive EPS that they produce under nutrient imbalances, leading to a lower compaction of the sludge bed in the secondary clarifiers. *Acinetobacter*, *Comamonas*, and *Pseudomonas* isolates from activated sludge can promote aggregation (Andersson et al. 2008). *Acinetobacter* spp. were reported as hydrophobic, bridging bacteria promoting microbial co-aggregation and floc formation in activated sludge plants (Malik and Kakii 2003, 2008; Phuong et al. 2009, 2012). However, bioaggregation is not restricted to a few microorganisms, and is widespread across the microbial tree of life. Among wastewater microbiomes, one of the main limitations for addressing the aggregation potential of specific microorganisms relates to the fact that only a few can be obtained in pure cultures.

Within the EPS produced by microorganisms in activated sludge and natural biofilms, amyloid adhesins pare abundant and lay a significant role for maintaining the internal adhesion of microcolonies inside flocs and for increasing the microbial surface hydrophobicity (Christiaens et al. 2022; Danielsen et al. 2017; Dueholm et al. 2010; Larsen et al. 2007, 2008). The expression of amyloid adhesins was detected from a diversity of bacterial populations inside classes of *Alphaproteobacteria*, *Betaproteobacteria*, and *Actinobacteria*. Denitrifiers like *Thauera*, *Zoogloea*, *Azoarcus*, *Aquaspirillum*, and *Pseudomonas* spp. as well as *Actinobacteria*-related PAOs show high potential to produce amyloids, whereas nitrifiers much less. Filamentous bacteria inside *Alpha*-, *Beta*- and *Gammaproteobacteria*, *Bacteroidetes*, *Chloroflexi* and *Actinobacteria* are other important producers of amyloid adhesins.

Microbial aggregation relates to cell surface charge and hydrophobicity, to EPS and amyloid adhesins, and to microbial populations that produce them.

2.5.3 Granular Methanogenic Sludge

The immobilization of bacteria is the key for modern, high-rate environmental biotechnologies (van Lier et al. 2020). Immobilization can be achieved not only on supports, either fixed (e.g., biotrickling filters) or free-floating (e.g., fluidized beds and MBBRs), but also without any carrier materials in densely packed granular sludge. Granular sludge particles are special cases of biofilms, that form without a substratum but that display biofilm characteristics (Hall-Stoodley et al. 2004).

The auto-immobilization of microorganisms in anaerobic digestion was a pioneering step in granular sludge technologies. Microbial self-aggregation was observed during the start-up of anaerobic up-flow filters (Lettinga et al. 1976; Young and McCarty 1969). It was actively researched and implemented to develop intensive processes for anaerobic wastewater treatment processes like the upflow anaerobic sludge blanket (UASB) and the expanded granular sludge blanket (EGSB) technologies (Lettinga and Hulshoff Pol 1991; Lettinga et al. 1979; Seghezzo et al. 1998; Tay and Yan 1996). In UASB and EGSB systems, volumetric organic loading rates (OLR_v) as high as 10–40 kg_{CODs} d^{-1} m^{-3} can be efficiently converted to methane gas.

Different physical, microbial and thermodynamic theories were developed to explain anaerobic granulation phenomena (Hulshoff Pol et al. 2004). Granulation is a natural process that proceeds in all systems where basic conditions for its occurrence are met, namely: soluble substrates, up-flow reactor operation, and HRT lower than the bacterial doubling time (van Lier et al. 2020).

Suspended solids in the seed sludge, like particulate (in)organic matter and bioaggregates, serve as colonization material. The application of dilution rates above the maximum growth rate and the gradual increase of the superficial liquid and gas velocities result in the wash-out of light and finely dispersed sludge, eventually leading to the formation of biofilms and granules that get retained in the process. Progressively, the SRT decouples from the HRT. Granulation is initiated by the adhesion of microorganisms to inert particles. The nuclei develop by internal and outer microbial growth into granules of a maximum size that depends on internal binding forces and external shear forces on particles. Detachment leads to the formation of new generations of nuclei from which novel granules are formed. Successions of particle growth and detachment lead to the maturation of the first generation of filamentous and voluminous aggregates into denser rod-type anaerobic granules as big as 5 mm, that can reside long in the process. Long retention of a concentrated and active granular biomass in the digester enables a high-rate anaerobic treatment (Skiadas et al. 2003).

Conditions favorable for dense anaerobic granulation stimulates the production of EPS and the preferential selection of microbial populations like the aceticlastic methanogens *Methanosaeta* (formerly referred to as *Methanothrix soehngenii* (Patel and Sprott 1990)) and *Methanosarcina* spp. (Dolfing et al. 1985; Grotenhuis et al. 1991a, 1992; Hofman-Bang et al. 2003; Hulshoff Pol et al. 2004). *Methanosaeta* is an important microbial player in anaerobic granule formation, forming filaments but also attaching to surfaces. UASB process conditions select for highly hydrophobic microbial populations, like *Methanosaeta*, that stabilize anaerobic granules. This was determined using the contact angle method developed to describe microbial adhesion (Grotenhuis et al. 1992; van Loosdrecht et al. 1987b). The initial filamentous bundles produced by this archaeal organism progressively transition towards dense rod-type granules under the increasing shear forces, microbial densification, EPS formation, and the colonization by acidogens and other methanogens. Under unfavorable shear conditions, *Methanosaeta* outgrows as filaments, leading to filamentous bulking and slow-settling granules (Hulshoff Pol et al. 2004). Managing the shear forces in such systems is important.

After maturation, the internal architecture of anaerobic granules is composed of different layers of trophic niches along internal gradients of substrates and products (Arcand et al. 1994; Hulshoff Pol et al. 2004; MacLeod et al. 1990; Pauss et al. 1990; Schmidt and Ahring 1996). Microbial compositions and spatial organization of methanogenic granules depend on substrates and their degradation kinetics (Fang 2000; Fukuzaki et al. 1995; Grotenhuis et al. 1991b; Lens et al. 1995).

2.6 Granular Sludge for a High-Rate Nutrient Removal

The success of anaerobic digestion using granular sludge opened avenues for other granular sludge processes under anoxic and aerobic conditions. Self-granulation is not restricted to methanogens. Anammox technologies using granular sludge made breakthrough for a fully chemolithautotrophic removal of nitrogen at low resource and energy expenditures from concentrated side streams and diluted main streams (De Clippeleir et al. 2013; Lotti et al. 2015; van der Star et al. 2007; Vlaeminck et al. 2009; Wett et al. 2015). AGS was developed as a possible new standard for full BNR from wastewater (Morgenroth et al. 1997; Pronk et al. 2020).

Process conditions favor microbial aggregation. An appropriate window of selection pressures applied by process design and operation leads to the formation of fast-settling AGS assembling microorganisms necessary for BNR.

Over the last 25 years, research went from the understanding of aerobic granulation phenomena to the design of conditions to select for BNR microorganisms in the process via the role of EPS in granulation, the effect of granule size on oxygen transfer and nitrogen removal, the microbial ecology of granules, the start-up of AGS reactors, the modelling of AGS systems, and the applications of AGS technologies and reactor configurations in engineering practice. The scale up of the technology at pilot and full scales went in parallel by involving a strong partnership between research institutions, the engineering sector, and water authorities. A scientific and engineering community assembled around granular sludge.

2.6.1 *Initial Observations and Investigations of Aerobic Granular Sludge*

Early observations and trials of aerobic granulation were made in the 1990s.

Auto-immobilization of activated sludge into fast-settling granules was observed in an aerobic up-flow sludge blanket (AUSB) pilot-plant fed with sewage wastewater (1.57 $kg_{BOD5}\ d^{-1}\ m_r^{-3}$) (Mishima and Nakamura 1991).

Aerobic granulation was further studied and controlled in a AUSB reactor fed with a synthetic wastewater composed of glucose, acetate and yeast extract in a step-wise increase of the OLR_v from 2.6 up to 7.2 $kg_{CODs}\ d^{-1}\ m^{-3}$ and by suppressing the overgrowth of filamentous organisms like *Sphaerotilus* or *Beggiatoa* spp. by applying high up-flow mixing velocities (Shin et al. 1992).

Aerobic granulation of digested sludge was achieved in a AUSB reactor equipped with an internal airlift and treating the effluent wastewater of a UASB digester by progressively increasing the organic (0.5–10.6 $kg_{COD}\ d^{-1}\ m^{-3}$) and nitrogen (0.04–0.70 $kg_N\ d^{-1}\ m^{-3}$) loads and by decreasing the HRT from 80 to 8.4 days over 2–4 months (Inamori et al. 1994). The effect of the internal recirculation of wastewater from the AUSB back to the UASB was tested on the granulation and nitrogen removal performances. Granule nuclei of 100 μm appeared after 1.5–5 weeks and evolved towards 2 mm granules after 6–10 weeks. Scanning electron microscopy (SEM) displayed different granulation mechanisms: (i) progressive growth of granular nuclei towards smooth granules; (ii) aggregation of granular nuclei in heterogeneous granular assemblies. Without internal recirculation of the wastewater, 97% of carbon and 20% of nitrogen were removed. The internal recirculation led to a nitrogen removal of 71% by denitrification in the UASB reactor, but in a decrease of 10% in the methane fraction in the biogas produced.

Lab-scale cultivations of AGS went forward with bubble-column SBR (Beun et al. 1999; Morgenroth et al. 1997). To stimulate granulation, operation conditions were transposed from the cultivation of granular methanogenic sludge (Dolfing 1986; Lettinga et al. 1980). Morgenroth et al. (1997) used a 31.4-L column with a height-to-diameter ratio (H/D) of 5 (height of 1 m, internal diameter of 20 cm). The reactor was pulse-fed with a synthetic wastewater composed of molasse carbohydrates (400 $mg_{COD}\ L^{-1}$, 1.2 $kg_{COD}\ m^{-3}\ d^{-1}$), and run aerobically under non-limiting supply of phosphorus and nitrogen (111 $mg_{COD}\ mg_P^{-1}$ and 18 $mg_{COD}\ mg_N^{-1}$). Calcium and magnesium divalent cations (Ca^{2+}, Mg^{2+}) were added to form precipitates acting as carrier materials for colonization, biofilm formation, and granulation. The beneficial effect of calcium precipitate was shown for aggregation (van der Hoek and Klapwijk 1987; Vogelaar et al. 2002). A fast-settling biomass was selected by short sedimentation (1 min) and effluent withdrawal (6 min) times. Aerobic granules with diameters of up to 7 mm (2.35 mm on average) exhibiting full COD removal were successfully cultivated over 4 months, after which the granule stability and reactor performances deteriorated by the proliferation of the filamentous yeast-like fungi *Geotrichum*.

In parallel, studies on microbial aggregation, biofilm formation, and filamentous bulking remedial helped better understand the triggers for the densification of activated sludge toward the development of aerobic granular sludge. The success of AGS arose from the examination of microbial aggregation phenomena from flocs to biofilms and granules (Beun et al. 2000; Martins et al. 2004c; Tijhuis et al. 1992). Studies on overcoming filamentous bulking have actually been important in AGS developments (Caluwé et al. 2022; Henriet et al. 2017; Martins et al. 2004b; Weissbrodt et al. 2012a).

2.6.2 *Aerobic Granulation Mechanisms*

The first AGS cultivations focused on getting dense fast-settling granules under aerobic conditions for COD removal. Aerobic granulation is driven by selection pressures like substrate gradients (feast-famine regimes) and wash-out conditions resulting from short HRTs and settling phases. Granulation is gradual process where flocs from the seed sludge progress toward compact aggregates, early-stage granules, and mature granules (Liu and Tay 2004; Liu et al. 2005f; Weissbrodt et al. 2013a). Impacts of process parameters on granulation were extensively studied and reviewed (Adav et al. 2008b; Lee et al. 2010; Liu et al. 2005e, 2009a; Liu and Tay 2004; Show et al. 2012; Winkler et al. 2018).

Wash-out conditions in bubble-column SBRs with short settling times (1–5 min), short withdrawal times (1–5 min), high volumetric exchange ratios (30–70%, typically 50%), and short HRTs below $1/\mu_{max,OHO}$ (typically 6 h) select for a fast-settling biomass and rapid granulation (Beun et al. 1999; Etterer and Wilderer 2001; Morgenroth et al. 1997; Qin et al. 2004; Tay et al. 2001b; Zhu et al. 2001). Slow-settling flocs are washed-out and fast-settling nuclei are retained.

CLSM highlighted different mechanisms of granule formation by (i) microcolony outgrowth from granule nuclei, (ii) aggregation of microcolonies in granular conglomerates, but also (iii) the differential developments of fast-growing and slow-growing organisms into smooth matrices and compact microcolonies, respectively, inside heterogeneous granules (Barr et al. 2010a; Weissbrodt et al. 2013a). Granulation mechanisms relate to both substrate gradients and microbial physiologies (de Graaff et al. 2020; Pronk et al. 2015a; Weissbrodt et al. 2013a).

2.6.3 *Physical Factors of Granulation*

Detailed analyses of process parameters impacting aerobic granulation were performed in a bubble column (1999). The application of feast-famine regimes, short HRTs, short settling times, and high shear stresses stimulated granulation. Feeding the influent as a pulse (fill-to-cycle time ratio of 0.01) was known to enhance the sludge settling (Chiesa et al. 1985). An HRT of 6.75 h was beneficial for granulation,

8 h were not sufficiently low to suppress the growth of the suspended biomass. The studied OLR_v (2.5–7.5 kg_{CODs} d^{-1} m^{-3}) did not directly impact the rate of granulation but mainly influenced the granule shape. Under full aeration, the higher OLR_v selected for fluffy aggregates overgrown by filamentous organisms. A settling time below 3 min induced the wash-out of the slow-settling flocs. Adapting the settling time between 1 and 3 min in function of the OLR_v helped control the accumulation of granules in the reactor. A higher amount of granular sludge was produced with higher OLR_v, resulting in hindered settling in the lower compartment of the bubble column. Higher settling times were required for a clear separation of the granular sludge below the effluent withdrawal point located at the half of the reactor height. An up-flow superficial gas velocity (SGV) of 0.041 m s^{-1} during aeration resulted in sufficient hydrodynamic shear stress for maintaining stable smooth granules even under high OLR_v. This enabled the detachment of filamentous structures from the granule surfaces. Lower SGVs of 0.014 and 0.020 m s^{-1} did not lead to stable granules. Higher OLR_v was therefore balanced by higher shear to form compact granules of 3.3 mm diameter on average.

Dense granules were formed in several other studies with SGVs higher than 0.010 m s^{-1}, typically between 0.025 and 0.045 m s^{-1}, whereas granulation was not successful below 0.008 m s^{-1} (Beun et al. 1999; Ebrahimi et al. 2010; Koh et al. 2009; Morgenroth et al. 1997; Tay et al. 2001a, 2003, 2004).

Like biofilms, cell hydrophobicity is an important factor for promoting granulation (Liu et al. 2003b, 2004a, b, c). A positive correlation was established between the imposed shear stress and the surface hydrophobicity of cells (Dulekgurgen et al. 2008; Liu et al. 2005b; Liu and Tay 2001a, b, 2002; Wang and Zheng 2008; Wang et al. 2008b; Zheng et al. 2009). A decrease of the surface negative charge from 0.20 to 0.02 meq g_{VSS}^{-1} and an increase of the relative hydrophobicity from 29 to 60% was measured during granulation (Li et al. 2006b). A high content of adenosine triphosphate (ATP) was measured in the biomass during granulation, confirming that non-limited cell growth and activity correlates with enhanced hydrophobic, adhesive, and agglomerative properties (Xiong and Liu 2010).

Feast-famine regimes promote granulation by generating substrate gradients and stimulating EPS production, cell surface hydrophobicity and adhesion. Feast-famine regimes were applied either by (i) 5 min pulse feeding prior to 3 h prolonged aeration (known as 'pulse feeding' regime), (ii) 1 h anaerobic feed followed by 2 h aeration (known as 'anaerobic feeding' regime), or (iii) intermittent feeding (Beun et al. 1999; de Kreuk and van Loosdrecht 2004; Liu and Tay 2008; Liu et al. 2007b; McSwain et al. 2004). A high content of adenosine triphosphate (ATP) was measured in the biomass during granulation, confirming that non-limited cell growth and activity correlates with enhanced hydrophobic, adhesive, and agglomerative properties (Xiong and Liu 2010).

Denitrification processes were established in granules by different strategies like (i) simultaneous nitrification and denitrification (SND) during aeration after anaerobic feeding (de Kreuk et al. 2005), (ii) alternating nitrification and denitrification (AND) by on/off aeration after anaerobic feeding (Lochmatter et al. 2013), or (iii) by alternating anoxic feast and aerobic starvation phases (Wan et al. 2009). The integration of denitrification in granules was favorable for their densification with a SGV as low as 0.006 m s^{-1} and settling time as high as 30 min (Wan et al. 2011). A new definition of granules was proposed as microbial aggregates that exhibit a cohesion higher than flocs and that is sufficiently high to resist turbulence and shear stress fluctuations, which enables them to grow to a size bigger than the turbulence micro-scale (Wan et al. 2011). A high SGV and short settling time is not requested for granulation when organisms producing compact biofilms or microcolonies are selected. Benefits of selecting for slow-growing organisms like PAOs was further demonstrated on granule stability (de Kreuk and van Loosdrecht 2004; Guimarães et al. 2018; Weissbrodt et al. 2013a). This is also valid for GAOs, nitrifiers, and anammox organisms among others.

2.6.4 Physical Characteristics of Granules for BNR

The physical characteristics of granules were reported from various studies.

Early-stage, smooth aerobic granules (< 50 days) displayed a size distribution between 1.1 and 6.5 mm (3.0 mm on average) in a bubble-column SBR fed with acetate and operated under wash-out conditions (Etterer and Wilderer 2001). The wet density of early-stage granules ranged between 1.038 and 1.050 g mL^{-1}. These values were close to bacterial cells of 1.040–1.100 g mL^{-1} (Woldringh et al. 1981) and activated sludge flocs of 1.010–1.056 g mL^{-1} (Dammel and Schroeder 1991; Schuler and Jang 2007). The granule porosity was 65–72%.

Granules settling velocities up to 72 m h^{-1} were measured, with an average at 38 m h^{-1}. From about 40 literature reports, the following ranges of settling parameters were published from early-stage to mature granules: mean granule diameter of 2.2 ± 1.7 mm (min–max: 0.2–10 mm), mean specific particle gravity of 1.042 ± 0.020 (1.017–1.082), mean settling velocity of 52 ± 25 m h^{-1} (8–116 m h^{-1}), mean SVI of 45 ± 26 mL g^{-1} (11–130 mL g^{-1}).

The effect of the temperature and salinity of wastewater were examined on the settling ability of mature aerobic granules (Winkler et al. 2012b). Slower settling velocities were measured with decreasing temperatures from 40 °C (60–140 m h^{-1}) to 5 °C (35–80 m h^{-1}) and with increasing salt concentrations from 5 g L^{-1} (50–110 m h^{-1}) to 40 g L^{-1} of NaCl (30–80 m h^{-1}), depending on the granule diameter.

Variations in wastewater density and viscosity, resulting, e.g., from the infiltration of saline water in sewers of coastal areas, from the discharge of industrial effluents, or from road de-icing in winter, should be considered for an optimal operation of AGS-SBRs (Pronk et al. 2014). Granulation was successfully obtained in seawater-adapted granular sludge (de Graaff et al. 2020).

2.6.5 Multiphase Flow Dynamics in AGS Sequencing Batch Reactors

The effect of multiphase flow dynamics (water, air, granules) on granulation was studied in details in a bubble-column AGS-SBR with a H/D of 10 (Zima 2008). In situ optical techniques by particle image and tracking velocimetry helped measure the flow dynamics (Zima et al. 2005, 2007).

The formation and structure of aerobic granules is strongly influenced by the (i) the magnitude of the mechanical forces induced by the three-phase flow dynamics, (ii) collisional forces, and (iii) microfauna at the surface of granules. Granulation only occurs under appropriate flow conditions. The shear stress, elongation stress, and stresses induced by fluctuating velocities influence granule formation.

Up-flow aeration is the main source of mechanical energy in bubble columns (Heijnen and Van't Riet 1984; Zima et al. 2007). The coalescent coarse bubble regime induces significant frictional forces on the fluidized biofilm particles. The velocity distribution in the multiphase flow shapes the granules. The relative velocity between the fluid and granules exerts a superficial stress and deforms of granules. A characteristic flow pattern occurs during aeration. A large vortex arises in the lower part of the bubble column and smaller eddies in the upper part.

The concentration of AGS impacts the velocity pattern. The presence of a less concentrated granular sludge at the top of the reactor leads to a higher fluid velocity in the upper part. The fluid velocity increases with increasing vertical coordinate, wall distance and up-flow aeration rate. The velocity of granules decreases with height. The amount and size of granules was also shown to affect the mass transfer of oxygen in the system during aeration (Weissbrodt et al. 2014d).

The hydrodynamic mechanical shear stress in bubble-column reactors is composed of a shear stress due to relative motion between particle and fluid, a normal stress due to pressure and velocity gradients, and a turbulent stress due to rapid velocity fluctuations (Zima-Kulisiewicz et al. 2008). The shear stress acting on granules is composed of normal (elongation) and tangential (shear) strains. In this reactor, an up-flow aeration rate of 4 L min^{-1} (SGV of 0.0105 m s^{-1}) provided the best conditions for the formation of spherical and compact granules. Higher up-flow aeration rates induced mechanical fatigue effects on aerobic granules.

Optimal granulation was obtained with flow elongation rates between 5 and 26 s^{-1} and with shear rates between 5 and 28 s^{-1}. A dimensionless elongation rate of 1 was measured in this bubble-column AGS-SBR, whereas conventional activated sludge tanks are characterized by a value of 0.2. Such high elongation rate prevented the growth of fluffy flocs, by breaking them down at an early stage.

The maximum shear stress measured in this AGS-SBR amounted to 0.015 Pa. It is far beyond the critical tangential stress of 10 Pa that can destroy granules. However, much higher shear stress can exist at the granule surface than observed in the bulk liquid phase, and thus the shear stress can impact on the surface structure of granules. The elongation and shear strains are also higher in the upper part of the column and decrease close to the reactor wall.

Collisional forces significantly contribute to the mechanical forces in the bubble column. Highest particle-particle (42.5×10^6 collisions s^{-1} m^{-3}) and particle-wall collision rates (6.0×10^3 collisions s^{-1} m^{-2}) occur in the bottom part of the reactor.

For an optimal granulation, homogeneous aeration should avoid dead zones in the flow pattern. The fraction of solids should amount to 10–20% of the overall liquid column. A sufficient height should allow an unconstrained settling of AGS.

Like for biofilm streamers (Taherzadeh et al. 2010), micro-flows induced by the beats of ciliates at the granule surface can significantly contribute to granulation and compaction through characteristic vortices.

Flow dynamics should be considered at different scales for managing their impact on granulation. The operation of full-scale AGS processes further results from the design of hydraulic flow regimes during anaerobic feeding and aeration (van Dijk et al. 2020, 2022).

2.6.6 *Importance of Extracellular Polymeric Substances in Granulation*

Different studies investigated the production and composition of EPS in AGS.

EPS levels were compared in two bubble-column SBRs operated with activated sludge and granular sludge under identical conditions (except different settling times of 10 and 10 min, respectively) (McSwain et al. 2005). Proteins were 50% more abundant than polysaccharides in the EPS of granules than flocs. According to CLSM with in situ staining of protein and glycoconjugates with a combination of fluorescent labeled probes and lectins, cells and exopolysaccharides were localized in the 200 μm outer layer of granules. The granule core mainly comprised proteins and cell lysis products.

Aerobic granules comprise an hydrophobic outer shell with high biomass density inside a matrix of poorly soluble and non-easily biodegradable exopolysaccharides, and an inner core with relatively low biomass density and filled with readily soluble and biodegradable exopolysaccharides (Wang et al. 2005b). The insoluble and non-biodegradable β-linked exopolysaccharides, such as cellulose or chitin, may sustain granular architecture. EPS contained in aerobic granules were composed of 40% biodegradable and 60% non-biodegradable fractions (Zhang et al. 2008). Dynamics of EPS production and consumption were analyzed during aerobic granulation, and over SBR cycles operated with intermittent feeding and aerobic phases (Wang et al. 2006c). EPS in granules were mainly produced during exponential phases. Around 50% of EPS were consumed during starvation. The consumed EPS were considered as carbon and energy source to sustain growth inside granules. The ratio of proteins-to-polysaccharides in EPS, the surface charge and hydrophobicity evolved dynamically with the periodic cycles of EPS production and consumption. The protein fraction of EPS mainly evolved during the formation of granules, becoming an important factor promoting granule cohesion (Li et al. 2008b; Zhang et al. 2007a, b). The distribution of EPS components (proteins, lipids, α- and β-D-glucopyranose polysaccharides) were mapped inside aerobic granules by CLSM with multiple fluorochrome stainings (Chen et al. 2007a, b). The core of acetate-fed granules consisted of proteins and β-polysaccharides, whereas cells and α-polysaccharides were present in the outer layers. The contribution of the individual EPS components on the granule structural cohesion was examined (Adav et al. 2008c). Selective enzymatic treatment of these macromolecules with proteinase K, lipase, α- and β-amylase revealed that the hydrolysis of β-polysaccharides leads to the disintegration of granules. The β-polysaccharides were considered to form the backbone that supported the structural integrity of granules by embedding other EPS components and cells.

Granules contain exopolymers that behave like alginates (Lin et al. 2008, 2010). These alginate-like exopolymers were more abundant in granules (310 $mg_{C6H7O6^-}\ g_{VSS}^{-1}$) than flocs (140 $mg_{C6H7O6^-}\ g_{VSS}^{-1}$). Sodium alignate extracted from granules comprised a higher ratio (1.18) of α-L-guluronate (G) to β-D-mannuronate acid residues (M) than sodium alginate extracted from the flocculent sludge (0.94). These exopolymers extracted form granules sampled from a pilot plant treating municipal wastewater were composed by 69 ± 9% of GG-blocks, 15 ± 2% of heteropolymeric MG-blocks, and 2 ± 1% of MM-blocks. The gelling property of these exopolymers was tested in a solution of $CaCl_2$. Compact spherical gels were with the exopolymers extracted from granules, whereas a suspension of tiny flocs was obtained with those extracted from flocs. The spherical tridimensional self-assembly of alginate molecules relies on interaction of G-blocks with divalent cations such as Ca^{2+}, leading to macromolecular cross-linking. Similar to commercial alginate used for cell encapsulations (Melvik and Dornish 2004), bacterial alginate rich in G-blocks were proposed as structural gel-forming polymer of granules that can provide inhabitancy and protection to microorganisms.

Properties and compositions of EPS in aerobic granules were further examined by selective enzymatic hydrolysis (Seviour et al. 2009b). Granules were mainly composed of proteins and α-polysaccharides. Hydrogel characteristics with reversible swelling phenomena were shown for the exopolymer. The EPS matrix can be weakened by environmental factors. Temperature > 50 °C, pH value > 10, and ionic strength ~ 1 mol L^{-1} NaCl equivalents were detrimental to macromolecular cohesion. Granule EPS exist as a gel at the pH of 6.0–8.5 of WWTP operations, whereas floc EPS do not (Seviour et al. 2009a). At pH between 5 and 9, granule EPS form a strong gel. A weaker gel was observed at pH 4–5, and sol–gel transition occurred at pH 9–12.

The strength of the gel of granule EPS was independent of the protein content. Polysaccharides and glycosides, such as proteoglycans, were important for EPS gelation. The granule EPS comprised three components, namely high-molecular-weight polysaccharide, medium-molecular-weight proteins and glycosides, and low-molecular-weight proteins and glycosides (Seviour et al. 2010b). Only the high-molecular-weight polysaccharide displayed gel-forming behavior of aerobic granules and were considered as a gelling agent for granulation. The structure of this high-molecular-weight polysaccharide called 'granulan' was elucidated by nuclear magnetic resonance (NMR) as a highly complex single heteropolysaccharide with a repeated sequence of α-galactose, β-mannose, β-glucosamine, N-acetyl-β-galactosamine, and 2-acetoamido-2-deoxy-α-galactopyranuronic acid (Seviour et al. 2010a). The synthesis of 'granulan' was however only induced under selection for "*Ca.* Competibacter", and has not been a key structural component for all types of AGS (Seviour et al. 2011).

Florescence lectin-binding analysis (FLBA) and CLSM highlighted a composition of granule EPS more complex than a tale of two exopolysaccharides (alginate, granulan). Glyconjugates screened with lectins displayed abundant aminosugars (like N-acetylglucosamine, N-acetylgalactosamine, and N-acetylneuraminic acids also known as sialic acids) over cross sections of BNR granules, besides sugar residues like α-glucose and α-mannose (Weissbrodt et al. 2013a). Different matrices of glycoconjugates were detected in the biofilm continuum and at the surfaces of microbial clusters, microcolonies, and cells. The abundance of these glycoconjugates varied with the predominant microorganisms enriched in the granules. Aminosugars like N-acetylglucosamine and sialic acids were abundant when "*Ca.* Accumulibacter" was dominant, while almost not detected when *Zoogloea* initially prevailed (Weissbrodt et al. 2014c). Sialic acids were further identified as interesting compounds (de Graaff et al. 2019). FLBA techniques were also used to address glycoconjugates in anaerobic granules exposed to high salinities in relation to microbial community compositions (Gagliano et al. 2018, 2020).

The characterization of EPS in environmental samples require strong analytical developments (Feng et al. 2019; Seviour et al. 2019). Investigations of EPS in granular sludge were further advanced over the last 5 years. Techniques for the extraction of exopolymers and analysis of their monomers were refined (Felz et al. 2016, 2019). Currently, structural are extracted by thermal (80 °C) and alkaline (pH = 10 with Na_2CO_3 at 9.5% w/v) conditions under continuous stirring. This enables a complete dissolution of the granule matrix. However, cell lysis can still occur, liberating intracellular compounds as well. The overall compositions of protein and sugars can be measured by modified colorimetric methods using mixtures of standards instead of single compounds. The monomeric compositions of sugars and aminosugars can be 'sequenced' by high-performance anion-exchange chromatography with pulsed amperometric detection (HPAE-PAD), similar to techniques used for characterizing the oligosaccharides in seaweed alginates (Ballance et al. 2005). These techniques were used to track the EPS compositions of aerobic and anammox granules under different conditions (Boleij et al. 2018, 2019, 2020; de Graaff et al. 2019; Pronk et al. 2017b; Schambeck et al. 2020), and to correlate the EPS mass fractions with the enrichment levels of organisms like "*Ca.* Accumulibacter" and their dynamics (Guimarães et al. 2016).

The turnover of EPS can now be examined with higher resolution by combining these techniques with other methods involving ^{13}C substrate labelling and mass spectrometry (Tomás-Martínez et al. 2022c). Mass spectrometry cannot only uncover the diversity and complexity of EPS, but also target specific compounds like nonulosonic acids (Kleikamp et al. 2020a) for better understanding their biosynthesis and biodegradation pathways (Tomás-Martínez et al. 2022a). Sialic acids are important components of the cell surface involved as recognition molecules, in pathogenicity and biofilm formation. This can further support the development of biotechnological applications from substances like sialic acids recovered from biofilms and granules (Tomás-Martínez et al. 2022b).

Advances on EPS of granules led to engineering developments for the recovery of exopolymers from AGS as biomaterials sustaining the civil, coating and textile industries (Lin et al. 2015). These exopolymers are commercialized under the name Kaumera Nereda® Gum (van der Roest et al. 2015). A first process for its production was recently installed at full scale (Dutch Water Sector 2019).

2.6.7 Slow-Settling Filamentous Bulking Granules, and Remedial Actions

Granular sludge represents the opposite morphology of bulking sludge (Pronk et al. 2020; van Loosdrecht et al. 2020b). Granules should form when conditions selecting for bulking sludge are minimized (Caluwé et al. 2022; de Graaff et al. 2020; Henriet et al. 2017; Martins et al. 2004c; Weissbrodt et al. 2013a). Nevertheless, different studies reported the deterioration of the settling properties of aerobic granules by filamentous bulking (Ebrahimi et al. 2010; Liu and Liu 2006; Weissbrodt et al. 2012a), in addition to the early reports (1997; Shin et al. 1992).

In flocculent activated sludge and granular sludge processes, filamentous bulking should be prevented to avoid a substantial wash-out of the biomass from the system. Filamentous outgrowth in AGS reactors can be triggered by different factors, such as long SRT, low substrate concentration in the bulk liquid phase, substrate gradients in the granular biofilms, DO and nutrient deficiency inside granules, as well as temperature shifts (Liu and Liu 2006). Under cyclic operation in SBRs, the various stresses can occur simultaneously and can become repetitive. This can lead to the progressive development of filamentous populations.

Compared to activated sludge flocs, granules have a compact biofilm structure with diameters varying between 0.2 and 5 mm. Diffusion limitations and gradients of substrates, nutrients and DO are therefore more pronounced over the depth of the granular biofilm (Liu and Liu 2006; Liu et al. 2005g). The magnitude of concentration gradients relies on the ratio between the rates of reaction and of mass transfer (Martins et al. 2004c). Mass transfer of soluble components depends on the diffusional processes, the concentration in the bulk liquid phase, and the size of the aggregates. As a result of diffusion, reaction and growth processes, a lower concentration of soluble chemical compounds is observed in the vicinity of flocs than in the bulk liquid phase. Mass transfer-limited regimes can lead to the proliferation of filamentous bacteria in the direction of the higher concentrations of the dissolved components present in the bulk liquid phase (Martins et al. 2003, 2004a), and to the formation of fluffy granules (Liu and Liu 2006). Under low substrate concentrations, outer filamentous organisms observe a higher concentration than floc-forming organisms present inside the floc. Due to their morphology and unidirectional mode of growth, filamentous organisms out-compete floc-formers under such conditions by growing outside aggregates.

Based on the differences in molecular diffusion coefficients of soluble components inside biofilms (acetate: 2.5×10^{-9} m^2 s^{-1}, DO: 1.67×10^{-9} m^2 s^{-1}, ammonium: 1.01×10^{-9} m^2 s^{-1}), bacteria present inside granules experience localized COD-to-nutrient ratios that are different the bulk liquid phase (Liu and Liu 2006). A minimum concentration of DO (2 mg_{O2} L^{-1}) was recommended for preventing filamentous outgrowth, as well as minimal concentrations of total inorganic nitrogen and orthophosphate between 1 and 2 mg L^{-1} in the effluent.

Carbon sources and OLR_v affect the balance between dense granules or unstable fluffy granules. High-energy carbohydrates and higher mesophilic temperatures (35 °C) were shown to select for filamentous organisms in AGS (Weber et al. 2007). The effect of the carbon source on the granular structure was investigated. Compact aerobic granules were obtained with an acetate feed pulsed with loading rates of up to 6 kg_{CODs} m_r^{-3} d^{-1} (Moy et al. 2002). Looser filamentous microstructures were obtained with a glucose carbon source but resulted in a reduced resistance towards diffusion of substrate into the aggregates. Extreme OLR_v between 9 and 15 kg_{CODs} m_r^{-3} d^{-1} were admitted in the glucose-fed reactor.

Despite the decrease in filamentous proliferation with acetate, filament organisms can still be observed with this substrate. Filamentous structures were presumed to act as backbones for the immobilization of microbial colonies (Martins et al. 2004c), and for the development of early stage granules (Liu and Liu 2006; Wang et al. 2006a).

The application of higher SGV typically above 0.020 m s^{-1} were shown to remediate outgrowth of filamentous organisms (McSwain Sturm and Irvine 2008; Tay et al. 2004). Higher hydrodynamic shear forces enable to detach filamentous structures from the surface of the aggregates, and to form smooth spherical granules. An air-lift mixing regime was proposed to be more energy-efficient for the application of localized shear with reduced up-flow aeration requirements, to this end (Beun et al. 2002). Filamentous organisms are nevertheless likely to remain inside the aggregates, and to develop filamentous outer structures as soon as the SGV will be set back to a lower value. Thus, although filaments could constitute skeletons for bacterial aggregation, filamentous bulking can remain a latent problem in the AGS system.

By analogy to activated sludge systems, the growth of filamentous organisms is not desired. Suppressing the conditions for their development is an important target (de Graaff et al. 2020; Henriet et al. 2017; Weissbrodt et al. 2013a). Their development should be specifically suppressed by operation using an anaerobic selector and applying substrate gradients across a SBR in feast-famine regimes and by supplying the wastewater in a plug-flow regime during the anaerobic feeding phase (Caluwé et al. 2022; de Kreuk et al. 2005; Guimarães et al. 2018; Henriet et al. 2016; Li et al. 2006a; Martins et al. 2004b; van Loosdrecht et al. 2008; Weissbrodt et al. 2013a). The success for a stable granulation is to make sure that the substrate consumption rate is lower than the substrate transport rate (de Graaff et al. 2020; Pronk et al. 2015a).

2.6.8 Full Biological Nutrient Removal in AGS-SBRs

In addition to the understanding of granulation phenomena, microorganisms with metabolisms necessary for the removal of carbon, nitrogen, and phosphorus should be selected in granular sludge.

Emphasis had been put on getting AGS with nitrification (Fang et al. 2009; Filali et al. 2012; Liu et al. 2007a; Tsuneda et al. 2003, 2006; Yang et al. 2003; Zhang et al. 2010), nitrogen removal (Adav et al. 2009a; Beun et al. 2001; Dangcong et al. 2004; Mosquera-Corral et al. 2005; Qin and Liu 2006; Qin et al. 2005; Wang et al. 2005a; Yang et al. 2003), phosphorus removal (Barr et al. 2010a; de Kreuk and van Loosdrecht 2004; Lin et al. 2003; Lou In et al. 2001; Zhu and Liu 1999), and combined nutrient removal capacities (Cassidy and Belia 2005; de Kreuk et al. 2005; Lemaire et al. 2008; Lu et al. 2007a; Yang et al. 2005; Yilmaz et al. 2008). Conversion of particulate organic matter is another point of attention for the operation of AGS systems with real wastewater (de Kreuk et al. 2010; Layer et al. 2020a; Pronk et al. 2015b; Toja Ortega et al. 2022; Wagner and da Costa 2013; Wagner et al. 2015b).

2.6.8.1 Nitrification in Granular Sludge

Nitrifying granules were cultivated in an aerobic up-flow fluidized bed reactor continuously fed with an inorganic wastewater and continuously aerated with a SGV of 0.009 m s^{-1}, by progressive decrease of the HRT from 5 days to 7.6 h in 1 year (Tsuneda et al. 2003). The flocculent seed sludge was taken from an activated sludge tank of a municipal WWTP and acclimated for 2 years to the synthetic inorganic wastewater. Compact 0.4 mm granules of nitrifiers were observed after 100 days. Nitrifying granules were cultivated in a bubble-column SBR (H/D = 14) started with a flocculent activated sludge taken from a municipal WWTP, fed with an inorganic synthetic wastewater, and operated with a 8 h HRT (Fang et al. 2009). Nitrifying granules with a diameter between 0.5 and 1.5 mm were obtained after 3 months, after decrease of the settling time stepwise from 20 to 5 min for preventing an excessive wash-out of inoculum. In both cases, no filamentous structures were detected.

Nitrifying granules of 0.3 mm were obtained after 120 days in a bubble-column SBR (H/D = 22) started with a flocculent activated sludge pre-adapted over 6 weeks, fed with a synthetic inorganic wastewater, and operated with a HRT of 12 h, a SGV of 0.018 m s^{-1}, and a progressive decrease in settling time from 20 to 8 min (Shi et al. 2010). An ammonium concentration of 100–250 $mg_{N\text{-}NH4}$ L^{-1} in the influent wastewater was optimal for cultivating nitrifying granules.

The effect of the co-presence of flocs and granules on the oxygen mass transfer and on the nitrification performance was investigated in a hybrid airlift AGS-SBR (Filali et al. 2012). Operation with a floc fraction in addition to granules efficiently reduced limitations in oxygen mass transfer and promoted nitrification.

The mass transfer of oxygen, nutrients, and soluble substrates in general is limited by their diffusion into the granular biofilms (Shi et al. 2009; van den Berg et al. 2022).

Nitrifying granules were permeable to molecules < 1000 Da by the interspersing channels that penetrated to a depth of 900 μm below the granule surface, and which facilitated the transport of substrates into the granule core.

2.6.8.2 Combined Nitrification and Denitrification in Granular Sludge

Nitrification and denitrification was achieved by alternance of aerobic and anoxic conditions (Adav et al. 2009a). Up to 80% N-removal was obtained by simultaneous nitrification and denitrification (SND) during the aeration phase of mature AGS-SBRs with granules bigger than 1 mm (Zhu et al. 2001). Stable EBPR was also maintained.

Fulfillment of SND in granules requires an aerobic zone for nitrification and an anoxic interior zone rich in C source for denitrification (de Kreuk et al. 2005). Because of the gradient of terminal electron acceptors in the biofilm, the granular architecture can be simplified as three successive aerobic, anoxic, and anaerobic redox zones (de Kreuk et al. 2005; Li et al. 2008c; Wei et al. 2012; Yuan and Gao 2010). The SND efficiency lies on an optimum between the oxygenation setpoint and the size distribution of aerobic granules. Granules should be sufficiently big to this end, and/or the DO concentration in the bulk liquid phase should be adapted to the size of granules to reduce the DO penetration into the biofilm. The evolution of the size of granules can lead to process disturbances (Toh et al. 2003). SND was obtained in mature granules of 2.5 mm (Beun et al. 2001). In mathematical model simulations, a DO of 40% was optimal for SND. In continuously fed systems, a stratification was obtained with fast-growing heterotrophs present in the outer layers of the biofilm and slow-growing autotrophs in the deeper layers. In discontinuously fed SBRs, the carbon source was present in temporarily high concentrations in the bulk liquid phase and penetrated completely into the granular biofilm, whereas DO was only present in the outer layer.

Two key conditions were pointed out for enabling simultaneous denitrification: presence of nitrate and availability of C source inside the biofilm (Beun et al. 2001). The presence of an anoxic zone inside granules was favored by decreasing the DO level in the bulk liquid phase. Under the same pulse feeding and aerobic starvation conditions, the decrease of the DO setpoint from 100 to 40% led to an improved N-removal (Mosquera-Corral et al. 2005). However, the granules were instable and disintegrated after the proliferation of filamentous bacteria at this lower DO concentration. PHAs stored under discontinuous feeding were used as C source for denitrification. Discontinuous feeding is therefore more efficient for denitrification in AGS than continuous feeding (Liu et al. 2005a). The COD/N ratio in the influent wastewater impacts on the SND process in AGS-SBRs (Wei et al. 2012). A COD/N ratio above 8.0 $g_{COD}\ g_N^{-1}$ was optimal for an efficient SND (92% N-removal), whereas a COD/N below 7.2 $g_{COD}\ g_N^{-1}$ was detrimental to SND.

The application of an anaerobic up-flow feeding phase of 60 min was not only successful for full BNR in a mature AGS-SBR but also for avoiding filamentous bulking (de Kreuk et al. 2005). Optimal SND was obtained with a DO maintained at

20% in the bulk, with granule diameters between 1.30 and 1.75 mm, under operation with OLR_v of 1.6 kg_{CODs} d^{-1} m^{-3} and NLR_v of 0.2 $kg_{N\text{-}NH4}$ d^{-1} m^{-3}. The anaerobic feeding regime is one key of AGS processes at full scale (Pronk et al. 2015b; van Dijk et al. 2018, 2022). Design of this phase can be supported by mathematical modelling in order to better understand the underlying phenomena (van Dijk et al. 2020; Weissbrodt et al. 2017).

SND has however been limited in different lab-scale and pilot studies (Layer et al. 2020b; Lochmatter et al. 2013). The implementation of alternation nitrification and denitrification (AND) by on/off aeration regimes can be a solution for an improved BNR in AGS. A specific attention should be given to N_2O in both SND and AND configurations (Lochmatter et al. 2014; van Dijk et al. 2021). Knowledge gained from partial nitritation and anammox systems highlighted that alternance of anoxic and aeration phases can lead to a bigger production of N_2O (Wunderlin et al. 2012).

2.6.8.3 Enhanced Biological Phosphorus Removal in Granular Sludge

EBPR was early obtained in A/O granular sludge systems (Lou In et al. 2001; Zhu and Liu 1999). Effect of ratios of nutrients in the influent wastewater, of the temperature and of the SRT were studied on granulation and P-removal. High COD/N (25 g_{COD} $g_N{}^{-1}$) and COD/P ratios (58 g_{COD} $g_P{}^{-1}$), 22 °C, and a low SRT (10 d) were optimal for cultivating granules removing phosphorus. A conventional stirred-tank SBR operated with flocculent activated sludge and under A/O conditions for full BNR from abattoir wastewater was converted into an AGS-SBR by decreasing the settling time from 60 to 2 min in 4 days (Cassidy and Belia 2005). Granules of 1.7 mm were obtained together with full BNR (97% COD-removal, 97% P-removal, 98% N-removal).

Selection for slow-growing PAOs and GAOs by alternating anaerobic feeding and aeration phases was efficient to select for EBPR and to stabilize compact granules (de Kreuk and van Loosdrecht 2004; Weissbrodt et al. 2013a). Long feeding periods were also more realistic and economically feasible for full-scale applications, where pumping rates cannot be applied as rapid as in the lab.

2.6.8.4 Nitrification, Denitrification, and Dephosphatation in Granular Sludge

Full BNR efficiencies were optimized by implementing sequences of anaerobic feeding and aerobic starvation phases, and by adapting the DO setpoint (de Kreuk et al. 2005). Efficient simultaneous BNR was obtained after decreasing the DO conditions from 100 to 20% O_2-saturation, namely 85–100% COD-removal, 74–94% P-removal, and 93–94% total N-removal with full nitrification. The lower DO concentration in the bulk liquid phase has resulted in a decreased oxygen penetration across the granular biofilms.

The formation of anoxic zones favored denitrification by clades of PAOs and GAOs (de Kreuk 2006; Yilmaz et al. 2008; Zeng et al. 2003a). Efficient simultaneous nitrification, denitrification and P-removal (SNDPR) was obtained in an SBR operated under A/O conditions for the treatment of putride abattoir wastewater with 85% COD-removal, 93% N-removal, 89% P-removal (Yilmaz et al. 2008). SNDPR in AGS-SBRs was achieved in other studies as well (Chen et al. 2009a; de Kreuk et al. 2007; Kishida et al. 2009; Lemaire et al. 2008; Liu et al. 2008b). SNDPR was achieved in A/O SBRs operated over wide ranges of psychrophilic (9–13 °C) to mesophilic temperatures (18–27 °C), and fed with a domestic wastewater complemented with acetate and propionate to reach a COD:N:P ratio of 360:60:6 (m/m/m) (Gao et al. 2010b).

Real-time control strategies based on the time derivatives of the on-line signals of electrical conductivity, pH and oxidation reduction potential (ORP) were used to adapt the length of each redox phase for an improved BNR in AGS-SBR (Gao et al. 2010c; Kishida et al. 2008). Electrical conductivity evolutions correlate nicely with the cycles of orthophosphate release and uptake along anaerobic-aerobic periods in lab-scale enrichment cultures notably, and can be used to derive the fraction of active PAOs and the EBPR potential of the sludge (Aguado et al. 2006; Maurer and Gujer 1995; Weissbrodt et al. 2014b). However, the electrical conductivity accounts for all ions evolving in the wastewater, and therefore correlations with orthophosphate will be noisier at full scale. Online monitoring of the liquid phase and the off-gas help control the performance of the AGS-SBRs (Baeten et al. 2021).

A 2-sludge and 3-stage system for full BNR from a nutrient-rich abattoir wastewater was developed with the combination of an A2/O granular sludge SBR and an aerated MBBR (Zhou et al. 2008a). The influent wastewater was fed to the SBR where the readily biodegradable COD was taken up and stored as PHAs by PAOs under anaerobic conditions. After a short settling phase, the supernatant was transferred into the MBBR for enhanced nitrification. The nitrified wastewater was sent back to the AGS-SBR where the PAOs have conducted SNDPR (81% N-removal, 94% P-removal). The treated effluent was used for irrigation.

Strategies for start-up and operation of single AGS-SBRs for full BNR were elucidated (de Kreuk et al. 2005; Lochmatter and Holliger 2014) and successfully applied at full scale (Pronk et al. 2015b). BNR can be improved not only by the application of AND but also by short-cut nitrogen removal via nitrite (Lochmatter et al. 2014).

2.6.8.5 Partial Nitrification and Anammox in Granular Sludge

Different studies investigated partial nitrification in AGS systems for combination with denitrification, anoxic dephosphatation, or anammox over the nitrite pathway, to save on aeration energy required for nitratation as well as on organic COD equivalents required for the reduction of nitrate into dinitrogen (Liu et al. 2008c; Lochmatter et al. 2014; Vazquez-Padin et al. 2010a; Zhang et al. 2006, 2011; Jin et al. 2008; Vazquez-Padin et al. 2009; Volcke et al. 2010).

A stable AGS conducting anoxic dephosphatation was cultivated with 90% removal of nitrogen and phosphorus (Zhang et al. 2006). Stable accumulation of nitrite was obtained in a nitrifying AGS-SBR fed with inorganic wastewater by increasing the NLR_v from 0.4 to 0.8 $kg_{N\text{-}NH}$ d^{-1} m^{-3}, and by maintaining the DO concentration in the bulk liquid phase between 2.0 and 3.5 mg_{O2} L^{-1} (Vazquez-Padin et al. 2010a). The AGS comprised granules of 1.9–2.9 mm and a low biofilm surface area to reactor volume ratio between 58 and 216 m^2 m^{-3}. These resistances to oxygen transfer were optimal for selecting for partial nitrification. A similar result was obtained in a nitrifying AGS-SBR treating the supernatant of an anaerobic digester, after increasing the NLR_v above 0.8 $kg_{N\text{-}NH}$ d^{-1} m^{-3} (Lopez-Palau et al. 2011b). When loading 1.1 $kg_{N\text{-}NH}$ d^{-1} m^{-3}, an effluent consisting of 50% ammonium and 50% nitrite was produced. An AGS process is therefore also suitable for the nitritation of putride process waters prior to anammox.

Partial nitritation and anammox (PN/A) was achieved inside biofilm, granule, and hybrid biofilm-floc or granule-floc processes from side stream systems to the main WWTP line (Hoekstra et al. 2019; Lackner et al. 2014; Laureni et al. 2016; Lotti et al. 2014; Weissbrodt et al. 2020b). Resistances to oxygen mass transfer into granules are optimal for partial nitrification by AOOs like *Nitrosomonas* spp. (Lochmatter et al. 2014; Lopez-Palau et al. 2011a; Vazquez-Padin et al. 2010a). Selection for AOOs over NOOs results from the control of dissolved oxygen in the bulk liquid phase and to the concentration of free ammonia that can inhibit the growth of NOOs (Lotti et al. 2014; Perez et al. 2009). Anammox organisms (AMOs) were integrated in an anaerobic-aerobic granular sludge process by segregation of biomass and preferential purge of the upper AGS bed fractions predominated by NOOs, after inoculation of AGS SBRs with anammox granules originating from a full-scale side stream system treating high-nitrogen-loaded centrates (Winkler et al. 2011b). Starting from a granular sludge dominated by chemolithoautotrophic AMOs, a chemoorganoheterotrophic anammox population of "*Candidatus* Brocadia fulgida" competing with DHOs for acetate was established in granules of an anoxic-aerobic AGS-SBR process after increasing the low influent COD/N ratio from 0.1 to 0.5 g_{COD} g_N^{-1} (Winkler et al. 2012d). In granular sludge, PN/A was further achieved by implementing an oxygen-limited autotrophic nitrification/denitrification (OLAND) (De Clippeleir et al. 2013; Vlaeminck et al. 2012). Similar to integrated fixed-film activated sludge processes (IFAS) (Macmanus et al. 2022; Veuillet et al. 2014), the development of hybrid granule-floc processes could provide additional possibilities to manage (short-cut) nitrogen removal processes (Filali et al. 2012). Hybrid processes lead to the partitioning of microbial niches on granules and flocs (Wei et al. 2021).

Initially the focus has mainly been on partial nitritation and anammox in different types of configurations with flocs, biofilms, and granules, and hybrid combinations of them from side-stream to main-stream applications in one-stage or two-stage processes (Christensson et al. 2013; Hoekstra et al. 2019; Joss et al. 2009; Lackner et al. 2014; Laureni et al. 2016; Perez et al. 2015; van der Star et al. 2007; Veuillet et al. 2014). However, the nitrification short-cut over nitrite and the suppression of NOB is not always easy to achieve (Han et al. 2015; Laureni et al. 2019; Pérez et al.

2014, 2020). Current efforts target a partial denitrification and anammox (Du et al. 2019; Le et al. 2019).

2.6.9 *Issues in the Start-Up of BNR Granular Sludge Systems After Seeding with Flocculent Activated Sludge*

The aforementioned studies demonstrated the great potential of AGS for full BNR. However, the start-up of AGS-SBR processes often took more than two months at lab scale for obtaining granules with nutrient removing activities after inoculation with flocculent activated sludge and imposing wash-out conditions.

Experimental and modeling studies highlighted that 75–100 days were required to obtain active PAOs in AGS, even though a seed sludge from a BNR-WWTP was used as inoculum and nitrification was initially inhibited with allylthiourea (ATU) to promote the growth of PAOs (de Kreuk et al. 2005; Ebrahimi et al. 2010). Full nitrification was only detected 100 days after stopping the supply of ATU. Fast-growing OHOs were modeled to outcompete PAOs and nitrifiers in early-stage granules independently from feeding and aeration regimes (Xavier et al. 2007). This led to questioning why active ANOs and PAOs are outcompeted during start-up under wash-out dynamics, which organisms are predominant during start-up, and how to reduce start-up period of AGS-SBRs?

Pre-cultivated anaerobic granules (Chen et al. 2009c; Dangcong et al. 2004; Hu et al. 2004; Lan et al. 2005; Linlin et al. 2005; Ruan et al. 2006) or aerobic granules (Lee et al. 2010; Liu et al. 2005c; Shen et al. 2007; Yuan et al. 2011; Zhang et al. 2005), as well as air-dried aerobic granules (Lee et al. 2010) were used as seed granular substrata for microbial colonization and reducing the start-up period of AGS reactors. However, the storage of pre-cultivated granules requires space and costs (Liu et al. 2011), and aerobic granules can lose structural integrity under prolonged storage (Adav et al. 2008a). For the start-up of new full-scale AGS installations, a strategy similar to the MBBR biofarm (Christensson et al. 2013) can be used by supplying granules from an already existing AGS plant.

Other strategies were also studied to reduce the start-up time. For instance, bioaugmenting the flocculent seed sludge with bacterial strains with high self-aggregation index (e.g., *Pseudomonas veronii* isolated from AGS) was proposed (Ivanov et al. 2006, 2008). However, although bioaugmentation strategies can be helpful for remediating a process issue on short term, the survival and stable activity of strains introduced in the system (i.e., the long-term effect of the bioaugmentation) is a common problem. Instead of an exogenous supply of microbial populations, the operational conditions should be better managed to select for the BNR microorganisms via microbial community engineering.

Low-loaded real domestic wastewater containing particulate substrates can lead to a more difficult granulation (de Kreuk and van Loosdrecht 2006; Derlon et al. 2016; Layer et al. 2020a). Low OLR_V can be insufficient for enabling rapid biomass growth in the reactor. Shortening the total SBR cycle time from 3 to 2 h can help increase the daily OLR_v from 1.0 to 1.5 $kg_{CODs}\ d^{-1}\ m^{-3}$. Granulation was enhanced in a pilot plant aiming at treating a real wastewater composed of domestic (40%) and industrial (60%) sewage by operating the system with a higher initial F/M ratio between 2.2 and 3.4 $kg_{CODs}\ d^{-1}\ kg_{CODx}{}^{-1}$ and a short settling time of 2 min (Liu et al. 2011). Efficient removal of COD (87%) and nitrogen (93%) was obtained after 50 days. A pilot plant fed with low-strength domestic wastewater (95–200 $mg_{COD}\ L^{-1}$) was operated with a high volume exchange ratio (75%) in order to increase the OLR_v from 0.6 to 1.0 $kg_{CODs}\ d^{-1}\ m^{-3}$, and to stimulate granule formation (Ni et al. 2009). At full scale, rain events dilute the influents. Shortening the SBR cycle and thus running more SBR cycles per day is an efficient action for promoting a stable AGS-SBR process (Pronk et al. 2015b). The operation of this full-scale installation in Garmerwolde in the Netherlands was designed with a sludge loading rate of 0.10 $kg_{COD}\ d^{-1}\ kg^{-1}$ TSS for an expected biomass concentration of 8 kg m^{-3}, and a volumetric loading rate of 1.5 $m^3\ d^{-1}\ m^{-3}$. While the SBRs were operated in with a total cycle of 6.5 h under dry weather, the cycle time was shortened to 3 h during rainy weather. Biomass concentrations of 6–8 $kg_{TSS}\ m^{-3}$ were achieved in reported full scale AGS processes (Ekholm et al. 2022; Pronk et al. 2015b), which is 2–3 times higher than conventional activated sludge.

An alternative start-up method was proposed by mixtures of flocculent activated sludge and crushed aerobic granules for reducing the time required to get aerobic granules with efficient BNR activities (Pijuan et al. 2011; Verawaty et al. 2012). This was achieved in a mechanically stirred 2-L A/O SBR with a H/D of 11, operated by stepwise decrease of the settling time from 20 to 1.5 min, gradual increase in volumetric exchange ratio from 12.5 to 50%, and thus progressive increase in organic (0.46–1.83 $kg_{CODs}\ d^{-1}\ m^{-3}$) and ammonium nitrogen volumetric loading rates (0.097–0.390 $kg_{N\text{-}NH4}\ d^{-1}\ m^{-3}$) from a putride abattoir wastewater. The shortest granulation time of 18 days was obtained with a 1:1 mixture of flocculent sludge and crushed AGS. Nitrogen removal was maintained at 90%, whereas phosphorus removal was hindered by excessive accumulation of nitrite after oxidation of the high loads of ammonium from the abattoir wastewater.

In pilot and full-scale installations operated on real wastewater, a start-up period of about 6–13 months was needed to obtain a substantial amount of granules over activated sludge flocs (Giesen et al. 2012; Liu et al. 2011; Pronk et al. 2015b; van der Roest and van Loosdrecht 2012). Limitations in pumping, aeration and energetic efficiencies are typically induced by the scale-up. However, a good effluent quality was obtained before granulation was complete (Pronk et al. 2015b).

For the translation of aerobic granulation from lab to pilot and full scales, one can question how to apply a sufficient shear stress and a hydraulic selection pressure for inducing the formation of dense aggregates and for preventing suspended growth? For SBR systems including simultaneous fill/draw phases by feeding the influent at the bottom and discharging the effluent at the top of the reactor, a high up-flow superficial liquid velocity (SLV) of up to 5 m h^{-1} was recommended during the start-up phase (van Haandel and van der Lubbe 2012). This should help the wash-out of suspended flocs with the effluent, and to selectively retain organisms and aggregates that settle well. As soon as granules develop and PAOs get established, the SLV can be reduced within a normal operating range of 2–3 m h^{-1}. In the aforementioned full-scale AGS system, the influent was pumped at an upward velocity of 3.0–3.3 m h^{-1} (Pronk et al. 2015b). This up-flow velocity should help select for fast-settling granules, control the suspended solids in the effluent, and prevent the mixing of the influent with the effluent (van Dijk et al. 2018, 2020; Weissbrodt et al. 2017).

2.6.10 Design of Granular Sludge Reactors for BNR

Granulation was mainly studied in bubble-column SBRs. Scaling-up the technology however leads to important changes of reactor designs, geometries, and volume factors. These can affect granulation, granule characteristics, BNR, and overall process performance. Technological adaptations need to be achieved to develop a robust granular sludge process at pilot and full scales (de Kreuk and van Loosdrecht 2006; Derlon et al. 2016; Ekholm et al. 2022; Guimarães et al. 2018; Pronk et al. 2015b).

Ideal bubble-column and air-lift SBRs were frequently used at lab and pilot scales with high H/D ratios, high up-flow gas supply, and short settling times for investigating granulation (Beun et al. 1999, 2002; Derlon et al. 2016; Kong et al. 2009; Layer et al. 2020a; Liu and Tay 2007; Morgenroth et al. 1997; Wagner et al. 2015a; Wilderer and McSwain 2004). These reactor designs together with the application of operation conditions selecting for a fast-settling biomass selected for aerobic granules in a relatively short start-up period. However, implementation of columns reactors at full scale can be limited by practical design considerations, and gradients along the column (Gujer 2008; Haringa 2022; Heijnen et al. 1991, 1993; McClure et al. 2016).

A 'spontaneous' formation of granules with good settleability (SVI < 80 mL g^{-1}) was reported from traditional stirred-tank SBRs with H/D ratios between < 1 and 2 and operated for BNR under alternated redox conditions. Gradual granulation of a flocculent activated sludge into granules of 0.3–0.5 mm was observed within one month in a stirred-tank SBR operated for aerobic COD-removal and nitrification, after decreasing the DO setpoint from 4 to 0.8 mg_{O2} L^{-1} (Dangcong et al. 1999). Granulation occurred in a stirred-tank SBR operated for biological pretreatment of an abattoir effluent, and for the production of a biomass with a low SVI as protein amendment for animals (de Villiers and Pretorius 2001). A biomass with a low SVI and composed of granules of 1.5 mm were formed when the DO was controlled to low values during the feeding phase, and notably when an anoxic feeding phase was achieved. The gradual formation of granules of 2.5–3.0 mm with an SVI < 40 mL $g_{VSS}{}^{-1}$ was noticed after decreasing the settling time from 30 to 15 min in a stirred-tank SBR operated for EBPR under A/O conditions (Dulekgurgen et al. 2003a, b). Stable and efficient BNR was obtained in this system (95% COD removal, 99% P-removal, 71% N-removal). The formation of an AGS was reported from a stirred-tank SBR operated for EBPR as well (You et al. 2008). Granules enriched with 70% of PAOs were obtained together with full EBPR. An AGS was cultivated in a stirred-tank SBR (H/D = 2.5, 100 rpm stirring) by imposing a 6 h HRT and a settling time as short as 1 min since reactor start-up (Mosquera-Corral et al. 2011). The SBR was operated with cycles of 3 min pulse feeding, 171 min aeration, 1 min settling, 3 min effluent withdrawal, and 2 min idle. The DO was maintained at 5 mg_{O2} L^{-1} during feast phases, and at 9 mg_{O2} L^{-1} during starvation phases. Granules were obtained with an OLR_v of 1.75 kg_{COD} d^{-1} m^{-3} whereas granulation was hindered when the OLR_v was set below 1 kg_{COD} d^{-1} m^{-3}. The granule size evolved from 0.5 mm after two weeks to 1.75 mm after < 3 months. COD was removed at 90% and nitrogen at 40%.

Operating AGS processes in traditional SBRs and with reduced aeration could lead to energetic and economic advantages for full-scale applications. Conventional EBPR systems from lab to full scales are prone to spontaneous granulation, because of the selection for PAOs that form compact microcolonies (de Kreuk and van Loosdrecht 2004; Larsen et al. 2006; Weissbrodt et al. 2013a, 2014b). Sludge densification has been observed in several full-scale installations in the world, often resulting in process control impairment for maintaining the sludge age, returning the sludge from the secondary clarifier into the activated sludge tank, and fluidizing the biomass by the aeration (Bangerter 2017; Bruce et al. 2014; Downing et al. 2022; Wei et al. 2020). EBPR and BNR activated sludge systems are often designed and operated in flow-through A2O configurations that resemble plug flows in a cascade of anaerobic, anoxic, and aerobic tanks. This can establish gradients of substrates across the plug flow that are favorable for granulation, on top of the anaerobic selector for PAOs. This natural granulation propensity is now desired for the upgrade and intensification of existing activated sludge installations into continuous-flow granulation processes (Cofré et al. 2018; Haaksman et al. 2020; Regmi et al. 2022; Strubbe et al. 2022).

AGS processes were achieved with other configurations as well. AGS was cultivated in aerated continuous-flow stirred-tank reactors (CSTR) with H/D ratio of 0.5–0.7, by imposing a HRT as short as 1 h and a settling velocity of 10 m h^{-1} by withdrawing the effluent through a half-submerged up-flow tube (diameter of 1.2–2.8 cm) (Morales et al. 2012). Granules with diameters of up to 7 mm were obtained. However, only 60% of the OLR_v (4.8–6.0 kg_{COD} d^{-1} m^{-3}) was removed and ammonium was not nitrified. Even though BNR was optimized in these systems, the possibility to convert already constructed systems into AGS reactors was revealed (Morales et al. 2012; Mosquera-Corral et al. 2011).

The AGS-SBR technology can also be combined with submerged membranes (Thanh et al. 2010). Membranes can deliver a treated effluent exempted from suspended solids and reclaim a water of high quality. A possible application can relate to the removal of antimicrobial resistance. One major limitation is the fouling of the microfiltration hollow-fiber membranes by EPS. Compared to operation with flocs, membrane fouling was significantly decreased when operated in combination with AGS (Li et al. 2005, 2007a; Tay et al. 2007; Zhou et al. 2007). The release of suspended solids measured in AGS-SBR effluents (Pratt et al. 2007; Schwarzenbeck et al. 2005; Yilmaz et al. 2008) was alleviated with the use of the membrane filtration modules (Thanh et al. 2008). FISH and fractionation measurements revealed that the secreted EPS were predominantly bound to granules (Yu et al. 2009a).

Only a low amount of free EPS remained in the supernatant, and membrane fouling was prevented. Nitrogen removal was achieved in these integrated granular sludge membrane systems by AND and SND (Thanh et al. 2010; Wang et al. 2008a). Full BNR was achieved by A/O sequences (Wang and Yu 2009a, b; Xu et al. 2010).

Additional process configurations have been used, like shaking SBR (Cai et al. 2004), annular gap reactor (Williams and de los Reyes III 2006), swimming bed reactor (Zhang et al. 2007c), and periodic biofilters (De Sanctis et al. 2010; Laconi et al. 2008). The use of hydrocyclones is further use to select for dense bioaggregates and granules, like in the deammonification process (DEAMON) expanding from side stream to mainstream processes (Regmi et al. 2022; Wett et al. 2007, 2013).

Overall, the AGS-SBR technology has successfully been implemented with the typical operation by anaerobic fill/draw phase followed by aeration, such as demonstrated by the more than 90 Nereda® WWTPs installed world-wide.

2.6.11 Practical Implications for Implementing the Granular Sludge Technology for BNR at Full Scale

At full-scale level, the application of granulation selection pressures is limited by pumping, aeration, energetical, and economical constrains.

A mathematical concept was developed for a rational design and to adjust the settling time, the effluent discharge time and the volume exchange ratio (or the depth of the effluent discharge port) in function of the minimum settling velocity setpoint (Liu et al. 2005e). At full-scale, a reactor design with low H/D ratios is more suitable for practical application and limiting pH gradients. A reactor design with simultaneous feeding at the bottom of the reactor and effluent discharge at the top is in addition efficient for decreasing pumping requirements (van Dijk et al. 2018, 2020; van Haandel and van der Lubbe 2012). The wash-out of activated sludge flocs is achieved during the fill/draw phase by fluidizing the sludge bed with a proper up-flow SLV. To select for fast-settling granules over flocs, the minimum settling velocity was recommended to be set at an initial value above the settling velocity of activated sludge flocs (3–5 m h^{-1}). An initial SLV of 5 m h^{-1} (van Haandel and van der Lubbe 2012) or above 8 m h^{-1} (Liu et al. 2005e) was recommended to induce the formation of fast-settling biomass. The extent of the SLV that can be applied depends on the influent pumping capacities. Full-scale AGS-SBRs are operated at a SLV of 3 m h^{-1} (Ekholm et al. 2022; Pronk et al. 2015b, 2020; van Dijk et al. 2018).

A SBR design with a H/D ratio of 0.4 and comprising a 1 h fill/draw phase (Table 2.1) was adopted for the world first AGS-WWTP installed at full scale in Epe, The Netherlands (Giesen et al. 2012; van der Roest and van Loosdrecht 2012). An overall continuous-flow operation at plant level was achieved by using 3 SBRs in parallel. This configuration was also tested at bench (Zhou et al. 2011a). Now 90 AGS WWTPs are installed in the world. The performance of some full scale plants have been described in literature (Ekholm et al. 2022; Pronk et al. 2015b). The design and operation criteria are summarized elsewhere (Pronk et al. 2020).

The high-rate AGS technology was applied for the removal of nutrients from different types of wastewater, ranging from low-strength domestic wastewater (de Kreuk and van Loosdrecht 2006; Derlon et al. 2016; Ekholm et al. 2022; Layer et al. 2020b; Pronk et al. 2015b, 2020; van der Roest et al. 2011; Wagner et al. 2015a; Wang et al. 2009a) to high-loaded food-industry wastewaters (Arrojo et al. 2004; Caluwé et al. 2022; Lopez-Palau et al. 2009; Schwarzenbeck et al. 2005; Wang et al. 2007) and putride process effluents from anaerobic digesters, landfills or abattoirs (Di Iaconi et al. 2009; Vogelaar et al. 2002; Yilmaz et al. 2008). Besides the treatment of nutrients, AGS applications were developed for the treatment of industrial chemical compounds such as petrochemical compounds, phenols, chlorophenols, anilines, nitrobenzene, MTBE, phthalates, and chelating agents (Adav et al. 2009c; Carucci et al. 2008; Fang et al. 2008; Jiang et al. 2004; Nancharaiah et al. 2006b; Wang and Zhou 2010; Winkler et al. 2018; Yu et al. 2009b; Zhu et al. 2008b). Biosorption of metals and biodecolorization of dying effluents have also been achieved with AGS (Caluwé et al. 2017; Gao et al. 2010a; Hailei et al. 2010; Liu et al. 2003a; Nancharaiah et al. 2006a). AGS reactors were able to remove some micropollutants such as endocrine disrupters (Balest et al. 2008), and have also been coupled with advanced oxidation processes for enhanced removal of xenobiotic organic compounds (Di Iaconi et al. 2006). The AGS technology was implemented for decentralized wastewater treatment (Li et al. 2006a), as well.

2.7 Microbial Ecology of Wastewater Treatment Systems

Microbial ecology is the key science underlying environmental biotechnology (Cerruti et al. 2021; Rittmann 2010; Verstraete et al. 2007). It aims at characterizing microbial communities from microorganisms to metabolisms, their network structures, and their dynamics. Microbial ecology principles when coupled with bioprocess engineering foster the management of microbial resources for environmental services. Since 'microbes do the job', microbial populations are present and (in)activated behind every bio-based process in biological wastewater treatment, like bioaggregation, EPS production, BNR, and resource capture. In environmental biotechnology, this links to the concepts of microbial resource management, microbial community engineering, and ecological engineering that are bound to principles of microbial ecology and thermodynamics (Heijnen et al. 2009; Kleerebezem and van Loosdrecht 2007; Lawson et al. 2019; Moralejo-Gárate et al. 2011; Rittmann 2006b; Verstraete et al. 2007; Weissbrodt et al. 2020a).

Examining microbial communities in function of operations and performances (Weissbrodt et al. 2014d; Wells et al. 2011) provide substantial knowledge on microbial selections and metabolic regulations for understanding and managing environmental biotechnologies like activated sludge and granular sludge systems. One important feature is the link between microbial compositions, identities, abundances, and

Table 2.1 Design parameters of the world first full-scale WWTP using granular sludge for full and high-rate biological nutrient removal installed in Epe, The Netherlands (Giesen et al. 2012; van der Roest and van Loosdrecht 2012)

Parameter	Units	Value
Wastewater		
Origin	–	Domestic and industrial (abattoir)
Maximum amount of treated wastewater	$m^3\ h^{-1}$	1500 (by rain weather)
	$m^3\ d^{-1}$	36,000
Person-equivalents (PE)	PE	59,000
Treatment objectives		
Primary treatment	–	2 mm screening Aerated sand and grease trap
Secondary treatment	–	Full BNR
Total nitrogen	$mg_N\ L^{-1}$	< 5
Total phosphorus	$mg_P\ L^{-1}$	< 0.3
SBR cycle		
Total SBR cycle duration	h	3
Anaerobic fill and draw	min	60
Aeration	min	110
Settling	min	10
Number of SBR cycles per day per SBR	cycles d^{-1} SBR^{-1}	8
Total number of SBR cycles per day per SBR	cycles d^{-1}	24
SBR design		
Number of SBRs in parallel	–	3
Height (H)	m	9
Diameter (D)	m	25
H/D ratio	–	0.36
Cross section area	m^2	491
SBR volume	m^3	4419
Volumetric exchange ratio	%	25–50
Biomass concentration	kg m^{-3}	10–20
Plant value		
Plant value	€	15 mio
Plant value per PE	€ PE^{-1}	254
Calculations		
Maximum amount of treated wastewater per SBR cycle	m^3 $cycle^{-1}$	1500

(continued)

Table 2.1 (continued)

Parameter	Units	Value
Maximum influent feeding flowrate	$m^3\ h^{-1}$	1500
Maximum SLV during feeding	$m\ h^{-1}$	3.1

Design criteria and operation parameters of the latest full-scale AGS installations are given elsewhere (Ekholm et al. 2022; Pronk et al. 2015b, 2020)

functions (Muyzer and Ramsing 1995; Nielsen et al. 2004; Wagner et al. 2006). Using microbial ecology principles, microbiomes can be engineered for controllable outputs (Bernstein 2019; Kleerebezem and van Loosdrecht 2007; Lawson et al. 2019; Lindemann et al. 2016; Scarborough et al. 2022). While being the biggest biotechnologies, WWTPs are the perfect systems for investigating microbial ecology questions (Daims et al. 2006).

The assessment of population shifts during process failures and recovery is of interest as well (Baldwin et al. 2019; Gunther et al. 2012; Weissbrodt et al. 2012a), leading to the development of reliability theory (Zilles et al. 2016) and engineering measures. Similar approaches are followed in medical sciences for designing therapies against microbially-induced diseases (Dethlefsen et al. 2007; Gonzalez et al. 2011a; Zemanick et al. 2011), or in agricultural systems for improving cultivation strategies (Brussaard et al. 2007; Holland and Coleman 1987; Johansson et al. 2004).

The development of environmental biotechnologies was closely related to the advances in molecular microbial ecology and ecophysiology methods, by analyzing the pool of informational macromolecules extracted from biomasses (i.e., nucleic acids and proteins besides lipids and polysaccharides). Modern, high-throughput, culture-independent molecular methods enable the transition from descriptive to predictive investigations and sustain the understanding and optimization of engineered mixed cultures (McMahon et al. 2007a; Nielsen and McMahon 2014; Weissbrodt et al. 2020a). Meta-omic methods like genome-centric metagenomics (using metagenome-assembled genomes, MAGs), metatranscriptomics and metaproteomics bring an unprecedented access to microorganisms and metabolic networks inside microbiomes (Albertsen et al. 2013b; Kleikamp et al. 2020b; Lawson et al. 2017; McDaniel et al. 2021; Narayanasamy et al. 2015; Oyserman et al. 2016b; Sampara et al. 2022; Singleton et al. 2021; Weissbrodt et al. 2020b; Ziels et al. 2018). The toolbox of bioanalytical methods can be eventually brought at treatment lines for a plant-wide systems microbiology for the wastewater industry (Cerruti et al. 2021).

2.7.1 Microbial Communities and Climax

The evolution of microbial communities is an important concept in activated sludge processes (Kucnerowicz and Verstraete 1983). The efficiency of activated sludge systems relies on the development of a functional microbial assembly. The evolution of microbial communities and their activities in activated sludge reactors relates to the concept of climax pattern (Krebs 1972): the microbial community adapts constantly to changes in environmental factors, and is characterized by permanent shifts of minor populations within the community. Process failures link to imbalances in the climax state of the microbiome. The climax of a biological wastewater treatment process is impacted both by deterministic factors (e.g., selection conditions) and probabilistic phenomena (e.g., immigration of microorganisms from the influent in the tank) (Cerruti et al. 2021; Curtis et al. 2003; Curtis and Sloan 2006; Frigon and Wells 2019; Ofiteru et al. 2010; Weissbrodt et al. 2020a).

The development of a microbial community was tracked by Kucnerowicz and Verstraete (1983) in aerobic reactors inoculated at lab bench with either activated sludge collected from a WWTP or microorganisms present in the non-filtered air supplied into the reactors. Even though an optimal state was obtained more rapidly with the activated sludge inoculum, functional optima were achieved on long term with both inoculation types. At least 2–4 sludge ages had to elapse for expecting a consistent response of the microbial community, and at least 10 sludge ages for obtaining an optimally adapted microbial community.

BNR activated sludge and granular sludge systems like are operated under highly dynamic conditions by integrating spatial or temporal sequences of anaerobic, anoxic, and aerobic conditions. The development of stable communities requires the acclimation to operational dynamics and variations in influent composition, loads and micro-environmental conditions. Addressing the links between microbial processes and process boundaries is an important target for research. Different frameworks have been developed to this end (Cerruti et al. 2021; de los Reyes III et al. 2015; Joyce et al. 2018; Park et al. 2010a; Vanwonterghem et al. 2014; Weissbrodt et al. 2014d).

2.7.2 Microbial Ecology and Its Analytical Toolbox for Environmental Biotechnologies

Microbial ecology established as a core science in the field of wastewater treatment. Besides being the largest biotechnologies, WWTPs are model systems for the investigation of microbial ecology questions (Daims et al. 2006). The selection, activities, and interactions of microbial niches is an important feature for robust BNR in mixed-culture processes in activated sludge, granular sludge, biofilm, and hybrid configurations. Microbial ecology evolved together with the continuous advances in bioanalytical methods for investigating microorganisms and their functions in

WWTPs (Amann et al. 1998; Wagner et al. 2002). Microbial ecology principles and analyses support the understanding and management of microbial processes in environmental biotechnologies (Rittmann 2006b; Verstraete 2007).

Knowledge gained on microbial interactions in complex communities is driven by the availability of methods for a fast analysis of microbial compositions and activities (Wuertz et al. 2004). More than 30 methods of microbial ecology are available to examine microorganisms and their metabolisms in microbiomes (Nielsen and McMahon 2014; Weissbrodt et al. 2020a). Ex situ and in situ molecular biology and ecophysiology methods can be applied for linking floc-granule-biofilm architectures, microbial community structure, and metabolic processes without the need for cultivation of target microorganisms (Wuertz et al. 2004).

Methods have been frequently reviewed along new analytical developments since the 1990s (Almstrand et al. 2013; Amann and Kuhl 1998; Amann et al. 1998; Cerruti et al. 2021; Christensen et al. 1999; Gilbride et al. 2006; Goel et al. 2011; Gu et al. 2011; Karst et al. 2016; McMahon et al. 2009; Muyzer and Ramsing 1995; Nielsen et al. 2016; Nielsen and McMahon 2014; Onuki et al. 2000; Weissbrodt et al. 2020a).

After the early developments to address phylogenies of microorganisms using the ribosomal RNA (rRNA) gene marker (Woese et al. 1985), several methods were developed to profile microbial communities by targeting the 16S rRNA gene pool of prokaryotic microorganisms (bacteria, archaea) and the 18S rRNA gene pool of eukaryotic microorganisms.

Non-spatial community profiling methods such as denaturing gradient gel electrophoresis (DGGE) (Muyzer et al. 1993) and terminal-restriction fragment length polymorphism (T-RFLP) (Liu et al. 1997) have been widely used to profile communities. These methods involve DNA extraction from the biomass, amplification of the 16S rRNA gene pool by polymerase chain reaction (PCR), and separation of operational taxonomic units (OTUs) by electrophoresis in a planar gel or in a gel capillary, respectively. DGGE, T-RFLP but also ARISA (automated ribosomal intergenic spacer analysis) are used to assess relative changes in microbial populations and their relative abundances as a function of time, environmental conditions, and process performances. Closest phylogenetic identities of microbial relatives are obtained by an additional sequencing of the OTUs, either by excising the DGGE gel bands or by clonal libraries (Mengoni et al. 2002; Rowan et al. 2003; Sekiguchi et al. 1998). Modern methods of amplicon sequencing, field guides and databases now help obtain community profiles with phylotypes and their relative abundances in single sequencing runs (Albertsen et al. 2015; Andersen et al. 2018; Dueholm et al. 2019; Weissbrodt et al. 2012b).

Since the early 2000s, quantitative real-time PCR (qPCR) (Agrawal et al. 2021; Calderón-Franco et al. 2022; Hall et al. 2002; Harms et al. 2003; Muurinen et al. 2022; O'Reilly et al. 2009; Pallares-Vega et al. 2021; Tsushima et al. 2007) and oligonucleotide microarrays (Brodie 2011; Herzberg and Elimelech 2008; Kelly et al. 2005; Lazazzera 2005; Loy et al. 2002; Yergeau et al. 2010) facilitate a rapid screening and quantification of populations and genes by targeting not only their 16S rRNA gene but also their functional genes, with high sensitivity. Reverse transcription and qPCR (RT-qPCR) is used to analyze the expression of genes, by extracting the RNA pool, enriching and reverse-transcribing the messenger RNA (mRNA) into complementary DNA (cDNA), and quantifying gene transcripts by qPCR (Park et al. 2015; Wu et al. 2016).

Fluorescence in situ hybridization (FISH) targets the rRNA of microorganisms with fluorescent-labelled oligonucleotide probes. After hybridization, the microorganisms can be detected by epifluorescence microscopy, CLSM, or flow cytometry (Daims et al. 2001; Gonzalez-Gil and Holliger 2011; Manz et al. 1992; Nielsen et al. 2004, 2016; Sekiguchi et al. 1999; Wagner et al. 2003; Weissbrodt et al. 2013a; Winkler et al. 2011b). FISH can be combined with ecophysiological methods like microautoradiography for an in situ structural and functional analysis of active microbial populations within flocs, biofilms, and granules (Gieseke et al. 2003; Nielsen and Nielsen 2002; Nielsen et al. 1999, 2016; Okabe et al. 1999). FISH has notably been used to early characterize filamentous populations from bulking systems.

For microautoradiography, biomasses are incubated with a substrate labelled with a radioactive isotope (e.g., ^{14}C, ^{3}H, $^{32/33}P$, ^{35}S). During growth, the radio-labelled substrate is metabolized and part of it is converted into anabolic products present in the cells. Since these products contain the radiolabel, they can be detected by microautoradiography or Raman spectroscopy, and localized in the specimen by overlay with the FISH and microscopy reflection signals (Fernando et al. 2019; Huang et al. 2007; Kong et al. 2004; Lee et al. 2002; Nielsen et al. 2003; Oshiki et al. 2010; Seviour et al. 2006; Wagner et al. 2006). Microautoradiography and Raman spectroscopy has notably been used to detect PAOs in WWTP biomasses.

The functionalities of microorganisms can further get resolved by stable isotope probing (SIP) (Dueholm et al. 2015; Dumont and Murrell 2005; Ginige et al. 2004; Mooshammer et al. 2021; Musat et al. 2012). Biomasses are incubated with a substrate labelled with a stable isotope (e.g., ^{13}C, ^{15}N) that get anabolized partly into the informational molecules of the biomass like DNA, RNA, and proteins. After extraction, the heavier fractions of the DNA or RNA that contain the stable isotopes can be separated by ultracentrifugation and sequenced to identify the microorganisms that metabolized the substrate (Gülay et al. 2019; McIlroy et al. 2014b; Zhang et al. 2020; Ziels et al. 2018). The heavier proteins/peptides can be detected and sequenced in a mass spectrometer (de Jonge et al. 2022; Kleiner et al. 2021; Macêdo et al. 2022). The stable isotopes can also be detected in situ by Raman spectroscopy, secondary ion mass spectrometry (SIMS), or nanoscale secondary ion mass spectrometry (nanoSIMS) (Eichorst et al. 2015; Koch et al. 2014; Musat et al. 2012; Renslow et al. 2016; Schreiber et al. 2016; Tao et al. 2019). SIP methods made breakthrough in the characterization of the metabolic versatility of nitrifiers.

Recent developments in sequencing, mass spectrometry, and bioinformatics (collectively known as 'meta-omics') enable the characterization of microbial and functional networks at high resolution. It opened the new era of systems microbiology (Nielsen et al. 2012b; Teitzel 2011). Translated at WWTP level, a plant-wide systems microbiology can be developed for the wastewater industry (Cerruti et al. 2021).

Metagenomics with high-throughput sequencing of the DNA pool of biomasses foster the characterization of microbial communities with high resolution on microorganisms, their relative abundance, and their genetic potential (Albertsen et al. 2011a, 2013b; Burke et al. 2011; Garcia Martin et al. 2006; Steele and Streit 2005). Genome-centric metagenomics can now be achieved by the binning of genomes from single lineages out of metagenomes, revisiting the microbial tree of life (Adler et al. 2022; Albertsen et al. 2013a; Palomo et al. 2018; Parks et al. 2017, 2018; Rubio-Rincon et al. 2019; Singleton et al. 2021).

Metatranscriptomic (mRNA) and metaproteomics (proteins) unravel the expressed functionalities of microorganisms according to their metabolic regulation from genes to mRNA and proteins (Abram et al. 2009; Hassa et al. 2018; Kleikamp et al. 2021, 2022; Lawson et al. 2015; Narayanasamy et al. 2015; Niederdorfer et al. 2021; Oyserman et al. 2016b; Thompson et al. 2008; Tomás-Martínez et al. 2021; Wilmes et al. 2008a, b; Yergeau et al. 2010). Metaproteomics can be used to analyze both the intracellular and extracellular protein pools of environmental samples. It is therefore useful for the analysis of not only microorganisms but also EPS.

Overall, the reunion of microbial ecology techniques allows for linking functions with identities (Nielsen et al. 2004; Wagner et al. 2006). These phylogenetic and functional signatures from microbiomes are important for disentangling microbial selection and metabolic processes in environmental biotechnologies. The design of WWTPs primarily results from the growth stoichiometries and kinetics of BNR microorganisms (Weissbrodt et al. 2020a). Analytical information from microbial communities opens the black boxes of activated sludge, granules, and biofilms to better understand how microorganisms establish their metabolic niches under the operational and environmental constraints and perturbations.

2.7.3 The Saga of PAOs in EBPR Processes

After extensive investigations with a battery of microbial ecology methods, the microbiome of EBPR processes is considered as one of the best studied model ecosystem in environmental biotechnology (Nielsen et al. 2011), enabling microbial resource management.

The microbiology challenge of the last 30 years was to identify the predominant organisms involved in EBPR. Despite several efforts, the major PAOs contributing to EBPR in full scale WWTPs could not be isolated. Some isolates exhibiting dephosphatating activities were obtained but were predominant in full scale plants. From the seventies to the early nineties, the genus *Acinetobacter* belonging to the family of *Moraxellaceae* inside *Gammaproteobacteria* was isolated from activated sludge and thought to undertake EBPR at full scale (Beacham et al. 1990; Deinema et al. 1980; Florentz and Hartemann 1984; Fuhs and Chen 1975; Streichan et al. 1990). *Acinetobacter* was therefore studied as a model EBPR bacterium (Bark et al. 1992; Deinema et al. 1985; Ohtake et al. 1985; van Groenestijn et al. 1989a; van Groenestijn and Deinema 1985; van Veen 1997).

The need for further investigations was stressed for unraveling other genera that significantly contribute to EBPR (Jenkins and Tandoi 1991). The development and the use of culture independent methods for the in situ characterization of community structures with rRNA targeted oligonucleotide FISH probes resulted in the conclusion that *Acinetobacter* spp. were only minor contributors of the EBPR community at full scale (Blackall et al. 1998; Bond et al. 1998; Helmer and Kunst 1998; Kampfer et al. 1996; Kortstee et al. 1994; Manz et al. 1992, 1994; Mudaly et al. 2001; van Loosdrecht et al. 1997c; Wagner et al. 1994b; Wedi et al. 1995).

The culture-dependent enumeration of *Acinetobacter* spp. therefore resulted in overestimations (Wagner et al. 1994a). Similar conclusions were obtained for other dephosphatating isolates like *Microlunatus phosphovorus* or *Lampropedia* spp. Nonetheless, insights on the diversity and novel strains of *Acinetobacter* in EBPR plants was obtained with the use of RAPD-PCR (random amplification of polymorphic DNA and PCR). It was suggested that the FISH probes that were used to detect *Acinetobacter* spp. in activated sludge could not cover all the strains, and that their role in EBPR should be re-examined (Carr et al. 2001).

The in situ rRNA-based probing methods revealed a population close to the genus *Rhodocyclus* (family of *Rhodocyclaceae* inside *Betaproteobacteria*[1]) as a dominant PAO (Blackall 2000; Blackall et al. 2002; Crocetti et al. 2000; Hesselmann et al. 1999). This population has so far not been isolated and obtained in pure culture and received the provisional name "*Candidatus* Accumulibacter phosphatis". The genus "*Ca.* Accumulibacter" is composed of several clades and sub-lineages (Albertsen et al. 2016; Camejo et al. 2016; McMahon et al. 2007b; Oyserman et al. 2016a; Rubio-Rincon et al. 2019). The difficulty to isolate key microorganisms involved in EBPR or in N-removal was related to the fact that the 16S rRNA sequence-based identification of a microorganism does not allow to infer its functional properties (Wagner et al. 2002). The rRNA approach was therefore complemented with ecophysiology methods for a functional assignment of PAOs and BNR transformations in general. Enrichment cultures and mathematical models also help unravel the metabolic properties of these organisms (Lopez-Vazquez et al. 2009b; Marques et al. 2017; Oehmen et al. 2007; Páez-Watson et al. 2023; Rubio-Rincon et al. 2019; Smolders et al. 1995; Weissbrodt et al. 2014b; Ziliani et al. 2023).

"*Ca.* Accumulibacter" was particularly involved in dephosphatation in bench-scale reactors fed with synthetic sewage like a VFA-based feed (Bond et al. 1999), but has not always been dominant in full-scale EBPR systems (Beer and Seviour 2006). According to 16 rRNA-targeted FISH and polyphosphate staining, "*Ca.* Accumulibacter" populations were present at 9–24% in full-scale EBPR biomass and formed a fraction between 40 and 60% of the guild of PAOs.

[1] The class of *Betaproteobacteria* was recently reclassified as an order within the class of *Gammaproteobacteria*. The tree of life was revised using a standardized bacterial taxonomy based on genome phylogeny made possible by the recovery of high-quality MAGs from metagenomes (Parks et al. 2017, 2018).

Dechloromonas spp. belong to the same family of *Rhodocyclaceae* as "*Ca.* Accumulibacter" and often gets selected under the same anaerobic-aerobic conditions. *Dechloromonas* and "*Ca.* Accumulibacter" are therefore often co-present in EBPR enrichment cultures fed with VFAs. *Dechloromonas* was recently shown to store polyphosphate. "*Ca.* Dechloromonas phosphatis" and "*Ca.* Dechloromonas phosphovora" are new members of the PAO guild (Petriglieri et al. 2020).

Tetrasphaera spp. (family of *Intrasporangiaceae* inside *Actinobacteria*) were proposed as other important PAOs in EBPR activated sludge and granular sludge plants (Daims et al. 2006; Ekholm et al. 2022; Eschenhagen et al. 2003; Gunther et al. 2009; Hanada et al. 2002; Kong et al. 2005; Liu et al. 2001; Maszenan et al. 2000; Nguyen et al. 2011; Nielsen et al. 2012a; Wang et al. 2009b). *Tetrasphaera* spp. are versatile fermenting bacteria able to ferment sugars and amino acids, and to store a variety of polymeric products (Fernando et al. 2019; Herbst et al. 2019; Xia et al. 2008). The anaerobic-aerobic dephosphatating metabolism of *Tetrasphaera* was linked to glycogen cycling in a genome-based metabolic model (Kristiansen et al. 2013). However, its metabolism(s) needs a closer look. Lately, the examination of high-quality MAGs of the *Tetrasphaera* relatives mainly detected in these EPBR systems led to their reclassified to "*Ca.* Phosphoribacter" (Singleton et al. 2022).

An additional PAO candidate of full-scale plants is "*Ca.* Halomonas phosphatis" (Nguyen et al. 2012). The diversity of members of the guild of PAOs is certainly not yet fully resolved (Stokholm-Bjerregaard et al. 2017). Future studies combining microbial ecology, omics, and ecophysiology methods will help unravel them and their effective contribution to EBPR.

2.7.4 A Holistic View on the BNR Microbiome

Difficulties in microbial isolations or enrichments also led researchers to consider an EBPR or BNR community as a whole ('meta-organism'). Instead of trying to define single organisms that do single jobs, microbial communities can be handled as pools of organisms that can undertake a function, and which develop in a continuous stochastic cycle of immigration, birth, and death (Curtis and Sloan 2006; Frigon and Wells 2019; Guo et al. 2019; Ofiteru et al. 2010).

A core microbiome is present in a large number of BNR plants, despite significant differences in wastewater composition, plant design and operation, such as exemplified in Denmark (Nielsen et al. 2010; Singleton et al. 2021) and world-wide (Dueholm et al. 2022; Wu et al. 2019). Clone libraries, quantitative FISH (qFISH), amplicon sequencing, and high-quality MAGs were successively used and combined to resolve the BNR microbiomes across latitudes. In Danish WWTPs, a limited number of core species represent around 80% of the bacterial communities. Major functional groups are filamentous bacteria (28%), denitrifiers (18%), PAOs (13%), nitrifiers (7%), other hydrolysers (3%), fermenters (3%) and GAOs (1%), and a remaining 27% fraction of 'other and unknown' bacteria. These flanking populations will be soon uncovered using MAGs.

Bacterial affiliates of the order *Rhodocyclales* form an important, widespread fraction of heterotrophic denitrifying organisms in full-scale BNR plants (Hesselsoe et al. 2009; Juretschko et al. 2002; Loy et al. 2005; Morgan-Sagastume et al. 2008; Rossello-Mora et al. 1995; Thomsen et al. 2007). The order *Rhodocyclales* is an assembly of physiologically versatile bacteria that contribute to removing anthropogenic compounds like C, N, and P. A total of 27 OTUs were grouped in the 7 dominant lineages affiliating with the family of *Rhodocyclaceae*, which can conduct nitrogen transformations with different carbon sources. These sub-groups are composed of the following genera: (i) *Azoarcus*, *Thauera* and *Denitromonas*; (ii) *Azovibrio*; (iii) *Zoogloea*; (iv) *Rhodocyclus*, *Propionivibrio* and "*Ca.* Accumulibacter"; (v) *Azospira*; (vi) *Dechloromonas*, *Ferribacterium*, *Quadricoccus* and *Azonexus*; and (vii) *Sterolibacterium*. These populations exhibit a high level of functional redundancy, notably in denitrification abilities. Their functional versatility in substrate usage is a factor governing the niche overlap and differentiation of this bacterial order. In addition, "*Ca.* Accumulibacter" and *Dechloromonas* belong to the PAO guild. Several clades of "*Ca.* Accumulibacter" can denitrify, by reducing different nitrogen oxides used as electron acceptors (Camejo et al. 2016; Gao et al. 2019; Rubio-Rincon et al. 2019).

Based on 16S rRNA-targeted microarray and sequencing data, specific FISH probes were designed to detect and quantify the presence of these important functional populations in activated sludge.

2.7.5 A Conceptual Model of the Microbial Ecosystem of BNR Processes

A comprehensive conceptual model was developed for the EBPR/BNR ecosystem, using knowledge gained on microbial guilds, taxa, and their ecophysiologies (Nielsen et al. 2010, 2012b).

In the model, the organic matter was fractionated into different components in relation to microbial substrate specificities, namely lipids, polysaccharides, VFAs and ethanol, nitrogen and phosphorus, proteins, and others compounds like nucleic acids. The model presented relationships between the guilds, the microbial populations, and the turnover of these various substrates under anaerobic, anoxic, and aerobic conditions. Information retrieved from the ecosystem model on guilds and populations is summarized in Table 2.2.

Under real conditions, EBPR dynamics impact the competition between the microbial populations, by acting on their physiologies like their respiratory capabilities, substrate uptake kinetics, accumulation of storage compounds, release of exoenzymes, sensitivity to starvation, and sensitivity to chemicals. Abiotic factors like pH, temperature, salinity, micronutrients, and the size of flocs further influence microbial competition. The ecosystem model can be used as practical information to manipulate populations and function in the EBPR process, based on knowledge

Table 2.2 Functional groups (guilds) and bacterial populations of the conceptual model of the EBPR ecosystem (Nielsen et al. 2010), detected in situ by rRNA-targeted FISH

Main guilds and populations of the EBPR/BNR ecosystem
Nitrifiers
AOOs: g_*Nitrosospira*, g_*Nitrosomonas* (s_*N. oligotropha*)
Other AOO: g_*Nitrosococcus*
NOOs: g_*Nitrospira* sublineage I and II
Denitrifiers
c_*Betaproteobacteria*[a]: g_*Thauera*, g_*Azoarcus*, g_*Zoogloea*, g_*Curvibacter* (or *Aquaspirrilum*), g_"*Ca.* Accumulibacter"
Other denitrifiers
c_*Betaproteobacteria*: g_*Dechloromonas*, f_*Comamonadaceae*
c_*Alphaproteobacteria*: f_*Rhodobacteraceae*, f_*Hyphomicrobiaceae*
c_*Gammaproteobacteria*: g_"*Ca.* Competibacter", g_*Pseudomonas*
g_*Paracoccus*, g_*Hyphomicrobium*, g_*Acidovorax*, g_*Pseudomonas*
PAOs
c_*Betaproteobacteria*: g_"*Ca.* Accumulibacter", g_*Decholoromomas*
c_*Actinobacteria*: g_*Tetrasphaera*[b]
GAOs
c_*Gammaproteobacteria*: g_"*Ca.* Competibacter"
c_*Alphaproteobacteria*: g_*Defluviicoccus*
Fermenters
c_*Actinobacteria*: g_*Tetrasphaera*
p_*Firmicutes*: g_*Streptococcus*
Hydrolyzers
Proteins hydrolyzers: g_*Curvibacter*, p_*Chloroflexi*, p_*TM7*[c], g_*Epiflobacter*
Polysaccharides hydrolyzers: g_*Tetrasphaera*
Lipids hydrolyzers: filamentous g_*Microthrix*, filamentous Mycolata group (g_*Gordonia*, g_ *Skermania*)
Others: p_*Bacteroidetes*
Filamentous organisms
p_*Chloroflexi*, p_*TM7*
c_*Betaproteobacteria*: g_*Curvibacter*
p_*Bacteroidetes*: g_*Haliscomenobacter*
c_*Actinobacteria*: g_*Microthrix*, Mycolata group (g_*Gordonia*, g_*Skermania*)
Other organisms

(continued)

Table 2.2 (continued)

Main guilds and populations of the EBPR/BNR ecosystem
Sulfate-reducing bacteria c_*Deltaproteobacteria*: f_*Desulfovibrionaceae*, f_*Desulfobacteriaceae*
Iron(III)-reducing bacteria c_*Deltaproteobacteria*: g_*Geobacter*

These organisms are abundant among the investigated 25 Danish EBPR plants (Mielczarek et al. 2013). The names of the bacterial populations are provided along phylogenetic levels, namely *p*_ phylum, *c*_ class, *o*_ order, *f*_ family, *g*_ genus, and *s*_ specie

AOOs aerobic ammonium-oxidizing organisms, *NOOs* aerobic nitrite-oxidizing organisms, *PAOs* polyphosphate-accumulating organism, *GAOs* glycogen-accumulating organisms

[a] The class of *Betaproteobacteria* were recently reclassified as an order of the class of *Gammaproteobacteria* based on the analysis of the thousands of MAGs available in public databases (Parks et al. 2017, 2018)

[b] *Tetrasphaera* affiliates detected in the sludge were recently reclassified as "*Ca.* Phosphoribacter" based on the annotation and comparison of high-quality MAGs (Singleton et al. 2022)

[c] The candidate division TM7 has recently been reappraised as candidate phylum "*Ca.* Saccharibacteria" (Albertsen et al. 2013a)

gained on factors that control their selection and metabolisms. The model can also address the relationships between the microbial and chemical compositions of flocs and their physicochemical properties.

Multiple interacting factors structure the guild of PAOs, e.g., the wastewater composition and operational conditions (He et al. 2008). The higher abundances of "*Ca.* Accumulibacter" measured in full scale EBPR plants were related to a lower COD/P ratio in the influent, and with flowcharts aiming at preventing the leakage of nitrate into anaerobic tanks (He et al. 2006).

Eukaryotic (protozoans, metazoans) and viral (phages) predators were included in the ecosystem model as factors regulating the microbial community structure. Other studies highlighted correlations between EBPR community compositions and the presence of phages. In enrichment cultures, "*Ca.* Accumulibacter" was prone to phage attack (Albertsen et al. 2011a; Barr et al. 2015; Khan et al. 2002; Kunin et al. 2008; Lee et al. 2004; Skennerton et al. 2011). Bacteriophage therapies were proposed as solutions to shape microbial communities, such as for remediating filamentous bulking and removing biofilms on membranes (Bhattacharjee et al. 2015; Kotay et al. 2011; Motlagh et al. 2016; Shivaram et al. 2023; Withey et al. 2005). Phages are also known to not only suppress but also propagate antimicrobial resistances in wastewater environments (Gunathilaka et al. 2017; Subirats et al. 2016).

Besides phages, eukaryotic predators like protozoans, nematodes, and oligochaetes can impact biofilm and granule structures by grazing the microorganisms (Aybar et al. 2019; Derlon et al. 2012; Klein et al. 2016). Interkingdom interactions between prokaryotes (bacteria and archaea), viruses (phages), and eukaryotic organisms can be harnessed to unravel ecological questions.

2.7.6 Selecting for Microorganisms with BNR Activities

Microbial community engineering relies on the selection of microorganisms with metabolisms necessary for BNR (Weissbrodt et al. 2020a).

Selecting for the right guilds and populations is not only important for conferring metabolic activities, but also for enabling a good solid-liquid separation and sludge dewatering. In activated sludge, the denitrifying *Zoogloea* and *Thauera* inside *Rhodocyclaceae* can produce large slimy colonies that trap water and hinder sludge compaction. Manipulating process parameters to either limit their EPS overproducing metabolism (e.g., by manipulating nutrient ratios) or select for other denitrifiers is required for an optimal plant operation.

The aforementioned ecosystem model supports the hypothesis of an intermediate disturbance where variations in, e.g., substrate composition, concentration levels or starvation times can maintain a high functional redundancy, stability and resiliency in the microbial community (Nielsen et al. 2010).

Robust wastewater treatment processes should be engineered using ecological principles of microbial selection. This requests a fundamental understanding of the metabolisms and the conditions that selects for the microorganisms of interest, such as for PAOs in EBPR systems.

2.7.7 Selecting for Polyphosphate-Accumulating Organisms Over Their Competitors

PAOs are selected in EBPR systems by alternance of anaerobic feast and aerobic starvation conditions (Comeau et al. 1986; Smolders et al. 1995). The deterioration of EBPR performances often results from the proliferation of undesired microorganisms that outcompete PAOs (van Loosdrecht et al. 2008; Wentzel et al. 2008). Different mechanisms and microorganisms can get involved. Most of the insights were gained from investigations of "*Ca.* Accumulibacter" which was used as model PAO over the last 25 years.

PAOs are outcompeted by OHOs when the soluble COD (e.g., VFAs) is only partially taken up anaerobically and leaks into aeration. Fast-growing OHOs like the *Zoogloea*, *Xanthomonas* and *Comamonas* floc-formers, or other filamentous organisms, scavenge the biodegradable organics aerobically, surpass the PAOs, and hinder sludge settling by viscous or filamentous bulking (Fang et al. 2002; Martins et al. 2004a; Montoya et al. 2008; van Loosdrecht et al. 2008). A similar mechanism can occur in anaerobic-anoxic processes, when DHOs outcompete PAOs when a large fraction of COD leaks from the anaerobic to the anoxic phase (Chuang et al. 1996). The design and control of an anaerobic selector is efficient to suppress the growth of these organisms and to preferentially select for PAOs. A full uptake of the biodegradable COD should be controlled under anaerobic conditions prior to contacting the wastewater with terminal electron acceptors like dissolved oxygen or nitrogen oxides.

GAOs compete with PAOs for the uptake of VFAs under anaerobic conditions. Specific combinations of nutrient ratios (i.e., COD/P) and environmental conditions (e.g., T, pH) impact the competition between PAOs and GAOs (Lopez-Vazquez et al. 2009b; Mino et al. 1994; Weissbrodt et al. 2013b, 2014b). PAO and GAO metabolisms are outlined in the previous Fig. 2.1. GAOs exhibit a similar A/O metabolism as PAOs, but do not store polyphosphate. The intracellular glycogen pool is used to provide energy and electrons for VFA uptake and PHA storage.

The competition between PAOs and GAOs has been widely studied from selective enrichment cultures in the lab to process scale under regional conditions (Brdjanovic et al. 1998a; Filipe and Daigger 1998; Lopez-Vazquez et al. 2009a, b; Mino 2000; Mino et al. 1994, 1998; Oehmen et al. 2006a, b, 2007; Qiu et al. 2022; Sabba et al. 2023; Schuler and Jenkins 2003; Smolders et al. 1994b, 1995; Weissbrodt et al. 2013b, 2014b, 2017; Winkler et al. 2011a; Zeng et al. 2003c; Law et al. 2016; Stokholm-Bjerregaard 2016). The factors governing their competition are summarised in the next section.

Like PAOs, the main GAOs were not isolated and cultivated in pure cultures. Culture-independent methods emended the guild of GAOs. GAOs exhibit a complex phylogenetic picture consisting of different morphotypes and phylotypes. "*Candidatus* Competibacter phosphatis" (inside *Gammaproteobacteria*) is a major GAO specie and considered as model organism in experimental investigations with enrichment cultures (Crocetti et al. 2002; Weissbrodt et al. 2014b; Zeng et al. 2003c). "*Ca.* Competibacter" is composed of seven phylogenetic clades (Serafim et al. 2006; Wong and Liu 2006). The family of *Competibacteraceae* was reappraised using MAGs, and FISH probes were refined to better target them (McIlroy et al. 2014a, 2015). *Defluviicoccus vanus* is an alphaproteobacterial GAO (Meyer et al. 2006). A tetrade-forming coccoid alphaproteobacterium was detected at 75% of the community of a non-EBPR sludge (Wilmes and Bond 2004). The GAO guild also comprises the betaproteobacterial genus *Azoarcus* (Serafim et al. 2006; Wong and Liu 2006).

When EBPR is not a treatment objective, GAOs can be selected for COD-removal and production of PHAs in mixed cultures (Johnson et al. 2009; Werker 2007). The selection of DGAOs in granular sludge can be interesting for densifying granules besides conferring them denitrification.

2.7.7.1 Factors Governing the Competitive Selection of PAOs and GAOs

The PAO-GAO competition in EBPR systems is a dynamic and complex process. It depends on the combined effect of different factors related to the wastewater composition (COD/P ratio, type of VFAs) and the pH and temperature of the system. The effects of the individual parameters are summarized hereafter.

COD/P ratio. The PAO-GAO metabolisms form a continuum along the COD/P ratio in the influent wastewater (Schuler and Jenkins 2003), such as summarized by Fig. 2.2. Extremely low COD/P ratios ($< 8\ g_{COD}\ g_P^{-1}$) are mostly used to enrich for highly active PAOs in research systems. However, phosphorus is supplied in excess and incompletely removed. A safety factor of 2–3 times the low COD/P ratio 8 $g_{COD}\ g_P^{-1}$ is recommended for low residual concentrations of orthophosphates in the treated effluent. COD/P ratios between 25 and 50 $g_{COD}\ g_P^{-1}$ are optimal for a safe operation of full-scale EBPR systems reaching a full P-removal. At a COD/P ratio near and above 100 $g_{COD}\ g_P^{-1}$, phosphorus limitation is high, and the GAO metabolism progressively replaces the PAO activity.

Carbon source. Both "*Ca.* Accumulibacter" and "*Ca.* Competibacter" mainly take up VFAs and store them as PHAs under anaerobic conditions. Alphaproteobacterial GAOs like *Defluviicoccus* also take up VFAs anaerobically. In a study investigating the impact of primary sludge fermentation on EBPR performances, PAOs were shown to take up the linear VFAs like acetate (C2, two-carbon chain) and propionate (C3) much faster than C4/C5 unbranched and branched VFAs (Moser-Engeler et al. 1998). The competition between the PAO "*Ca.* Accumulibacter" and

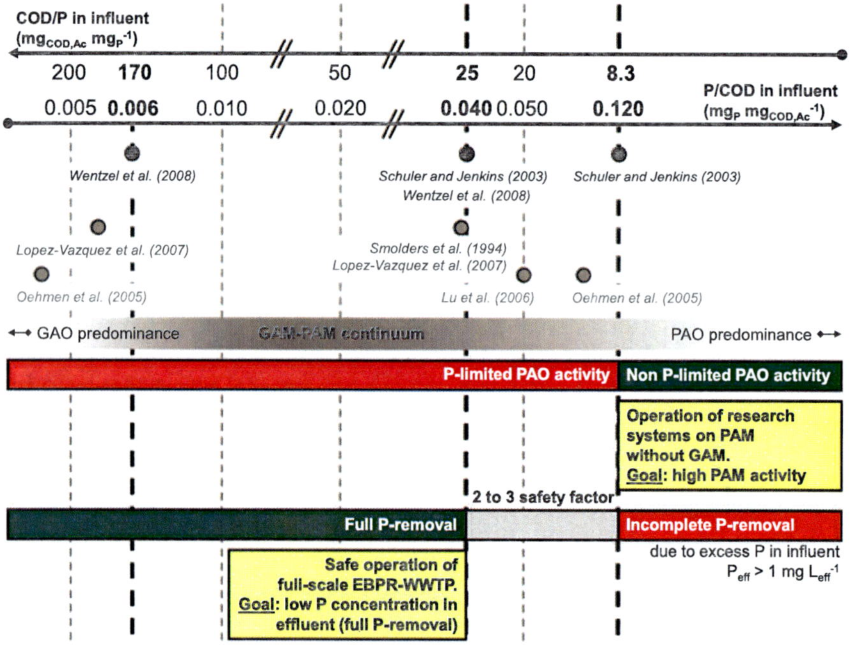

Fig. 2.2 Continuum of the PAO-GAO metabolism along the COD/P ratio of the influent wastewater. Adapted from Schuler and Jenkins (2003). *PAM* polyphosphate-accumulating metabolism, *GAM* glycogen-accumulating metabolisms. PAOs are preferentially selected by low COD/P ratios, i.e. in the presence of orthophosphate in the bulk. GAOs are preferentially selected by high COD/P ratios, i.e. under P-limitation. Bench-scale enrichment cultures of PAOs and GAOs are typically operated at COD/P ratios below 20 $g_{COD}\ g_P^{-1}$ (P in excess) and above 200 $g_{COD}\ g_P^{-1}$ (depleted P), respectively. Safe operation of full-scale plants designed for and enhanced biological phosphorus removal (EBPR) targets low residual concentrations of orthophosphate; this is typically achieved under normal operation conditions between 25 and 100 $g_{COD}\ g_P^{-1}$ in the influent wastewater

the GAOs "*Ca.* Competibacter" and *Defluviicoccus* is regulated by the ape of VFAs present in the medium (Filipe et al. 2001b, d; Lopez-Vazquez et al. 2009b; Oehmen et al. 2007; Weissbrodt et al. 2013b, 2014b). At pH 7.0 and 20 °C, "*Ca.* Accumulibacter" takes up acetate and propionate at an identical maximum specific uptake rate under anaerobic conditions ($q_{VFA,An,max}$) of 0.20 C-mmol$_{VFA}$ h^{-1} C-mmol$_X{}^{-1}$. "*Ca.* Competibacter" selectively takes up acetate as fast as "*Ca.* Accumulibacter" (0.20), but propionate much slower (0.01). *Defluviicoccus* takes up propionate at a similar rate as "*Ca.* Accumulibacter" (0.20) and acetate at a lower rate (0.10). PAOs can take up acetate and propionate at the same rate. GAOs are slower to respond to VFA switches or mixtures. PAOs are selected with mixtures of acetate and propionate in ratios of 50:50 or 75:25 in COD equivalents, or with periodical alternance of acetate and propionate in the feed (e.g., changing substrate after 1 sludge age) (Lu et al. 2006). Depending on the type of VFA, PAOs and GAOs produce different fractions of PHAs at different yields, like poly-β-hydroxybutyrate (PHB), poly-β-hydroxyvalerate (PHV), and poly-β-hydroxymethylvalerate (PH2MV) (Lopez-Vazquez et al. 2009b). "*Ca.* Accumulibacter" condenses acetate into PHB, and propionate into PHV and PH2MV. GAOs condense acetate and propionate into mixtures of PHB, PHV and PH2MV.

The versatile *Tetrasphaera*-like PAOs ferment sugars and amino acids. It can hydrolyze polysaccharides like starch into glucose by the secretion of amylases, ferment glucose into VFAs, and grow under anaerobic conditions. *Tetrasphaera* clade I can consume acetate, whereas clade II cannot. *Tetrasphera* and "*Ca.* Acumulibacter" can form an interesting cross-feeding association (Adler and Holliger 2020; Weissbrodt et al. 2014d). "*Ca.* Accumulibacter" was recently suspected to also metabolise glucose anaerobically (Ziliani et al. 2023).

pH. The extent of hydrolysis of the phosphoanhydrid bonds of intracellular polyphosphate by PAOs and the release of orthophosphate in the bulk medium are proportional to the pH of the cultivation medium (Smolders et al. 1994a). Under anaerobic conditions, more ATP is required for the transport of VFAs across the cell membrane when a decreasing pH gradient exists between the external and internal sides of the cell membrane. The yield of orthophosphate released to acetate taken up under anaerobic conditions ($Y_{P/C,An}$) therefore varies linearly from 0.25 to 0.75 mmol$_{P\text{–}PO4}$ C-mmol$_{Ac}{}^{-1}$ from pH 5.5 to 8.5. pH values above 7.25 are detrimental to GAOs, because these organisms do not have luxury pools of energetic polyphosphate that can be used to compensate the pH gradient, and to buffer the higher energy demand for VFA uptake (Lopez-Vazquez et al. 2009b; Oehmen et al. 2005). Under aerobic conditions, a pH value of 7.25–8.0 is more beneficial for the growth of PAOs on PHA stocks, and for P-uptake (Filipe et al. 2001a, c).

Temperature. It affects the anaerobic uptake of VFAs by PAOs and GAO, and their aerobic biomass production rate. The global temperature effect is a resultant of an equilibrium reached among the different A/O metabolic conversions (Lopez-Vazquez et al. 2009a, b). The net production of PAO and GAO biomass is a function of the temperature dependency of the biomass production rate, and on the PHAs available for growth. The competition between PAO-GAO metabolisms is also a continuum along temperature. Below 10 °C, the hydrolysis of glycogen under anaerobic conditions is partly inhibited. Both PAO and GAO metabolisms are affected. PAOs are however advantaged by using internal energetic pools of polyphosphate. Temperatures between 10 and 20 °C affect the anaerobic acetate uptake rate and aerobic biomass production rate of GAOs and select for PAOs. At temperatures above 20 °C, GAOs anaerobically take up acetate at higher rate than PAOs. GAOs are thus selected between 25 and 30 °C, even though their aerobic biomass production rate remains lower than PAOs. From 15 to 30 °C, the anaerobic and aerobic stoichiometries of PAOs and GAOs are insensitive to temperature changes. From 30 to 40 °C, the oxidative phosphorylation process gets less efficient for both functional groups. Even though kinetic expressions are usually enhanced at higher temperature, this results in decreased energy production, yields, and biomass growth aerobically. The SRT should be increased from 8 days at 20 °C to 16 days at 30 °C, to conserve a sufficient biomass concentration in the reactor. Strong increases in anaerobic maintenance requirements were measured at temperatures of 35–40 °C, whereas temperatures above 40 °C immediately affects the aerobic stoichiometries of both organisms.

Salinity. The salinity of the wastewater impacts the anaerobic and aerobic metabolisms of PAOs and GAOs (Welles et al. 2014, 2015a). Both PAOs and GAOs are affected under anaerobic conditions by high salinity levels (6% vs. 0.02% m/v as NaCl). PAOs are more sensitive to increasing salinity. The maximum biomass specific rates of acetate uptake by PAOs decrease by as high as 71% and GAOs by 41%, after an increase in salinity from 0 to 1%. All aerobic metabolic processes of PAOs are drastically affected by elevated salinities.

Aeration. The DO concentration and the aerobic HRT affect the PAO-GAO competition (Carvalheira et al. 2014). Because of a higher affinity for oxygen, "*Ca.* Accumulibacter" is advantaged over "*Ca.* Competibacter" at low DO levels. An increased aerobic HRT at a DO of 2 $mg_{O2}\ L^{-1}$ unfavorably sustained GAOs over PAOs and hamper EBPR. PAO selection and EBPR should benefit from an operation at low aeration, on top of energy savings.

2.7.7.2 Can PAO Clades Be or Behave Like GAOs?

The metabolisms of PAOs and GAOs are both based on the cycling of PHAs and glycogen, while PAOs can in addition use polyphosphate as energy source. It was postulated that PAOs with depleted pools of inorganic polyphosphate, e.g., under phosphorus limitation in the feed, can behave as GAOs by involving glycogen as main energy source under anaerobic conditions (Zhou et al. 2008b). This initially went against previous investigations on PAOs (Brdjanovic et al. 1998b), and was scientific debated, reappraised, and studied (Lopez-Vazquez et al. 2008a).

The "*Ca.* Accumulibacter" PAO clades I and II can adopt a GAO metabolism (Welles et al. 2015b). Under long-term operation without polyphosphate limitation for acetate uptake, clade I displayed a typical PAO metabolism ($Y_{P/C,An} = 0.64$ P-mol C-mol^{-1}), while clade II exhibited a mixed PAO and GAO metabolism ($Y_{P/C,An}$

= 0.22 P-mol C-mol^{-1}). Under batch conditions, the activity of both clades gradually shifted toward a GAO metabolism along the decrease in intracellular polyphosphate. Clade II outcompeted clade I under such limiting conditions, with a four times higher rate of acetate uptake. A novel framework was proposed on the metabolic flexibility of PAO clades. PAOS are therefore very interesting for molecular and metabolic research (Páez-Watson et al. 2023).

2.7.7.3 Measuring and Linking PAO and GAO Abundances to Operational Conditions

A rapid method was developed for quantifying the ratio of PAOs and GAOs in activated sludges (Lopez-Vazquez et al. 2007), using standardized anaerobic metabolic batch tests based on the yield of phosphorus release to acetate uptake under anaerobic conditions ($Y_{P/C,An}$) and FISH.

At full scale, correlations were drawn between the operational parameters, occurrence of PAOs and GAOs, and performance of seven EBPR plants in winter (12 °C) (Lopez-Vazquez et al. 2008b). FISH measurements and anaerobic, anoxic and aerobic metabolic batch tests were conducted. At this low temperature, GAOs could not compete with PAOs. "*Ca.* Accumulibacter" accounted for 5.7–16.4% of the bacterial community, "*Ca.* Competibacter" for 0.4–3.2%, and *Defluviicoccus* was barely detected (< 0.1%).

The occurrence of "*Ca.* Competibacter" was correlated positively with the amount of organic matter in the influent. The relative abundance of "*Ca.* Accumulibacter" correlated with the concentration of total nitrogen in the influent, the pH value in the anaerobic zones, and the rates of acetate uptake and P-release.

"*Ca.* Accumulibacter" was favored in configurations including denitrification stages. The denitrification process is known to produce alkalinity. Internal recirculation flows between anoxic and anaerobic tanks can recirculate alkalinity and stimulate VFA uptake by PAOs at higher pH values.

2.7.7.4 Denitrifying PAOs and GAOs

Denitrification was positively correlated with the occurrence of PAOs with denitrifying capacities. Denitrifying PAOs (DPAOs) affiliating with "*Ca.* Accumulibacter" (Carvalho et al. 2007; Chuang et al. 1996; Janssen and van der Roest 1996; Kong et al. 2004; Kuba et al. 1993, 1996b, 1997; Meinhold et al. 1999; Oehmen et al. 2010a; Zeng et al. 2003b) and denitrifying GAOs (DGAOs) affiliating with "*Ca.* Competibacter" and *Defluviicoccus* (Oehmen et al. 2010c; Zeng et al. 2003d) were detected in the bacterial community of BNR activated sludge processes, and contributing to N-removal.

P- and N-removal by DPAOs was achieved in biofilm systems (Brandt et al. 2002). In the presence of the nitrate or nitrite electron acceptors, DPAOs and DGAOs exhibit similar metabolic activities as with oxygen, but with a 30–40% lower energy production efficiency (de Kreuk et al. 2007; Kuba et al. 1996a; Oehmen et al. 2010c; Vargas et al. 2011). In contrast to separate P- and N-removal, the denitrifying dephosphatation by DPAOs leads to savings in the organic e-donor, in aeration energy, and in excess sludge production (Bortone et al. 1999; Kuba et al. 1996b). Under unfavorable operation of BNR systems, the accumulation of free nitrous acid inhibits the uptake of orthophosphate and the respiration rate of PAOs (Saito et al. 2004; Zhou et al. 2011b).

2.7.7.5 Microbial Community Engineering to Select for PAOs and EBPR at Full Scale

The microbial ecology knowledge gained on the PAO-GAO competition from full scale WWTPs, experimental lab work, and mathematical modelling can be used to develop strategies for minimizing the growth of GAOs and improving the operation of wastewater treatment systems aiming at conducting EBPR (Daigger and Nolasco 1995; Lemaire et al. 2008; Lemos et al. 2008; Lopez-Vazquez et al. 2008b; Manga et al. 2001; Oehmen et al. 2005; Stokholm-Bjerregaard et al. 2017).

One engineering question remains on the application of the optimal conditions at full-scale. Temperature is difficult to regulate. PAO selection can be achieved by acting on the composition of the soluble COD. A sludge pre-fermentation step directing acetate or propionate formation may be implemented upstream to amend the influent with the required amount of the target VFA that are preferentially stored by PAOs under anaerobic conditions (Thomas et al. 2003; Thomas 2008). The implementation of a side stream process for the hydrolysis and fermentation of the return sludge was favorable to control GAOs and select for PAOs (Stokholm-Bjerregaard 2016). Integrating denitrification producing alkalinity could favor PAOs at a pH > 7.0 (Lopez-Vazquez et al. 2008b).

As *Tetrasphaera*-related PAO seem to grow on a different substrate than "*Ca.* Accumulibacter" and "*Ca.* Competibacter", the classical PAO-GAO competition might be suppressed by selecting for *Tetrasphaera* at the expense of the two traditional "*Ca.* Accumulibacter" and "*Ca.* Competibacter" competitors (Daims et al. 2006; Kong et al. 2005). *Tetrasphaera* was shown to be selected under the same conditions as GAOs (Weissbrodt et al. 2013b, 2014b). Therefore, a closer look at the interactions between these different PAO and GAO genera is of interest.

Process engineering for robust EBPR performances should be driven by microbial community engineering.

2.7.8 *Resolving Molecular and Metabolic Signatures of PAOs and GAOs*

Ecophysiological knowledge is not sufficient for managing EBPR installations (Beer and Seviour 2006). The processes are mainly designed and controlled based on stoichiometry and kinetics.

This knowledge can be complemented by a deeper understanding of molecular signatures and metabolic regulations to open the microbial black box. Understanding the microorganisms and their metabolisms in their environment is important. This can go by studying the expression functional genes, their translation into proteins, their activation into enzymes, and the metabolic fluxes in the PAO and GAO metabolisms, as well as their metabolic flexibility and resource allocations in the anaerobic-aerobic turnover of PHAs, glycogen, polyphosphate, and biomass (da Silva et al. 2020; Páez-Watson et al. 2023).

2.7.8.1 High-Throughput Methods for Analyzing the Metabolisms of EBPR Lineages

Metagenomes and metaproteomes were relatively early obtained from EBPR consortia and resulted in an improved knowledge of functionalities. It provided an inventory of the metabolic potential of "*Ca.* Accumulibacter" and other EBPR lineages (Albertsen et al. 2011a; Garcia Martin et al. 2006; Wilmes et al. 2008b).

Genome-centric metagenomics, metatranscriptomics, metaproteomics, and metabolomics can be combined to investigate the metabolic regulation of EBPR populations under dynamic anaerobic-aerobic conditions. Functional expression and metabolic fluxes can be tracked in response to factors like temperature, pH, electron/carbon sources, electron acceptors. High-quality MAGs from targeted PAOs and GAOs will further lead to the formulation of lineage-specific metabolic models (Barr et al. 2015; Skennerton et al. 2015).

Screening for specific functional genes, enzymes, and metabolic pathways involved in the metabolism of PAOs helped correlate the evolution of bulk chemical species with the activities of EBPR activated sludges (Blackall et al. 2002; Burow et al. 2008a, b; He et al. 2006, 2007; He and McMahon 2011; Keasling et al. 2000; McMahon et al. 2002a, 2007b; Pramanik et al. 1999; Wilmes and Bond 2004; Wilmes et al. 2008b; da Silva et al. 2020; Páez-Watson et al. 2023).

Differences in the abundance of protein variants associated with the different clades of "*Ca.* Accumulibacter" can be a reflection of functional partitioning within the population (Wilmes et al. 2008a). This highlights the apparent importance of genetic diversity in maintaining a stable process performance.

2.7.8.2 Functional Markers of the Polyphosphate Metabolism and Beyond

Polyphosphate is formed primarily from the reversible enzyme polyphosphate kinase (PPK) or from some analogues, while two primary types of enzymes can degrade polyphosphate: (i) PPK acting in reversed mode, and (ii) the exopolyphosphatase (PPX) (Pramanik et al. 1999). Degradation by PPK yields ATP, while degradation by PPX yields orthophosphate residues. An EBPR metabolic-flux model was used to identify that the degradation of polyphosphate occurred by PPK and barely by PPX in a bench scale EBPR SBR. PPK was targeted as a predominant enzyme involved in EBPR (Gavigan et al. 1999; He et al. 2007; McMahon et al. 2002a, 2007b).

The polyphosphate enzymatic model was refined, and two kinds of PPK were highlighted (McMahon et al. 2002b; Wilmes et al. 2008a). PPK1 produces polyphosphate at the expense of ATP under aerobic conditions and is unable to hydrolyze polyphosphate under anaerobic conditions. PPK2 produces polyphosphate at the expense of GTP under aerobic conditions and degrades polyphosphate for producing GTP under anaerobic conditions. However, PPK were detected in diverse organisms, and are probably not specific to PAOs.

Further research should be conducted to determine if other enzymes can be targeted for specifically describing the anaerobic depolymerisation of polyphosphate into orthophosphate residues within PAO cells. The conceptual model of Wilmes et al. (2008a) still suggested the involvement of PPX in the anaerobic degradation of polyphosphate. Early studies on polyphosphate-degrading activated sludge revealed that in addition to polyphosphatase, polyphosphate:AMP phosphotransferase and adenylate kinase were active in the degradation of polyphosphate, and correlated with the metabolism of PAOs (van Groenestijn et al. 1987, 1989b). Other literature reports on the molecular biology of the polyphosphate metabolism are available here (Ahn et al. 2006; Chavez et al. 2009; Hernandez et al. 2008; Lee et al. 2006; Lindner et al. 2009; Nesmeyanova 2000).

Besides polyphosphate, additional research on the PAO metabolism and enzymes highlighted that PAOs synthesize PHAs using acetoacetyl coenzyme A reductase driven by NADH (instead of NADPH initially assumed) (da Silva et al. 2020). It confers them with a metabolic flexibility for maintaining their metabolic activity under substrate variations, high carbon conservation, and low energetic costs. PHA accumulation can therefore be considered as not only a stress response but also a fermentation product. This leads to new research on resource allocation in the PAO metabolism and on their propensity for growth under anaerobic-aerobic alternance and other substrates that these organisms can consume (Páez-Watson et al. 2023; Ziliani et al. 2023).

2.7.8.3 PAO Clades of the "*Ca.* Accumulibacter" Lineage

On top of the analysis of the phylogenetic diversity, the functional diversity can associate microbial populations with a function in EBPR systems.

Degenerated primers targeting *ppk* genes were designed to follow their diversity over time in EBPR systems, and to differentiate clades of "*Ca.* Accumulibacter" (McMahon et al. 2002a, b, 2007b). The *ppk1* gene is used genetic marker to study the structure of EBPR communities with high phylogenetic resolution. Analyzing the relative abundances of "*Ca.* Accumulibacter" by targeting *ppk1* led to similar results a with the 16S rRNA gene (He et al. 2007). *ppk1* analyses enabled to structure the "*Ca.* Accumulibacter" lineage in at least 5 different *ppk1* clades (IA, IIA, IIB, IIC, IID) forming ecologically distinct ecotypes. These were close to the 4 clades obtained with the 16S rRNA gene (IA, IIA, IIB, IIC) in full-scale EBPR plants. EBPR sludges cultivated under laboratory conditions have exhibited mainly clades I and IIA.

Full scale operation conditions with more complex feeds and temporal fluctuations create more niches for closely related groups. Clade-specific *ppk1*-targeting PCR primers were designed to analyze the relative distributions and abundances of the five clades among full-scale EBPR and non-EBPR plants in addition to the full relative abundance of the genus "*Ca.* Accumulibacter".

In a lab-scale EBPR reactor, qPCR and ARISA showed that the "*Ca.* Accumulibacter" clades IA and IIA shifted their abundances several times without disturbing the P-removal performance (He et al. 2010a). The different clades provided functional redundancy and increased the robustness of the EBPR systems.

With metatranscriptomic array analyses and RT-qPCR, the transcription of functional genes involved in the metabolism of "*Ca.* Accumulibacter" were linked to the alternated anaerobic feast and aerobic starvation conditions imposed in a bench-scale EBPR SBR (He et al. 2010b; He and McMahon 2011). Dynamic patterns in genetic expression were detected during EBPR cycles. The total RNA yield profile correlated with the typical A/O profile of orthophosphate in the bulk. Among others, the *ppk1* gene was functional in both the degradation polyphosphate and its synthesis (i.e., active in reverse mode as well).

Genome-based phylogenetic trees computed with high-quality MAGs retrieved from "*Ca.* Accumulibacter" lineages resulted in similar taxonomic pictures as obtained with *ppk1*. Genome annotations helped further resolve the metabolic traits of these clades (Camejo et al. 2016; Rubio-Rincon et al. 2019). Enrichment cultures, metabolic experiments, and high-quality MAGs allowed for disentangling the metabolic capabilities of "*Ca.* Accumulibacter" notably in relation to denitrification. It was notably shown that the "*Ca.* Accumulibacter delftensis" affiliating with clade IC was not able to denitrify using nitrate to remove phosphorus (Rubio-Rincon et al. 2019). The annotation of its MAG highlighted that this organism does not contain the nitrate reductase (*nar*) that is essential to drive a nitrate-dependent denitrification catabolism.

2.8 Microbial Ecology of AGS Systems

The AGS science investigates across physical, chemical, and biological phenomena within the process boundaries (Winkler et al. 2018). Resolving the selection microorganisms, their role in granulation, and the establishment of their metabolisms for BNR is an important consideration in parallel to the development, scale-up, installation, and operation of the AGS technology under local constrains.

The use of microbial ecology principles is central for shaping a robust granular sludge and intensifying the BNR process. It is also useful for unraveling failures and developing process strategies. A multilevel integration should be adopted for managing the physical, chemical, microbial and metabolic properties of AGS biofilms. These levels of investigation are interconnected and impact on each other back-and-forth.

Microbial ecology knowledge gained from the last 100 years of research and practice on activated sludge, biofilm, and other granular sludge reactors (Jenkins and Wanner 2014; Sheik et al. 2014) can support investigations of AGS systems. Granules lie on a continuum from flocs to biofilms (Weissbrodt et al. 2014a). Based on their metrics and internal architectures, granules differ from flocs and resemble biofilms. Based on their mobility, granules are similar to flocs. Inside the continuum of flocs-granules-biofilms, granules combine the benefits of biofilms and flocs. In a simple definition: granules are large bioaggregates that behave like biofilms and that are mobile like flocs.

Granular sludge nonetheless displays its heterogeneous specificity since not a biomass as homogeneous as flocculent activated sludge and not a biofilm as homogeneous as a planar biofilm system. This heterogeneity can lead to specific microbial selection phenomena and metabolic niche establishment inside granules. These mechanisms rely on granule metrics, particle size distribution, internal architectures, and physical-chemical factors like mass transfer patterns and limitations of dissolved compounds, as well as biomass detachment and retention.

2.8.1 Out-Selecting Filamentous Populations and Selecting Floc-Forming BNR Microorganisms in Granules by Managing Selective Pressures

Analyses of microbial communities of aerobic granules were conducted by 16S rRNA gene fingerprinting by DGGE, T-RFLP, cloning-sequencing, amplicon sequencing, metagenomics, and metaproteomics, as well as light microscopy, qFISH or cryosectioning and FISH-CLSM (Adler and Holliger 2020; Ali et al. 2019; Caluwé et al. 2022; Ebrahimi et al. 2010; Etterer 2006; Guimarães et al. 2018; Henriet et al. 2017; Kleikamp et al. 2022; Li et al. 2008a; Liu and Liu 2006; Lochmatter and Holliger 2014; Morgenroth et al. 1997; Shin et al. 1992; Szabo et al. 2017; Wagner et al.

2015b; Weissbrodt et al. 2012a; Adav et al. 2009b; Juang et al. 2009; Williams and de los Reyes III 2006).

Like for activated sludge systems, the AGS microbiology early focused on the detection and identification of filamentous organisms that proliferate to the outer sphere of granules, induce filamentous bulking, impair the granule settling, and lead to process failures. Depending on the carbon source used, filamentous prokaryotes or eukaryotes are obtained. Filamentous bacteria like *Sphaerotilus*, *Beggiatoa*, *Leptothrix*, *Thiothrix* spp. and the filamentous Eikelboom Type 0411 (closely related to *Runella slithyformis*) (e.g., on acetate) or filamentous yeast-like fungi like *Geotrichum* sp. (e.g., on glucose or COD-mixtures of acetate and glucose 1:3) were detected when granulation under wash-out dynamics was hampered by filamentous bulking. When filamentous bulking was prevented during start-up, early-stage granules were often composed of the floc-forming and EPS-producing *Zoogloea* and *Thauera* spp. (> 50% of the community), accompanied by flanking populations of *Pseudomonas*, *Acinetobacter*, *Alcaligenes*, *Flavobacterium*, and *Chryseobacterium*, as well as *Comamonadaceae*, *Sphingomonadaceae* and *Cytophagales* relatives among others. Selection for *Zoogloea* mainly occurred under operation with pulse feeding of a VFA-based medium followed by aeration with OLR_v of in general 1.5–6 but also as high as 16.7 kg_{CODs} d^{-1} m^{-3}. Its proliferation was also detected in anaerobic-aerobic granular sludge SBRs, when the VFAs were incompletely converted during the anaerobic feeding phase. Temperature impacted the competition between floc formers and filamentous bacteria in anaerobic-aerobic SBRs. The latter were more abundant at 30 °C than at 20 °C.

During start-up, BNR was frequently hampered, leading to primarily COD removal only (Ebrahimi et al. 2010; Weissbrodt et al. 2012a). Ammonium was not nitrified, and phosphorus was not removed. Over the first 2–3 months, even though a full BNR activated sludge was used as inoculum, nitrifiers, PAOs and GAOs were frequently outcompeted. The growth of OHOs was sustained by a low uptake of VFAs during the anaerobic phase (e.g., max 25%) and their leakage into aeration. Managing the anaerobic selector was the key to maintain a BNR active biomass during the start-up of granulation processes, selecting for and re-equilibrating the relative abundance of PAOs and nitrifiers (Lochmatter and Holliger 2014; Weissbrodt et al. 2013a). Stepwise increase of the OLR_v and adaptation of the anaerobic contact time in function of the residual biomass concentration and accumulation were important keys.

Managing the selective sludge discharge is an important aspect for granulation and microbial selection in granular sludge processes (Guimarães et al. 2018; Henriet et al. 2016; Li and Li 2009; Lochmatter and Holliger 2014; van Dijk et al. 2018, 2020; Weissbrodt et al. 2013b; Winkler et al. 2011a, b). Applying a fixed short settling period in the SBR cycle (e.g., between 1 and 5 min) or progressively decreasing the settling time (e.g., from 30 to 3 min) to avoid wash-out and purging a fraction of the mixed liquor before the end of the aeration phase or selectively removing top or bottom fractions of the AGS sludge bed can lead to different efficiencies in granulation and in microbial selection. A selective removal of the top fractions of the sludge bed was used to out-select GAOs and maintain a granular sludge enriched in

PAOs (located in dense granules at the bottom of the sludge bed) at high temperature (Winkler et al. 2011a). A similar approach was used to select for anammox bacteria forming dense colonies in granular sludge while out-selecting NOB mostly present in the slower-settling upper bed fraction (Winkler et al. 2011b). A good molecular characterization of the gradients of microbial niches not only inside granules but also over the height of the settled AGS bed is important to this end. These gradients can be system specific (Weissbrodt et al. 2013b).

A battery of microscopy and FISH analyses can reveal the composition of the microfauna embedded in the EPS of granules, like bacteria, protozoans, and fungi and their contribution to granulation (Weber et al. 2007). Colonization of ciliates like *Epistylis* by bacteria was proposed to initiate granule formation. Surface outgrowth of rotifers further resulted in the deterioration and disintegration of aerobic granules, when cultivated at a low SGV (Li et al. 2007b). Granules developed under higher shear exhibited higher resistance to grazing organisms.

Additional studies investigated the microbial communities of AGS reactors operated for the removal of industrial organic compounds such as phenol, chlorophenols, fluorophenols, nitrophenols, chloranilines and petrochemical compounds, and the impacts of their loadings (Caluwé et al. 2022; Chen et al. 2009b; Duque et al. 2015; Jemaat et al. 2013; Jiang et al. 2004; Liu et al. 2008a; Zhu et al. 2008a).

Conditions to select both for granules, BNR microorganisms, and a high-rate BNR are important for robust AGS processes. The combination of microbial ecology methods with process engineering is efficient to manage selective pressures.

2.8.2 Competition of PAOs and GAOs in Granular Sludge

Alternate anaerobic-aerobic (A/O) conditions are beneficial for remediating the growth of filamentous organisms and for selecting slow-growing PAOs that stabilize the compactness of granules and remove phosphorus (de Kreuk and van Loosdrecht 2004; Guimarães et al. 2018; Henriet et al. 2016; Weissbrodt et al. 2013a; Winkler et al. 2012c). When P-removal is not an objective, the presence of slow-growing GAOs is also beneficial for stabilizing granules.

Conditions selecting for PAOs and GAOs were studied in granular sludge systems (Barr et al. 2010a; de Kreuk and van Loosdrecht 2004; Weissbrodt et al. 2013b; Winkler et al. 2011a). Previous knowledge from selection mechanisms in activated sludge provided a strong basis. PAOs were predominant at 15–20 °C in mature AGS SBRs fed with an influent synthetic wastewater composed of 20 $g_{COD}\ g_P^{-1}$. GAOs have proliferated over PAOs under phosphate depletion in the influent (de Kreuk and van Loosdrecht 2004), or after a step-wise switch in temperature from 20 to 30 °C (Ebrahimi et al. 2010). Operating the reactor under phosphorus-limiting conditions has led to a predominance of PAOs over GAOs.

Winkler et al. (2011a) have demonstrated that PAOs can be preferentially selected over GAOs even at higher mesophilic temperatures by directed sludge removal from the upper part of the heterogeneous settled AGS bed. In this study, fast-settling granules present in the lower part of the AGS bed were predominantly composed of PAOs, whereas GAOs were present in the upper granules. This segregation was linked to a higher density of granules containing polyphosphate storage compounds.

Gonzalez-Gil and Holliger (2011) have analyzed the bacterial community of aerobic granules cultivated on either acetate or propionate carbon sources for EBPR in bubble-column AGS-SBRs operated without fixing the sludge age. Whereas floc-forming *Zoogloea-* and filamentous *Thiothrix*-affiliated OTUs were abundantly present in the early-stage granules (50 days), the propionate carbon source has selected for more stable community and EBPR. GAOs have never been able to compete with PAOs on this propionate carbon source, whereas GAOs have outcompeted PAOs when PAO cells present in granules cultivated on acetate have presumably been saturated with polyphosphate after 200 days. The conduction of a polyphosphate-stripping operation has enabled to recover the PAO population and the EBPR activity.

By applying T-RFLP and cloning-sequencing methods combining the traditional strategy targeting 16S rRNA genes with the functional *ppk1*-targeted approach developed by McMahon et al. (2007b), the same authors have detected that propionate selected for the "*Ca.* Accumulibacter" clade IIA. With acetate, transient shifts between "*Ca.* Accumulibacter" clade I and clade II were observed before clade IIA became predominant. The authors have concluded that the type of "*Ca.* Accumulibacter" lineage determines the robustness of the EBPR process in AGS.

Acidovorax- and *Herpetosiphon*-related populations have in addition been detected in the same study as important flanking populations in aerobic granules. *Acidovorax* spp. inside the betaproteobacterial *Comamonadaceae* can reduce nitrate with acetate and propionate VFA, as well as with PHA carbon sources originating from cell lysis. The *Herpetosiphon* spp. were described as a non-bulking filamentous population inside *Chloroflexi* that can use EPS produced by other organisms or originating from cell lysis as well.

At viral level, Barr et al. (2010b) have correlated the deterioration of the structure of granular biofilms and EBPR performances in an A/O AGS-SBR with the proliferation of bacteriophages, and their specific infection into the enriched "*Ca.* Accumulibacter"-affiliates. This phenomenon illustrates the ecological hypothesis on 'killing-the-winner', which aims at explaining the dynamics of microbial communities under predation and parasitic pressures (Shapiro and Kushmaro 2011; Winter et al. 2010). For microbial enrichment cultures, it formulates that the predominant organism will mainly be affected by adverse operational and micro-environmental conditions.

2.8.3 Favoring Aerobic-Anoxic Gradients for PAOs, GAOs, Nitrifiers and Denitrifiers Inside BNR Granules

The integration of microorganisms with nitrification and denitrification activities is important for BNR in granular sludge. Establishing a cooperative consortium of nitrifiers and denitrifiers inside AGS is important. Nitrification and denitrification can be combined by simultaneous or alternating processes during the aeration phase of AGS SBRs (de Kreuk et al. 2005; Layer et al. 2020b; Lochmatter et al. 2013). In fully granulated systems, granules are engineered to maintain aerobic and anoxic biovolumes favorable to nitrifiers and denitrifiers, respectively, in function of the gradients of dissolved oxygen and nitrogen oxides in their biofilm (Mosquera-Corral et al. 2005). At full scale and in hybrid granule-flocs sludge, AOOs, NOOs, DHOs, and also AMOs can stay at different locations between and inside granules and flocs. Like for hybrid biofilm-floc systems, a good consideration of microbial segregation processes is important to control the (out-)selection of (un)desired populations (De Clippeleir et al. 2013; Hoekstra et al. 2019; Laureni et al. 2019; Lotti et al. 2014; Perez et al. 2014; Winkler et al. 2011b).

The use of qPCR, FISH and CLSM helped unravel the relative abundances of nitrifiers, denitrifiers, and anammox bacteria and their localization inside granules (Sun et al. 2006; Tsuneda et al. 2003; Winkler et al. 2012a, 2018). In purely nitrifying granules, AOOs like *Nitrosomonas* and *Nitrosococcus* spp. dominated the community (as high as 75–85%) and were located at the granule surface because of their growth rate higher than NOOs (Fang et al. 2009; Shi et al. 2010; Tsuneda et al. 2003). AOOs formed 10–20 μm clusters located at a depth of 100 μm. NOOs like *Nitrobacter* and *Nitrospira* spp. were detected in lower relative abundances (5–10%) and located either close to the granule surface or in the deeper layers. Even though no organic C source was fed in these systems, OHOs could still be detected, growing on cell lysis products. Slow-growing nitrifiers stimulated granule compactness, and no filamentous organisms were detected. In reactors operated for COD-removal and nitrification, a balance between ANOs and OHOs was achieved in mature aerobic granules by adapting the COD/N ratio in the influent wastewater (Yang et al. 2003, 2004). Decreasing the COD/N ratio from 20 to 3 $g_{COD}\ g_N^{-1}$ led to lower predominance of heterotrophs and higher relative abundance of nitrifiers. The effect of oxygen mass transfer resistances on the nitrifying performances was investigated in a hybrid flocs-granules AGS-SBR (Filali et al. 2012). AOOs were present in equal amounts in granules and in flocs, whereas NOOs were preferentially located in granules. Difference in the SRTs of flocs (2 days) and granules (up to 260 days), and in the growth rates of AOOs and NOOs, selected for the faster-growing AOOs in flocs. Inside granules, AOOs were located near the granule surface. NOOs were present around the internal core, and their growth was enabled by oxygen and nutrient transport by the interspersing channels. An overabundance of NOOs (like *Nitrobacter* and *Nitrospira* spp.) over AOOs was detected in AGS (Weissbrodt et al. 2013b, 2014d; Winkler et al. 2012a). The prevalence of NOOs can be explained by different reasons: (i) some populations of *Nitrospira* were recently identified as comammox organisms (CMOs) completely oxidizing ammonium into nitrate; (ii) the back-supply of nitrite by denitrification in a nitrite oxidation/nitrate reduction loop ('nitrite loop') can further feed NOOs; (iii) some NOB populations like *Nitrobacter* can grow mixotrophically by acetate-dependent dissimilatory nitrate reduction (DNRA), leading to a 'ping-pong effect' in the production and supply of nitrite and nitrate (Freitag et al. 1987).

In processes involving SND during aeration, the location of the autotrophic biomass impacted the net removal of nitrogen (Beun et al. 2001; Jang et al. 2003). The distribution of nitrifiers in granules was influenced by the DO concentration in the bulk liquid phase. Nitrifiers were present in the outer aerobic layers (e.g., in the first 300 μm from the granule surface), while acetate was partly stored anoxically as PHAs by heterotrophs in deeper layers. With AND, nitrifiers and denitrifiers were both located near the granule surface (Adav et al. 2009a). Adaptation of lower DO concentrations in the bulk liquid phase results in a decreased oxygen penetration into the inner part of granules, and to the formation of anoxic zones within the biofilm architectures. Anoxic biovolumes favor the establishment of denitrifying clades of PAOs and GAOs in the microbial ecosystem of granules in SBRs operated for SNDPR (de Kreuk 2006; Yilmaz et al. 2008; Zeng et al. 2003a). Some clades of "*Ca.* Accumulibacter" can remove phosphorus by denitrification. At process level, this results in lower demand in carbon source. DPAOs were also detected in granules cultivated under anaerobic/aerobic/anoxic (A/O/A) conditions (Kishida et al. 2006). SND was also obtained in AGS maintained under phosphorus-limitation and enriched in GAOs (Wang et al. 2006b). If EBPR is an objective, the presence of GAOs in an AGS system is not desired since decreasing the potential of SNDPR by hampering dephosphatation (Lemaire et al. 2008). Efficient SNDPR was obtained with an AGS community dominated by $48 \pm 18\%$ of PAOs and $26 \pm 8\%$ of GAOs forming a mean PAO/GAO ratio of 1.9 ± 0.5. Full nitrification occurred with low abundances of nitrifiers (< 2–3%) (Ebrahimi et al. 2010; Lemaire et al. 2008; Lochmatter et al. 2013; Weissbrodt et al. 2014d).

One important challenge seeks to identify microbial populations within the functional guilds involved for BNR in AGS. Classical and modern methods of microbial ecology can be integrated in the engineering context. Microbial community composition of nitrifying-denitrifying granules were measured under different conditions like SND and AND and with different types of synthetic, municipal and industrial wastewater (Dobbeleers et al. 2017; Adav et al. 2010; Adler and Holliger 2020; Ali et al. 2019; Ekholm et al. 2022; Guimarães et al. 2018; Layer et al. 2019; Lochmatter et al. 2013; Szabo et al. 2017; Weissbrodt et al. 2013b, 2014d). Besides the aforementioned populations of PAOs, GAOs, AOOs and NOOs, the communities were composed of *Xanthomonadaceae* (e.g., genus *Thermomonas*), *Comamonadaceae* (*Acidovorax*), *Pseudomonadaceae* (*Pseudomonas*), *Rhodocyclaceae* (*Azoarcus*), *Moraxellaceae* (*Acinetobacter*), *Alcaligenaceae* (*Alcaligenes*), *Hyphomicrobiaceae* (*Hyphomicrobium* and *Devosia*), as well as the *Bacteroidetes* families of *Saprospiraceae* and *Porphyromonadaceae*. Among these populations, several are facultative aerobic chemoorganoheterotrophic organisms that display a potential for denitrification.

In the ecosystem model of anaerobic-aerobic BNR granular sludge (Weissbrodt et al. 2014d, 2020a; Winkler et al. 2018) (also refer to Fig. 12.1 in the conclusion Chap. 12 of this book), one important question remains on how these potential denitrifiers obtain their organic electron donor for the reduction of nitrogen oxides. Microorganisms able to anaerobically store compounds present in the wastewater as other intracellular stocks can be more diverse than the known PAOs and GAOs. Since some PAOs and GAOs can denitrify, other denitrifiers could be PAOs or GAOs as well. Besides unravelling the storage physiology of these diversified denitrifiers under anaerobic conditions, it will be important to also uncover their effective contribution to nitrogen removal in AGS.

The localisation and denitrification involvement of PAOs and GAOs were studied across DO gradients in granules. PAOs were detected both at the surface and in the inner parts at depth > 100 μm, whereas GAOs were detected near the granule surface (Kishida et al. 2006). The synthesis of PHV in GAOs was postulated to be more efficient aerobically, explaining the presence of GAOs in the oxic layers. A strong positive correlation was obtained between the PAO/GAO ratio and the oxygen profile inside granules (Lemaire et al. 2008). "*Ca.* Accumulibacter" PAOs dominated in the first 200 μm (PAO/GAO > 1), and GAOs from 200 μm inwards (PAO/GAO < 1) where DO was $< 0.05\ \mathrm{mg_{O2}\ L^{-1}}$. This trend was however not always observed for small granules < 500 μm. Additional measurements revealed that PHAs were not depleted by the end of aeration in populations located in the central anoxic zone. "*Ca.* Competibacter" was therefore presumed conduct denitrification in the system. These granules cultivated on a synthetic influent were subsequently fed with a nutrient-rich real abattoir wastewater (Yilmaz et al. 2008). Under these high nutrient loads, "*Ca.* Accumulibacter" (41%) progressively outcompeted "*Ca.* Competibacter" (n.d.). Concomitant dephosphatation and nitrite reduction was reported, with a presumable involvement of "*Ca.* Accumulibacter" for denitrification over nitrite. The nitrite shunt helped achieving an efficient N-removal in parallel to P-removal in AGS (Dobbeleers et al. 2017; Lochmatter et al. 2014). Managing N_2O emissions in SND and AND processes is another endpoint of microbial resource management (Dobbeleers et al. 2018; Lochmatter et al. 2014; van Dijk et al. 2021). The localization of AOOs and PAOs was further studied in SNDPR (Gao et al. 2010b, c). AOOs (12%) were located near the granule surface, and PAOs (40%) were present in inner parts. Three different clades of PAOs were detected in different relative abundances according to their affinity with the electron acceptors present. The overall PAO guild of the granules comprised 14% of PAOs using oxygen, 74% of PAOs used nitrate, and 12% used nitrite.

No unique conclusion converged from studies on the localization of ecological niches of PAOs and GAOs in granules, and on their predominant involvement in denitrification. DO can be supplied into the granules core not only by diffusion from the granule surface but also by transport by the interspersing channels. The distribution of redox local microenvironments and microbial populations across granules can be more complex than an onion-layered stratification that is conventionally used for rationalization of the granule ecosystem in mathematical models (de Kreuk et al. 2005, 2007; Winkler et al. 2018). At actual state of knowledge, an adequate balance of PAO and GAOs as well as ANOs guilds should be maintained for an efficient BNR in AGS systems. Other DHOs can co-exist with DPAOs and DGAOs in the granule structure. Their organic substrate can either originate from a not-yet uncovered anaerobic storage capability, or by thriving on products formed by cell lysis or by the hydrolysis of particulate organic substrates that leak to aeration. This can arise when the anaerobic contact time is not sufficient to promote hydrolysis of particulate substrates (X_s) into dissolved fractions (S_s), their fermentation as volatile fatty acids (S_{VFA}) and storage as PHAs (X_{PHA}) by PAOs and GAOs, such as exemplified in the Eawag Bio-P module of Activated Sludge Models (Koch et al. 2000b; Siegrist et al. 2002), and observed in AGS systems operated with real wastewater (Derlon et al. 2016; Layer et al. 2020a; Wagner et al. 2015b).

2.8.4 *Toward an Ecological Engineering of Granular Sludge Using Principles of Microbial Ecology*

Microbial ecology principles are essential for managing the microbial resource and engineering the microbial community of BNR granular sludge systems. Microbial ecology methods provide high resolution on the selection of microorganisms, their metabolisms, and their distributions inside bioaggregates in function of process conditions. The elaboration of conceptual ecosystem models is powerful to rationalize the complex microbial and metabolic network of microbiomes like BNR activated sludge and granular sludge processes (Nielsen et al. 2010, 2012b; Weissbrodt et al. 2014d, 2020a; Winkler et al. 2018). Microbial ecology principles, methods, and ecosystem models are used to formulate and investigate research questions on targeted microbial processes.

Microbial communities of AGS are composed of microbial guilds and populations that are commonly found in activated sludge or biofilms systems (Alloul et al. 2021; Ebrahimi et al. 2010; Li and Li 2009; Li et al. 2008a; Liu et al. 2009b; Weissbrodt et al. 2013a, 2014d; Winkler et al. 2018). The same microorganisms do the job, but granule metrics, gradients, and mobility lead to specific microbial distribution patterns in granular sludge that require a specific consideration for the design and operation of AGS SBRs for BNR. Microbial resource management fosters the stability of physical, chemical, and biological processes in granular sludge systems.

2.9 Mathematical Modelling of Activated Sludge, Biofilm, and Granular Sludge Systems

Computational workflows enabled the development mathematical models for an in-depth understanding of microbial processes in environmental engineering systems using activated sludge, biofilms, and granular sludge. Mathematical models can help integrate knowledge gained across length and time scales from experimental observations and predict the occurrence and extent of phenomena underlying the overall performance of a process. Model supports not only research investigations but also education (Morgenroth et al. 2002; Weissbrodt et al. 2023).

2.9.1 Mathematical Modelling of Activated Sludge Systems

The IWA Task Group on Mathematical Modelling for Design and Operation of Biological Wastewater Treatment developed the well-known Activated Sludge Models (ASM) Nos. 1–3 (Henze et al. 2000; van Loosdrecht et al. 2020a). ASM1 to ASM3 are reference modelling approaches for simulating microbial processes involved in biological wastewater treatment systems, understanding their interactions, and predicting BNR performances at reactor scale.

The principal features of ASM1–3 are given in Table 2.3. The Petersen or Gujer matrix of biological processes provides a compact and easy readable form (Weissbrodt et al. 2023). The ASM model formulation comprises: (i) a stoichiometric matrix comprising the stoichiometric description of the microbial processes of interest involving soluble (S) and particulate (X) model compounds; (ii) a composition matrix aiming at describing the COD, elemental, and charge compositions of all compounds for material balance purposes; and (iii) a biokinetic model with the related process rates. Briefly, each process rate is expressed as a multiplication of the biomass specific maximum transformation rate by the saturation and inhibition switching functions relative to the materials involved (so-called Monod terms) and the concentration of the target biomass population.

Metabolic models were developed besides ASM1–3 to assess mechanisms of microbial competition in activated sludge. A metabolic model was developed and calibrated to predict selections of the "*Ca.* Accumulibacter" PAO and the gammaproteobacterial "*Ca.* Competibacter" and alphaproteobacterial *Defluviicoccus* GAOs in function of the VFA composition, pH, temperature, and their combined effects (Lopez-Vazquez et al. 2009b). Mathematical relations describing the pH and temperature dependencies of stoichiometric coefficients and kinetic functions were developed for PAOs and GAOs. This model was extended with the microbial processes related to the different clades of the "*Ca.* Accumulibacter" and "*Ca.* Competibacter" lineages and their involvement in denitrification processes (Oehmen et al. 2010b, c).

Several one-dimensional (1-D) dynamic mathematical models were implemented in the AQUASIM software (Reichert 1994; Reichert et al. 1995). It provides a structured and user-friendly implementation interface composed of 4 boxes (variables, processes, compartments, links), and in which the differential equations were included at the root for different kinds of reactor compartments (e.g., mixed, biofilm, advective–diffusive). Wastewater treatment process configurations can be implemented by a sequence of reactor compartments interconnected with advective or diffusive links, and bifurcations.

Table 2.3 Simplified description of activated sludge models No. 1–2–3 (Henze et al. 2000) and the underlying COD degradation pathways

ASM No. and processes	COD degradation pathways[a]
ASM1 (1987) Nitrification Hydrolysis of particulate COD (Ox, Ax) Growth of OHO on soluble substrate (Ox, Ax) Biomass decay to hydrolysable particulate COD, and undegradable particulate compounds	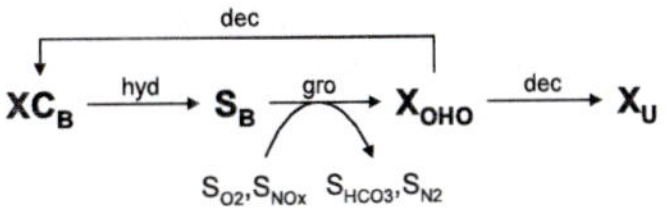
ASM2 (1991) *ASM1 + additional EBPR PAO processes* Storage of PHA by PAO with P-release (An) Storage of PP by PAO (Ox) Growth of PAO on PHA (Ox) Lysis of PAO Lysis of PP Lysis of PHA *ASM2d (1994)* *ASM2 + additional EBPR DPAO processes* Storage of PP by PAO (Ax) Growth of PAO on PHA (Ax)	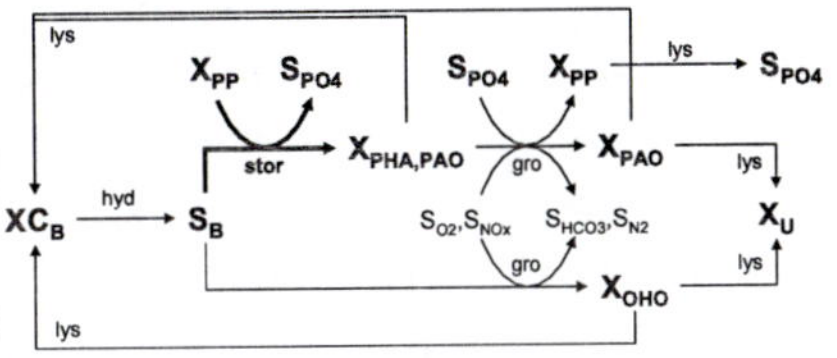
ASM3 (1998) Nitrification (independent from redox conditions) Hydrolysis particulate COD Storage of soluble COD (Ox, Ax) Growth of OHO on storage products (Ox, Ax) Endogenous respiration of biomass, and of storage products (Ox, Ax)	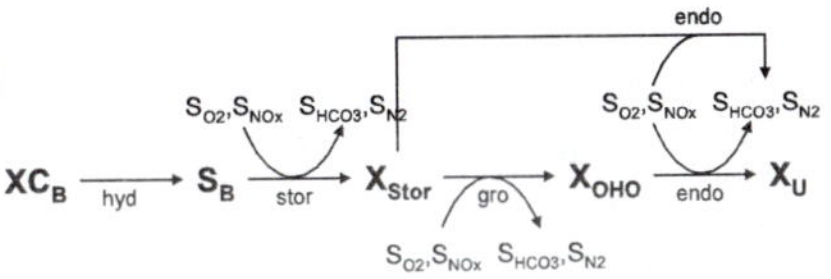
Eawag BioP module for ASM3 Storage of PHA by PAO (An) Storage of PP by PAO (Ox, Ax) Growth of PAO on stored PHA (Ox, Ax) Endogenous respiration of PAO biomass (Ox, Ax) Endogenous respiration of PHA (Ox, Ax) Lysis of polyphosphate (Ox, Ax)	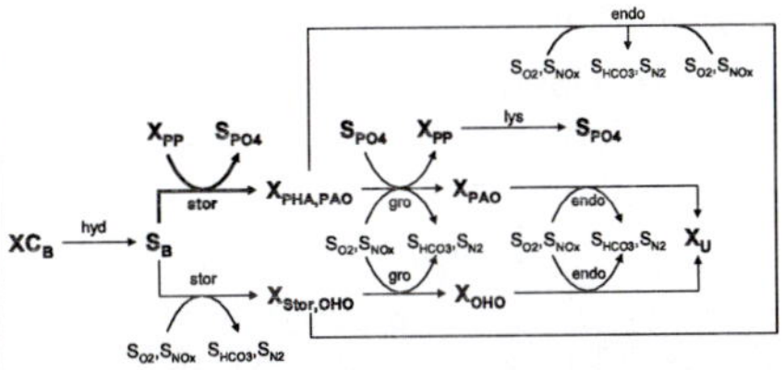

[a] The main differences between ASM1 to ASM3 rely on the conceptual models developed for the degradation pathway of chemical oxygen demand (COD). These conceptual models were compiled here as simplified metabolic flow-schemes for compact understanding. The nomenclature was adapted from the *New framework for standardized notation in wastewater treatment modelling* (Corominas et al. 2010). Processes: *hyd* hydrolysis, *gro* growth, *dec* decay, *lys* lysis, *stor* storage, *endo* endogenous respiration. Conditions: *Ox* aerobic (oxygen as terminal electron acceptor), *Ax* anoxic (nitrite and nitrate as terminal electron acceptors), *An* anaerobic (no terminal electron acceptor in the bulk). Materials: XC_B particulate and colloidal biodegradable COD, S_B biodegradable dissolved substrate, S_{O2} dissolved oxygen, S_{NOx} dissolve nitrite and nitrate, S_{HCO3} dissolved bicarbonate, S_{N2} dissolved dinitrogen, X_{OHO} OHO biomass, X_U undegradable particulate materials, X_{PP} intracellular stocks of inorganic polyphosphate, S_{PO4} dissolved orthophosphate, $X_{PHA,PAO}$ intracellular stock of PHA in the PAO biomass, X_{PAO} PAO biomass, $X_{Stor,OHO}$ intracellular storage compounds of OHOs

2.9.2 Modelling Biofilm Systems Across Length and Time Scales

Mathematical models are important to investigate biofilm systems (Morgenroth 2020; Van Loosdrecht et al. 2002; Wanner et al. 2006; Wanner and Reichert 1996). Since the early eighties, numerous approaches of biofilm modelling were developed. Biofilm models integrate phenomena across different length and time scales (Morgenroth and Milferstedt 2009). One-dimensional (1-D) to two-dimensional (2-D) and three-dimensional (3-D) model structures can be established under steady-state or dynamic conditions depending on the specific research questions (Noguera and Morgenroth 2004). The objectives of the modelling study should be made explicit before selecting a particular modelling approach and developing the model (Morgenroth 2020).

Based on the available literature, the biofilm modelling research can be divided along interconnected axes focusing on:

- **Substrate mass transfer and conversion kinetics in biofilms** (Beyenal and Lewandowski 2005; Eberl et al. 2000; Flora et al. 1993; Horn and Hempel 1998; Morgenroth et al. 2000a; Olivieri et al. 2011; Picioreanu et al. 2000b, 2016; Reino et al. 2016; Rittmann and Dovantzis 1983; Stewart 2003);
- **Formation of biofilm structures and their mechanical properties** (Aspa et al. 2011; Bishop 1997; Eberl et al. 2004; Ebrahimi et al. 2005; Hermanowicz 1998; Laspidou and Rittmann 2004; Pavissich et al. 2014; Picioreanu et al. 2000a, 2007, 2018; Prades et al. 2020; Reichert and Wanner 1997; Van Loosdrecht et al. 2002; Xavier et al. 2004);
- **Formation and role of EPS** (Horn et al. 2001; Janus and Ulanicki 2010; Kreft and Wimpenny 2001; Kuehn et al. 2001; Laspidou and Rittmann 2002; Xavier et al. 2005a; Xu et al. 2021);
- **Detachment phenomena** (Alpkvist and Klapper 2007; De Bivar Xavier et al. 2005; Delavar and Wang 2021; Elenter et al. 2007; Morgenroth and Wilderer 2000; Peyton and Characklis 1992; Picioreanu et al. 2001; Stewart 1993; Tierra et al. 2015);
- **Biofilm microbial processes** involved in nutrient removal (Beg and Chaudhry 1999; Brockmann and Morgenroth 2010; Falkentoft et al. 2000; Hubaux et al. 2015; Kermani et al. 2009; Park et al. 2010b; Sabba et al. 2015; Vannecke et al. 2015), in anaerobic digestion (Batstone et al. 2006; Buffiere et al. 1995; Cooke et al. 1999; Fuentes et al. 2008; Suidan et al. 1994), and in wastewater transformations occurring in sewer systems (Huisman and Gujer 2002; Hvitved-Jacobsen et al. 1998; Jiang et al. 2009; McCall et al. 2016; Nielsen et al. 1992);
- **Structure and dynamics of biofilm microbial communities** (Battin et al. 2007; Bishop and Rittmann 1995; Fagerlind et al. 2012; Gujer 1987; Jayathilake et al. 2017; Lardon et al. 2011; Lu et al. 2007b; Massoudieh et al. 2010; Morgenroth and Milferstedt 2009; Noguera and Picioreanu 2004; Noguera et al. 1999b; Picioreanu et al. 2004; Rittmann et al. 2002; van Loosdrecht et al. 1995; Xavier et al. 2005b);
- **Social biofilm models** (Xavier and Foster 2007) used to demonstrate that EPS-producers have a strong competitive advantage for pushing their lineages up

to better oxygenation conditions while harming non EPS-producers, whereas biofilms have often previously been considered as a cooperative mode of growth in a protective EPS matrix;

- **Development of biofilm technologies** such as biotrickling filters (Popat and Deshusses 2011; Vanhooren et al. 2000; Vayenas et al. 1997), rotating biological contactors (Dutta et al. 2007; Huilinir et al. 2010; Koch et al. 2000a), biofilters (Falkentoft et al. 2000; Hodge and Devinny 1995; Miller and Allen 2005), three-phase fluidized bed reactors (Boaventura and Rodrigues 1988; Boessmann et al. 2003; Nicolella et al. 1999; Suidan 1986; Toumi et al. 2008; Tsuneda et al. 2002), moving bed reactors (Alpkvist et al. 2007; Boltz et al. 2009; Mannina et al. 2011; Plattes et al. 2007), sequencing batch biofilm reactors (Morgenroth and Wilderer 1998; Zinatizadeh et al. 2011), anaerobic granular sludge reactors (Batstone et al. 2004; Buffiere et al. 1998; Fuentes et al. 2009a; Picioreanu et al. 2005; Tartakovsky and Guiot 1997), and aerobic granular sludge reactors (de Kreuk et al. 2007; Dong et al. 2012; Matsumoto et al. 2010; Ni and Yu 2010a; Vazquez-Padin et al. 2010b; Volcke et al. 2012; Xavier et al. 2007);
- **Implementation of biofilm models for practitioners** (Boltz et al. 2010, 2011; Daigger 2011; Moelants et al. 2010; Morgenroth et al. 2000b; Rittmann et al. 2018; Takcs et al. 2007), and as teaching supports (Morgenroth et al. 2002; Shiflet and Shiflet 2010).

2.9.3 Mathematical Modelling of BNR Granular Sludge Systems

Mathematical models were developed in the field of the AGS science for a deeper understanding of physical phenomena (e.g., settling behavior of aerobic granules), mass transfer of substrates and degradation kinetics inside granules, BNR processes and efficiencies, and microbial competition and localization across chemical gradients. The different types of models were reviewed (Ni and Yu 2010a). Modelling concepts were substantially developed in other PhD theses for uncovering AGS from granulation principles to full scale operation management (Baeten 2020; Ni 2013; van Dijk 2022).

Most of the dynamic mathematical models published were developed for understanding and optimizing BNR in AGS systems. AGS-SBRs are operated at long sludge ages typically between 15 and 70 days. Model simulations are thus efficient to save long experimental times, such as required to reach steady-states after at least 3–5 sludge ages (Liu et al. 2009a). Models requires calibration with experimental data though. The biofilm compartment of AQUASIM (Wanner and Morgenroth 2004; Wanner and Reichert 1996) was used for implementing several of 1-D AGS models which provide a good information on the distribution of microbial populations and BNR performances in granular systems.

Advanced models were used to understand granulation phenomena and their management at pilot and full scales (Baeten et al. 2018; Derlon et al. 2022; van Dijk et al. 2018, 2020, 2022). Models were efficient to aggregate the knowledge gained from granular sludge across redox conditions like anaerobic granules, anammox granules, and aerobic granules (Baeten et al. 2019).

2.9.3.1 Modelling the Settling Behavior of BNR Granular Aggregates

A mathematical model was developed to describe the settling behavior of granular sludges (Liu et al. 2005d). The application of an adequate settling velocity was shown to enable a good sludge granulation. An analytical solution was obtained for the zone settling velocity as a function of the particle diameter, the SVI, and the biomass concentration (Liu et al. 2005e). Based on this relation, a critical settling velocity setpoint can be calculated, and a proper reactor design determined, for enabling activated sludge granulation.

An analytical solution was proposed for the granule settling velocity in function of the temperature and salinity of the bulk liquid phase (Winkler et al. 2012b). This relation can be used for predicting the settleability of granular sludge in function of variations in temperature or salinity in the influent wastewater.

A dynamic mathematical model was used to show that the selective sludge discharge of activated sludge flocs was crucial for granulation (Li and Li 2009). Small and loose flocs have a better substrate uptake capacity than larger and denser precursors of granules. The selective removal of the flocculent competitors leads to preferential uptake of substrates by granule nuclei, and to sludge granulation. Selective discharge of small and slow-settling flocs can be achieved by purging the bulk liquid phase after 1–5 min biomass settling. Unselective discharge of fractions of mixed liquors at the end of the aeration phase did not lead to granulation. Sludge discharge ratios of 10–30% were applied in both cases, preventing extensive biomass wash-out.

Modelling highlighted that a higher substrate concentration and a higher biomass retention have promoted the formation of EPS, soluble microbial products and intracellular storage products in aerobic granules (Ni and Yu 2010b).

2.9.3.2 Modelling BNR Conversion Processes in Granular Sludge Biofilms

A dynamic 1-D mathematical model was developed for simulating the SND conversion processes (Beun et al. 2001). This helped obtain knowledge on the location of ANOs and DHOs in granules, and for predicting the efficiency of nitrogen conversion processes. ASM3 was used to describe the stoichiometries and kinetics of the microbial processes. A DO of 40% was optimal for SND in the AGS-SBR. At this DO, ANOs were localized in the aerobic outer layer of aerobic granules, whereas DHOs were present from the surface to the central anoxic zone where they conducted denitrification.

Similar results were obtained in a model investigating the simultaneous growth of ANOs and DHOs in aerobic granules (Ni et al. 2008). The level of COD and ammonium concentrations in the bulk liquid phase has significantly impacted on the balance of DHOs and ANOs in granules. AOOs and NOOs were preferentially selected at low COD/N ratios.

A 1-D dynamic model was used to describe the kinetics of BNR microbial processes inside AGS (de Kreuk et al. 2007). Effects of DO concentration, temperature, granule diameter, sludge loading rate, and SBR cycle configuration were addressed on BNR performances. Model simulations pointed out the DO concentration and penetration depth as important parameters for SNDPR in AGS.

The SND performances of an AGS-SBR were modelled in function of the COD/N ratio in the influent wastewater (Vazquez-Padin et al. 2010b). The ASM platform was extended with simultaneous growth and storage of the soluble COD by DHOs, and with nitrite as intermediate compound. With a COD/N ratio as low as 1.25 $g_{CODs}\ g_N^{-1}$, denitrification was hampered and DHOs were only located in the aerobic layer at the granule surface. Denitrification was efficient with a COD/N ratio of 5.5 $g_{CODs}\ g_N^{-1}$ and DHOs were present up to anoxic layers as well. ANOs were not abundant (5%) at the COD/N ratio of 5.5 $g_{CODs}\ g_N^{-1}$. ANOs increased up to 30% and 100% when decreasing the COD/N ratio to 1.25 and 0 $g_{CODs}\ g_N^{-1}$, respectively. At the lowest COD/N ratio, AOOs were twice more abundant than NOOs. At the highest COD/N ratio, AOOs were 5.3 times more abundant. The over-abundance of AOOs linked with nitrite accumulation, due to the three times lower oxidation rate of NOOs. The fraction of inert particulate material amounted to 60 and 80% of TSS at high and low COD/N ratios, respectively. A higher fraction of PHAs was stored with the highest COD/N ratio. The definition of different biomass densities for DHOs (150 $kg_{VSS}\ m^{-3}$ of biomass) and ANOs (350 $kg_{VSS}\ m^{-3}$ of biomass), and the introduction of a decreasing porosity profile across the granule depth were key for a good description of experimental results.

A mathematical model was used to describe COD and N removals in an airlift AGS-SBR (Wan 2009). A heterogeneous biomass distribution was obtained once granules reached a size > 400 μm. Gradients in microbial activity and oxygen profile increased across the biofilm depth and led to the formation of a central anoxic zone. The presence of nitrates enabled growth in the center of granules, as well as a higher PHA storage capacity. The concentration in active biomass was higher under such conditions, than in aerobic granules fully penetrated by oxygen. When developing in the biofilm deepness, the active biomass is better protected from detachment. Integrating a description of the rate of loss of active biomass in function of the detachment and the sludge discharge ratio can improve the model.

Wan's model was complemented for the description of nitrification phenomena in a hybrid granules-flocs reactor (Filali 2011). A surface detachment process and a purge of excess flocculent sludge was implemented. According to the model, AOOs and NOOs in this hybrid system were both present in granules, whereas the nitrifying community of flocs was predominated by the faster-growing AOOs.

The operation of AGS-SBR was optimized for COD- and N-removal from dairy wastewater with the help of a dynamic modelling (Wichern et al. 2008). This model integrated the diffusion and substrate limitation processes inside granular biofilms. With this model, the thickness of aerobic layers of 2.5 mm granules amounted between 65 and 95 μm in function of the DO concentration in the bulk liquid phase (0–9 mg_{O2} L^{-1}). Optimal N-removal from dairy wastewater was obtained at an ideal DO concentration of 5 mg_{O2} L^{-1}, and with an OLR_v between 4.5 and 9.0 kg_{CODs} d^{-1} m^{-3} with a fraction of readily biodegradable COD of 20%.

2.9.3.3 A Multiscale Model for BNR Granular Sludge

A multiscale model was developed for AGS-SBRs (Xavier et al. 2007). At the macro scale, a 1-D model described the evolution of the concentrations of solutes in the bulk liquid phase. At the micro scale, the spatial arrangement of OHOs, AOOs, NOOs and PAOs inside granules was modelled in 2-D by individual-based modelling with process kinetics specific to bacterial populations. Short-term and long-term dynamics of solutes and populations were predicted by this model.

This model pointed out the predominance of OHOs and the out-selection of PAOs during the start-up of AGS-SBRs. OHOs are sustained by the leakage of readily biodegradable COD into aeration. The model identified that N-removal was to a higher extent achieved in AGS systems by AND rather than SND.

2.9.3.4 Unraveling PAO-GAO Competition in Granules by Modelling

Different mathematical models are available for understanding and managing processes in AGS. Like activated sludge (Lopez-Vazquez et al. 2009b; Oehmen et al. 2010c), the competitive selection of PAOs and GAOs can be addressed in granular sludge using simple metabolic models.

This can be done either using models integrating the full SBR cycle or using models that focus on specific processes that mainly govern the PAO-GAO competition. For instance, a 1-D dynamic model combining hydraulic transport phenomena and microbial processes was developed to understand the role the anaerobic feeding phase and the underlying conditions (flow rate, T, pH, among others) on the preferential uptake of VFAs by PAOs and GAOs (Weissbrodt et al. 2017). The model is useful to support the design of granular sludge bed geometries and anaerobic feeding phases to manage the anaerobic selector of AGS SBRs.

2.10 Situation Analysis of the Wastewater Engineering and Molecular Biology Research

A scientometric method (Larsen and von Ins 2010) was used at the start of this state-of-the-art review to address the degree of scientific maturity of wastewater engineering (including the AGS technology) and molecular biology methods by comparing the growth rates of scientific publications. The following trends can be delineated from Fig. 2.3.

Scientific publishing reaches a continuous annual growth rate of 5–9% in the cumulative number of records over the years when the target research field has reached a mature state (Larsen and von Ins 2010; Price 1963). This is exemplified for publications recorded per general fields such as natural sciences, technology, biology, chemistry, physics, or electronics.

Similar trends were obtained for wastewater engineering and biology (Fig. 2.3a). Mature technologies such as activated sludge, anaerobic digestion, biofilm, and up-flow anaerobic sludge blanket (UASB) processes have reached an average slope of 0.03 log $year^{-1}$ in the number of articles recorded in the Scopus[2] on-line database of literature references over the years. This corresponds to an annual publication growth rate of 7.2% and to a publication doubling time of about 10 years.

The AGS science is getting close to this rate. The logarithmic cumulated curve of publications is inflecting towards linear tendency with an annual turnover around 7%. This inflection point matches with the on-going active installation of AGS WWTPs at full scale; with already 90 Nereda® plants world-wide (Giesen et al. 2013, 2015; Pronk et al. 2017a, 2020; van der Roest et al. 2011).[3] The AGS technology is becoming an important player in the environmental engineering sector. Recent innovation on the extraction of exopolymer biomaterials from granular sludge with the first Kaumera Nereda® Gum processing plant already installed in Zutphen in the Netherlands (Dutch Water Sector 2019; van der Roest et al. 2015) will establish the technology in the circular economy (Weissbrodt 2018).

Environmental biotechnology developed closely together with microbial ecology (Fig. 2.3b). The integration of microbial ecology principles and methods is the key for the development of environmental biotechnologies with microbial community engineering concepts. The scientometric analysis was further used to address the maturity level of molecular biology methods used over the last 30 years to characterize microbiomes from natural to engineered environments. More than 30 microbial ecology methods are available depending on research questions to address microorganisms and their metabolisms in microbiomes (Nielsen and McMahon 2014; van Loosdrecht et al. 2016; Weissbrodt et al. 2020a).

The traditional fingerprinting methods like denaturing gradient gel electrophoresis (DGGE) (Muyzer et al. 1993) and terminal-fragment length polymorphism (T-RFLP) (Liu et al. 1997; Weissbrodt et al. 2012b) are getting replaced by the sequencing

---

[2] Scopus is Elsevier's abstract and citation database www.scopus.com.

[3] Royal HaskoningDHV (status 2022) https://nereda.royalhaskoningdhv.com/.

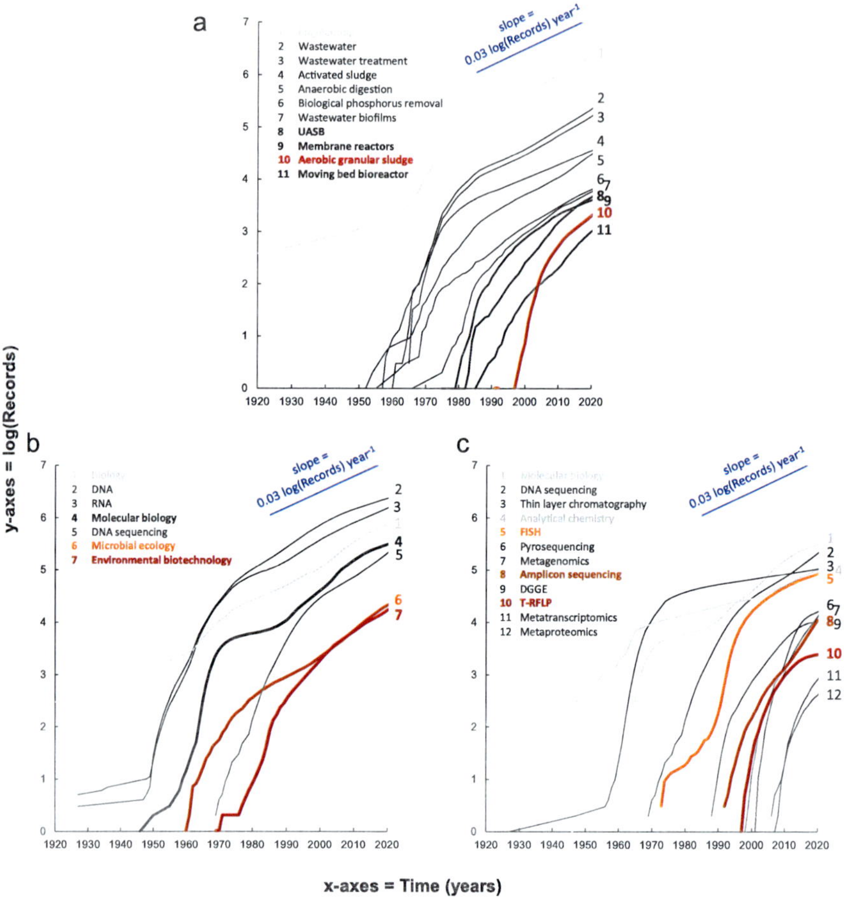

Fig. 2.3 Scientometric analysis of one century of publication rates in the fields of engineering and wastewater treatment (**a**), environmental biotechnology and microbial ecology (**b**), and analytical chemistry and molecular biology (**c**). Cumulative profiles of literature records are given on a logarithmic scale. An average annual publication growth rate of 0.03 log(Records) y^{-1} was computed along the linear tendencies at steady-state. This corresponds to a publication doubling time of about 10 years. When expressed as a percentage (7.2%), the average publication rate matches with the traditional annual rates of 5–9% described in literature (Larsen and von Ins 2010; Price 1963) for publication records of in natural sciences and engineering. An interesting trend is the concomitant growth of the aerobic granular sludge technology and microbial community fingerprinting techniques of like T-RFLP now replaced by amplicon sequencing. *Data source* Scopus®, the largest abstract and citation database of peer-reviewed literature

and mass spectrometry era. Amplicon sequencing methods are getting widely used to measure microbial community compositions (obtaining relative abundances and taxonomic affiliations in one single sequencing run) and track microbial selection phenomena in mixed-culture biotechnologies (Albertsen et al. 2015; Karst et al. 2016; Pinto and Raskin 2012; Weissbrodt et al. 2020a).

Amplicon sequencing is nicely complemented by microscopy and fluorescence in situ hybridization (FISH) to localize and semi-quantify the abundances of microorganisms in wastewater biomasses (Nielsen et al. 2009, 2016; Wagner et al. 1994a, 1995, 1998; Yilmaz and Noguera 2004). FISH can be combined with ecophysiology methods to localize microorganisms with specific anabolic properties (Nielsen and Nielsen 2005; Wagner and Haider 2012; Wagner et al. 2006). Quantitative polymerase chain reaction (qPCR) increases the sensitivity in detection of microorganisms and their functional genes (Agrawal et al. 2021; Auerbach et al. 2007; Calderón-Franco et al. 2021; Hu et al. 2010; Lebuhn et al. 2004; Wu et al. 2016).

Systems microbiology methods are getting implemented to obtain higher resolution on microorganisms and the regulation of their metabolic functions, involving multi meta-omics methods (Cerruti et al. 2021; McDaniel et al. 2021; Narayanasamy et al. 2015; Rodríguez et al. 2015). This integrates genome-centric metagenomics, metatranscriptomics, and metaproteomics to unravel microbial functionalities out of the pool of informational molecules comprised in biomass (DNA, RNA, proteins) (Albertsen et al. 2013b; Hassa et al. 2018; Kleikamp et al. 2020b; Oyserman et al. 2016b; Singleton et al. 2021; Wilmes et al. 2008a, 2015).

Collectively, the scientometric analysis given in Fig. 2.3 highlights the connection between the breakthrough developments in environmental biotechnology and microbial ecology principles and methods. The aerobic granular sludge science is an excellent example of integration of new-generation engineering concepts with systems microbiology methods to better understand microorganisms, conversions, and products inside their process boundaries.

2.11 Conclusion

This state-of-the-art review revealed the complexity of AGS. The following main conclusions can be outlined:

- The AGS technology is highly attractive for the design of intensified and flexible SBR processes toward high-rate BNR from used water streams. This technology is an excellent fit along paradigm shifts in the wastewater sector toward achieving sustainability on top of environmental and public health protection.
- Because of its multiple redox characteristics across length and time scales over process and biofilm boundaries for full BNR, AGS can be referred to as BNR granular sludge in a more integrated consideration.

- The three-phase AGS systems are highly heterogeneous processes. They rely on mobile and fast-settling biofilms composed of complex EPS matrices that embed a network of microorganisms and metabolisms inside their microbiome.
- Granules are biofilm ecosystems as mobile as flocs. Their complex outer structure, internal architecture, microbial niche partitioning, and metabolic distributions and interactions depend on mass transfer phenomena across their biofilm.
- Managing the microbial resource in AGS systems requires a detailed understanding of selection, competition, and interaction phenomena inside the granular sludge microbiome. Ecosystem models can help disentangle the microbial complexity, and to examine metabolic functionalities of microorganisms.
- Multi-level mechanistic studies are required to investigate the impact of operational factors, environmental conditions, and their variations as basis for the development of microbial community engineering strategies in AGS systems.
- An ecological engineering of AGS processes integrates principles of process engineering, environmental biotechnology, biofilm engineering and science, biomaterial science, microbiology, microscopy, microbial ecology, molecular biology and meta-omics, and mathematical modelling.

References

Abram F, Gunnigle E, O'Flaherty V (2009) Optimisation of protein extraction and 2-DE for metaproteomics of microbial communities from anaerobic wastewater treatment biofilms. Electrophoresis 30(23):4149–4151

Adav SS, Lee DJ, Lai JY (2008a) Proteolytic activity in stored aerobic granular sludge and structural integrity. Bioresour Technol 100(1):68–73

Adav SS, Lee DJ, Show KY, Tay JH (2008b) Aerobic granular sludge: recent advances. Biotechnol Adv 26(5):411–423

Adav SS, Lee DJ, Tay JH (2008c) Extracellular polymeric substances and structural stability of aerobic granule. Water Res 42(6–7):1644–1650

Adav SS, Lee DJ, Lai JY (2009a) Biological nitrification-denitrification with alternating oxic and anoxic operations using aerobic granules. Appl Microbiol Biotechnol 84(6):1181–1189

Adav SS, Lee DJ, Lai JY (2009b) Functional consortium from aerobic granules under high organic loading rates. Bioresour Technol 100(14):3465–3470

Adav SS, Lee DJ, Lai JY (2009c) Treating chemical industries influent using aerobic granular sludge: recent development. J Taiwan Inst Chem Eng 40(3):333–336

Adav SS, Lee DJ, Lai JY (2010) Microbial community of acetate utilizing denitrifiers in aerobic granules. Appl Microbiol Biotechnol 85(3):753–762

Adler A, Holliger C (2020) Multistability and reversibility of aerobic granular sludge microbial communities upon changes from simple to complex synthetic wastewater and back. Front Microbiol 11:574361

Adler A, Poirier S, Pagni M, Maillard J, Holliger C (2022) Disentangle genus microdiversity within a complex microbial community by using a multi-distance long-read binning method: example of *Candidatus* Accumulibacter. Environ Microbiol 24(4):2136–2156

Agrawal S, Weissbrodt DG, Annavajhala M, Jensen MM, Arroyo JMC, Wells G, Chandran K, Vlaeminck SE, Terada A, Smets BF, Lackner S (2021) Time to act–assessing variations in qPCR analyses in biological nitrogen removal with examples from partial nitritation/anammox systems. Water Res 190:116604

Aguado D, Montoya T, Ferrer J, Seco A (2006) Relating ions concentration variations to conductivity variations in a sequencing batch reactor operated for enhanced biological phosphorus removal. Environ Model Softw 21(6):845–851

Aguilar C, Vlamakis H, Losick R, Kolter R (2007) Thinking about *Bacillus subtilis* as a multicellular organism. Curr Opin Microbiol 10(6):638–643

Ahn J, Lee M, Kwon H (2006) Changes in respiratory quinone profiles of enhanced biological phosphorus removal activated sludge under different influent phosphorus/carbon ratio conditions. Bioprocess Biosyst Eng 29(3):143–148

Akbari A, Wang Z, He P, Wang D, Lee J, Han IL, Li G, Gu AZ (2021) Unrevealed roles of polyphosphate-accumulating microorganisms. Microb Biotechnol 14(1):82–87

Albertsen M, Hansen LBS, Saunders AM, Nielsen PH, Nielsen KL (2011a) A metagenome of a full-scale microbial community carrying out enhanced biological phosphorus removal. ISME J 6(6):1094–1106

Albertsen M, Sorensen DK, Stensballe A, Nielsen KL, Nielsen PH (2011b) Identification of exopolymers in mixed microbial biofilms: a metagenomic approach. In: Qi Z (ed) IWA biofilm conference: 2011 processes in biofilms. Tongji University, Shanghai, China

Albertsen M, Hugenholtz P, Skarshewski A, Nielsen KL, Tyson GW, Nielsen PH (2013a) Genome sequences of rare, uncultured bacteria obtained by differential coverage binning of multiple metagenomes. Nat Biotechnol 31(6):533–538

Albertsen M, Saunders AM, Nielsen KL, Nielsen PH (2013b) Metagenomes obtained by 'deep sequencing'—what do they tell about the enhanced biological phosphorus removal communities? Water Sci Technol 68(9):1959–1968

Albertsen M, Karst SM, Ziegler AS, Kirkegaard RH, Nielsen PH (2015) Back to basics—the influence of DNA extraction and primer choice on phylogenetic analysis of activated sludge communities. PLoS ONE 10(7):e0132783

Albertsen M, McIlroy SJ, Stokholm-Bjerregaard M, Karst SM, Nielsen PH (2016) "*Candidatus* Propionivibrio aalborgensis": a novel glycogen accumulating organism abundant in full-scale enhanced biological phosphorus removal plants. Front Microbiol 7:1033

Ali M, Wang Z, Salam KW, Hari AR, Pronk M, van Loosdrecht MCM, Saikaly PE (2019) Importance of species sorting and immigration on the bacterial assembly of different-sized aggregates in a full-scale aerobic granular sludge plant. Environ Sci Technol 53(14):8291–8301

Alldredge AL, Silver MW (1988) Characteristics, dynamics and significance of marine snow. Prog Oceanogr 20(1):41–82

Allison DG (2003) The biofilm matrix. Biofouling 19(2):139–150

Allison DG, Gilbert P (1995) Modification by surface association of antimicrobial susceptibility of bacterial populations. J Ind Microbiol 15(4):311–317

Alloul A, Cerruti M, Adamczyk D, Weissbrodt DG, Vlaeminck SE (2021) Operational strategies to selectively produce purple bacteria for microbial protein in raceway reactors. Environ Sci Technol 55(12):8278–8286

Almstrand R, Daims H, Persson F, Sörensson F, Hermansson M (2013) New methods for analysis of spatial distribution and coaggregation of microbial populations in complex biofilms. Appl Environ Microbiol 79(19):5978–5987

Alpkvist E, Klapper I (2007) Description of mechanical response including detachment using a novel particle model of biofilm/flow interaction. Water Sci Technol 55(8–9):265–273

Alpkvist E, Picioreanu C, Van Loosdrecht MCM, Heyden A (2006) Three-dimensional biofilm model with individual cells and continuum EPS matrix. Biotechnol Bioeng 94(5):961–979

Alpkvist E, Bengtsson J, Overgaard NC, Christensson M, Heyden A (2007) Simulation of nitrification of municipal wastewater in a moving bed biofilm process: a bottom-up approach based on a 2D-continuum model for growth and detachment. Water Sci Technol 55(8–9):247–255

Amann R, Kuhl M (1998) In situ methods for assessment of microorganisms and their activities. Curr Opin Microbiol 1(3):352–358

Amann R, Lemmer H, Wagner M (1998) Monitoring the community structure of wastewater treatment plants: a comparison of old and new techniques. FEMS Microbiol Ecol 25(3):205–215

Aminov RI (2011) Horizontal gene exchange in environmental microbiota. Front Microbiol 2:158
Andersen KS, Kirkegaard RH, Karst SM, Albertsen M (2018) ampvis2: an R package to analyse and visualise 16S rRNA amplicon data. bioRxiv 299537
Andersson S, Kuttuva Rajarao G, Land CJ, Dalhammar G (2008) Biofilm formation and interactions of bacterial strains found in wastewater treatment systems. FEMS Microbiol Lett 283(1):83–90
Andreottola G, Damiani E, Foladori P, Nardelli P, Ragazzi M (2004) Treatment of mountain refuge wastewater by fixed and moving bed biofilm systems. Water Sci Technol 48(11–12):169–177
Aqeel H, Weissbrodt D, Cerruti M, Wolfaardt GM, Wilen B-M, Liss SN (2019) Drivers of bioaggregation from flocs to biofilms and granular sludge. Environ Sci Water Res Technol 5:2072–2089
Arcand Y, Chavarie C, Guiot SR (1994) Dynamic modelling of the population distribution in the anaerobic granular biofilm. Water Sci Technol 30(12):63–73
Arrojo B, Mosquera-Corral A, Garrido JM, Mendez R (2004) Aerobic granulation with industrial wastewater in sequencing batch reactors. Water Res 38(14–15):3389–3399
Arvin E, Harremoes P (1990) Concepts and models for biofilm reactor performance. Water Sci Technol 22(1–2):171–192
Aspa Y, Debenest G, Quintard M (2011) Effective dispersion in channelled biofilms. Int J Environ Waste Manage 7(1–2):112–131
Auerbach EA, Seyfried EE, McMahon KD (2007) Tetracycline resistance genes in activated sludge wastewater treatment plants. Water Res 41(5):1143–1151
Aybar M, Perez-Calleja P, Li M, Pavissich JP, Nerenberg R (2019) Predation creates unique void layer in membrane-aerated biofilms. Water Res 149:232–242
Baeten JE (2020) Wastewater treatment with aerobic granular sludge: challenges and opportunities for modelling and off gas analyses. Universiteit Gent, Belgium
Baeten JE, van Loosdrecht MCM, Volcke EIP (2018) Modelling aerobic granular sludge reactors through apparent half-saturation coefficients. Water Res 146:134–145
Baeten JE, Batstone DJ, Schraa OJ, van Loosdrecht MCM, Volcke EIP (2019) Modelling anaerobic, aerobic and partial nitritation-anammox granular sludge reactors—a review. Water Res 149:322–341
Baeten JE, van Dijk EJH, Pronk M, van Loosdrecht MCM, Volcke EIP (2021) Potential of off-gas analyses for sequentially operated reactors demonstrated on full-scale aerobic granular sludge technology. Sci Total Environ 787:147651
Balcázar JL, Subirats J, Borrego CM (2015) The role of biofilms as environmental reservoirs of antibiotic resistance. Front Microbiol 6:1216
Baldwin SA, Taylor JC, Ziels R (2019) Genome-resolved metagenomics links microbial dynamics to failure and recovery of a bioreactor removing nitrate and selenate from mine-influenced water. Biochem Eng J 151:107297
Balest L, Lopez A, Mascolo G, Di Iaconi C (2008) Removal of endocrine disrupter compounds from municipal wastewater using an aerobic granular biomass reactor. Biochem Eng J 41(3):288–294
Ballance S, Holtan S, Aarstad OA, Sikorski P, Skjåk-Braek G, Christensen BE (2005) Application of high-performance anion-exchange chromatography with pulsed amperometric detection and statistical analysis to study oligosaccharide distributions—a complementary method to investigate the structure and some properties of alginates. J Chromatogr A 1093(1–2):59–68
Bangerter B (2017) Abwasserkennzahlen ARA Thunersee. ARA Thunersee, Uetendorf, p 27
Bark K, Sponner A, Kampfer P, Grund S, Dott W (1992) Differences in polyphosphate accumulation and phosphate adsorption by *Acinetobacter* isolates from wastewater producing polyphosphate: AMP phosphotransferase. Water Res 26(10):1379–1388
Barnard JL, Abraham K (2006) Key features of successful BNR operation. Water Sci Technol 53(12):1–9
Barnard JL, Steichen MT (2006) Where is biological nutrient removal going now? Water Sci Technol 53(3):155–164
Barr JJ, Cook AE, Bond PL (2010a) Granule formation mechanisms within an aerobic wastewater system for phosphorus removal. Appl Environ Microbiol 76(22):7588–7597

Barr JJ, Slater FR, Fukushima T, Bond PL (2010b) Evidence for bacteriophage activity causing community and performance changes in a phosphorus-removal activated sludge. FEMS Microbiol Ecol 74(3):631–642

Barr JJ, Dutilh BE, Skennerton CT, Fukushima T, Hastie ML, Gorman JJ, Tyson GW, Bond PL (2015) Metagenomic and metaproteomic analyses of *Accumulibacter phosphatis*-enriched floccular and granular biofilm. Environ Microbiol 18(1):273–287

Basuvaraj M, Fein J, Liss SN (2015) Protein and polysaccharide content of tightly and loosely bound extracellular polymeric substances and the development of a granular activated sludge floc. Water Res 82:104–117

Batstone DJ, Keller J, Blackall LL (2004) The influence of substrate kinetics on the microbial community structure in granular anaerobic biomass. Water Res 38(6):1390–1404

Batstone DJ, Picioreanu C, van Loosdrecht MCM (2006) Multidimensional modelling to investigate interspecies hydrogen transfer in anaerobic biofilms. Water Res 40(16):3099–3108

Battin TJ, Kaplan LA, Newbold JD, Hansen CME (2003) Contributions of microbial biofilms to ecosystem processes in stream mesocosms. Nature 426(6965):439–442

Battin TJ, Sloan WT, Kjelleberg S, Daims H, Head IM, Curtis TP, Eberl L (2007) Microbial landscapes: new paths to biofilm research. Nat Rev Microbiol 5(1):76–81

Beacham AM, Seviour RJ, Lindrea KC, Livingston I (1990) Genospecies diversity of *Acinetobacter* isolates obtained from a biological nutrient removal pilot plant of a modified UCT configuration. Water Res 24(1):23–29

Beer M, Seviour RJ (2006) Gene cassette-associated sequences from phosphorus and non-phosphorus removing microbial communities in aerobic:anaerobic sequencing batch reactors. Water Sci Technol 54(1):55–61

Beg SA, Chaudhry MAS (1999) A review of mathematical modelling of biofilm processes: advances in modelling of selected biofilm processes. Int J Environ Stud 56(3):285–312

Bernstein HC (2019) Reconciling ecological and engineering design principles for building microbiomes. mSystems 4(3):e00106–19

Beun JJ, Hendriks A, van Loosdrecht MCM, Morgenroth E, Wilderer PA, Heijnen JJ (1999) Aerobic granulation in a sequencing batch reactor. Water Res 33(10):2283–2290

Beun JJ, van Loosdrecht MCM, Heijnen JJ (2000) Aerobic granulation. Water Sci Technol 41(4–5):41–48

Beun JJ, Heijnen JJ, van Loosdrecht MCM (2001) N-removal in a granular sludge sequencing batch airlift reactor. Biotechnol Bioeng 75(1):82–92

Beun JJ, van Loosdrecht MCM, Heijnen JJ (2002) Aerobic granulation in a sequencing batch airlift reactor. Water Res 36(3):702–712

Beyenal H, Lewandowski Z (2005) Modeling mass transport and microbial activity in stratified biofilms. Chem Eng Sci 60(15):4337–4348

Bhattacharjee AS, Choi J, Motlagh AM, Mukherji ST, Goel R (2015) Bacteriophage therapy for membrane biofouling in membrane bioreactors and antibiotic-resistant bacterial biofilms. Biotechnol Bioeng 112(8):1644–1654

Bishop PL (1997) Biofilm structure and kinetics. Water Sci Technol 36(1):287–294

Bishop PL (2007) The role of biofilms in water reclamation and reuse. Water Sci Technol 55(1):19–26

Bishop PL, Rittmann BE (1995) Modelling heterogeneity in biofilms: report of the discussion session. Water Sci Technol 32(8):263–265

Blackall LL (2000) A summary of recent microbial discoveries in biological nutrient removal from wastewater. Australas Biotechnol 10(3):29–32

Blackall LL, Burrell PC, Gwilliam H, Bradford D, Bond PL, Hugenholtz P (1998) The use of 16S rDNA clone libraries to describe the microbial diversity of activated sludge communities. Water Sci Technol 37(4–5):451–454

Blackall LL, Crocetti GR, Saunders AM, Bond PL (2002) A review and update of the microbiology of enhanced biological phosphorus removal in wastewater treatment plants. Anton Van Leeuw Int J Gen Mol Microbiol 81(1–4):681–691

Boaventura RA, Rodrigues AE (1988) Consecutive reactions in fluidized-bed biological reactors: modeling and experimental study of wastewater denitrification. Chem Eng Sci 43(10):2715–2728

Boelee NC, Temmink H, Janssen M, Buisman CJN, Wijffels RH (2011) Nitrogen and phosphorus removal from municipal wastewater effluent using microalgal biofilms. Water Res 45(18):5925–5933

Boessmann M, Staudt C, Neu TR, Horn H, Hempel DC (2003) Investigation and modeling of growth, structure and oxygen penetration in particle supported biofilms. Chem Eng Technol 26(2):219–222

Boleij M, Pabst M, Neu TR, Van Loosdrecht MCM, Lin Y (2018) Identification of glycoproteins isolated from extracellular polymeric substances of full-scale anammox granular sludge. Environ Sci Technol 52(22):13127–13135

Boleij M, Seviour T, Wong LL, van Loosdrecht MCM, Lin Y (2019) Solubilization and characterization of extracellular proteins from anammox granular sludge. Water Res 164:114952

Boleij M, Kleikamp H, Pabst M, Neu TR, van Loosdrecht MCM, Lin Y (2020) Decorating the anammox house: sialic acids and sulfated glycosaminoglycans in the extracellular polymeric substances of anammox granular sludge. Environ Sci Technol 54(8):5218–5226

Boles BR, Thoendel M, Singh PK (2004) Self-generated diversity produces "insurance effects" in biofilm communities. Proc Natl Acad Sci USA 101(47):16630–16635

Boltz JP, Johnson BR, Daigger GT, Sandino J (2009) Modeling integrated fixed-film activated sludge and moving-bed biofilm reactor systems I: mathematical treatment and model development. Water Environ Res 81(6):555–575

Boltz JP, Morgenroth E, Sen D (2010) Mathematical modelling of biofilms and biofilm reactors for engineering design. Water Sci Technol 62(8):1821–1836

Boltz JP, Morgenroth E, Brockmann D, Bott C, Gellner WJ, Vanrolleghem PA (2011) Systematic evaluation of biofilm models for engineering practice: components and critical assumptions. Water Sci Technol 64(4):930–944

Boltz JP, Smets BF, Rittmann BE, Van Loosdrecht MCM, Morgenroth E, Daigger GT (2017) From biofilm ecology to reactors: a focused review. Water Sci Technol 75(8):1753–1760

Bond PL, Keller J, Blackall LL (1998) Characterisation of enhanced biological phosphorus removal activated sludges with dissimilar phosphorus removal performances. Water Sci Technol 37(4–5):567–571

Bond PL, Erhart R, Wagner M, Keller J, Blackall LL (1999) Identification of some of the major groups of bacteria in efficient and nonefficient biological phosphorus removal activated sludge systems. Appl Environ Microbiol 65(9):4077–4084

Bortone G, Libelli SM, Tilche A, Wanner J (1999) Anoxic phosphate uptake in the dephanox process. Water Sci Technol 40(4–5):177–185

Bos R, van der Mei HC, Busscher HJ (1999) Physico-chemistry of initial microbial adhesive interactions—its mechanisms and methods for study. FEMS Microbiol Rev 23(2):179–230

Bossier P, Verstraete W (1996) Triggers for microbial aggregation in activated sludge? Appl Microbiol Biotechnol 45(1–2):1–6

Brandt D, Sieker C, Hegemann W (2002) Combined denitrification and excess biological phosphorus removal in discontinuous operated biofilm systems. Water Sci Technol 46(4–5):193–200

Brdjanovic D, Logemann S, Van Loosdrecht MCM, Hooijmans CM, Alaerts GJ, Heijnen JJ (1998a) Influence of temperature on biological phosphorus removal: process and molecular ecological studies. Water Res 32(4):1035–1048

Brdjanovic D, van Loosdrecht MCM, Hooijmans CM, Mino T, Alaerts GJ, Heijnen JJ (1998b) Effect of polyphosphate limitation on the anaerobic metabolism of phosphorus-accumulating microorganisms. Appl Microbiol Biotechnol 50(2):273–276

Brockmann D, Morgenroth E (2010) Evaluating operating conditions for outcompeting nitrite oxidizers and maintaining partial nitrification in biofilm systems using biofilm modeling and Monte Carlo filtering. Water Res 44(6):1995–2009

Brodie EL (2011) Phylogenetic microarrays (phylochips) for analysis of complex microbial communities. In: de Bruijn FJ (ed) Handbook of molecular microbial ecology I: metagenomics and complementary approaches. Wiley, Hoboken, NJ, pp 521–532

Bruce SCR, Downing L, Young M, Nerenberg R (2014) Floc or granule? Evidence of granulation in a continuous flow system. Proc Water Environ Fed 19:2891–2897

Bruckner S, Mosch HU (2012) Choosing the right lifestyle: adhesion and development in *Saccharomyces cerevisiae*. FEMS Microbiol Rev 36(1):25–58

Brussaard L, de Ruiter PC, Brown GG (2007) Soil biodiversity for agricultural sustainability. Agr Ecosyst Environ 121(3):233–244

Bucs S, Farhat N, Kruithof JC, Picioreanu C, van Loosdrecht MCM, Vrouwenvelder JS (2018) Review on strategies for biofouling mitigation in spiral wound membrane systems. Desalination 434:189–197

Buffiere P, Steyer JP, Fonade C, Moletta R (1995) Comprehensive modeling of methanogenic biofilms in fluidized bed systems: mass transfer limitations and multisubstrate aspects. Biotechnol Bioeng 48(6):725–736

Buffiere P, Steyer JP, Fonade C, Moletta R (1998) Modeling and experiments on the influence of biofilm size and mass transfer in a fluidized bed reactor for anaerobic digestion. Water Res 32(3):657–668

Burke C, Steinberg P, Rusch D, Kjelleberg S, Thomas T (2011) Bacterial community assembly based on functional genes rather than species. Proc Nat Acad Sci USA 108(34):14288–14293

Burow LC, Mabbett AN, Blackall LL (2008a) Anaerobic glyoxylate cycle activity during simultaneous utilization of glycogen and acetate in uncultured Accumulibacter enriched in enhanced biological phosphorus removal communities. ISME J 2(10):1040–1051

Burow LC, Mabbett AN, McEwan AG, Bond PL, Blackall LL (2008b) Bioenergetic models for acetate and phosphate transport in bacteria important in enhanced biological phosphorus removal. Environ Microbiol 10(1):87–98

Cabau-Peinado O, Straathof AJJ, Jourdin L (2021) A general model for biofilm-driven microbial electrosynthesis of carboxylates from CO_2. Front Microbiol 12:669218

Cai C, Xu F, Liu J, Zhu N, Cai W (2004) Cultivation of aerobic granules in a sequential batch shaking reactor. Environ Technol 25:937–944

Calderón-Franco D, van Loosdrecht MCM, Abeel T, Weissbrodt DG (2021) Free-floating extracellular DNA: systematic profiling of mobile genetic elements and antibiotic resistance from wastewater. Water Res 189:116592

Calderón-Franco D, Sarelse R, Christou S, Pronk M, van Loosdrecht MCM, Abeel T, Weissbrodt DG (2022) Metagenomic profiling and transfer dynamics of antibiotic resistance determinants in a full-scale granular sludge wastewater treatment plant. Water Res 219:118571

Caluwé M, Dobbeleers T, D'aes J, Miele S, Akkermans V, Daens D, Geuens L, Kiekens F, Blust R, Dries J (2017) Formation of aerobic granular sludge during the treatment of petrochemical wastewater. Bioresour Technol 238:559–567

Caluwé M, Goossens K, Seguel Suazo K, Tsertou E, Dries J (2022) Granulation strategies applied to industrial wastewater treatment: from lab to full-scale. Water Sci Technol 85(9):2761–2771

Camejo PY, Owen BR, Martirano J, Ma J, Kapoor V, Santo Domingo J, McMahon KD, Noguera DR (2016) *Candidatus* Accumulibacter phosphatis clades enriched under cyclic anaerobic and microaerobic conditions simultaneously use different electron acceptors. Water Res 102:125–137

Carr E, Eason H, Feng S, Hoogenraad A, Croome R, Soddel J, Lindrea K, Seviour R (2001) RAPD-PCR typing of *Acinetobacter* isolates from activated sludge systems designed to remove phosphorus microbiologically. J Appl Microbiol 90(3):309–319

Carucci A, Milia S, De Gioannis G, Piredda M (2008) Acetate-fed aerobic granular sludge for the degradation of chlorinated phenols. Water Sci Technol 58(2):309–315

Carvalheira M, Oehmen A, Carvalho G, Eusebio M, Reis MA (2014) The impact of aeration on the competition between polyphosphate accumulating organisms and glycogen accumulating organisms. Water Res 66:296–307

Carvalho G, Lemos PC, Oehmen A, Reis MAM (2007) Denitrifying phosphorus removal: linking the process performance with the microbial community structure. Water Res 41(19):4383–4396
Cassidy DP, Belia E (2005) Nitrogen and phosphorus removal from an abattoir wastewater in a SBR with aerobic granular sludge. Water Res 39(19):4817–4823
Cerruti M, Guo B, Delatolla R, de Jonge N, Hommes de Vos van Steenwijk A, Kadota P, Lawson CE, Mao T, Oosterkamp MJ, Sabba F, Stokholm-Bjerregaard M, Watson I, Frigon D, Weissbrodt DG (2021) Plant-wide systems microbiology for the wastewater industry. Environ Sci Water Res Technol 7(10):1687–1706
Characklis WG (1973) Attached microbial growths: I. Attachment and growth. Water Res 7(8):1113–1127
Chavez FP, Mauriaca C, Jerez CA (2009) Constitutive and regulated expression vectors to construct polyphosphate deficient bacteria. BMC Res Notes 2:50
Chen M-Y, Lee D-J, Tay J-H, Show K-Y (2007a) Staining of extracellular polymeric substances and cells in bioaggregates. Appl Microbiol Biotechnol 75:467–474
Chen MY, Lee DJ, Tay JH (2007b) Distribution of extracellular polymeric substances in aerobic granules. Appl Microbiol Biotechnol 73(6):1463–1469
Chen RN, Gao JF, Guo JQ, Su K, Zhang Q (2009a) Simultaneous nitrogen and phosphorus removal by aerobic granular sludge at normal and low temperatures. Huanjing Kexue/Environ Sci 30(10):2995–3001
Chen YC, Chen D, Peng LC, Fu SY, Zhan HY (2009b) The microorganism community of pentachlorophenol (PCP)-degrading coupled granules. Water Sci Technol 59(5):987–994
Chen YC, Lin CJ, Chen HL, Fu SY, Zhan HY (2009c) Cultivation of biogranules in a continuous flow reactor at low dissolved oxygen. Water Air Soil Pollut Focus 9(3–4):213–221
Chen G, van Loosdrecht MCM, Ekama GA, Brdjanovic D (2020) Biological wastewater treatment: principles, modelling and design, 2nd edn. IWA Publishing, London
Chiesa SC, Irvine RL, Manning JF Jr (1985) Feast/famine growth environments and activated sludge population selection. Biotechnol Bioeng 27(5):562–568
Christensen BE (1989) The role of extracellular polysaccharides in biofilms. J Biotechnol 10(3–4):181–201
Christensen BB, Sternberg C, Andersen JB, Palmer RJ Jr, Nielsen AT, Givskov M, Molin S (1999) Molecular tools for study of biofilm physiology. Methods Enzymol 310:20–42
Christenson L, Sims R (2011) Production and harvesting of microalgae for wastewater treatment, biofuels, and bioproducts. Biotechnol Adv 29(6):686–702
Christensson M, Ekström S, Chan AA, Le Vaillant E, Lemaire R (2013) Experience from start-ups of the first ANITA Mox plants. Water Sci Technol 67(12):2677–2684
Christiaens A-S, Van Steenkiste M, Rummens K, Smets I (2022) Amyloid adhesin production in activated sludge is enhanced in lab-scale sequencing batch reactors: feeding regime impacts microbial community and amyloid distribution. Water Res X 17:100162
Chuang SH, Ouyang CF, Wang YB (1996) Kinetic competition between phosphorus release and denitrification on sludge under anoxic condition. Water Res 30(12):2961–2968
Close K, Marques R, Carvalho VCF, Freitas EB, Reis MAM, Carvalho G, Oehmen A (2021) The storage compounds associated with Tetrasphaera PAO metabolism and the relationship between diversity and P removal. Water Res 204:117621
Cofré C, Campos JL, Valenzuela-Heredia D, Pavissich JP, Camus N, Belmonte M, Pedrouso A, Carrera P, Mosquera-Corral A, Val del Río A (2018) Novel system configuration with activated sludge like-geometry to develop aerobic granular biomass under continuous flow. Bioresour Technol 267:778–781
Comeau Y, Hall KJ, Hancock REW, Oldham WK (1986) Biochemical model for enhanced biological phosphorus removal. Water Res 20(12):1511–1521
Cooke AJ, Rowe RK, Rittmann BE, Fleming IR (1999) Modeling biochemically driven mineral precipitation in anaerobic biofilms. Water Sci Technol 39:57–64

Corominas L, Rieger L, Takacs I, Ekama G, Hauduc H, Vanrolleghem PA, Oehmen A, Gernaey KV, van Loosdrecht MCM, Comeau Y (2010) New framework for standardized notation in wastewater treatment modelling. Water Sci Technol 61:841–857

Costerton JW (1999a) Introduction to biofilm. Int J Antimicrob Agents 11(3–4):217–221

Costerton JW (1999b) The role of bacterial exopolysaccharides in nature and disease. J Ind Microbiol Biotechnol 22(4–5):551–563

Costerton B (2004) Microbial ecology comes of age and joins the general ecology community. Proc Natl Acad Sci USA 101(49):16983–16984

Costerton JW (2007) The biofilm primer. Springer, New York

Costerton JW, Geesey GG, Cheng KJ (1978) How bacteria stick. Sci Am 238(1):86–95

Costerton JW, Irvin RT, Cheng KJ (1981) The bacterial glycocalyx in nature and disease. Annu Rev Microbiol 35:299–324

Costerton JW, Lewandowski Z, Caldwell DE, Korber DR, Lappin-Scott HM (1995) Microbial biofilms. Annu Rev Microbiol 49:711–745

Crocetti GR, Hugenholtz P, Bond PL, Schuler A, Keller J, Jenkins D, Blackall LL (2000) Identification of polyphosphate-accumulating organisms and design of 16S rRNA-directed probes for their detection and quantitation. Appl Environ Microbiol 66(3):1175–1182

Crocetti GR, Banfield JF, Keller J, Bond PL, Blackall LL (2002) Glycogen-accumulating organisms in laboratory-scale and full-scale wastewater treatment processes. Microbiology 148:3353–3364

Curtis TP, Sloan WT (2006) Towards the design of diversity: stochastic models for community assembly in wastewater treatment plants. Water Sci Technol 54(1):227–236

Curtis TP, Head IM, Graham DW (2003) Theoretical ecology for engineering biology. Environ Sci Technol 37(3):64A–70A

da Silva LG, Gamez KO, Gomes JC, Akkermans K, Welles L, Abbas B, Loosdrecht MCMV, Wahl SA (2020) Revealing the metabolic flexibility of "Candidatus Accumulibacter phosphatis" through redox cofactor analysis and metabolic network modeling. Appl Environ Microbiol 86(24):1–17

Daigger GT (2011) A practitioner's perspective on the uses and future developments for wastewater treatment modelling. Water Sci Technol 63(3):516–526

Daigger GT, Nolasco D (1995) Evaluation and design of full-scale wastewater treatment plants using biological process models. Water Sci Technol 31(2):245–255

Daims H, Purkhold U, Bjerrum L, Arnold E, Wilderer PA, Wagner M (2001) Nitrification in sequencing biofilm batch reactors: lessons from molecular approaches. Water Sci Technol 43(3):9–18

Daims H, Taylor MW, Wagner M (2006) Wastewater treatment: a model system for microbial ecology. Trends Biotechnol 24(11):483–489

Daims H, Lebedeva EV, Pjevac P, Han P, Herbold C, Albertsen M, Jehmlich N, Palatinszky M, Vierheilig J, Bulaev A, Kirkegaard RH, Bergen MV, Rattei T, Bendinger B, Nielsen PH, Wagner M (2015) Complete nitrification by *Nitrospira* bacteria. Nature 528(7583):504–509

Dammel EE, Schroeder ED (1991) Density of activated sludge solids. Water Res 25(7):841–846

Dangcong P, Bernet N, Delgenes JP, Moletta R (1999) Aerobic granular sludge—a case report. Water Res 33:890–893

Dangcong P, Yi W, Hao W, Xiaochang W (2004) Biological denitrification in a sequencing batch reactor. Water Sci Technol 50(10):67–72

Danielsen HN, Hansen SH, Herbst FA, Kjeldal H, Stensballe A, Nielsen PH, Dueholm MS (2017) Direct identification of functional amyloid proteins by label-free quantitative mass spectrometry. Biomolecules 7(3):58

Davies DG, Parsek MR, Pearson JP, Iglewski BH, Costerton JW, Greenberg EP (1998) The involvement of cell-to-cell signals in the development of a bacterial biofilm. Science 280(5361):295–298

de Beer D, Schramm A (1999) Micro-environments and mass transfer phenomena in biofilms studied with microsensors. Water Sci Technol 39(7):173–178

de Beer D, Schramm A, Santegoeds CM, Kuhl M (1997) A nitrite microsensor for profiling environmental biofilms. Appl Environ Microbiol 63(3):973–977

De Bivar Xavier J, Picioreanu C, Van Loosdrecht MCM (2005) A general description of detachment for multidimensional modelling of biofilms. Biotechnol Bioeng 91(6):651–669
De Clippeleir H, Vlaeminck SE, De Wilde F, Daeninck K, Mosquera M, Boeckx P, Verstraete W, Boon N (2013) One-stage partial nitritation/anammox at 15 C on pretreated sewage: feasibility demonstration at lab-scale. Appl Microbiol Biotechnol 97(23):10199–10210
de Graaff DR, Felz S, Neu TR, Pronk M, van Loosdrecht MCM, Lin Y (2019) Sialic acids in the extracellular polymeric substances of seawater-adapted aerobic granular sludge. Water Res 155:343–351
de Graaff DR, van Loosdrecht MCM, Pronk M (2020) Stable granulation of seawater-adapted aerobic granular sludge with filamentous *Thiothrix* bacteria. Water Res 175:115683
de Jonge N, Poulsen JS, Vechi NT, Kofoed MVW, Nielsen JL (2022) Wood-Ljungdahl pathway utilisation during in situ H(2) biomethanation. Sci Total Environ 806(Pt 3):151254
de Kreuk MK (2006) Aerobic granular sludge, scaling up a new technology. PhD thesis, Delft University of Technology
de Kreuk MK, van Loosdrecht MCM (2004) Selection of slow growing organisms as a means for improving aerobic granular sludge stability. Water Sci Technol 49(11–12):9–17
de Kreuk MK, van Loosdrecht MCM (2006) Formation of aerobic granules with domestic sewage. J Environ Eng ASCE 132(6):694–697
de Kreuk M, Heijnen JJ, van Loosdrecht MCM (2005) Simultaneous COD, nitrogen, and phosphate removal by aerobic granular sludge. Biotechnol Bioeng 90(6):761–769
de Kreuk MK, Picioreanu C, Hosseini M, Xavier JB, van Loosdrecht MCM (2007) Kinetic model of a granular sludge SBR: influences on nutrient removal. Biotechnol Bioeng 97(4):801–815
de Kreuk MK, Kishida N, Tsuneda S, van Loosdrecht MCM (2010) Behavior of polymeric substrates in an aerobic granular sludge system. Water Res 44(20):5929–5938
de los Reyes FL III, Weaver JE, Wang L (2015) A methodological framework for linking bioreactor function to microbial communities andenvironmental conditions. Curr Opin Biotechnol 33:112–118
De Oliveira DMP, Forde BM, Kidd TJ, Harris PNA, Schembri MA, Beatson SA, Paterson DL, Walker MJ (2020) Antimicrobial resistance in ESKAPE pathogens. Clin Microbiol Rev 33(3):e00181-19
De Sanctis M, Di Iaconi C, Lopez A, Rossetti S (2010) Granular biomass structure and population dynamics in sequencing batch biofilter granular reactor (SBBGR). Bioresour Technol 101(7):2152–2158
de Villiers GH, Pretorius WA (2001) Abattoir effluent treatment and protein production. Water Sci Technol 43(11):243–250
Decho AW, Gutierrez T (2017) Microbial extracellular polymeric substances (EPSs) in ocean systems. Front Microbiol 8:922
Deinema MH, Habets LHA, Scholten J (1980) The accumulation of polyphosphate in *Acinetobacter* spp. FEMS Microbiol Lett 9(4):275–279
Deinema MH, van Loosdrecht M, Scholten A (1985) Some physiological characteristics of *Acinetobacter* spp. accumulating large amounts of phosphate. Water Sci Technol 17(11–12):119–125
Delavar MA, Wang J (2021) Lattice Boltzmann method in modeling biofilm formation, growth and detachment. Sustainability 13(14):7968
Derjaguin BV, Churaev NV, Muller VM (1987) The Derjaguin–Landau–Verwey–Overbeek (DLVO) theory of stability of lyophobic colloids. In: Derjaguin BV, Churaev NV, Muller VM (eds) Surface forces. Springer US, Boston, MA, pp 293–310
Derlon N, Peter-Varbanets M, Scheidegger A, Pronk W, Morgenroth E (2012) Predation influences the structure of biofilm developed on ultrafiltration membranes. Water Res 46(10):3323–3333
Derlon N, Mimoso J, Klein T, Koetzsch S, Morgenroth E (2014) Presence of biofilms on ultrafiltration membrane surfaces increases the quality of permeate produced during ultra-low pressure gravity-driven membrane filtration. Water Res 60:164–173

Derlon N, Wagner J, da Costa RHR, Morgenroth E (2016) Formation of aerobic granules for the treatment of real and low-strength municipal wastewater using a sequencing batch reactor operated at constant volume. Water Res 105:341–350

Derlon N, Garcia Villodres M, Kovács R, Brison A, Layer M, Takács I, Morgenroth E (2022) Modelling of aerobic granular sludge reactors: the importance of hydrodynamic regimes, selective sludge removal and gradients. Water Sci Technol 86(3):410–431

Desbos G, Rogalla F, Sibony J, Bourbigot MM (1990) Biofiltration as a compact technique for small wastewater treatment plants. Water Sci Technol 22(3–4):145–152

Desmond P, Best JP, Morgenroth E, Derlon N (2018) Linking composition of extracellular polymeric substances (EPS) to the physical structure and hydraulic resistance of membrane biofilms. Water Res 132:211–221

Dethlefsen L, McFall-Ngai M, Relman DA (2007) An ecological and evolutionary perspective on human-microbe mutualism and disease. Nature 449(7164):811–818

Di Iaconi C, Ramadori R, Lopez A (2006) Combined biological and chemical degradation for treating a mature municipal landfill leachate. Biochem Eng J 31(2):118–124

Di Iaconi C, Del Moro G, Pagao M, Ramadori R (2009) Municipal landfill leachate treatment by SBBGR technology. IJEWM 4(3–4):422–432

Dobbeleers T, D'aes J, Miele S, Caluwé M, Akkermans V, Daens D, Geuens L, Dries J (2017) Aeration control strategies to stimulate simultaneous nitrification-denitrification via nitrite during the formation of aerobic granular sludge. Appl Microbiol Biotechnol 101(17):6829–6839

Dobbeleers T, Caluwé M, Daens D, Geuens L, Dries J (2018) Evaluation of two start-up strategies to obtain nitrogen removal via nitrite and examination of the nitrous oxide emissions for different nitritation levels during the treatment of slaughterhouse wastewater. J Chem Technol Biotechnol 93(2):569–576

Dolfing J (1986) Granulation in UASB reactors. Water Sci Technol 18(12):15–25

Dolfing J, Griffioen A, van Neerven ARW, Zevenhuizen LPTM (1985) Chemical and bacteriological composition of granular methanogenic sludge. Can J Microbiol 31(8):744–750

Dong F, Zhang HM, Yang FL (2012) Modeling formation of aerobic granule and influence of hydrodynamic shear forces on granule diameter. Huanjing Kexue/Environ Sci 33(1):181–190

Downing L, Redmond E, Avila I (2022) When density is desirable. Water Online, September 9, 2022

Du R, Peng Y, Ji J, Shi L, Gao R, Li X (2019) Partial denitrification providing nitrite: opportunities of extending application for anammox. Environ Int 131:105001

Dueholm MS, Petersen SV, Sønderkær M, Larsen P, Christiansen G, Hein KL, Enghild JJ, Nielsen JL, Nielsen KL, Nielsen PH, Otzen DE (2010) Functional amyloid in Pseudomonas. Mol Microbiol 77(4):1009–1020

Dueholm MS, Marques IG, Karst SM, D'Imperio S, Tale VP, Lewis D, Nielsen PH, Nielsen JL (2015) Survival and activity of individual bioaugmentation strains. Bioresour Technol 186:192–199

Dueholm MS, Andersen KS, Petriglieri F, McIlroy SJ, Nierychlo M, Petersen JF, Kristensen JM, Yashiro E, Karst SM, Albertsen M, Nielsen PH (2019) Generation of Comprehensive Ecosystem-Specific Reference Databases with Species-Level Resolution by High-Throughput Full-Length 16S rRNA Gene Sequencing and Automated Taxonomy Assignment (AutoTax). mBio 1(5):e01557–20

Dueholm MKD, Nierychlo M, Andersen KS, Rudkjøbing V, Knutsson S, Arriaga S, Bakke R, Boon N, Bux F, Christensson M, Chua ASM, Curtis TP, Cytryn E, Erijman L, Etchebehere C, Fatta-Kassinos D, Frigon D, Garcia-Chaves MC, Gu AZ, Horn H, Jenkins D, Kreuzinger N, Kumari S, Lanham A, Law Y, Leiknes TO, Morgenroth E, Muszyński A, Petrovski S, Pijuan M, Pillai SB, Reis MAM, Rong Q, Rossetti S, Seviour R, Tooker N, Vainio P, van Loosdrecht M, Vikraman R, Wanner J, Weissbrodt D, Wen X, Zhang T, Nielsen PH, Albertsen M, Nielsen PH, Mi DASGC (2022) MiDAS 4: a global catalogue of full-length 16S rRNA gene sequences and taxonomy for studies of bacterial communities in wastewater treatment plants. Nat Commun 13(1):1908

Dueholm MKD, Besteman M, Zeuner EJ, Riisgaard-Jensen M, Nielsen ME, Vestergaard SZ, Heidelbach S, Bekker NS, Nielsen PH (2023) Genetic potential for exopolysaccharide synthesis in activated sludge bacteria uncovered by genome-resolved metagenomics. Water Res 229:119485

Dulekgurgen E, Ovez S, Artan N, Orhon D (2003a) Enhanced biological phosphate removal by granular sludge in a sequencing batch reactor. Biotechnol Lett 25(9):687–693

Dulekgurgen E, Yesiladali K, Ovez S, Tamerler C, Artan N, Orhon D (2003b) Conventional morphological and functional evaluation of the microbial populations in a sequencing batch reactor performing EBPR. J Environ Sci Health Part A 38(8):1499–1515

Dulekgurgen E, Artan N, Orhon D, Wilderer PA (2008) How does shear affect aggregation in granular sludge sequencing batch reactors? Relations between shear, hydrophobicity, and extracellular polymeric substances. Water Sci Technol 58(2):267–276

Dumont MG, Murrell JC (2005) Stable isotope probing—linking microbial identity to function. Nat Rev Microbiol 3(6):499–504

Dunne WM Jr (2002) Bacterial adhesion: seen any good biofilms lately? Clin Microbiol Rev 15(2):155–166

Duque AF, Bessa VS, Castro PML (2015) Characterization of the bacterial communities of aerobic granules in a 2-fluorophenol degrading process. Biotechnol Rep 5:98–104

Dutch Water Sector (2019) World's first waste water treatment plant to produce biopolymer. Kaumera Water & Technology, Netherlands

Dutta S, Hoffmann E, Hahn HH (2007) Study of rotating biological contactor performance in wastewater treatment using multi-culture biofilm model. Water Sci Technol 55(8–9):345–353

Eberl HJ, Picioreanu C, Heijnen JJ, Van Loosdrecht MCM (2000) Three-dimensional numerical study on the correlation of spatial structure, hydrodynamic conditions, and mass transfer and conversion in biofilms. Chem Eng Sci 55(24):6209–6222

Eberl HJ, van Loosdrecht MCM, Morgenroth E, Noguera DR, Perez J, Picioreanu C, Rittmann BE, Schwarz AO, Wanner O (2004) Modelling a spatially heterogeneous biofilm and the bulk fluid: selected results from benchmark problem 2 (BM2). Water Sci Technol 49(11–12):155–162

Ebrahimi S, Picioreanu C, Xavier JB, Kleerebezem R, Kreutzer M, Kapteijn F, Moulijn JA, Van Loosdrecht MCM (2005) Biofilm growth pattern in honeycomb monolith packings: effect of shear rate and substrate transport limitations. Catal Today 105(3–4):448–454

Ebrahimi S, Gabus S, Rohrbach-Brandt E, Hosseini M, Rossi P, Maillard J, Holliger C (2010) Performance and microbial community composition dynamics of aerobic granular sludge from sequencing batch bubble column reactors operated at 20°C, 30°C, and 35°C. Appl Microbiol Biotechnol 87(4):1555–1568

Eichorst SA, Strasser F, Woyke T, Schintlmeister A, Wagner M, Woebken D (2015) Advancements in the application of NanoSIMS and Raman microspectroscopy to investigate the activity of microbial cells in soils. FEMS Microbiol Ecol 91(10):fiv106

Eigentler L, Davidson FA, Stanley-Wall NR (2022) Mechanisms driving spatial distribution of residents in colony biofilms: an interdisciplinary perspective. Open Biol 12(12):220194

Eikelboom DH (1975) Filamentous organisms observed in activated sludge. Water Res 9(4):365–388

Eikelboom DH (2000) Process control of activated sludge plants by microscopic investigation. IWA Publishing, London

Ekholm J, Persson F, de Blois M, Modin O, Pronk M, van Loosdrecht MCM, Suarez C, Gustavsson DJI, Wilén B-M (2022) Full-scale aerobic granular sludge for municipal wastewater treatment—granule formation, microbial succession, and process performance. Environ Sci Water Res Technol 8(12):3138–3154

Elenter D, Milferstedt K, Zhang W, Hausner M, Morgenroth E (2007) Influence of detachment on substrate removal and microbial ecology in a heterotrophic/autotrophic biofilm. Water Res 41(20):4657–4671

Elias S, Banin E (2012) Multi-species biofilms: living with friendly neighbors. FEMS Microbiol Rev 36(5):990–1004

Eschenhagen M, Schuppler M, Roske I (2003) Molecular characterization of the microbial community structure in two activated sludge systems for the advanced treatment of domestic effluents. Water Res 37(13):3224–3232

Etterer TJ (2006) Formation, structure and function of aerobic granular sludge. PhD thesis, Technische Universität München

Etterer T, Wilderer PA (2001) Generation and properties of aerobic granular sludge. Water Sci Technol 43:19–26

Fagerlind MG, Webb JS, Barraud N, McDougald D, Jansson A, Nilsson P, Harlen M, Kjelleberg S, Rice SA (2012) Dynamic modelling of cell death during biofilm development. J Theor Biol 295:23–36

Falkentoft CM, Harremoes P, Mosbaek H, Wilderer P (2000) Combined denitrification and phosphorus removal in a biofilter. Water Sci Technol 41(4–5):493–501

Fang HHP (2000) Microbial distribution in UASB granules and its resulting effects. Water Sci Technol 42(12):201–208

Fang HHP, Zhang T, Liu Y (2002) Characterization of an acetate-degrading sludge without intracellular accumulation of polyphosphate and glycogen. Water Res 36(13):3211–3218

Fang F, Zhu R, Zhang L, Chen J (2008) Biodegradation of MTBE with ethanol co-substrate using an SBR and microbial community structure. Acta Sci Circum 28(11):2206–2212

Fang F, Ni BJ, Li XY, Sheng GP, Yu HQ (2009) Kinetic analysis on the two-step processes of AOB and NOB in aerobic nitrifying granules. Appl Microbiol Biotechnol 83(6):1–11

Felz S, Al-Zuhairy S, Aarstad OA, van Loosdrecht MCM, Lin YM (2016) Extraction of structural extracellular polymeric substances from aerobic granular sludge. J Vis Exp 115:e54534

Felz S, Vermeulen P, van Loosdrecht MCM, Lin YM (2019) Chemical characterization methods for the analysis of structural extracellular polymeric substances (EPS). Water Res 157:201–208

Feng C, Lotti T, Lin Y, Malpei F (2019) Extracellular polymeric substances extraction and recovery from anammox granules: evaluation of methods and protocol development. Chem Eng J 374:112–122

Fernando EY, McIlroy SJ, Nierychlo M, Herbst F-A, Petriglieri F, Schmid MC, Wagner M, Nielsen JL, Nielsen PH (2019) Resolving the individual contribution of key microbial populations to enhanced biological phosphorus removal with Raman–FISH. ISME J 13(8):1933–1946

Filali A (2011) Analyse et modélisation du traitement de l'azote dans un procédé de granulation aérobie hybride. PhD thesis, INSA

Filali A, Bessiere Y, Sperandio M (2012) Effects of oxygen concentration on the nitrifying activity of an aerobic hybrid granular sludge reactor. Water Sci Technol 65(2):289–295

Filipe CDM, Daigger GT (1998) Development of a revised metabolic model for the growth of phosphorus-accumulating organisms. Water Environ Res 70(1):67–79

Filipe CDM, Daigger GT, Grady CPL (2001a) Effects of pH on the rates of aerobic metabolism of phosphate-accumulating and glycogen-accumulating organisms. Water Environ Res 73(2):213–222

Filipe CDM, Daigger GT, Grady CPL (2001b) A metabolic model for acetate uptake under anaerobic conditions by glycogen accumulating organisms: stoichiometry, kinetics, and the effect of pH. Biotechnol Bioeng 76(1):17–31

Filipe CDM, Daigger GT, Grady CPL (2001c) pH as a key factor in the competition between glycogen-accumulating organisms and phosphorus-accumulating organisms. Water Environ Res 73(2):223–232

Filipe CDM, Daigger GT, Grady CPL (2001d) Stoichiometry and kinetics of acetate uptake under anaerobic conditions by an enriched culture of phosphorus-accumulating organisms at different pHs. Biotechnol Bioeng 76(1):32–43

Finstein MS (1967) Growth and flocculation in a Zoogloea culture. Appl Microbiol 15(4):962–963

Flemming HC (1993) Biofilms and environmental protection. Water Sci Technol 27(7–8):1–10

Flemming HC (2011) The perfect slime. Colloids Surf B 86(2):251–259

Flemming HC, Wuertz S (2019) Bacteria and archaea on earth and their abundance in biofilms. Nat Rev Microbiol 17(4):247–260

Flemming HC, Neu TR, Wozniak DJ (2007) The EPS matrix: the "house of biofilm cells." J Bacteriol 189(22):7945–7947
Flemming HC, Wingender J, Szewzyk U, Steinberg P, Rice SA, Kjelleberg S (2016) Biofilms: an emergent form of bacterial life. Nat Rev Microbiol 14(9):563–575
Flora JRV, Suidan MT, Biswas P, Sayles GD (1993) Modeling substrate transport into biofilms: role of multiple ions and pH effects. J Environ Eng 119(5):908–930
Florentz M, Hartemann P (1984) Screening for phosphate accumulating bacteria isolated from activated sludge. Environ Technol Lett 5(10):457–463
Fowler SJ, Palomo A, Dechesne A, Mines PD, Smets BF (2018) Comammox *Nitrospira* are abundant ammonia oxidizers in diverse groundwater-fed rapid sand filter communities. Environ Microbiol 20(3):1002–1015
Freitag A, Rudert M, Bock E (1987) Growth of *Nitrobacter* by dissimilatoric nitrate reduction. FEMS Microbiol Lett 48(1–2):105–109
Friedman BA, Dugan PR (1968) Identification of *Zoogloea* species and the relationship to zoogloeal matrix and floc formation. J Bacteriol 95(5):1903–1909
Frigon D, Wells G (2019) Microbial immigration in wastewater treatment systems: analytical considerations and process implications. Curr Opin Biotechnol 57:151–159
Frolund B, Griebe T, Nielsen PH (1995) Enzymatic activity in the activated-sludge floc matrix. Appl Microbiol Biotechnol 43(4):755–761
Frolund B, Palmgren R, Keiding K, Nielsen PH (1996) Extraction of extracellular polymers from activated sludge using a cation exchange resin. Water Res 30(8):1749–1758
Fuentes M, Scenna NJ, Aguirre PA, Mussati MC (2008) Application of two anaerobic digestion models to biofilm systems. Biochem Eng J 38(2):259–269
Fuentes M, Aguirre PA, Scenna NJ (2009a) Heterogeneous anaerobic biofilm reactor models application to UASB, EGSB and AFB reactors. Comput Aided Chem Eng 27:297–302
Fuentes M, Mussati MC, Scenna NJ, Aguirre PA (2009b) Global modeling and simulation of a three-phase fluidized bed bioreactor. Comput Chem Eng 33(1):359–370
Fuhs GW, Chen M (1975) Microbiological basis of phosphate removal in the activated sludge process for the treatment of wastewater. Microb Ecol 2(2):119–138
Fukuzaki S, Nishio N, Nagai S (1995) High rate performance and characterization of granular methanogenic sludges in upflow anaerobic sludge blanket reactors fed with various defined substrates. J Ferment Bioeng 79(4):354–359
Gagliano MC, Neu TR, Kuhlicke U, Sudmalis D, Temmink H, Plugge CM (2018) EPS glycoconjugate profiles shift as adaptive response in anaerobic microbial granulation at high salinity. Front Microbiol 9:1423
Gagliano MC, Sudmalis D, Pei R, Temmink H, Plugge CM (2020) Microbial community drivers in anaerobic granulation at high salinity. Front Microbiol 11:235
Gajda I, Greenman J, Melhuish C, Ieropoulos I (2015) Self-sustainable electricity production from algae grown in a microbial fuel cell system. Biomass Bioenergy 82:87–93
Gao J, Zhang Q, Su K, Chen R, Peng Y (2010a) Biosorption of acid yellow 17 from aqueous solution by non-living aerobic granular sludge. J Hazard Mater 174(1–3):215–225
Gao JF, Chen RN, Su K, Zhang Q, Peng YZ (2010b) Formation and reaction mechanism of simultaneous nitrogen and phosphorus removal by aerobic granular sludge. Huanjing Kexue/Environ Sci 31(4):1021–1029
Gao JF, Chen RN, Su K, Zhang Q, Peng YZ (2010c) Real time control of simultaneous nitrogen and phosphorus removal by aerobic granular sludge. Zhongguo Huanjing Kexue/China Environ Sci 30(2):180–185
Gao H, Mao Y, Zhao X, Liu WT, Zhang T, Wells G (2019) Genome-centric metagenomics resolves microbial diversity and prevalent truncated denitrification pathways in a denitrifying PAO-enriched bioprocess. Water Res 155:275–287
Garcia Martin H, Ivanova N, Kunin V, Warnecke F, Barry KW, McHardy AC, Yeates C, He S, Salamov AA, Szeto E, Dalin E, Putnam NH, Shapiro HJ, Pangilinan JL, Rigoutsos I, Kyrpides

NC, Blackall LL, McMahon KD, Hugenholtz P (2006) Metagenomic analysis of two enhanced biological phosphorus removal (EBPR) sludge communities. Nat Biotechnol 24(10):1263–1269

Gavigan J-A, Marshall LM, Dobson ADW (1999) Regulation of polyphosphate kinase gene expression in *Acinetobacter baumannii* 252. Microbiology 145(10):2931–2937

Geesey GG, Richardson WT, Yeomans HG, Irvin RT, Costerton JW (1977) Microscopic examination of natural sessile bacterial populations from an alpine stream. Can J Microbiol 23(12):1733–1736

Ghigo JM (2001) Natural conjugative plasmids induce bacterial biofilm development. Nature 412(6845):442–445

Gieseke A, Purkhold U, Wagner M, Amann R, Schramm A (2001) Community structure and activity dynamics of nitrifying bacteria in a phosphate-removing biofilm. Appl Environ Microbiol 67(3):1351–1362

Gieseke A, Bjerrum L, Wagner M, Amann R (2003) Structure and activity of multiple nitrifying bacterial populations co-existing in a biofilm. Environ Microbiol 5(5):355–369

Giesen A, Niermans R, van Loosdrecht MCM (2012) Aerobic granular biomass: the new standard for domestic and industrial wastewater treatment? Water 21(4):28–30

Giesen A, de Bruin LMM, Niermans RP, van der Roest HF (2013) Advancements in the application of aerobic granular biomass technology for sustainable treatment of wastewater. Water Pract Technol 8(1):47–54

Giesen A, van Loosdrecht M, Robertson S, de Bruin B (2015) Aerobic granular biomass technology: further innovation, system development and design optimisation. Proc Water Environ Fed 2015(16):1897–1917

Gilbride KA, Lee DY, Beaudette LA (2006) Molecular techniques in wastewater: understanding microbial communities, detecting pathogens, and real-time process control. J Microbiol Methods 66(1):1–20

Ginige MP, Hugenholtz P, Daims H, Wagner M, Keller J, Blackall LL (2004) Use of stable-isotope probing, full-cycle rRNA analysis, and fluorescence in situ hybridization-microautoradiography to study a methanol-fed denitrifying microbial community. Appl Environ Microbiol 70(1):588–596

Glancer M, Solian V, Matic V (2003) Granular gains. Water 21:31

Goel R, Kotay SM, Butler CS, Torres CI, Mahendra S (2011) Molecular biological methods in environmental engineering. Water Environ Res 83(10):927–955

Gonzalez A, Stombaugh J, Lozupone C, Turnbaugh PJ, Gordon JI, Knight R (2011a) The mind-body-microbial continuum. Dialogues Clin Neurosci 13(1):55–62

Gonzalez BC, Spinola ALG, Lamon AW, Araujo JC, Campos JR (2011b) The use of microsensors to study the role of the loading rate and surface velocity on the growth and the composition of nitrifying biofilms. Water Sci Technol 64(8):1607–1613

Gonzalez-Gil G, Holliger C (2011) Dynamics of microbial community structure of and enhanced biological phosphorus removal by aerobic granules cultivated on propionate or acetate. Appl Environ Microbiol 77(22):8041–8051

Grandclément C, Tannières M, Moréra S, Dessaux Y, Faure D (2015) Quorum quenching: role in nature and applied developments. FEMS Microbiol Rev 40(1):86–116

Grotenhuis JTC, Smit M, Lammeren AAM, Stams AJM, Zehnder AJB (1991a) Localization and quantification of extracellular polymers in methanogenic granular sludge. Appl Microbiol Biotechnol 36(1):115–119

Grotenhuis JTC, Smit M, Plugge CM, Yuansheng X, Van Lammeren AAM, Stams AJM, Zehnder AJB (1991b) Bacteriological composition and structure of granular sludge adapted to different substrates. Appl Environ Microbiol 57(7):1942–1949

Grotenhuis JTC, Plugge CM, Stams AJM, Zehnder AJB (1992) Hydrophobicities and electrophoretic mobilities of anaerobic bacterial isolates from methanogenic granular sludge. Appl Environ Microbiol 58(3):1054–1056

Gu AZ, Nerenberg R, Sturm BM, Chul P, Goel R (2011) Molecular methods in biological systems. Water Environ Res 82(10):908–930

Guimarães LB, Gubser NR, Lin Y, Pronk M, Welles L, Albertsen M, Daudt GC, Geleijnse MAA, da Costa RHR, Nielsen PH, van Loosdrecht MCM, Weissbrodt DG (2016) Exopolysaccharides biorefining from used water: an enterprise in the microbiome of granular sludge. In: van Loosdrecht et al (eds) 13th IWA leading edge conference on water and wastewater technologies: evaluating impacts of innovation. International Water Association, Jerez de la Frontera

Guimarães LB, Wagner J, Akaboci TRV, Daudt GC, Nielsen PH, van Loosdrecht MCM, Weissbrodt DG, da Costa RHR (2018) Elucidating performance failures in use of granular sludge for nutrient removal from domestic wastewater in a warm coastal climate region. Environ Technol 41(15):1896–1911

Gujer W (1987) The significance of segregation of biomass in biofilms. Water Sci Technol 19(3–4):495–503

Gujer W (2008) Systems analysis for water technology. Springer Verlag, Berlin, Heidelberg

Gülay A, Fowler SJ, Tatari K, Thamdrup B, Albrechtsen H-J, Al-Soud WA, Sørensen SJ, Smets BF (2019) DNA- and RNA-SIP reveal *Nitrospira* spp. as key drivers of nitrification in groundwater-fed biofilters. mBio 10(6):e01870–19

Gunathilaka GU, Tahlan V, Mafiz AI, Polur M, Zhang Y (2017) Phages in urban wastewater have the potential to disseminate antibiotic resistance. Int J Antimicrob Agents 50(5):678–683

Gunther S, Trutnau M, Kleinsteuber S, Hause G, Bley T, Roske I, Harms H, Muller S (2009) Dynamics of polyphosphate-accumulating bacteria in wastewater treatment plant microbial communities detected via DAPI (4′,6′-diamidino-2-phenylindole) and tetracycline labeling. Appl Environ Microbiol 75(7):2111–2121

Gunther S, Koch C, Hubschmann T, Roske I, Muller RA, Bley T, Harms H, Muller S (2012) Correlation of community dynamics and process parameters as a tool for the prediction of the stability of wastewater treatment. Environ Sci Technol 46(1):84–92

Guo B, Liu C, Gibson C, Frigon D (2019) Wastewater microbial community structure and functional traits change over short timescales. Sci Total Environ 662:779–785

Haaksman VA, Mirghorayshi M, van Loosdrecht MCM, Pronk M (2020) Impact of aerobic availability of readily biodegradable COD on morphological stability of aerobic granular sludge. Water Res 187:116402

Hailei W, Ping L, Guosheng L, Xin L, Jianming Y (2010) Rapid biodecolourization of eriochrome black T wastewater by bioaugmented aerobic granules cultivated through a specific method. Enzyme Microb Technol 47(1–2):37–43

Hall CW, Mah TF (2017) Molecular mechanisms of biofilm-based antibiotic resistance and tolerance in pathogenic bacteria. FEMS Microbiol Rev 41(3):276–301

Hall SJ, Hugenholtz P, Siyambalapitiya N, Keller J, Blackall LL (2002) The development and use of real-time PCR for the quantification of nitrifiers in activated sludge. Water Sci Technol 46(1–2):267–272

Hall-Stoodley L, Costerton JW, Stoodley P (2004) Bacterial biofilms: from the natural environment to infectious diseases. Nat Rev Microbiol 2(2):95–108

Han M, De Clippeleir H, Al-Omari A, Stewart H, Wett B, Vlaeminck SE, Bott C, Murthy S (2015) NOB out-selection in mainstream deammonification: a resilience evaluation. In: WEF (ed) Water environment federation's annual technical exhibition and conference (WEFTEC), Chicago, IL

Hanada S, Liu WT, Shintani T, Kamagata Y, Nakamura K (2002) *Tetrasphaera elongata* sp. nov., a polyphosphate-accumulating bacterium isolated from activated sludge. Int J Syst Evol Microbiol 52(3):883–887

Haringa C (2022) An analysis of organism lifelines in an industrial bioreactor using Lattice-Boltzmann CFD. Eng Life Sci 23(1):e2100159

Harms G, Layton AC, Dionisi HM, Gregory IR, Garrett VM, Hawkins SA, Robinson KG, Sayler GS (2003) Real-time PCR quantification of nitrifying bacteria in a municipal wastewater treatment plant. Environ Sci Technol 37(2):343–351

Hassa J, Maus I, Off S, Puhler A, Scherer P, Klocke M, Schluter A (2018) Metagenome, metatranscriptome, and metaproteome approaches unraveled compositions and functional relationships of microbial communities residing in biogas plants. Appl Microbiol Biotechnol 102(12):5045–5063

Hausner M, Wuertz S (1999) High rates of conjugation in bacterial biofilms as determined by quantitative in situ analysis. Appl Environ Microbiol 65(8):3710–3713

He S, McMahon KD (2011) "*Candidatus* Accumulibacter" gene expression in response to dynamic EBPR conditions. ISME J 5(2):329–340

He S, Gu AZ, McMahon KD (2006) Fine-scale differences between *Accumulibacter*-like bacteria in enhanced biological phosphorus removal activated sludge. Water Sci Technol 54(1):111–117

He S, Gall DL, McMahon KD (2007) "*Candidatus* Accumulibacter" population structure in enhanced biological phosphorus removal sludges as revealed by polyphosphate kinase genes. Appl Environ Microbiol 73(18):5865–5874

He S, Gu AZ, McMahon KD (2008) Progress toward understanding the distribution of *Accumulibacter* among full-scale enhanced biological phosphorus removal systems. Microb Ecol 55(2):229–236

He S, Bishop FI, McMahon KD (2010a) Bacterial community and "*Candidates* Accumulibacter" population dynamics in laboratory-scale enhanced biological phosphorus removal reactors. Appl Environ Microbiol 76(16):5479–5487

He S, Kunin V, Haynes M, Martin HG, Ivanova N, Rohwer F, Hugenholtz P, McMahon KD (2010b) Metatranscriptomic array analysis of "*Candidatus* Accumulibacter phosphatis"-enriched enhanced biological phosphorus removal sludge. Environ Microbiol 12(5):1205–1217

Heijnen JJ, Van't Riet K (1984) Mass transfer, mixing and heat transfer phenomena in low viscosity bubble column reactors. Chem Eng J 28(2)

Heijnen JJ, Mulder A, Weltevrede R, Hols J, Van Leeuwen HLJM (1991) Large scale anaerobic-aerobic treatment of complex industrial waste water using biofilm reactors. Water Sci Technol 23(7–9):1427–1436

Heijnen JJ, Van Loosdrecht MCM, Mulder A, Tijhuis L (1992) Formation of biofilms in a biofilm air-lift suspension reactor. Water Sci Technol 26(3–4):647–654

Heijnen JJ, van Loosdrecht MCM, Mulder R, Weltevrede R, Mulder A (1993) Development and scale-up of an aerobic biofilm air-lift suspension reactor. Water Sci Technol 27(5–6):253–261

Heijnen JJ, Kleerebezem R, Flickinger MC (2009) Bioenergetics of microbial growth. Encyclopedia of industrial biotechnology. Wiley

Helmer C, Kunst S (1998) Low temperature effects on phosphorus release and uptake by microorganisms in EBPR plants. Water Sci Technol 37(4–5):531–539

Henrici AT (1933) Studies of freshwater bacteria. J Bacteriol 25(3):277–287

Henriet O, Meunier C, Henry P, Mahillon J (2016) Improving phosphorus removal in aerobic granular sludge processes through selective microbial management. Bioresour Technol 211:298–306

Henriet O, Meunier C, Henry P, Mahillon J (2017) Filamentous bulking caused by *Thiothrix* species is efficiently controlled in full-scale wastewater treatment plants by implementing a sludge densification strategy. Sci Rep 7(1):1430

Henze M, Harremoes P (1983) Anaerobic treatment of wastewater in fixed film reactors—a literature review. Water Sci Technol 15(8–9):1–101

Henze M, Gujer W, Mino T, van Loosdrecht MCM (2000) Activated sludge models ASM1, ASM2, ASM2d and ASM3. IWA Publishing, London

Henze M, van Loosdrecht MCM, Ekama GA, Brdjanovic D (2008) Biological wastewater treatment: principles, modelling and design. IWA Publishing, London

Herbst FA, Dueholm MS, Wimmer R, Nielsen PH (2019) The proteome of *Tetrasphaera elongata* is adapted to changing conditions in wastewater treatment plants. Proteomes 7(2):16

Hermanowicz SW (1998) A model of two-dimensional biofilm morphology. Water Sci Technol 37(4–5):219–222

Hernandez M, Mohn W, Martinez E, Rost E, Alvarez A, Alvarez H (2008) Biosynthesis of storage compounds by *Rhodococcus jostii* RHA1 and global identification of genes involved in their metabolism. BMC Genomics 9(1):600

Herzberg M, Elimelech M (2008) Physiology and genetic traits of reverse osmosis membrane biofilms: a case study with *Pseudomonas aeruginosa*. ISME J 2(2):180–194

Hesselmann RPX, Werlen C, Hahn D, van der Meer JR, Zehnder AJB (1999) Enrichment, phylogenetic analysis and detection of a bacterium that performs enhanced biological phosphate removal in activated sludge. Syst Appl Microbiol 22(3):454–465

Hesselsoe M, Fureder S, Schloter M, Bodrossy L, Iversen N, Roslev P, Nielsen PH, Wagner M, Loy A (2009) Isotope array analysis of *Rhodocyclales* uncovers functional redundancy and versatility in an activated sludge. ISME J 3(12):1349–1364

Hodge DS, Devinny JS (1995) Modeling removal of air contaminants by biofiltration. J Environ Eng 121(1):21–32

Hoekstra M, Geilvoet SP, Hendrickx TLG, van ErpTaalman Kip CS, Kleerebezem R, van Loosdrecht MCM (2019) Towards mainstream anammox: lessons learned from pilot-scale research at WWTP Dokhaven. Environ Technol 40(13):1721–1733

Hofman-Bang J, Zheng D, Westermann P, Ahring BK, Raskin L (2003) Molecular ecology of anaerobic reactor systems. Adv Biochem Eng Biotechnol 81:151–203

Holland EA, Coleman DC (1987) Litter placement effects on microbial and organic matter dynamics in an agroecosystem. Ecology 68(2):425–433

Horn H (1994) Dynamics of a nitrifying bacteria population in a biofilm controlled by an oxygen microelectrode. Water Sci Technol 29(10–11):69–76

Horn H, Hempel DC (1998) Modeling mass transfer and substrate utilization in the boundary layer of biofilm systems. Water Sci Technol 37(4–5):139–147

Horn H, Neu TR, Wulkow M (2001) Modelling the structure and function of extracellular polymeric substances in biofilms with new numerical techniques. Water Sci Technol 43(6):121–127

Hu LL, Wang JL, Wen XH, Yang N, Qian Y (2004) Cultivation of aerobic granular sludge in SBR by seeding anaerobic granular sludge. Huanjing Kexue/Environ Sci 25(4):74

Hu BL, Zheng P, Tang CJ, Chen JW, van der Biezen E, Zhang L, Ni BJ, Jetten MSM, Yan J, Yu HQ, Kartal B (2010) Identification and quantification of anammox bacteria in eight nitrogen removal reactors. Water Res 44(17):5014–5020

Huang WE, Stoecker K, Griffiths R, Newbold L, Daims H, Whiteley AS, Wagner M (2007) Raman–FISH: combining stable-isotope Raman spectroscopy and fluorescence in situ hybridization for the single cell analysis of identity and function. Environ Microbiol 9(8):1878–1889

Hubaux N, Wells G, Morgenroth E (2015) Impact of coexistence of flocs and biofilm on performance of combined nitritation-anammox granular sludge reactors. Water Res 68:127–139

Huilinir C, Romero R, Munoz C, Bornhardt C, Roeckel M, Antileo C (2010) Dynamic modeling of partial nitrification in a rotating disk biofilm reactor: calibration, validation and simulation. Biochem Eng J 52(1):7–18

Huisman JL, Gujer W (2002) Modelling wastewater transformation in sewers based on ASM3. Water Sci Technol 45(6):51–60

Huisman IL, Krebs P, Gujer W (2003) Integral and unified model for the sewer and wastewater treatment plant focusing on transformations. Water Sci Technol 47(12):65–71

Hulshoff Pol LW, De Castro Lopes SI, Lettinga G, Lens PNL (2004) Anaerobic sludge granulation. Water Res 38(6):1376–1389

Hvitved-Jacobsen T, Vollertsen J, Nielsen PH (1998) A process and model concept for microbial wastewater transformations in gravity sewers. Water Sci Technol 37(1):233–241

Inamori Y, Kong HN, Nakanishi H, Sudo R (1994) Granulation property in USB-aerobic biofilter recirculation treatment process. Doboku Gakkai Rombun-Hokokushu/Proc Jpn Soc Civ Eng 503:139–147

Ivanov V, Wang XH, Tay STL, Tay JH (2006) Bioaugmentation and enhanced formation of microbial granules used in aerobic wastewater treatment. Appl Microbiol Biotechnol 70(3):374–381

Ivanov V, Wang X-H, Stabnikova O (2008) Starter culture of *Pseudomonas veronii* strain B for aerobic granulation. World J Microbiol Biotechnol 24(4):533–539

Iwai S, Oshino Y, Tsukada T (1990) Design and operation of small wastewater treatment plants by the microbial film process. Water Sci Technol 22(3–4):139–144

Jafari M, Vanoppen M, van Agtmaal JMC, Cornelissen ER, Vrouwenvelder JS, Verliefde A, van Loosdrecht MCM, Picioreanu C (2020) Cost of fouling in full-scale reverse osmosis and nanofiltration installations in the Netherlands. Desalination 500:114865

James GA, Beaudette L, Costerton JW (1995) Interspecies bacterial interactions in biofilms. J Ind Microbiol 15(4):257–262

Jang A, Yoon YH, Kim IS, Kim KS, Bishop PL (2003) Characterization and evaluation of aerobic granules in sequencing batch reactor. J Biotechnol 105(1–2):71–82

Janssen P, van der Roest H (1996) Winning combination. Water Qual Int 1996(9–10):12–17

Janus T, Ulanicki B (2010) Modelling SMP and EPS formation and degradation kinetics with an extended ASM3 model. Desalination 261(1):117–125

Jayathilake PG, Gupta P, Li B, Madsen C, Oyebamiji O, González-Cabaleiro R, Rushton S, Bridgens B, Swailes D, Allen B, McGough AS, Zuliani P, Ofiteru ID, Wilkinson D, Chen J, Curtis T (2017) A mechanistic individual-based model of microbial communities. PLoS ONE 12(8):e0181965

Jemaat Z, Suarez-Ojeda ME, Perez J, Carrera J (2013) Simultaneous nitritation and p-nitrophenol removal using aerobic granular biomass in a continuous airlift reactor. Bioresour Technol 150:307–313

Jenkins D, Richard MG (1985) The causes and control of activated-sludge bulking. Tappi J 68(12):73–76

Jenkins D, Tandoi V (1991) The applied microbiology of enhanced biological phosphate removal accomplishments and needs. Water Res 25(12):1471–1478

Jenkins D, Wanner J (eds) (2014) Activated sludge—100 years and counting. IWA Publishing, London

Jenkinson HF, Lappin-Scott HM (2001) Biofilms adhere to stay. Trends Microbiol 9(1):9–10

Jetten MSM, Strous M, van de Pas-Schoonen KT, Schalk J, van Dongen UGJM, van De Graaf AA, Logemann S, Muyzer G, van Loosdrecht MCM, Kuenen JG (1998) The anaerobic oxidation of ammonium. FEMS Microbiol Rev 22(5):421–437

Jiang HL, Tay JH, Maszenan AM, Tay STL (2004) Bacterial diversity and function of aerobic granules engineered in a sequencing batch reactor for phenol degradation. Appl Environ Microbiol 70(11):6767–6775

Jiang F, Leung DH, Li S, Chen GH, Okabe S, van Loosdrecht MCM (2009) A biofilm model for prediction of pollutant transformation in sewers. Water Res 43(13):3187–3198

Jin RC, Zheng P, Mahmood Q, Zhang L (2008) Performance of a nitrifying airlift reactor using granular sludge. Sep Purif Technol 63(3):670–675

Johansson JF, Paul LR, Finlay RD (2004) Microbial interactions in the mycorrhizosphere and their significance for sustainable agriculture. FEMS Microbiol Ecol 48(1):1–13

Johnson K, Jiang Y, Kleerebezem R, Muyzer G, Van Loosdrecht MCM (2009) Enrichment of a mixed bacterial culture with a high polyhydroxyalkanoate storage capacity. Biomacromolecules 10(4):670–676

Joss A, Salzgeber D, Eugster J, König R, Rottermann K, Burger S, Fabijan P, Leumann S, Mohn J, Siegrist HR (2009) Full-scale nitrogen removal from digester liquid with partial nitritation and anammox in one SBR. Environ Sci Technol 43(14):5301–5306

Joyce A, Ijaz UZ, Nzeteu C, Vaughan A, Shirran SL, Botting CH, Quince C, O'Flaherty V, Abram F (2018) Linking microbial community structure and function during the acidified anaerobic digestion of grass. Front Microbiol 9:540

Juang YC, Adav SS, Lee DJ, Lai JY (2009) Biodiversity in aerobic granule membrane bioreactor at high organic loading rates. Appl Microbiol Biotechnol 85(2):1–6

Juretschko S, Loy A, Lehner A, Wagner M (2002) The microbial community composition of a nitrifying-denitrifying activated sludge from an industrial sewage treatment plant analyzed by the full-cycle rRNA approach. Syst Appl Microbiol 25(1):84–99

Kaballo HP, Zhao Y, Wilderer PA (1995) Elimination of p-chlorophenol in biofilm reactors—a comparative study of continuous flow and sequenced batch operation. Water Sci Technol 31(1):51–60

Kampfer P, Erhart R, Beimfohr C, Bahringer J, Wagner M, Amann R (1996) Characterization of bacterial communities from activated sludge: culture-dependent numerical identification versus in situ identification using group- and genus-specific rRNA-targeted oligonucleotide probes. Microb Ecol 32(2):101–121

Karst SM, Albertsen M, Kirkegaard RH, Dueholm MS, Nielsen PH (2016) Molecular methods. In: van Loosdrecht MCM, Nielsen PH, Lopez-Vazquez CM, Brdjanovic D (eds) Experimental methods in wastewater treatment. IWA Publishing, London, pp 285–323

Kartal B, Kuenen JG, van Loosdrecht MC (2010) Engineering. Sewage treatment with anammox. Science 328(5979):702–703

Keasling JD, Van Dien SJ, Trelstad P, Renninger N, McMahon K (2000) Application of polyphosphate metabolism to environmental and biotechnological problems. Biochem (Mosc) 65(3):324–331

Kelly JJ, Siripong S, McCormack J, Janus LR, Urakawa H, El Fantroussi S, Noble PA, Sappelsa L, Rittmann BE, Stahl DA (2005) DNA microarray detection of nitrifying bacterial 16S rRNA in wastewater treatment plant samples. Water Res 39(14):3229–3238

Kermani M, Bina B, Movahedian H, Amin MM, Nikaeen M (2009) Biological phosphorus and nitrogen removal from wastewater using moving bed biofilm process. Iran J Biotechnol 7(1):19–27

Khan MA, Satoh H, Mino T, Katayama H, Kurisu F, Matsuo T (2002) Bacteriophage-host interaction in the enhanced biological phosphate removing activated sludge system. Water Sci Technol 46(1–2):39–43

Kishida N, Kim J, Tsuneda S, Sudo R (2006) Anaerobic/oxic/anoxic granular sludge process as an effective nutrient removal process utilizing denitrifying polyphosphate-accumulating organisms. Water Res 40(12):2303–2310

Kishida N, Tsuneda S, Sakakibara Y, Kim JH, Sudo R (2008) Real-time control strategy for simultaneous nitrogen and phosphorus removal using aerobic granular sludge. Water Sci Technol 58(2):445–450

Kishida N, Tsuneda S, Kim JH, Sudo R (2009) Simultaneous nitrogen and phosphorus removal from high-strength industrial wastewater using aerobic granular sludge. J Environ Eng 135(3):153–158

Kjelleberg S, Hermansson M (1984) Starvation-induced effects on bacterial surface characteristics. Appl Environ Microbiol 48(3):497–503

Kleerebezem R, van Loosdrecht MC (2007) Mixed culture biotechnology for bioenergy production. Curr Opin Biotechnol 18(3):207–212

Kleikamp HBC, Lin Y, McMillan DGG, Geelhoed JS, Naus-Wiezer SNH, van Baarlen P, Saha C, Louwen R, Sorokin DY, van Loosdrecht MCM, Pabst M (2020a) Tackling the chemical diversity of microbial nonulosonic acids—a universal large-scale survey approach. Chem Sci 11:3074

Kleikamp HBC, Pronk M, Tugui C, da Silva LG, Abbas B, Lin YM, van Loosdrecht MCM, Pabst M (2020b) Quantitative profiling of microbial communities by *de novo* metaproteomics. bioRxiv 2020.08.16.252924

Kleikamp HBC, Pronk M, Tugui C, Guedes da Silva L, Abbas B, Lin YM, van Loosdrecht MCM, Pabst M (2021) Database-independent *de novo* metaproteomics of complex microbial communities. Cell Syst 12(5):375–383.e5

Kleikamp HBC, Grouzdev D, Schaasberg P, van Valderen R, van der Zwaan R, van de Wijgaart R, Lin Y, Abbas B, Pronk M, van Loosdrecht MCM, Pabst M (2022) Comparative metaproteomics demonstrates different views on the complex granular sludge microbiome. bioRxiv 2022.03.07.483319

Klein T, Zihlmann D, Derlon N, Isaacson C, Szivák I, Weissbrodt DG, Pronk W (2016) Biological control of biofilms on membranes by metazoans. Water Res 88:20–29

Kleiner M, Kouris A, Jensen M, Liu Y, McCalder J, Strous M (2021) Ultra-sensitive protein-SIP to quantify activity and substrate uptake in microbiomes with stable isotopes. Microbiome 11:24
Koch G, Egli K, Van Der Meer JR, Siegrist H (2000a) Mathematical modeling of autotrophic denitrification in a nitrifying biofilm of a rotating biological contactor. Water Sci Technol 41(4–5):191–198
Koch G, Kuhni M, Gujer W, Siegrist H (2000b) Calibration and validation of activated sludge model no. 3 for Swiss municipal wastewater. Water Res 34(14):3580–3590
Koch H, Galushko A, Albertsen M, Schintlmeister A, Gruber-Dorninger C, Lücker S, Pelletier E, Le Paslier D, Spieck E, Richter A, Nielsen PH, Wagner M, Daims H (2014) Growth of nitrite-oxidizing bacteria by aerobic hydrogen oxidation. Science 345(6200):1052–1054
Kofoed MVW, Nielsen DA, Revsbech NP, Schramm A (2012) Fluorescence in situ hybridization (FISH) detection of nitrite reductase transcripts (nirS mRNA) in *Pseudomonas stutzeri* biofilms relative to a microscale oxygen gradient. Syst Appl Microbiol
Koh KY, Kueh KH, Loh KT, Leong HJ, Chua ASM, Hashim MA (2009) Effect of seeding sludge type and hydrodynamic shear force on the aerobic sludge granulation in sequencing batch airlift reactors. Asia-Pac J Chem Eng 4(5):826–831
Koh KS, Matz C, Tan CH, Le HL, Rice SA, Marshall DJ, Steinberg PD, Kjelleberg S (2012) Minimal increase in genetic diversity enhances predation resistance. Mol Ecol 21(7):1741–1753
Kolter R, Losick R (1998) One for all and all for one. Science 280(5361):226–227
Kong YH, Nielsen JL, Nielsen PH (2004) Microautoradiographic study of *Rhodocyclus*-related polyphosphate accumulating bacteria in full-scale enhanced biological phosphorus removal plants. Appl Environ Microbiol 70(9):5383–5390
Kong Y, Nielsen JL, Nielsen PH (2005) Identity and ecophysiology of uncultured actinobacterial polyphosphate-accumulating organisms in full-scale enhanced biological phosphorus removal plants. Appl Environ Microbiol 71(7):4076–4085
Kong Y, Liu YQ, Tay JH, Wong FS, Zhu J (2009) Aerobic granulation in sequencing batch reactors with different reactor height/diameter ratios. Enzyme Microb Technol 45(5):379–383
Kortstee GJ, Appeldoorn KJ, Bonting CF, van Niel EW, van Veen HW (1994) Biology of polyphosphate-accumulating bacteria involved in enhanced biological phosphorus removal. FEMS Microbiol Rev 15(2–3):137–153
Kotay SM, Datta T, Choi J, Goel R (2011) Biocontrol of biomass bulking caused by *Haliscomenobacter hydrossis* using a newly isolated lytic bacteriophage. Water Res 45(2):694–704
Krebs CJ (1972) Ecology: the experimental analysis of distribution and abundance. Harper & Row, London
Kreft JU, Wimpenny JWT (2001) Effect of EPS on biofilm structure and function as revealed by an individual-based model of biofilm growth. Water Sci Technol 43(6):135–141
Kreft JU, Picioreanu C, Wimpenny JWT, van Loosdrecht MCM (2001) Individual-based modelling of biofilms. Microbiology 147(11):2897–2912
Kristiansen R, Nguyen HTT, Saunders AM, Nielsen JL, Wimmer R, Le VQ, McIlroy SJ, Petrovski S, Seviour RJ, Calteau A, Nielsen KL, Nielsen PH (2013) A metabolic model for members of the genus *Tetrasphaera* involved in enhanced biological phosphorus removal. ISME J 7(3):543–554
Kuba T, Smolders G, van Loosdrecht MCM, Heijnen JJ (1993) Biological phosphorus removal from wastewater by anaerobic-anoxic sequencing batch reactor. Water Sci Technol 27(5–6):241–252
Kuba T, Murnleitner E, van Loosdrecht MCM, Heijnen JJ (1996a) A metabolic model for biological phosphorus removal by denitrifying organisms. Biotechnol Bioeng 52(6):685–695
Kuba T, van Loosdrecht MCM, Heijnen JJ (1996b) Phosphorus and nitrogen removal with minimal COD requirement by integration of denitrifying dephosphatation and nitrification in a two-sludge system. Water Res 30(7):1702–1710
Kuba T, van Loosdrecht MCM, Heijnen JJ (1997) Biological dephosphatation by activated sludge under denitrifying conditions: pH influence and occurrence of denitrifying dephosphatation in a full-scale waste water treatment plant. Water Sci Technol 36(12):75–82
Kucnerowicz F, Verstraete W (1983) Evolution of microbial communities in the activated sludge process. Water Res 17(10):1275–1279

Kuehn M, Mehl M, Hausner M, Bungartz HJ, Wuertz S (2001) Time-resolved study of biofilm architecture and transport processes using experimental and simulation techniques: the role of EPS. Water Sci Technol 43(6):143–150

Kuhl M, Jorgensen BB (1992) Microsensor measurements of sulfate reduction and sulfide oxidation in compact microbial communities of aerobic biofilms. Appl Environ Microbiol 58(4):1164–1174

Kunin V, He S, Warnecke F, Peterson SB, Garcia Martin H, Haynes M, Ivanova N, Blackall LL, Breitbart M, Rohwer F, McMahon KD, Hugenholtz P (2008) A bacterial metapopulation adapts locally to phage predation despite global dispersal. Genome Res 18(2):293–297

Lacamp B, Hansen F, Penillard P, Rogalla F (1993) Wastewater nutrient removal with advanced biofilm reactors. Water Sci Technol 27(5–6):263–276

Lackner S, Gilbert EM, Vlaeminck SE, Joss A, Horn H, van Loosdrecht MCM (2014) Full-scale partial nitritation/anammox experiences—an application survey. Water Res 55:292–303

Laconi CD, Moro GD, Lopez A, De Sanctis M, Ramadori R (2008) Municipal wastewater treatment by a periodic biofilter with granular biomass. Water Sci Technol 58(12):2395–2401

Lan HX, Chen YC, Chen ZH, Chen R (2005) Cultivation and characters of aerobic granules for pentachlorophenol (PCP) degradation under microaerobic condition. J Environ Sci 17(3):506–510

Lardon LA, Merkey BV, Martins S, Dotsch A, Picioreanu C, Kreft JU, Smets BF (2011) iDynoMiCS: next-generation individual-based modelling of biofilms. Environ Microbiol 13(9):2416–2434

Larsen P, von Ins M (2010) The rate of growth in scientific publication and the decline in coverage provided by Science Citation Index. Scientometrics 84(3):575–603

Larsen P, Eriksen PS, Lou MA, Thomsen TR, Kong YH, Nielsen JL, Nielsen PH (2006) Floc-forming properties of polyphosphate accumulating organisms in activated sludge. Water Sci Technol 54(1):257–265

Larsen P, Nielsen JL, Dueholm MS, Wetzel R, Otzen D, Nielsen PH (2007) Amyloid adhesins are abundant in natural biofilms. Environ Microbiol 9(12):3077–3090

Larsen P, Nielsen JL, Otzen D, Nielsen PH (2008) Amyloid-like adhesins produced by floc-forming and filamentous bacteria in activated sludge. Appl Environ Microbiol 74(5):1517–1526

Laspidou CS, Rittmann BE (2002) Non-steady state modeling of extracellular polymeric substances, soluble microbial products, and active and inert biomass. Water Res 36(8):1983–1992

Laspidou CS, Rittmann BE (2004) Modeling the development of biofilm density including active bacteria, inert biomass, and extracellular polymeric substances. Water Res 38(14–15):3349–3361

Lau AO, Strom PF, Jenkins D (1984) The competitive growth of floc-forming and filamentous bacteria: a model for activated sludge bulking. J Water Pollut Control Fed 56(1):52–61

Laureni M, Falås P, Robin O, Wick A, Weissbrodt DG, Nielsen JL, Ternes T, Morgenroth E, Joss A (2016) Mainstream partial nitritation and anammox: long-term process stability and effluent quality at low temperatures. Water Res 101:628–639

Laureni M, Weissbrodt DG, Villez K, Robin O, de Jonge N, Rosenthal A, Wells G, Nielsen JL, Morgenroth E, Joss A (2019) Biomass segregation between biofilm and flocs improves the control of nitrite-oxidizing bacteria in mainstream partial nitritation and anammox processes. Water Res 154:104–116

Law Y, Kirkegaard RH, Cokro AA, Liu X, Arumugam K, Xie C, Stokholm-Bjerregaard M, Drautz-Moses DI, Nielsen PH, Wuertz S, Williams RB (2016) Integrative microbial community analysis reveals full-scale enhanced biological phosphorus removal under tropical conditions. Sci Rep 6:25719

Lawrence JR, Wolfaardt GM, Korber DR (1994) Determination of diffusion coefficients in biofilms by confocal laser microscopy. Appl Environ Microbiol 60(4):1166–1173

Lawrence JR, Korber DR, Wolfaardt GM (1996) Heterogeneity of natural biofilm communities. Cells Mater 6(1–3):175–191

Lawrence JR, Swerhone GDW, Kuhlicke U, Neu TR (2007) In situ evidence for microdomains in the polymer matrix of bacterial microcolonies. Can J Microbiol 53(3):450–458

Lawson CE, Wu SJ, Bhattacharjee A, McMahon KD, Goel R, Noguera DR (2015) Ecogenomics reveals distributed metabolic networks in suspended and attached growth anammox bioreactors. In: WEF (ed) Water environment federation's annual technical exhibition and conference (WEFTEC). Water Environment Federation, Chicago

Lawson CE, Wu S, Bhattacharjee AS, Hamilton JJ, McMahon KD, Goel R, Noguera DR (2017) Metabolic network analysis reveals microbial community interactions in anammox granules. Nat Commun 8:15416

Lawson CE, Harcombe WR, Hatzenpichler R, Lindemann SR, Löffler FE, O'Malley MA, García Martín H, Pfleger BF, Raskin L, Venturelli OS, Weissbrodt DG, Noguera DR, McMahon KD (2019) Common principles and best practices for engineering microbiomes. Nat Rev Microbiol 17:725–741

Layer M, Adler A, Reynaert E, Hernandez A, Pagni M, Morgenroth E, Holliger C, Derlon N (2019) Organic substrate diffusibility governs microbial community composition, nutrient removal performance and kinetics of granulation of aerobic granular sludge. Water Res X 4:100033

Layer M, Bock K, Ranzinger F, Horn H, Morgenroth E, Derlon N (2020a) Particulate substrate retention in plug-flow and fully-mixed conditions during operation of aerobic granular sludge systems. Water Res X 9:100075

Layer M, Villodres MG, Hernandez A, Reynaert E, Morgenroth E, Derlon N (2020b) Limited simultaneous nitrification-denitrification (SND) in aerobic granular sludge systems treating municipal wastewater: mechanisms and practical implications. Water Res X 7:100048

Lazazzera BA (2005) Lessons from DNA microarray analysis: the gene expression profile of biofilms. Curr Opin Microbiol 8(2):222–227

Le T, Peng B, Su C, Massoudieh A, Torrents A, Al-Omari A, Murthy S, Wett B, Chandran K, deBarbadillo C, Bott C, De Clippeleir H (2019) Nitrate residual as a key parameter to efficiently control partial denitrification coupling with anammox. Water Environ Res 91(11):1455–1465

Lebuhn M, Effenberger M, Garcés G, Gronauer A, Wilderer PA (2004) Evaluating real-time PCR for the quantification of distinct pathogens and indicator organisms in environmental samples. Water Sci Technol 50(1):263–270

Lee SP, Kwon OS, Sinskey AJ (1996) Localization of genes involved in exopolysaccharide biosynthesis in *Zoogloea ramigera* 115SLR. J Microbiol Biotechnol 6(5):321–325

Lee N, La Cour Jansen J, Aspegren H, Henze M, Nielsen PH, Wagner M (2002) Population dynamics in wastewater treatment plants with enhanced biological phosphorus removal operated with and without nitrogen removal. Water Sci Technol 46:163–170

Lee SH, Satoh H, Katayama H, Mino T (2004) Isolation, physiological characterization of bacteriophages from enhanced biological phosphorous removal activated sludge and their putative role. J Microbiol Biotechnol 14(4):730–736

Lee SJ, Lee YS, Lee YC, Choi YL (2006) Molecular characterization of polyphosphate (PolyP) operon from *Serratia marcescens*. J Basic Microbiol 46(2):108–115

Lee D-J, Chen Y-Y, Show K-Y, Whiteley CG, Tay J-H (2010) Advances in aerobic granule formation and granule stability in the course of storage and reactor operation. Biotechnol Adv 28(6):919–934

Lee WH, Lee JH, Choi WH, Hosni AA, Papautsky I, Bishop PL (2011) Needle-type environmental microsensors: design, construction and uses of microelectrodes and multi-analyte MEMS sensor arrays. Meas Sci Technol 22(4):1–22

Legner M, McMillen DR, Cvitkovitch DG (2019) Role of dilution rate and nutrient availability in the formation of microbial biofilms. Front Microbiol 10:916

Lemaire R, Yuan Z, Blackall LL, Crocetti GR (2008) Microbial distribution of *Accumulibacter* spp. and *Competibacter* spp. in aerobic granules from a lab-scale biological nutrient removal system. Environ Microbiol 10(2):354–363

Lemos PC, Viana C, Salgueiro EN, Ramos AM, Crespo JPSG, Reis MAM (1998) Effect of carbon source on the formation of polyhydroxyalkanoates (PHA) by a phosphate-accumulating mixed culture. Enzyme Microb Technol 22(8):662–671

Lemos P, Levantesi C, Serafim L, Rossetti S, Reis M, Tandoi V (2008) Microbial characterisation of polyhydroxyalkanoates storing populations selected under different operating conditions using a cell-sorting RT-PCR approach. Appl Microbiol Biotechnol 78(2):351–360
Lens P, De Beer D, Cronenberg C, Ottengraf S, Verstraete W (1995) The use of microsensors to determine population distributions in UASB aggregates. Water Sci Technol 31(1):273–280
Lettinga G, Hulshoff Pol LW (1991) USAB-process design for various types of wastewaters. Water Sci Technol 24(8):87–107
Lettinga G, Van Der Sar J, Van Der Ben J (1976) Anaerobic treatment of sugar beet waste water (Dutch). H2O 9(2):38–43
Lettinga G, van Velsen AFM, de Zeeuw W, Hobma SW (1979) Feasibility of the upflow anaerobic sludge blanket (UASB)-process. SAE Preprints 35–45
Lettinga G, van Velsen AFM, Hobma SW, Dezeeuw W, Klapwijk A (1980) Use of the upflow sludge blanket (Usb) reactor concept for biological wastewater-treatment, especially for anaerobic treatment. Biotechnol Bioeng 22(4):699–734
Li AJ, Li XY (2009) Selective sludge discharge as the determining factor in SBR aerobic granulation: numerical modelling and experimental verification. Water Res 43(14):3387–3396
Li X, Gao F, Hua Z, Du G, Chen J (2005) Treatment of synthetic wastewater by a novel MBR with granular sludge developed for controlling membrane fouling. Sep Purif Technol 46(1–2):19–25
Li ZH, Kuba T, Kusuda T (2006a) Aerobic granular sludge: a promising technology for decentralised wastewater treatment. Water Sci Technol 53(9):79–85
Li ZH, Kuba T, Kusuda T (2006b) The influence of starvation phase on the properties and the development of aerobic granules. Enzyme Microb Technol 38(5):670–674
Li X, Li Y, Liu H, Hua Z, Du G, Chen J (2007a) Characteristics of aerobic biogranules from membrane bioreactor system. J Membr Sci 287(2):294–299
Li ZH, Kuba T, Kusuda T, Wang XC (2007b) Effect of rotifers on the stability of aerobic granules. Environ Technol 28:235–242
Li AJ, Yang SF, Li XY, Gu JD (2008a) Microbial population dynamics during aerobic sludge granulation at different organic loading rates. Water Res 42(13):3552–3560
Li XF, Li YJ, Liu H, Hua ZZ, Du GC, Chen J (2008b) Correlation between extracellular polymeric substances and aerobic biogranulation in membrane bioreactor. Sep Purif Technol 59(1):26–33
Li Y, Liu Y, Shen L, Chen F (2008c) DO diffusion profile in aerobic granule and its microbiological implications. Enzyme Microb Technol 43(4–5):349–354
Liao BQ, Allen DG, Droppo IG, Leppard GG, Liss SN (2001) Surface properties of sludge and their role in bioflocculation and settleability. Water Res 35(2):339–350
Lin YM, Liu Y, Tay JH (2003) Development and characteristics of phosphorus-accumulating microbial granules in sequencing batch reactors. Appl Microbiol Biotechnol 62(4):430–435
Lin YM, Wang L, Chi ZM, Liu XY (2008) Bacterial alginate role in aerobic granular bio-particles formation and settleability improvement. Sep Sci Technol 43:1642–1652
Lin Y, de Kreuk M, van Loosdrecht MCM, Adin A (2010) Characterization of alginate-like exopolysaccharides isolated from aerobic granular sludge in pilot-plant. Water Res 44(11):3355–3364
Lin YM, Nierop KGJ, Girbal-Neuhauser E, Adriaanse M, van Loosdrecht MCM (2015) Sustainable polysaccharide-based biomaterial recovered from waste aerobic granular sludge as a surface coating material. Sustain Mater Technol 4:24–29
Lindemann SR, Bernstein HC, Song H-S, Fredrickson JK, Fields MW, Shou W, Johnson DR, Beliaev AS (2016) Engineering microbial consortia for controllable outputs. ISME J 10(9):2077–84
Lindner SN, Knebel S, Wesseling H, Schoberth SM, Wendisch VF (2009) Exopolyphosphatases PPX1 and PPX2 from *Corynebacterium glutamicum*. Appl Environ Microbiol 75(10):3161–3170
Linlin H, Jianlong W, Xianghua W, Yi Q (2005) The formation and characteristics of aerobic granules in sequencing batch reactor (SBR) by seeding anaerobic granules. Process Biochem 40(1):5–11

Liss SN, Droppo IG, Flannigan DT, Leppard GG (1996) Floc architecture in wastewater and natural riverine systems. Environ Sci Technol 30(2):680–686

Liss SN, Liao BQ, Droppo IG, Allen DG, Leppard GG (2002) Effect of solids retention time on floc structure. Water Sci Technol 46(1–2):431–438

Liu Y, Liu Q-S (2006) Causes and control of filamentous growth in aerobic granular sludge sequencing batch reactors. Biotechnol Adv 24(1):115–127

Liu Y, Tay JH (2001a) Detachment forces and their influence on the structure and metabolic behaviour of biofilms. World J Microbiol Biotechnol 17(2):111–117

Liu Y, Tay JH (2001b) Metabolic response of biofilm to shear stress in fixed-film culture. J Appl Microbiol 90(3):337–342

Liu Y, Tay J-H (2002) The essential role of hydrodynamic shear force in the formation of biofilm and granular sludge. Water Res 36(7):1653–1665

Liu Y, Tay JH (2004) State of the art of biogranulation technology for wastewater treatment. Biotechnol Adv 22(7):533–563

Liu YQ, Tay JH (2007) Cultivation of aerobic granules in a bubble column and an airlift reactor with divided draft tubes at low aeration rate. Biochem Eng J 34(1):1–7

Liu YQ, Tay JH (2008) Influence of starvation time on formation and stability of aerobic granules in sequencing batch reactors. Bioresour Technol 99(5):980–985

Liu WT, Marsh TL, Cheng H, Forney LJ (1997) Characterization of microbial diversity by determining terminal restriction fragment length polymorphisms of genes encoding 16S rRNA. Appl Environ Microbiol 63(11):4516–4522

Liu WT, Nielsen AT, Wu JH, Tsai CS, Matsuo Y, Molin S (2001) In situ identification of polyphosphate- and polyhydroxyalkanoate-accumulating traits for microbial populations in a biological phosphorus removal process. Environ Microbiol 3(2):110–122

Liu Y, Xu H, Yang SF, Tay JH (2003a) A general model for biosorption of Cd^{2+}, Cu^{2+} and Zn^{2+} by aerobic granules. J Biotechnol 102:233–239

Liu Y, Yang SF, Liu QS, Tay JH (2003b) The role of cell hydrophobicity in the formation of aerobic granules. Curr Microbiol 46(4):270–274

Liu Y, Yang SF, Qin L, Tay JH (2004a) A thermodynamic interpretation of cell hydrophobicity in aerobic granulation. Appl Microbiol Biotechnol 64(3):410–415

Liu Y, Yang SF, Tay JH, Liu QS, Qin L, Li Y (2004b) Cell hydrophobicity is a triggering force of biogranulation. Enzyme Microb Technol 34(5):371–379

Liu YQ, Liu Y, Tay JH (2004c) The effects of extracellular polymeric substances on the formation and stability of biogranules. Appl Microbiol Biotechnol 65(2):143–148

Liu Q, Hu X, Wang J (2005a) Performance characteristics for nitrogen removal in SBR by aerobic granules. Chin J Chem Eng 13(5):669–672

Liu QS, Liu Y, Tay JH, Kuan YS (2005b) Responses of sludge flocs to shear strength. Process Biochem 40(10):3213–3217

Liu QS, Liu Y, Tay STL, Show KY, Ivanov V, Benjamin M, Tay JH (2005c) Startup of pilot-scale aerobic granular sludge reactor by stored granules. Environ Technol 26(12):1363–1369

Liu Y, Wang Z-W, Liu Y-Q, Qin L, Tay J-H (2005d) A generalized model for settling velocity of aerobic granular sludge. Biotechnol Prog 21(2):621–626

Liu Y, Wang Z-W, Tay J-H (2005e) A unified theory for upscaling aerobic granular sludge sequencing batch reactors. Biotechnol Adv 23(5):335–344

Liu Y, Wang ZW, Qin L, Liu YQ, Tay JH (2005f) Selection pressure-driven aerobic granulation in a sequencing batch reactor. Appl Microbiol Biotechnol 67(1):26–32

Liu YQ, Liu Y, Tay JH (2005g) Relationship between size and mass transfer resistance in aerobic granules. Lett Appl Microbiol 40:312–315

Liu YQ, Moy BYP, Tay JH (2007a) COD removal and nitrification of low-strength domestic wastewater in aerobic granular sludge sequencing batch reactors. Enzyme Microb Technol 42(1):23–28

Liu YQ, Wu WW, Tay JH, Wang JL (2007b) Starvation is not a prerequisite for the formation of aerobic granules. Appl Microbiol Biotechnol 76(1):211–216

Liu H, Li G, Li X, Chen J (2008a) Molecular characterization of bacterial community in aerobic granular sludge stressed by pentachlorophenol. J Environ Sci 20(10):1243–1249

Liu XY, Zhao HM, Peng DC, Sui XJ (2008b) Denitrifying phosphate uptake of biological phosphorous removal granular sludge in SBR. Huanjing Kexue/Environ Sci 29(8):2254–2259

Liu YQ, Wu WW, Tay JH, Wang JL (2008c) Formation and long-term stability of nitrifying granules in a sequencing batch reactor. Bioresour Technol 99(9):3919–3922

Liu X-W, Yu H-Q, Ni B-J, Sheng G-P (2009a) Characterization, modeling and application of aerobic granular sludge for wastewater treatment. Adv Biochem Eng Biotechnol 113:275–303

Liu YJ, Wang XC, Yuan HL (2009b) Characterization of microbial communities in a fluidized-pellet-bed bioreactor for wastewater treatment. Desalination 249(1):445–452

Liu YQ, Kong Y, Tay JH, Zhu J (2011) Enhancement of start-up of pilot-scale granular SBR fed with real wastewater. Sep Purif Technol 82(1):190–196

Lochmatter S, Holliger C (2014) Optimization of operation conditions for the startup of aerobic granular sludge reactors biologically removing carbon, nitrogen, and phosphorous. Water Res 59:58–70

Lochmatter S, Gonzalez-Gil G, Holliger C (2013) Optimized aeration strategies for nitrogen and phosphorus removal with aerobic granular sludge. Water Res 47(16):6187–6197

Lochmatter S, Maillard J, Holliger C (2014) Nitrogen removal over nitrite by aeration control in aerobic granular sludge sequencing batch reactors. Int J Environ Res Public Health 11(7):6955

Lopez-Palau S, Dosta J, Mata-Alvarez J (2009) Start-up of an aerobic granular sequencing batch reactor for the treatment of winery wastewater. Water Sci Technol 60(4):1049–1054

Lopez-Palau S, Dosta J, Pericas A, Mata-Alvarez J (2011a) Partial nitrification of sludge reject water using suspended and granular biomass. J Chem Technol Biotechnol 86(12):1480–1487

Lopez-Palau S, Pericas A, Dosta J, Mata-Alvarez J (2011b) Partial nitrification of sludge reject water by means of aerobic granulation. Water Sci Technol 64(9):1906–1912

Lopez-Vazquez CM, Hooijmans CM, Brdjanovic D, Gijzen HJ, van Loosdrecht MCM (2007) A practical method for quantification of phosphorus- and glycogen-accumulating organism populations in activated sludge systems. Water Environ Res 79(13):2487–2498

Lopez-Vazquez CM, Brdjanovic D, van Loosdrecht MC (2008a) Comment on "Could polyphosphate-accumulating organisms (PAOs) be glycogen-accumulating organisms (GAOs)?" by Zhou Y, Pijuan M, Zeng R, Lu Huabing, Yuan Z Water Res 42(13):3561–3562

Lopez-Vazquez CM, Hooijmans CM, Brdjanovic D, Gijzen HJ, van Loosdrecht MCM (2008b) Factors affecting the microbial populations at full-scale enhanced biological phosphorus removal (EBPR) wastewater treatment plants in The Netherlands. Water Res 42(10–11):2349–2360

Lopez-Vazquez CM, Hooijmans CM, Brdjanovic D, Gijzen HJ, van Loosdrecht MCM (2009a) Temperature effects on glycogen accumulating organisms. Water Res 43(11):2852–2864

Lopez-Vazquez CM, Oehmen A, Hooijmans CM, Brdjanovic D, Gijzen HJ, Yuan Z, van Loosdrecht MCM (2009b) Modeling the PAO-GAO competition: effects of carbon source, pH and temperature. Water Res 43(2):450–462

Lopez-Vazquez CM, Wentzel MC, Comeau Y, Ekama GA, van Loosdrecht MCM, Brdjanovic D, Oehmen A (2020) Enhanced biological phosphorus removal. In: Chen G, van Loosdrecht MCM, Ekama GA, Brdjanovic D (eds) Biological wastewater treatment: principles, modelling and design, 2nd edn. IWA Publishing, London, pp 239–326

Lotti T, Kleerebezem R, Hu Z, Kartal B, Jetten MS, van Loosdrecht MC (2014) Simultaneous partial nitritation and anammox at low temperature with granular sludge. Water Res 66:111–121

Lotti T, Kleerebezem R, Hu Z, Kartal B, de Kreuk MK, van Erp Taalman Kip C, Kruit J, Hendrickx TL, van Loosdrecht MC (2015) Pilot-scale evaluation of anammox-based mainstream nitrogen removal from municipal wastewater. Environ Technol 36(9–12):1167–1177

Lou In C, Zhang X, Zhang Y, Zhu J (2001) Aerobic sludge granulation and biological phosphorus removal in different operating conditions of SBR. Huanjing Kexue/Environ Sci 22(2):87–90

Loy A, Lehner A, Lee N, Adamczyk J, Meier H, Ernst J, Schleifer K-H, Wagner M (2002) Oligonucleotide microarray for 16S rRNA gene-based detection of all recognized lineages of sulfate-reducing prokaryotes in the environment. Appl Environ Microbiol 68(10):5064–5081

Loy A, Schulz C, Lucker S, Schopfer-Wendels A, Stoecker K, Baranyi C, Lehner A, Wagner M (2005) 16S rRNA gene-based oligonucleotide microarray for environmental monitoring of the betaproteobacterial order *Rhodocyclales*. Appl Environ Microbiol 71(3):1373–1386

Lu H, Oehmen A, Virdis B, Keller J, Yuan Z (2006) Obtaining highly enriched cultures of "*Candidatus* Accumulibacter phosphates" through alternating carbon sources. Water Res 40(20):3838–3848

Lu S, Ji M, Wang JF, Wei YJ (2007a) Simultaneous phosphorus and nitrogen removal of domestic sewage with aerobic granular sludge SBR. Huanjing Kexue/Environ Sci 28(8):1687–1692

Lu T, Saikaly PE, Oerther DB (2007b) Modelling the competition of planktonic and sessile aerobic heterotrophs for complementary nutrients in biofilm reactor. Water Sci Technol 55(8–9):227–235

Macêdo WV, Poulsen JS, Zaiat M, Nielsen JL (2022) Proteogenomics identification of TBBPA degraders in anaerobic bioreactor. Environ Pollut 310:119786

MacLeod FA, Guiot SR, Costerton JW (1990) Layered structure of bacterial aggregates produced in an upflow anaerobic sludge bed and filter reactor. Appl Environ Microbiol 56(6):1598–1607

Macmanus J, Long C, Klaus S, Parsons M, Chandran K, De Clippeleir H, Bott C (2022) Nitrogen removal capacity and carbon demand requirements of partial denitrification/anammox MBBR and IFAS processes. Water Environ Res 94(8):e10766

Madsen JS, Burmølle M, Hansen LH, Sørensen SJ (2012) The interconnection between biofilm formation and horizontal gene transfer. FEMS Immunol Med Microbiol 65(2):183–195

Malik A, Kakii K (2003) Pair-dependent co-aggregation behavior of non-flocculating sludge bacteria. Biotechnol Lett 25(12):981–986

Malik A, Kakii K (2008) Novel coaggregating microbial consortium: testing strength for field applications. Bioresour Technol 99(11):4627–4634

Manga J, Ferrer J, Garcia-Usach F, Seco A (2001) A modification to the activated sludge model no. 2 based on the competition between phosphorus-accumulating organisms and glycogen-accumulating organisms. Water Sci Technol 43(11):161–171

Mannina G, Trapani DD, Viviani G, Odegaard H (2011) Modelling and dynamic simulation of hybrid moving bed biofilm reactors: model concepts and application to a pilot plant. Biochem Eng J 56(1–2):23–36

Manz W, Amann R, Ludwig W, Wagner M, Schleifer KH (1992) Phylogenetic oligodeoxynucleotide probes for the major subclasses of proteobacteria: problems and solutions. Syst Appl Microbiol 15(4):593–600

Manz W, Wagner M, Amann R, Schleifer KH (1994) In situ characterization of the microbial consortia active in two wastewater treatment plants. Water Res 28(8):1715–1723

Marques R, Santos J, Nguyen H, Carvalho G, Noronha JP, Nielsen PH, Reis MAM, Oehmen A (2017) Metabolism and ecological niche of *Tetrasphaera* and *Ca.* Accumulibacter in enhanced biological phosphorus removal. Water Res 122:159–171

Marshall KC, Goodman AE (1994) Effects of adhesion on microbial cell physiology. Colloids Surf B 2(1–3):1–7

Martins AMP, Heijnen JJ, van Loosdrecht MCM (2003) Effect of dissolved oxygen concentration on sludge settleability. Appl Microbiol Biotechnol 62(5–6):586–593

Martins AMP, Heijnen JJ, van Loosdrecht MCM (2004a) Bulking sludge in biological nutrient removal systems. Biotechnol Bioeng 86(2):125–135

Martins AMP, Pagilla K, Heijnen JJ, van Loosdrecht MCM (2004b) Filamentous bulking sludge—a critical review. Water Res 38(4):793–817

Martins AMP, Picioreanu C, Heijnen JJ, van Loosdrecht MCM (2004c) Three-dimensional dual-morphotype species modeling of activated sludge flocs. Environ Sci Technol 38(21):5632–5641

Martins AMP, Karahan O, van Loosdrecht MCM (2011) Effect of polymeric substrate on sludge settleability. Water Res 45(1):263–273

Massoudieh A, Crain C, Lambertini E, Nelson KE, Barkouki T, L'Amoreaux P, Loge FJ, Ginn TR (2010) Kinetics of conjugative gene transfer on surfaces in granular porous media. J Contam Hydrol 112(1–4):91–102

Maszenan AM, Seviour RJ, Patel BKC, Schumann P, Burghardt J, Tokiwa Y, Stratton HM (2000) Three isolates of novel polyphosphate-accumulating Gram-positive cocci, obtained from activated sludge, belong to a new genus, *Tetrasphaera* gen. nov., and description of two new species, *Tetrasphaera japonica* sp. nov. and *Tetrasphaera australiensis* sp. nov. Int J Syst Evol Microbiol 50(2):593–603

Matsumoto S, Katoku M, Saeki G, Terada A, Aoi Y, Tsuneda S, Picioreanu C, Van Loosdrecht MCM (2010) Microbial community structure in autotrophic nitrifying granules characterized by experimental and simulation analyses. Environ Microbiol 12(1):192–206

Maurer M, Gujer W (1995) Monitoring of microbial phosphorus release in batch experiments using electric conductivity. Water Res 29(11):2613–2617

McCall A-K, Scheidegger A, Madry MM, Steuer AE, Weissbrodt DG, Vanrolleghem PA, Kraemer T, Morgenroth E, Ort C (2016) Influence of different sewer biofilms on transformation rates of drugs. Environ Sci Technol 50(24):13351–13360

McCann KS (2000) The diversity-stability. Nature 405(6783):228–233

McClure DD, Kavanagh JM, Fletcher DF, Barton GW (2016) Characterizing bubble column bioreactor performance using computational fluid dynamics. Chem Eng Sci 144:58–74

McDaniel EA, Wahl SA, Si I, Pinto A, Ziels R, Nielsen PH, McMahon KD, Williams RBH (2021) Prospects for multi-omics in the microbial ecology of water engineering. Water Res 205:117608

McIlroy SJ, Albertsen M, Andresen EK, Saunders AM, Kristiansen R, Stokholm-Bjerregaard M, Nielsen KL, Nielsen PH (2014a) '*Candidatus* Competibacter'-lineage genomes retrieved from metagenomes reveal functional metabolic diversity. ISME J 8(3):613–624

McIlroy SJ, Starnawska A, Starnawski P, Saunders AM, Nierychlo M, Nielsen PH, Nielsen JL (2014b) Identification of active denitrifiers in full-scale nutrient removal wastewater treatment systems. Environ Microbiol 18(1):50–64

McIlroy SJ, Nittami T, Kanai E, Fukuda J, Saunders AM, Nielsen PH (2015) Re-appraisal of the phylogeny and fluorescence *in situ* hybridization probes for the analysis of the *Competibacteraceae* in wastewater treatment systems. Environ Microbiol Rep 7(2):166–174

McMahon KD, Dojka MA, Pace NR, Jenkins D, Keasling JD (2002a) Polyphosphate kinase from activated sludge performing enhanced biological phosphorus removal. Appl Environ Microbiol 68(10):4971–4978

McMahon KD, Jenkins D, Keasling JD (2002b) Polyphosphate kinase genes from activated sludge carrying out enhanced biological phosphorus removal. Water Sci Technol 46(1–2):155–162

McMahon KD, Garcia Martin H, Hugenholtz P (2007a) Integrating ecology into biotechnology. Curr Opin Biotechnol 18:287–292

McMahon KD, Yilmaz S, He S, Gall DL, Jenkins D, Keasling JD (2007b) Polyphosphate kinase genes from full-scale activated sludge plants. Appl Microbiol Biotechnol 77(1):167–173

McMahon KD, Gu AZ, Nerenberg R, Sturm BM (2009) Molecular methods in biological systems. Water Environ Res 81(10):986–1002

McSwain BS, Irvine RL, Wilderer PA (2004) The effect of intermittent feeding on aerobic granule structure. Water Sci Technol 49(11–12):19–25

McSwain BS, Irvine RL, Hausner M, Wilderer PA (2005) Composition and distribution of extracellular polymeric substances in aerobic flocs and granular sludge. Appl Environ Microbiol 71(2):1051–1057

McSwain Sturm BS, Irvine RL (2008) Dissolved oxygen as key parameter to aerobic granule formations. Water Sci Technol 58(4):781–787

Medipally SR, Yusoff FM, Banerjee S, Shariff M (2015) Microalgae as sustainable renewable energyfeedstock for biofuel production. BioMed Res Int 2015:519513

Meinhold J, Filipe CDM, Daigger GT, Isaacs S (1999) Characterization of the denitrifying fraction of phosphate accumulating organisms in biological phosphate removal. Water Sci Technol 39(1):31–42

Melvik JE, Dornish M (2004) Alginate as carrier for cell immobilisation. In: Nedovic V, Willaert R (eds) Fundamentals of cell immobilisation biotechnology. Kluwer Academic Publishers, Dordrecht, pp 33–51

Mengoni A, Grassi E, Bazzicalupo M (2002) Cloning method for taxonomic interpretation of T-RFLP patterns. Biotechniques 33(5):990–992

Meyer RL, Saunders AM, Blackall LL (2006) Putative glycogen-accumulating organisms belonging to the *Alphaproteobacteria* identified through rRNA-based stable isotope probing. Microbiology 152:419–429

Mielczarek AT, Saunders AM, Larsen P, Albertsen M, Stevenson M, Nielsen JL, Nielsen PH (2013) The microbial database for Danish wastewater treatment plants with nutrient removal (MiDas-DK)—a tool for understanding activated sludge population dynamics and community stability. Water Sci Technol 67(11):2519–2526

Miller MJ, Allen DG (2005) Modelling transport and degradation of hydrophobic pollutants in biofilter biofilms. Chem Eng J 113(2–3):197–204

Mino T (2000) Microbial selection of polyphosphate-accumulating bacteria in activated sludge wastewater treatment processes for enhanced biological phosphate removal. Biochem (Mosc) 65(3):341–348

Mino T, Satoh H, Matsuo T (1994) Metabolisms of different bacterial populations in enhanced biological phosphate removal processes. Water Sci Technol 29(7):67–70

Mino T, van Loosdrecht MCM, Heijnen JJ (1998) Microbiology and biochemistry of the enhanced biological phosphate removal process. Water Res 32(11):3193–3207

Mishima K, Nakamura M (1991) Self-immobilization of aerobic activated sludge—a pilot study of the aerobic upflow sludge blanket process in municipal sewage treatment. Water Sci Technol 23(4–6):981–990

Moelants N, Smets IY, Van Den Broeck R, Lambert N, Liers S, Declerck P, Vanysacker L, Van Impe JF (2010) Towards a low complexity carbon removal model for the optimal design of compact decentralised wastewater treatment systems. Water Sci Technol 61(6):1579–1588

Molin S, Tolker-Nielsen T (2003) Gene transfer occurs with enhanced efficiency in biofilms and induces enhanced stabilisation of the biofilm structure. Curr Opin Biotechnol 14(3):255–261

Montoya T, Borrás L, Aguado D, Ferrer J, Seco A (2008) Detection and prevention of enhanced biological phosphorus removal deterioration caused by Zoogloea overabundance. Environ Technol 29(1):35–42

Mooshammer M, Kitzinger K, Schintlmeister A, Ahmerkamp S, Nielsen JL, Nielsen PH, Wagner M (2021) Flow-through stable isotope probing (flow-SIP) minimizes cross-feeding in complex microbial communities. ISME J 15(1):348–353

Moralejo-Gárate H, Mar'Atusalihat E, Kleerebezem R, Van Loosdrecht MCM (2011) Microbial community engineering for biopolymer production from glycerol. Appl Microbiol Biotechnol 92(3):631–639

Morales N, Figueroa M, Mosquera-Corral A, Campos JL, Mendez R (2012) Aerobic granular-type biomass development in a continuous stirred tank reactor. Sep Purif Technol 89:199–205

Morgan-Sagastume F, Nielsen JL, Nielsen PH (2008) Substrate-dependent denitrification of abundant probe-defined denitrifying bacteria in activated sludge. FEMS Microbiol Ecol 66(2):447–461

Morgenroth E (2020) Modelling biofilms. In: Chen GH, van Loosdrecht MCM, Ekama GA, Brdjanovic D (eds) Biological wastewater treatment: principles, modelling and design, 2nd edn. IWA Publishing, London

Morgenroth E, Wilderer PA (1998) Modeling of enhanced biological phosphorus removal in a sequencing batch biofilm reactor. Water Sci Technol 37(4–5):583–587

Morgenroth E, Wilderer PA (1999) Controlled biomass removal—the key parameter to achieve enhanced biological phosphorus removal in biofilm systems. Water Sci Technol 39(7):33–40

Morgenroth E, Wilderer PA (2000) Influence of detachment mechanisms on competition in biofilms. Water Res 34(2):417–426

Morgenroth E, Milferstedt K (2009) Biofilm engineering: linking biofilm development at different length and time scales. Rev Environ Sci Biotechnol 8(3):203–208

Morgenroth E, Sherden T, van Loosdrecht MCM, Heijnen JJ, Wilderer PA (1997) Aerobic granular sludge in a sequencing batch reactor. Water Res 31(12):3191–3194

Morgenroth E, Eberl H, Van Loosdrecht MCM (2000a) Evaluating 3-D and 1-D mathematical models for mass transport in heterogeneous biofilms. Water Sci Technol 41(4–5):347–356

Morgenroth E, Van Loosdrecht MCM, Wanner O (2000b) Biofilm models for the practitioner. Water Sci Technol 41(4–5):509–512

Morgenroth E, Arvin E, Vanrolleghem P (2002) The use of mathematical models in teaching wastewater treatment engineering. Water Sci Technol 45(6):229–233

Morse GK, Brett SW, Guy JA, Lester JN (1998) Review: phosphorus removal and recovery technologies. Sci Total Environ 212(1):69–81

Moser-Engeler R, Udert KM, Wild D, Siegrist H (1998) Products from primary sludge fermentation and their suitability for nutrient removal. Water Sci Technol 38(1):265–273

Mosquera-Corral A, de Kreuk MK, Heijnen JJ, van Loosdrecht MCM (2005) Effects of oxygen concentration on N-removal in an aerobic granular sludge reactor. Water Res 39(12):2676–2686

Mosquera-Corral A, Arrojo B, Figueroa M, Campos JL, Mendez R (2011) Aerobic granulation in a mechanical stirred SBR: treatment of low organic loads. Water Sci Technol 64(1):155–161

Motlagh AM, Bhattacharjee AS, Goel R (2016) Biofilm control with natural and genetically-modified phages. World J Microbiol Biotechnol 32(4):67

Moy BYP, Tay JH, Toh SK, Liu Y, Tay STL (2002) High organic loading influences the physical characteristics of aerobic sludge granules. Lett Appl Microbiol 34(6):407–412

Mudaly DD, Atkinson BW, Bux F (2001) 16S rRNA in situ probing for the determination of the family level community structure implicated in enhanced biological nutrient removal. Water Sci Technol 43:91–98

Murgel GA, Lion LW, Acheson C, Shuler ML, Emerson D, Ghiorse WC (1991) Experimental apparatus for selection of adherent microorganisms under stringent growth conditions. Appl Environ Microbiol 57(7):1987–1996

Musat N, Foster R, Vagner T, Adam B, Kuypers MMM (2012) Detecting metabolic activities in single cells, with emphasis on nanoSIMS. FEMS Microbiol Rev 36(2):486–511

Muurinen J, Muziasari WI, Hultman J, Pärnänen K, Narita V, Lyra C, Fadlillah LN, Rizki LP, Nurmi W, Tiedje JM, Dwiprahasto I, Hadi P, Virta MPJ (2022) Antibiotic resistomes and microbiomes in the surface water along the code river in Indonesia reflect drainage basin anthropogenic activities. Environ Sci Technol 56(21):14994–15006

Muyzer G, Ramsing NB (1995) Molecular methods to study the organization of microbial communities. Water Sci Technol 32:1–9

Muyzer G, De Waal EC, Uitterlinden AG (1993) Profiling of complex microbial populations by denaturing gradient gel electrophoresis analysis of polymerase chain reaction-amplified genes coding for 16S rRNA. Appl Environ Microbiol 59(3):695–700

Nagarajan K, Ni C, Lu T (2022) Agent-based modeling of microbial communities. ACS Synth Biol 11(11):3564–3574

Nancharaiah YV, Joshi HM, Mohan TVK, Venugopalan VP, Narasimhan SV (2006a) Aerobic granular biomass: a novel biomaterial for efficient uranium removal. Curr Sci 91(4):503–509

Nancharaiah YV, Schwarzenbeck N, Mohan TVK, Narasimhan SV, Wilderer PA, Venugopalan VP (2006b) Biodegradation of nitrilotriacetic acid (NTA) and ferric-NTA complex by aerobic microbial granules. Water Res 40(8):1539–1546

Narayanasamy S, Muller EEL, Sheik AR, Wilmes P (2015) Integrated omics for the identification of key functionalities in biological wastewater treatment microbial communities. Microb Biotechnol 8(3):363–368

Navada S, Vadstein O, Gaumet F, Tveten AK, Spanu C, Mikkelsen Ø, Kolarevic J (2020) Biofilms remember: osmotic stress priming as a microbial management strategy for improving salinity acclimation in nitrifying biofilms. Water Res 176:115732

Nerenberg R (2016) The membrane-biofilm reactor (MBfR) as a counter-diffusional biofilm process. Curr Opin Biotechnol 38:131–136

Nesmeyanova MA (2000) Polyphosphates and enzymes of polyphosphate metabolism in *Escherichia coli*. Biochem Mosc 65(3):309–314

Neu TR, Lawrence JR (1999) Lectin-binding analysis in biofilm systems. Methods Enzymol 310:145–152

Nguyen HTT, Le VQ, Hansen AA, Nielsen JL, Nielsen PH (2011) High diversity and abundance of putative polyphosphate-accumulating *Tetrasphaera*-related bacteria in activated sludge systems. FEMS Microbiol Ecol 76(2):256–267

Nguyen HTT, Nielsen JL, Nielsen PH (2012) "*Candidatus* Halomonas phosphatis", a novel polyphosphate-accumulating organism in full-scale enhanced biological phosphorus removal plants. Environ Microbiol 14(10):2826–2837

Ni B-J (2013) Formation, characterization and mathematical modeling of the aerobic granular sludge, 1st edn. Springer-Verlag, Berlin, Heidelberg

Ni BJ, Yu HQ (2010a) Mathematical modeling of aerobic granular sludge: a review. Biotechnol Adv 28(6):895–909

Ni BJ, Yu HQ (2010b) Modeling and simulation of the formation and utilization of microbial products in aerobic granular sludge. AIChE J 56(2):546–559

Ni BJ, Yu HQ, Sun YJ (2008) Modeling simultaneous autotrophic and heterotrophic growth in aerobic granules. Water Res 42(6–7):1583–1594

Ni BJ, Xie WM, Liu SG, Yu HQ, Wang YZ, Wang G, Dai XL (2009) Granulation of activated sludge in a pilot-scale sequencing batch reactor for the treatment of low-strength municipal wastewater. Water Res 43(3):751–761

Nicolella C, van Loosdrecht MCM, Heijnen JJ (1999) Identification of mass transfer parameters in three-phase biofilm reactors. Chem Eng Sci 54(15–16):3143–3152

Nicolella C, Zolezzi M, Rabino M, Furfaro M, Rovatti M (2005) Development of particle-based biofilms for degradation of xenobiotic organic compounds. Water Res 39(12):2495–2504

Niederdorfer R, Hausherr D, Palomo A, Wei J, Magyar P, Smets BF, Joss A, Bürgmann H (2021) Temperature modulates stress response in mainstream anammox reactors. Commun Biol 4(1):23

Nielsen PH, McMahon KD (2014) Microbiology and microbial ecology of the activated sludge process. In: Jenkins D, Wanner J (eds) Activated sludge—100 years and counting. IWA Publishing, London, p 464

Nielsen JL, Nielsen PH (2002) Quantification of functional groups in activated sludge by microautoradiography. Water Sci Technol 46(1–2):389–395

Nielsen PH, Nielsen JL (2005) Microautoradiography: recent advances within the studies of the ecophysiology of bacteria in biofilms. Water Sci Technol 52:187–194

Nielsen PH, Raunkjaer K, Norsker NH, Jensen NA, Hvitved-Jacobsen T (1992) Transformation of wastewater in sewer systems—a review. Water Sci Technol 25(6):17–31

Nielsen PH, Jahn A, Palmgren R (1997) Conceptual model for production and composition of exopolymers in biofilms. Water Sci Technol 36(1):11–19

Nielsen PH, Andreasen K, Lee N, Wagner M (1999) Use of microautoradiography and fluorescent in situ hybridization for characterization of microbial activity in activated sludge. Water Sci Technol 39(1):1–9

Nielsen JL, Christensen D, Kloppenborg M, Nielsen PH (2003) Quantification of cell-specific substrate uptake by probe-defined bacteria under in situ conditions by microautoradiography and fluorescence in situ hybridization. Environ Microbiol 5(3):202–211

Nielsen PH, Thomsen TR, Nielsen JL (2004) Bacterial composition of activated sludge—importance for floc and sludge properties. Water Sci Technol 49(10):51–58

Nielsen PH, Daims H, Lemmer H (2009) FISH handbook for biological wastewater treatment—identification and quantification of microorganisms in activated sludge and biofilms by FISH, 1st edn. IWA Publishing, London

Nielsen PH, Mielczarek AT, Kragelund C, Nielsen JL, Saunders AM, Kong Y, Hansen AA, Vollertsen J (2010) A conceptual ecosystem model of microbial communities in enhanced biological phosphorus removal plants. Water Res 44(17):5070–5088

Nielsen PH, Saunders AM, Hansen AA, Larsen P, Nielsen JL (2011) Microbial communities involved in enhanced biological phosphorus removal from wastewater—a model system in environmental biotechnology. Curr Opin Biotechnol

Nielsen JL, Nguyen H, Meyer RL, Nielsen PH (2012a) Identification of glucose-fermenting bacteria in a full-scale enhanced biological phosphorus removal plant by stable isotope probing. Microbiology 158(7):1818–1825
Nielsen PH, Saunders AM, Hansen AA, Larsen P, Nielsen JL (2012b) Microbial communities involved in enhanced biological phosphorus removal from wastewater—a model system in environmental biotechnology. Curr Opin Biotechnol 23(3):452–459
Nielsen JL, Seviour RJ, Nielsen PH (2016) Microscopy. In: van Loosdrecht MCM, Nielsen PH, Lopez-Vazquez CM, Brdjanovic D (eds) Experimental methods in wastewater treatment. IWA Publishing, London, pp 263–284
Noguera DR, Morgenroth E (2004) Introduction to the IWA task group on biofilm modeling. Water Sci Technol 49(11–12):131–136
Noguera DR, Picioreanu C (2004) Results from the multi-species benchmark problem 3 (BM3) using two-dimensional models. Water Sci Technol 49(11–12):169–176
Noguera DR, Okabe S, Picioreanu C (1999a) Biofilm modeling: present status and future directions. Water Sci Technol 39:273–278
Noguera DR, Pizarro G, Stahl DA, Rittmann BE (1999b) Simulation of multispecies biofilm development in three dimensions. Water Sci Technol 39(7):123–130
O'Reilly J, Lee C, Collins G, Chinalia F, Mahony T, O'Flaherty V (2009) Quantitative and qualitative analysis of methanogenic communities in mesophilically and psychrophilically cultivated anaerobic granular biofilims. Water Res 43(14):3365–3374
O'Toole GA (2004) Jekyll or hide? Nature 432(7018):680–681
O'Toole G, Kaplan HB, Kolter R (2000) Biofilm formation as microbial development. Annu Rev Microbiol 54:49–79
Odegaard H (2006) Innovations in wastewater treatment: the moving bed biofilm process. Water Sci Technol 53(9):17–33
Odegaard H, Storhaug R (1990) Small wastewater treatment plants in Norway. Water Sci Technol 22(3–4):33–40
Odegaard H, Rusten B, Westrum T (1994) A new moving bed biofilm reactor—applications and results. Water Sci Technol 29(10–11):157–165
Oehmen A, Vives MT, Lu H, Yuan Z, Keller J (2005) The effect of pH on the competition between polyphosphate-accumulating organisms and glycogen-accumulating organisms. Water Res 39(15):3727–3737
Oehmen A, Saunders AM, Vives MT, Yuan Z, Keller J (2006a) Competition between polyphosphate and glycogen accumulating organisms in enhanced biological phosphorus removal systems with acetate and propionate as carbon sources. J Biotechnol 123(1):22–32
Oehmen A, Zeng RJ, Saunders AM, Blackall LL, Keller J, Yuan Z (2006b) Anaerobic and aerobic metabolism of glycogen-accumulating organisms selected with propionate as the sole carbon source. Microbiology 152(9):2767–2778
Oehmen A, Lemos PC, Carvalho G, Yuan Z, Keller J, Blackall LL, Reis MAM (2007) Advances in enhanced biological phosphorus removal: from micro to macro scale. Water Res 41:2271–2300
Oehmen A, Carvalho G, Freitas F, Reis MAM (2010a) Assessing the abundance and activity of denitrifying polyphosphate accumulating organisms through molecular and chemical techniques. Water Sci Technol 61(8):2061–2068
Oehmen A, Carvalho G, Lopez-Vazquez CM, van Loosdrecht MCM, Reis MAM (2010b) Incorporating microbial ecology into the metabolic modelling of polyphosphate accumulating organisms and glycogen accumulating organisms. Water Res 44(17):4992–5004
Oehmen A, Lopez-Vazquez CM, Carvalho G, Reis MAM, van Loosdrecht MCM (2010c) Modelling the population dynamics and metabolic diversity of organisms relevant in anaerobic/anoxic/aerobic enhanced biological phosphorus removal processes. Water Res 44(15):4473–4486
Ofiteru ID, Lunn M, Curtis TP, Wells GF, Criddle CS, Francis CA, Sloan WT (2010) Combined niche and neutral effects in a microbial wastewater treatment community. Proc Natl Acad Sci USA 107(35):15345–15350

Ohandja DG, Stuckey DC (2006) Development of a membrane-aerated biofilm reactor to completely mineralise perchloroethylene in wastewaters. J Chem Technol Biotechnol 81(11):1736–1744

Ohtake H, Takahashi K, Tsuzuki Y, Toda K (1985) Uptake and release of phosphate by a pure culture of *Acinetobacter calcoaceticus*. Water Res 19(12):1587–1594

Okabe S, Yasuda T, Watanabe Y (1997) Uptake and release of inert fluorescence particles by mixed population biofilms. Biotechnol Bioeng 53(5):459–469

Okabe S, Satoh H, Watanabe Y (1999) In situ analysis of nitrifying biofilms as determined by in situ hybridization and the use of microelectrodes. Appl Environ Microbiol 65(7):3182–3191

Okshevsky M, Meyer RL (2015) The role of extracellular DNA in the establishment, maintenance and perpetuation of bacterial biofilms. Crit Rev Microbiol 41(3):341–352

Olivieri G, Russo ME, Marzocchella A, Salatino P (2011) Modeling of an aerobic biofilm reactor with double-limiting substrate kinetics: bifurcational and dynamical analysis. Biotechnol Prog 27(6):1599–1613

Onuki M, Satoh H, Mino T, Matsuo T (2000) Application of molecular methods to microbial community analysis of activated sludge. Water Sci Technol 42:17–22

Orhon D, Sekoulov I, Dulkadiroglu H (2002) Innovative technologies for wastewater treatment in coastal tourist areas. Water Sci Technol 46(8):67–74

Oshiki M, Satoh H, Mino T (2010) Acetate uptake by PHA-accumulating and non-PHA-accumulating organisms in activated sludge from an aerobic sequencing batch reactor fed with acetate. Water Sci Technol J Int Assoc Water Pollut Res 62(1):8–14

Oyserman BO, Moya F, Lawson CE, Garcia AL, Vogt M, Heffernen M, Noguera DR, McMahon KD (2016a) Ancestral genome reconstruction identifies the evolutionary basis for trait acquisition in polyphosphate accumulating bacteria. ISME J 10(12):2931–2945

Oyserman BO, Noguera DR, Del Rio TG, Tringe SG, McMahon KD (2016b) Metatranscriptomic insights on gene expression and regulatory controls in *Candidatus* Accumulibacter phosphatis. ISME J 10(4):810–822

Páez-Watson T, van Loosdrecht MCM, Wahl SA (2023) Predicting the impact of temperature on metabolic fluxes using resource allocation modelling: application to polyphosphate accumulating organisms. Water Res 228(Pt A):119365

Pallares-Vega R, Hernandez Leal L, Fletcher BN, Vias-Torres E, van Loosdrecht MCM, Weissbrodt DG, Schmitt H (2021) Annual dynamics of antimicrobials and resistance determinants in flocculent and aerobic granular sludge treatment systems. Water Res 190:116752

Palm JC, Jenkins D, Parker DS (1980) Relationship between organic loading, dissolved oxygen concentration and sludge settleability in the completely-mixed activated sludge process. J Water Pollut Control Fed 52(10):2484–2506

Palomo A, Pedersen AG, Fowler SJ, Dechesne A, Sicheritz-Pontén T, Smets BF (2018) Comparative genomics sheds light on niche differentiation and the evolutionary history of comammox *Nitrospira*. ISME J 12(7):1779–1793

Park H, Rosenthal A, Ramalingam K, Fillos J, Chandran K (2010a) Linking community profiles, gene expression and N-removal in anammox bioreactors treating municipal anaerobic digestion reject water. Environ Sci Technol 44(16):6110–6116

Park S, Bae W, Rittmann BE (2010b) Multi-species nitrifying biofilm model (MSNBM) including free ammonia and free nitrous acid inhibition and oxygen limitation. Biotechnol Bioeng 105(6):1115–1130

Park H, Sundar S, Ma Y, Chandran K (2015) Differentiation in the microbial ecology and activity of suspended and attached bacteria in a nitritation-anammox process. Biotechnol Bioeng 112(2):272–279

Parks DH, Rinke C, Chuvochina M, Chaumeil P-A, Woodcroft BJ, Evans PN, Hugenholtz P, Tyson GW (2017) Recovery of nearly 8,000 metagenome-assembled genomes substantially expands the tree of life. Nat Microbiol 2(11):1533–1542

Parks DH, Chuvochina M, Waite DW, Rinke C, Skarshewski A, Chaumeil PA, Hugenholtz P (2018) A standardized bacterial taxonomy based on genome phylogeny substantially revises the tree of life. Nat Biotechnol 36(10):996–1004

Patel GB, Sprott GD (1990) *Methanosaeta concilii* gen. nov., sp. nov. ("*Methanothrix concilii*") and *Methanosaeta thermoacetophila* nom. rev., comb. nov. Int J Syst Bacteriol 40(1):79–82
Pauss A, Samson R, Guiot S (1990) Thermodynamic evidence of trophic microniches in methanogenic granular sludge-bed reactors. Appl Microbiol Biotechnol 33(1):88–92
Pavissich JP, Aybar M, Martin KJ, Nerenberg R (2014) A methodology to assess the effects of biofilm roughness on substrate fluxes using image analysis, substrate profiling, and mathematical modelling. Water Sci Technol 69(9):1932–1941
Perez J, Picioreanu C, van Loosdrecht M (2005) Modeling biofilm and floc diffusion processes based on analytical solution of reaction-diffusion equations. Water Res 39(7):1311–1323
Perez J, Costa E, Kreft JU (2009) Conditions for partial nitrification in biofilm reactors and a kinetic explanation. Biotechnol Bioeng 103(2):282–295
Perez J, Lotti T, Kleerebezem R, Picioreanu C, van Loosdrecht MC (2014) Outcompeting nitrite-oxidizing bacteria in single-stage nitrogen removal in sewage treatment plants: a model-based study. Water Res 66:208–218
Perez J, Isanta E, Carrera J (2015) Would a two-stage N-removal be a suitable technology to implement at full scale the use of anammox for sewage treatment? Water Sci Technol 72(6):858–864
Pérez J, Laureni M, van Loosdrecht MCM, Persson F, Gustavsson DJI (2020) The role of the external mass transfer resistance in nitrite oxidizing bacteria repression in biofilm-based partial nitritation/anammox reactors. Water Res 186:116348
Peterson SB, Warnecke F, Madejska J, McMahon KD, Hugenholtz P (2008) Environmental distribution and population biology of *Candidatus* Accumulibacter, a primary agent of biological phosphorus removal. Environ Microbiol 10(10):2692–2703
Petriglieri F, Singleton C, Peces M, Petersen JF, Nierychlo M, Nielsen PH (2020) "*Candidatus* Dechloromonas phosphatis" and "*Candidatus* Dechloromonas phosphovora", two novel polyphosphate accumulating organisms abundant in wastewater treatment systems. ISME J 15(12):3605-3614
Peyton BM, Characklis WG (1992) Kinetics of biofilm detachment. Water Sci Technol 26(9–11):1995–1998
Phuong K, Kakii K, Nikata T (2009) Intergeneric coaggregation of non-flocculating *Acinetobacter* spp. isolates with other sludge-constituting bacteria. J Biosci Bioeng 107(4):394–400
Phuong K, Hanazaki S, Kakii K, Nikata T (2012) Involvement of *Acinetobacter* sp. in the floc-formation in activated sludge process. J Biotechnol 157(4):505–511
Picioreanu C, van Loosdrecht MCM, Heijnen JJ (2000a) Effect of diffusive and convective substrate transport on biofilm structure formation: a two-dimensional modeling study. Biotechnol Bioeng 69(5):504–515
Picioreanu C, van Loosdrecht MCM, Heijnen JJ (2000b) A theoretical study on the effect of surface roughness on mass transport and transformation in biofilms. Biotechnol Bioeng 68(4):355–369
Picioreanu C, van Loosdrecht MCM, Heijnen JJ (2001) Two-dimensional model of biofilm detachment caused by internal stress from liquid flow. Biotechnol Bioeng 72(2):205–218
Picioreanu C, Kreft JU, van Loosdrecht MCM (2004) Particle-based multidimensional multispecies biofilm model. Appl Environ Microbiol 70(5):3024–3040
Picioreanu C, Batstone DJ, van Loosdrecht MCM (2005) Multidimensional modelling of anaerobic granules. Water Sci Technol 52(1–2):501–507
Picioreanu C, Kreft JU, Klausen M, Haagensen JAJ, Tolker-Nielsen T, Molin S (2007) Microbial motility involvement in biofilm structure formation—a 3D modelling study. Water Sci Technol 55:337–343
Picioreanu C, Pérez J, van Loosdrecht MCM (2016) Impact of cell cluster size on apparent half-saturation coefficients for oxygen in nitrifying sludge and biofilms. Water Res 106:371–382
Picioreanu C, Blauert F, Horn H, Wagner M (2018) Determination of mechanical properties of biofilms by modelling the deformation measured using optical coherence tomography. Water Res 145:588–598

Pijuan M, Guisasola A, Baeza JA, Carrera J, Casas C, Lafuente J (2005) Aerobic phosphorus release linked to acetate uptake: influence of PAO intracellular storage compounds. Biochem Eng J 26(2–3):184–190

Pijuan M, Werner U, Yuan Z (2011) Reducing the startup time of aerobic granular sludge reactors through seeding floccular sludge with crushed aerobic granules. Water Res 45(16):5075–5083

Pinto AJ, Raskin L (2012) PCR biases distort bacterial and archaeal community structure in pyrosequencing datasets. PLoS ONE 7(8):e43093

Plattes M, Fiorelli D, Gille S, Girard C, Henry E, Minette F, O'Nagy O, Schosseler PM (2007) Modelling and dynamic simulation of a pilot-scale moving bed bioreactor for the treatment of municipal wastewater: model concepts and the use of respirometry for the estimation of kinetic parameters. Water Sci Technol 55(8–9):309–316

Poortinga AT, Bos R, Norde W, Busscher HJ (2002) Electric double layer interactions in bacterial adhesion to surfaces. Surf Sci Rep 47(1):1–32

Popat SC, Deshusses MA (2011) Kinetics and inhibition of reductive dechlorination of trichloroethene, cis-1,2-dichloroethene and vinyl chloride in a continuously fed anaerobic biofilm reactor. Environ Sci Technol 45(4):1569–1578

Prades L, Fabbri S, Dorado AD, Gamisans X, Stoodley P, Picioreanu C (2020) Computational and experimental investigation of biofilm disruption dynamics induced by high-velocity gas jet impingement. mBio 11(1):e02813–19

Pramanik J, Trelstad PL, Schuler AJ, Jenkins D, Keasling JD (1999) Development and validation of a flux-based stoichiometric model for enhanced biological phosphorus removal metabolism. Water Res 33(2):462–476

Pratt S, Tan M, Gapes D, Shilton A (2007) Development and examination of a granular nitrogen-fixing wastewater treatment system. Process Biochem 42(5):863–872

Price DJdS (1963) Little science, big science. The impact of the explosive growth of science on the wolrd we live in. Columbia University Press, New York

Pronk M, Bassin JP, de Kreuk MK, Kleerebezem R, van Loosdrecht MC (2014) Evaluating the main and side effects of high salinity on aerobic granular sludge. Appl Microbiol Biotechnol 98(3):1339–1348

Pronk M, Abbas B, Al-Zuhairy SH, Kraan R, Kleerebezem R, van Loosdrecht MC (2015a) Effect and behaviour of different substrates in relation to the formation of aerobic granular sludge. Appl Microbiol Biotechnol 99(12):5257–5268

Pronk M, de Kreuk MK, de Bruin B, Kamminga P, Kleerebezem R, van Loosdrecht MCM (2015b) Full scale performance of the aerobic granular sludge process for sewage treatment. Water Res 84:207–217

Pronk M, Giesen A, Thompson A, Robertson S, van Loosdrecht M (2017a) Aerobic granular biomass technology: advancements in design, applications and further developments. Water Pract Technol 12(4):987–996

Pronk M, Neu TR, van Loosdrecht MCM, Lin YM (2017b) The acid soluble extracellular polymeric substance of aerobic granular sludge dominated by *Defluviicoccus* sp. Water Res 122:148–158

Pronk M, van Dijk EJH, van Loosdrecht MCM (2020) Aerobic granular sludge. In: Chen GH, van Loosdrecht MCM, Ekama GA, Brdjanovic D (eds) Biological wastewater treatment: principles, modelling and design, 2nd edn. IWA Publishing, London

Qin L, Liu Y (2006) Aerobic granulation for organic carbon and nitrogen removal in alternating aerobic-anaerobic sequencing batch reactor. Chemosphere 63(6):926–933

Qin L, Tay JH, Liu Y (2004) Selection pressure is a driving force of aerobic granulation in sequencing batch reactors. Process Biochem 39(5):579–584

Qin L, Liu Y, Tay JH (2005) Denitrification on poly-b-hydroxybutyrate in microbial granular sludge sequencing batch reactor. Water Res 39(8):1503–1510

Qiu G, Law Y, Zuniga-Montanez R, Deng X, Lu Y, Roy S, Thi SS, Hoon HY, Nguyen TQN, Eganathan K, Liu X, Nielsen PH, Williams RBH, Wuertz S (2022) Global warming readiness: feasibility of enhanced biological phosphorus removal at 35 °C. Water Res 216:118301

Rabaey K, Verstraete W (2005) Microbial fuel cells: novel biotechnology for energy generation. Trends Biotechnol 23(6):291–298
Regmi P, DeBarbadillo C, Weissbrodt DG (2017) Biofilm reactor technology and design. In: Krause TL et al (eds) Design of water resource recovery facilities—MOP 8. WEF manual of practice no. 8, ASCE manuals and reports on engineering practice no. 76, 6th edn. Water Environment Federation, American Society of Civil Engineers and Environmental and Water Resources Institute, McGraw-Hill Education, Alexandria VA, Reston VA, New York
Regmi P, Sturm B, Hiripitiyage D, Keller N, Murthy S, Jimenez J (2022) Combining continuous flow aerobic granulation using an external selector and carbon-efficient nutrient removal with AvN control in a full-scale simultaneous nitrification-denitrification process. Water Res 210:117991
Reichert P (1994) AQUASIM—a tool for simulation and data analysis of aquatic systems. Water Sci Technol 30(2 pt 2):21–30
Reichert P, Wanner O (1997) Movement of solids in biofilms: significance of liquid phase transport. Water Sci Technol 36(1):321–328
Reichert P, Von Schulthess R, Wild D (1995) The use of AQUASIM for estimating parameters of activated sludge models. Water Sci Technol 31(2):135–147
Reino C, Suárez-Ojeda ME, Pérez J, Carrera J (2016) Kinetic and microbiological characterization of aerobic granules performing partial nitritation of a low-strength wastewater at 10 °C. Water Res 101:147–156
Renslow RS, Lindemann SR, Cole JK, Zhu Z, Anderton CR (2016) Quantifying element incorporation in multispecies biofilms using nanoscale secondary ion mass spectrometry image analysis. Biointerphases 11(2):02a322
Rittmann BE (2006a) The membrane biofilm reactor: the natural partnership of membranes and biofilm. Water Sci Technol 53(3):219–225
Rittmann BE (2006b) Microbial ecology to manage processes in environmental biotechnology. Trends Biotechnol 24(6):261–266
Rittmann BE (2007) Where are we with biofilms now? Where are we going? Water Sci Technol 55(8–9):1–7
Rittmann BE (2010) Environmental biotechnology in water and wastewater treatment. J Environ Eng 136(4):348–353
Rittmann BE, Dovantzis K (1983) Dual limitation of biofilm kinetics. Water Res 17(12):1727–1734
Rittmann BE, McCarty PL (1980) Model of steady-state-biofilm kinetics. Biotechnol Bioeng 22(11):2343–2357
Rittmann BE, Stilwell D, Ohashi A (2002) The transient-state, multiple-species biofilm model for biofiltration processes. Water Res 36(9):2342–2356
Rittmann BE, Boltz JP, Brockmann D, Daigger GT, Morgenroth E, Sørensen KH, Takács I, van Loosdrecht M, Vanrolleghem PA (2018) A framework for good biofilm reactor modeling practice (GBRMP). Water Sci Technol 77(5–6):1149–1164
Rodríguez E, García-Encina P, Stams AM, Maphosa F, Sousa D (2015) Meta-omics approaches to understand and improve wastewater treatment systems. Rev Environ Sci Bio 14(3):385–406
Roeselers G, Loosdrecht MCMV, Muyzer G (2008) Phototrophic biofilms and their potential applications. J Appl Phycol 20(3):227–235
Rogalla F, Roudon G, Sibony J, Blondeau F (1992) Minimising nuisances by covering compact sewage treatment plants. Water Sci Technol 25(4–5):363–374
Rossello-Mora RA, Wagner M, Amann R, Schleifer KH (1995) The abundance of *Zoogloea ramigera* in sewage treatment plants. Appl Environ Microbiol 61(2):702–707
Rossi F, De Philippis R (2015) Role of cyanobacterial exopolysaccharides in phototrophic biofilms and in complex microbial mats. Life 5(2):1218–1238
Rowan AK, Snape JR, Fearnside D, Barer MR, Curtis TP, Head IM (2003) Composition and diversity of ammonia-oxidising bacterial communities in wastewater treatment reactors of different design treating identical wastewater. FEMS Microbiol Ecol 43(2):195–206

Ruan W, Hua Z, Chen J (2006) Simultaneous nitrification and denitrification in an aerobic reactor with granular sludge originating from an upflow anaerobic sludge bed reactor. Water Environ Res 78(8):792–796

Rubio-Rincon FJ, Weissbrodt DG, Lopez-Vazquez CM, Welles L, Abbas B, Albertsen M, Nielsen PH, van Loosdrecht MCM, Brdjanovic D (2019) "*Candidatus* Accumulibacter delftensis": a clade IC novel polyphosphate-accumulating organism without denitrifying activity on nitrate. Water Res 161:136–151

Sabba F, Picioreanu C, Perez J, Nerenberg R (2015) Hydroxylamine diffusion can enhance N_2O emissions in nitrifying biofilms: a modeling study. Environ Sci Technol 49(3):1486–1494

Sabba F, Farmer M, Jia Z, Di Capua F, Dunlap P, Barnard J, Qin CD, Kozak JA, Wells G, Downing L (2023) Impact of operational strategies on a sidestream enhanced biological phosphorus removal (S2EBPR) reactor in a carbon limited wastewater plant. Sci Total Environ 857(Pt 1):159280

Saia SM, Sullivan PJ, Regan JM, Carrick HJ, Buda AR, Locke NA, Walter MT (2017) Evidence for polyphosphate accumulating organism (PAO)-mediated phosphorus cycling in stream biofilms under alternating aerobic/anaerobic conditions. Freshw Sci 36(2):284–296

Saia SM, Carrick HJ, Buda AR, Regan JM, Walter MT (2021) Critical review of polyphosphate and polyphosphate accumulating organisms for agricultural water quality management. Environ Sci Technol 55(5):2722–2742

Saito T, Brdjanovic D, van Loosdrecht MCM (2004) Effect of nitrite on phosphate uptake by phosphate accumulating organisms. Water Res 38(17):3760–3768

Sam T, Le Roes-Hill M, Hoosain N, Welz PJ (2022) Strategies for controlling filamentous bulking in activated sludge wastewater treatment plants: the old and the new. Water (Switzerland) 14(20):3223

Sampara P, Luo Y, Lin X, Ziels RM (2022) Integrating genome-resolved metagenomics with trait-based process modeling to determine biokinetics of distinct nitrifying communities within activated sludge. Environ Sci Technol 56(16):11670–11682

Santegoeds CM, Damgaard LR, Hesselink G, Zopfi J, Lens P, Muyzer G, De Beer D (1999) Distribution of sulfate-reducing and methanogenic bacteria in anaerobic aggregates determined by microsensor and molecular analyses. Appl Environ Microbiol 65(10):4618–4629

Satoh H, Ramey WD, Koch FA, Oldham WK, Mino T, Matsuo T (1996) Anaerobic substrate uptake by the enhanced biological phosphorus removal activated sludge treating real sewage. Water Sci Technol 34(1–2):9–16

Sauer K, Camper AK (2001) Characterization of phenotypic changes in *Pseudomonas putida* in response to surface-associated growth. J Bacteriol 183(22):6579–6589

Scarborough MJ, Lawson CE, DeCola AC, Gois IM (2022) Microbiomes for sustainable biomanufacturing. Curr Opin Microbiol 65:8–14

Schambeck CM, Magnus BS, de Souza LCR, Leite WRM, Derlon N, Guimaraes LB, da Costa RHR (2020) Biopolymers recovery: dynamics and characterization of alginate-like exopolymers in an aerobic granular sludge system treating municipal wastewater without sludge inoculum. J Environ Manage 263:110394

Schmidt JE, Ahring BK (1996) Granular sludge formation in upflow anaerobic sludge blanket (UASB) reactors. Biotechnol Bioeng 49(3):229–246

Schramm A, De Beer D, Wagner M, Amann R (1998) Identification and activities in situ of *Nitrosospira* and *Nitrospira* spp. as dominant populations in a nitrifying fluidized bed reactor. Appl Environ Microbiol 64(9):3480–3485

Schreiber F, Littmann S, Lavik G, Escrig S, Meibom A, Kuypers MMM, Ackermann M (2016) Phenotypic heterogeneity driven by nutrient limitation promotes growth in fluctuating environments. Nat Microbiol 1(6):16055

Schreier HJ, Mirzoyan N, Saito K (2010) Microbial diversity of biological filters in recirculating aquaculture systems. Curr Opin Biotechnol 21(3):318–325

Schuler AJ, Jang H (2007) Causes of variable biomass density and its effects on settleability in full-scale biological wastewater treatment systems. Environ Sci Technol 41(5):1675–1681

Schuler AJ, Jenkins D (2003) Enhanced biological phosphorus removal from wastewater by biomass with different phosphorus contents, part I: experimental results and comparison with metabolic models. Water Environ Res 75(6):485–498

Schwarzenbeck N, Borges JM, Wilderer PA (2005) Treatment of dairy effluents in an aerobic granular sludge sequencing batch reactor. Appl Microbiol Biotechnol 66(6):711–718

Seghezzo L, Zeeman G, van Lier JB, Hamelers HVM, Lettinga G (1998) A review: the anaerobic treatment of sewage in UASB and EGSB reactors. Bioresour Technol 65(3):175–190

Sekiguchi Y, Kamagata Y, Syutsubo K, Ohashi A, Harada H, Nakamura K (1998) Phylogenetic diversity of mesophilic and thermophilic granular sludges determined by 16S rRNA gene analysis. Microbiology 144(9):2655–2665

Sekiguchi Y, Kamagata Y, Nakamura K, Ohashi A, Harada H (1999) Fluorescence in situ hybridization using 16S rRNA-targeted oligonucleotides reveals localization of methanogens and selected uncultured bacteria in mesophilic and thermophilic sludge granules. Appl Environ Microbiol 65(3):1280–1288

Serafim LS, Lemos PC, Rossetti S, Levantesi C, Tandoi V, Reis MAM (2006) Microbial community analysis with a high PHA storage capacity. Water Sci Technol 54(1):183–188

Seviour EM, Eales K, Izzard L, Beer M, Carr EL, Seviour RJ (2006) The in situ physiology of "*Nostocoida limicola*" II, a filamentous bacterial morphotype in bulking activated sludge, using fluorescence in situ hybridization an microautoradiography. Water Sci Technol 54:47–53

Seviour T, Pijuan M, Nicholson T, Keller J, Yuan Z (2009a) Gel-forming exopolysaccharides explain basic differences between structures of aerobic sludge granules and floccular sludges. Water Res 43(18):4469–4478

Seviour T, Pijuan M, Nicholson T, Keller J, Yuan Z (2009b) Understanding the properties of aerobic sludge granules as hydrogels. Biotechnol Bioeng 102(5):1483–1493

Seviour T, Donose BC, Pijuan M, Yuan Z (2010a) Purification and conformational analysis of a key exopolysaccharide component of mixed culture aerobic sludge granules. Environ Sci Technol 44(12):4729–4734

Seviour T, Lambert LK, Pijuan M, Yuan Z (2010b) Structural determination of a key exopolysaccharide in mixed culture aerobic sludge granules using NMR spectroscopy. Environ Sci Technol 44(23):8964–8970

Seviour TW, Lambert LK, Pijuan M, Yuan Z (2011) Selectively inducing the synthesis of a key structural exopolysaccharide in aerobic granules by enriching for "*Candidatus* Competibacter phosphatis". Appl Microbiol Biotechnol 92(6):1297–1305

Seviour T, Derlon N, Dueholm MS, Flemming H-C, Girbal-Neuhauser E, Horn H, Kjelleberg S, van Loosdrecht MCM, Lotti T, Malpei MF, Nerenberg R, Neu TR, Paul E, Yu H, Lin Y (2019) Extracellular polymeric substances of biofilms: suffering from an identity crisis. Water Res 151:1–7

Shapiro JA (1988) Bacteria as multicellular organisms. Sci Am 256:82–89

Shapiro JA (1998) Thinking about bacterial populations as multicellular organisms. Annu Rev Microbiol 52:81–104

Shapiro OH, Kushmaro A (2011) Bacteriophage ecology in environmental biotechnology processes. Curr Opin Biotechnol 22(3):449–455

Sheik AR, Muller EEL, Wilmes P (2014) A hundred years of activated sludge: time for a rethink. Front Microbiol 5:47

Shen XX, Li XM, Yang Q, Zeng GM, Xu WX, Liao Q, Zheng Y (2007) Acceleration of the formation of aerobic granules in SBR by inoculating mature aerobic granules. Huanjing Kexue/Environ Sci 28(11):2467–2472

Shi XY, Yu HQ, Sun YJ, Huang X (2009) Characteristics of aerobic granules rich in autotrophic ammonium-oxidizing bacteria in a sequencing batch reactor. Chem Eng J 147(2–3):102–109

Shi XY, Sheng GP, Li XY, Yu HQ (2010) Operation of a sequencing batch reactor for cultivating autotrophic nitrifying granules. Bioresour Technol 101(9):2960–2964

Shiflet AB, Shiflet GW (2010) Simulating the formation of biofilms in an undergraduate modeling course. Procedia Comput Sci 1:895–901

Shin HS, Lim KH, Park HS (1992) Effect of shear stress on granulation in oxygen aerobic upflow sludge bed reactors. Water Sci Technol 26(3–4):601–605

Shivaram KB, Bhatt P, Applegate B, Simsek H (2023) Bacteriophage-based biocontrol technology to enhance the efficiency of wastewater treatment and reduce targeted bacterial biofilms. Sci Total Environ 862:160723

Show KY, Lee DJ, Tay JH (2012) Aerobic granulation: advances and challenges. Appl Biochem Biotechnol 167(6):1–19

Siegrist H, Rieger L, Koch G, Kuhni M, Gujer W (2002) The EAWAG bio-P module for activated sludge model no. 3. Water Sci Technol 45(6):61–76

Simon M, Grossart HP, Schweitzer B, Ploug H (2002) Microbial ecology of organic aggregates in aquatic ecosystems. Aquat Microb Ecol 28(2):175–211

Singh PK, Schaefer AL, Parsek MR, Moninger TO, Welsh MJ, Greenberg EP (2000) Quorum-sensing signals indicate that cystic fibrosis lungs are infected with bacterial biofilms. Nature 407(6805):762–764

Singleton CM, Petriglieri F, Kristensen JM, Kirkegaard RH, Michaelsen TY, Andersen MH, Kondrotaite Z, Karst SM, Dueholm MS, Nielsen PH, Albertsen M (2021) Connecting structure to function with the recovery of over 1000 high-quality metagenome-assembled genomes from activated sludge using long-read sequencing. Nat Commun 12(1):2009

Singleton CM, Petriglieri F, Wasmund K, Nierychlo M, Kondrotaite Z, Petersen JF, Peces M, Dueholm MS, Wagner M, Nielsen PH (2022) The novel genus, '*Candidatus* Phosphoribacter', previously identified as *Tetrasphaera*, is the dominant polyphosphate accumulating lineage in EBPR wastewater treatment plants worldwide. ISME J 16(6):1605–1616

Skennerton CT, Angly FE, Breitbart M, Bragg L, He S, McMahon KD, Hugenholtz P, Tyson GW (2011) Phage encoded H-NS: a potential achilles heel in the bacterial defence system. PLoS ONE 6(5)

Skennerton CT, Barr JJ, Slater FR, Bond PL, Tyson GW (2015) Expanding our view of genomic diversity in *Candidatus* Accumulibacter clades. Environ Microbiol 17(5):1574–1585

Skiadas IV, Gavala HN, Schmidt JE, Ahring BK (2003) Anaerobic granular sludge and biofilm reactors. Adv Biochem Eng Biotechnol 82:35–67

Smolders GJF, van der Meij J, van Loosdrecht MCM, Heijnen JJ (1994a) Model of the anaerobic metabolism of the biological phosphorus removal process: stoichiometry and pH influence. Biotechnol Bioeng 43(6):461–470

Smolders GJF, van Loosdrecht MCM, Heijnen JJ (1994b) pH: keyfactor in the biological phosphorus removal process. Water Sci Technol 29(7):71–74

Smolders GJF, van der Meij J, van Loosdrecht MCM, Heijnen JJ (1995) A structured metabolic model for anaerobic and aerobic stoichiometry and kinetics of the biological phosphorus removal process. Biotechnol Bioeng 47(3):277–287

Sørensen KH, Morgenroth E (2020) Biofilm reactors. In: Chen GH, van Loosdrecht MCM, Ekama GA, Brdjanovic D (eds) Biological wastewater treatment: principles, modelling and design, 2nd edn. IWA Publishing, London

Sorensen SJ, Bailey M, Hansen LH, Kroer N, Wuertz S (2005) Studying plasmid horizontal transfer in situ: a critical review. Nat Rev Microbiol 3(9):700–710

Staudt C, Horn H, Hempel DC, Neu TR (2004) Volumetric measurements of bacterial cells and extracellular polymeric substance glycoconjugates in biofilms. Biotechnol Bioeng 88(5):585–592

Steele HL, Streit WR (2005) Metagenomics: advances in ecology and biotechnology. FEMS Microbiol Lett 247(2):105–111

Stewart PS (1993) A model of biofilm detachment. Biotechnol Bioeng 41(1):111–117

Stewart PS (2003) Diffusion in biofilms. J Bacteriol 185(5):1485–1491

Stewart PS, Franklin MJ (2008) Physiological heterogeneity in biofilms. Nat Rev Microbiol 6(3):199–210

Stokholm-Bjerregaard M (2016) Control of GAOs in wastewater treatment plants with enhanced biological phosphorus removal. PhD thesis, Aalborg University

Stokholm-Bjerregaard M, McIlroy SJ, Nierychlo M, Karst SM, Albertsen M, Nielsen PH (2017) A critical assessment of the microorganisms proposed to be important to enhanced biological phosphorus removal in full-scale wastewater treatment systems. Front Microbiol 8:718

Stoodley P, DeBeer D, Lewandowski Z (1994) Liquid flow in biofilm systems. Appl Environ Microbiol 60(8):2711–2716

Streichan M, Golecki JR, Schon G (1990) Polyphosphate-accumulating bacteria from sewage plants with different processes for biological phosphorus removal. FEMS Microbiol Ecol 73(2):113–124

Strubbe L, Pennewaerde M, Baeten JE, Volcke EIP (2022) Continuous aerobic granular sludge plants: better settling versus diffusion limitation. Chem Eng J 428(4):131427

Subirats J, Sànchez-Melsió A, Borrego CM, Balcázar JL, Simonet P (2016) Metagenomic analysis reveals that bacteriophages are reservoirs of antibiotic resistance genes. Int J Antimicrob Agents 48(2):163–167

Suidan MT (1986) Performance of deep biofilm reactors. J Environ Eng 112(1):78–93

Suidan MT, Flora JRV, Biswas P, Sayles GD (1994) Optimization modelling of anaerobic biofilm reactors. Water Sci Technol 30(12):347–355

Sun YJ, Zuo JE, Yang Y, Lu YQ, Xing W, Bu DH (2006) Community structure of nitrification bacteria in aerobic short-cut nitrification granule. Huanjing Kexue/Environ Sci 27(9):1858–1861

Sutherland IW (2001) Biofilm exopolysaccharides: a strong and sticky framework. Microbiology 147(1):3–9

Szabo E, Liebana R, Hermansson M, Modin O, Persson F, Wilen BM (2017) Microbial population dynamics and ecosystem functions of anoxic/aerobic granular sludge in sequencing batch reactors operated at different organic loading rates. Front Microbiol 8:770

Taherzadeh D, Picioreanu C, Kuttler U, Simone A, Wall WA, Horn H (2010) Computational study of the drag and oscillatory movement of biofilm streamers in fast flows. Biotechnol Bioeng 105(3):600–610

Takcs I, Bye CM, Chapman K, Dold PL, Fairlamb PM, Jones RM (2007) A biofilm model for engineering design. Water Sci Technol 55:329–336

Tao Y, Huang X, Gao D, Wang X, Chen C, Liang H, van Loosdrecht MCM (2019) NanoSIMS reveals unusual enrichment of acetate and propionate by an anammox consortium dominated by *Jettenia asiatica*. Water Res 159:223–232

Tartakovsky B, Guiot SR (1997) Modeling and analysis of layered stationary anaerobic granular biofilms. Biotechnol Bioeng 54(2):122–130

Tay JH, Yan YG (1996) Influence of substrate concentration on microbial selection and granulation during start-up of upflow anaerobic sludge blanket reactors. Water Environ Res 68(7):1140–1150

Tay JH, Liu QS, Liu Y (2001a) The effects of shear force on the formation, structure and metabolism of aerobic granules. Appl Microbiol Biotechnol 57(1–2):227–233

Tay JH, Liu QS, Liu Y (2001b) Microscopic observation of aerobic granulation in sequential aerobic sludge blanket reactor. J Appl Microbiol 91:168–175

Tay JH, Pan S, Tay STL, Ivanov V, Liu Y (2003) The effect of organic loading rate on the aerobic granulation: the development of shear force theory. Water Sci Technol 47(11):235–240

Tay JH, Liu QS, Liu Y (2004) The effect of upflow air velocity on the structure of aerobic granules cultivated in a sequencing batch reactor. Water Sci Technol 49(11–12):35–40

Tay JH, Yang P, Zhuang WQ, Tay STL, Pan ZH (2007) Reactor performance and membrane filtration in aerobic granular sludge membrane bioreactor. J Membr Sci 304(1–2):24–32

Teitzel G (2011) Microbiology goes big: microbial systems biology. Trends Microbiol 19(10):471

Thanh BX, Visvanathan C, Sperandio M, Aim RB (2008) Fouling characterization in aerobic granulation coupled baffled membrane separation unit. J Membr Sci 318(1–2):334–339

Thanh BX, Sperandio M, Guigui C, Aim RB, Wan J, Visvanathan C (2010) Coupling sequencing batch airlift reactor (SBAR) and membrane filtration: influence of nitrate removal on sludge characteristics, effluent quality and filterability. Desalination 250(2):850–854

Thomas MP (2008) The secret to achieving reliable biological phosphorus removal. Water Sci Technol 58(6):1231–1236

Thomas M, Wright P, Blackall L, Urbain V, Keller J (2003) Optimisation of Noosa BNR plant to improve performance and reduce operating costs. Water Sci Technol 47(12):141–148

Thompson MR, Chourey K, Froelich JM, Erickson BK, VerBerkmoes NC, Hettich RL (2008) Experimental approach for deep proteome measurements from small-scale microbial biomass samples. Anal Chem 80(24):9517–9525

Thomsen TR, Kong Y, Nielsen PH (2007) Ecophysiology of abundant denitrifying bacteria in activated sludge. FEMS Microbiol Ecol 60(3):370–382

Tierra G, Pavissich JP, Nerenberg R, Xu Z, Alber MS (2015) Multicomponent model of deformation and detachment of a biofilm under fluid flow. J R Soc Interface 12(106):20150045

Tijhuis L, Van Loosdrecht MCM, Heijnen JJ (1992) Nitrification with biofilms on small suspended particles in airlift reactors. Water Sci Technol 26(9–11):2207–2211

Tijhuis L, Vanloosdrecht MCM, Heijnen JJ (1994) Formation and growth of heterotrophic aerobic biofilms on small suspended particles in airlift reactors. Biotechnol Bioeng 44(5):595–608

Toh SK, Tay JH, Moy BYP, Ivanov V, Tay STL (2003) Size-effect on the physical characteristics of the aerobic granule in a SBR. Appl Microbiol Biotechnol 60(6):687–695

Toja Ortega S, van den Berg L, Pronk M, de Kreuk MK (2022) Hydrolysis capacity of different sized granules in a full-scale aerobic granular sludge (AGS) reactor. Water Res X 16:100151

Tomás-Martínez S, Kleikamp HBC, Neu TR, Pabst M, Weissbrodt DG, van Loosdrecht MCM, Lin Y (2021) Production of nonulosonic acids in the extracellular polymeric substances of "*Candidatus* Accumulibacter phosphatis". Appl Microbiol Biotechnol 105(8):3327–3338

Tomás-Martínez S, Chen LM, Neu TR, Weissbrodt DG, Van Loosdrecht MCM, Lin Y (2022a) Catabolism of sialic acids in an environmental microbial community. FEMS Microbiol Ecol 98(5):fiac047

Tomás-Martínez S, Chen LM, Pabst M, Weissbrodt DG, van Loosdrecht MCM, Lin Y (2022b) Enrichment and application of extracellular nonulosonic acids containing polymers of Accumulibacter. Appl Microbiol Biotechnol 107(2-3):931–941

Tomás-Martínez S, Zwolsman EJ, Merlier F, Pabst M, Lin Y, van Loosdrecht MCM, Weissbrodt DG (2022c) Turnover of the extracellular polymeric matrix in an EBPR microbial community. Appl Microbiol Biotechnol 107(5-6):1997–2009

Torley VJ (2007) The anatomy of a minimal mind. PhD thesis, University of Melbourne

Toumi LB, Fedailaine M, Allia K (2008) Modelling three-phase fluidized bed bioreactor for wastewater treatment. Int J Chem Reactor Eng 6(1):1–16

Tsuneda S, Auresenia J, Morise T, Hirata A (2002) Dynamic modeling and simulation of a three-phase fluidized bed batch process for wastewater treatment. Process Biochem 38(4):599–604

Tsuneda S, Nagano T, Hoshino T, Ejiri Y, Noda N, Hirata A (2003) Characterization of nitrifying granules produced in an aerobic upflow fluidized bed reactor. Water Res 37(20):4965–4973

Tsuneda S, Ogiwara M, Ejiri Y, Hirata A (2006) High-rate nitrification using aerobic granular sludge. Water Sci Technol 53(3):147–154

Tsushima I, Kindaichi T, Okabe S (2007) Quantification of anaerobic ammonium-oxidizing bacteria in enrichment cultures by real-time PCR. Water Res 41(4):785–794

Unz RF, Farrah SR (1976) Exopolymer production and flocculation by Zoogloea mp6. Appl Environ Microbiol 31(4):623–626

Vallina SM, Martinez-Garcia R, Smith SL, Bonachela JA (2019) Models in microbial ecology. In: Encyclopedia of microbiology, pp 211–246

van den Berg L, Kirkland CM, Seymour JD, Codd SL, van Loosdrecht MCM, de Kreuk MK (2020) Heterogeneous diffusion in aerobic granular sludge. Biotechnol Bioeng 117(12):3809–3819

van den Berg L, van Loosdrecht MCM, de Kreuk MK (2021) How to measure diffusion coefficients in biofilms: a critical analysis. Biotechnol Bioeng 118(3):1273–1285

van den Berg L, Toja Ortega S, van Loosdrecht MCM, de Kreuk MK (2022) Diffusion of soluble organic substrates in aerobic granular sludge: effect of molecular weight. Water Res X 16:100148

van der Hoek JP, Klapwijk A (1987) Nitrate removal from ground water. Water Res 21(8):989–997 van der Roest HF, van Loosdrecht MCM (2012) Water purification, the new standard: purely based on character. Delft outlook—magazine of Delft University of Technology. TU Delft, Delft, pp 6–11

van der Roest HF, de Bruin LMM, Gademan G, Coelho F (2011) Towards sustainable waste water treatment with Dutch Nereda® technology. Water Pract Technol 6(3):59

van der Roest H, van Loosdrecht M, Langkamp EJ, Uijterlinde C (2015) Recovery and reuse of alginate from granular Nereda sludge. Water 21:48

van der Star WRL, Abma WR, Blommers D, Mulder J-W, Tokutomi T, Strous M, Picioreanu C, van Loosdrecht MCM (2007) Startup of reactors for anoxic ammonium oxidation: experiences from the first full-scale anammox reactor in Rotterdam. Water Res 41(18):4149–4163

van Dijk EJH (2022) Principles of the full-scale aerobic granular sludge process. Delft University of Technology, The Netherlands

van Dijk EJH, Pronk M, van Loosdrecht MCM (2018) Controlling effluent suspended solids in the aerobic granular sludge process. Water Res 147:50–59

van Dijk EJH, Pronk M, van Loosdrecht MCM (2020) A settling model for full-scale aerobic granular sludge. Water Res 186:116135

van Dijk EJH, van Loosdrecht MCM, Pronk M (2021) Nitrous oxide emission from full-scale municipal aerobic granular sludge. Water Res 198:117159

van Dijk EJH, Haaksman VA, van Loosdrecht MCM, Pronk M (2022) On the mechanisms for aerobic granulation—model based evaluation. Water Res 216:118365

van Groenestijn JW, Deinema MH (1985) Phosphate uptake and release by *Acinetobacter* strain 210A. Anton Van Leeuw 51(5–6):581–582

van Groenestijn JW, Deinema MH, Zehnder AJB (1987) ATP production from polyphosphate in *Acinetobacter* strain 210A. Arch Microbiol 148(1):14–19

van Groenestijn J, Zuidema M, van de Worp JJM, Deinema MH, Zehnder AJB (1989a) Influence of environmental parameters on polyphosphate accumulation in *Acinetobacter* sp. Anton Van Leeuw Int J Gen Mol Microbiol 55(1):67–82

van Groenestijn JW, Bentvelsen MMA, Deinema MH, Zehnder AJB (1989b) Polyphosphate-degrading enzymes in *Acinetobacter* spp. and activated sludge. Appl Environ Microbiol 55(1):219–223

van Haandel AC, van der Lubbe JGM (2012) Handbook of biological wastewater treatment, design and optimisation of activated sludge systems, 2nd edn. IWA Publishing, London

Van Hulle SWH, Vandeweyer HJP, Meesschaert BD, Vanrolleghem PA, Dejans P, Dumoulin A (2010) Engineering aspects and practical application of autotrophic nitrogen removal from nitrogen rich streams. Chem Eng J 162(1):1–20

van Kessel MA, Speth DR, Albertsen M, Nielsen PH, Op den Camp HJ, Kartal B, Jetten MS, Lucker S (2015) Complete nitrification by a single microorganism. Nature 528(7583):555–559

van Lier JB, Mahmoud N, Zeeman G (2020) Anaerobic wastewater treatment. In: Chen G, van Loosdrecht MCM, Ekama GA, Brdjanovic D (eds) Biological wastewater treatment: principles, modelling and design, 2nd edn. IWA Publishing, London, pp 701–756

van Loosdrecht MCM, Heijnen SJ (1993) Biofilm bioreactors for waste-water treatment. Trends Biotechnol 11(4):117–121

van Loosdrecht MC, Lyklema J, Norde W, Schraa G, Zehnder AJ (1987a) Electrophoretic mobility and hydrophobicity as a measured to predict the initial steps of bacterial adhesion. Appl Environ Microbiol 53(8):1898–1901

van Loosdrecht MC, Lyklema J, Norde W, Schraa G, Zehnder AJ (1987b) The role of bacterial cell wall hydrophobicity in adhesion. Appl Environ Microbiol 53(8):1893–1897

van Loosdrecht MCM, Lyklema J, Norde W, Zehnder AJB (1989) Bacterial adhesion: a physio-chemical approach. Microb Ecol 17(1):1–15

van Loosdrecht MCM, Lyklema J, Norde W, Zehnder AJB (1990) Influence of interfaces on microbial activity. Microbiol Rev 54(1):75–87

van Loosdrecht MCM, Tijhuis L, Wijdieks AMS, Heijnen JJ (1995) Population distribution in aerobic biofilms on small suspended particles. Water Sci Technol 31(1):163–171 van Loosdrecht MCM, Hooijmans CM, Brdjanovic D, Heijnen JJ (1997a) Biological phosphate removal processes. Appl Microbiol Biotechnol 48(3):289–296

van Loosdrecht MCM, Picioreanu C, Heijnen JJ (1997b) A more unifying hypothesis for biofilm structures. FEMS Microbiol Ecol 24(2):181–183

van Loosdrecht MCM, Smolders GJ, Kuba T, Heijnen JJ (1997c) Metabolism of micro-organisms responsible for enhanced biological phosphorus removal from wastewater. Use of dynamic enrichment cultures. Anton Van Leeuw Int J Gen Mol Microbiol 71(1–2):109–116

Van Loosdrecht MCM, Heijnen JJ, Eberl H, Kreft J, Picioreanu C (2002) Mathematical modelling of biofilm structures. Anton Van Leeuw Int J Gen Mol Microbiol 81(1–4):245–256

van Loosdrecht MCM, Martins AM, Ekama GA (2008) Bulking sludge. In: Henze M, van Loosdrecht MCM, Ekama GA, Brdjanovic D (eds) Biological wastewater treatment: principles, modelling and design. IWA Publishing, London, pp 291–308

van Loosdrecht MCM, Nielsen PH, Lopez-Vazquez CM, Brdjanovic D (2016) Experimental methods in wastewater treatment. IWA Publishing, London

van Loosdrecht MCM, Ekama GA, Lopez Vazquez CM, Meijer SCF, Hooijmans CM, Brdjanovic D (2020a) Modelling activated sludge processes. In: Chen G, van Loosdrecht MCM, Ekama GA, Brdjanovic D (eds) Biological wastewater treatment: principles, modelling and design, 2nd edn. IWA Publishing, London, pp 613–665

van Loosdrecht MCM, Martins AM, Ekama GA (2020b) Bulking sludge. In: Chen G, van Loosdrecht MCM, Ekama GA, Brdjanovic D (eds) Biological wastewater treatment: principles, modelling and design, 2nd edn. IWA Publishing, London, pp 475–496

van Niekerk AM, Jenkins D, Richard MG (1987) The competitive growth of *Zoogloea ramigera* and type 021N in activated sludge and pure culture—a model for low F:M bulking. J Water Pollut Control Fed 59(5):262–273

van Veen HW (1997) Phosphate transport in prokaryotes: molecules, mediators and mechanisms. Anton Van Leeuw 72(4):299–315

Vanhooren H, Demey D, Vannijvel I, Vanrolleghem PA (2000) Monitoring and modelling an industrial trickling filter using on-line off-gas analysis and respirometry. Water Sci Technol 41:139–148

Vannecke TP, Wells G, Hubaux N, Morgenroth E, Volcke EI (2015) Considering microbial and aggregate heterogeneity in biofilm reactor models: how far do we need to go? Water Sci Technol 72(10):1692–1699

Vanwonterghem I, Jensen PD, Ho DP, Batstone DJ, Tyson GW (2014) Linking microbial community structure, interactions and function in anaerobic digesters using new molecular techniques. Curr Opin Biotechnol 27:55–64

Vargas M, Guisasola A, Artigues A, Casas C, Baeza JA (2011) Comparison of a nitrite-based anaerobic-anoxic EBPR system with propionate or acetate as electron donors. Process Biochem 46(3):714–720

Vayenas DV, Pavlou S, Lyberatos G (1997) Transient modeling of trickling filters for biological ammonia removal. Environ Model Assess 2(3):221–226

Vazquez-Padin J, Fernandez I, Figueroa M, Mosquera-Corral A, Campos JL, Mendez R (2009) Applications of anammox based processes to treat anaerobic digester supernatant at room temperature. Bioresour Technol 100(12):2988–2994

Vazquez-Padin JR, Figueroa M, Campos JL, Mosquera-Corral A, Mendez R (2010a) Nitrifying granular systems: a suitable technology to obtain stable partial nitrification at room temperature. Sep Purif Technol 74(2):178–186

Vazquez-Padin JR, Mosquera-Corral A, Campos JL, Mendez R, Carrera J, Perez J (2010b) Modelling aerobic granular SBR at variable COD/N ratios including accurate description of total solids concentration. Biochem Eng J 49(2):173–184

Verawaty M, Pijuan M, Yuan Z, Bond PL (2012) Determining the mechanisms for aerobic granulation from mixed seed of floccular and crushed granules in activated sludge wastewater treatment. Water Res 46(3):761–771

Verstraete W (2007) Microbial ecology and environmental biotechnology. ISME J 1(1):4–8

Verstraete W, Wittebolle L, Heylen K, Vanparys B, de Vos P, van de Wiele T, Boon N (2007) Microbial resource management: the road to go for environmental biotechnology. Eng Life Sci 7(2):117–126

Veuillet F, Lacroix S, Bausseron A, Gonidec E, Ochoa J, Christensson M, Lemaire R (2014) Integrated fixed-film activated sludge ANITA™ Mox process—a new perspective for advanced nitrogen removal. Water Sci Technol 69(5):915–922

Vlaeminck SE, Terada A, Smets BF, van der Linden D, Boon N, Verstraete W, Carballa M (2009) Nitrogen removal from digested black water by one-stage partial nitritation and anammox. Environ Sci Technol 43(13):5035–5041

Vlaeminck SE, De Clippeleir H, Verstraete W (2012) Microbial resource management of one-stage partial nitritation/anammox. Microb Biotechnol 5(3):433–448

Vogelaar JCT, van der Wal F, Lettinga G (2002) A new post-treatment system for anaerobic effluents containing a high Ca^{2+} content. Biotechnol Lett 24:1981–1986

Volcke EIP, Picioreanu C, De Baets B, Van Loosdrecht MCM (2010) Effect of granule size on autotrophic nitrogen removal in a granular sludge reactor. Environ Technol 31(11):1271–1280

Volcke EIP, Picioreanu C, De Baets B, van Loosdrecht MCM (2012) The granule size distribution in an anammox-based granular sludge reactor affects the conversion—implications for modeling. Biotechnol Bioeng 109(7):1629–1636

Wagner J, da Costa RHR (2013) Aerobic granulation in a sequencing batch reactor using real domestic wastewater. J Environ Eng (US) 139(11):1391–1396

Wagner M, Haider S (2012) New trends in fluorescence in situ hybridization for identification and functional analyses of microbes. Curr Opin Biotechnol 23(1):96–102

Wagner M, Amann R, Lemmer H, Manz W, Schleifer KH (1994a) Probing activated sludge with fluorescently labeled rRNA targeted oligonucleotides. Water Sci Technol 29(7):15–23

Wagner M, Erhart R, Manz W, Amann R, Lemmer H, Wedi D, Schleifer KH (1994b) Development of an rRNA-targeted oligonucleotide probe specific for the genus *Acinetobacter* and its application for in situ monitoring in activated sludge. Appl Environ Microbiol 60(3):792–800

Wagner M, Rath G, Amann R, Koops HP, Schleifer KH (1995) In situ identification of ammonia-oxidizing bacteria. Syst Appl Microbiol 18(2):251–264

Wagner M, Noguera DR, Juretschko S, Rath G, Koops HP, Schleifer KH (1998) Combining fluorescent in situ hybridization (FISH) with cultivation and mathematical modeling to study population structure and function of ammonia-oxidizing bacteria in activated sludge. Water Sci Technol 37(4–5):441–449

Wagner M, Loy A, Nogueira R, Purkhold U, Lee N, Daims H (2002) Microbial community composition and function in wastewater treatment plants. Anton Van Leeuw Int J Gen Mol Microbiol 81(1–4):665–680

Wagner M, Hornt M, Daims H (2003) Fluorescence in situ hybridisation for the identification and characterisation of prokaryotes. Curr Opin Microbiol 6(3):302–309

Wagner M, Nielsen PH, Loy A, Nielsen JL, Daims H (2006) Linking microbial community structure with function: fluorescence in situ hybridization-microautoradiography and isotope arrays. Curr Opin Biotechnol 17(1):83–91

Wagner J, Guimaraes LB, Akaboci TR, Costa RH (2015a) Aerobic granular sludge technology and nitrogen removal for domestic wastewater treatment. Water Sci Technol 71(7):1040–1046

Wagner J, Weissbrodt DG, Manguin V, Ribeiro da Costa RH, Morgenroth E, Derlon N (2015b) Effect of particulate organic substrate on aerobic granulation and operating conditions of sequencing batch reactors. Water Res 85:158–166

Wan J (2009) Interaction entre l'élimination des polluants azotés et la formation des granules aérobies en réacteur biologique séquencé. PhD thesis, INSA

Wan J, Bessiere Y, Sperandio M (2009) Alternating anoxic feast/aerobic famine condition for improving granular sludge formation in sequencing batch airlift reactor at reduced aeration rate. Water Res 43(20):5097–5108

Wan J, Mozo I, Filali A, Line A, Bessiere Y, Sperandio M (2011) Evolution of bioaggregate strength during aerobic granular sludge formation. Biochem Eng J 58–59(1):69–78

Wang YL, Yu SL (2009a) Comparative performance between a novel aerobic granular sludge membrane bioreactor and a conventional activated floc sludge membrane bioreactor. In: 3rd international conference on bioinformatics and biomedical engineering, iCBBE 2009
Wang YL, Yu SL (2009b) Simultaneous COD, nitrogen and phosphate removal in an aerobic granular sludge membrane bioreactor. In: 3rd international conference on bioinformatics and biomedical engineering, iCBBE 2009
Wang C, Zheng XY (2008) Effect of shear stress on morphology, structure and microbial activity of aerobic granules. Huanjing Kexue/Environ Sci 29(8):2235–2241
Wang DZ, Zhou LX (2010) Cultivation of aerobic granular sludge and characterization of nitrobenzene-degrading. Huanjing Kexue/Environ Sci 31(1):147–152
Wang F, Yang FL, Liu YH, Zhang XW (2005a) Cultivation of aerobic granules for simultaneous nitrification and denitrification by seeding different inoculated sludge. J Environ Sci 17(2):268–270
Wang ZW, Liu Y, Tay JH (2005b) Distribution of EPS and cell surface hydrophobicity in aerobic granules. Appl Microbiol Biotechnol 69(4):469–473
Wang HL, Yu GL, Liu GS, Pan F (2006a) A new way to cultivate aerobic granules in the process of papermaking wastewater treatment. Biochem Eng J 28(1):99–103
Wang JF, Wang X, Ji M, Liu WH, Yang ZY (2006b) Intracellular storage polymer driven simultaneous nitrification and denitrification of GAOs granular sludge. Huanjing Kexue/Environ Sci 27(3):473–477
Wang Z, Liu L, Yao J, Cai W (2006c) Effects of extracellular polymeric substances on aerobic granulation in sequencing batch reactors. Chemosphere 63(10):1728–1735
Wang SG, Liu XW, Gong WX, Gao BY, Zhang DH, Yu HQ (2007) Aerobic granulation with brewery wastewater in a sequencing batch reactor. Bioresour Technol 98(11):2142–2147
Wang J, Wang X, Zhao Z, Li J (2008a) Organics and nitrogen removal and sludge stability in aerobic granular sludge membrane bioreactor. Appl Microbiol Biotechnol 79(4):679–685
Wang YF, Zhang HM, Wang XH, Yang FL (2008b) Effects of aeration intensity on characteristics of aerobic granules in sequencing batch airlift reactor (SBAR). Huanjing Kexue/Environ Sci 29(6):1598–1603
Wang SG, Gai LH, Zhao LJ, Fan MH, Gong WX, Gao BY, Ma Y (2009a) Aerobic granules for low-strength wastewater treatment: formation, structure, and microbial community. J Chem Technol Biotechnol 84(7):1015–1020
Wang X, Zhang K, Ren N, Li N, Ren L (2009b) Monitoring microbial community structure and succession of an A/O SBR during start-up period using PCR-DGGE. J Environ Sci 21(2):223–228
Wanner O, Morgenroth E (2004) Biofilm modeling with AQUASIM. Water Sci Technol 49:137–144
Wanner O, Reichert P (1996) Mathematical modeling of mixed-culture biofilms. Biotechnol Bioeng 49(2):172–184
Wanner O, Eberl HJ, Morgenroth E, Noguera DR, Picioreanu C, Rittmann BE, van Loosdrecht MCM (2006) Mathematical modeling of biofilms. IWA Publishing, London
Watnick P, Kolter R (2000) Biofilm, city of microbes. J Bacteriol 182(10):2675–2679
Weber SD, Ludwig W, Schleifer KH, Fried J (2007) Microbial composition and structure of aerobic granular sewage biofilms. Appl Environ Microbiol 73(19):6233–6240
Wedi D, Wagner M, Amann R, Erhart R, Lemmer H, Wilderer PA (1995) In-situ proof of the secondary role of *Acinetobacter* on enhanced biological phosphate removal. Gas Wasser Abwasser 136(13):S17–S23
Wei D, Qiao Z, Zhang Y, Hao L, Si W, Du B, Wei Q (2012) Effect of COD/N ratio on cultivation of aerobic granular sludge in a pilot-scale sequencing batch reactor. Appl Microbiol Biotechnol
Wei SP, Stensel HD, Nguyen Quoc B, Stahl DA, Huang X, Lee PH, Winkler MKH (2020) Flocs in disguise? High granule abundance found in continuous-flow activated sludge treatment plants. Water Res 179:115865

Wei SP, Stensel HD, Ziels RM, Herrera S, Lee PH, Winkler MKH (2021) Partitioning of nutrient removal contribution between granules and flocs in a hybrid granular activated sludge system. Water Res 203:117514
Weissbrodt DG (2018) StaRRE—stations de récupération des ressources de l'eau. Aqua Gas 1:20–24
Weissbrodt DG (2022) Microbial ecology principles for urban circular economies—polyphosphate-accumulating organisms: nutrient cycling and beyond. In: ISME (ed) 18th international symposium on microbial ecology (ISME18), Lausanne
Weissbrodt DG, Lochmatter S, Ebrahimi S, Rossi P, Maillard J, Holliger C (2012a) Bacterial selection during the formation of early-stage aerobic granules in wastewater treatment systems operated under wash-out dynamics. Front Microbiol 3:332
Weissbrodt DG, Shani N, Sinclair L, Lefebvre G, Rossi P, Maillard J, Rougemont J, Holliger C (2012b) PyroTRF-ID: a novel bioinformatics methodology for the affiliation of terminal-restriction fragments using 16S rRNA gene pyrosequencing data. BMC Microbiol 12:306
Weissbrodt DG, Neu TR, Kuhlicke U, Rappaz Y, Holliger C (2013a) Assessment of bacterial and structural dynamics in aerobic granular biofilms. Front Microbiol 4:175
Weissbrodt DG, Schneiter GS, Fürbringer JM, Holliger C (2013b) Identification of trigger factors selecting for polyphosphate- and glycogen-accumulating organisms in aerobic granular sludge sequencing batch reactors. Water Res 47(19):7006–7018
Weissbrodt DG, Derlon N, Morgenroth E, Holliger C (2014a) A consolidated approach of flocculent and granular sludge systems under the perspective of bacterial resource management. Proc Water Environ Fed 2014(19):5008–5009
Weissbrodt DG, Maillard J, Brovelli A, Chabrelie A, May J, Holliger C (2014b) Multilevel correlations in the biological phosphorus removal process: from bacterial enrichment to conductivity-based metabolic batch tests and polyphosphatase assays. Biotechnol Bioeng 111(12):2421–2435
Weissbrodt DG, Neu TR, Derlon N, Szivák I, Kuhlicke U, Holliger C, Morgenroth E (2014c) Fluorescence lectin-binding analysis reveals the complexity of extracellular glycoconjugate matrices in aerobic granular sludge biofilms. In: Flemming et al (eds) IWA EPS conference—the perfect slime—nature, properties, regulation and dynamics of EPS. University of Duisburg-Essen, Germany
Weissbrodt DG, Shani N, Holliger C (2014d) Linking bacterial population dynamics and nutrient removal in the granular sludge biofilm ecosystem engineered for wastewater treatment. FEMS Microbiol Ecol 88(3):579–595
Weissbrodt DG, Holliger C, Morgenroth E (2017) Modeling hydraulic transport and anaerobic uptake by PAOs and GAOs during wastewater feeding in EBPR granular sludge reactors. Biotechnol Bioeng 114(8):1688–1702
Weissbrodt DG, Laureni M, van Loosdrecht MCM, Comeau Y (2020a) Basic microbiology and metabolism. In: Chen GH, van Loosdrecht MCM, Ekama GA, Brdjanovic D (eds) Biological wastewater treatment: principles, modelling and design, 2nd edn. IWA Publishing, London
Weissbrodt DG, Wells GF, Laureni M, Agrawal S, Goel R, Russo G, Men Y, Johnson D, Christensson M, Lackner S, Joss A, Nielsen JL, Bürgmann H, Morgenroth E (2020b) Systems microbiology and engineering of aerobic-anaerobic ammonium oxidation. ChemRxiv 12243077
Weissbrodt DG, Laureni M, van Loosdrecht MCM, Comeau Y (2023) Examples and exercises of basic microbiology and metabolism. In: Lopez-Vazquez C, Volcke E, Wu D, van Loosdrecht MCM, Brdjanovic D, Chen GH (eds) Biological wastewater treatment: principles, modelling and design: examples and exercises, 1st edn. IWA Publishing, London
Welles L, Lopez-Vazquez CM, Hooijmans CM, van Loosdrecht MC, Brdjanovic D (2014) Impact of salinity on the anaerobic metabolism of phosphate-accumulating organisms (PAO) and glycogen-accumulating organisms (GAO). Appl Microbiol Biotechnol 98(17):7609–7622
Welles L, Lopez-Vazquez CM, Hooijmans CM, van Loosdrecht MC, Brdjanovic D (2015a) Impact of salinity on the aerobic metabolism of phosphate-accumulating organisms. Appl Microbiol Biotechnol 99(8):3659–3672

Welles L, Tian WD, Saad S, Abbas B, Lopez-Vazquez CM, Hooijmans CM, van Loosdrecht MC, Brdjanovic D (2015b) *Accumulibacter* clades type I and II performing kinetically different glycogen-accumulating organisms metabolisms for anaerobic substrate uptake. Water Res 83:354–366

Wells GF, Park HD, Eggleston B, Francis CA, Criddle CS (2011) Fine-scale bacterial community dynamics and the taxa-time relationship within a full-scale activated sludge bioreactor. Water Res 45(17):5476–5488

Wentzel MC, Comeau Y, Ekama GA, van Loosdrecht MCM, Brdjanovic D (2008) Enhanced biological phosphorus removal. In: Henze M, van Loosdrecht MCM, Ekama GA, Brdjanovic D (eds) Biological wastewater treatment: principles, modelling and design. IWA Publishing, London, pp 155–220

Werker A (2007) The next generation of wastewater and sludge treatment: biorefineries producing biopolymer. Water 21:30–31

West SA, Griffin AS, Gardner A, Diggle SP (2006) Social evolution theory for microorganisms. Nat Rev Microbiol 4(8):597–607

Wett B, Murthy S, Takács I, Hell M, Bowden G, Deur A, O'Shaughnessy M (2007) Key parameters for control of DEMON deammonification process. Water Pract 1(5):1–11

Wett B, Omari A, Podmirseg SM, Han M, Akintayo O, Gómez Brandón M, Murthy S, Bott C, Hell M, Takács I, Nyhuis G, O'Shaughnessy M (2013) Going for mainstream deammonification from bench to full scale for maximized resource efficiency. Water Sci Technol 68(2):283–289

Wett B, Podmirseg SM, Gomez-Brandon M, Hell M, Nyhuis G, Bott C, Murthy S (2015) Expanding DEMON sidestream deammonification technology towards mainstream application. Water Environ Res 87(12):2084–2089

Wichern M, Lübken M, Horn H (2008) Optimizing sequencing batch reactor (SBR) reactor operation for treatment of dairy wastewater with aerobic granular sludge. Water Sci Technol 58:1199–1206

Wilderer PA, Characklis WG (1989) Structure and function of biofilms. In: Characklis WG, Wilderer PA (eds) Structure and function of biofilms. Wiley, New York, pp 5–17

Wilderer PA, McSwain BS (2004) The SBR and its biofilm application potentials. Water Sci Technol 50(10):1–10

Wilen BM, Lund Nielsen J, Keiding K, Nielsen PH (2000) Influence of microbial activity on the stability of activated sludge flocs. Colloids Surf B 18(2):145–156

Williams JC, de los Reyes FL III (2006) Microbial community structure of activated sludge during aerobic granulation in an annular gap bioreactor. Water Sci Technol 54(1):139–146

Williamson K, McCarty PL (1976) A model of substrate utilization by bacterial films. J Water Pollut Control Fed 48(1):9–24

Wilmes P, Bond PL (2004) The application of two-dimensional polyacrylamide gel electrophoresis and downstream analyses to a mixed community of prokaryotic microorganisms. Environ Microbiol 6(9):911–920

Wilmes P, Andersson AF, Lefsrud MG, Wexler M, Shah M, Zhang B, Hettich RL, Bond PL, VerBerkmoes NC, Banfield JF (2008a) Community proteogenomics highlights microbial strain-variant protein expression within activated sludge performing enhanced biological phosphorus removal. ISME J 2(8):853–864

Wilmes P, Wexler M, Bond PL (2008b) Metaproteomics provides functional insight into activated sludge wastewater treatment. PLoS ONE 3(3):e1778

Wilmes P, Heintz-Buschart A, Bond PL (2015) A decade of metaproteomics: where we stand and what the future holds. Proteomics 15(20):3409–3417

Wimpenny JWT, Colasanti R (1997a) A more unifying hypothesis for biofilm structures—a reply. FEMS Microbiol Ecol 24(2):185–186

Wimpenny JWT, Colasanti R (1997b) A unifying hypothesis for the structure of microbial biofilms based on cellular automaton models. FEMS Microbiol Ecol 22(1):1–16

Wimpenny J, Poole RK (2009) Microbial metropolis. Adv Microb Physiol 56:29–84

Winkler MKH, Bassin JP, Kleerebezem R, de Bruin LMM, van den Brand TPH, van Loosdrecht MCM (2011a) Selective sludge removal in a segregated aerobic granular biomass system as a strategy to control PAO-GAO competition at high temperatures. Water Res 45(11):3291–3299
Winkler MKH, Kleerebezem R, Kuenen JG, Yang J, van Loosdrecht MCM (2011b) Segregation of biomass in cyclic anaerobic/aerobic granular sludge allows the enrichment of anaerobic ammonium oxidizing bacteria at low temperatures. Environ Sci Technol 45(17):7330–7337
Winkler MKH, Bassin JP, Kleerebezem R, Sorokin DY, van Loosdrecht MCM (2012a) Unravelling the reasons for disproportion in the ratio of AOB and NOB in aerobic granular sludge. Appl Microbiol Biotechnol 94(6):1657–1666
Winkler MKH, Bassin JP, Kleerebezem R, van der Lans RGJM, van Loosdrecht MCM (2012b) Temperature and salt effect on settling velocity in granular sludge technology. Water Res
Winkler MKH, Kleerebezem R, Strous M, Chandran K, van Loosdrecht MCM (2012c) Factors influencing the density of aerobic granular sludge. Appl Microbiol Biotechnol 1–10
Winkler MKH, Kleerebezem R, van Loosdrecht MCM (2012d) Integration of anammox into the aerobic granular sludge process for main stream wastewater treatment at ambient temperatures. Water Res 46(1):136–144
Winkler MKH, Meunier C, Henriet O, Mahillon J, Suarez-Ojeda ME, Del Moro G, De Sanctis M, Di Iaconi C, Weissbrodt DG (2018) An integrative review of granular sludge for the biological removal of nutrients and of recalcitrant organic matter from wastewater. Chem Eng J 336:489–502
Winter C, Bouvier T, Weinbauer MG, Thingstad TF (2010) Trade-offs between competition and defense specialists among unicellular planktonic organisms: the "killing the winner" hypothesis revisited. Microbiol Mol Biol Rev 74(1):42–57
Withey S, Cartmell E, Avery LM, Stephenson T (2005) Bacteriophages—potential for application in wastewater treatment processes. Sci Total Environ 339(1–3):1–18
Woese CR, Stackebrandt E, Macke TJ, Fox GE (1985) A phylogenetic definition of the major eubacterial taxa. Syst Appl Microbiol 6(2):143–151
Woldringh CL, Binnerts JS, Mans A (1981) Variation in *Escherichia coli* buoyant density measured in Percoll gradients. J Bacteriol 148(1):58–63
Wong MT, Liu WT (2006) Microbial succession of glycogen accumulating organisms in an anaerobic-aerobic membrane bioreactor with no phosphorus removal. Water Sci Technol 54(1):29–37
Wu S, Bhattacharjee AS, Weissbrodt DG, Morgenroth E, Goel R (2016) Effect of short term external perturbations on bacterial ecology and activities in a partial nitritation and anammox reactor. Bioresour Technol 219:527–535
Wu L, Ning D, Zhang B, Li Y, Zhang P, Shan X, Zhang Q, Brown M, Li Z, Van Nostrand JD, Ling F, Xiao N, Zhang Y, Vierheilig J, Wells GF, Yang Y, Deng Y, Tu Q, Wang A, Zhang T, He Z, Keller J, Nielsen PH, Alvarez PJJ, Criddle CS, Wagner M, Tiedje JM, He Q, Curtis TP, Stahl DA, Alvarez-Cohen L, Rittmann BE, Wen X, Zhou J, Acevedo D, Agullo-Barcelo M, Andersen GL, de Araujo JC, Boehnke K, Bond P, Bott CB, Bovio P, Brewster RK, Bux F, Cabezas A, Cabrol L, Chen S, Etchebehere C, Ford A, Frigon D, Sanabria J, Griffin JS, Gu AZ, Habagil M, Hale L, Hardeman SD, Harmon M, Horn H, Hu Z, Jauffur S, Johnson DR, Keucken A, Kumari S, Leal CD, Lebrun LA, Lee J, Lee M, Lee ZMP, Li M, Li X, Liu Y, Luthy RG, Mendonça-Hagler LC, de Menezes FGR, Meyers AJ, Mohebbi A, Oehmen A, Palmer A, Parameswaran P, Park J, Patsch D, Reginatto V, de los Reyes FL III, Noyola A, Rossetti S, Sidhu J, Sloan WT, Smith K, de Sousa OV, Stephens K, Tian R, Tooker NB, De los Cobos Vasconcelos D, Wakelin S, Wang B, Weaver JE, West S, Wilmes P, Woo SG, Wu JH, Wu L, Xi C, Xu M, Yan T, Yang M, Young M, Yue H, Zhang Q, Zhang W, Zhang Y, Zhou H, Global Water Microbiome C (2019) Global diversity and biogeography of bacterial communities in wastewater treatment plants. Nat Microbiol 4(7):1183–1195
Wuertz S, Okabe S, Hausner M (2004) Microbial communities and their interactions in biofilm systems: an overview. Water Sci Technol 49(11–12):327–336

Wunderlin P, Mohn J, Joss A, Emmenegger L, Siegrist H (2012) Mechanisms of N_2O production in biological wastewater treatment under nitrifying and denitrifying conditions. Water Res 46(4):1027–1037
Xavier JB, Foster KR (2007) Cooperation and conflict in microbial biofilms. Proc Natl Acad Sci USA 104(3):876–881
Xavier JB, Picioreanu C, van Loosdrecht MCM (2004) Assessment of three-dimensional biofilm models through direct comparison with confocal microscopy imaging. Water Sci Technol 49(11–12):177–185
Xavier JB, Picioreanu C, Abdul Rani S, van Loosdrecht MCM, Stewart PS (2005a) Biofilm-control strategies based on enzymic disruption of the extracellular polymeric substance matrix—a modelling study. Microbiology 151(12):3817–3832
Xavier JB, Picioreanu C, van Loosdrecht MCM (2005b) A framework for multidimensional modelling of activity and structure of multispecies biofilms. Environ Microbiol 7(8):1085–1103
Xavier JB, de Kreuk MK, Picioreanu C, van Loosdrecht MCM (2007) Multi-scale individual-based model of microbial and bioconversion dynamics in aerobic granular sludge. Environ Sci Technol 41(18):6410–6417
Xia Y, Kong Y, Nielsen PH (2008) In situ detection of starch-hydrolyzing microorganisms in activated sludge. FEMS Microbiol Ecol 66(2):462–471
Xiong Y, Liu Y (2010) Involvement of ATP and autoinducer-2 in aerobic granulation. Biotechnol Bioeng 105(1):51–58
Xu LR, Huang D, Li XN, Zhu JR (2010) Characteristics of operational performance and membrane fouling in a sidestream membrane sequencing batch reactor with aerobic granule. Huanjing Kexue/Environ Sci 31(3):750–755
Xu R-Z, Cao J-S, Feng G, Luo J-Y, Wu Y, Ni B-J, Fang F (2021) Modeling molecular structure and behavior of microbial extracellular polymeric substances through interacting-particle reaction dynamics. Chem Eng J Adv 8:100154
Yang SF, Tay JH, Liu Y (2003) A novel granular sludge sequencing batch reactor for removal of organic and nitrogen from wastewater. J Biotechnol 106(1):77–86
Yang SF, Tay JH, Liu Y (2004) Respirometric activities of heterotrophic and nitrifying populations in aerobic granules developed at different substrate N/COD ratios. Curr Microbiol 49(1):42–46
Yang GJ, Li XM, Zeng GM, Xie S, Yang Q (2005) Study on simultaneous phosphorus and nitrogen removal through aerobic granular sludge. Hunan Daxue Xuebao/j Hunan Univ Nat Sci 32(3):97–100
Yergeau E, Lawrence JR, Waiser MJ, Korber DR, Greer CW (2010) Metatranscriptomic analysis of the response of river biofilms to pharmaceutical products, using anonymous DNA microarrays. Appl Environ Microbiol 76(16):5432–5439
Yilmaz LS, Noguera DR (2004) Mechanistic approach to the problem of hybridization efficiency in fluorescent in situ hybridization. Appl Environ Microbiol 70(12):7126–7139
Yilmaz G, Lemaire R, Keller J, Yuan Z (2008) Simultaneous nitrification, denitrification, and phosphorus removal from nutrient-rich industrial wastewater using granular sludge. Biotechnol Bioeng 100(3):529–541
You Y, Peng Y, Yuan ZG, Li XY, Peng YZ (2008) Cultivation and characteristic of aerobic granular sludge enriched by phosphorus accumulating organisms. Huanjing Kexue/Environ Sci 29(8):2242–2248
Young JC, McCarty PL (1969) The anaerobic filter for waste treatment. J Water Pollut Control Fed 41(5):R160–R173
Young B, Banihashemi B, Forrest D, Kennedy K, Stintzi A, Delatolla R (2016) Meso and micro-scale response of post carbon removal nitrifying MBBR biofilm across carrier type and loading. Water Res 91:235–243
Yu T, de la Rosa C, Lu R (2004) Microsensor measurement of oxygen concentration in biofilms: from one dimension to three dimensions. Water Sci Technol 49(11–12):353–358
Yu GH, Juang YC, Lee DJ, He PJ, Shao LM (2009a) Filterability and extracellular polymeric substances of aerobic granules for AGMBR process. J Taiwan Inst Chem Eng 40(4):479–483

Yu HY, Yao L, Ye ZF (2009b) Aerobic granulation for dimethyl phthalate biodegradation in a sequencing batch reactor. Huanjing Kexue/Environ Sci 30(9):2661–2666
Yuan X, Gao D (2010) Effect of dissolved oxygen on nitrogen removal and process control in aerobic granular sludge reactor. J Hazard Mater 178(1–3):1041–1045
Yuan X, Gao D, Liang H (2011) Reactivation characteristics of stored aerobic granular sludge using different operational strategies. Appl Microbiol Biotechnol 94(5):1365–1374
Zemanick ET, Sagel SD, Harris JK (2011) The airway microbiome in cystic fibrosis and implications for treatment. Curr Opin Pediatr 23(3):319–324
Zeng RJ, Lemaire R, Yuan Z, Keller J (2003a) Simultaneous nitrification, denitrification, and phosphorus removal in a lab-scale sequencing batch reactor. Biotechnol Bioeng 84(2):170–178
Zeng RJ, Saunders AM, Yuan Z, Blackall LL, Keller J (2003b) Identification and comparison of aerobic and denitrifying polyphosphate-accumulating organisms. Biotechnol Bioeng 83(2):140–148
Zeng RJ, Van Loosdrecht MCM, Yuan Z, Keller J (2003c) Metabolic model for glycogen-accumulating organisms in anaerobic/aerobic activated sludge systems. Biotechnol Bioeng 81(1):92–105
Zeng RJ, Yuan Z, Keller J (2003d) Enrichment of denitrifying glycogen-accumulating organisms in anaerobic/anoxic activated sludge system. Biotechnol Bioeng 81(4):397–404
Zhang LL, Zhang B, Huang YF, Cai WM (2005) Re-activation characteristics of preserved aerobic granular sludge. J Environ Sci 17(4):655–658
Zhang XL, Wang L, Wang ZY (2006) Experiment study on anoxic phosphate accumulation with nitrite. Huanjing Kexue/Environ Sci 27(5):930–934
Zhang L, Feng X, Zhu N, Chen J (2007a) Role of extracellular protein in the formation and stability of aerobic granules. Enzyme Microb Technol 41(5):551–557
Zhang LL, Chen X, Chen JM, Cai WM (2007b) Role mechanism of extracellular polymeric substances in the formation of aerobic granular sludge. Huanjing Kexue/Environ Sci 28(4):795–799
Zhang Y, Wang YS, Bai YH, Chen C, Lu J, Zhang J (2007c) Characteristics of novel wastewater treatment technology by swimming bed combined with aerobic granular sludge. Huanjing Kexue/Environ Sci 28(10):2249–2254
Zhang YX, Ji M, Li C, Wang XD, Wang SH (2008) Biodegradability of extracellular polymeric substances (EPS) produced by aerobic granules. Huanjing Kexue/Environ Sci 29(11):3124–3127
Zhang ZJ, Wu WW, Wang JL (2010) Granulation of completely autotrophic nitrifying sludge in sequencing batch reactor. Huanjing Kexue/Environ Sci 31(1):140–146
Zhang SH, Xie K, Hua YM, Liu YZ (2011) Coupling of anammox and shortcut denitrifying dephosphatation in a single reactor. Fresenius Environ Bull 20(6A):1564–1569
Zhang S, Zhang Z, Xia S, Ding N, Long X, Wang J, Chen M, Ye C, Chen S (2020) Combined genome-centric metagenomics and stable isotope probing unveils the microbial pathways of aerobic methane oxidation coupled to denitrification process under hypoxic conditions. Bioresour Technol 318:124043
Zheng X, Chen W, Zhu N, Li X (2009) Effect of shear stress on the cultivation and characteristics of aerobic granules. In: Chou KC (ed) 3rd international conference on bioinformatics and biomedical engineering, Beijing
Zhou J, Fl Y, Fg M, An P, Wang D (2007) Comparison of membrane fouling during short-term filtration of aerobic granular sludge and activated sludge. J Environ Sci 19(11):1281–1286
Zhou Y, Pijuan M, Yuan Z (2008a) Development of a 2-sludge, 3-stage system for nitrogen and phosphorous removal from nutrient-rich wastewater using granular sludge and biofilms. Water Res 42(12):3207–3217
Zhou Y, Pijuan M, Zeng RJ, Lu H, Yuan Z (2008b) Could polyphosphate-accumulating organisms (PAOs) be glycogenaccumulating organisms (GAOs)? Water Res 42(10–11):2361–2368
Zhou Y, Pijuan M, Oehmen A, Yuan Z (2010) The source of reducing power in the anaerobic metabolism of polyphosphate accumulating organisms (PAOs)—a mini-review. Water Sci Technol 61(7):1653–1662

Zhou J, Wei S, Li J, He M, Hille A, Horn H (2011a) Aerobic granulation in a modified continuous flow system. In: Chou KC (ed) 5th international conference on bioinformatics and biomedical engineering, Wuhan

Zhou Y, Oehmen A, Lim M, Vadivelu V, Ng WJ (2011b) The role of nitrite and free nitrous acid (FNA) in wastewater treatment plants. Water Res 45(15):4672–4682

Zhu J, Liu C (1999) Cultivation and physico-chemical characteristics of granular activated sludge in alternation of anaerobic/aerobic process. Huanjing Kexue/Environ Sci 20(2):38–41

Zhu J, Etterer T, Wilderer PA (2001) Generation and characteristics of granular activated sludge (Herstellung und Eigenschaften von granuliertem Belebtschlamm). GWF Wasser-Abwasser 142(10):698–702

Zhu L, Xu X, Luo W, Cao D, Yang Y (2008a) Formation and microbial community analysis of chloroanilines-degrading aerobic granules in the sequencing airlift bioreactor. J Appl Microbiol 104:152–160

Zhu L, Xu X, Luo W, Tian Z, Lin H, Zhang N (2008b) A comparative study on the formation and characterization of aerobic 4-chloroaniline-degrading granules in SBR and SABR. Appl Microbiol Biotechnol 79:867–874

Zhu IX, Getting T, Bruce D (2010) Review of biologically active filters in drinking water applications. J Am Water Works Assoc 102(12):67–77

Ziels RM, Sousa DZ, Stensel HD, Beck DAC (2018) DNA-SIP based genome-centric metagenomics identifies key long-chain fatty acid-degrading populations in anaerobic digesters with different feeding frequencies. ISME J 12(1):112–123

Ziliani A, Bovio-Winkler P, Cabezas A, Etchebehere C, Garcia HA, López-Vázquez CM, Brdjanovic D, van Loosdrecht MCM, Rubio-Rincón FJ (2023) Putative metabolism of *Ca.* Accumulibacter via the utilization of glucose. Water Res 229:119446

Zilles JL, Rodríguez LF, Bartolerio NA, Kent AD (2016) Microbial community modeling using reliability theory. ISME J 10(8):1809–1814

Zima BE (2008) Impact of fluid dynamic effects on granular activated sludge. PhD thesis, Universität Erlangen-Nürnberg

Zima BE, Kowalczyk W, Hartmann H, Delgado A (2005) Influence of velocity distribution in the multiphase flow on the building of aerobic granules in a sequencing batch reactor (SBR). Proc Appl Math Mech 5(1):603–604

Zima BE, Diez L, Kowalczyk W, Delgado A (2007) Sequencing batch reactor (SBR) as optimal method for production of granular activated sludge (GAS)—fluid dynamic investigations. Water Sci Technol 55(8–9):151–158

Zima-Kulisiewicz BE, Diez L, Kowalczyk W, Hartmann C, Delgado A (2008) Biofluid mechanical investigations in sequencing batch reactor (SBR). Chem Eng Sci 63(3):599–608

Zinatizadeh AAL, Mansouri Y, Akhbari A, Pashaei S (2011) Biological treatment of a synthetic dairy wastewater in a sequencing batch biofilm reactor: statistical modeling using optimization using response surface methodology. Chem Ind Chem Eng Q 17(4):485–495

Chapter 3
Research Questions and Scientific Overview

It is the best of times for biofilm research.
(Battin et al. 2007)

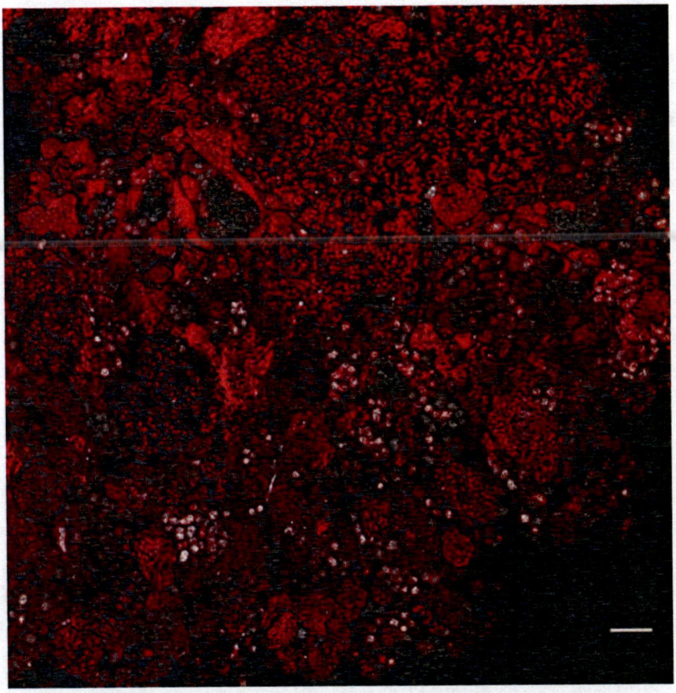

Granular sludge microbiome (scale bar = 10 μm)

D. G. Weissbrodt, *Engineering Granular Microbiomes*, Springer Theses,
https://doi.org/10.1007/978-3-031-41009-3_3

3.1 Motivation and Scope of This Scientific Research

Engineered microbial and biofilm systems involve the interaction of different length and time scales (Morgenroth and Milferstedt 2009). In a holistic vision, the process (i.e. macro scale), biofilm (i.e. meso scale), microbial community (i.e. micro scale), and molecular (i.e. genetic scale) boundaries are closely interconnected.

Fundamental understanding of the microbial ecology of aerobic granular sludge (AGS) systems and the underlying principles of microbial selection is essential to apprehend the mechanisms of granular biofilm formation at the bacterial community level and the associated effects on the process performance. Systematic profiling of the bacterial community of AGS is needed to assess the impact of selection pressures on the granular biofilm properties and on the bacterial compositions under both transient and steady-state conditions. Understanding the mechanisms of bacterial selection during the transition from activated sludge flocs to granular biofilms is required to optimize the conditions for the cultivation of dense fast-settling granules with tailored metabolic activities. Insights into the mechanisms of bacterial competition at reactor steady-state allow for the design of favorable operational conditions for a robust biological nutrient removal (BNR) on long term. Gaining knowledge on the bacterial population dynamics under transient unsteady conditions, such as failure events, is crucial to trigger a fast recovery of high removal efficiencies in a resilient environmental biotechnology process. Biofilms are the 'house of cells' (Flemming et al. 2007). One objective further addressed the impact of the granular biofilm barrier to mass transfer of solutes on bacterial competition. Multilevel correlations were addressed from the macro-scale to the micro- and molecular scales in a systems approach.

The core microbiome of wastewater treatment systems operated for enhanced biological phosphorus removal (EBPR) has been characterized with high resolution on its complexity, leading to the formulation of a conceptual model of the EBPR ecosystem (Nielsen et al. 2010). Most of studies targeting bacterial community structures in BNR systems have been conducted in conventional systems using flocculent activated sludge. Only reduced knowledge has been acquired on the behavior of bacterial communities of AGS biofilm systems operated for BNR. Highlighting the core bacterial microbiome of BNR AGS and its comparison with conventional activated sludge is key for the subsequent understanding of microbial selection mechanisms. The delineation of a conceptual model of the AGS ecosystem can also sustain investigations of metabolic functions of targeted populations.

The competition of polyphosphate- (PAOs) and glycogen-accumulating organisms (GAOs) is an important aspect of EBPR, and has been widely studied for activated sludge systems (Oehmen et al. 2010). These guilds that grow slower than ordinary heterotrophic organisms (OHOs) have also been highlighted as key for the formation of stable granular aggregates (de Kreuk and van Loosdrecht 2004). Specific selection and long-term maintenance of PAOs is crucial for a stable BNR process. EBPR may on the other hand not be a treatment objective such as for the handling high-loaded effluents of food industry. GAOs could play in such a specific context a favorable role for maintaining the granule structure, and for recovering high-value biopolymers such as poly-β-hydroxyalkanoates (PHAs) from wastewater (Johnson et al. 2009). A substantial work is required to specifically assess the selection and competition mechanisms of PAOs and GAOs in granular sludge processes. Since "microbes do the job", understanding the conditions that promote the growth of either PAOs or GAOs in AGS systems, and that suppress the proliferation of competitors, is key for being able to tailor optimal operation conditions for a maximized process efficiency.

This research follows up on microbial ecology studies that have highlighted the importance of considering bacterial communities as a whole. It in addition aimed to target the behavior of predominant populations of engineering interest inside the complexity of the bacterial community continuum. The key findings of this research trigger the proposition of strategies for the optimization of bacterial selection as a prerequisite for an enhanced BNR performance in AGS-SBRs.

3.2 Research Questions and Scientific Overview

The central research question of this scientific research aimed to understand:

Which factors do trigger bacterial selection during the formation of granular biofilms and during the steady-state operation of sequencing batch reactors using granular sludge for high-rate biological nutrient removal?

A set of following twelve research questions was delineated in eight research chapter that was addressed (A) in a method development approach, (B) during granule formation, and (C) under operation of AGS-SBRs for BNR with mature granules. Disciplines of environmental bioprocess engineering, molecular microbial ecology, bioinformatics, confocal laser scanning microscopy, numerical ecology, and mathematical modelling were integrated to these ends.

(A) ***In a Method Development Approach:***

(i) **Can a flexible bubble-column reactor be designed for AGS studies conducted at bench scale toward the bacterial community level?**

Granules are efficiently cultivated at lab scale using bubble-column reactors. Fundamental investigations of microbial community systems require the mastering of operation conditions. Biological systems are impacted by temperature, therefore temperature control is required. Double-wall column reactors constructed as one single piece with inert glass material do however not offer flexibility as soon as the piece is manufactured. A flexible column reactor design enabling the achievement of variable reactor dimensions was conceived to investigate AGS research questions at bench scale (Chap. 4).

(ii) **Can a wet-lab and dry-lab molecular methodology of microbial ecology be developed for a high-throughput profiling of bacterial communities**

with phylogenetic identification in the rational engineering context? Can classical and new-generation workflows be combined to this end?

Classical 16S rRNA gene-based fingerprinting methods of molecular biology such as terminal-restriction fragment length polymorphism (T-RFLP) have been widely applied for the analysis of bacterial community structures and dynamics in open mixed-culture biological systems. However, affiliation of 16S rRNA gene fragments that form operational taxonomic units (OTUs) via cloning and sequencing is time-consuming, and only leads to the identification of a restricted number of OTUs. Microbial ecology methods have progressively been complemented over the last 5 years with massive sequencing technologies for the investigation of microbiomes with high throughput and resolution. However, rational integration of massive sequencing is required for meaningfulness in the engineering context. Time-series sequencing investigations of bacterial community dynamics on long term is related to substantial costs and substantial amount of data generated. A bioinformatics methodology was therefore conceived to harness the rational and high-resolution advantages of the two types of classical fingerprinting and new-generation sequencing, respectively. The PyroTRF-ID workflow was developed to combine datasets from wet-lab analyses of T-RFLP and new-generation amplicon sequencing of identical samples (Weissbrodt et al. 2012b/Chap. 5).

(iii) **Can an on-line metabolic test and a simple enzymatic assay be developed to rapidly assess the fraction of active PAOs and dephosphatation potential of activated and granular sludges?**

PAOs are crucial for EBPR and have been highlighted as possible key microorganisms for stable granulation. Solving their competition with GAOs is of paramount importance. Investigations were conducted here to assess multi-level correlations in their metabolisms from reactor scale to bacterial community, functional gene diversity, and enzymatic activity (Weissbrodt et al. 2014b/Chap. 6). This study mainly focused on the anaerobic metabolism, since PAOs and GAOs compete for the uptake of the carbon source under such conditions. The anaerobic phase was therefore considered to govern the selection of these organisms. Enrichments of PAOs and GAOs were required to tackle this objective. Start-up conditions were therefore optimized to cultivate stable enrichments. Standardized anaerobic metabolic batch tests were conducted with defined mixtures of fractions of PAO- and GAO-enrichments following previous work (Lopez-Vazquez et al. 2007), and were supplemented by on-line electrical conductivity measurements as rapid measurement of fractions of active PAOs present in sludge. Mathematical modelling with the geochemical software PHREEQC was performed to highlight the metabolic contributions of PAOs and GAOs to electrical conductivity profiles. Under anaerobic conditions PAOs hydrolyze their intracellular stocks of inorganic polyphosphate by involving polyphosphatase-like enzymes. A polyphosphatase assay was evaluated for the analysis of the dephosphatation potential of sludge. Similar to rates of electrical conductivity evolution, the enzymatic response was correlated to the fraction of PAOs present in the sludge.

These rapid assays can be attractive for fast and low-cost assessment of PAO fractions and EBPR properties of sludge. The diversity of functional genes encoding for exopolyphosphatase, an enzyme putatively involved in polyphosphate hydrolysis under anaerobic conditions, was in addition attempted by degenerate polymerase chain reaction (PCR).

(B) ***During the Formation of Granules:***

(iv) **Which bacterial populations are predominantly selected during granulation under wash-out conditions?**

Different architectures of granular aggregate can arise from granulation process operated under wash-out dynamics, namely fast-settling dense granules and slow-settling fluffy granules. Methanogenic granular sludge is known to display distinct compact or loose phenotypic structures exhibited by the same predominant phylotype depending on conditions (van Lier et al. 2008). Microbial ecology was investigated here to identify the relationships between bacterial relatives selected under wash-out conditions and the formation of either dense fast-settling granules or loose filamentous bulking granules (Weissbrodt et al. 2012a/Chap. 7).

(v) **Can the deterioration in biomass settling and BNR performances that is commonly observed during start-up of AGS-SBRs under wash-out conditions be explained by preferential selection of unfavorable specific organisms?**

Deterioration of sludge settling properties and of BNR performances has commonly been reported during start-up of AGS-SBRs under wash-out conditions, although these are traditionally applied to select for a fast-settling granular biomass. Microbial ecology and process engineering were combined here to determine whether such deficiencies are explained by preferential selection of unfavorable organisms under wash-out conditions (Weissbrodt et al. 2012a/Chap. 7).

(vi) **How can shifts in predominant populations during transitions from flocs to early-stage and mature granules be explained by the operational conditions?**

Bacterial community dynamics were then followed in details from inoculation with flocculent activated sludge to formation of early-stage granules, and maturation of granules. The objective was to assess correlations between shifts in predominant populations and intrinsic process factors related to operation under wash-out dynamic conditions (Weissbrodt et al. 2013a/Chap. 8).

(vii) **Is the granulation mechanism restricted to single bacterial populations? If not, does this mechanism depend on the predominant population involved?**

Granulation further spontaneously occurred in this study in the stirred-tank SBRs operated at steady-state to enrich for PAOs and for GAOs. Confocal laser scanning microscopy (CLSM), fluorescence lectin-binding analysis (FLBA), fluorescence in situ hybridization (FISH), and T-RFLP analyses were combined to assess the structural and bacterial dynamics during granulation in these systems. This analyses were conducted by comparison to granulation phenomena underlying operation under wash-out conditions in the bubble-column reactor. The aim was (*i*) to determine whether single organisms are responsible for granulation, or whether different populations can form granules, and (*ii*) to assess the mechanism of granular biofilm formation according to the predominant organism(s) involved and their physiological traits (Weissbrodt et al. 2013a/Chap. 8).

(viii) **What conditions do select for PAOs in early-stage granules?**

Conditions that select for PAOs in granules were investigated based on the microbial ecology and reactor performance datasets collected during operations under wash-out dynamics in the bubble-column SBR and under steady-state conditions in the PAO-enrichment (Weissbrodt et al. 2013a/Chap. 8).

(C) ***Under Operation of AGS-SBRs for BNR with Mature Granules:***

(ix) **Do fluctuations in operation variables impact on bacterial community structures and BNR performances? Can this be investigated in a multivariate numerical framework?**

Wastewater treatment systems are commonly operated on the long run under daily, weekly, and seasonal variations in operation variables such as the composition, physicochemical characteristics, and loading of the influent wastewater. The impacts of fluctuations in operation variables on BNR performances and bacterial community structures were assessed in two SBRs operated at steady state for full BNR with mature AGS (Weissbrodt et al. 2014c/Chap. 9). A multivariate numerical approach involving hierarchical clustering, ordination, multiple factor, statistical, and correlation analyses was applied to assess relationships between operation variables, BNR performances, and bacterial community behavior. Understanding was also generated on the impact of the particle size distribution of size on oxygen mass transfer and aerobic processes of nitrification and dephosphatation in AGS.

(x) **Can the bacterial microbiome of granular sludge be rationalized in a conceptual ecosystem model? Which key phylotypes do compose the core community, and how can they preferentially be selected for an efficient BNR?**

Information on bacterial community structures are optimally gained from dynamic states. The multivariate numerical approach developed above was also used for fine-scale delineation of the structure of the bacterial continuum of BNR AGS (Weissbrodt et al. 2014c/Chap. 9). A conceptual ecosystem model was built to rationalize the high-resolution phylogenetic information gained by 16S rRNA gene-based amplicon sequencing, as basis for functional metabolic analyses.

(xi) **Which main factors do trigger the competitive selection of PAOs and GAOs in AGS? Can a multifactorial experimental design be efficiently used for screening factors and effect?**

The PAO/GAO competition has extensively been studied in flocculent activated sludge systems. However, knowledge on this competition in AGS-SBRs is restricted to only few studies conducted on only single parameters. The effects of main factors governing the PAO/GAO competition in mature BNR granular sludge were investigated under steady-state operational conditions by screening within a set of six parameters selected based on previous knowledge (Weissbrodt et al. 2013b/Chap. 10). A Plackett–Burman multifactorial experimental design was used to identify major factors in a reduced number of experiments.

(xii) **Can a mathematical model provide information on hydraulic transport and PAO/GAO competition for the carbon source during up-flow feeding of wastewater across settled beds of AGS? Can this model be used to design proper anaerobic plug-flow feeding conditions as a mean for preferential selection of PAOs in AGS-SBRs?**

System analysis and mathematical modelling was applied to investigate the hydraulic transport and diffusional phenomena occurring during up-flow feeding of wastewater across settled beds of AGS. Residence time distribution experiments were conducted to calibrate a hydraulic transport model. The hydraulic transport model was combined with stoichiometric and biokinetic formulations of PAO- and GAO-metabolisms described by Lopez-Vazquez et al. (2009). This model was then used to investigate the mechanisms of PAO/GAO competition for the uptake of the carbon source during anaerobic conditions achieved plug-flow feeding phases, and their dependencies to temperature and pH (Weissbrodt et al. 2017/Chap. 11). Simulations were run to assess the impact of these parameters on the volumetric rates of acetate uptake by the two guilds under anaerobic conditions. Bed geometries for optimal loading and full anaerobic C-uptake were defined based under different pH and temperature conditions.

Outlook: How can the bacterial resource be optimally managed to engineer granules and trigger efficient nutrient removal in AGS systems?

The extended insights gained in this fundamental scientific research on the phylogenetic signatures and mechanisms of bacterial selection in AGS were harnessed to define applied strategies for an optimal management of the bacterial resource in AGS systems designed for BNR (Weissbrodt et al. 2014a; Weissbrodt and Holliger 2014; Winkler et al. 2018/Chap. 12). Similarities and differences between the microbial diversity of AGS and activated sludge were also discussed. The integration of the present research in the state of the art of the AGS technology will sustain the design of efficient process conditions for engineering bacterial communities in AGS. Directions for future investigations are finally proposed. Overall, sound microbial ecology findings result from the investigation of these engineering systems by combining environmental biotechnology and community systems microbiology in an interdisciplinary approach of environmental life science engineering.

References

Battin TJ, Sloan WT, Kjelleberg S, Daims H, Head IM, Curtis TP, Eberl L (2007) Microbial landscapes: new paths to biofilm research. Nat Rev Microbiol 5(1):76–81

de Kreuk MK, van Loosdrecht MCM (2004) Selection of slow growing organisms as a means for improving aerobic granular sludge stability. Water Sci Technol 49(11–12):9–17

Flemming HC, Neu TR, Wozniak DJ (2007) The EPS matrix: the "house of biofilm cells." J Bacteriol 189(22):7945–7947

Johnson K, Kleerebezem R, Van Loosdrecht MCM (2009) Model-based data evaluation of polyhydroxybutyrate producing mixed microbial cultures in aerobic sequencing batch and fed-batch reactors. Biotechnol Bioeng 104(1):50–67

Lopez-Vazquez CM, Hooijmans CM, Brdjanovic D, Gijzen HJ, van Loosdrecht MCM (2007) A practical method for quantification of phosphorus- and glycogen-accumulating organism populations in activated sludge systems. Water Environ Res 79(13):2487–2498

Lopez-Vazquez CM, Oehmen A, Hooijmans CM, Brdjanovic D, Gijzen HJ, Yuan Z, van Loosdrecht MCM (2009) Modeling the PAO-GAO competition: effects of carbon source, pH and temperature. Water Res 43(2):450–462

Morgenroth E, Milferstedt K (2009) Biofilm engineering: linking biofilm development at different length and time scales. Rev Environ Sci Biotechnol 8(3):203–208

Nielsen PH, Mielczarek AT, Kragelund C, Nielsen JL, Saunders AM, Kong Y, Hansen AA, Vollertsen J (2010) A conceptual ecosystem model of microbial communities in enhanced biological phosphorus removal plants. Water Res 44(17):5070–5088

Oehmen A, Carvalho G, Lopez-Vazquez CM, van Loosdrecht MCM, Reis MAM (2010) Incorporating microbial ecology into the metabolic modelling of polyphosphate accumulating organisms and glycogen accumulating organisms. Water Res 44(17):4992–5004

van Lier JB, Mahmoud N, Zeeman G (2008) Anaerobic wastewater treatment. In: Henze M, van Loosdrecht MCM, Ekama GA, Brdjanovic D (eds) Biological wastewater treatment: principles, modelling and design. IWA Publishing, London, pp 415–456

Weissbrodt DG, Holliger C (2014) Towards management of the bacterial resource for nutrient removal in granular sludge biofilm systems. In: Fatone et al (eds) Keynote lecture at the 2nd IWA specialized conference on ecotechnologies for wastewater treatment, University of Verona, Italy

Weissbrodt DG, Lochmatter S, Ebrahimi S, Rossi P, Maillard J, Holliger C (2012a) Bacterial selection during the formation of early-stage aerobic granules in wastewater treatment systems operated under wash-out dynamics. Front Microbiol 3:332

Weissbrodt DG, Shani N, Sinclair L, Lefebvre G, Rossi P, Maillard J, Rougemont J, Holliger C (2012b) PyroTRF-ID: a novel bioinformatics methodology for the affiliation of terminal-restriction fragments using 16S rRNA gene pyrosequencing data. BMC Microbiol 12:306

Weissbrodt DG, Neu TR, Kuhlicke U, Rappaz Y, Holliger C (2013a) Assessment of bacterial and structural dynamics in aerobic granular biofilms. Front Microbiol 4:175

Weissbrodt DG, Schneiter GS, Fürbringer JM, Holliger C (2013b) Identification of trigger factors selecting for polyphosphate- and glycogen-accumulating organisms in aerobic granular sludge sequencing batch reactors. Water Res 47(19):7006–7018

Weissbrodt DG, Derlon N, Wagner J, Holliger C, Morgenroth E (2014a) A consolidated approach of flocculent and granular sludge systems under the perspective of bacterial resource management. In: WEF (ed) Knowledge development forum "Floc versus granules—the ultimate match" of the Water Environment Federation's annual technical exhibition and conference (WEFTEC), New Orleans, LA, USA

Weissbrodt DG, Maillard J, Brovelli A, Chabrelie A, May J, Holliger C (2014b) Multilevel correlations in the biological phosphorus removal process: from bacterial enrichment to conductivity-based metabolic batch tests and polyphosphatase assays. Biotechnol Bioeng 111(12):2421–2435

Weissbrodt DG, Shani N, Holliger C (2014c) Linking bacterial population dynamics and nutrient removal in the granular sludge biofilm ecosystem engineered for wastewater treatment. FEMS Microbiol Ecol 88(3):579–595

Weissbrodt DG, Holliger C, Morgenroth E (2017) Modeling hydraulic transport and anaerobic uptake by PAOs and GAOs during wastewater feeding in EBPR granular sludge reactors. Biotechnol Bioeng 114(8):1688–1702

Winkler MKH, Meunier C, Henriet O, Mahillon J, Suarez-Ojeda ME, Del Moro G, De Sanctis M, Di Iaconi C, Weissbrodt DG (2018) An integrative review of granular sludge for the biological removal of nutrients and of recalcitrant organic matter from wastewater. Chem Eng J 336:489–502

Chapter 4
Infrastructure and Flexible Bioreactor Design for Experimental Research with Granular Sludge

Chemical engineering is an amalgam of engineering and art; biotechnology is an amalgam of engineering, art and affection.
(Ebrahimi 2005)

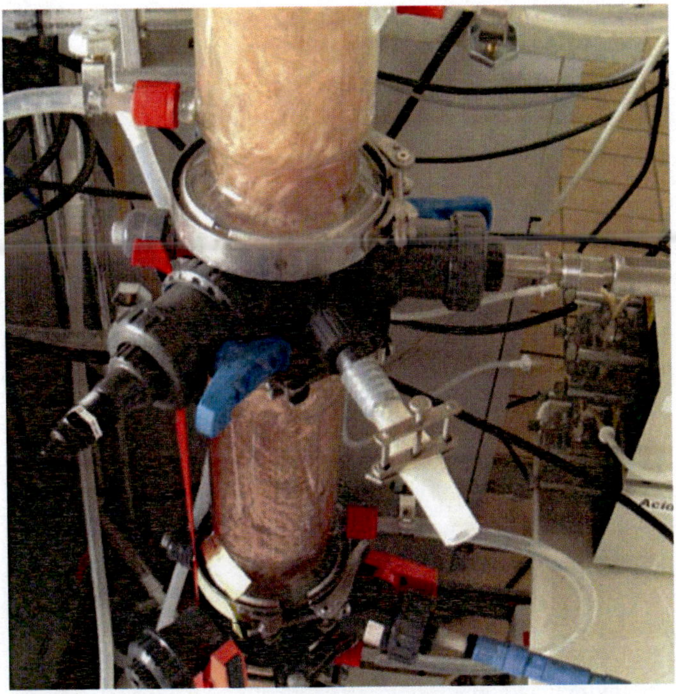

Bioreactor design

Supplementary Information The online version contains supplementary material available at https://doi.org/10.1007/978-3-031-41009-3_4.

D. G. Weissbrodt, *Engineering Granular Microbiomes*, Springer Theses,
https://doi.org/10.1007/978-3-031-41009-3_4

4.1 Introduction

Aerobic granular sludge (AGS) is advantageous for high-rate simultaneous biological nutrient removal, and secondary clarification in single sequencing batch reactors (SBR). Robust AGS-SBR operation relies on the maintenance of a high fraction of granular sludge in the system, and to prevent high proliferation of flocculent sludge (Liu and Tay 2012). Selection and maintenance of key bacterial populations that metabolize carbonaceous, nitrogenous and phosphorous nutrients is also required for efficient wastewater treatment.

The formation of AGS from flocculent activated sludge has been driven by the application of operation conditions selecting for a fast-settling biomass (Morgenroth et al. 1997; Beun et al. 1999). According to Liu and Tay (2012), growth kinetics of flocculent sludge exceed AGS ones. The fraction of flocculent sludge should be controlled at a low level either by increasing the growth of granular sludge or by selective discharge of flocs in order to promote granulation. Activated sludge flocs typically consist of a diameter of 5–500 μm, a specific gravity of 1.010–1.020 (up to 1.060 in dephosphatation processes), a settling velocity of 17–27 m day^{-1}, and a sludge volume index of < 80 to > 150 mL $g_{TSS}{}^{-1}$ (Dammel and Schroeder 1991; Grady et al. 1999; Schuler et al. 2001). Aerobic granules exhibit larger settling velocities than activated sludge flocs. Over about 40 studies referenced in literature, the following granules characteristics have been published from early-stage to mature AGS: granule diameter of 2.2 ± 1.7 mm (min–max 0.2–10 mm), specific particle gravity of 1.042 ± 0.020 (1.017–1.082), settling velocity of 52 ± 25 m h^{-1} (8–116 m h^{-1}), and sludge volume index (SVI) of 45 ± 26 mL $g_{TSS}{}^{-1}$ (11–130 mL $g_{VSS}{}^{-1}$). Winkler et al. (2012) have shown that the settling velocity of mature aerobic granules is a function of temperature (from 35–80 m h^{-1} at 5 °C to 60–140 m h^{-1} at 40 °C) and salt concentration of the wastewater (from 50–110 m h^{-1} at 5 g L^{-1} to 30–80 m h^{-1} at 40 g L^{-1}). Based on the settling characteristics of granules and flocs, granulation has been favored in lab-scale bubble-column SBRs designed with high height-to-diameter ratios (H/D) and operated with short settling times (2–6 min), and effluent withdrawal times (5 min) (Beun et al. 1999; Liu and Tay 2004; de Kreuk et al. 2005b). Additional combination with short hydraulic retention times (HRT; 6 h) and with up-flow superficial gas velocities (SGV) of 0.010–0.040 m s^{-1} has resulted in an increase in the bacterial hydrophobicity and in the formation of dense aerobic granules (Morgenroth et al. 1997; Beun et al. 1999; Liu and Tay 2002; Zima et al. 2007; Dulekgurgen et al. 2008).

The operation of activated sludge and AGS reactors in the SBR mode has been related to definite advantages (Wilderer and McSwain 2004). SBRs provide flexibility in reactor operation. The number of phases and their respective lengths in the SBR cycle can be tailored and adapted at every moment to dedicated selection of robust microbial communities, variations in influent loads, nutrient treatment requirements, and biomass settling objectives. The combination of SBR mode with an automated supervisory control and data acquisition (SCADA) interface can provide operators with an optimal remote control strategy. SBRs are operated with a sequence of batches that allow bacteria to work at their maximum growth rate and that lead to faster conversion rates of pollutants. Contrary to continuous-flow reactors where the influent is diluted in the reactor volume, SBRs work at initial highest concentrations that favor high-rate kinetics, and usually prevent the proliferation of filamentous bacteria. Such conditions lead to an increased bacterial hydrophobicity as well, and formation of a fast-settling biomass. "Spontaneous" granulation has been observed in traditional SBR systems (Dangcong et al. 1999; Beun et al. 2000; Wilderer and McSwain 2004; Mosquera-Corral et al. 2011).

Design, installation and optimization of a full SBR infrastructure was required upfront to the completion of the experimental research. A fleet of 8 bubble-column SBRs and 2 stirred-tank SBRs was developed for conducting granulation experiments and studies on bacterial competition in AGS, and cultivating enrichments of polyphosphate- (PAO) and glycogen-accumulating organisms (GAO), respectively. The progressive developments conducted on bubble-column and stirred-tank reactor designs are presented here, from simple and low-cost single-wall reactors to novel and highly flexible segmented double-wall reactors, together with the SCADA infrastructure. The practicability of bubble-column SBR designs is discussed along with troubleshooting during operation of new bioprocess SBR infrastructures, efficiency of the implementation of an influent de-oxygenation unit, and the use of an anaerobic buffer tank to foster an anaerobic selector in practice. Benefits from the flexibility of SBR processes are demonstrated.

4.2 Material and Methods

4.2.1 Bubble-Column Reactor Designs

Starting from the reactor designs used by Beun et al. (1999) and de Kreuk et al. (2005a) with H/D ratios between 15 and 30, five different bubble-column SBR designs were developed stepwise in function of target research objectives (Table 4.1).

4.2.1.1 Single-Wall PVC Designs

The simple and low-cost Design I was used for first investigations on granulation and on the underlying bacterial community dynamics (Weissbrodt et al. 2012/Chap. 7). A similar Design II with more holes for sensors and sampling was used in the second part of this study for detailed assessment of the impact of wash-out dynamics on bacterial selection during early-stage granulation, of shifts in predominant populations during maturation of AGS (Weissbrodt et al. 2013a/Chap. 8), and of the impact of the volatile fatty acid (VFA) composition on the PAO-GAO competition (Weissbrodt et al. 2013b/Chap. 10). On-line data acquisition of pH, dissolved oxygen (DO) and electrical conductivity was required for detailed monitoring. Design III comprised a

Table 4.1 Sequencing-batch reactor designs used in this research

Design parameters	Units	Design I	Design II	Design III	Design IV	Design V	Design VI
Reactor type[a]		BC	BC	BC	BC	BC	ST
Mixing[a]		Up-flow	Up-flow	Up-flow	Up-flow	Up-flow	Mechanical
Column type[b]		Single	Single	Single	Single	Segmented	–
Wall[c]		Single	Single	Single	Double	Double	Double
Material[c]		PVC	PVC	PVC	Glass	Glass	Glass
Temperature control[c]		No	No	No	Yes	Yes	Yes
Bottom[d]		Flat	Flat	Flat	Rounded	Rounded	Flat
D[e]	(mm)	52	52	80	55	59	130
H_{tot}[e]	(cm)	130	145	110	143	96–140	23
H_{tot}/D[e]	(–)	25	28	14	26	15–23	2
V_{tot}[f]	(L)	2.8	3.1	5.5	3.4	2.6–3.8	3.0
V[f]	(L)	2.1	2.5	4.75	2.5	1.75–3.2	2.0–2.5
Gas headspace volume[f]	(L)	0.7	0.6	0.8	0.9	0.6–0.9	0.6
Gas headspace volume fraction[f]	(%)	24	19	14	26	16–33	18
Gas headspace length[f]	(cm)	31	27	16	38	38–34	4
Gas headspace recirculation[f]		No	No	No	No/Yes	Yes	No
Secondary clarifier[g]		No	Yes	No	No/Yes	No/Yes	Yes

(continued)

Table 4.1 (continued)

Design parameters	Units	Design I	Design II	Design III	Design IV	Design V	Design VI
Influent de-oxygenation[h]		No	No	No	No/Yes	Yes	No
SBR automation[i]		TR	TR	TR	TR/SCADA	SCADA	SCADA
On-line data acquisition		No	Yes	Yes	No/Yes	Yes	Yes
Number of reactors	(–)	2	1	1	2	2	2
Used in Chapter		7	7, 8, 10	10	7, 9, 10	9, 10	6, 8

[a] AGS studies were mainly operated in bubble-column (*BC*) SBRs, whereas enrichments of PAO and GAO were cultivated in stirred-tank (*ST*) SBRs. Mixing was performed in bubble-columns with up-flow aeration, and in stirred-tank reactors with mechanical agitation

[b] The columns consisted either of one single piece with fixed volume, height and positions for sensors and effluent withdrawal, or of interchangeable segments allowing for flexible volumes, heights and positions

[c] The columns were constructed either with single-wall of polyvinylchloride (PVC) tubes, or with double-wall glass pieces enabling temperature control

[d] The bottom of the reactors was either flat or rounded. This parameter did apparently not impact on granulation

[e] Reactors with height-to-diameter (*H/D*) ratios of 15–30 have commonly been used in granulation studies

[f] Total (V_{tot}) and working (V) reactor volumes ranged between 2.8–5.5 L and 1.75–4.75, respectively. The gas headspaces volumes amounted to 0.6–1.1 L and to 14–40% of V_{tot}. Larger headspace volumes and lengths are preferred to protect gas headspace recirculation pumps against direct humidity

[g] Secondary clarifiers designed together with Lochmatter (2013) were used to measure the amount of biomass present in SBR effluents

[h] The tap water fed to dilute the concentrated nutrient media was de-oxygenated either by sonication or by stripping with dinitrogen gas in order to prevent introduction of DO in anaerobic feeding phases

[i] The SBR cycles were operated either with arrays of time relays (*TR*), or via full SCADA interfaces

4.75-L PVC column and was used to cultivate and maintain a high amount of fresh mature AGS for the inoculation of multifactorial experiments run to screen for main effects of operation parameters on bacterial community dynamics (Weissbrodt et al. 2013b/Chap. 10).

4.2.1.2 Double-Wall Glass Designs

Experiments were run with double-wall glass bubble-column reactor (Design IV) to achieve additional research objectives that required temperature control (Weissbrodt et al. 2012, 2013b, 2014b/Chaps. 7, 9 and 10). This design was used to investigate granulation at 20 and 30 °C, AGS maturation with acetate and propionate at 20 °C, and bacterial community dynamics at 20 °C under fluctuations of operation variables. A full SCADA interface was used for the latter objective for pH, DO and conductivity monitoring, and for DO regulation at target setpoints. A novel Design V comprising double-wall glass segments of different heights and short single wall PVC segments for probing and sampling was conceived for full flexibility in reactor construction in order to meet with target research objectives (Fig. 4.1). The PVC segments were equipped with open/close connections for fast insertion and removal of sensors at any time during SBR cycles. This was convenient for regular cleaning of sensors from surface biofilms that can lead to disturbances in measuring and control processes. Contrary to glass, the PVC material is advantageous for drilling as many holes as required for probing and sampling purposes, and to drill additional holes if required during the project. The double-wall segments were made in glass because of its transparency, its inertness, and its higher thermal conductivity (1.05 W m^{-1} K^{-1}) than PVC (0.19 W m^{-1} K^{-1}). Temperature control is therefore more efficient with glass material. This design was conceived in the years when plastic materials have been stressed to leach organic trace compounds that can exerce toxic effects in aquatic media (Howdeshell et al. 2003; Oehlmann et al. 2009; Santhi et al. 2012). Glass was therefore chosen in order to protect the biological reactor systems. The reactors were operated with volume exchange ratios of 50%. The influent was fed at the bottom of the bubble-columns through the settled sludge blanket. Gas was sparged at the bottom of the reactor by a cannula inserted from the top, and equipped with a porous plastic tube.

4.2.2 Stirred-Tank Reactor Design

Two stirred-tank SBRs were installed for cultivating enrichments of PAO and GAO in flocculent activated sludge. Each SBR consisted of a 3-L double-wall glass stirred-tank and of H/D ratios of 2, equipped with two 45-mm Rushton impellers and L-type gas aerators (Applikon Biotechnology, The Netherlands). The drain pipe was adapted with a self-built rounded anchor in order to prevent biomass losses from the settled sludge blanket during effluent withdrawal. The mechanical agitation system was

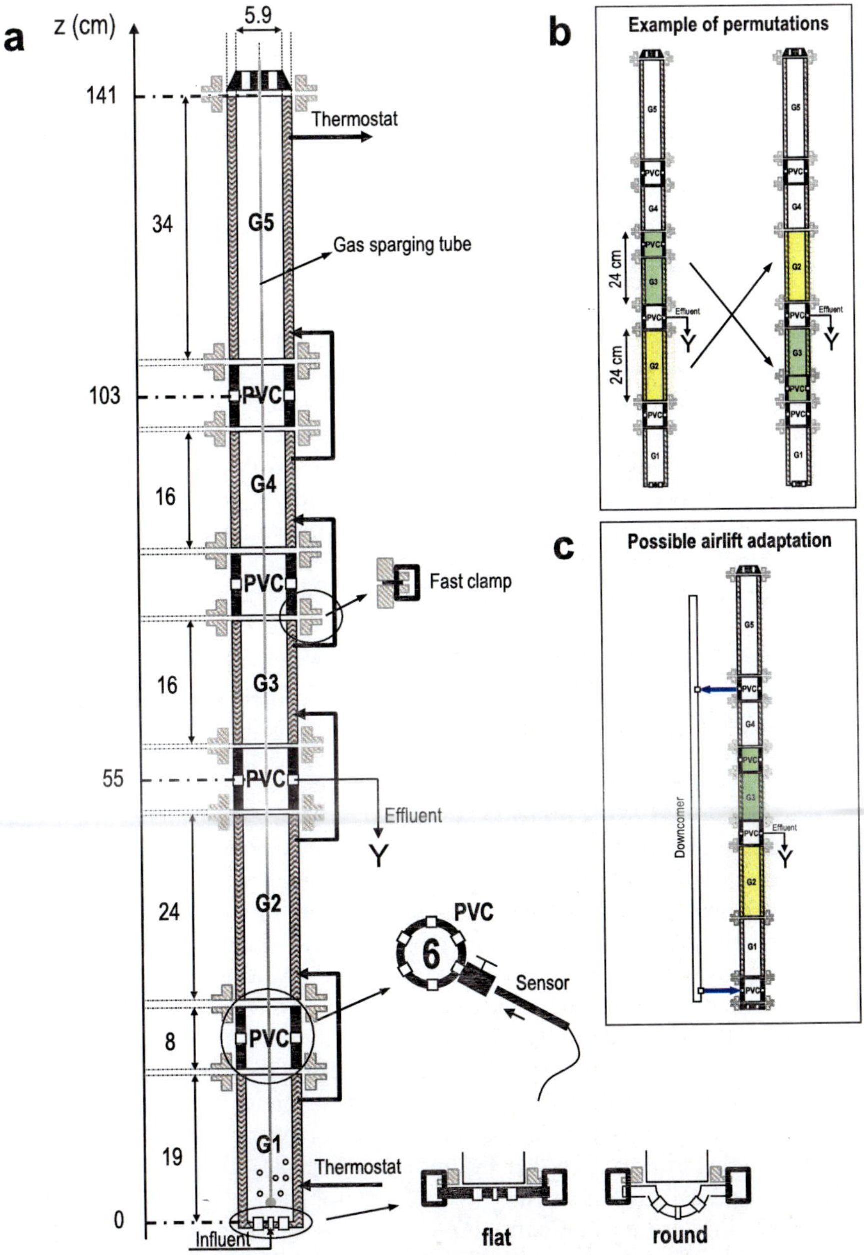

Fig. 4.1 Scheme of the novel bubble-column reactor Design V conceived for high flexibility in lab-scale experimental research on AGS (**a**). This design comprises double-wall glass segments of different heights (G1–G5), and single-wall PVC segments with 6 holes for introduction of sensors (3 open/close valves) and for connection of effluent pump, acid/base pumps, and sampling device (3 holes). Flat and round bottom plates were available if one aimed at investigating the effect of reactor geometry on the granulation process. Different reactor configurations, heights and volumes could be conceived by re-arrangement of the different types of segments (**b**). This design enabled connection of a downcomer if one aimed at operating external airlift reactor operation (**c**)

modified with a RZR 2051 control stirrer (Heidolph Instruments, Germany) and a self-made long axis for a robust long-term continuous SBR operation (> 300 days). The influent wastewater was fed from the top of the SBRs with volume exchange ratios of 50%.

4.2.3 Implementation of Sequencing Batch Reactor Operations

The SBRs cycles consisted of different kinds of sequences depending on research objectives, namely (i) pulse (6 min) or anaerobic (60 min) feeding, aeration, settling, and withdrawal for granulation studies (Weissbrodt et al. 2012, 2013a, 2014b/ Chaps. 7, 8 and 9), (ii) N_2-flush, pulse feeding, N_2-flush, anaerobic batch, aeration, settling, and withdrawal for PAO/GAO enrichments (Weissbrodt et al. 2013a, 2014a/Chaps. 6 and 8), and (iii) N_2-flush, pulse feeding, N_2-flush, anaerobic batch, oxic or anoxic starvation, settling, and withdrawal for multifactorial experiments on mature granular sludge (Weissbrodt et al. 2013b/Chap. 10).

The cycles of the bubble-column SBRs built with Designs I-III were implemented with arrays of multifunctional time delay relays C55 (Comat, Switzerland). Such arrays of relays are convenient for fast and low-cost implementation without computational requirements, but do not enable process control loops. When required, on-line data acquisition was performed with an Ecograph C paperless recorder (Endress+Hauser, E+H, Switzerland), or with a U12 data acquisition device (Labjack, USA) connected to a data-logging DAQFactory computer interface (Azeotech, USA). For reactor Designs IV–VI, a full SCADA interface was developed for full monitoring and control of SBR cycles. Coding was done in DAQFactory installed on a standard laptop computer (3 GHz Intel i7 CPU, Microsoft Windows XP operating system). Coding and optimization in DAQFactory are relatively user-friendly, but nevertheless require time. The SBR phases were programmed as a cascade of interconnected individual programmed entities, which enabled flexibility in the SBR operation. This modus operandi provided higher degrees of freedom for manual prolongation, shortening or stopping of a particular phase at any time during SBR cycles. Every order was doubled with a delay of 0.2 s in order to prevent losses in the network. All computers, Wago, Siemens and Fieldpoint modules were connected to an uninterruptible power supply (UPS) APC Smart-UPS1400 (American Power Conversion, USA) providing emergency power and buffering dropouts.

For bubble-column SBRs, the orders were supplied via an Object-linking-and-embedding-for-Process-Control (OPC) server comprising an Ethernet TCP/IP Programmable Fieldbus Controller 750-841 10/100 Mbit/s and a Binary Spacer Module (Wago-I/O-System750, Wago, Switzerland). Plug-in relays LZX PT270024 (Siemens, Germany) were actuated in functions of the programmed orders. For stirred-tank SBRs, only a limited number of four FP-relays FP-RLY-420 were available per reactor on a programmable logic controller (PLC) FieldPoint P-2000

Ethernet Controller Module (National Instruments, USA). Intermediary multifunctional time delay relays C55 were thus used to multiply actuation possibilities. The first FP-relay was activating a time delay relay switching from dinitrogen supply to influent wastewater supply. The second FP-relay was activating mechanical stirring, and was enabling pH regulation after 30 s of initial mixing via a second time delay relay. The third FP-relay was activating air supply. The fourth FP-relay was activating the effluent pump for mixed liquor purge and for effluent withdrawal. The FieldPoint PLC module was in addition composed of a FP-AI-110 analog input array for data acquisition.

4.2.4 On-Line Sensors and Amplifiers

DO was recorded with 12 mm × 12 cm (bubble columns) or 12 mm × 22 cm (stirred tanks) membrane sensors Ingold (Mettler Toledo, Switzerland) or COS21-1K0 (E+H), connected to Liquisys M DO amplifiers (E+H). Regular cleaning of DO sensors from surface biofilms once per week was required for accurate measurements and regulation.

pH was recorded in bubble columns with 12 mm × 12 cm liquid electrolyte glass electrodes InLab Routine Pro (Mettler Toledo) or ion-sensitive field-effect transistor (ISFET) electrodes CPS471-2ESB1 (E+H), and in stirred tanks with long 12 mm × 22 cm gel electrolyte electrodes S8/225 (Mettler Toledo). The pH signals were recorded with Liquisys M pH-redox amplifiers (E+H). Glass electrodes were advantageous for their high measuring surface area, but required the regular addition of liquid electrolyte. ISFET electrodes did not require addition of electrolyte and could be placed in every direction, but comprised an only small measuring surface area that required to be cleaned twice a week from surface biofilm. Glass electrodes were mainly used in reactors with fixed sensor holes, whereas ISFET electrodes were only used in reactors equipped with open/close valves for fast maintenance. Regulation of pH was actuated by the amplifiers that comprised proportional-integral controllers.

Electrical conductivity was recorded with 12 mm × 12 cm Conductivity High Purity Water Cells with 2-electrodes system Type 202922 (10/cm cell constant, max 200 mS/cm) connected to ecoTRANS Lf03 amplifiers (Jumo, Germany). Long 12-mm electrical conductivity sensors were not available on the market. The same electrodes as the ones used for bubble columns were inserted below cover plate of the stirred tanks with a self-built muff-arm to reach the bottom half liquid phases.

4.2.5 Influent De-oxygenation Unit

Granular sludge SBRs operated for full BNR are optimally operated with a slow up-flow feeding regime of influent wastewater under anaerobic conditions across the settled bed of granules (Weissbrodt et al. 2017/Chap. 11). This feeding regime has previously been called "anaerobic feeding" (de Kreuk and van Loosdrecht 2004).

For this purpose, a de-oxygenation unit was designed and optimized to remove residual DO present in the tap water used to dilute the cultivation media (Additional File 4.2a in the Supplementary Information). This unit consisted of PVC column sparged with dinitrogen gas to strip DO from the liquid phase during reactor feeding (controlled via SCADA).

4.3 Results and Discussion

4.3.1 Practicability of Bubble-Column SBR Designs

Different single-wall PVC and double-wall glass bubble-column designs were used in function of target research objectives (Table 4.1). The pros and cons of these systems is presented hereafter, in terms of impact of reactor material and flexibility.

PVC designs I, II and III were convenient for fast, home-made, and low-cost process implementation. Their main disadvantage was that thermostatization and SCADA could not be implemented.

Double-wall glass columns were efficient for temperature control. The main disadvantages of glass were related to high cost and to the need of external glassmaker competencies for manufacturing the columns. This resulted in extended time delays (6 months) for the installation of the reactor systems. Glass is also breakable, and handling must be conducted with precautions. A relatively low number of sensors holes can be manufactured in glass columns. Once the glass piece is manufactured, additional modifications and holes can hardly be done.

The segmented double-wall glass column Design V (Fig. 4.1) and the full SCADA infrastructure enabled high flexibility for tailoring installation schemes to target research objectives. Small PVC sections with open/close valves for sensors displayed high practicability. Maintenance and calibration of sensors was possible at any moment during SBR cycles. Optimal control strategy on dissolved oxygen (DO) was enabled as well.

4.3.2 *Troubleshooting During Operation of New Bioprocess SBR Infrastructures*

The operation of newly installed SBR infrastructures for bioprocess research required organisation, precaution, pragmatic handling, and patience. This was in agreement with Ebrahimi (2005) who reported that biotechnology is an amalgam of engineering, art, and affection.

The Additional File 4.1 in the Supplementary Information summarizes main issues related to the handling of biological SBRs over periods of long term continuous operations of typically 1–2 years. Although it is tempting to reduce them as Murphy's Law effect, improved communication on experienced experimental issues is beneficial for principal investigators, lab mates and for the scientific community, and particularly for technology transfer.

Communicating risks is probably less appealing than communicating success stories. However, energy savings in optimization and troubleshooting can obviously result in an increased number of research successes. Catabolic energy can be saved for the production of new articles.

Operation of SBR infrastructures was new in the lab where the research project was conducted, and therefore energy was spent for process design, construction, optimization, and troubleshooting. It was thus intended here to provide information on optimization together with research advances.

4.3.3 *Efficiency of the Influent De-oxygenation Unit*

Depending on feeding conditions, DO can be present in the influent and might hamper anaerobic conversions during the plug-flow feeding phase. The installation and use of the influent de-oxygenation pre-column was efficient to strip the residual DO present in the tap water and to lead to full anaerobic conditions during the feeding phase of granular sludge SBRs operated under anaerobic slow up-flow feeding regime (Weissbrodt et al. 2012, 2013a, 2013b, 2014b).

DO stripping equipment is mostly useful for fundamental studies at bench scale, but can nonetheless hardly be conceived for implementation in practice at pilot or full scales. In such cases, residual DO is transported with the influent wastewater into the reactor and most probably rapidly depleted along the first sections of the settled bed of granular sludge, while the Mathematical modelling computations went in this direction (Weissbrodt et al. 2017/Chap. 11). Some results are preliminarily presented here shortly. Simulations run at worst case scenario with 9 $mg_{O2}\ L_{Inf}^{-1}$ in the influent showed that the presence of DO had a negligible effect on anaerobic conversions of acetate uptake. In the simulated case of a 30-cm bed of mature AGS comprising 1.75 $g_{TSS}\ cm^{-1}$ of biomass along the bed height and of a feeding condition with at least 300 $mg_{COD}\ L_{Inf}^{-1}$, DO was rapidly depleted in the first 2 cm. This meant that the 93% of the remaining height of the bed was fully anaerobic. By assuming a growth yield of aerobic ordinary heterotrophic organisms (OHOs) on acetate of 0.64 g_{CODx} $g_{COD,Ac}{}^{-1}$ (Siegrist et al. 2002), 25 $mg_{COD,Ac}$ L^{-1} of acetate were removed with 9 mg_{O2} L^{-1}. Depending on the loading conditions, this may or may not lead to deteriorated anaerobic conversions in the bed, i.e. in the case of low-loaded (< 200 mg_{COD} $L_{Inf}{}^{-1}$) and high-loaded (> 300 mg_{COD} $L_{Inf}{}^{-1}$) pre-settled wastewater.

4.3.4 The Use of an Anaerobic Buffer Tank in Practice

Such as described above, de-oxygenation stripping units in practice are hardly applicable. The use of an anaerobic buffer tank could nevertheless be very opportune to remove residual dissolved oxygen from the influent wastewater prior to feeding to the reactor.

Such technical measure can in addition be interestingly used (*i*) to equalize the flow rates of wastewater and (*ii*) to pre-acidify the wastewater, namely the transformation of particulate organic matter (X_S) into dissolved organics (S_s) and their fermentation into volatile fatty acids (S_{VFA}) such as acetate and propionate alkanoates that can preferentially be taken up by polyphosphate-accumulating organisms (PAOs) under anaerobic slow feeding conditions across the settled bed of granular sludge. The hydraulic retention time in the buffer tank can easily be mastered to adequately promote the production of VFAs.

An anaerobic buffer tank can therefore constitute a very simple and low-cost measure to favour the selection of PAOs and the establishment of EBPR in AGS-SBRs operated for nutrient removal from the complex matrix of real municipal wastewater. An implementation example is provided in Additional File 4.2b, c in the Supplementary Information.

4.4 Conclusions

A pragmatic engineering approach was presented for the design, analysis, and modelling of bubble-column SBR systems used in AGS research, and led to the following conclusions:

- A novel bubble-column SBR design consisting of double-wall segments and sensor holder segments was efficient for high flexibility in experimental AGS research.
- A full SCADA interface was optimal for enhanced process control and monitoring, and for detailed and real-time data acquisition on microbial activities.

Acknowledgements Jean-Pierre Kradolfer, Sirous Ebrahimi (EPFL), and Jonathan May (master student from AgroSup Dijon, France) for collaboration on the design and construction of SBRs. Marc Deront, Simon Taillard, and Graciela Gonzalez-Gil (EPFL) for collaboration on the design and implementation of the SCADA interface. Yoan Rappaz and Guillaume Schneiter (EPFL) for excellent assistance in reactor operation.

Supplementary Information

Additional File 4.1 Description of main issues during the operation of biological SBRs.

Additional File 4.2 Examples of a lab-scale unit for de-oxygenating the influent wastewater, and of an anaerobic buffer tank to pre-acidify the influent wastewater by hydrolysis and fermentation of particulate organic matter into dissolved VFA prior to feeding into the AGS-SBR.

References

Beun JJ, Hendriks A, van Loosdrecht MCM, Morgenroth E, Wilderer PA, Heijnen JJ (1999) Aerobic granulation in a sequencing batch reactor. Water Res 33(10):2283–2290

Beun JJ, van Loosdrecht MCM, Heijnen JJ (2000) Aerobic granulation. Water Sci Technol 41(4–5):41–48

Dammel EE, Schroeder ED (1991) Density of activated sludge solids. Water Res 25(7):841–846

Dangcong P, Bernet N, Delgenes JP, Moletta R (1999) Aerobic granular sludge—a case report. Water Res 33:890–893

de Kreuk MK, van Loosdrecht MCM (2004) Selection of slow growing organisms as a means for improving aerobic granular sludge stability. Water Sci Technol 49(11–12):9–17

de Kreuk MK, Heijnen JJ, van Loosdrecht MCM (2005a) Simultaneous COD, nitrogen, and phosphate removal by aerobic granular sludge. Biotechnol Bioeng 90(6):761–769

de Kreuk MK, Pronk M, van Loosdrecht MCM (2005b) Formation of aerobic granules and conversion processes in an aerobic granular sludge reactor at moderate and low temperatures. Water Res 39(18):4476–4484

Dulekgurgen E, Artan N, Orhon D, Wilderer PA (2008) How does shear affect aggregation in granular sludge sequencing batch reactors? Relations between shear, hydrophobicity, and extracellular polymeric substances. Water Sci Technol 58(2):267–276

Ebrahimi S (2005) Biotreatment of SO_2 and H_2S contaminated gas streams. Ph.D. thesis, Delft University of Technology

Grady CPLJ, Daigger GT, Lim HC (1999) Biological wastewater treatment, 2nd edn. Marcel Dekker, New York

Howdeshell KL, Peterman PH, Judy BM, Taylor JA, Orazio CE, Ruhlen RL, vom Saal FS, Welshons WV (2003) Bisphenol A is released from used polycarbonate animal cages into water at room temperature. Environ Health Perspect 111(9):1180–1187

Liu Y, Tay J-H (2002) The essential role of hydrodynamic shear force in the formation of biofilm and granular sludge. Water Res 36(7):1653–1665

Liu Y, Tay J-H (2004) State of the art of biogranulation technology for wastewater treatment. Biotechnol Adv 22(7):533–563
Liu Y-Q, Tay J-H (2012) The competition between flocculent sludge and aerobic granules during the long-term operation period of granular sludge sequencing batch reactor. Environ Technol 1–8
Lochmatter S (2013) Optimization of reactor startup and nitrogen removal of aerobic granular sludge systems. PhD thesis No. 5879, Ecole Polytechnique Fédérale de Lausanne
Morgenroth E, Sherden T, van Loosdrecht MCM, Heijnen JJ, Wilderer PA (1997) Aerobic granular sludge in a sequencing batch reactor. Water Res 31(12):3191–3194
Mosquera-Corral A, Arrojo B, Figueroa M, Campos JL, Mendez R (2011) Aerobic granulation in a mechanical stirred SBR: treatment of low organic loads. Water Sci Technol 64(1):155–161
Oehlmann J, Schulte-Oehlmann U, Kloas W, Jagnytsch O, Lutz I, Kusk KO, Wollenberger L, Santos EM, Paull GC, VanLook KJW, Tyler CR (2009) A critical analysis of the biological impacts of plasticizers on wildlife. Philos Trans Roy Soc B Biol Sci 364(1526):2047–2062
Santhi VA, Sakai N, Ahmad ED, Mustafa AM (2012) Occurrence of bisphenol A in surface water, drinking water and plasma from Malaysia with exposure assessment from consumption of drinking water. Sci Total Environ 427–428:332–338
Schuler AJ, Jenkins D, Ronen P (2001) Microbial storage products, biomass density, and setting properties of enhanced biological phosphorus removal activated sludge. Water Sci Technol 43(1):173–180
Siegrist H, Rieger L, Koch G, Kuhni M, Gujer W (2002) The EAWAG Bio-P module for activated sludge model No. 3. Water Sci Technol 45 (6):61–76
Weissbrodt DG, Lochmatter S, Ebrahimi S, Rossi P, Maillard J, Holliger C (2012) Bacterial selection during the formation of early-stage aerobic granules in wastewater treatment systems operated under wash-out dynamics. Front Microbiol 3:332
Weissbrodt DG, Neu TR, Kuhlicke U, Rappaz Y, Holliger C (2013a) Assessment of bacterial and structural dynamics in aerobic granular biofilms. Front Microbiol 4:175
Weissbrodt DG, Schneiter GS, Fürbringer JM, Holliger C (2013b) Identification of trigger factors selecting for polyphosphate- and glycogen-accumulating organisms in aerobic granular sludge sequencing batch reactors. Water Res 47(19):7006–7018
Weissbrodt DG, Maillard J, Brovelli A, Chabrelie A, May J, Holliger C (2014a) Multilevel correlations in the biological phosphorus removal process: from bacterial enrichment to conductivity-based metabolic batch tests and polyphosphatase assays. Biotechnol Bioeng 111(12):2421–2435
Weissbrodt DG, Shani N, Holliger C (2014b) Linking bacterial population dynamics and nutrient removal in the granular sludge biofilm ecosystem engineered for wastewater treatment. FEMS Microbiol Ecol 88(3):579–595
Weissbrodt DG, Holliger C, Morgenroth E (2017) Modeling hydraulic transport and anaerobic uptake by PAOs and GAOs during wastewater feeding in EBPR granular sludge reactors. Biotechnol Bioeng 114(8):1688–1702
Wilderer PA, McSwain BS (2004) The SBR and its biofilm application potentials. Water Sci Technol 50(10):1–10
Winkler MKH, Bassin JP, Kleerebezem R, van der Lans RGJM, van Loosdrecht MCM (2012) Temperature and salt effects on settling velocity in granular sludge technology. Water Res 46(12):3897–3902
Zima BE, Diez L, Kowalczyk W, Delgado A (2007) Sequencing batch reactor (SBR) as optimal method for production of granular activated sludge (GAS)—fluid dynamic investigations. Water Sci Technol 55(8–9):151–158

Chapter 5
PyroTRF-ID: A Bioinformatics Methodology for Profiling Microbiomes with T-RLFP and Amplicon Sequencing Data

It is an exciting time to be a molecular ecologist. Comprehensive evaluation of microbial diversity in almost any environment is now possible.
Adapted from Glenn (2011) and Sun et al. (2011)

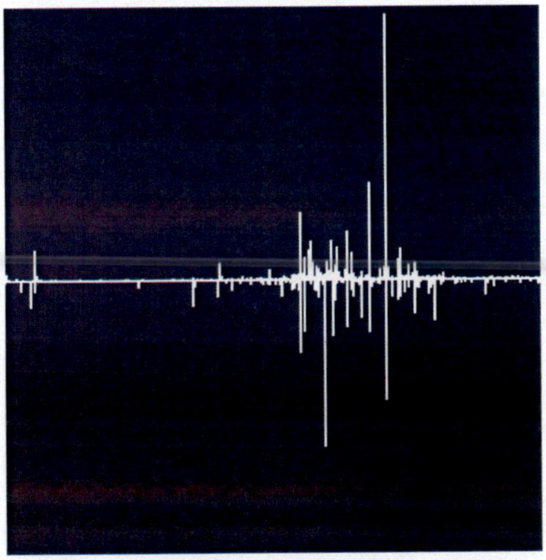

Profiling microbiomes

The content of this chapter was published in a modified version in: Weissbrodt DG, Shani N, Sinclair L, Lefebvre G, Rossi P, Maillard J, Rougemont J, Holliger C (2012) PyroTRF-ID: a novel bioinformatics methodology for the affiliation of terminal-restriction fragments using 16S rRNA gene pyrosequencing data. BMC Microbiol 12:306. https://doi.org/10.1186/1471-2180-12-306. Permission was granted to reuse the figure materials (© 2012 BioMed Central Ltd., Springer Nature).

Supplementary Information The online version contains supplementary material available at https://doi.org/10.1007/978-3-031-41009-3_5.

D. G. Weissbrodt, *Engineering Granular Microbiomes*, Springer Theses,
https://doi.org/10.1007/978-3-031-41009-3_5

5.1 Introduction

Molecular microbial ecology has become an important discipline in natural and medical sciences. Research on the structure, dynamics and evolution of microbial communities in environmental, human, and engineered systems provides substantial scientific knowledge for understanding the underlying microbial processes, for predicting their behavior, and for controlling, favoring, or suppressing target populations (Kent et al. 2007; Mazzola 2004).

Different analytical methods have been successively developed for the assessment of microbial communities via profiling or metagenomic approaches (Gu et al. 2011). Terminal-restriction fragment length polymorphism (T-RFLP) analysis has been widely used over the last decade for culture-independent assessment of complex microbial community structures (Marsh 1999; Schutte et al. 2008). Standardized, robust, and highly reproducible T-RFLP has become the method of choice for community fingerprinting since its automation in capillary electrophoresis devices has enabled the simultaneous analysis of numerous samples at relatively low cost (Militsopoulou et al. 2002; Rossi et al. 2009; Thies 2007). Cloning and sequencing methods have been optimized in parallel for taxonomic affiliation of terminal-restriction fragments (T-RF) (Grant and Ogilvie 2004; Mengoni et al. 2002). This approach however remains time-consuming and often leads to only partial characterization of the apparent microbial diversity (Mao et al. 2011). On the other hand, next-generation sequencing (NGS) technologies have recently been applied for comprehensive high-throughput analyses of microbiomes with reduced sequencing costs (Petrosino et al. 2009; Roesch et al. 2007; Ronaghi 2001; Sun et al. 2011; Wommack et al. 2008) and high reproducibility (Pilloni et al. 2012). Metagenomics projects have however generated novel requirements in resource and expertise for generating, processing, and interpreting large datasets (Desai et al. 2012; Edwards 2008; Glenn 2011; Kunin et al. 2008; Rodriguez-Ezpeleta et al. 2012; Trombetti et al. 2007). Overall, 'omics' technologies challenge the field of bioinformatics to design tailored computing solutions for enhanced production of scientific knowledge from massive datasets. While NGS techniques tend to progressively replace the traditional combination of T-RFLP and cloning-sequencing, recent studies have demonstrated the benefits of using both techniques to complement each other (Camarinha-Silva et al. 2012; Collins and Rocap 2007; Hume et al. 2011; Jakobsson et al. 2010; Mushegian et al. 2011). The combination of routine T-RFLP and NGS strategies could offer an efficient trade-off between laboratory efforts required for the in-depth analysis of bacterial communities and the financial and infrastructural costs related to datasets processing.

If T-RFLP and NGS are meant to be used concomitantly for the investigation of microbial systems, one key objective is to link T-RFs to phylotypes. In parallel to early developments of T-RFLP methods, several computational procedures have been proposed to predict T-RF sizes and to phylogenetically affiliate T-RFs. For instance, TAP T-RFLP (Marsh et al. 2000), TRiFLe (Junier et al. 2008) and T-RFPred (Fernandez-Guerra et al. 2010) have been developed to perform in silico digestion of datasets of 16S rRNA gene sequences, originating mostly from clone libraries or reference public databases. REPK (Collins and Rocap 2007) has been designed to screen for single and combinations of restriction enzymes for the optimization of T-RFLP profiles, and to design experimental strategies. All these programs do not involve comparison of in silico profiles with experimental data.

In the current study, we propose a novel bioinformatics methodology, called PyroTRF-ID, to assign phylogenetic affiliations to experimental T-RFs by coupling pyrosequencing and T-RFLP datasets obtained from the same biological samples. A recent study showing that natural bacterial community structures analyzed with both techniques were very similar (Pilloni et al. 2012) strengthened the here adopted conceptual approach. The methodological objectives were to generate digital T-RFLP (dT-RFLP) profiles from full pyrosequencing datasets, to cross-correlate them to the experimental T-RFLP (eT-RFLP) profiles, and to affiliate eT-RFs to closest bacterial relatives, in a fully automated procedure. The effects of different processing algorithms are discussed. An additional functionality was developed to assess the impact of restriction enzymes on resolution and representativeness of T-RFLP profiles. Validation was conducted with high- and low-complexity bacterial communities. This dual methodology was meant to process single DNA extracts in T-RFLP and pyrosequencing with similar PCR conditions, and therefore aimed to preserve the original microbial complexity of the investigated samples.

5.2 Material and Methods

5.2.1 Biological Samples

Two different biological systems were used for analytical procedure validation. The first set comprised ten groundwater (GRW) samples from two different chloroethene-contaminated aquifers that have been previously described by Aeppli et al. (2010) and Shani (2012). The second set consisted of five aerobic granular sludge (AGS) biofilm samples from anaerobic-aerobic sequencing batch reactors operated for full biological nutrient removal from an acetate-based synthetic wastewater. The AGS system has been described previously (Weissbrodt et al. 2012a) and displayed a lower bacterial community complexity (richness of 42 ± 6 eT-RFs, Shannon's H′ diversity of 2.5 ± 0.2) than the GRW samples (richness of 67 ± 15 eT-RFs, Shannon's H′ diversity of 3.3 ± 0.5).

5.2.2 DNA Extraction

GRW samples were filtered through 0.2-µm autoclaved polycarbonate membranes (Isopore™ Membrane Filters, Millipore) with a mobile filtration system (Filter

Funnel Manifolds, Pall Corporation). DNA was extracted using the PowerSoil™ DNA Extraction Kit (Mo-Bio Laboratories, Inc.) following the manufacturer instructions, except that the samples were processed in a bead-beater (Fastprep FP120, Bio101) at 4.5 m s^{-1} for 30 s after the addition of solution C1.

DNA from AGS samples was extracted with the automated Maxwell 16 Tissue DNA Purification System (Promega, Duebendorf, Switzerland) according to manufacturer's instructions with following modifications. An aliquot of 100 mg of ground granular sludge was preliminarily digested during 1 h at 37 °C in 500 μL of a solution composed of 5 mg mL^{-1} lysozyme in TE buffer (10 mM Tris-HCl, 0.1 mM EDTA, pH 7.5). The DNA extracts were resuspended in 300 μL of TE buffer.

All extracted DNA samples were quantified with the ND-1000 Nanodrop® spectrophoto-meter (Thermo Fisher Scientific, USA) and stored at −20 °C until analysis.

5.2.3 Experimental T-RFLP

The eT-RFLP analysis of the GRW series was done according to Rossi et al. (2009) with following modifications: (i) 30 μL PCR reactions contained 3 μL 10× Y buffer, 2.4 μL 10 mM dNTPs, 1.5 μL of each primer at 10 μM, 6 μL 5× enhancer P solution, 1.5 U PeqGold Taq polymerase (Peqlab), and 0.2 ng μL^{-1} template DNA (final concentration), completed with autoclaved and UV-treated Milli-Q water (Millipore, USA); (ii) for each DNA extract, PCR amplification was carried out in triplicate. Samples from the AGS series were analyzed by eT-RFLP according to Ebrahimi et al. (2010) with following modifications: (i) Go Taq polymerase (Promega, Switzerland) was used for PCR amplification; (ii) forward primer was FAM-labeled; (iii) the PCR program was modified to increase the initial denaturation to 10 min, the cycle denaturation step to 1 min, and 30 cycles of amplification. All PCRs were carried out using the labeled forward primer 8f (FAM-5′-AGAGTTTGATCMTGGCTCAG-3′) and the reverse primer 518r (5′-ATTACCGCGGCTGCTGG-3′). For details, refer to Weissbrodt et al. (2012a).

The resulting eT-RFLP profiles were generated between 50 and 500 bp as described in (2009). The eT-RFLP profiles were aligned using the Treeflap crosstab macro (Rees et al. 2004) and expressed as relative contributions of operational taxonomic units (OTUs). For GRW samples which exhibited numerous low abundant OTUs, the final bacterial community datasets were constructed as follows: multivariate Ruzicka dissimilarities were computed between replicates of eT-RFLP profiles with R (R Development Core Team 2008) and the additional package Vegan (Oksanen et al. 2009); the profile at the centroid (i.e. displaying the lowest dissimilarity with its replicates) was selected for each sample to build the final community profiles. For AGS samples which were characterized by less complex communities, triplicates were periodically measured and resulted in a mean relative standard coefficient of 6% over the analytical method.

5.2.4 *Cloning and Sequencing*

Clone libraries were constructed with the 16S rRNA gene pool amplified from DNA samples using the same PCR procedures as described in the eT-RFLP method but with an unlabeled 8f primer.

The PCR products were purified with the purification kit Montage® PCR Centrifugal Filter Devices (Millipore, USA), ligated into pGEM®-T Easy vector (Promega, USA) and transformed into *E. coli* XL1-Blue competent cells (Agilent Technologies, USA).

The eT-RFLP procedure was applied on isolated colonies in order to screen for the dominant eT-RFs obtained previously by eT-RFLP on the entire 16S rRNA gene pool.

The 16S rRNA gene was then amplified from selected colonies using PCR with primers T7 and SP6 (Promega, USA) and purified as described above.

A sequencing reaction was carried out on each purified PCR product as described in Regeard et al. (2004). Sequences were aligned in BioEdit (Hall 1999), and primer sequences were removed. Sequences were analyzed for chimeras using Bellerophon (Huber et al. 2004), and dT-RFs of selected clones were produced by in silico digestion using TRiFLe (Junier et al. 2008) for comparison with eT-RFs.

5.2.5 *High-Throughput Amplicon Sequencing*

A total of 15 biological samples were analyzed using bacterial tag encoded FLX amplicon pyrosequencing analysis. A first set of DNA extracts from GRW and AGS samples were sent for sequencing to Research and Testing Laboratory LLC (Lubbock, TX, USA). The samples underwent partial amplification of the V1–V3 region of the 16S rRNA gene by PCR with unlabeled 8f and 518r primers, secondary PCR with tagged fusion primers for FLX amplicon sequencing, emulsion-based clonal amplification (emPCR), and GS FLX sequencing targeting at least 3’000 reads with the 454 GS-FLX Titanium Genome Sequencing System technology (Roche, Switzerland). The whole sample preparation protocol has been made available by the company in the publication of Sun et al. (2011). This series refers, in the present study, to the low reads amount pyrosequencing procedure (LowRA). The DNA extract of one AGS sample was analyzed in triplicate through the whole analytical method from pyrosequencing (LowRA) to PyroTRF-ID analysis.

A second set of amplicons from different GRW samples was analyzed by GATC Biotech AG (Konstanz, Germany) following an analog procedure but targeting at least 10'000 reads (referred to as the high reads amount method, HighRA, hereafter). The A- and B-adapters for sequencing with the Roche technology were ligated to the ends of the DNA fragments. The samples were run on a 2% agarose gel with TAE buffer and the band in a size range of 700–900 bp, 450–650 bp, or 100–500 bp, respectively, was excised and column purified. After concentration measurement the differently tagged libraries were pooled. The three resulting library pools were immobilized onto DNA capture beads and the amplicon-beads obtained were amplified through emPCR according to the manufacturer's recommendations. Following amplification, the emulsion was chemically broken and the beads carrying the amplified DNA library were recovered and washed by filtration. Each pool was sequenced on a quarter GS FLX Pico-Titer plate device with GS FLX Titanium XLR70 chemistry on a GS FLX+ Instrument. The GS FLX System Software Version 2.6 was used and the GS FLX produced the sequence data as Standard Flowgram Format (SFF) file containing flowgrams for each read with basecalls and per-base quality scores.

5.2.6 *Development of the PyroTRF-ID Bioinformatics Methodology*

The PyroTRF-ID bioinformatics methodology for identification of T-RFs from pyrosequenc-ing datasets was coded in Python for compatibility with the *BioLinux* open software strategy (Field et al. 2006). PyroTRF-ID jobs were submitted to the Vital-IT high performance computing center (HPCC) of the Swiss Institute of Bioinformatics (Switzerland). All documentation needed for implementing the methodology is available in open access.[1] The flowchart description of PyroTRF-ID is depicted in Fig. 5.1, and computational parameters are described hereafter.

5.2.6.1 Input Files: Combining Amplicon Sequencing and T-RFLP Datasets

Input 454 tag-encoded pyrosequencing datasets were used either in raw standard flowgram (.sff), or as pre-denoised fasta format (.fasta) as presented below. Input eT-RFLP datasets were provided in coma-separated-values format (.csv).

[1] Electronic web link: http://bbcf.epfl.ch/PyroTRF-ID/.

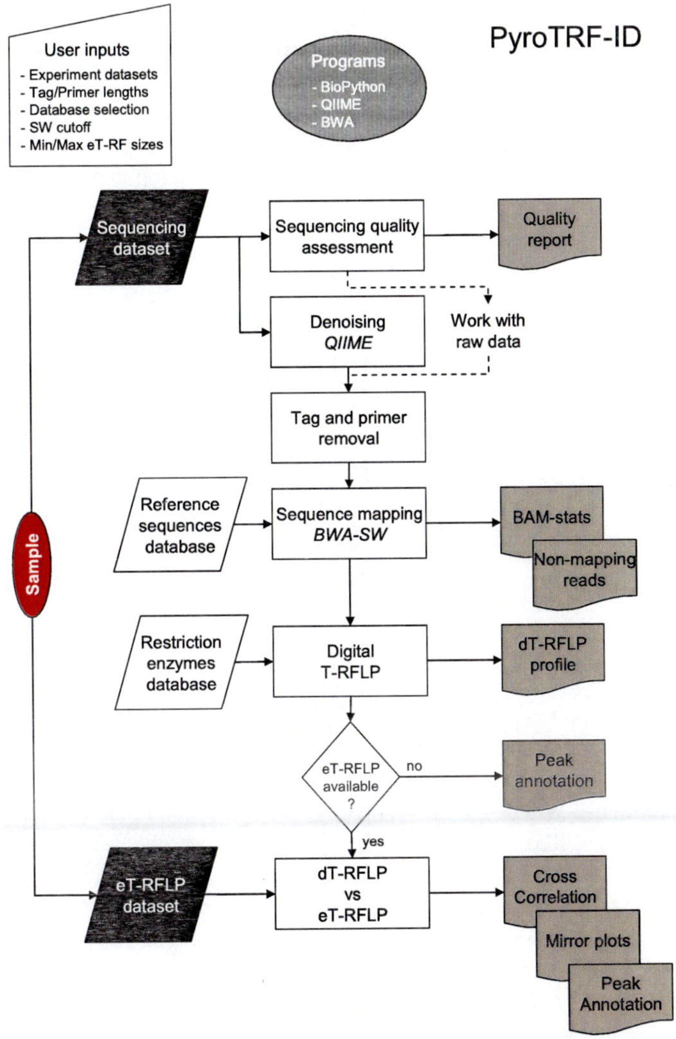

Fig. 5.1 Data workflow in the PyroTRF-ID bioinformatics methodology. Experimental pyrosequencing and T-RFLP input datasets (*black parallelograms*), reference input databases (*white parallelograms*), data processing (*white rectangles*), output files (*grey sheets*)

5.2.6.2 Denoising of Sequencing Datasets

Sequence denoising was integrated in the PyroTRF-ID workflow but this feature can be disabled by the user. It requires the independent installation of the QIIME software (Caporaso et al. 2010) to decompose and denoise the .sff files containing the whole pyrosequencing information into .sff.txt, .fasta and .qual files. Briefly, the script *split_libraries.py* was used first to remove tags and primers. Sequences were then filtered based on two criteria: (i) a sequence length ranging from the minimum (default value of 300 bp) and maximum 500-bp amplicon length, and (ii) a PHRED sequencing quality score above 20 according to Ewing and Green (1998).

Denoising for the removal of classical 454 pyrosequencing flowgram errors such as homopolymers (Balzer et al. 2011; Quince et al. 2009) was carried out with the script *denoise_wrapper.py*. Denoised sequences were processed using the script *inflate_denoiser_output.py* in order to generate clusters of sequences with at least 97% identity as conventionally used in the microbial ecology community (Reeder and Knight 2010). Based on computation of statistical distance matrices, one representative sequence (centroid) was selected for each cluster. With this procedure, a new file was created containing cluster centroids inflated according to the original cluster sizes as well as non-clustering sequences (singletons). The denoising step on the HPCC typically lasted approximately 13 h and 5 h for HighRA and LowRA datasets, respectively.

5.2.6.3 Mapping

Mapping of sequences was performed using the Burrows-Wheeler Aligner's Smith-Waterman (BWA-SW) alignment algorithm (Li and Durbin 2010) against the Greengenes database (McDonald et al. 2012). The SW score was used as mapping quality criterion (Smith and Waterman 1981; Wilson et al. 2000). It can be set by the user according to research needs. Sequences with SW scores below 150 were removed from the pipeline. SW cutoffs have typically been used in a range between 100 and 250 (House et al. 2003; Smit et al. 2003). This score can be adapted to reduce the probability of mismatches. SW scores normalized by sequence length were computed to allow comparison between sequences of various lengths. Two files were generated consecutive to mapping. The first one provided general mapping statistics for each sample. The second one provided the list of unmapped sequences, which were removed from the PyroTRF-ID pipeline.

5.2.6.4 Generation of dT-RFLP Profiles

Sequences that passed through all previous steps of the procedure were digested in silico using the restriction enzyme *Hae*III which was selected from the Bio.Restriction BioPython database. The dT-RFLP profiles were generated for each sample considering both the size of the dT-RFs and their relative abundance in the sample. Sequences containing no restriction site were discarded. A raw dT-RFLP profile plot was generated as output file. Different restriction enzymes can be tested in the PyroTRF-ID workflow for the optimization of dT-RFLP profiles. This is particularly convenient for designing new eT-RFLP approaches. Such screening can be performed on the pyrosequencing datasets without requirements of eT-RFLP data as input file.

5.2.6.5 Comparison of eT-RFLP and dT-RFLP Profiles

In order to allow comparison with eT-RFLP profiles, T-RFs below 50 bp were removed, and a second set of dT-RFLP profiles was generated. To overcome any possible discrepancy between experimental and in silico T-RFLP (Junier et al. 2008), PyroTRF-ID evaluated the most probable drift between e- and dT-RFLP profiles by computing the cross-correlation of the two. A plot showing the results of the cross-correlation was generated in order to help the user assessing the optimal shift to apply for aligning both profiles. By default, PyroTRF-ID corrected the dT-RFLP profile based on the drift with the highest cross-correlation. However, the user can optionally define a specific shift to apply. After shifting the dT-RFLP data, a mirror plot was generated allowing visual comparison of the dT-RFLP and eT-RFLP profiles.

5.2.6.6 Assignment of Affiliation to dT-RFs

Peak annotation files were generated in comma-separated-values format (.csv), listing all digitally obtained T-RFs within each dT-RFLP profile, together with their original and shifted lengths. Closest phylogenetic affiliations were provided together with the number of reads and their relative contribution to the T-RF, as well as with the absolute and normalized SW mapping scores, and the Genbank code of each reference sequence. When eT-RFLP data were not provided in the workflow, the peak annotation file was directly obtained after dT-RFLP processing without removing dT-RFs below 50 bp and without indication of T-RF shift.

5.2.7 Optimization and Testing of PyroTRF-ID

The initial testing and validation steps were carried out with the 17 pyrosequencing datasets originating from the two environments. The impact of the data processing steps of the PyroTRF-ID pipeline was assessed using two samples (GRW01 and AGS01). Three different combinations of algorithms were tested for the processing of sequences (Table 5.1), and their respective impact on the final dT-RFLP profiles was compared by calculating richness and Shannon's H′ diversity indices. The aim was to optimize the cross-correlation between dT-RFLP and the corresponding eT-RFLP profiles. The optimal standardized PyroTRF-ID procedure was selected based on this

Table 5.1 Combinations of algorithms tested for dT-RFLP profiling of pyrosequencing datasets in PyroTRF-ID

Pyrosequencing data processing procedure	Processing algorithms				
	PHRED-filtering[a]	Sequence length cut-off	Denoising	Filtering by SW mapping score[b]	Restriction of sequences[c]
(1) Standard dT-RFLP[d]	> 20[e]	> 300	Yes	> 150[f]	Yes
(2) Filtered dT-RFLP[e]	> 20	> 300	No	> 150	Yes
(3) Raw dT-RFLP[d]	> 20	> 300	No	No (0)[g]	Yes

[a] PHRED score $= -10 \log P_{error}$ with $P_{error} = 10^{-PHRED/10}$ as the probability that a base was called incorrectly. For all trials, the raw pyrosequencing datasets were systematically filtered according to the PHRED quality score. Only sequences with a related PHRED score above 20 were conserved. This corresponds to a P_{error} of 1/100

[b] A SW mapping score of 150 was set as cutoff. In the case when sequences were preliminarily denoised, it was nevertheless observed that no denoised sequence was rejected at the mapping stage. Processing without filtering by the SW mapping score was done by setting a cutoff of 0

[c] The processed sequences were digested in silico with the restriction enzyme

[d] The first combination with denoising was defined as the standard PyroTRF-ID procedure

[e] In the second combination, only a filtering method at the mapping stage was considered

[f] In the third combination, raw datasets of sequences obtained after PHRED-filtering of the pyrosequencing datasets were digested without post-processing

assessment. The optimal procedure was then applied for the comparison of PyroTRF-ID results obtained from groundwater and wastewater environments. Finally, restriction enzymes commonly used in T-RFLP analyses of bacterial communities (*Alu*I, *Hha*I, *Msp*I, *Rsa*I, *Taq*I, and *Hae*III) were selected for comparison of profiling resolutions. Visual observation, richness and diversity indices, as well as density plots were used to analyze the distributions of T-RFs along the e- and dT-RFLP profiles.

5.3 Results

5.3.1 Pyrosequencing Quality Control and Read Length Limitation

The principal quality outputs given by PyroTRF-ID are presented in Additional File 5.1 in the Supporting Information for the low throughput (LowRA) and high throughput (HighRA) pyrosequencing methods used in this study. On average, 6'380 and 32'480 reads were obtained for each method, respectively. Filtering based on the PHRED quality criterion allowed discarding low quality sequences. Most of the remaining sequences had a length below 400–450 bp (Additional File 5.1a). For the LowRA and HighRA methods, the median number of reads (800 and 2'750) was related to a PHRED score of 30 and 35, respectively, and more than 99% of reads were related to a PHRED score above 20 (Additional File 5.1b). Only reads longer than 300 bp were conserved for subsequent in silico digestion, because including short sequences in the dT-RFLP profiles may have altered the relative proportions of T-RFs to eT-RFLP profiles. Pyrosequencing datasets obtained with the HighRA method were predominantly composed of short reads below 300 bp (69% of a total of 24'810 reads in the example presented, Additional File 5.1c). However, 7'641 reads (31%) of high quality sequences were still available for PyroTRF-ID analysis, which was even larger than the number of high quality sequences remaining with the LowRA method (2'804 reads, 47%).

5.3.2 *Effect of Denoising and Mapping Procedures*

Denoising of pyrosequencing datasets was performed in order to correct for classical 454 analytical errors including the above-mentioned cut-off values: a minimum PHRED quality score of 20, as well as minimum and maximum sequence lengths of 300 and 500 bp, respectively. The denoising process generated a subset of representative sequences harboring at least 3% dissimilarity to each other. This amounted to 17 ± 1% and 43 ± 9% of the number of reads present in the raw datasets obtained with the HighRA and LowRA methods, respectively.

After denoising, the mapping process was the time-limiting step in the PyroTRF-ID pipeline. Twenty minutes were required for mapping the largest datasets against the Greengenes database. Discarding sequences shorter than 300 bp did not lead to a reduced number of detected bacterial phylotypes (Additional File 5.2). Bacterial community composi-tions obtained both without and with minimum sequence length cut-off exhibited high correspondences with determination coefficients of R^2 between 0.80 and 0.99 depending on the sample type and the reference database used for mapping (Greengenes and RDP). Within the sets of identified phlyotypes, sequences affiliated to *Geobacter* sp. displayed the highest difference in relative abundance (18%), resulting from a high proportion of short reads below 200 bp in the dataset GRW01.

After PHRED-filtering, the remaining raw sequences had maximum lengths of 450 bp and therefore the maximal SW mapping scores amounted to around 450. The distributions of the absolute and normalized SW scores are provided in Additional File 5.3, and are compared to the distribution of the sequence identity score, usually used for phylogenetic affiliation of sequences. These two scoring methods are conceptually different, since nucleotide positions and gaps are taken into account in the computation of SW scores. The median absolute and normalized SW scores amounted to 270 and 0.736, respectively. The relative number of bacterial affiliations obtained with normalized SW scores higher than 0.600 and 0.900 amounted to 89% and 37%, respectively. A total of 81% of the affiliations up to the genus level were related to a sequence identity score of 100%, and 91% with an identity score above 97%. The normalized SW scores obtained for the predominant affiliations presented in Tables 5.2 and 5.3 were comprised between 0.630 and 1.000, and are related to sequence identity scores above 97%.

5.3.3 Generation of Digital T-RFLP Profiles

The dT-RFLP profiles were successfully generated with the standard PyroTRF-ID procedure (Table 5.1) from denoised bacterial pyrosequencing datasets of the GRW and the AGS sample series (Additional File 5.4). With *Hae*III, 165 ± 29 and 87 ± 11 T-RFs were present in the dT-RFLP profiles of the GRW and AGS series, respectively. For all samples, only a reduced number of dT-RFs above 400 bp were obtained because of the low pyrosequencing quality at sequence lengths between 400 and 500 bp.

An additional feature of PyroTRF-ID is the generation of dT-RFLP profiles with any restriction enzyme. Here, profiles were obtained with five additional restriction enzymes and compared. Profiles of GRW samples were on average 2.3 times richer than ones of AGS samples, and each restriction enzyme generated characteristic dT-RFLP features regardless of the sample complexity (Fig. 5.2 and Additional File 5.4). *Hae*III provided dT-RFLP profiles with the highest richness. The use of this enzyme resulted in the generation of dT-RFs stacked mainly between 200 and 300 bp. Highest diversities in dT-RFLP profiles were obtained with *Msp*I and *Rsa*I, respectively. Digestion with *Msp*I resulted in the most homogeneous distributions of dT-RFs up to approximately 300 bp. With the exception of *Hha*I, endonucleases did not produce numerous dT-RFs in the second half of the profiles, and cumulative curves flattened off. With *Hha*I, the cumulative curves increased step-wise. *Rsa*I resulted in dT-RFLP profiles of homogeneous distributions of dT-RFs for GRW samples, but lower diversity than *Hae*III, *Alu*I, *Msp*I, and *Hha*I. *Taq*I always provided profiles with the lowest richness and diversity.

5.3.4 Comparison of Digital and Experimental T-RFLP Profiles

Mirror plots generated by PyroTRF-ID computed with raw and denoised pyrosequencing datasets obtained from a complex bacterial community (GRW01) are presented in Fig. 5.3. Further examples of mirror plots are available in Additional File 5.5. Digital profiles generated from raw pyrosequencing datasets displayed Gaussian distributions around the most dominant dT-RFs of neighbor peaks (Fig. 5.3a)

Table 5.2 Phylogenetic annotation of identified T-RFs

eTRF[a] (bp)	dTRF[a] (bp)	dTRF shifted[b] (bp)	Counts[c] (–)	Relative contribution to T-RF[d] (%)	Phylogenetic affiliation[e]	Reference GenBank accession number[f]	Absolute SW mapping score[g] (–)	Normalized SW mapping score[h] (–)
Aerobic granular sludge biofilms from wastewater treatment reactors								
n.a (32)[i]	39	34	550 (276) (128)	70.6 (35.0) (16.0)	*F: Xanthomonadaceae* *(G: Thermomonas)* *(G: Pseudoxanthomonas)*	GQ396926 (EU834762) (EU834761)	386 (452) (385)	0.960 (0.983) (0.955)
			112	14.3	*O: Flavobacteriales*	AY468464	434	1.000
			46	5.9	*F: Rhodobacteraceae*	AY212706	448	1.000
			37	4.8	*S: Rhodocyclus tenuis*	AB200295	363	0.917
			18	2.3	*O: Sphingobacteriales*	GU454872	394	0.990
			5	0.6	*C: Gammaproteobacteria*	AY098896	403	0.906
			4	0.5	*O: Rhizobiales*	EU429497	360	0.981
			4	0.5	*O: Myxococcales*	DQ228369	302	0.765
			1	0.1	*O: Bacteroidales*	EU104248	180	0.636
194	198	193	10	90.9	*G: Acidovorax*	AJ864847	384	1.000
			1	9.1	*F: Xanthomonadaceae*	EF027004	303	0.819
214	219	214	769	99.6	*S: Rhodocyclus tenuis*	AB200295	371	0.949
			1	0.1	*G: Methyloversatilis*	DQ066958	368	0.958
			1	0.1	*G: Dechloromonas*	DQ413103	321	0.988
			1	0.1	*G: Nitrosomonas*	EU937892	278	0.753
220	225	220	50 (31)	92.6 (57.0)	*O: Rhizobiales* *(G: Aminobacter)*	NR025302	448	1.000
			2	3.7	*S: Rhodocyclus tenuis*	AB200295	206	0.703

(continued)

Table 5.2 (continued)

eTRF[a] (bp)	dTRF[a] (bp)	dTRF shifted[b] (bp)	Counts[c] (–)	Relative contribution to T-RF[d] (%)	Phylogenetic affiliation[e]	Reference GenBank accession number[f]	Absolute SW mapping score[g] (–)	Normalized SW mapping score[h] (–)
			1	1.9	*F: Hyphomonadaceae*	AF236001	229	0.636
			1	1.9	*P: Firmicutes*	DQ413080	284	1.000
216	221	216	10	34.5	*S: Rhodocyclus tenuis*	AF502230	296	0.773
			8	27.6	*G: Nitrosomonas*	GU183579	364	0.948
			6	20.7	*C: Anaerolineae*	EU104216	202	0.598
			3	10.3	*G: Methyloversatilis*	CU922545	360	0.909
			1	3.4	*G: Aminobacter*	L20802	281	0.829
			1	3.4	*G: Dechloromonas*	DQ413103	273	0.898
223	228	223	44	72.1	*F: Intrasporangiaceae (G: Tetrasphaera)*	AF255629	373	0.961
			15	24.6	*F: Hyphomonadaceae*	AF236001	298	0.674
			1	1.6	*F: Microbacteriaceae*	GQ009478	228	0.544
			1	1.6	*O: Acidimicrobiales*	GQ009478	153	0.447
239	243	238	275	98.9	*C: Gammaproteobacteria*	EU529737	446	0.982
			2	0.7	*G: Leptospira*	AB476706	350	0.926
			1	0.4	*P: Armatimonadetes*	EU332819	275	0.846
249	253	249	9	100.0	*S: Rhodocyclus tenuis*	AB200295	228	0.752
255	258	253	7	100.0	*O: Sphingobacteriales*	FJ793188	355	0.989
260	263	258	16	94.1	*G: Nitrospira*	GQ487996	389	0.982
			1	5.9	*O: Sphingobacteriales*	FJ536916	251	0.640
260	264	259	38	97.4	*O: Sphingobacteriales*	EU104185	267	0.706

(continued)

Table 5.2 (continued)

eTRF[a] (bp)	dTRF[a] (bp)	dTRF shifted[b] (bp)	Counts[c] (–)	Relative contribution to T-RF[d] (%)	Phylogenetic affiliation[e]	Reference GenBank accession number[f]	Absolute SW mapping score[g] (–)	Normalized SW mapping score[h] (–)
			1	2.6	*G: Nitrospira*	GQ487996	319	0.788
297	302	297	26	100.0	*G: Herpetosiphon*	NC009972	339	0.867
307	311	306	38	97.4	*P: Armatimonadetes*	CU921283	218	0.472
			1	2.6	*O: Sphingobacteriales*	EU104210	196	0.525
321	323	318	17	100.0	*G: Cytophaga*	EU104191	367	0.968
393	397	392	33	100.0	*G: Bdellovibrio*	CU466777	262	0.663
Groundwater samples from chloroethene-contaminated aquifers								
63	69	64	93	85.3	F: Methylococcaceae	AB354618	432	0.915
			14	12.8	F: Crenotrichaceae	GU454947	290	0.816
			1	0.9	F: Ectothiorhodospiraceae	AM902494	168	0.542
			1	0.9	P: candidate phylum OP3	GQ356152	187	0.488
165	168	163	143	100.0	G: Dehalococcoides	EF059529	448	0.953
190	193	191	12	54.6	F: Desulfobulbaceae	AJ389624	379	0.945
			4	13.6	F: Sphingomonadaceae	AY785128	263	0.555
			2	9.1	F: Erythrobacteraceae	DQ811848	343	0.771
			2	9.1	C: Alphaproteobacteria	AY921822	337	0.926
			1	4.6	F: Rhodospirillaceae	AY625147	294	0.679
			1	4.6	F: Rhodobiaceae	AB374390	328	0.877
198	201	196	140	98.6	G: Desulfovibrio	FJ810587	473	1.000
			2	1.4	F: Comamonadaceae	FN428768	311	0.814

(continued)

Table 5.2 (continued)

eTRF[a] (bp)	dTRF[a] (bp)	dTRF shifted[b] (bp)	Counts[c] (–)	Relative contribution to T-RF[d] (%)	Phylogenetic affiliation[e]	Reference GenBank accession number[f]	Absolute SW mapping score[g] (–)	Normalized SW mapping score[h] (–)
210	214	209	233	98.3	F: Dehalococcoidaceae	EU679418	262	0.665
			2	0.8	O: Burkhorderiales	AM777991	367	0.927
			1	0.4	F: Spirochaetaceae	EU073764	295	0.848
			1	0.4	P: candidate phylum TM7	DQ404736	277	0.723
216	221	216	1010	90.9	F: Gallionellaceae	EU802012	353	0.869
			94	8.5	G: Rhodoferax	DQ628925	369	0.920
			3	0.3	G: Methylotenera	AY212692	291	0.744
			1	0.1	G: Methyloversatilis	GQ340363	296	0.765
			1	0.1	F: Clostridiaceae	AJ863357	338	0.833
			1	0.1	C: Anaerolineae	AB179693	229	0.511
			1	0.1	C: Actinobacteria	EU644115	372	0.907
243	247	243	389	99.7	F: Dehalococcoidaceae	EU679418	255	0.631
			1	0.3	F: Anaerolinaceae	AB447642	253	0.806

[a] Experimental (eT-RF) and digital T-RFs (dT-RF)
[b] Digital T-RF obtained after having shifted the digital dataset with the most probable average cross-correlation lag
[c] Number of reads of the target phylotype that contribute to the T-RF
[d] Diverse bacterial affiliates can contribute to the same T-RF
[e] Phylogenetic affiliation of the T-RF (K: kingdom, P: phylum, C: class, O: order, F: family, G: genus, S: species). Only the last identified phylogenetic branch is presented here
[f] GenBank accession numbers provided by Greengenes for reference sequences
[g] Best SW mapping score obtained. SW scores consider nucleotide positions and gaps. The highest SW mapping score that can be obtained for a read is the length of the read itself
[h] SW mapping score normalized by the read length, as an estimation of the percentage of identity
[I] After having observed the presence of the dT-RF 34 bp, we returned to the raw eT-RFLP data and found an important eT-RF at 32 bp. However, Rossi et al. (2009) considered that T-RFs below 50 bp are inconsistent and lacks of precision in sizing. This peak was therefore initially not taken into account in the original eT-RFLP profiles

Table 5.3 T-RF diversity for single phylogenetic descriptions

Phylogenetic affiliation	dTRF (bp)	dTRF shifted[a] (bp)	Counts[b] (–)	Relative contribution to T-RF[c] (%)	Reference OTU[d]	Reference GenBank accession number[e]	Absolute SW mapping score[f] (–)	Normalized SW mapping score[g] (–)
Flocculent and aerobic granular sludge samples from wastewater treatment systems								
Rhodocyclus tenuis	39	34	37	4.8	3160	AB200295	363	0.917
	199	194	1	25.0	3160	AB200295	248	0.648
	205	200	3	100.0	3160	AF204247	314	0.858
	210	205	1	100.0	3160	AF204247	211	0.699
	218	213	11	91.7	3160	AB200295	356	0.942
	219	214	769	99.6	3160	AB200295	371	0.949
	220	215	6	37.5	3160	AF502230	318	0.817
	221	216	1	7.7	3160	AF502230	276	0.865
	225	220	2	3.7	3160	AB200295	206	0.703
	252	247	3	100.0	3160	AB200295	305	0.762
	253	248	9	100.0	3160	AB200295	228	0.752
	257	252	1	20.0	3160	AF502230	241	0.660
Groundwater samples from aquifers contaminated with chloroethenes								
Dehalococcoides spp.	166	161	1	100.0	1368	EF059529	290	0.775
	168	163	143	100.0	1368	EF059529	241	0.717
	169	164	2	100.0	1368	EF059529	331	0.768
	170	165	2	100.0	1368	EF059529	241	0.693
	171	166	1	50.0	1368	EF059529	303	0.783

(continued)

Table 5.3 (continued)

Phylogenetic affiliation	dTRF (bp)	dTRF shifted[a] (bp)	Counts[b] (–)	Relative contribution to T-RF[c] (%)	Reference OTU[d]	Reference GenBank accession number[e]	Absolute SW mapping score[f] (–)	Normalized SW mapping score[g] (–)
	173	168	1	100.0	1368	EF059529	241	0.717
	176	171	1	100.0	1369	DQ833317	211	0.687
	179	174	1	100.0	1369	DQ833317	193	0.629
	188	183	4	66.7	1369	DQ833340	464	0.947

[a] Digital T-RF obtained after having shifted the digital dataset with the most probable average cross-correlation lag
[b] Number of reads of the target phylotype that contribute to the T-RF
[c] Diverse bacterial affiliates can contribute to the same T-RF
[d] Reference OTU from the Greengenes public database obtained after mapping
[e] GenBank accession numbers provided by Greengenes for reference sequences
[f] Best SW mapping score obtained
[g] SW mapping score normalized by the read length

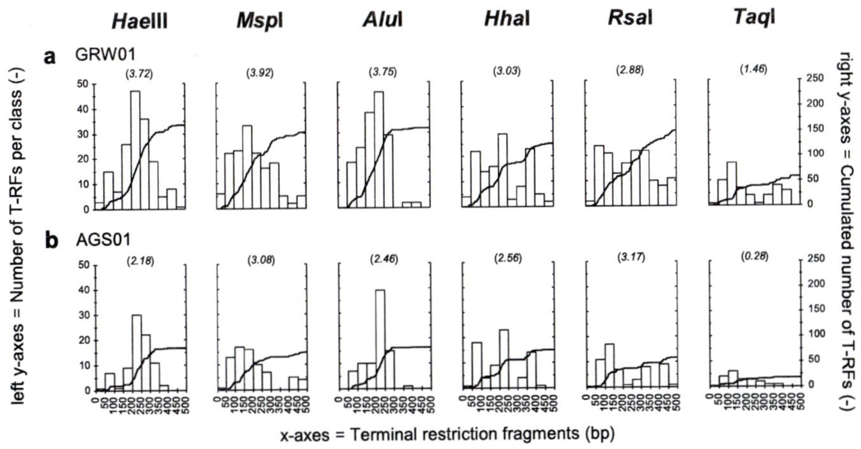

Fig. 5.2 Density plots displaying the repartition of T-RFs along the 0–500 bp domain with different endonucleases (*Hae*III, *Alu*I, *Msp*I, *Hha*I, *Rsa*I and *Taq*I) tested on pyrosequencing datasets GRW01 (**a**) and AGS01 (**b**). Histograms represent the number of T-RFs produced per class of 50 bp (*left y-axes*). Thick black lines represent the cumulated number of T-RFs over the 500-bp fingerprints (*right y-axes*). The total cumulated number of T-RFs corresponds to the richness index. Shannon's diversity indices are given in brackets

which exhibited identical bacterial affiliations (data not shown). This feature was attributed to errors of the 454 pyrosequencing analysis. Denoised dT-RFLP profiles displayed enhanced relative abundances of dominant peaks and had a higher cross-correlation with eT-RFLP profiles (Fig. 5.3). By selecting representative sequences (so-called centroids) for clusters containing reads sharing at least 97% identity, in the QIIME denoising process, all neighbor peaks were integrated in the dominant dT-RFs resulting from the centroid sequences. Cross-correlations between dT-RFLP and eT-RFLP profiles issued from sample GRW01 increased from 0.43 to 0.62 after denoising of the pyrosequencing data.

The dT-RFLP profiles exhibited a drift of 4–6 bp compared to eT-RFLP profiles. From in silico restriction of a library of 150 clones obtained in our laboratory using the TRiFLe software (Junier et al. 2008), we confirmed that in silico T-RFs were, on average, 4 ± 1 bp (min 3 bp–max 6 bp) longer than the experimental ones (data not shown). After correcting with an optimal shift (Additional File 5.6), maximum cross-correlation coefficients between denoised dT-RFLP and eT-RFLP profiles ranged from 0.55 ± 0.14 and 0.67 ± 0.05 for the GRW samples (HighRA and LowRA method, respectively) to 0.82 ± 0.10 for the AGS samples (LowRA method) (Table 5.4).

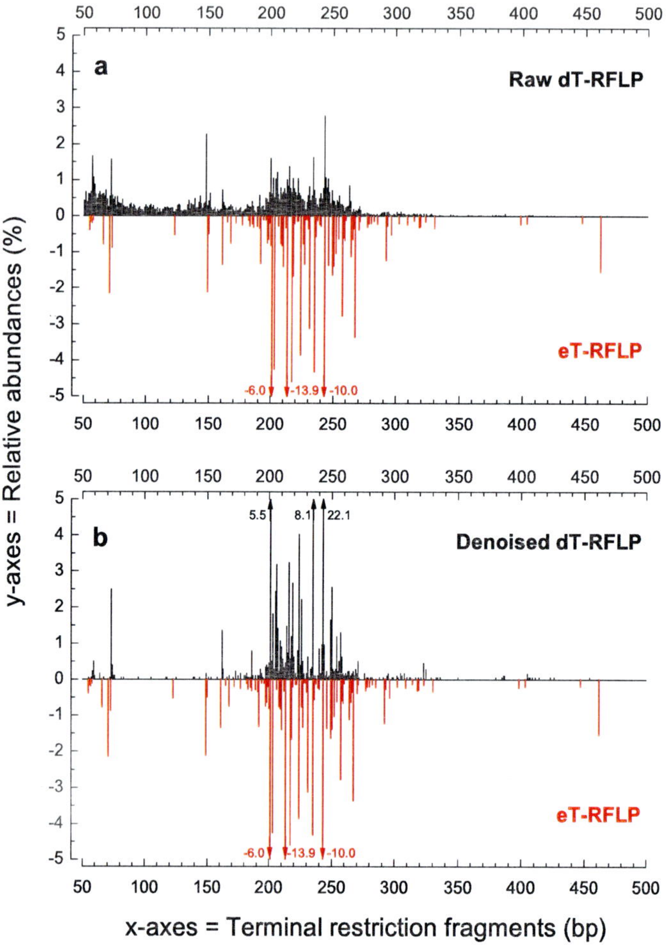

Fig. 5.3 Mirror plot displaying the cross-correlation between digital and experimental T-RFLP profiles. This mirror plot was generated for the complex bacterial community of the groundwater sample GRW01. Comparison of mirror plots constructed with raw (**a**) and denoised sequences (**b**). Relative abundances are displayed up to 5% absolute values. For those T-RFs exceeding these limits, the actual relative abundance is displayed next to the peak

5.3.5 Impact of Sequence Processing Steps, Pyrosequencing Methods and Sample Types

Indices of richness (number of T-RFs) and diversity (number of T-RFs and distributions of abundances) were used to evaluate the impacts of data processing steps, pyrosequencing methods and sample types on the structure of the final dT-RFLP profiles (Fig. 5.4). The changes of the indices were considered positive if they approached the indices determined for eT-RFLP profiles. The raw dT-RFLP profiles

Table 5.4 Cross-correlations between experimental and standard digital T-RFLP profiles

Samples	Optimal cross-correlation lag between digital and experimental T-RFLP profiles[a] (bp)	Maximum cross-correlation coefficient at optimal lag[b] (–)	Total number of experimental T-RFs per profile (–)	Number of experimental T-RFs affiliated with digital T-RFs[c] (–)	Percentage of experimental T-RFs affiliated with digital T-RFs[c] (%)
Groundwater					
GRW01[d]	– 4	0.62	88	58	66
GRW02[d]	– 5	0.69	50	23	46
GRW03[d]	– 4	0.44	76	62	82
GRW04[d]	– 5	0.71	44	24	44
GRW05[d]	– 5	0.35	75	56	75
GRW06[d]	– 6	0.51	87	70	81
Avg ± stdev (min–max)	– 5 ± 1 – (4–6)	0.55 ± 0.14 (0.35–0.71)	70 ± 19 (44–88)	49 ± 20 (23–70)	67 ± 14 (44–82)
GRW07[e]	– 6	0.70	57	17	30
GRW08[e]	– 4	0.59	54	43	80
GRW09[e]	– 4	0.69	71	66	93
GRW10[e]	– 5	0.68	70	22	31
Avg ± stdev (min–max)	– 5 ± 1 – (4–6)	0.67 ± 0.05 (0.59–0.70)	59 ± 11 (44–71)	34 ± 20 (17–66)	59 ± 33 (30–93)
Aerobic granular sludge					
AGS01[e]	– 5	0.75	48	31	65
AGS02[e, f]	– 5	0.90	38	22	58
AGS03[e, f]	– 5	0.90	38	19	50
AGS04[e]	– 5	0.72	52	24	46

(continued)

Table 5.4 (continued)

Samples	Optimal cross-correlation lag between digital and experimental T-RFLP profiles[a] (bp)	Maximum cross-correlation coefficient at optimal lag[b] (–)	Total number of experimental T-RFs per profile (–)	Number of experimental T-RFs affiliated with digital T-RFs[c] (–)	Percentage of experimental T-RFs affiliated with digital T-RFs[c] (%)
AGS05[e]	– 4	0.67	43	29	67
AGS06[e, f]	– 5	0.91	38	19	50
AGS07[e]	– 5	0.80	38	31	82
Avg ± stdev (min–max)	– 5 ± 0 – (4–5)	0.82 ± 0.10 (0.67–0.91)	42 ± 6 (38–52)	25 ± 5 (19–31)	61 ± 12 (46–82)

[a] Shift leading to optimal matching of the digital to the experimental T-RFLP profile
[b] Maximum cross-correlation coefficients obtained after matching of the digital to the experimental T-RFLP profile
[c] Number and percentage of experimental T-RFs having corresponding digital T-RFs
[d] Samples GRW01-06 were pyrosequenced with the HighRA method
[e] Samples GRW07-10 and AGS01-07 were pyrosequenced with the LowRA method
[f] Samples AGS02, AGS03, and AGS06 are triplicates from the same DNA extract

were composed of 2.4- to 7.4-times more T-RFs than the eT-RFLP profiles. Denoising resulted in a decrease of richness and diversity. The ratios of richness and diversity between standard dT-RFLP and eT-RFLP profiles amounted to 2.5 ± 0.6 and 1.0 ± 0.3, respectively, for high-complexity samples (GRW), and to 2.1 ± 0.5 and 0.8 ± 0.2, respectively, for low-complexity samples (AGS). The raw dT-RFLP profiles of the groundwater samples GRW01-GRW06, which were sequenced with the HighRA method, were composed of 4 to 7.4-times more T-RFs than their respective eT-RFLP profiles. Groundwater samples GRW07-GRW10 sequenced with the LowRA method displayed ratios of raw dT-RFs to eT-RFs which were between 2.4 and 5.2. After denoising, both sets of groundwater-related dT-RFLP profiles exhibited similar richness and diversity and were closer to indices of eT-RFLP profiles than raw dT-RFLP profiles (Fig. 5.4).

The DNA extract of one AGS sample was analyzed in triplicate from pyrosequencing to PyroTRF-ID. The resulting standard dT-RFLP profiles contained 94 ± 10 T-RFs, and exhibited very close diversity indices of 1.48 ± 0.03. In comparison, denoised profiles of all AGS samples collected over 50 days contained similar numbers of T-RFs (84 ± 9) but exhibited quite different diversity indices of 2.12 ± 0.48. There was also very little variation in the cross-correlation coefficients (0.90 ± 0.01) between the dT-RFLP profiles and the corresponding eT-RFLP profile. All three denoised T-RFLP profiles exhibited similar structures, and affiliations were the same for T-RFs that could be identified.

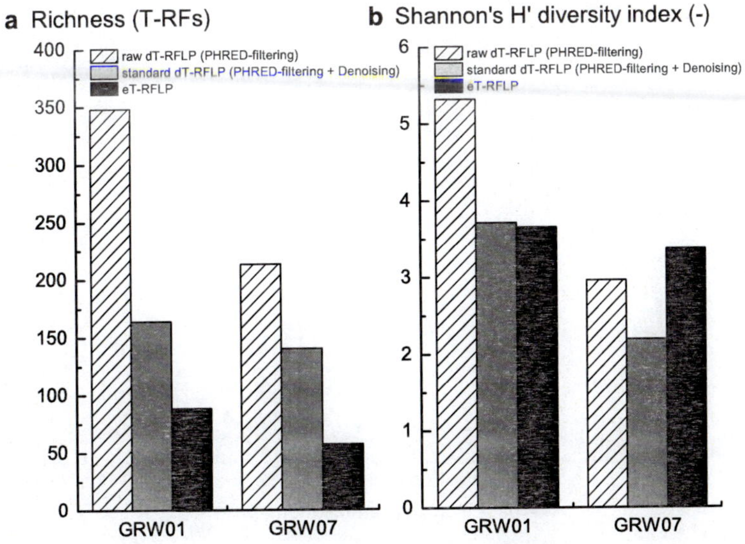

Fig. 5.4 Assessment of the impact of data processing on dT-RFLP profiles, and comparison with eT-RFLP profiles. Richness and Shannon's H′ diversity indices were calculated in a way to quantify the impact of the pyrosequencing data processing parameters on the resulting dT-RFLP profiles. Two examples are given for samples pyrosequenced with the HighRA (GRW01) and LowRA methods (GRW07)

5.3.6 Efficiency of Phylogenetic Affiliation of T-RFs

Comprehensive phylogenetic information was provided by PyroTRF-ID for each dT-RF, as exemplified in Table 5.2. Depending on the sample type, between 45 and 60% of all dT-RFs were affiliated with a unique bacterial phylotype (Fig. 5.5). The other dT-RFs were affiliated with two or more phylotypes, showing different contribution patterns. In such cases, a single phylotype was usually clearly predominating with a relative contribution ranging from 50 to 99%. However, for some T-RFs no clear dominant phylotype emerged (e.g. eT-RF 216 in AGS samples, Table 5.2).

Some reference sequences were sometimes represented by several T-RFs (Table 5.3). For instance, in AGS01, six dT-RFs (34, 194, 213, 214, 220, 247 bp) were affiliated to the same reference sequence of *Rhodocyclus tenuis* (accession number AB200295), with shifted T-RF 214 being predominant (769 of 844 reads). The *Dehalococcoides* sp. affiliation in sample GRW05 was related to eight T-RFs, with shifted T-RF 163 being predominant (143 of 156 reads). To investigate this phenomenon, reads resulting in different *Dehalococcoides*-affiliated T-RFs were retrieved from the pyrosequencing dataset and aligned with ClustalX (Additional File 5.7). This analysis showed that the multiple T-RF sizes observed were due to reads harboring insertions or deletions of nucleotides before the first *Hae*III restriction site or to nucleotide modifications within *Hae*III sites.

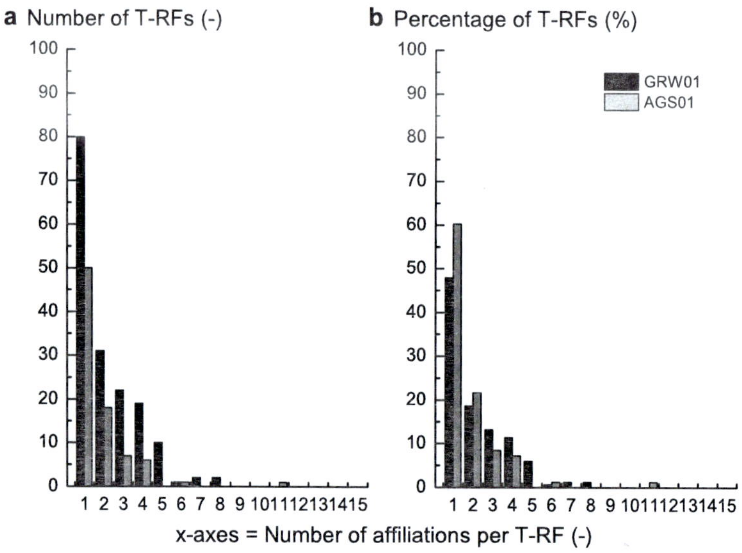

Fig. 5.5 Amount of bacterial affiliations contributing to T-RFs. The absolute (**a**) and relative numbers (**b**) of T-RFs that comprised one to several bacterial affiliations is given for the samples GRW01 and AGS01

5.4 Discussion

5.4.1 Advantages and Novelties of the PyroTRF-ID Bioinformatics Methodology

This study describes the development of the PyroTRF-ID bioinformatics methodology for the analysis of microbial community structures, and its application on low- and high-complexity environments. PyroTRF-ID can be seen as the core of a high-throughput methodology for assessing microbial community structures and their dynamics combining NGS technologies and more traditional community fingerprinting techniques such as T-RFLP. More than just predicting the most probable T-RF size of target phylotypes, PyroTRF-ID allows the generation of dT-RFLP profiles from 16S rRNA gene pyrosequencing datasets and the identification of experimental T-RFs by comparing dT-RFLP to eT-RFLP profiles constructed from the same DNA samples.

At the initial stage of the assessment of a microbial community, PyroTRF-ID can be used for the design of an eT-RFLP procedure adapted to a given microbial community through digital screening of restriction enzymes. In contrast to previous studies involving in silico restriction of artificial microbial communities compiled from selected reference sequences from public or cloning-sequencing databases (Collins and Rocap 2007; Fernandez-Guerra et al. 2010; Marsh et al. 2000), PyroTRF-ID works on sample-based pyrosequencing datasets. This requires the pyrosequencing of a limited number of initial samples. The number of T-RFs, the homogeneity in their distribution, and the number of phylotypes contributing to T-RFs should be used as criteria for the choice of the best suited enzyme. Combination of pyrosequencing and eT-RFLP datasets obtained on the same initial set of samples enables the beginning of the study of new microbial systems with knowledge on T-RFs affiliation. The length of T-RFs and their sequences are directly representative of the investigated sample rather than inferred from existing databases. In this sense, the complexity of the original environment is accurately investigated. For all types of low- and high-complexity environments assessed in this study, *Hae*III, *Alu*I and *Msp*I were good candidates for the generation of rich and diverse dT-RFLP profiles.

Subsequently, eT-RFLP can be used as a routine method to assess the dynamics of the stuctures of microbial communites, avoiding the need for systematic pyrosequencing analyses. We suggest that pyrosequencing should be applied at selected time intervals or on representative samples to ensure that the T-RFs still display the same phylogenetic composition. Combining T-RFLP and pyrosequencing is particularly adapted for the temporal follow-up of a microbial system, taking advantage of the relative low costs of T-RFLP and its convenience for routine assessment of microbial community structures, and of the power of pyrosequencing for assessing the composition of these communities. PyroTRF-ID has already been used for the study of bacterial communities involved in start-up of aerobic granular sludge systems (Weissbrodt et al. 2012a) and in natural attenuation of chloroethene-contaminated aquifers (Shani 2012).

5.4.2 Performance Assessment and Limitations of PyroTRF-ID

Classical 454 pyrosequencing errors, such as inaccurate resolving of homopolymers and single base insertions (Huse et al. 2007), were expected to impact the quality of dT-RFLP profiles by overestimating the number of dT-RFs present (Kunin et al. 2010; Niu et al. 2010). The use of a denoising procedure based on the analysis of rank-abundance distributions (Reeder and Knight 2010) was a prerequisite to minimize pyrosequencing errors and to generate dT-RFLP profiles approaching the structure of eT-RFLP profiles, as assessed by the improved cross-correlation coefficients. Filtering pyrosequencing reads with the SW mapping score threshold only slightly reduced overestimations. In addition, this filtering approach does not specifically remove reads based on their intrinsic quality but rather on similarities with existing sequences from the database, hence reducing the complexity of the studied bacterial community to what is already known (Gilbert et al. 2007; Huse et al. 2007). When denoising was applied, the use of a SW mapping score threshold did not improve the shape of dT-RFLP profiles. Whereas small-size reads were more abundant in the HighRA pyrosequencing datasets. The pyrosequencing method and the initial amount of reads did not impact the final PyroTRF-ID output. Only the level of complexity of the bacterial communities of the ecosystems could have explained the differences in richness among T-RFLP profiles.

Clipping the low-quality end parts of sequences is an option to improve sequence quality but it is quite improbable that it has an impact on the outcome of the taxon assignment and the creation of dT-RFLP profile. When PyroTRF-ID is run with the "–qiime" option, quality trimming is done using the protocol proposed in QIIME (Caporaso et al. 2010) and its online tutorial.[2] This includes the amplicon noise procedure that is efficient in correcting for sequencing errors, PCR single base substitutions, and PCR chimeras (Quince et al. 2011). Even if some wrong base calls remain in the consensus sequences after this, they should not affect the assignment to taxon as the BWA aligner can account for mismatches. It should not influence the dT-RFLP profile either since a mismatch outside of the enzyme cleavage site does not affect the length of the fragment produced. As the fragment length is determined by counting the number of base pairs before the enzyme cleavage site and that the BWA aligner does not necessarily use the whole sequence when selecting a match, clipping the low-quality ends of sequences would probably have no measurable effect.

Discrepancies of 0–7 bp between the size of in silico predicted T-RFs and eT-RFs have previously been reported (2008; 2001). In the present study, an average discrepancy of 4–6 bp was observed between dT-RFLP and eT-RFLP profiles. This drift was confirmed by comparison of in silico and experimental digestion of 150 clones from a clone library. To overcome the bias induced by the experimental drift, we introduced the calculation of a cross-correlation between dT-RFLP and eT-RFLP profiles. The entire dT-RFLP profile was shifted by the number of base pairs enabling

[2] Electronic web link: http://qiime.org/tutorials/denoising_454_data.html.

better fitting to the corresponding eT-RFLP profile. It is known that the drift is not constant across the T-RFs but rather depends on the true T-RF length, on its purine content, and on its secondary structure (Bukovska et al. 2010; Kaplan and Kitts 2003; Kitts 2001). Mirror plots sometimes displayed a 1-bp difference between eT-RFs and dT-RFs. It was crucial for the user to visually inspect the mirror plots prior to semi-manually assigning phylotypes to eT-RFs. The approach adopted here consisted of selecting eT-RFs to identify prior to checking their alignment with dT-RFs. In order to overcome manual inspection, a shift could be computed for each single dT-RF in relation with its sequence composition and theoretical secondary structure (2010). However, the standard deviation associated with this method is still higher than 1 bp. Shifting each single dT-RF based on this function was therefore not expected to improve the alignment accuracy. If at a later stage an improved method for calculating drift for single dT-RFs will be available, it could replace our approach combining a shift of the whole profile, cross-correlation calculation between dT-RFLP and eT-RFLP profiles, and manual inspection. Though user interpretation can introduce a subjective step, final manual processing of T-RFLP profiles can remain the only way to resolve T-RF alignment problems (Kitts 2001). We nevertheless suggest that selected samples of the investigated system should pass through PyroTRF-ID in triplicates in order to validate the optimal drift determined in the cross-correlation analysis.

Following the standard PyroTRF-ID procedure, high level of correspondence was obtained between dT-RFLP and eT-RFLP profiles. Over all samples, $63 \pm 18\%$ of all eT-RFs could be affiliated with a corresponding dT-RF. Correspondence between dT-RFs and eT-RFs was relatively obvious for high abundance T-RFs, in contrast to low abundance dT-RFs. Numerous low abundance dT-RFs were present in dT-RFLP profiles but absent in eT-RFLP profiles. Conversely, eT-RFs were sometimes lacking a corresponding dT-RF. This mainly occurred in profiles generated using pyrosequencing datasets with an initially low amount of reads exceeding 400 bp. The lower proportion of long reads was associated with a decreasing probability of finding a restriction site in the final portion of the sequences. For eT-RFs near 500 bp, incomplete enzymatic restriction could explain that undigested amplicons were detected in the electrophoresis runs (Clement et al. 1998; Osborn et al. 2000). These features, however, do not explain missing dT-RFs, which sometimes occurred in the initial portion of the dT-RFLP profile. Egert and Friedrich (2003) have attributed the presence of ‘pseudo T-RFs’ to undigested single stranded DNA amplicons, and have cleared them by cleaving amplicons with single-strand-specific mung bean nuclease. An interesting possibility to increase considerably the number of long reads would be to use bidirectional reads as used by Pilloni et al. for the characterization of tar-oil-degrading microbial communities (Pilloni et al. 2011).

The majority of dT-RFs were affiliated to several phylotypes, revealing the underlying phylogenetic complexity, which was in agreement with Kitts (2001). PyroTRF-ID enabled assessing the relative contributions of each phylotype, and determining the most abundant ones. In most cases, one phylotype clearly displayed the highest number of reads for one dT-RF. However, for some dT-RFs several phylotypes contributed almost equally to the total number of reads. Although problematic while aiming at identifying T-RFs, this information is of primary importance if PyroTRF-ID is intended to be used for designing the most adapted T-RFLP procedure for the study of a particular bacterial community. Finally, as exemplified by Additional File 5.7, the reference mapping database can have an impact on the identification of T-RFs. A fraction of 35–45% of the reads was unassigned during mapping in MG-RAST with the Greengenes database, while only 3–5% was unassigned with RDP. This aspect stresses the need of standardized databases and microbiome dataset processing approaches in the microbial ecology field.

5.4.3 Comparison of Community Compositions Obtained with PyroTRF-ID and MG-RAST

The full compositions of the "bacteriomes" of biomass samples collected in the *Accumulibacter* and *Competibacter* enrichments (Weissbrodt et al. 2014a/Chap. 6) and in the AGS-SBRs (Weissbrodt et al. 2012a, 2014b/Chaps. 7 and 9) were analyzed in MG-RAST, in parallel to eTRFLP and PyroTRF-ID analyses. Surprisingly, the comparison of these data revealed some divergences between some predominant phylotypes and abundances obtained. For instance, in the sample AGS01 described in the present chapter and corresponding to sample X066 of mature AGS in Chap. 6, important abundances of the genera *Zoogloea*, *Nitrococcus*, and *Chryseobacterium* were resulting from the analysis in MG-RAST (Additional File 5.7). However, these organisms were almost not detected in this sample according to eTRFLP analysis and PyroTRF-ID processing of the same pyrosequencing dataset, with only two denoised centroid sequences affiliating with *Zoogloea* spp. The combined 16S rRNA targeted FISH and CLSM analyses conducted on cross-sections of mature granules confirmed that *Zoogloea* spp. were outcompeted (Weissbrodt et al. 2013/Chap. 5).

MG-RAST and PyroTRF-ID workflows are based on different algorithms and cutoff values, notably in the mapping procedures. MG-RAST relies on the definition of cutoff values according to a percentage of identity (e.g. > 97%) conventionally used in the microbial ecology science for the definition of "species". Mapping in PyroTRF-ID is performed according to the BWA-SW algorithm that also takes sequence gaps into account. This method is typically used since the nineties by bioinformaticians active in life sciences. It seems that the divergences between the results obtained with the two methods for the same pyroseqeuncing dataset possibly rely on the differences in the implemented algorithms.

Overall, and according to the numerous jobs submitted to processing workflows available in open-access, there is an urgent need for harmonizing and cross-validating bioinformatics approaches at international level (e.g. standardization of mapping algorithms and reference sequence databases, suppressing black box effects at user level). A common language should be found between microbiologists and bioinformaticians, such as exemplified by the current different uses of identity scores and SW-scores. The constitution of a task group within the International Society for Microbial Ecology (ISME) might be considered to these ends.

5.5 Conclusions

This study presented the successful development of the PyroTRF-ID bioinformatics methodology for high-throughput generation of digital T-RFLP profiles from massive sequencing datasets and for assigning phylotypes to eT-RFs based on pyrosequences obtained from the same samples. In addition, this study leads to the following conclusions:

- The combination of pyrosequencing and eT-RFLP data directly obtained from the same samples was a powerful characteristic of the PyroTRF-ID methodology, enabling generation of dT-RFLP profiles that integrate the whole complexity of microbiomes of interest.
- The LowRA and HighRA 454 pyrosequencing method did not impact on the final results of the PyroTRF-ID procedure.
- As in any new generation sequencing analysis, denoising was a crucial step in the 454 pyrosequencing dataset processing pipeline in order to generate representative digital fingerprints.
- The PyroTRF-ID workflow could be applied to the screening of restriction enzymes for the optimization of favorably distributed eT-RFLP profiles by considering the entire underlying microbial communities. *Hae*III, *Msp*I and *Alu*I were good candidates for T-RFLP profiling with high richness and diversity indices.
- The PyroTRF-ID methodology was validated with different samples from low- and high-complexity environments, and could be implemented in a broad spectrum of biological samples in environmental to medical applications with optimized laboratory and computational costs. This methodology is probably not restricted to pyrosequencing datasets, and could be, after some modifications, applied to datasets obtained with any kind of sequencing techniques.

Acknowledgements Lucas Sinclair, Grégory Lefebvre, and Jacques Rougemont from the Bioinformatics and Biostatistics Core Facility at EPFL for collaboration on the implementation of the bioinformatics procedure, as well as Julien Delafontaine for the PyroTRF-ID website. Ioannis Xenarios for support and access to the Vital-IT HPCC of the Swiss Institute of Bioinformatics (Lausanne, Switzerland). Scot E. Dowd, Yan Sun, Lars Koenig and at Research and Testing Laboratory (Lubbock, Texas, USA), Timothy M. Vogel, Sébastien Cecillon and the Environmental Microbial Genomics Group at Ecole Centrale de Lyon (France), and GATC Biotech (Konstanz, Germany) for pyrosequencing analyses and advice.

Appendix: Phylogenetic Affiliations Obtained with PyroTRF-ID

Description of Biological Samples Processed

Phylogenetic affiliations of operational taxonomic units (OTUs) were identified in a combination of terminal-restriction fragment length polymorphism (T-RFLP) and amplicon sequencing methods targeting the v1-v3 hypervariable region of the bacterial 16S rRNA gene pool. T-RFLP (Rossi et al. 2009) was used as low-cost analysis for routine measurement of bacterial community structures and population dynamics with high temporal resolution with regular sampling of biomass over the time course of environmental biotechnology processes. Amplicon sequencing was conducted on a reduced number of representative samples selected for bacterial tag-encoded FLX amplicon pyrosequencing (bTEFAP, Sun et al. 2011) for analysis of community compositions with high phylogenetic resolution. The dry-lab PyroTRF-ID workflow (Weissbrodt et al. 2012b) was used to rationally combine T-RFLP and bTEFAP datasets to generate OTU affiliations and to sustain the graphing of time series of bacterial community profiles (Table 5.5).

Table 5.5 Description of the 10 representative biomass samples selected for amplicon sequencing (bTEFAP, > 3000 reads per sample) and used to calibrate the bacterial affiliations of the majority of OTUs from the broad and rational T-RFLP profiles generated with high time resolution along experimental campaigns, using PyroTRF-ID (Weissbrodt et al. 2012b)

Sample	SBR	Day	T (°C)	Experiment	Stage	Chapters
PAO109[c]	Stirred-tank PAO	109	17	Enrichment of PAOs	Steady-state	6, 8
GAO398[c]	Stirred-tank GAO	398	30	Enrichment of GAOs	Steady-state	6, 8
BC002[a]	AGS SBR ambient	2	23 ± 2	Inoculation	Flocculent sludge	7, 8
BC059[a]	AGS SBR ambient	59	23 ± 2	Start-up of granulation	Early-stage AGS	7, 8
BC-II[b]	AGS SBR ambient		20	Maturation of granules	Mature AGS	8, 9
XP06[d]	AGS SBR-20	6	20	Fluctuating variables	Excess sludge	9
X066[b, d]	AGS SBR-20	66	20	Fluctuating variables	Mature AGS	9
X084[d]	AGS SBR-20	84	20	Fluctuating variables	Mature AGS	9

(continued)

Table 5.5 (continued)

Sample	SBR	Day	T (°C)	Experiment	Stage	Chapters
X114[d]	AGS SBR-20	114	20	Fluctuating variables	Mature AGS	9
O125[d]	AGS SBR-25	125	25	Fluctuating variables	Mature AGS	9

[a] BC002 and BC059 were taken during start-up from flocs to early-stage granules
[b] BC-II and X066 are identical but were named differently according to objectives of Chaps. 5 and 6. The biomass sample was taken in the aerobic granular sludge (AGS) SBR-20 66 days after starting the experiments under fluctuations in operation variables with fresh pre-cultivated mature AGS older than 500 days
[c] PAO109 and GAO398 were taken in stirred-tank SBRs for PAO and GAOs
[d] X066, X084, and X114, and O125 were taken from the AGS SBR-20 and SBR-25, respectively, operated with mature AGS under fluctuations in operation variables. Sample XP06 was taken from excess AGS purge from the bottom of SBR-20 6 days after having started the experiment with the fresh pre-cultivated and acclimatized mature AGS older than 500 days

Predominant Bacterial Phylotypes and Corresponding OTUs

See Table 5.6.

Table 5.6 Predominant bacterial relatives detected across the 16S rRNA gene-targeted amplicon sequencing datasets of all 10 representative biomass samples, and listed in descending numerical order of their total read counts

Broader affiliation Phylum → Class → Order	Closest affiliation Order → Family → Genus	Total read counts (–)	Fraction of total reads (%)	T-RF or OUT (bp)
O: Rhodocyclales	*G: Rhodocyclus*	8944	22.4	214
O: Rhodocyclales	*G: Zoogloea*	5050	12.6	195
C: Gammaproteobacteria		4267	10.7	238
C: Acidobacteria	*F: Acidobacteriaceae*	4009	10.0	209
O: Xanthomonadales	*G: Thermomonas*	2772	6.9	32
C: Gammaproteobacteria		1181	3.0	239
C: Alphaproteobacteria	*F: Rhodospirillaceae*	1099	2.8	178
O: Xanthomonadales	*G: Pseudoxanthomonas*	930	2.3	32
O: Actinomycetales	*G: Tetrasphaera*	575	1.4	223
O: Rhodocyclales	*G: Rhodocyclus*	492	1.2	213
O: Rhizobiales	*G: Aminobacter*	474	1.2	220
P: Armatimonadetes		427	1.1	209
C: Gammaprotebacteria		412	1.0	298

(continued)

Table 5.6 (continued)

Broader affiliation Phylum → Class → Order	Closest affiliation Order → Family → Genus	Total read counts (–)	Fraction of total reads (%)	T-RF or OUT (bp)
O: Sphingobacteriales		385	1.0	256
O: Actinomycetales	*F: Intrasporangiaceae*	377	0.9	223
O: Rhodocyclales	*G: Dechloromonas*	352	0.9	214
C: Gammaproteobacteria	*F: Xanthomonadaceae*	302	0.8	32
O: Burkholderiales	*G: Paucibacter*	283	0.7	213
O: Flavobacteriales	*F: Cryomorphaceae*	274	0.7	32
C: Gammaproteobacteria	*O: Xanthomonadaceae*	257	0.6	32
O: Flavobacteriales	*G: Sejongia*	251	0.6	32
C: Alphaproteobacteria	*F: Rhodospirillaceae*	244	0.6	194
O: Sphingobacteriales		233	0.6	253
O: Rhodobacterales	*F: Hyphomonadaceae*	231	0.6	222
O: Xanthomonadales	*G: Dokdonella*	221	0.6	32
P: Chloroflexi	*G: Herpetosiphon*	221	0.6	298
O: Rhodobacterales	*G: Thioclava*	202	0.5	32
O: Rhodocyclales	*G: Rhodocyclus*	198	0.5	32
P: Bacteroidetes	*F: Flavobacteriaceae*	188	0.5	32
O: Sphingomonadales	*G: Sphingomonas*	175	0.4	288
O: Burkholderiales	*G: Acidovorax*	168	0.4	193
P: Chloroflexi	*G: Herpetosiphon*	158	0.4	297
O: Rhodocyclales	*G: Rhodocyclus*	155	0.4	215
C: Betaproteobacteria		150	0.4	214
O: Rhodobacterales	*F: Hyphomonadaceae*	135	0.3	224
C: Gammaproteobacteria		124	0.3	303
O: Rhodobacterales	*G: Rhodobacter*	122	0.3	32
C: Gammaproteobacteria		116	0.3	239–248
O: Actinomycetales	*F: Intrasporangiaceae*	100	0.3	228
O: Sphingobacteriales	*G: Cytophaga*	100	0.3	321
O: Burkholderiales	*G: Rhodoferax*	92	0.2	214
C: Gammaproteobacteria		80	0.2	237
P: Armatimonadetes		80	0.2	307
C: Deltaproteobacteria	*G: Bdellovibrio*	77	0.2	392
O: Burkholderiales		72	0.2	212
O: Nitrospirales	*G: Nitrospira*	71	0.2	258

(continued)

Table 5.6 (continued)

Broader affiliation Phylum → Class → Order	Closest affiliation Order → Family → Genus	Total read counts (–)	Fraction of total reads (%)	T-RF or OUT (bp)
O: Rhizobiales	*G: Bradyrhizobium*	71	0.2	286
P: candidate phylum TM7		65	0.2	198
O: Rhodocyclales	*G: Methyloversatilis*	63	0.2	216
C: Gammaproteobacteria		58	0.1	240
O: Nitrospirales	*G: Nitrospira*	53	0.1	260

The corresponding terminal-restriction fragments (T-RFs, restriction with *Hae*III endonuclease) referred to as operational taxonomic units (OTUs) were obtained after dry-lab processing of the sequencing datasets and experimental T-RFLP profiles obtained for the same biological samples in PyroTRF-ID (Weissbrodt et al. 2012b). The phylogenetic affiliations are first given at broader level from Phylum → Class → Order and then with more resolution on the deepest lineage identified from Order → Family → Genus after mapping against the Greengenes database (McDonald et al. 2012)

Predominant OTUs and Corresponding Bacterial Phylotypes

See Table 5.7.

Table 5.7 Predominant terminal-restriction fragments (T-RFs, restriction with *Hae*III endonuclease) here referred to as operational taxonomic units (OTUs) detected across all experimental T-RFLP datasets, and listed in numerical order

T-RF or OTU (bp)	Broader affiliation Phylum → Class → Order	Closest affiliation Order → Family → Genus	Total read counts (–)	Fraction of total reads (%)
32	*O: Xanthomonadales*	*G: Thermomonas*	2772	6.9
32	*O: Xanthomonadales*	*G: Pseudoxanthomonas*	930	2.3
32	*C: Gammaproteobacteria*	*F: Xanthomonadaceae*	302	0.8
32	*O: Flavobacteriales*	*F: Cryomorphaceae*	274	0.7
32	*C: Gammaproteobacteria*	*O: Xanthomonadaceae*	257	0.6
32	*O: Flavobacteriales*	*G: Sejongia*	251	0.6
32	*O: Xanthomonadales*	*G: Dokdonella*	221	0.6
32	*O: Rhodobacterales*	*G: Thioclava*	202	0.5
32	*O: Rhodocyclales*	*G: Rhodocyclus*	198	0.5
32	*P: Bacteroidetes*	*F: Flavobacteriaceae*	188	0.5
32	*O: Rhodobacterales*	*G: Rhodobacter*	122	0.3
32	*C: Alphaproteobacteria*	*F: Rhodobacteraceae*	32	0.1

(continued)

Table 5.7 (continued)

T-RF or OTU (bp)	Broader affiliation Phylum → Class → Order	Closest affiliation Order → Family → Genus	Total read counts (–)	Fraction of total reads (%)
32	*O: Flavobacteriales*	*G: Kaistella*	28	0.1
32	*O: Sphingobacteriales*	*F: Saprospiraceae*	27	0.1
32	*O: Rhizobiales*	*G: Methylocella*	26	0.1
32	*O: Sphingobacteriales*	*G: Cytophaga*	23	0.1
62	*O: Actinomycetales*	*G: Tessaracoccus*	22	0.1
72	*O: Rhodocyclales*	*G: Zoogloea*	29	0.1
178	*C: Alphaproteobacteria*	*F: Rhodospirillaceae*	1099	2.8
185	*O: Rhizobiales*		22	0.1
187	*O: Rhodobacterales*	*F: Hyphomonadaceae*	48	0.1
190	*O: Rhizobiales*		37	0.1
193	*O: Burkholderiales*	*G: Acidovorax*	168	0.4
193	*O: Burkholderiales*	*G: Simplicispira*	25	0.1
194	*C: Alphaproteobacteria*	*F: Rhodospirillaceae*	244	0.6
195	*O: Rhodocyclales*	*G: Zoogloea*	5050	12.6
195	*O: Xanthomonadales*	*G: Pseudoxanthomonas*	38	0.1
195	*O: Burkholderiales*	*F: Comamonadaceae*	30	0.1
198	*P: candidate phylum TM7*		65	0.2
206	*O: Sphingobacteriales*		32	0.1
208	*O: Burkholderiales*		22	0.1
209	*C: Acidobacteria*	*F: Acidobacteriaceae*	4009	10.0
209	*P: Armatimonadetes*		427	1.1
210	*C: Acidobacteria*	*F: Acidobacteriaceae*	23	0.1
212	*O: Burkholderiales*		72	0.2
212	*O: Burkholderiales*	*G: Mitsuaria*	31	0.1
212	*O: Rhodocyclales*	*F: Rhodocyclaceae*	21	0.1
212	*O: Burkholderiales*	*F: Comamonadaceae*	20	0.1
213	*O: Rhodocyclales*	*G: Rhodocyclus*	492	1.2
213	*O: Burkholderiales*	*G: Paucibacter*	283	0.7
214	*O: Rhodocyclales*	*G: Rhodocyclus*	8944	22.4
214	*O: Rhodocyclales*	*G: Dechloromonas*	352	0.9
214	*C: Betaproteobacteria*		150	0.4
214	*O: Burkholderiales*	*G: Rhodoferax*	92	0.2
214	*O: Rhodocyclales*	*G: Methyloversatilis*	41	0.1
215	*O: Rhodocyclales*	*G: Rhodocyclus*	155	0.4

(continued)

Table 5.7 (continued)

T-RF or OTU (bp)	Broader affiliation Phylum → Class → Order	Closest affiliation Order → Family → Genus	Total read counts (–)	Fraction of total reads (%)
216	*O: Rhodocyclales*	*G: Methyloversatilis*	63	0.2
216	*O: Nitrosomonadales*	*G: Nitrosomonas*	32	0.1
216	*O: Rhodocyclales*	*G: Rhodocyclus*	27	0.1
220	*O: Rhizobiales*	*G: Aminobacter*	474	1.2
222	*O: Rhodobacterales*	*F: Hyphomonadaceae*	231	0.6
223	*O: Actinomycetales*	*G: Tetrasphaera*	575	1.4
223	*O: Actinomycetales*	*F: Intrasporangiaceae*	377	0.9
223	*O: Rhodobacterales*	*F: Hyphomonadaceae*	40	0.1
224	*O: Rhodobacterales*	*F: Hyphomonadaceae*	135	0.3
228	*O: Actinomycetales*	*F: Intrasporangiaceae*	100	0.3
232	*C: Alphaproteobacteria*	*F: Rhodospirillaceae*	47	0.1
233	*P: candidate phylum TM7*		26	0.1
237	*C: Gammaproteobacteria*		80	0.2
238	*C: Gammaproteobacteria*		4267	10.7
239	*C: Gammaproteobacteria*		1181	3.0
239–248	*C: Gammaproteobacteria*		116	0.3
240	*C: Gammaproteobacteria*		58	0.1
248	*O: Rhodocyclales*	*G: Rhodocyclus*	43	0.1
250	*O: Pseudomonadales*	*G: Acinetobacter*	35	0.1
252	*O: Sphingobacteriales*		23	0.1
253	*O: Sphingobacteriales*		233	0.6
253	*O: Ignavibacteriales*	*F: Ignavibacteriaceae*	28	0.1
254	*O: Sphingobacteriales*		21	0.1
255	*O: Sphingobacteriales*		45	0.1
256	*O: Sphingobacteriales*		385	1.0
257	*O: Nitrospirales*	*G: Nitrospira*	39	0.1
257	*O: Sphingobacteriales*		26	0.1
258	*O: Nitrospirales*	*G: Nitrospira*	71	0.2
259	*O: Sphingobacteriales*		38	0.1
260	*O: Nitrospirales*	*G: Nitrospira*	53	0.1
260	*O: Sphingobacteriales*		47	0.1
262	*O: Sphingobacteriales*		23	0.1
286	*O: Rhizobiales*	*G: Bradyrhizobium*	71	0.2
288	*O: Sphingomonadales*	*G: Sphingomonas*	175	0.4
297	*P: Chloroflexi*	*G: Herpetosiphon*	158	0.4

(continued)

Table 5.7 (continued)

T-RF or OTU (bp)	Broader affiliation Phylum → Class → Order	Closest affiliation Order → Family → Genus	Total read counts (–)	Fraction of total reads (%)
297	*O: Anaerolineae*		22	0.1
298	*C: Gammaprotebacteria*		412	1.0
298	*P: Chloroflexi*	*G: Herpetosiphon*	221	0.6
303	*C: Gammaproteobacteria*		124	0.3
304	*C: Gammaproteobacteria*		28	0.1
306	*P: Armatimonadetes*		45	0.1
306	*P: Armatimonadetes*		38	0.1
307	*P: Armatimonadetes*		80	0.2
318	*O: Sphingobacteriales*	*G: Cytophaga*	34	0.1
321	*O: Sphingobacteriales*	*G: Cytophaga*	100	0.3
321	*O: Sphingobacteriales*		23	0.1
364	*O: Sphingobacteriales*		38	0.1
392	*C: Deltaproteobacteria*	*G: Bdellovibrio*	77	0.2
393	*C: Deltaproteobacteria*	*G: Bdellovibrio*	26	0.1

The corresponding bacterial relatives are summarized across the 16S rRNA gene-targeted amplicon sequencing datasets obtained for all 10 representative biomass samples after dry-lab processing in PyroTRF-ID (Weissbrodt et al. 2012b). The phylogenetic affiliations are first given at broader level from Phylum → Class → Order and then with more resolution on the deepest relative identified in this lineage from Order → Family → Genus after mapping against Greengenes (McDonald et al. 2012)

All Bacterial Phylotypes and Their Corresponding OTUs

See Table 5.8.

Table 5.8 List of all bacterial relatives detected across the 16S rRNA gene-targeted amplicon sequencing datasets of all 10 representative biomass samples, and classified in alphabetical order of their broader phylogenetic level

Broader affiliation Phylum → Class → Order	Closest affiliation Order → Family → Genus	Total read counts (–)	Fraction of total reads (%)	T-RF or OTU (bp)
C: Acidobacteria	*F: Acidobacteriaceae*	6	0.0	178
C: Acidobacteria	*F: Acidobacteriaceae*	18	0.0	200
C: Acidobacteria	*F: Acidobacteriaceae*	17	0.0	208

(continued)

Table 5.8 (continued)

Broader affiliation Phylum → Class → Order	Closest affiliation Order → Family → Genus	Total read counts (–)	Fraction of total reads (%)	T-RF or OTU (bp)
C: Acidobacteria	*F: Acidobacteriaceae*	4009	10.0	209
C: Acidobacteria	*F: Acidobacteriaceae*	23	0.1	210
C: Acidobacteria	*F: Candidatus* Solibacter	3	0.0	211
C: Acidobacteria	*F: Holophagaceae*	3	0.0	250
C: Actinobacteria	*F: Microthrixaceae*	6	0.0	228
C: Alphaproteobacteria		4	0.0	188
C: Alphaproteobacteria	*F: Hyphomicrobiaceae*	4	0.0	202
C: Alphaproteobacteria	*F: Hyphomonadaceae*	2	0.0	184
C: Alphaproteobacteria	*F: Rhodobacteraceae*	32	0.1	32
C: Alphaproteobacteria	*F: Rhodospirillaceae*	1099	2.8	178
C: Alphaproteobacteria	*F: Rhodospirillaceae*	4	0.0	186
C: Alphaproteobacteria	*F: Rhodospirillaceae*	244	0.6	194
C: Alphaproteobacteria	*F: Rhodospirillaceae*	6	0.0	195
C: Alphaproteobacteria	*F: Rhodospirillaceae*	3	0.0	229
C: Alphaproteobacteria	*F: Rhodospirillaceae*	47	0.1	232
C: Alphaproteobacteria	*F: Rhodospirillaceae*	17	0.0	290
C: Alphaproteobacteria	*F: Rhodospirillaceae*	5	0.0	294
C: Alphaproteobacteria	*F: Sphingomonadaceae*	9	0.0	222
C: Alphaproteobacteria	*G: Rhodobacter*	8	0.0	32
C: Alphaproteobacteria	*G: Sphingomonas*	7	0.0	66
C: Alphaproteobacteria	*O: Rhodospirillales*	5	0.0	196
C: Anaerolineae		3	0.0	176
C: Anaerolineae		10	0.0	215
C: Anaerolineae		4	0.0	216
C: Anaerolineae		3	0.0	293
C: Anaerolineae		4	0.0	298
C: Anaerolineae		6	0.0	302
C: Armatimonadetes		8	0.0	155
C: Betaproteobacteria		150	0.4	214
C: Betaproteobacteria		5	0.0	248
C: Betaproteobacteria	*F: Rhodocyclaceae*	3	0.0	196
C: Deltaproteobacteria	*G: Bdellovibrio*	77	0.2	392
C: Deltaproteobacteria	*G: Bdellovibrio*	26	0.1	393
C: Deltaproteobacteria	*G: Bdellovibrio*	3	0.0	400
C: Deltaproteobacteria	*O: Myxococcales*	2	0.0	196

(continued)

Table 5.8 (continued)

Broader affiliation Phylum → Class → Order	Closest affiliation Order → Family → Genus	Total read counts (–)	Fraction of total reads (%)	T-RF or OTU (bp)
C: Flavobacteria		10	0.0	281
C: Gammaprotebacteria		412	1.0	298
C: Gammaprotebacteria		8	0.0	304
C: Gammaproteobacteria		28	0.1	304
C: Gammaproteobacteria		11	0.0	67
C: Gammaproteobacteria		12	0.0	68
C: Gammaproteobacteria		6	0.0	204
C: Gammaproteobacteria		8	0.0	235
C: Gammaproteobacteria		80	0.2	237
C: Gammaproteobacteria		4267	10.7	238
C: Gammaproteobacteria		1181	3.0	239
C: Gammaproteobacteria		58	0.1	240
C: Gammaproteobacteria		6	0.0	242
C: Gammaproteobacteria		124	0.3	303
C: Gammaproteobacteria		8	0.0	386
C: Gammaproteobacteria		116	0.3	239–248
C: Gammaproteobacteria	*F: Xanthomonadaceae*	302	0.8	32
C: Gammaproteobacteria	*F: Xanthomonadaceae*	7	0.0	195
C: Gammaproteobacteria	*G: Legionella*	2	0.0	202
C: Gammaproteobacteria	*G: Rhodanobacter*	2	0.0	250
C: Gammaproteobacteria	*G: Thermomonas*	7	0.0	195
C: Gammaproteobacteria	*G:Pseudoxanthomonas*	7	0.0	408
C: Gammaproteobacteria	*O: Xanthomonadales*	257	0.6	32
C: Gammaproteobacteria	*O: Xanthomonaales*	10	0.0	72
C: Gammaproteobacteria	*O: Xanthomonadales*	3	0.0	196
C: Gammaproteobacteria	*O: Xanthomonadales*	4	0.0	249
O: Acidimicrobiales	*F: Microthrixaceae*	4	0.0	189
O: Acidobacteriales		6	0.0	210
O: Actinomycetales		3	0.0	216
O: Actinomycetales		15	0.0	217
O: Actinomycetales	*F: Intrasporangiaceae*	377	0.9	223
O: Actinomycetales	*F: Intrasporangiaceae*	19	0.0	224
O: Actinomycetales	*F: Intrasporangiaceae*	2	0.0	225
O: Actinomycetales	*F: Intrasporangiaceae*	100	0.3	228
O: Actinomycetales	*F: Microbacteriaceae*	13	0.0	227
O: Actinomycetales	*F: Nocardiaceae*	8	0.0	63

(continued)

Table 5.8 (continued)

Broader affiliation Phylum → Class → Order	Closest affiliation Order → Family → Genus	Total read counts (–)	Fraction of total reads (%)	T-RF or OTU (bp)
O: Actinomycetales	*G: Propionicimonas*	11	0.0	223
O: Actinomycetales	*G: Tessaracoccus*	22	0.1	62
O: Actinomycetales	*G: Tetrasphaera*	575	1.4	223
O: Anaerolineae		22	0.1	297
O: Burkholderiales		22	0.1	208
O: Burkholderiales		72	0.2	212
O: Burkholderiales	*F: Comamonadaceae*	7	0.0	64
O: Burkholderiales	*F: Comamonadaceae*	3	0.0	188
O: Burkholderiales	*F: Comamonadaceae*	30	0.1	195
O: Burkholderiales	*F: Comamonadaceae*	9	0.0	196
O: Burkholderiales	*F: Comamonadaceae*	7	0.0	211
O: Burkholderiales	*F: Comamonadaceae*	20	0.1	212
O: Burkholderiales	*F: Comamonadaceae*	17	0.0	213
O: Burkholderiales	*F: Comamonadaceae*	2	0.0	215
O: Burkholderiales	*F: Comamonadaceae*	7	0.0	216
O: Burkholderiales	*G: Acidovorax*	18	0.0	190
O: Burkholderiales	*G: Acidovorax*	168	0.4	193
O: Burkholderiales	*G: Acidovorax*	4	0.0	194
O: Burkholderiales	*G: Alicycliphilus*	1	0.0	195
O: Burkholderiales	*G: Aquamonas*	12	0.0	195
O: Burkholderiales	*G: Aquamonas*	11	0.0	214
O: Burkholderiales	*G: Delftia*	13	0.0	193
O: Burkholderiales	*G: Hydrogenophaga*	4	0.0	64
O: Burkholderiales	*G: Hylemonella*	6	0.0	212
O: Burkholderiales	*G: Ideonella*	2	0.0	217
O: Burkholderiales	*G: Mitsuaria*	31	0.1	212
O: Burkholderiales	*G: Paucibacter*	283	0.7	213
O: Burkholderiales	*G: Rhodoferax*	6	0.0	208
O: Burkholderiales	*G: Rhodoferax*	92	0.2	214
O: Burkholderiales	*G: Rhodoferax*	4	0.0	245
O: Burkholderiales	*G: Simplicispira*	25	0.1	193
O: Burkholderiales	*G: Sphaerotilus*	1	0.0	208
O: Burkholderiales	*G: Xenophilus*	10	0.0	212
O: Chromatiales	*F: Sinobacteraceae*	2	0.0	195

(continued)

Table 5.8 (continued)

Broader affiliation Phylum → Class → Order	Closest affiliation Order → Family → Genus	Total read counts (–)	Fraction of total reads (%)	T-RF or OTU (bp)
O: Clostridiales	*G: Ruminococcus*	1	0.0	294
O: Desulfuromonadales		3	0.0	224
O: Flavobacteriales	*F: Cryomorphaceae*	274	0.7	32
O: Flavobacteriales	*F: Flavobacteriaceae*	8	0.0	32
O: Flavobacteriales	*G: Flavobacterium*	4	0.0	32
O: Flavobacteriales	*G: Flavobacterium*	1	0.0	398
O: Flavobacteriales	*G: Kaistella*	28	0.1	32
O: Flavobacteriales	*G: Sejongia*	251	0.6	32
O: Hydrogenophilales	*G: Thiobacillus*	3	0.0	212
O: Hydrogenophilales	*G: Thiobacillus*	15	0.0	216
O: Ignavibacteriales	*F: Ignavibacteriaceae*	28	0.1	253
O: Myxococcales		5	0.0	198
O: Myxococcales	*F: Haliangiaceae*	5	0.0	32
O: Myxococcales	*F: Haliangiaceae*	4	0.0	71
O: Nitrosomonadales	*G: Nitrosomonas*	1	0.0	214
O: Nitrosomonadales	*G: Nitrosomonas*	2	0.0	215
O: Nitrosomonadales	*G: Nitrosomonas*	32	0.1	216
O: Nitrosomonadales	*G: Nitrosomonas*	2	0.0	217
O: Nitrospirales	*G: Nitrospira*	39	0.1	257
O: Nitrospirales	*G: Nitrospira*	71	0.2	258
O: Nitrospirales	*G: Nitrospira*	1	0.0	259
O: Nitrospirales	*G: Nitrospira*	53	0.1	260
O: Nitrospirales	*G: Nitrospira*	8	0.0	261
O: Nitrospirales	*G: Nitrospira*	10	0.0	326
O: Phycisphaerales		4	0.0	58
O: Phycisphaerales		10	0.0	220
O: Phycisphaerales		2	0.0	233
O: Pirellulales		2	0.0	227
O: Planctomycetales	*G: Planctomyces*	8	0.0	404
O: Pseudomonadales	*G: Acinetobacter*	18	0.0	195
O: Pseudomonadales	*G: Acinetobacter*	35	0.1	250
O: Rhizobiales		22	0.1	185
O: Rhizobiales		10	0.0	186
O: Rhizobiales		18	0.0	189
O: Rhizobiales		37	0.1	190

(continued)

Table 5.8 (continued)

Broader affiliation Phylum → Class → Order	Closest affiliation Order → Family → Genus	Total read counts (–)	Fraction of total reads (%)	T-RF or OTU (bp)
O: Rhizobiales		6	0.0	220
O: Rhizobiales	*F: Bradyrhizobiaceae*	5	0.0	186
O: Rhizobiales	*F: Bradyrhizobiaceae*	7	0.0	187
O: Rhizobiales	*F: Bradyrhizobiaceae*	10	0.0	188
O: Rhizobiales	*F: Bradyrhizobiaceae*	14	0.0	190
O: Rhizobiales	*F: Bradyrhizobiaceae*	12	0.0	290
O: Rhizobiales	*F: Hyphomicrobiaceae*	9	0.0	32
O: Rhizobiales	*G: Aminobacter*	10	0.0	219
O: Rhizobiales	*G: Aminobacter*	474	1.2	220
O: Rhizobiales	*G: Aminobacter*	10	0.0	283
O: Rhizobiales	*G: Aminobacter*	7	0.0	371
O: Rhizobiales	*G: Bradyrhizobium*	71	0.2	286
O: Rhizobiales	*G: Devosia*	2	0.0	217
O: Rhizobiales	*G: Devosia*	4	0.0	220
O: Rhizobiales	*G: Devosia*	6	0.0	286
O: Rhizobiales	*G: Hyphomicrobium*	6	0.0	32
O: Rhizobiales	*G: Methylocella*	26	0.1	32
O: Rhizobiales	*G: Methylocystis*	11	0.0	184
O: Rhizobiales	*G: Methylosinus*	16	0.0	186
O: Rhizobiales	*G: Mezorhizobium*	6	0.0	220
O: Rhizobiales	*G: Rhodoplanes*	2	0.0	186
O: Rhodobacterales		1	0.0	185
O: Rhodobacterales	*F: Hyphomonadaceae*	9	0.0	185
O: Rhodobacterales	*F: Hyphomonadaceae*	48	0.1	187
O: Rhodobacterales	*F: Hyphomonadaceae*	8	0.0	188
O: Rhodobacterales	*F: Hyphomonadaceae*	3	0.0	211
O: Rhodobacterales	*F: Hyphomonadaceae*	9	0.0	221
O: Rhodobacterales	*F: Hyphomonadaceae*	231	0.6	222
O: Rhodobacterales	*F: Hyphomonadaceae*	40	0.1	223
O: Rhodobacterales	*F: Hyphomonadaceae*	135	0.3	224
O: Rhodobacterales	*G: Rhodobacter*	122	0.3	32
O: Rhodobacterales	*G: Thioclava*	202	0.5	32
O: Rhodocyclales	*F: Rhodocyclaceae*	6	0.0	211
O: Rhodocyclales	*F: Rhodocyclaceae*	21	0.1	212
O: Rhodocyclales	*F: Rhodocyclaceae*	7	0.0	216

(continued)

Table 5.8 (continued)

Broader affiliation Phylum → Class → Order	Closest affiliation Order → Family → Genus	Total read counts (–)	Fraction of total reads (%)	T-RF or OTU (bp)
O: Rhodocyclales	*G: Dechloromonas*	2	0.0	195
O: Rhodocyclales	*G: Dechloromonas*	1	0.0	212
O: Rhodocyclales	*G: Dechloromonas*	1	0.0	213
O: Rhodocyclales	*G: Dechloromonas*	352	0.9	214
O: Rhodocyclales	*G: Dechloromonas*	2	0.0	215
O: Rhodocyclales	*G: Dechloromonas*	12	0.0	399
O: Rhodocyclales	*G: Methyloversatilis*	2	0.0	195
O: Rhodocyclales	*G: Methyloversatilis*	3	0.0	196
O: Rhodocyclales	*G: Methyloversatilis*	1	0.0	213
O: Rhodocyclales	*G: Methyloversatilis*	41	0.1	214
O: Rhodocyclales	*G: Methyloversatilis*	18	0.0	215
O: Rhodocyclales	*G: Methyloversatilis*	63	0.2	216
O: Rhodocyclales	*G: Rhodocyclus*	198	0.5	32
O: Rhodocyclales	*G: Rhodocyclus*	4	0.0	195
O: Rhodocyclales	*G: Rhodocyclus*	6	0.0	200
O: Rhodocyclales	*G: Rhodocyclus*	2	0.0	202
O: Rhodocyclales	*G: Rhodocyclus*	7	0.0	204
O: Rhodocyclales	*G: Rhodocyclus*	12	0.0	205
O: Rhodocyclales	*G: Rhodocyclus*	6	0.0	207
O: Rhodocyclales	*G: Rhodocyclus*	13	0.0	212
O: Rhodocyclales	*G: Rhodocyclus*	492	1.2	213
O: Rhodocyclales	*G: Rhodocyclus*	8944	22.4	214
O: Rhodocyclales	*G: Rhodocyclus*	155	0.4	215
O: Rhodocyclales	*G: Rhodocyclus*	27	0.1	216
O: Rhodocyclales	*G: Rhodocyclus*	5	0.0	217
O: Rhodocyclales	*G: Rhodocyclus*	4	0.0	219
O: Rhodocyclales	*G: Rhodocyclus*	3	0.0	247
O: Rhodocyclales	*G: Rhodocyclus*	43	0.1	248
O: Rhodocyclales	*G: Rhodocyclus*	12	0.0	249
O: Rhodocyclales	*G: Rhodocyclus*	1	0.0	307
O: Rhodocyclales	*G: Thauera*	13	0.0	72
O: Rhodocyclales	*G: Thauera*	12	0.0	217
O: Rhodocyclales	*G: Zoogloea*	5050	12.6	195
O: Rhodocyclales	*G: Zoogloea*	12	0.0	214
O: Rhodocyclales	*G: Zoogloea*	29	0.1	72

(continued)

Table 5.8 (continued)

Broader affiliation Phylum → Class → Order	Closest affiliation Order → Family → Genus	Total read counts (–)	Fraction of total reads (%)	T-RF or OTU (bp)
O: Rhodospirillales	*F: Rhodospirillaceae*	3	0.0	289
O: Rhodospirillales	*F: Rhodospirillaceae*	4	0.0	290
O: Roseiflexales		3	0.0	233
O: Sphingobacteriales		7	0.0	257
O: Sphingobacteriales		4	0.0	260
O: Sphingobacteriales		8	0.0	32
O: Sphingobacteriales		32	0.1	206
O: Sphingobacteriales		2	0.0	250
O: Sphingobacteriales		23	0.1	252
O: Sphingobacteriales		233	0.6	253
O: Sphingobacteriales		21	0.1	254
O: Sphingobacteriales		45	0.1	255
O: Sphingobacteriales		385	1.0	256
O: Sphingobacteriales		26	0.1	257
O: Sphingobacteriales		10	0.0	258
O: Sphingobacteriales		38	0.1	259
O: Sphingobacteriales		47	0.1	260
O: Sphingobacteriales		5	0.0	261
O: Sphingobacteriales		23	0.1	262
O: Sphingobacteriales		11	0.0	263
O: Sphingobacteriales		4	0.0	309
O: Sphingobacteriales		23	0.1	321
O: Sphingobacteriales		7	0.0	322
O: Sphingobacteriales		17	0.0	323
O: Sphingobacteriales		4	0.0	325
O: Sphingobacteriales		38	0.1	364
O: Sphingobacteriales		1	0.0	404
O: Sphingobacteriales	*F: Flexibacteraceae*	8	0.0	303
O: Sphingobacteriales	*F: Saprospiraceae*	27	0.1	32
O: Sphingobacteriales	*G: Cytophaga*	23	0.1	32
O: Sphingobacteriales	*G: Cytophaga*	2	0.0	251
O: Sphingobacteriales	*G: Cytophaga*	34	0.1	318
O: Sphingobacteriales	*G: Cytophaga*	100	0.3	321
O: Sphingobacteriales	*G: Niabella*	2	0.0	253

(continued)

Table 5.8 (continued)

Broader affiliation Phylum → Class → Order	Closest affiliation Order → Family → Genus	Total read counts (–)	Fraction of total reads (%)	T-RF or OTU (bp)
O: Sphingobacteriales	*G: Niabella*	4	0.0	254
O: Sphingobacteriales	*G: Spirosoma*	5	0.0	32
O: Sphingomonadales		18	0.0	222
O: Sphingomonadales	*F: Sphingomonadaceae*	5	0.0	288
O: Sphingomonadales	*G: Novosphingobium*	4	0.0	222
O: Sphingomonadales	*G: Sphingobium*	4	0.0	289
O: Sphingomonadales	*G: Sphingomonas*	175	0.4	288
O: Sphingomonadales	*G: Sphingopyxis*	12	0.0	288
O: Sphingomonadales	*G: Sphingopyxis*	10	0.0	289
O: Sphingomonadales	*G: Sphingosinicella*	14	0.0	288
O: Spirochaetales	*G: Spirochaeta*	3	0.0	251
O: Thiotrichales	*G: Thiothrix*	3	0.0	264
O: Verrucomicrobiales	*G: Verrucomicrobium*	2	0.0	222
O: Xanthomonadales		5	0.0	193
O: Xanthomonadales	*F: Xanthomonadaceae*	8	0.0	201
O: Xanthomonadales	*G: Dokdonella*	221	0.6	32
O: Xanthomonadales	*G: Pseudoxanthomonas*	930	2.3	32
O: Xanthomonadales	*G: Pseudoxanthomonas*	3	0.0	183
O: Xanthomonadales	*G: Pseudoxanthomonas*	38	0.1	195
O: Xanthomonadales	*G: Pseudoxanthomonas*	13	0.0	200
O: Xanthomonadales	*G: Stenotrophonomas*	12	0.0	32
O: Xanthomonadales	*G: Thermomonas*	2772	6.9	32
O: Xanthomonadales	*G: Thermomonas*	5	0.0	72
O: Xanthomonadales	*G: Thermomonas*	8	0.0	194
O: Xanthomonadales	*G: Thermomonas*	5	0.0	195
O: Xanthomonadales	*G: Thermomonas*	3	0.0	197
O: Xanthomonadales	*G: Thermomonas*	7	0.0	201
P: Acidobacteria		4	0.0	180
P: Acidobacteria	*G: Candidatus* Solibacter	3	0.0	265
P: Armatimonadetes		45	0.1	306
P: Armatimonadetes		2	0.0	58
P: Armatimonadetes		427	1.1	209
P: Armatimonadetes		15	0.0	211

(continued)

Table 5.8 (continued)

Broader affiliation Phylum → Class → Order	Closest affiliation Order → Family → Genus	Total read counts (–)	Fraction of total reads (%)	T-RF or OTU (bp)
P: Armatimonadetes		13	0.0	238
P: Armatimonadetes		38	0.1	306
P: Armatimonadetes		80	0.2	307
P: Bacteroidetes	*F: Flavobacteriaceae*	188	0.5	32
P: candidate phylum TM7		10	0.0	62
P: candidate phylum TM7		65	0.2	198
P: candidate phylum TM7		3	0.0	200
P: candidate phylum TM7		26	0.1	233
P: candidate phylum TM7		2	0.0	234
P: candidate phylum TM7		5	0.0	257
P: Chlorobi		2	0.0	57
P: Chlorobi		7	0.0	402
P: Chloroflexi		10	0.0	201
P: Chloroflexi	*C: Anaerolineae*	18	0.0	215
P: Chloroflexi	*C: Anaerolineae*	10	0.0	216
P: Chloroflexi	*C: Anaerolineae*	19	0.0	270
P: Chloroflexi	*G: Caldilinea*	9	0.0	215
P: Chloroflexi	*G: Herpetosiphon*	12	0.0	294
P: Chloroflexi	*G: Herpetosiphon*	11	0.0	295
P: Chloroflexi	*G: Herpetosiphon*	158	0.4	297
P: Chloroflexi	*G: Herpetosiphon*	221	0.6	298
P: Chloroflexi	*G: Herpetosiphon*	6	0.0	314
P: Firmicutes	*G: Trichococcus*	5	0.0	210
P: Gemmatimonadetes	*G: Gemmatimonas*	6	0.0	58
P: Spirochaetes	*F: Leptospiraceae*	10	0.0	277
P: Spirochaetes	*F: Spirochaetaceae*	3	0.0	179
P: Verrucomicrobia	*F: Opitutaceae*	2	0.0	176

The corresponding terminal-restriction fragments (T-RFs, restriction with *Hae*III endonuclease) here referred to as operational taxonomic units (OTUs) were obtained after dry-lab processing of the sequencing datasets and experimental T-RFLP profiles obtained for the same biological samples in PyroTRF-ID (Weissbrodt et al. 2012b). The phylogenetic affiliations are first given at broader level from Phylum → Class → Order and then with more resolution on the deepest relative identified in this lineage from Order → Family → Genus after mapping against the Greengenes database (McDonald et al. 2012)

All OTUs and Their Corresponding Bacterial Phylotypes

See Table 5.9.

Table 5.9 List of all predominant terminal-restriction fragments (T-RFs, restriction with *Hae*III endonuclease) here referred to as operational taxonomic units (OTUs) detected across all experimental T-RFLP datasets, and listed in numerical order

T-RF or OTU (bp)	Biomass origin[a]	Biological sample	Fraction of T-RF (%)[b]	Read counts (–)[c]	Broader affiliation Phylum → Class → Order	Closest affiliation Order → Family → Genus	GenBank accession number	Absolute SW mapping score (–)[d]	Normalized SW mapping score (–)[e]
32	AGS 20 °C	X066	35	276	*O: Xanthomonadales*	*G: Thermomonas*	EU834762	425	0.562
32	AGS 20 °C	X066			*O: Xanthomonadales*	*G: Thermomonas*	AB355702	425	0.775
32	AGS 20 °C	X066	16	128	*O: Xanthomonadales*	*G: Pseudoxanthomonas*	AY512829	385	0.753
32	AGS 20 °C	X066	14	111	*C: Gammaproteobacteria*	*O: Xanthomonadaceae*	AY212636	386	0.897
32	AGS 20 °C	X066	11	88	*O: Flavobacteriales*	*G: Sejongia*	AY468464	434	0.655
32	AGS 20 °C	X066	5	37	*O: Rhodocyclales*	*G: Rhodocyclus*	AF502230	363	0.851
32	AGS 20 °C	X066	4	32	*O: Xanthomonadales*	*G: Dokdonella*	AM981200	349	0.918
32	AGS 20 °C	X066	4	29	*O: Rhodobacterales*	*G: Thioclava*	CU919741	401	0.965
32	AGS 20 °C	X066	1	17	*O: Rhodobacterales*	*G: Rhodobacter*	AY212706	448	0.965
32	AGS 20 °C	X066	1	11	*O: Flavobacteriales*	*F: Cryomorphaceae*	EU803886	320	0.617
32	AGS 20 °C	X066	0.9	7	*O: Flavobacteriales*	*G: Kaistella*	AF502204	339	0.407
32	AGS 20 °C	X066	0.4	3	*O: Xanthomonadales*	*G: Stenotrophonomas*	AY259519	322	0.861
32	AGS 20 °C	X084	43	348	*O: Xanthomonadales*	*G: Pseudoxanthomonas*	AY512829	385	0.964
32	AGS 20 °C	X084			*O: Xanthomonadales*	*G: Pseudoxanthomonas*	EU834761	385	0.865
32	AGS 20 °C	X084	13	107	*O: Xanthomonadales*	*G: Thermomonas*	EU834762	418	0.722
32	AGS 20 °C	X084			*O: Xanthomonadales*	*G: Thermomonas*	DQ376561	418	0.915
32	AGS 20 °C	X084	10	85	*O: Xanthomonadales*	*G: Dokdonella*	FJ660523	391	0.912

(continued)

Table 5.9 (continued)

T-RF or OTU (bp)	Biomass origin[a]	Biological sample	Fraction of T-RF (%)[b]	Read counts (–)[c]	Broader affiliation Phylum → Class → Order	Closest affiliation Order → Family → Genus	GenBank accession number	Absolute SW mapping score (–)[d]	Normalized SW mapping score (–)[e]
32	AGS 20 °C	X084			*C: Xanthomonadales*	*G: Dokdonella*	AM981200	391	0.943
32	AGS 20 °C	X084	9	71	*C: Rhodocyclales*	*G: Rhodocyclus*	AF502227	367	0.982
32	AGS 20 °C	X084	5	39	*C: Rhodobacterales*	*G: Thioclava*	CU919741	426	1
32	AGS 20 °C	X084	5	39	*C: Rhodobacterales*	*G: Rhodobacter*	AY212706	418	0.902
32	AGS 20 °C	X084	5	38	*C: Flavobacteriales*	*F: Cryomorphaceae*	EU803886	305	0.902
32	AGS 20 °C	X084	3	21	*C: Flavobacteriales*	*G: Sejongia*	AY468464	369	0.651
32	AGS 20 °C	X084	2	14	*C: Gammaproteobacteria*	*O: Xanthomonadaceae*	AY212636	385	0.738
32	AGS 20 °C	X084	1	9	*C: Sphingobacteriales*	*F: Saprospiraceae*	DQ856550	257	0.529
32	AGS 20 °C	X084	1	8	*C: Flavobacteriales*	*F: Flavobacteriaceae*	AY682382	221	0.738
32	AGS 20 °C	X084	0.7	6	*C: Sphingobacteriales*	*G: Cytophaga*	FJ516870	262	0.738
32	AGS 20 °C	X084	0.7	6	*O: Rhizobiales*	*G: Hyphomicrobium*	GQ264377	298	0.791
32	AGS 20 °C	X084	0.6	5	*O: Myxococcales*	*F: Haliangiaceae*	AB179520	283	0.784
32	AGS 20 °C	X084	0.5	4	*O: Flavobacteriales*	*G: Flavobacterium*	AJ440981	294	0.919
32	AGS 20 °C	X084	0.1	1	*O: Xanthomonadales*	*G: Stenotrophonomas*	AY259519	225	0.746
32	AGS 20 °C	X114	32	149	*O: Xanthomonadales*	*G: Thermomonas*	EU834762	373	0.935
32	AGS 20 °C	X114			*O: Xanthomonadales*	*G: Thermomonas*	DQ376561	373	0.937
32	AGS 20 °C	X114	28	128	*O: Xanthomonadales*	*G: Pseudoxanthomonas*	EU834761	390	0.88
32	AGS 20 °C	X114			*O: Xanthomonadales*	*G: Pseudoxanthomonas*	AY512829	390	0.92
32	AGS 20 °C	X114	10	46	*O: Flavobacteriales*	*F: Cryomorphaceae*	EU803886	301	0.674

(continued)

Table 5.9 (continued)

T-RF or OTU (bp)	Biomass origin[a]	Biological sample	Fraction of T-RF (%)[b]	Read counts (–)[c]	Broader affiliation Phylum → Class → Order	Closest affiliation Order → Family → Genus	GenBank accession number	Absolute SW mapping score (–)[d]	Normalized SW mapping score (–)[e]
32	AGS 20 °C	X114	9	40	*O: Xanthomonadales*	*G: Dokdonella*	FM213040	390	0.78
32	AGS 20 °C	X114			*O: Xanthomonadales*	*G: Dokdonella*	AM981200	390	0.508
32	AGS 20 °C	X114	5	21	*C: Gammaproteobacteria*	*O: Xanthomonadaceae*	GQ396926	375	0.503
32	AGS 20 °C	X114	5	21	*O: Rhodobacterales*	*G: Thioclava*	CU919741	396	0.882
32	AGS 20 °C	X114	2	11	*O: Flavobacteriales*	*G: Sejongia*	AY468464	346	0.755
32	AGS 20 °C	X114	2	8	*C: Alphaproteobacteria*	*G: Rhodobacter*	AY212710	314	0.727
32	AGS 20 °C	X114	3	2	*O: Rhodocyclales*	*G: Rhodocyclus*	AF502227	356	0.727
32	AGS 23 ± 2 °C	BC-II	35	276	*O: Xanthomonadales*	*G: Thermomonas*	EU834762	425	0.794
32	AGS 23 ± 2 °C	BC-II			*O: Xanthomonadales*	*G: Thermomonas*	AB355702	425	0.886
32	AGS 23 ± 2 °C	BC-II	16	128	*O: Xanthomonadales*	*G: Pseudoxanthomonas*	AY512829	385	0.831
32	AGS 23 ± 2 °C	BC-II	14	111	*C: Gammaproteobacteria*	*O: Xanthomonadaceae*	AY212636	386	0.51
32	AGS 23 ± 2 °C	BC-II	11	88	*O: Flavobacteriales*	*G: Sejongia*	AY468464	434	0.93
32	AGS 23 ± 2 °C	BC-II	5	37	*O: Rhodocyclales*	*G: Rhodocyclus*	AF502230	363	0.566
32	AGS 23 ± 2 °C	BC-II	4	32	*O: Xanthomonadales*	*G: Dokdonella*	AM981200	349	0.752
32	AGS 23 ± 2 °C	BC-II	3.7	29	*O: Rhodobacterales*	*G: Thioclava*	CU919741	401	0.663
32	AGS 23 ± 2 °C	BC-II	1.2	17	*O: Rhodobacterales*	*G: Rhodobacter*	AY212706	448	0.663
32	AGS 23 ± 2 °C	BC-II	1.3	11	*O: Flavobacteriales*	*F: Cryomorphaceae*	EU803886	320	0.612
32	AGS 23 ± 2 °C	BC-II	0.9	7	*O: Flavobacteriales*	*G: Kaistella*	AF502204	339	0.613
32	AGS 23 ± 2 °C	BC-II	0.4	3	*O: Xanthomonadales*	*G: Stenotrophonomas*	AY259519	322	0.81

(continued)

Table 5.9 (continued)

T-RF or OTU (bp)	Biomass origin[a]	Biological sample	Fraction of T-RF (%)[b]	Read counts (–)[c]	Broader affiliation Phylum → Class → Order	Closest affiliation Order → Family → Genus	GenBank accession number	Absolute SW mapping score (–)[d]	Normalized SW mapping score (–)[e]
32	AGS 25 °C	O125	53	402	*O: Xanthomonadales*	*G: Thermomonas*	EU834762	455	0.481
32	AGS 25 °C	O125			*O: Xanthomonadales*	*G: Thermomonas*	DQ376561	455	0.87
32	AGS 25 °C	O125	20	151	*O: Flavobacteriales*	*F: Cryomorphaceae*	EU803886	324	0.819
32	AGS 25 °C	O125	4	27	*O: Rhodobacterales*	*G: Rhodobacter*	CU920503	432	0.87
32	AGS 25 °C	O125	3	26	*O: Flavobacteriales*	*G: Sejongia*	AY468464	388	0.565
32	AGS 25 °C	O125	3	26	*O: Rhodobacterales*	*G: Thioclava*	CU919741	408	0.733
32	AGS 25 °C	O125	3	19	*O: Xanthomonadales*	*G: Pseudoxanthomonas*	AY550263	371	0.565
32	AGS 25 °C	O125	2	18	*O: Sphingobacteriales*	*F: Saprospiraceae*	GU454872	347	0.667
32	AGS 25 °C	O125	2	15	*O: Xanthomonadales*	*G: Dokdonella*	FM213023	338	0.667
32	AGS 25 °C	O125	2	12	*O: Rhodocyclales*	*G: Rhodocyclus*	AF204247	307	0.73
32	AGS 25 °C	O125	1	7	*O: Flavobacteriales*	*G: Kaistella*	AF527584	336	0.844
32	AGS purge	XP06	76	1562	*O: Xanthomonadales*	*G: Thermomonas*	EU834762	452	0.979
32	AGS purge	XP06			*O: Xanthomonadales*	*G: Thermomonas*	DQ376561	452	0.951
32	AGS purge	XP06	9	179	*O: Xanthomonadales*	*G: Pseudoxanthomonas*	EU834761	356	0.909
32	AGS purge	XP06			*O: Xanthomonadales*	*G: Pseudoxanthomonas*	AY512829	356	0.719
32	AGS purge	XP06	3	64	*C: Gammaproteobacteria*	*F: Xanthomonadaceae*	GQ396926	377	0.86
32	AGS purge	XP06	2	39	*O: Rhodocyclales*	*G: Rhodocyclus*	AF204247	333	0.883
32	AGS purge	XP06	0.9	18	*O: Rhodobacterales*	*G: Thioclava*	CU919741	415	0.862
32	AGS purge	XP06	0.8	17	*O: Xanthomonadales*	*G: Dokdonella*	AM981200	346	0.982

(continued)

Table 5.9 (continued)

T-RF or OTU (bp)	Biomass origin[a]	Biological sample	Fraction of T-RF (%)[b]	Read counts (–)[c]	Broader affiliation Phylum → Class → Order	Closest affiliation Order → Family → Genus	GenBank accession number	Absolute SW mapping score (–)[d]	Normalized SW mapping score (–)[e]
32	AGS purge	XP06	0.8	17	*O: Flavobacteriales*	*G: Sejongia*	AY468464	372	0.982
32	AGS purge	XP06	0.8	17	*O: Flavobacteriales*	*F: Cryomorphaceae*	EU803886	291	0.917
32	AGS purge	XP06	0.8	17	*O: Rhodobacterales*	*G: Rhodobacter*	AY212710	369	0.661
32	AGS purge	XP06	0.8	17	*O: Sphingobacteriales*	*G: Cytophaga*	FJ529937	260	0.883
32	AGS purge	XP06	0.4	8	*O: Sphingobacteriales*		GQ340105	332	0.795
32	AGS purge	XP06	0.2	7	*O: Flavobacteriales*	*G: Kaistella*	AF502204	358	0.634
32	AGS purge	XP06	0.2	5	*O: Xanthomonadales*	*G: Stenotrophonomas*	AY259519	275	0.974
32	GAO 28 °C	GAO398	24	40	*O: Rhodobacterales*	*G: Thioclava*	AB079681	373	0.982
32	GAO 28 °C	GAO398	18	30	*C: Gammaproteobacteria*	*F: Xanthomonadaceae*	EU834776	342	0.982
32	GAO 28 °C	GAO398	17	28	*P: Bacteroidetes*	*F: Flavobacteriaceae*	GQ089203	299	0.985
32	GAO 28 °C	GAO398	16	26	*O: Rhizobiales*	*G: Methylocella*	CP001280	344	0.985
32	GAO 28 °C	GAO398	5	9	*O: Rhizobiales*	*F: Hyphomicrobiaceae*	AM411913	367	0.972
32	GAO 28 °C	GAO398	3	5	*O: Rhodobacterales*	*G: Rhodobacter*	FN428770	289	1
32	GAO 28 °C	GAO398	3	5	*O: Sphingobacteriales*	*G: Spirosoma*	EU370956	351	1
32	PAO 17 °C	PAO109	44	208	*C: Gammaproteobacteria*	*F: Xanthomonadaceae*	EU834761	407	0.616
32	PAO 17 °C	PAO109	34	160	*P: Bacteroidetes*	*F: Flavobacteriaceae*	AF502206	395	0.697
32	PAO 17 °C	PAO109	7	32	*C: Alphaproteobacteria*	*F: Rhodobacteraceae*	CU924686	434	0.701
57	AGS 25 °C	O125	100	2	*P: Chlorobi*		AF445728	201	0.686
58	PAO 17 °C	PAO109	50	6	*P: Gemmatimonadetes*	*G: Gemmatimonas*	AP009153	368	0.797

(continued)

Table 5.9 (continued)

T-RF or OTU (bp)	Biomass origin[a]	Biological sample	Fraction of T-RF (%)[b]	Read counts (–)[c]	Broader affiliation Phylum → Class → Order	Closest affiliation Order → Family → Genus	GenBank accession number	Absolute SW mapping score (–)[d]	Normalized SW mapping score (–)[e]
58	PAO 17 °C	PAO109	33	4	*O: Phycisphaerales*		AY957928	165	0.84
58	PAO 17 °C	PAO109	17	2	*P: Armatimonadetes*		GQ263883	157	0.816
62	Flocs	BC002	67	22	*O: Actinomycetales*	*G: Tessaracoccus*	GQ097568	380	0.608
62	Flocs	BC002	30	10	*P: candidate phylum TM7*		EU104134	373	0.721
63	GAO 28 °C	GAO398	100	8	*O: Actinomycetales*	*F: Nocardiaceae*	AF210769	299	0.924
64	AGS 20 °C	X114	50	4	*O: Burkholderiales*	*F: Comamonadaceae*	NR028717	338	0.875
64	AGS 25 °C	O125	100	3	*O: Burkholderiales*	*F: Comamonadaceae*	NR028717	338	0.943
64	AGS purge	XP06	100	2	*O: Burkholderiales*	*G: Hydrogenophaga*	AF078770	339	0.859
64	PAO 17 °C	PAO109	100	2	*O: Burkholderiales*	*G: Hydrogenophaga*	AB300163	362	0.832
66	AGS 20 °C	X084	100	2	*C: Alphaproteobacteria*	*G: Sphingomonas*	D16146	395	0.906
66	AGS 20 °C	X114	100	5	*C: Alphaproteobacteria*	*G: Sphingomonas*	DQ532310	357	0.786
67	AGS 20 °C	X114	100	6	*C: Gammaproteobacteria*		FJ356055	376	0.795
67	AGS purge	XP06	100	5	*C: Gammaproteobacteria*		FJ356050	347	0.96
68	AGS 20 °C	X084	100	12	*C: Gammaproteobacteria*		FJ356055	351	0.96
71	AGS 25 °C	O125	100	2	*O: Myxococcales*	*F: Haliangiaceae*	FM253578	330	0.982
71	AGS purge	XP06	100	2	*O: Myxococcales*	*F: Haliangiaceae*	GQ264118	352	0.982
72	AGS early	BC059	100	11	*O: Rhodocyclales*	*G: Zoogloea*	AJ011506	373	0.959
72	AGS 20 °C	X084	100	2	*C: Gammaproteobacteria*	*O: Xanthomonadaceae*	FN563156	373	0.915
72	AGS 25 °C	O125	100	5	*O: Xanthomonadales*	*G: Thermomonas*	AF508105	268	0.803

(continued)

Table 5.9 (continued)

T-RF or OTU (bp)	Biomass origin[a]	Biological sample	Fraction of T-RF (%)[b]	Read counts (–)[c]	Broader affiliation Phylum → Class → Order	Closest affiliation Order → Family → Genus	GenBank accession number	Absolute SW mapping score (–)[d]	Normalized SW mapping score (–)[e]
72	AGS purge	XP06	100	8	*C: Gammaproteobacteria*	*O: Xanthomonadaceae*	FN436156	314	0.719
72	Flocs	BC002	40	18	*O: Rhodocyclales*	*G: Zoogloea*	AJ011506	376	0.982
72	Flocs	BC002	35	13	*O: Rhodocyclales*	*G: Thauera*	AY945909	348	0.988
155	AGS 20 °C	X084	89	8	*C: Armatimonadetes*		GQ264378	195	0.664
176	AGS 25 °C	O125	100	2	*C: Anaerolineae*		AY592661	195	0.955
176	AGS purge	XP06	100	1	*C: Anaerolineae*		AY921672	175	0.961
176	PAO 17 °C	PAO109	100	2	*P: Verrucomicrobia*	*F: Opitutaceae*	EF018451	299	0.979
178	GAO 28 °C	GAO398	99.5	1099	*C: Alphaproteobacteria*	*F: Rhodospirillaceae*	AY351639	397	0.961
178	GAO 28 °C	GAO398	0.5	6	*C: Acidobacteria*	*F: Acidobacteriaceae*	AY326572	257	0.934
179	AGS purge	XP06	100	3	*P: Spirochaetes*	*F: Spirochaetaceae*	FM178812	247	0.972
180	Flocs	BC002	100	4	*P: Acidobacteria*		GQ396818	323	0.9
183	PAO 17 °C	PAO109	100	3	*O: Xanthomonadales*	*G: Pseudoxanthomonas*	DQ376568	229	0.787
184	AGS 20 °C	X084	100	2	*C: Alphaproteobacteria*	*F: Hyphomonadaceae*	DQ376579	303	0.818
184	GAO 28 °C	GAO398	100	11	*O: Rhizobiales*	*G: Methylocystis*	AJ868421	289	0.818
185	AGS early	BC059	100	14	*O: Rhizobiales*		AF502218	409	0.635
185	Flocs	BC002	89	8	*O: Rhizobiales*		FJ719048	296	0.946
185	Flocs	BC002	11	1	*O: Rhodobacterales*		FJ719099	318	0.946
185	PAO 17 °C	PAO109	100	9	*O: Rhodobacterales*	*F: Hyphomonadaceae*	AJ617876	371	0.923
186	AGS 20 °C	X086	100	3	*O: Rhizobiales*		FJ230932	433	0.644

(continued)

Table 5.9 (continued)

T-RF or OTU (bp)	Biomass origin[a]	Biological sample	Fraction of T-RF (%)[b]	Read counts (–)[c]	Broader affiliation Phylum → Class → Order	Closest affiliation Order → Family → Genus	GenBank accession number	Absolute SW mapping score (–)[d]	Normalized SW mapping score (–)[e]
186	AGS 25 °C	O125	85	12	*O: Rhizobiales*	*G: Methylosinus*	AJ458477	375	0.854
186	AGS purge	XP06	80	4	*O: Rhizobiales*	*G: Methylosinus*	AJ458477	318	0.977
186	GAO 28 °C	GAO398	45	5	*O: Rhizobiales*	*F: Bradyrhizobiaceae*	AY345540	327	0.944
186	GAO 28 °C	GAO398	36	4	*C: Alphaproteobacteria*	*F: Rhodospirillaceae*	AY351640	317	0.949
186	GAO 28 °C	GAO398	18	2	*O: Rhizobiales*	*G: Rhodoplanes*	CU918797	346	0.655
186	PAO 17 °C	PAO109	100	7	*O: Rhizobiales*		FM209362	300	0.928
187	GAO 28 °C	GAO398	87	48	*O: Rhodobacterales*	*F: Hyphomonadaceae*	AF236001	293	0.901
187	GAO 28 °C	GAO398	13	7	*O: Rhizobiales*	*F: Bradyrhizobiaceae*	AF502220	368	0.913
188	AGS 25 °C	O125	100	4	*C: Alphaproteobacteria*		EF520416	227	0.944
188	PAO 17 °C	PAO109	40	10	*O: Rhizobiales*	*F: Bradyrhizobiaceae*	AJ300771	397	0.961
188	PAO 17 °C	PAO109	32	8	*O: Rhodobacterales*	*F: Hyphomonadaceae*	AF236001	304	0.883
188	PAO 17 °C	PAO109	12	3	*O: Burkholderiales*	*F: Comamonadaceae*	DQ413087	375	0.869
189	AGS 20 °C	X084	81	13	*O: Rhizobiales*		CU926125	363	0.94
189	AGS 25 °C	O125	67	4	*O: Acidimicrobiales*	*F: Microthrixaceae*	CU925307	379	0.922
189	AGS 25 °C	O125	33	2	*O: Rhizobiales*		CU926125	363	0.881
189	AGS purge	XP06	75	3	*O: Rhizobiales*		CU926125	367	0.925
190	Flocs	BC002	49	20	*O: Rhizobiales*		CU918969	404	0.979
190	Flocs	BC002	44	18	*O: Burkholderiales*	*G: Acidovorax*	EU539550	356	0.979
190	GAO 28 °C	GAO398	52	17	*O: Rhizobiales*		GU455290	354	0.895

(continued)

Table 5.9 (continued)

T-RF or OTU (bp)	Biomass origin[a]	Biological sample	Fraction of T-RF (%)[b]	Read counts (–)[c]	Broader affiliation Phylum → Class → Order	Closest affiliation Order → Family → Genus	GenBank accession number	Absolute SW mapping score (–)[d]	Normalized SW mapping score (–)[e]
190	GAO 28 °C	GAO398	42	14	*O: Rhizobiales*	*F: Bradyrhizobiaceae*	AF208509	351	0.853
193	AGS early	BC059	94	72	*O: Burkholderiales*	*G: Acidovorax*	AJ864847	383	0.895
193	AGS early	BC059	6	5	*O: Xanthomonadales*		EU662583	343	0.944
193	AGS 20 °C	X066	91	10	*O: Burkholderiales*	*G: Acidovorax*	AJ864847	384	0.936
193	AGS 20 °C	X084	100	9	*O: Burkholderiales*	*G: Acidovorax*	AJ864847	367	0.923
193	AGS 20 °C	X114	75	6	*O: Burkholderiales*	*G: Acidovorax*	AJ864847	376	1
193	AGS 23 ± 2 °C	BC-II	91	10	*O: Burkholderiales*	*G: Acidovorax*	AJ864847	384	1
193	AGS 25 °C	O125	64	9	*O: Burkholderiales*	*G: Acidovorax*	AJ864847	372	1
193	AGS 25 °C	O125			*O: Burkholderiales*	*G: Acidovorax*	DQ814244	372	0.95
193	AGS purge	XP06	100	8	*O: Burkholderiales*	*G: Acidovorax*	AJ864847	371	1
193	Flocs	BC002	30	22	*O: Burkholderiales*	*G: Acidovorax*	EU375646	363	0.956
193	Flocs	BC002	28	21	*O: Burkholderiales*	*G: Simplicispira*	AJ505861	377	0.982
193	Flocs	BC002	17	13	*O: Burkholderiales*	*G: Delftia*	EF515241	355	0.954
193	PAO 17 °C	PAO109	58	22	*O: Burkholderiales*	*G: Acidovorax*	EU037281	361	0.939
193	PAO 17 °C	PAO109	10	4	*O: Burkholderiales*	*G: Simplicispira*	AJ505861	352	0.818
194	AGS 20 °C	X084	82	9	*C: Alphaproteobacteria*	*F: Rhodospirillaceae*	AF527585	350	0.919
194	AGS 20 °C	X084	18	2	*O: Xanthomonadales*	*G: Thermomonas*	DQ376561	260	0.919
194	AGS 20 °C	X114	100	3	*C: Alphaproteobacteria*	*F: Rhodospirillaceae*	EU834764	325	0.951
194	AGS 25 °C	O125	50	4	*O: Burkholderiales*	*G: Acidovorax*	AF078767	356	0.92

(continued)

Table 5.9 (continued)

T-RF or OTU (bp)	Biomass origin[a]	Biological sample	Fraction of T-RF (%)[b]	Read counts (–)[c]	Broader affiliation Phylum → Class → Order	Closest affiliation Order → Family → Genus	GenBank accession number	Absolute SW mapping score (–)[d]	Normalized SW mapping score (–)[e]
194	AGS 25 °C	O125	25	2	*O: Xanthomonadales*	*G: Thermomonas*	AF508105	247	0.916
194	AGS 25 °C	O125	13	1	*C: Alphaproteobacteria*	*F: Rhodospirillaceae*	EU834764	301	0.887
194	AGS purge	XP06	76	16	*C: Alphaproteobacteria*	*F: Rhodospirillaceae*	AF527585	408	0.929
194	AGS purge	XP06	19	4	*O: Xanthomonadales*	*G: Thermomonas*	AF508105	304	0.758
194	GAO 28 °C	GAO398	100	215	*C: Alphaproteobacteria*	*F: Rhodospirillaceae*	AF527585	420	0.758
195	AGS early	BC059	100	4793	*O: Rhodocyclales*	*G: Zoogloea*	EU639144	402	0.826
195	AGS 20 °C	X066	40	6	*O: Burkholderiales*	*G: Aquamonas*	DQ337068	361	0.699
195	AGS 20 °C	X066	33	5	*O: Xanthomonadales*	*G: Pseudoxanthomonas*	GU122958	288	0.931
195	AGS 20 °C	X066	20	3	*C: Alphaproteobacteria*	*F: Rhodospirillaceae*	AF527585	333	0.948
195	AGS 20 °C	X066	7	1	*O: Rhodocyclales*	*G: Zoogloea*	EU834814	303	0.885
195	AGS 20 °C	X084	65	28	*O: Xanthomonadales*	*G: Pseudoxanthomonas*	GU122958	327	0.619
195	AGS 20 °C	X084	16	7	*C: Gammaproteobacteria*	*F: Xanthomonadaceae*	AM696987	328	0.969
195	AGS 20 °C	X084	5	2	*O: Rhodocyclales*	*G: Rhodocyclus*	AF502230	245	0.936
195	AGS 20 °C	X084	5	2	*O: Chromatiales*	*F: Sinobacteraceae*	FJ902650	329	0.979
195	AGS 20 °C	X084	2	1	*O: Rhodocyclales*	*G: Zoogloea*	DQ413150	374	0.812
195	AGS 20 °C	X084	2	1	*O: Burkholderiales*	*G: Alicycliphilus*	FJ657757	306	0.814
195	AGS 20 °C	X114	36	5	*O: Xanthomonadales*	*G: Thermomonas*	DQ376561	249	0.906
195	AGS 20 °C	X114	28	4	*O: Rhodocyclales*	*G: Zoogloea*	DQ413151	383	0.379
195	AGS 20 °C	X114	14	2	*O: Rhodocyclales*	*G: Rhodocyclus*	AB200295	286	0.779

(continued)

Table 5.9 (continued)

T-RF or OTU (bp)	Biomass origin[a]	Biological sample	Fraction of T-RF (%)[b]	Read counts (–)[c]	Broader affiliation Phylum → Class → Order	Closest affiliation Order → Family → Genus	GenBank accession number	Absolute SW mapping score (–)[d]	Normalized SW mapping score (–)[e]
195	AGS 23 ± 2 °C	BC-II	40	6	*O: Burkholderiales*	*G: Aquamonas*	DQ337068	361	0.79
195	AGS 23 ± 2 °C	BC-II	33	5	*O: Xanthomonadales*	*G: Pseudoxanthomonas*	GU122958	288	0.779
195	AGS 23 ± 2 °C	BC-II	20	3	*C: Alphaproteobacteria*	*F: Rhodospirillaceae*	AF527585	333	0.77
195	AGS 23 ± 2 °C	BC-II	7	1	*O: Rhodocyclales*	*G: Zoogloea*	EU834814	303	0.768
195	AGS 25 °C	O125	78	19	*O: Rhodocyclales*	*G: Zoogloea*	EU639144	383	0.762
195	AGS 25 °C	O125	8	2	*O: Rhodocyclales*	*G: Dechloromonas*	AB240507	299	0.567
195	AGS purge	XP06	70	30	*O: Burkholderiales*	*F: Comamonadaceae*	GQ023492	367	0.705
195	AGS purge	XP06	15	7	*C: Gammaproteobacteria*	*G: Thermomonas*	DQ376561	305	0.759
195	AGS purge	XP06	3	2	*O: Rhodocyclales*	*G: Methyloversatilis*	FJ810571	300	0.921
195	AGS purge	XP06			*O: Rhodocyclales*	*G: Methyloversatilis*		300	0.911
195	Flocs	BC002	84	231	*O: Rhodocyclales*	*G: Zoogloea*	AF234684	385	0.921
195	Flocs	BC002	7	18	*O: Pseudomonadales*	*G: Acinetobacter*	GQ073520	337	0.92
196	AGS 20 °C	X084	45	9	*O: Burkholderiales*	*F: Comamonadaceae*	GQ023492	342	1
196	AGS 20 °C	X084	15	3	*O: Rhodocyclales*	*G: Methyloversatilis*	FJ810571	301	1
196	AGS 20 °C	X084	15	3	*C: Alphaproteobacteria*	*O: Rhodospirillales*	DQ066972	248	1
196	AGS 20 °C	X084	10	2	*C: Deltaproteobacteria*	*O: Myxococcales*	AF280857	192	0.946
196	AGS 20 °C	X114	100	3	*C: Gammaproteobacteria*	*O: Xanthomonadaceae*	DQ376561	249	1
196	AGS 25 °C	O125	60	3	*C: Betaproteobacteria*	*F: Rhodocyclaceae*	AB200295	328	0.944
196	AGS purge	XP06	50	2	*C: Alphaproteobacteria*	*O: Rhodospirillales*	DQ066972	287	0.435

(continued)

Table 5.9 (continued)

T-RF or OTU (bp)	Biomass origin[a]	Biological sample	Fraction of T-RF (%)[b]	Read counts (–)[c]	Broader affiliation Phylum → Class → Order	Closest affiliation Order → Family → Genus	GenBank accession number	Absolute SW mapping score (–)[d]	Normalized SW mapping score (–)[e]
197	AGS 25 °C	O125	75	3	*O: Xanthomonadales*	*G: Thermomonas*	EU071527	296	0.871
198	Flocs	BC002	94	65	*P: candidate phylum TM7*		DQ640696	367	0.977
198	PAO 17 °C	PAO109	83	5	*O: Myxococcales*		DQ088736	265	0.977
200	AGS 20 °C	X066	100	3	*O: Rhodocyclales*	*G: Rhodocyclus*	AF204247	314	0.923
200	AGS 20 °C	X084	100	13	*O: Xanthomonadales*	*G: Pseudoxanthomonas*	AJ306770	287	0.755
200	AGS 20 °C	X084			*O: Xanthomonadales*	*G: Pseudoxanthomonas*	AB008507	287	0.833
200	AGS 23 ± 2 °C	BC-II	100	3	*O: Rhodocyclales*	*G: Rhodocyclus*	AF204247	314	0.882
200	GAO 28 °C	GAO398	86	18	*C: Acidobacteria*	*F: Acidobacteriaceae*	FJ466403	320	0.753
200	GAO 28 °C	GAO398	14	3	*P: candidate phylum TM7*		EU135398	251	0.788
201	AGS purge	XP06	78	7	*O: Xanthomonadales*	*G: Thermomonas*	EF219043	347	0.788
201	Flocs	BC002	48	10	*P: Chloroflexi*		CU927307	443	0.909
201	Flocs	BC002	38	8	*O: Xanthomonadales*	*F: Xanthomonadaceae*	FJ612198	360	0.917
202	AGS 20 °C	X084	66	2	*C: Gammaproteobacteria*	*G: Legionella*	Z49739	316	0.948
202	AGS 25 °C	O125	80	4	*C: Alphaproteobacteria*	*F: Hyphomicrobiaceae*	EF018886	165	0.835
202	PAO 17 °C	PAO109	100	2	*O: Rhodocyclales*	*G: Rhodocyclus*	AF204247	315	0.808
204	AGS 20 °C	X084	67	6	*C: Gammaproteobacteria*		AY098896	263	0.764
204	AGS 20 °C	X114	100	5	*O: Rhodocyclales*	*G: Rhodocyclus*	AF502227	305	0.764
204	AGS 25 °C	O125	67	2	*O: Rhodocyclales*	*G: Rhodocyclus*	AB276369	313	0.944
205	AGS 20 °C	X084	100	1	*O: Rhodocyclales*	*G: Rhodocyclus*	AF204247	265	0.944

(continued)

Table 5.9 (continued)

T-RF or OTU (bp)	Biomass origin[a]	Biological sample	Fraction of T-RF (%)[b]	Read counts (–)[c]	Broader affiliation Phylum → Class → Order	Closest affiliation Order → Family → Genus	GenBank accession number	Absolute SW mapping score (–)[d]	Normalized SW mapping score (–)[e]
205	AGS 25 °C	O125	100	11	*O: Rhodocyclales*	*G: Rhodocyclus*	AF314417	221	0.982
206	AGS 20 °C	X066	100	6	*O: Sphingobacteriales*		EF019693	387	0.983
206	AGS 23 ± 2 °C	BC-II	100	6	*O: Sphingobacteriales*		EF019693	387	0.982
206	AGS 25 °C	O125	85	11	*O: Sphingobacteriales*		EF019693	385	0.954
206	AGS purge	XP06	90	9	*O: Sphingobacteriales*		EF019693	348	0.881
207	AGS 20 °C	X084	100	6	*O: Rhodocyclales*	*G: Rhodocyclus*	EU834760	257	0.881
208	Flocs	BC002	30	6	*O: Burkholderiales*	*G: Rhodoferax*	AB154311	351	0.788
208	Flocs	BC002	10	1	*O: Burkholderiales*	*G: Sphaerotilus*	AB087568	374	0.924
208	GAO 28 °C	GAO398	55	22	*O: Burkholderiales*		EU491323	364	0.788
208	GAO 28 °C	GAO398	43	17	*C: Acidobacteria*	*F: Acidobacteriaceae*	EU445233	293	0.785
209	GAO 28 °C	GAO398	90	4009	*C: Acidobacteria*	*F: Acidobacteriaceae*	AF200696	418	0.956
209	GAO 28 °C	GAO398	10	427	*P: Armatimonadetes*		GQ264378	193	0.92
210	Flocs	BC002	42	6	*O: Acidobacteriales*		FJ230900	403	0.885
210	Flocs	BC002	35	5	*P: Firmicutes*	*G: Trichococcus*	EU234209	295	0.702
210	GAO 28 °C	GAO398	72	23	*C: Acidobacteria*	*F: Acidobacteriaceae*	AF200696	277	0.935
211	AGS early	BC059	54	7	*O: Burkholderiales*	*F: Comamonadaceae*	EF540425	345	0.935
211	AGS early	BC059	46	6	*O: Rhodocyclales*	*F: Rhodocyclaceae*	DQ088735	343	0.747
211	GAO 28 °C	GAO398	79	15	*P: Armatimonadetes*		GQ264378	189	0.448
211	GAO 28 °C	GAO398	9	3	*C: Acidobacteria*	*F: Candidatus* Solibacter	DQ404599	342	0.468

(continued)

Table 5.9 (continued)

T-RF or OTU (bp)	Biomass origin[a]	Biological sample	Fraction of T-RF (%)[b]	Read counts (–)[c]	Broader affiliation Phylum → Class → Order	Closest affiliation Order → Family → Genus	GenBank accession number	Absolute SW mapping score (–)[d]	Normalized SW mapping score (–)[e]
211	GAO 28 °C	GAO398	9	3	*O: Rhodobacterales*	*F: Hyphomonadaceae*	EU770258	166	0.625
212	AGS 20 °C	X084	95	21	*O: Burkholderiales*		DQ066963	373	0.981
212	AGS 20 °C	X084	5	1	*O: Rhodocyclales*	*G: Dechloromonas*	EF632559	292	0.522
212	AGS 20 °C	X114	100	4	*O: Burkholderiales*		DQ066963	373	0.988
212	AGS 25 °C	O125	100	20	*O: Burkholderiales*	*F: Comamonadaceae*	EU499692	376	0.883
212	AGS purge	XP06	83	6	*O: Burkholderiales*	*G: Hylemonella*	CU917922	364	0.757
212	GAO 28 °C	GAO398	78	21	*O: Rhodocyclales*	*F: Rhodocyclaceae*	DQ664245	431	0.969
212	GAO 28 °C	GAO398	11	3	*O: Hydrogenophilales*	*G: Thiobacillus*	AY955087	199	0.723
212	PAO 17 °C	PAO109	33	47	*O: Burkholderiales*		DQ066963	376	0.967
212	PAO 17 °C	PAO109	22	31	*O: Burkholderiales*	*G: Mitsuaria*	AB240354	376	1
212	PAO 17 °C	PAO109	11	13	*O: Rhodocyclales*	*G: Rhodocyclus*	EF565152	356	1
212	PAO 17 °C	PAO109			*O: Rhodocyclales*	*G: Rhodocyclus*	AB276369	356	0.984
212	PAO 17 °C	PAO109	7	10	*O: Burkholderiales*	*G: Xenophilus*	EF125952	307	0.962
213	AGS 20 °C	X066	92	11	*O: Rhodocyclales*	*G: Rhodocyclus*	AF502230	356	0.881
213	AGS 20 °C	X084	73	11	*O: Rhodocyclales*	*G: Rhodocyclus*	AB200295	359	0.943
213	AGS 20 °C	X084	7	1	*O: Rhodocyclales*	*G: Dechloromonas*	DQ413103	326	0.811
213	AGS 20 °C	X084	7	1	*O: Rhodocyclales*	*G: Methyloversatilis*	GQ340363	314	0.882
213	AGS 20 °C	X114	98	41	*O: Rhodocyclales*	*G: Rhodocyclus*	AB200295	358	0.932
213	AGS 23 ± 2 °C	BC-II	92	11	*O: Rhodocyclales*	*G: Rhodocyclus*	AF502230	356	0.9

(continued)

Table 5.9 (continued)

T-RF or OTU (bp)	Biomass origin[a]	Biological sample	Fraction of T-RF (%)[b]	Read counts (–)[c]	Broader affiliation Phylum → Class → Order	Closest affiliation Order → Family → Genus	GenBank accession number	Absolute SW mapping score (–)[d]	Normalized SW mapping score (–)[e]
213	AGS 25 °C	O125	100	14	*O: Rhodocyclales*	*G: Rhodocyclus*	AF502230	327	0.897
213	AGS purge	XP06	100	404	*O: Rhodocyclales*	*G: Rhodocyclus*	EF565156	375	0.923
213	Flocs	BC002	73	17	*O: Burkholderiales*	*F: Comamonadaceae*	AY662010	365	0.88
213	PAO 17 °C	PAO109	98	283	*O: Burkholderiales*	*G: Paucibacter*	FJ535228	330	0.979
214	AGS early	BC059	88	14	*O: Rhodocyclales*	*G: Dechloromonas*	DQ413103	381	0.789
214	AGS early	BC059	12	2	*O: Rhodocyclales*	*G: Rhodocyclus*	AF502224	372	1
214	AGS 20 °C	X066	99.6	769	*O: Rhodocyclales*	*G: Rhodocyclus*	AB200295	371	0.936
214	AGS 20 °C	X066			*O: Rhodocyclales*	*G: Rhodocyclus*	AF502230	371	0.925
214	AGS 20 °C	X066	0.1	1	*O: Rhodocyclales*	*G: Dechloromonas*	DQ413103	321	1
214	AGS 20 °C	X066	0.1	1	*O: Rhodocyclales*	*G: Methyloversatilis*	DQ066958	368	0.942
214	AGS 20 °C	X084	99.8	3909	*O: Rhodocyclales*	*G: Rhodocyclus*	AF502227	372	0.984
214	AGS 20 °C	X084			*O: Rhodocyclales*	*G: Rhodocyclus*	EF565156	372	1
214	AGS 20 °C	X084	0.1	5	*O: Rhodocyclales*	*G: Methyloversatilis*	AF204251	354	0.792
214	AGS 20 °C	X114	99.5	772	*O: Rhodocyclales*	*G: Rhodocyclus*	AF502227	379	0.86
214	AGS 20 °C	X114			*O: Rhodocyclales*	*G: Rhodocyclus*	EF565156	379	0.878
214	AGS 23 ± 2 °C	BC-II	99.6	769	*O: Rhodocyclales*	*G: Rhodocyclus*	AB200295	371	0.611
214	AGS 23 ± 2 °C	BC-II			*O: Rhodocyclales*	*G: Rhodocyclus*	AF502230	371	0.859
214	AGS 23 ± 2 °C	BC-II	0.1	1	*O: Rhodocyclales*	*G: Dechloromonas*	DQ413103	321	0.859
214	AGS 23 ± 2 °C	BC-II	0.1	1	*O: Rhodocyclales*	*G: Methyloversatilis*	DQ066958	368	1

(continued)

Table 5.9 (continued)

T-RF or OTU (bp)	Biomass origin[a]	Biological sample	Fraction of T-RF (%)[b]	Read counts (–)[c]	Broader affiliation Phylum → Class → Order	Closest affiliation Order → Family → Genus	GenBank accession number	Absolute SW mapping score (–)[d]	Normalized SW mapping score (–)[e]
214	AGS 23 ± 2 °C	BC-II	0.1	1	*O: Nitrosomonadales*	*G: Nitrosomonas*	EU937892	278	0.853
214	AGS 25 °C	O125	99.5	1349	*O: Rhodocyclales*	*G: Rhodocyclus*	AF502227	379	0.885
214	AGS 25 °C	O125			*O: Rhodocyclales*	*G: Rhodocyclus*	EF565156	379	0.814
214	AGS 25 °C	O125	0.4	5	*O: Rhodocyclales*	*G: Methyloversatilis*	DQ066958	375	0.804
214	AGS 25 °C	O125	0.1	2	*O: Rhodocyclales*	*G: Dechloromonas*	DQ413103	268	0.691
214	AGS purge	XP06	97	116	*O: Rhodocyclales*	*G: Rhodocyclus*	AB200295	371	0.839
214	AGS purge	XP06	3	3	*O: Rhodocyclales*	*G: Dechloromonas*	DQ413103	366	0.83
214	Flocs	BC002	48	136	*O: Rhodocyclales*	*G: Dechloromonas*	AY064177	382	0.978
214	Flocs	BC002	16	45	*O: Burkholderiales*	*G: Rhodoferax*	AB452981	366	0.929
214	Flocs	BC002	10	29	*O: Rhodocyclales*	*G: Methyloversatilis*	AY436796	364	0.989
214	Flocs	BC002	4	12	*O: Rhodocyclales*	*G: Zoogloea*	DQ413172	344	0.941
214	Flocs	BC002	4	11	*O: Burkholderiales*	*G: Aquamonas*	DQ521469	366	0.961
214	Flocs	BC002	1	4	*O: Rhodocyclales*	*G: Rhodocyclus*	EF565151	368	0.757
214	PAO 17 °C	PAO109	74	1254	*O: Rhodocyclales*	*G: Rhodocyclus*	AB200295	422	0.776
214	PAO 17 °C	PAO109			*O: Rhodocyclales*	*G: Rhodocyclus*	AF502230	422	0.401
214	PAO 17 °C	PAO109	12	195	*O: Rhodocyclales*	*G: Dechloromonas*	AY032611	320	0.696
214	PAO 17 °C	PAO109	9	150	*C: Betaproteobacteria*		CU924494	409	0.638
214	PAO 17 °C	PAO109	3	47	*O: Burkholderiales*	*G: Rhodoferax*	FM955857	340	0.638
215	AGS 20 °C	X066	31	5	*O: Rhodocyclales*	*G: Rhodocyclus*	AF502230	337	0.692

(continued)

Table 5.9 (continued)

T-RF or OTU (bp)	Biomass origin[a]	Biological sample	Fraction of T-RF (%)[b]	Read counts (–)[c]	Broader affiliation Phylum → Class → Order	Closest affiliation Order → Family → Genus	GenBank accession number	Absolute SW mapping score (–)[d]	Normalized SW mapping score (–)[e]
215	AGS 20 °C	X066	13	2	*O: Rhodocyclales*	***G: Methyloversatilis***	GQ340363	298	0.674
215	AGS 20 °C	X066	6	1	*O: Nitrosomonadales*	***G: Nitrosomonas***	GQ396862	278	0.674
215	AGS 20 °C	X084	87	66	*O: Rhodocyclales*	***G: Rhodocyclus***	AB200295	364	0.614
215	AGS 20 °C	X084	12	9	*P: Chloroflexi*	***C: Anaerolineae***	X84576	163	0.625
215	AGS 23 ± 2 °C	BC-II	31	5	*O: Rhodocyclales*	***G: Rhodocyclus***	AF502230	337	1
215	AGS 23 ± 2 °C	BC-II	13	2	*O: Rhodocyclales*	***G: Methyloversatilis***	GQ340363	298	0.658
215	AGS 23 ± 2 °C	BC-II	6	1	*O: Nitrosomonadales*	***G: Nitrosomonas***	GQ396862	278	1
215	AGS 25 °C	O125	77	30	*O: Rhodocyclales*	***G: Rhodocyclus***	AB200295	348	1
215	AGS 25 °C	O125	23	9	*P: Chloroflexi*	***C: Anaerolineae***	EU104216	261	0.965
215	AGS purge	XP06	55	21	*O: Rhodocyclales*	***G: Rhodocyclus***	AB200295	368	0.932
215	AGS purge	XP06	21	8	*C: Anaerolineae*		EU104216	287	0.973
215	Flocs	BC002	44	14	*O: Rhodocyclales*	***G: Methyloversatilis***	DQ066958	349	0.829
215	Flocs	BC002	28	9	*P: Chloroflexi*	***G: Caldilinea***	CU917747	279	0.962
215	Flocs	BC002	6	2	*O: Rhodocyclales*	***G: Rhodocyclus***	FJ719063	320	0.973
215	Flocs	BC002	3	1	*O: Rhodocyclales*	***G: Dechloromonas***	AY062126	306	0.995
215	PAO 17 °C	PAO109	84	26	*O: Rhodocyclales*	***G: Rhodocyclus***	EF590005	366	1
215	PAO 17 °C	PAO109	6.5	2	*O: Burkholderiales*	***F: Comamonadaceae***	EU180529	375	0.941
215	PAO 17 °C	PAO109	6.5	2	*C: Anaerolineae*		EU862289	322	0.724
215	PAO 17 °C	PAO109	3	1	*O: Rhodocyclales*	***G: Dechloromonas***	AY064177	304	0.947

(continued)

Table 5.9 (continued)

T-RF or OTU (bp)	Biomass origin[a]	Biological sample	Fraction of T-RF (%)[b]	Read counts (–)[c]	Broader affiliation Phylum → Class → Order	Closest affiliation Order → Family → Genus	GenBank accession number	Absolute SW mapping score (–)[d]	Normalized SW mapping score (–)[e]
216	AGS 20 °C	X066	62	8	*O: Rhodocyclales*	*G: Methyloversatilis*	CU922545	361	0.881
216	AGS 20 °C	X066	15	2	*C: Anaerolineae*		EU104216	269	0.938
216	AGS 20 °C	X066	8	1	*O: Rhodocyclales*	*G: Rhodocyclus*	AF502230	276	0.938
216	AGS 20 °C	X066	8	1	*O: Rhodocyclales*	*F: Rhodocyclaceae*	GU454920	347	0.857
216	AGS 20 °C	X084	35	10	*O: Rhodocyclales*	*G: Rhodocyclus*	AF502230	296	0.964
216	AGS 20 °C	X084	28	8	*O: Nitrosomonadales*	*G: Nitrosomonas*	GU183579	364	0.959
216	AGS 20 °C	X084	21	6	*P: Chloroflexi*	*C: Anaerolineae*	EU104216	267	0.977
216	AGS 20 °C	X084	10	3	*O: Rhodocyclales*	*G: Methyloversatilis*	CU922545	360	0.89
216	AGS 20 °C	X114	67	2	*O: Rhodocyclales*	*G: Methyloversatilis*	CU922545	310	0.988
216	AGS 20 °C	X114	33	1	*O: Nitrosomonadales*	*G: Nitrosomonas*	EU937892	325	0.707
216	AGS 23 ± 2 °C	BC-II	62	8	*O: Rhodocyclales*	*G: Methyloversatilis*	CU922545	361	0.703
216	AGS 23 ± 2 °C	BC-II	15	2	*C: Anaerolineae*		EU104216	269	0.988
216	AGS 23 ± 2 °C	BC-II	8	1	*O: Rhodocyclales*	*G: Rhodocyclus*	AF502230	276	1
216	AGS 23 ± 2 °C	BC-II	8	1	*O: Rhodocyclales*	*F: Rhodocyclaceae*	GU454920	347	0.82
216	AGS 25 °C	O125	43	12	*O: Nitrosomonadales*	*G: Nitrosomonas*	GU183579	355	0.916
216	AGS 25 °C	O125	14	4	*O: Rhodocyclales*	*G: Rhodocyclus*	DQ856536	343	0.922
216	AGS 25 °C	O125	14	4	*P: Chloroflexi*	*C: Anaerolineae*	EU104216	265	0.949
216	AGS 25 °C	O125	11	3	*O: Actinomycetales*		DQ372707	239	0.949
216	AGS purge	XP06	74	42	*O: Rhodocyclales*	*G: Methyloversatilis*	CU922545	345	0.904

(continued)

Table 5.9 (continued)

T-RF or OTU (bp)	Biomass origin[a]	Biological sample	Fraction of T-RF (%)[b]	Read counts (–)[c]	Broader affiliation Phylum → Class → Order	Closest affiliation Order → Family → Genus	GenBank accession number	Absolute SW mapping score (–)[d]	Normalized SW mapping score (–)[e]
216	AGS purge	XP06	18	10	*O: Nitrosomonadales*	*G: Nitrosomonas*	AF272422	325	0.921
216	AGS purge	XP06	8	5	*O: Rhodocyclales*	*G: Rhodocyclus*	EF565156	357	0.958
216	AGS purge	XP06			*O: Rhodocyclales*	*G: Rhodocyclus*	DQ856536	357	0.958
216	Flocs	BC002	35	5	*O: Rhodocyclales*	*F: Rhodocyclaceae*	NR029035	311	0.957
216	Flocs	BC002	21	3	*O: Burkholderiales*	*F: Comamonadaceae*	EU180529	273	0.958
216	Flocs	BC002	7	1	*O: Nitrosomonadales*	*G: Nitrosomonas*	EU937892	348	0.941
216	GAO 28 °C	GAO398	100	15	*O: Hydrogenophilales*	*G: Thiobacillus*	FM212998	400	0.928
216	PAO 17 °C	PAO109	46	6	*O: Rhodocyclales*	*G: Rhodocyclus*	CU920179	352	0.914
216	PAO 17 °C	PAO109	31	4	*O: Burkholderiales*	*F: Comamonadaceae*	AB475015	331	0.914
217	AGS 20 °C	X066	33	1	*O: Nitrosomonadales*	*G: Nitrosomonas*	EU937892	295	0.914
217	AGS 20 °C	X066	33	1	*O: Rhizobiales*	*G: Devosia*	AF236010	298	0.914
217	AGS 20 °C	X066	33	1	*O: Burkholderiales*	*G: Ideonella*	GQ472390	207	0.909
217	AGS 23 ± 2 °C	BC-II	33	1	*O: Nitrosomonadales*	*G: Nitrosomonas*	EU937892	295	0.928
217	AGS 23 ± 2 °C	BC-II	33	1	*O: Rhizobiales*	*G: Devosia*	AF236010	298	0.887
217	AGS 23 ± 2 °C	BC-II	33	1	*O: Burkholderiales*	*G: Ideonella*	GQ472390	207	0.917
217	Flocs	BC002	46	15	*O: Actinomycetales*		AF513101	385	0.783
217	Flocs	BC002	36	12	*O: Rhodocyclales*	*G: Thauera*	AM084110	386	0.917
217	PAO 17 °C	PAO109	100	5	*O: Rhodocyclales*	*G: Rhodocyclus*	AY913841	344	0.941
219	AGS 20 °C	X084	66	10	*O: Rhizobiales*	*G: Aminobacter*	AF034798	371	0.92

(continued)

Table 5.9 (continued)

T-RF or OTU (bp)	Biomass origin[a]	Biological sample	Fraction of T-RF (%)[b]	Read counts (–)[c]	Broader affiliation Phylum → Class → Order	Closest affiliation Order → Family → Genus	GenBank accession number	Absolute SW mapping score (–)[d]	Normalized SW mapping score (–)[e]
219	AGS 20 °C	X084	27	4	*O: Rhodocyclales*	*G: Rhodocyclus*	AF502231	322	0.936
220	AGS 20 °C	X066	89	48	*O: Rhizobiales*	*G: Aminobacter*	NR025302	448	0.638
220	AGS 20 °C	X084	93	181	*O: Rhizobiales*	*G: Aminobacter*	NR025302	420	0.753
220	AGS 20 °C	X084	3	6	*O: Rhizobiales*		AY307924	384	0.858
220	AGS 20 °C	X084	2	4	*O: Rhizobiales*	*G: Devosia*	EU881088	373	0.858
220	AGS 20 °C	X114	100	40	*O: Rhizobiales*	*G: Aminobacter*	NR025302	429	0.812
220	AGS 23 ± 2 °C	BC-II	89	48	*O: Rhizobiales*	*G: Aminobacter*	NR025302	448	0.822
220	AGS 25 °C	O125	91	109	*O: Rhizobiales*	*G: Aminobacter*	NR025302	425	0.809
220	AGS 25 °C	O125	3	3	*O: Rhizobiales*	*G: Mezorhizobium*	AJ278249	329	0.597
220	AGS purge	XP06	96	44	*O: Rhizobiales*	*G: Aminobacter*	NR025302	414	0.803
220	GAO 28 °C	GAO398	75	3	*O: Rhizobiales*	*G: Mezorhizobium*	GQ221761	356	0.669
220	PAO 17 °C	PAO109	67	10	*O: Phycisphaerales*		AY957928	191	0.944
220	PAO 17 °C	PAO109	27	4	*O: Rhizobiales*	*G: Aminobacter*	NR025302	333	0.944
221	AGS 20 °C	X114	100	9	*O: Rhodobacterales*	*F: Hyphomonadaceae*	AF236001	312	0.942
222	AGS 20 °C	X084	64	18	*O: Rhodobacterales*	*F: Hyphomonadaceae*	AF236001	306	0.852
222	AGS 20 °C	X084	32	9	*C: Alphaproteobacteria*	*F: Sphingomonadaceae*	GQ500778	346	0.942
222	AGS 20 °C	X084	100	2	*O: Verrucomicrobiales*	*G: Verrucomicrobium*	EU703281	222	0.932
222	GAO 28 °C	GAO398	100	18	*O: Sphingomonadales*		AJ746092	379	0.932
222	PAO 17 °C	PAO109	98	213	*O: Rhodobacterales*	*F: Hyphomonadaceae*	AF236001	284	0.949

(continued)

Table 5.9 (continued)

T-RF or OTU (bp)	Biomass origin[a]	Biological sample	Fraction of T-RF (%)[b]	Read counts (–)[c]	Broader affiliation Phylum → Class → Order	Closest affiliation Order → Family → Genus	GenBank accession number	Absolute SW mapping score (–)[d]	Normalized SW mapping score (–)[e]
222	PAO 17 °C	PAO109			*O: Rhodobacterales*	*F: Hyphomonadaceae*	AJ227809	284	0.981
222	PAO 17 °C	PAO109	2	4	*O: Sphingomonadales*	*G: Novosphingobium*	CU919596	367	0.923
223	AGS early	BC059	100	23	*O: Actinomycetales*	*G: Tetrasphaera*	AF527583	371	0.949
223	AGS 20 °C	X066	72	44	*O: Actinomycetales*	*F: Intrasporangiaceae*	AF255629	373	0.949
223	AGS 20 °C	X066	25	15	*O: Rhodobacterales*	*F: Hyphomonadaceae*	AF236001	298	0.957
223	AGS 20 °C	X084	97	93	*O: Actinomycetales*	*F: Intrasporangiaceae*	AF387308	365	0.957
223	AGS 20 °C	X114	86	31	*O: Actinomycetales*	*F: Intrasporangiaceae*	AF255629	381	0.898
223	AGS 23 ± 2 °C	BC-II	72	44	*O: Actinomycetales*	*F: Intrasporangiaceae*	AF255629	373	0.898
223	AGS 23 ± 2 °C	BC-II	25	15	*O: Rhodobacterales*	*F: Hyphomonadaceae*	AF236001	298	0.949
223	AGS 25 °C	O125	99	82	*O: Actinomycetales*	*F: Intrasporangiaceae*	AF255629	380	0.949
223	AGS purge	XP06	86	83	*O: Actinomycetales*	*F: Intrasporangiaceae*	AF255629	360	0.935
223	AGS purge	XP06	10	10	*O: Rhodobacterales*	*F: Hyphomonadaceae*	AJ227808	275	0.935
223	Flocs	BC002	99	545	*O: Actinomycetales*	*G: Tetrasphaera*	AF255629	374	0.962
223	GAO 28 °C	GAO398	55	11	*O: Actinomycetales*	*G: Propionicimonas*	FM178834	393	0.962
223	GAO 28 °C	GAO398	35	7	*O: Actinomycetales*	*G: Tetrasphaera*	DQ007320	309	0.923
224	AGS early	BC059	96	135	*O: Rhodobacterales*	*F: Hyphomonadaceae*	AF236001	285	0.938
224	AGS 20 °C	X084	100	2	*O: Actinomycetales*	*F: Intrasporangiaceae*	AF255629	269	0.864
224	AGS 25 °C	O125	100	1	*O: Actinomycetales*	*F: Intrasporangiaceae*	AF255629	218	0.744
224	PAO 17 °C	PAO109	84	16	*O: Actinomycetales*	*F: Intrasporangiaceae*	AF255629	365	0.853

(continued)

Table 5.9 (continued)

T-RF or OTU (bp)	Biomass origin[a]	Biological sample	Fraction of T-RF (%)[b]	Read counts (–)[c]	Broader affiliation Phylum → Class → Order	Closest affiliation Order → Family → Genus	GenBank accession number	Absolute SW mapping score (–)[d]	Normalized SW mapping score (–)[e]
224	PAO 17 °C	PAO109	16	3	*O: Desulfuromonadales*		DQ309326	172	0.864
225	AGS 20 °C	X084	100	2	*O: Actinomycetales*	*F: Intrasporangiaceae*	AF387311	254	0.936
227	AGS 20 °C	X114	100	2	*O: Pirellulales*		GQ263926	359	0.944
227	PAO 17 °C	PAO109	100	13	*O: Actinomycetales*	*F: Microbacteriaceae*	CU922723	381	0.865
228	Flocs	BC002	88	50	*O: Actinomycetales*	*F: Intrasporangiaceae*	AF513091	382	0.847
228	Flocs	BC002	88	50	*O: Actinomycetales*	*F: Intrasporangiaceae*	AF513091	382	0.946
228	Flocs	BC002	5	3	*C: Actinobacteria*	*F: Microthrixaceae*	CU917839	388	0.865
228	Flocs	BC002	5	3	*C: Actinobacteria*	*F: Microthrixaceae*	CU917839	388	0.773
229	AGS 20 °C	X084	100	3	*C: Alphaproteobacteria*	*F: Rhodospirillaceae*	DQ066972	207	0.918
232	AGS 20 °C	X066	71	5	*C: Alphaproteobacteria*	*F: Rhodospirillaceae*	DQ066972	259	0.918
232	AGS 20 °C	X084	100	22	*C: Alphaproteobacteria*	*F: Rhodospirillaceae*	DQ066972	259	0.843
232	AGS 20 °C	X114	100	2	*C: Alphaproteobacteria*	*F: Rhodospirillaceae*	DQ066972	330	0.819
232	AGS 23 ± 2 °C	BC-II	71	5	*C: Alphaproteobacteria*	*F: Rhodospirillaceae*	DQ066972	259	0.762
232	AGS 25 °C	O125	100	3	*C: Alphaproteobacteria*	*F: Rhodospirillaceae*	EU133337	249	0.837
232	AGS purge	XP06	100	10	*C: Alphaproteobacteria*	*F: Rhodospirillaceae*	DQ066972	291	0.762
233	AGS 20 °C	X084	100	2	*O: Roseiflexales*		EU589291	202	0.703
233	AGS 25 °C	O125	100	1	*O: Roseiflexales*		EU589291	173	0.789
233	Flocs	BC002	87	26	*P: candidate phylum TM7*		FJ534960	271	0.775
233	Flocs	BC002	7	2	*O: Phycisphaerales*		FJ612210	283	0.429

(continued)

Table 5.9 (continued)

T-RF or OTU (bp)	Biomass origin[a]	Biological sample	Fraction of T-RF (%)[b]	Read counts (–)[c]	Broader affiliation Phylum → Class → Order	Closest affiliation Order → Family → Genus	GenBank accession number	Absolute SW mapping score (–)[d]	Normalized SW mapping score (–)[e]
234	AGS 25 °C	O125	67	2	*P: candidate phylum TM7*		FJ534960	183	0.926
235	GAO 28 °C	GAO398	100	8	*C: Gammaproteobacteria*		FJ356048	276	1
237	AGS 23 ± 2 °C	BC-II	94	17	*C: Gammaproteobacteria*		AF361092	324	0.99
237	GAO 28 °C	GAO398	97	63	*C: Gammaproteobacteria*		EU529737	395	0.854
238	AGS 20 °C	X066	94	17	*C: Gammaproteobacteria*		AF361092	324	0.953
238	AGS 20 °C	X114	82	9	*C: Gammaproteobacteria*		AF361092	369	0.985
238	AGS 23 ± 2 °C	BC-II	99	272	*C: Gammaproteobacteria*		FJ356056	446	0.953
238	AGS 23 ± 2 °C	BC-II			*C: Gammaproteobacteria*		AF361096	446	0.791
238	AGS 25 °C	O125	95	35	*C: Gammaproteobacteria*		AF361091	248	0.99
238	AGS purge	XP06	96	24	*C: Gammaproteobacteria*		AF361092	204	0.901
238	GAO 28 °C	GAO398	100	3910	*C: Gammaproteobacteria*		AY098896	431	0.962
238	PAO 17 °C	PAO109	93	13	*P: Armatimonadetes*		DQ975217	213	0.942
239	AGS 20 °C	X066	99	272	*C: Gammaproteobacteria*		FJ356056	446	0.967
239	AGS 20 °C	X066			*C: Gammaproteobacteria*		AF361096	446	1
239	AGS 20 °C	X084	100	358	*C: Gammaproteobacteria*		FJ356056	465	0.932
239	AGS 20 °C	X084			*C: Gammaproteobacteria*		EU529737	465	1
239	AGS 20 °C	X114	100	162	*C: Gammaproteobacteria*		AY098896	459	0.466
239	AGS 25 °C	O125	99.4	157	*C: Gammaproteobacteria*		FJ356056	454	0.575
239	AGS purge	XP06	99	232	*C: Gammaproteobacteria*		FJ356056	473	0.997

(continued)

Table 5.9 (continued)

T-RF or OTU (bp)	Biomass origin[a]	Biological sample	Fraction of T-RF (%)[b]	Read counts (–)[c]	Broader affiliation Phylum → Class → Order	Closest affiliation Order → Family → Genus	GenBank accession number	Absolute SW mapping score (–)[d]	Normalized SW mapping score (–)[e]
239	AGS purge	XP06			*C: Gammaproteobacteria*		EU529737	473	0.859
240	AGS 20 °C	X084	100	22	*C: Gammaproteobacteria*		AF361092	267	0.968
240	AGS 25 °C	O125	100	26	*C: Gammaproteobacteria*		AF361091	273	0.896
240	AGS purge	XP06	100	10	*C: Gammaproteobacteria*		AF361092	253	0.99
242	AGS 25 °C	O125	86	6	*C: Gammaproteobacteria*		AF361096	304	0.896
245	AGS 25 °C	O125	66	4	*O: Burkholderiales*	*G: Rhodoferax*	EU499692	273	0.888
247	AGS 23 ± 2 °C	BC-II	100	3	*O: Rhodocyclales*	*G: Rhodocyclus*	AB200295	305	0.905
248	AGS 20 °C	X066	100	3	*O: Rhodocyclales*	*G: Rhodocyclus*	AB200295	305	0.905
248	AGS 20 °C	X084	66	5	*C: Betaproteobacteria*		EU834771	218	0.928
248	AGS 20 °C	X084	44	4	*O: Rhodocyclales*	*G: Rhodocyclus*	AB200295	249	0.84
248	AGS 20 °C	X114	100	2	*O: Rhodocyclales*	*G: Rhodocyclus*	AB200295	281	0.928
248	AGS 25 °C	O125	100	34	*O: Rhodocyclales*	*G: Rhodocyclus*	EF565156	323	0.989
249	AGS 20 °C	X084	92	12	*O: Rhodocyclales*	*G: Rhodocyclus*	AB200295	258	0.654
249	AGS 20 °C	X114	57	4	*C: Gammaproteobacteria*	*O: Xanthomonadaceae*	EF540404	336	0.989
250	AGS 20 °C	X084	60	3	*C: Acidobacteria*	*F: Holophagaceae*	AM180889	424	0.885
250	AGS 25 °C	O125	66	2	*O: Sphingobacteriales*		CU466731	361	0.842
250	Flocs	BC002	92	35	*O: Pseudomonadales*	*G: Acinetobacter*	EU467673	415	0.893
250	GAO 28 °C	GAO398	100	2	*C: Gammaproteobacteria*	*G: Rhodanobacter*	FJ536898	356	0.664
251	AGS 25 °C	O125	66	2	*O: Sphingobacteriales*	*G: Cytophaga*	EU104191	151	0.787

(continued)

Table 5.9 (continued)

T-RF or OTU (bp)	Biomass origin[a]	Biological sample	Fraction of T-RF (%)[b]	Read counts (–)[c]	Broader affiliation Phylum → Class → Order	Closest affiliation Order → Family → Genus	GenBank accession number	Absolute SW mapping score (–)[d]	Normalized SW mapping score (–)[e]
251	PAO 17 °C	PAO109	100	3	*O: Spirochaetales*	*G: Spirochaeta*	AJ565434	300	0.834
252	AGS 20 °C	X066	80	4	*O: Sphingobacteriales*		DQ984594	349	0.951
252	AGS 23 ± 2 °C	BC-II	80	4	*O: Sphingobacteriales*		DQ984594	349	0.681
252	AGS 25 °C	O125	100	11	*O: Sphingobacteriales*		DQ984594	330	0.71
252	Flocs	BC002	80	4	*O: Sphingobacteriales*		FJ793188	295	0.742
253	AGS 20 °C	X066	100	7	*O: Sphingobacteriales*		AM411964	355	0.959
253	AGS 23 ± 2 °C	BC-II	100	7	*O: Sphingobacteriales*		AM411964	355	0.914
253	AGS 25 °C	O125	100	2	*O: Sphingobacteriales*	*G: Niabella*	DQ457019	262	0.876
253	Flocs	BC002	88	14	*O: Ignavibacteriales*	*F: Ignavibacteriaceae*	AB186808	462	0.914
253	Flocs	BC002	88	14	*O: Ignavibacteriales*	*F: Ignavibacteriaceae*	AB186808	462	0.921
253	GAO 28 °C	GAO398	100	219	*O: Sphingobacteriales*		DQ984594	304	0.909
254	AGS 20 °C	X084	100	2	*O: Sphingobacteriales*		AB179522	325	0.706
254	AGS 20 °C	X114	100	2	*O: Sphingobacteriales*		GQ488016	258	0.878
254	AGS 25 °C	O125	100	4	*O: Sphingobacteriales*	*G: Niabella*	DQ457019	274	0.728
254	PAO 17 °C	PAO109	95	17	*O: Sphingobacteriales*		DQ413096	346	0.706
255	AGS 20 °C	X114	100	6	*O: Sphingobacteriales*		EU104041	274	0.723
255	AGS 25 °C	O125	100	4	*O: Sphingobacteriales*		EU104272	328	0.586
255	GAO 28 °C	GAO398	100	19	*O: Sphingobacteriales*		EF018676	287	0.871
255	PAO 17 °C	PAO109	100	16	*O: Sphingobacteriales*		FM872812	331	0.726

(continued)

Table 5.9 (continued)

T-RF or OTU (bp)	Biomass origin[a]	Biological sample	Fraction of T-RF (%)[b]	Read counts (–)[c]	Broader affiliation Phylum → Class → Order	Closest affiliation Order → Family → Genus	GenBank accession number	Absolute SW mapping score (–)[d]	Normalized SW mapping score (–)[e]
256	AGS 20 °C	X084	100	7	*O: Sphingobacteriales*		EU104041	337	0.957
256	AGS 25 °C	O125	100	49	*O: Sphingobacteriales*		GQ263449	333	0.739
256	AGS purge	XP06	75	3	*O: Sphingobacteriales*		EU283377	328	0.676
256	GAO 28 °C	GAO398	100	326	*O: Sphingobacteriales*		GQ396989	294	0.848
257	AGS 20 °C	X066	100	7	*O: Sphingobacteriales*		EU283377	267	0.815
257	AGS 20 °C	X084	100	39	*O: Nitrospirales*	*G: Nitrospira*	GQ487996	373	0.779
257	AGS 20 °C	X084			*O: Nitrospirales*	*G: Nitrospira*	AF314422	373	0.581
257	AGS 23 ± 2 °C	BC-II	100	7	*O: Sphingobacteriales*		EU283377	267	0.838
257	Flocs	BC002	58	7	*O: Sphingobacteriales*		EF562554	380	0.658
257	Flocs	BC002	42	5	*P: candidate phylum TM7*		AB200304	283	0.863
257	PAO 17 °C	PAO109	100	12	*O: Sphingobacteriales*		EU104313	345	0.786
258	AGS 20 °C	X066	93	16	*O: Nitrospirales*	*G: Nitrospira*	AF314422	389	0.679
258	AGS 20 °C	X114	92	12	*O: Nitrospirales*	*G: Nitrospira*	GQ487996	371	0.703
258	AGS 23 ± 2 °C	BC-II	93	16	*O: Nitrospirales*	*G: Nitrospira*	AF314422	389	0.487
258	AGS 25 °C	O125	83	19	*O: Nitrospirales*	*G: Nitrospira*	GQ487996	402	0.525
258	AGS 25 °C	O125	17	4	*O: Sphingobacteriales*		DQ202140	338	0.525
258	AGS purge	XP06	100	8	*O: Nitrospirales*	*G: Nitrospira*	GQ487996	341	0.919
258	AGS purge	XP06			*O: Nitrospirales*	*G: Nitrospira*	AF314422	341	0.936
258	PAO 17 °C	PAO109	100	6	*O: Sphingobacteriales*		EU104041	341	0.936

(continued)

Table 5.9 (continued)

T-RF or OTU (bp)	Biomass origin[a]	Biological sample	Fraction of T-RF (%)[b]	Read counts (–)[c]	Broader affiliation Phylum → Class → Order	Closest affiliation Order → Family → Genus	GenBank accession number	Absolute SW mapping score (–)[d]	Normalized SW mapping score (–)[e]
259	AGS 23 ± 2 °C	BC-II	97	38	*O: Sphingobacteriales*		EU104185	267	0.975
259	AGS 23 ± 2 °C	BC-II		1	*O: Nitrospirales*	*G: Nitrospira*	GQ487996	319	0.975
260	AGS early	BC059	100	3	*O: Sphingobacteriales*		AY302128	354	0.596
260	AGS 20 °C	X066	97	38	*O: Sphingobacteriales*		EU104185	267	0.667
260	AGS 20 °C	X066		1	*O: Nitrospirales*	*G: Nitrospira*	GQ487996	319	0.687
260	AGS 20 °C	X084	67	2	*O: Nitrospirales*	*G: Nitrospira*	GQ487996	310	0.931
260	AGS 20 °C	X114	100	10	*O: Nitrospirales*	*G: Nitrospira*	AF314422	324	0.92
260	AGS purge	XP06	96	24	*O: Nitrospirales*	*G: Nitrospira*	GQ487996	368	0.972
260	Flocs	BC002	76	16	*O: Nitrospirales*	*G: Nitrospira*	AF314422	366	0.979
260	Flocs	BC002	19	4	*O: Sphingobacteriales*		FJ660602	334	0.785
260	PAO 17 °C	PAO109	100	6	*O: Sphingobacteriales*		EU104185	214	0.868
261	AGS 20 °C	X114	100	5	*O: Sphingobacteriales*		EU104185	248	0.884
261	AGS 25 °C	O125	83	5	*O: Nitrospirales*	*G: Nitrospira*	GQ487996	345	0.966
261	AGS purge	XP06	100	3	*O: Nitrospirales*	*G: Nitrospira*	GQ487996	273	0.923
262	AGS 20 °C	X084	95	21	*O: Sphingobacteriales*		EU431801	222	0.9
262	AGS purge	XP06	100	2	*O: Sphingobacteriales*		EU104041	244	0.949
263	AGS purge	XP06	100	11	*O: Sphingobacteriales*		EU104185	275	0.951
264	AGS early	BC059	100	3	*O: Thiotrichales*	*G: Thiothrix*	L79963	334	0.896
265	AGS 20 °C	X084	100	3	*P: Acidobacteria*	*G: Candidatus* Solibacter	AY212581	243	0.933

(continued)

Table 5.9 (continued)

T-RF or OTU (bp)	Biomass origin[a]	Biological sample	Fraction of T-RF (%)[b]	Read counts (–)[c]	Broader affiliation Phylum → Class → Order	Closest affiliation Order → Family → Genus	GenBank accession number	Absolute SW mapping score (–)[d]	Normalized SW mapping score (–)[e]
270	AGS 20 °C	X084	100	19	*P: Chloroflexi*	*C: Anaerolineae*	AB117714	166	0.625
277	AGS 20 °C	X066	100	3	*P: Spirochaetes*	*F: Leptospiraceae*	AY293856	364	0.868
277	AGS 23 ± 2 °C	BC-II	100	3	*P: Spirochaetes*	*F: Leptospiraceae*	AY293856	364	0.923
277	PAO 17 °C	PAO109	100	4	*P: Spirochaetes*	*F: Leptospiraceae*	AY293856	381	0.881
281	AGS 20 °C	X066	100	4	*C: Flavobacteria*		CU925607	321	0.906
281	AGS 23 ± 2 °C	BC-II	100	4	*C: Flavobacteria*		CU925607	321	0.881
281	AGS 25 °C	O125	100	2	*C: Flavobacteria*		CU925607	236	0.98
283	AGS 20 °C	X084	100	9	*O: Rhizobiales*	*G: Aminobacter*	AF034798	312	0.98
283	AGS 25 °C	O125	100	1	*O: Rhizobiales*	*G: Aminobacter*	AF034798	251	0.982
286	AGS 20 °C	X084	100	4	*O: Rhizobiales*	*G: Devosia*	CU926373	339	0.982
286	GAO 28 °C	GAO398	99	71	*O: Rhizobiales*	*G: Bradyrhizobium*	FJ192733	275	0.908
286	PAO 17 °C	PAO109	100	2	*O: Rhizobiales*	*G: Devosia*	CU926373	319	0.955
288	AGS 20 °C	X084	75	9	*O: Sphingomonadales*	*G: Sphingosinicella*	EF363041	408	0.946
288	AGS 20 °C	X084	83	5	*O: Sphingomonadales*	*F: Sphingomonadaceae*	FJ946577	423	0.955
288	AGS 20 °C	X084	16	2	*O: Sphingomonadales*	*G: Sphingopyxis*	EF534729	342	0.965
288	AGS purge	XP06	100	4	*O: Sphingomonadales*	*G: Sphingopyxis*	AF532188	335	0.965
288	GAO 28 °C	GAO398	100	175	*O: Sphingomonadales*	*G: Sphingomonas*	EU133552	395	0.98
288	PAO 17 °C	PAO109	46	6	*O: Sphingomonadales*	*G: Sphingopyxis*	AJ416410	395	0.98
288	PAO 17 °C	PAO109	39	5	*O: Sphingomonadales*	*G: Sphingosinicella*	EF363041	351	0.959
289	Flocs	BC002	57	4	*O: Sphingomonadales*	*G: Sphingobium*	AB040739	277	0.959

(continued)

Table 5.9 (continued)

T-RF or OTU (bp)	Biomass origin[a]	Biological sample	Fraction of T-RF (%)[b]	Read counts (–)[c]	Broader affiliation Phylum → Class → Order	Closest affiliation Order → Family → Genus	GenBank accession number	Absolute SW mapping score (–)[d]	Normalized SW mapping score (–)[e]
289	Flocs	BC002	43	3	*O: Rhodospirillales*	*F: Rhodospirillaceae*	EU864465	350	0.571
289	PAO 17 °C	PAO109	100	10	*O: Sphingomonadales*	*G: Sphingopyxis*	EF424392	394	0.742
290	AGS 20 °C	X066	100	1	*O: Rhodospirillales*	*F: Rhodospirillaceae*	AM935307	358	0.742
290	AGS 20 °C	X084	100	3	*O: Rhizobiales*	*F: Bradyrhizobiaceae*	CU919831	338	0.777
290	AGS 23 ± 2 °C	BC-II	100	1	*O: Rhodospirillales*	*F: Rhodospirillaceae*	AM935307	358	0.872
290	AGS 25 °C	O125	60	3	*O: Rhizobiales*	*F: Bradyrhizobiaceae*	CU919831	374	0.872
290	AGS 25 °C	O125	40	2	*O: Rhodospirillales*	*F: Rhodospirillaceae*	AM411928	315	0.839
290	AGS purge	XP06	100	6	*O: Rhizobiales*	*F: Bradyrhizobiaceae*	CU919831	336	0.839
290	GAO 28 °C	GAO398	100	17	*C: Alphaproteobacteria*	*F: Rhodospirillaceae*	AM411928	365	0.818
293	AGS 20 °C	X114	100	3	*C: Anaerolineae*		EF020035	304	0.846
294	AGS 20 °C	X084	100	5	*C: Alphaproteobacteria*	*F: Rhodospirillaceae*	DQ066972	314	0.983
294	AGS 25 °C	O125	100	12	*P: Chloroflexi*	*G: Herpetosiphon*	AB079642	364	0.983
294	Flocs	BC002	100	1	*O: Clostridiales*	*G: Ruminococcus*	DQ796981	289	0.962
295	PAO 17 °C	PAO109	100	11	*P: Chloroflexi*	*G: Herpetosiphon*	CP000875	271	0.962
297	AGS 20 °C	X084	100	26	*P: Chloroflexi*	*G: Herpetosiphon*	NC009972	339	0.983
297	AGS 20 °C	X114	100	3	*P: Chloroflexi*	*G: Herpetosiphon*	CP000875	296	0.983
297	AGS 25 °C	O125	59	22	*O: Anaerolineae*		EF020035	361	0.97
297	AGS 25 °C	O125	41	15	*P: Chloroflexi*	*G: Herpetosiphon*	CP000875	320	0.97
297	PAO 17 °C	PAO109	100	114	*P: Chloroflexi*	*G: Herpetosiphon*	NC009972	394	0.961

(continued)

Table 5.9 (continued)

T-RF or OTU (bp)	Biomass origin[a]	Biological sample	Fraction of T-RF (%)[b]	Read counts (–)[c]	Broader affiliation Phylum → Class → Order	Closest affiliation Order → Family → Genus	GenBank accession number	Absolute SW mapping score (–)[d]	Normalized SW mapping score (–)[e]
298	AGS 20 °C	X066	88	7	*C: Gammaprotebacteria*		AB255053	209	0.961
298	AGS 20 °C	X084	100	4	*C: Anaerolineae*		EF020035	256	0.968
298	AGS 23 ± 2 °C	BC-II	88	7	*C: Gammaprotebacteria*		AB255053	209	0.968
298	AGS 25 °C	O125	100	398	*C: Gammaprotebacteria*		AF361096	338	0.705
298	PAO 17 °C	PAO109	99.5	221	*P: Chloroflexi*	*G: Herpetosiphon*	CP000875	369	0.635
302	Flocs	BC002	75	6	*C: Anaerolineae*		EU332818	310	0.672
303	GAO 28 °C	GAO398	100	123	*C: Gammaproteobacteria*		AY098896	310	0.786
303	PAO 17 °C	PAO109	89	8	*O: Sphingobacteriales*	*F: Flexibacteraceae*	AY854022	239	0.677
303	PAO 17 °C	PAO109	11	1	*C: Gammaproteobacteria*		FJ356049	304	0.779
304	AGS 20 °C	X114	100	6	*C: Gammaprotebacteria*		AF361092	246	0.911
304	AGS 20 °C	X114			*C: Gammaprotebacteria*		AB255053	246	0.807
304	AGS purge	XP06	67	2	*C: Gammaprotebacteria*		AF361096	323	0.739
304	Flocs	BC002	93	28	*C: Gammaproteobacteria*		FJ356049	383	0.417
306	AGS 23 ± 2 °C	BC-II	97	38	*P: Armatimonadetes*		CU921283	218	0.376
306	PAO 17 °C	PAO109	100	45	*P: Armatimonadetes*		CU921283	196	0.41
307	AGS 20 °C	X066	97	38	*P: Armatimonadetes*		CU921283	218	0.512
307	AGS 20 °C	X084	97	33	*P: Armatimonadetes*		CU921283	188	0.552
307	AGS 20 °C	X114	100	1	*O: Rhodocyclales*	*G: Rhodocyclus*	FJ623276	162	0.472

(continued)

Table 5.9 (continued)

T-RF or OTU (bp)	Biomass origin[a]	Biological sample	Fraction of T-RF (%)[b]	Read counts (–)[c]	Broader affiliation Phylum → Class → Order	Closest affiliation Order → Family → Genus	GenBank accession number	Absolute SW mapping score (–)[d]	Normalized SW mapping score (–)[e]
307	AGS 25 °C	O125	100	4	*P: Armatimonadetes*		CU921283	165	0.359
307	AGS purge	XP06	100	5	*P: Armatimonadetes*		CU921283	168	0.472
309	AGS 25 °C	O125	100	4	*O: Sphingobacteriales*		AF502211	332	0.404
314	PAO 17 °C	PAO109	100	6	*P: Chloroflexi*	*G: Herpetosiphon*	NC009972	259	0.462
318	AGS 20 °C	X066	100	17	*O: Sphingobacteriales*	*G: Cytophaga*	EU104191	196	0.799
318	AGS 23 ± 2 °C	BC-II	100	17	*O: Sphingobacteriales*	*G: Cytophaga*	EU104191	196	0.956
321	AGS 20 °C	X084	100	17	*O: Sphingobacteriales*	*G: Cytophaga*	EU104191	292	0.912
321	AGS 20 °C	X084			*O: Sphingobacteriales*	*G: Cytophaga*	DQ499309	292	0.87
321	AGS 20 °C	X114	100	45	*O: Sphingobacteriales*	*G: Cytophaga*	EU104191	313	0.693
321	AGS 20 °C	X114			*O: Sphingobacteriales*	*G: Cytophaga*	DQ499309	313	0.666
321	AGS 25 °C	O125	100	15	*O: Sphingobacteriales*	*G: Cytophaga*	EU104191	350	0.55
321	AGS purge	XP06	100	23	*O: Sphingobacteriales*	*G: Cytophaga*	EU104191	220	0.663
321	GAO 28 °C	GAO398	100	10	*O: Sphingobacteriales*		DQ984594	281	0.672
321	PAO 17 °C	PAO109	93	13	*O: Sphingobacteriales*		AF368190	261	0.782
322	AGS 20 °C	X114	100	7	*O: Sphingobacteriales*		FM866271	352	1
323	AGS 20 °C	X084	100	17	*O: Sphingobacteriales*		FM866271	312	0.779
325	Flocs	BC002	100	4	*O: Sphingobacteriales*		GQ396974	299	0.422
326	AGS 25 °C	O125	33	3	*O: Nitrospirales*	*G: Nitrospira*	EU937868	372	0.789
326	AGS 25 °C	O125			*O: Nitrospirales*	*G: Nitrospira*	AB252938	372	0.792

(continued)

Table 5.9 (continued)

T-RF or OTU (bp)	Biomass origin[a]	Biological sample	Fraction of T-RF (%)[b]	Read counts (–)[c]	Broader affiliation Phylum → Class → Order	Closest affiliation Order → Family → Genus	GenBank accession number	Absolute SW mapping score (–)[d]	Normalized SW mapping score (–)[e]
326	AGS purge	XP06	100	7	*O: Nitrospirales*	*G: Nitrospira*	GQ487996	340	0.398
364	AGS purge	XP06	100	38	*O: Sphingobacteriales*		EU803667	215	0.73
371	AGS purge	XP06	100	7	*O: Rhizobiales*	*G: Aminobacter*	NR025302	317	0.897
386	AGS 20 °C	X114	100	6	*C: Gammaproteobacteria*		AY098896	305	0.763
386	AGS purge	XP06	100	2	*C: Gammaproteobacteria*		AF361091	236	0.779
392	AGS 20 °C	X066	100	33	*C: Deltaproteobacteria*	*G: Bdellovibrio*	CU466777	262	0.851
392	AGS 20 °C	X084	100	11	*C: Deltaproteobacteria*	*G: Bdellovibrio*	CU466777	233	0.867
392	AGS 23 ± 2 °C	BC-II	100	33	*C: Deltaproteobacteria*	*G: Bdellovibrio*	CU466777	262	0.804
393	PAO 17 °C	PAO109	100	26	*C: Deltaproteobacteria*	*G: Bdellovibrio*	CU466777	273	0.795
398	AGS 25 °C	O125	100	1	*O: Flavobacteriales*	*G: Flavobacterium*	EU703431	293	0.685
399	AGS early	BC059	100	12	*O: Rhodocyclales*	*G: Dechloromonas*	EF632559	378	1
400	AGS 25 °C	O125	100	3	*C: Deltaproteobacteria*	*G: Bdellovibrio*	EU431712	324	0.944
402	AGS purge	XP06	100	7	*P: Chlorobi*		EF632929	309	0.963
404	AGS 25 °C	O125	100	1	*O: Sphingobacteriales*		AB504930	335	0.828

(continued)

Table 5.9 (continued)

T-RF or OTU (bp)	Biomass origin[a]	Biological sample	Fraction of T-RF (%)[b]	Read counts (–)[c]	Broader affiliation Phylum → Class → Order	Closest affiliation Order → Family → Genus	GenBank accession number	Absolute SW mapping score (–)[d]	Normalized SW mapping score (–)[e]
404	GAO 28 °C	GAO398	100	8	*O: Planctomycetales*	*G: Planctomyces*	CU926004	209	0.963
408	GAO 28 °C	GAO398	100	7	*C: Gammaproteobacteria*	*G:Pseudoxanthomonas*	DQ984530	372	0.567

The corresponding bacterial relatives are summarized across the 16S rRNA gene-targeted amplicon sequencing datasets obtained for all 10 representative biomass samples after dry-lab processing in PyroTRF-ID (Weissbrodt et al. 2012b). The phylogenetic affiliations are first given at broader level from Phylum → Class → Order and then with more resolution on the deepest relative identified in this lineage from Order → Family → Genus after mapping against Greengenes (McDonald et al. 2012)

[a] Bioreactor environment from which the biological samples were collected: *AGS* aerobic granular sludge, *AGS early* AGS collected during the early stage of granule formation, *AGS purge* AGS collected in the excess sludge purged out of the reactor to maintain a specific sludge age, *Flocs* flocculent activated sludge inoculated from a full-scale BNR wastewater treatment plant, *GAO* enrichment of glycogen-accumulating organisms, *PAO* enrichment of polyphosphate-accumulating organisms. Indication is given on temperatures maintained in bulk liquid phases

[b] Different populations can contribute to the same T-RF. The relative contribution of the bacterial population of interest to the T-RF is provided for each biological sample

[c] Total number of reads affiliating with the bacterial phylotype of interest in each sample

[d] The absolute Smith-Waterman (*SW*) score was used as criterion of mapping quality against the Greengenes database of reference sequences. It corresponds to the read length minus penalties allocated for gaps and nucleotide mismatches

[e] The normalized Smith-Waterman score corresponds to the absolute SW score divided by the read length, and provides an indication of the extent of the correspondence between the read and the reference sequence from the database. It is only an indicative value that can definitely not be considered as an identity factor (see Chap. 2 on PyroTRF-ID; Weissbrodt et al. 2012b)

Supplementary Information

Additional File 5.1 Quality plots generated for samples pyrosequenced with LowRA (> 3'000 reads) and HighRA methods (> 10'000 reads).

Additional File 5.2 Assessment of mapping performances with pyrosequencing datasets denoised without (0–500 bp) and with (300–500 bp) minimal read length cutoff.

Additional File 5.3 Comparison of the distributions of the SW mapping score and of the traditional identity score used by microbial ecologists in the field of environmental sciences for phylogenetic affiliation of sequences.

Additional File 5.4 Full digital T-RFLP profiles.

Additional File 5.5 Mirror plots obtained with raw (left) and with denoised (right) pyrosequencing datasets.

Additional File 5.6 Cross-correlation and optimal lag between denoised dT-RFLP and eT-RFLP profiles.

Additional File 5.7 Alignment of sequences mapping with the same reference sequence with identical accession number in the Greengenes database, and resulting in different digital T-RFs.

References

Aeppli C, Hofstetter TB, Amaral HIF, Kipfer R, Schwarzenbach RP, Berg M (2010) Quantifying in situ transformation rates of chlorinated ethenes by combining compound-specific stable isotope analysis, groundwater dating, and carbon isotope mass balances. Environ Sci Technol 44(10):3705–3711

Balzer S, Malde K, Jonassen I (2011) Systematic exploration of error sources in pyrosequencing flowgram data. Bioinformatics 27(13):i304–i309

Bukovska P, Jelinkova M, Hrselova H, Sykorova Z, Gryndler M (2010) Terminal restriction fragment length measurement errors are affected mainly by fragment length, G plus C nucleotide content and secondary structure melting point. J Microbiol Methods 82(3):223–228

Camarinha-Silva A, Wos-Oxley ML, Jauregui R, Becker K, Pieper DH (2012) Validating T-RFLP as a sensitive and high-throughput approach to assess bacterial diversity patterns in human anterior nares. FEMS Microbiol Ecol 79(1):98–108

Caporaso JG, Kuczynski J, Stombaugh J, Bittinger K, Bushman FD, Costello EK, Fierer N, Pena AG, Goodrich JK, Gordon JI, Huttley GA, Kelley ST, Knights D, Koenig JE, Ley RE, Lozupone CA, McDonald D, Muegge BD, Pirrung M, Reeder J, Sevinsky JR, Turnbaugh PJ, Walters WA, Widmann J, Yatsunenko T, Zaneveld J, Knight R (2010) QIIME allows analysis of high-throughput community sequencing data. Nat Methods 7(5):335–336

Clement BG, Kehl LE, DeBord KL, Kitts CL (1998) Terminal restriction fragment patterns (TRFPs), a rapid, PCR-based method for the comparison of complex bacterial communities. J Microbiol Methods 31(3):135–142

Cole JR, Wang Q, Cardenas E, Fish J, Chai B, Farris RJ, Kulam-Syed-Mohideen AS, McGarrell DM, Marsh T, Garrity GM, Tiedje JM (2009) The Ribosomal Database Project: improved alignments and new tools for rRNA analysis. Nucleic Acids Res 37:D141–D145

Collins RE, Rocap G (2007) REPK: an analytical web server to select restriction endonucleases for terminal restriction fragment length polymorphism analysis. Nucleic Acids Res 35(2):W58–W62

Desai N, Antonopoulos D, Gilbert JA, Glass EM, Meyer F (2012) From genomics to metagenomics. Curr Opin Biotechnol 23(1):72–76

Ebrahimi S, Gabus S, Rohrbach-Brandt E, Hosseini M, Rossi P, Maillard J, Holliger C (2010) Performance and microbial community composition dynamics of aerobic granular sludge from sequencing batch bubble column reactors operated at 20 °C, 30 °C, and 35 °C. Appl Microbiol Biotechnol 87:1555–1568

Edwards RA (2008) The smallest cells pose the biggest problems: high-performance computing and the analysis of metagenome sequence data. J Phys Conf Ser 125:012050

Egert M, Friedrich MW (2003) Formation of pseudo-terminal restriction fragments, a PCR-related bias affecting terminal restriction fragment length polymorphism analysis of microbial community structure. Appl Environ Microbiol 69(5):2555–2562

Ewing B, Green P (1998) Base-calling of automated sequencer traces using phred. II. Error probabilities. Genome Res 8(3):186–194

Fernandez-Guerra A, Buchan A, Mou X, Casamayor EO, Gonzalez JM (2010) T-RFPred: a nucleotide sequence size prediction tool for microbial community description based on terminal-restriction fragment length polymorphism chromatograms. BMC Microbiol 10:262

Field D, Tiwari B, Booth T, Houten S, Swan D, Bertrand N, Thurston M (2006) Open software for biologists: from famine to feast. Nat Biotechnol 24(7):801–803

Gilbert MTP, Binladen J, Miller W, Wiuf C, Willerslev E, Poinar H, Carlson JE, Leebens-Mack JH, Schuster SC (2007) Recharacterization of ancient DNA miscoding lesions: insights in the era of sequencing-by-synthesis. Nucleic Acids Res 35(1):1–10

Glenn TC (2011) Field guide to next-generation DNA sequencers. Mol Ecol Resour 11(5):759–769

Grant A, Ogilvie LA (2004) Name that microbe: rapid identification of taxa responsible for individual fragments in fingerprints of microbial community structure. Mol Ecol Notes 4(1):133–136

Gu AZ, Nerenberg R, Sturm BM, Chul P, Goel R (2011) Molecular methods in biological systems. Water Environ Res 82(10):908–930

Hall TA (1999) BioEdit: a user-friendly biological sequence alignment editor and analysis program for Windows 95/98/NT. Nucleic Acids Symp Ser 41:95–98

House CH, Runnegar B, Fitz-Gibbon ST (2003) Geobiological analysis using whole genome-based tree building applied to the Bacteria, Archaea, and Eukarya. Geobiology 1:15–26

Huber T, Faulkner G, Hugenholtz P (2004) Bellerophon: a program to detect chimeric sequences in multiple sequence alignments. Bioinformatics 20(14):2317–2319

Hume ME, Barbosa NA, Dowd SE, Sakomura NK, Nalian AG, Martynova-Van Kley A, Oviedo-Rondon EO (2011) Use of pyrosequencing and denaturing gradient gel electrophoresis to examine the effects of probiotics and essential oil blends on digestive microflora in broilers under mixed Eimeria infection. Foodborne Pathog Dis 8(11):1159–1167

Huse SM, Huber JA, Morrison HG, Sogin ML, Welch DM (2007) Accuracy and quality of massively parallel DNA pyrosequencing. Genome Biol 8(7):R143

Jakobsson HE, Jernberg C, Andersson AF, Sjolund-Karlsson M, Jansson JK, Engstrand L (2010) Short-term antibiotic treatment has differing long-term impacts on the human throat and gut microbiome. PLoS One 5(3):e9836

Junier P, Junier T, Witzel KP (2008) TRiFLe, a program for in silico terminal restriction fragment length polymorphism analysis with user-defined sequence sets. Appl Environ Microbiol 74(20):6452–6456

Kaplan CW, Kitts CL (2003) Variation between observed and true terminal restriction fragment length is dependent on true TRF length and purine content. J Microbiol Methods 54(1):121–125

Kent AD, Yannarell AC, Rusak JA, Triplett EW, McMahon KD (2007) Synchrony in aquatic microbial community dynamics. ISME J 1(1):38–47

Kitts CL (2001) Terminal restriction fragment patterns: a tool for comparing microbial communities and assessing community dynamics. Curr Issues Intest Microbiol 2(1):17–25

Kunin V, Copeland A, Lapidus A, Mavromatis K, Hugenholtz P (2008) A bioinformatician's guide to metagenomics. Microbiol Mol Biol Rev 72(4):557–578

Kunin V, Engelbrektson A, Ochman H, Hugenholtz P (2010) Wrinkles in the rare biosphere: pyrosequencing errors can lead to artificial inflation of diversity estimates. Environ Microbiol 12(1):118–123

Li H, Durbin R (2010) Fast and accurate long-read alignment with Burrows-Wheeler transform. Bioinformatics 26(5):589–595

Mao Y, Yannarell AC, Mackie RI (2011) Changes in N-transforming archaea and bacteria in soil during the establishment of bioenergy crops. PLoS One 6(9):e24750

Marsh TL (1999) Terminal restriction fragment length polymorphism (T-RFLP): an emerging method for characterizing diversity among homologous populations of amplification products. Curr Opin Microbiol 2(3):323–327

Marsh TL, Saxman P, Cole J, Tiedje J (2000) Terminal restriction fragment length polymorphism analysis program, a web-based research tool for microbial community analysis. Appl Environ Microbiol 66(8):3616–3620

Mazzola M (2004) Assessment and management of soil microbial community structure for disease suppression. Annu Rev Phytopathol 42(1):35–59

McDonald D, Price MN, Goodrich J, Nawrocki EP, DeSantis TZ, Probst A, Andersen GL, Knight R, Hugenholtz P (2012) An improved greengenes taxonomy with explicit ranks for ecological and evolutionary analyses of bacteria and archaea. ISME J 6(3):610–618

Mengoni A, Grassi E, Bazzicalupo M (2002) Cloning method for taxonomic interpretation of T-RFLP patterns. Biotechniques 33(5):990–992

Meyer F, Paarmann D, D'Souza M, Olson R, Glass EM, Kubal M, Paczian T, Rodriguez A, Stevens R, Wilke A, Wilkening J, Edwards RA (2008) The metagenomics RAST server—a public resource for the automatic phylogenetic and functional analysis of metagenomes. BMC Bioinform 9:386

Militsopoulou M, Lamari FN, Hjerpe A, Karamanos NK (2002) Adaption of a fragment analysis technique to an automated high-throughput multicapillary electrophoresis device for the precise qualitative and quantitative characterization of microbial communities. Electrophoresis 23(7–8):1070–1079

Mushegian AA, Peterson CN, Baker CCM, Pringle A (2011) Bacterial diversity across individual lichens. Appl Environ Microbiol 77(12):4249–4252

Niu B, Fu L, Sun S, Li W (2010) Artificial and natural duplicates in pyrosequencing reads of metagenomic data. BMC Bioinform 11(1):187

Oksanen J, Kindt R, Legendre P, O'Hara B, Simpson GL, Solymos P, Stevens MHH, Wagner H (2009) Vegan: community ecology package. R package version 1.15-4. R Foundation for Statistical Computing, Vienna, Austria

Osborn AM, Moore ERB, Timmis KN (2000) An evaluation of terminal-restriction fragment length polymorphism (T-RFLP) analysis for the study of microbial community structure and dynamics. Environ Microbiol 2(1):39–50

Parks DH, Beiko RG (2010) Identifying biologically relevant differences between metagenomic communities. Bioinformatics 26(6):715–721

Petrosino JF, Highlander S, Luna RA, Gibbs RA, Versalovic J (2009) Metagenomic pyrosequencing and microbial identification. Clin Chem 55(5):856–866

Pilloni G, von Netzer F, Engel M, Lueders T (2011) Electron acceptor-dependent identification of key anaerobic toluene degraders at a tar-oil-contaminated aquifer by Pyro-SIP. FEMS Microbiol Ecol 78(1):165–175

Pilloni G, Granitsiotis MS, Engel M, Lueders T (2012) Testing the limits of 454 pyrotag sequencing: reproducibility, quantitative assessment and comparison to T-RFLP fingerprinting of aquifer microbes. PLoS One 7(7):e40467

Quince C, Lanzen A, Curtis TP, Davenport RJ, Hall N, Head IM, Read LF, Sloan WT (2009) Accurate determination of microbial diversity from 454 pyrosequencing data. Nat Methods 6(9):639–641

Quince C, Lanzen A, Davenport RJ, Turnbaugh PJ (2011) Removing noise from pyrosequenced amplicons. BMC Bioinform 12:38

R Development Core Team (2008) R: a language and environment for statistical computing. R Foundation for Statistical Computing, Vienna, Austria. http://cran.r-project.org/

Reeder J, Knight R (2010) Rapidly denoising pyrosequencing amplicon reads by exploiting rank-abundance distributions. Nat Methods 7(9):668–669

Rees G, Baldwin D, Watson G, Perryman S, Nielsen D (2004) Ordination and significance testing of microbial community composition derived from terminal restriction fragment length polymorphisms: application of multivariate statistics. Antonie Van Leeuwenhoek 86(4):339–347

Regeard C, Maillard J, Holliger C (2004) Development of degenerate and specific PCR primers for the detection and isolation of known and putative chloroethene reductive dehalogenase genes. J Microbiol Methods 56(1):107–118

Rodriguez-Ezpeleta N, Hackenberg M, Aransay AM (2012) Bioinformatics for high throughput sequencing. Springer, New York

Roesch LFW, Fulthorpe RR, Riva A, Casella G, Hadwin AKM, Kent AD, Daroub SH, Camargo FAO, Farmerie WG, Triplett EW (2007) Pyrosequencing enumerates and contrasts soil microbial diversity. ISME J 1(4):283–290

Ronaghi M (2001) Pyrosequencing sheds light on DNA sequencing. Genome Res 11(1):3–11

Rossi P, Gillet F, Rohrbach E, Diaby N, Holliger C (2009) Statistical assessment of variability of terminal restriction fragment length polymorphism analysis applied to complex microbial communities. Appl Environ Microbiol 75(22):7268–7270

Schutte UME, Abdo Z, Bent SJ, Shyu C, Williams CJ, Pierson JD, Forney LJ (2008) Advances in the use of terminal restriction fragment length polymorphism (T-RFLP) analysis of 16S rRNA genes to characterize microbial communities. Appl Microbiol Biotechnol 80(3):365–380

Shani N (2012) Assessing the bacterial ecology of organohalide respiration for the design of bioremediation strategies. Ph.D. thesis, Ecole Polytechnique Fédérale de Lausanne

Smit AFA, Hubley R, Green P (2003) RepeatMasker. Institute for Systems Biology, Seattle, USA. http://www.repeatmasker.org

Smith TF, Waterman MS (1981) Identification of common molecular subsequences. J Mol Biol 147(1):195–197

Sun Y, Wolcott RD, Dowd SE (2011) Tag-encoded FLX amplicon pyrosequencing for the elucidation of microbial and functional gene diversity in any environment. Methods Mol Biol 733:129–141

Thies JE (2007) Soil microbial community analysis using terminal restriction fragment length polymorphisms. Soil Sci Soc Am J 71(2):579–591

Trombetti GA, Bonnal RJP, Rizzi E, De Bellis G, Milanesi L (2007) Data handling strategies for high throughput pyrosequencers. BMC Bioinform 8(1):S22

Weissbrodt DG, Lochmatter S, Ebrahimi S, Rossi P, Maillard J, Holliger C (2012a) Bacterial selection during the formation of early-stage aerobic granules in wastewater treatment systems operated under wash-out dynamics. Front Microbiol 3:332

Weissbrodt DG, Shani N, Sinclair L, Lefebvre G, Rossi P, Maillard J, Rougemont J, Holliger C (2012b) PyroTRF-ID: a novel bioinformatics methodology for the affiliation of terminal-restriction fragments using 16S rRNA gene pyrosequencing data. BMC Microbiol 12:306

Weissbrodt DG, Neu TR, Kuhlicke U, Rappaz Y, Holliger C (2013) Assessment of bacterial and structural dynamics in aerobic granular biofilms. Front Microbiol 4:175

Weissbrodt DG, Maillard J, Brovelli A, Chabrelie A, May J, Holliger C (2014a) Multilevel correlations in the biological phosphorus removal process: from bacterial enrichment to conductivity-based metabolic batch tests and polyphosphatase assays. Biotechnol Bioeng 111(12):2421–2435

Weissbrodt DG, Shani N, Holliger C (2014b) Linking bacterial population dynamics and nutrient removal in the granular sludge biofilm ecosystem engineered for wastewater treatment. FEMS Microbiol Ecol 88(3):579–595

Wilson CA, Kreychman J, Gerstein M (2000) Assessing annotation transfer for genomics: quantifying the relations between protein sequence, structure and function through traditional and probabilistic scores. J Mol Biol 297(1):233–249

Wommack KE, Bhavsar J, Ravel J (2008) Metagenomics: read length matters. Appl Environ Microbiol 74(5):1453–1463

Chapter 6
Multilevel Correlations in the Metabolism of Polyphosphate-Accumulating Organisms

From Enrichment to Electrical-Conductivity-Based Metabolic Batch Tests and Polyphosphatase Assays

Inorganic polyphosphate: making a forgotten polymer unforgettable.
The primary energy currency of living systems is the phosphoanhydride bond.
Adapted from Kornberg (1995) and Keasling et al. (2000)

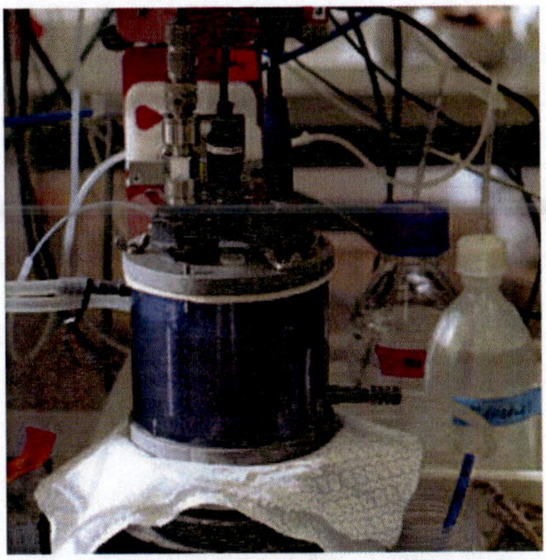

At-line metabolic testing

The content of this chapter was published in a modified version in: Weissbrodt DG, Maillard J, Brovelli A., Chabrelie A, May J, Holliger C (2014) Multilevel correlations in the biological phosphorus removal process: From bacterial enrichment to conductivity-based metabolic batch tests and polyphosphatase assays. Biotechnol Bioeng 111:2421–2435. https://doi.org/10.1002/bit.25320. Permission was granted to reuse the figure materials (© 2014 Wiley Periodicals, Inc.).

Supplementary Information The online version contains supplementary material available at https://doi.org/10.1007/978-3-031-41009-3_6.

D. G. Weissbrodt, *Engineering Granular Microbiomes*, Springer Theses,
https://doi.org/10.1007/978-3-031-41009-3_6

6.1 Introduction

Enhanced biological phosphorus removal (EBPR) from wastewater relies on proper management of the bacterial resource towards preferential selection of active polyphosphate-accumulating organisms (PAOs) (Comeau et al. 1987; Bond et al. 1995; van Loosdrecht et al. 1997; Mino et al. 1998; Hesselmann et al. 2000; Blackall et al. 2002; McMahon et al. 2010; Nielsen et al. 2012b). PAOs exhibit a complex metabolism that has grabed the attention of environmental engineers and microbiologists since more than 20 years. It involves uptake and condensation of volatile fatty acids (VFA) as poly-β-hydroxyalkanoates (PHAs) under anaerobic conditions, sustained by hydrolysis of intracellular stocks of inorganic polyphosphate (PP) and glycogen (Smolders et al. 1995b; Mino et al. 1998). PAOs aerobically grow on PHAs, and replenish PP stocks by uptake and polymerization of orthophosphate residues (Pi) present in the medium. The uncoupled supply of VFA as electron donors and terminal electron acceptors such as dissolved oxygen (DO), nitrite, and nitrate, by alternation of anaerobic feast and aerobic or anoxic starvation phases selects for PAOs over fast-growing ordinary heterotrophic organisms (OHO) (Wentzel et al. 2008). Glycogen-accumulating organisms (GAOs), which display similar anaerobic-aerobic metabolism but without uptake of phosphorus, can however compete for VFA (Mino et al. 1994; Satoh et al. 1994; Bond et al. 1999; Mino 2000; Crocetti et al. 2002; Zeng et al. 2003; Oehmen et al. 2007; Lopez-Vazquez et al. 2008).

Numerous studies have investigated the PAOs metabolism and the PAO/GAO competition for optimizing EBPR systems (Smolders et al. 1995b; Filipe et al. 2001; Meyer et al. 2003; Schuler and Jenkins 2003; Zeng et al. 2003; Lopez-Vazquez et al. 2009b; Oehmen et al. 2010a). Conditions of slight alkaline pH (7.25–7.50), temperatures below 20 °C, non-limiting phosphorus in the wastewater (20–40 $g_{COD}\ g_P^{-1}$), and mixtures of acetate and propionate have selected for the betaproteobacterial PAOs "*Candidatus* Accumulibacter phosphatis". On the contrary, slight acidic pH (6.5), higher mesophilic temperatures (30 °C), limiting phosphorus content (100–200 $g_{COD}\ g_P^{-1}$), and single VFA have selected for the gamma- and alphaproteobacterial GAOs "*Candidatus* Competibacter phosphatis" and *Defluviicoccus vanus*, respectively.

Different methods have been used to characterize EBPR from reactor to bacterial community, metabolic, and genetic levels, making this biotechnological application one of the best described microbial ecosystem (Albertsen et al. 2012; Nielsen et al. 2012b). Since electrical conductivity (hereafter referred to as conductivity) positively correlates with anaerobic-aerobic Pi-cycling (Maurer and Gujer 1995), online conductivity profiles have been used for real-time control of EBPR sequencing batch reactors (SBR) (Serralta et al. 2004; Aguado et al. 2006). Relative abundances of active PAOs and GAOs populations have been measured by 16S rRNA-targeted fluorescence in situ hybridization (FISH) (Bond et al. 1999; Hesselmann et al. 1999; Crocetti et al. 2000, 2002; Meyer et al. 2006), and have been correlated with the yield of Pi-release to acetate-uptake under anaerobic conditions ($Y_{PO4/Ac}$) (Lopez-Vazquez et al. 2007a). EBPR microbiomes and metagenomes have recently been investigated with the advent of massive sequencing techniques (Garcia Martin et al. 2006; Wilmes et al. 2008; Albertsen et al. 2012). Investigations have also focused on metabolic fluxes (Satoh et al. 1992; Maurer et al. 1997; Pramanik et al. 1999), functional gene diversity and transcription (Gavigan et al. 1999; Keasling et al. 2000; McMahon et al. 2002; Garcia Martin et al. 2006; He et al. 2010b; He and McMahon 2011b), and enzymes involved (Wilmes and Bond 2004; Burow et al. 2008; Wilmes et al. 2008). PP synthesis is catalyzed by polyphosphate kinases (PPK, EC 2.7.4.1) that are widely conserved within prokaryotes, namely PPK1 and PPK2 homologs, acting as polyphosphate:adenosine (ADP) and guanosine diphosphate (GDP) phosphotransferases (Ahn and Kornberg 1990; Zhang et al. 2002; McMahon et al. 2007; Wilmes et al. 2008). PP hydrolysis yields ATP or GTP when catalyzed by PPK acting in reverse mode, and Pi when catalyzed by polyphosphatases such as exopolyphoshatases (PPX, EC 3.6.1.11). Polyphosphate:AMP phosphotransferase (EC 2.7.4.B2) and adenylate kinase (EC 2.7.4.3) have also been shown to be involved (van Groenestijn et al. 1989). PPK1 is specific to ADP and ATP, whereas PPK2 is nucleotide-aspecific. PP utilization by PPK2 occurs at a rate 75- and 240-fold greater than PP synthesis by PPK2 and PPK1, respectively (Rao et al. 2009). The *ppk* genes have been targeted for analysis of "*Ca.* Accumulibacter" clades (McMahon et al. 2007; He et al. 2010a; Slater et al. 2010; Gonzalez-Gil and Holliger 2011). The *ppx* genes have on the contrary barely been described.

Multilevel correlations in the PAOs metabolism were investigated from reactor to enzymatic scales (Fig. 6.1). Since the cultivation of PAO- and GAO-enrichments has not been straightforward, reactor start-up conditions were first optimized together with monitoring of bacterial community dynamics. Core bacterial microbiomes were analyzed by pyrosequencing. Correlations were assessed between conductivity evolution, polyphosphatase activity, and PAO fractions in anaerobic metabolic batches conducted with PAO/GAO mixtures (Fig. 6.2). A mathematical model was implemented in PHREEQC to describe metabolic activities of PAO/GAO mixtures and the conductivity evolution under anaerobic conditions. Degenerated PCR was attempted to describe the diversity of *ppx* genes of EBPR sludges.

6.2 Material and Methods

6.2.1 Cultivation of PAO- and GAO-Enrichments

Hierarchical clustering and principal component analysis (PCA) of 35 different operation conditions reported in literature (Additional File 6.1 in Supplementary Information) was performed in R (R-Development-Core-Team 2008) equipped with the vegan package (Oksanen et al. 2009) to identify triggers of PAOs and GAOs selection, and to define initial composition of cultivation media (Table 6.1).

The 2.5-L double-wall glass stirred-tank PAO-SBR and GAO-SBR (Applikon Biotechnology, The Netherlands) were inoculated with 3 g_{VSS} L^{-1} of activated sludge

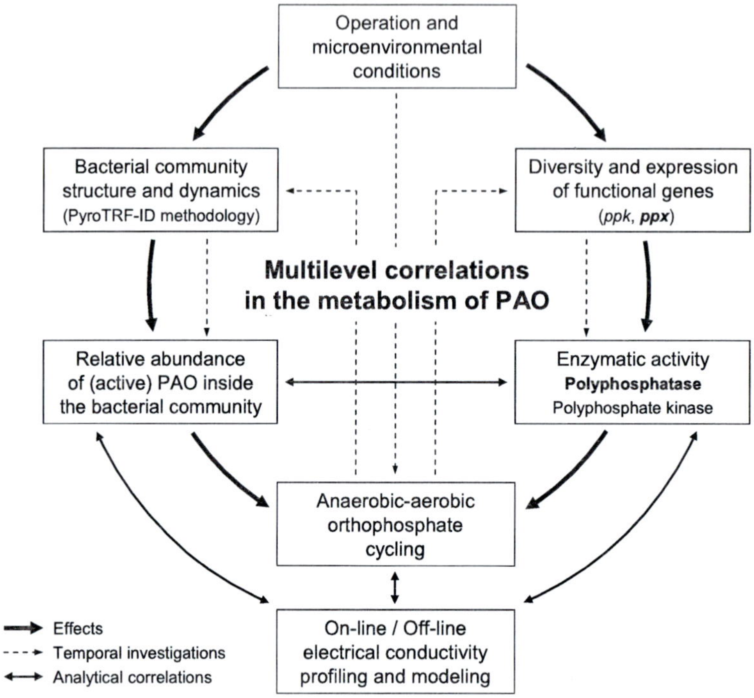

Fig. 6.1 Conceptual representation of the multilevel correlations in the metabolism of PAOs. The anaerobic-aerobic orthophosphate-cycling performances of an EBPR reactor can be assessed by on-line and off-line measurements of conductivity profiles. This activity results from the presence and the expression of functional *ppx* and *ppk* genes coding for exopolyphosphatases and polyphosphate kinases, respectively. Since many studies have already focused on *ppk* genes, *ppx* genes were targeted here with the aim to characterize their diversity in EBPR sludges. Gene expression was not investigated here. On a metabolic point of view, special attention was given to the overall activity of the synthesized pool of proteins. Since the hydrolysis of energetic polyphosphate under anaerobic conditions relies on a complex biochemistry, the developed polyphosphatase enzymatic assay targeted the pool of enzymes present in cell extracts of activated sludges. Since the volumetric rates of orthophosphate-cycling measured in EBPR reactors rely on the abundance of (active) PAOs inside the underlying bacterial community, analytical methods based on conductivity evolution under anaerobic conditions and polyphosphatase activity were developed for fast assessment of relative abundances of PAOs. Since the efficiency of EBPR systems also rely on the structure of the bacterial microbiome, bacterial community compositions and dynamics in a lab-scale enrichment culture of "*Ca.* Accumulibacter" were assessed by combining T-RFLP and pyrosequencing via the PyroTRF-ID methodology (Weissbrodt et al. 2012b/Chap. 5)

from a full-scale EBPR-WWTP (ARA Thunersee, Switzerland). The SBR cycles were adapted from Lopez-Vazquez et al. (2009b), and comprised N_2-flush (7 min), pulse feeding (7.3 min), N_2-flush (5 min), anaerobic, aerobic and settling phases (for timing see below), and withdrawal (5 min; 50% volume exchange ratio). Mixed liquors were stirred at 300 rpm during anaerobic and aerobic phases. Nitrification was inhibited by addition of allyl-*N*-thiourea. The sludge retention times (SRT) were

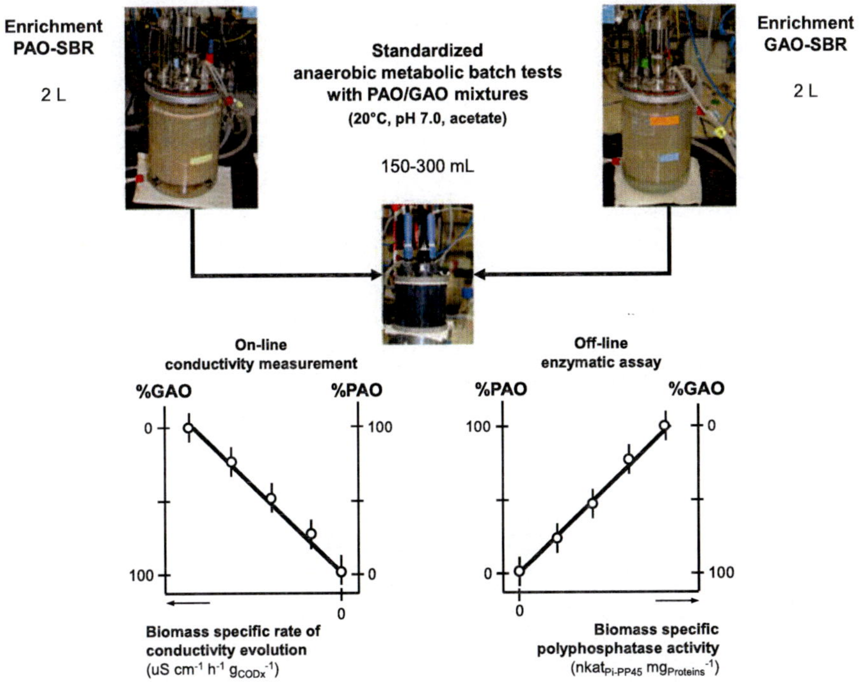

Fig. 6.2 Analytical concept for the determination of PAO fractions and EBPR potential of sludge from conductivity-based anaerobic metabolic batch tests and polyphosphatase activity assays

controlled by automated purge of excess sludge from aerated mixed liquors. Stepwise optimizations in operation parameters were compiled in Table 6.2.

The PAO-SBR was optimally operated at 17 °C, pH 7.0–8.0, with 12-h hydraulic retention time (HRT) and 8-days SRT at steady state, propionate, and 9 $g_{CODs}\ g_{P\text{-}PO4}{}^{-1}$ in the influent wastewater. Enhanced anaerobic propionate uptake and Pi-cycling activities were ensured by stepwise adaptation of the volumetric organic loading rate (OLR) from 15 to 200 mg_{CODs} cycle^{-1} $L_R{}^{-1}$ in 12 days, and by proper control of the anaerobic and aerobic contact times (3–5 h) based on on-line conductivity profiles. Since a fast-settling biomass formed after 30 days, the settling time was decreased from 60 to 10 min to save cycle time, and to prevent prolonged endogenous respiration. The GAO-SBR was operated at 30 °C, pH 6.5 ± 0.2, with 12-h HRT, acetate, and 185 $g_{COD}\ g_{P\text{-}PO4}{}^{-1}$. The SRT was doubled after 100 days from 8 to 16 days according to Lopez-Vazquez et al. (2009a) in order to sustain the growth of GAOs. The OLR, anaerobic phase length, and settling time were fixed at 200 mg_{CODs} cycle^{-1} $L_R{}^{-1}$, 3 h, and 60 min right from start-up, respectively. In both reactors, pH was regulated with HCl and NaOH (1 mol L^{-1}), and 3.5 ± 0.5 mg_{O2} L^{-1} were maintained during aeration by on/off control. DO, pH, temperature, and conductivity were recorded on-line.

Table 6.1 Optimal composition of nutritive media used for the cultivation of stable PAO- and GAO-enrichments

Compound	CAS No.	Molecular formula	Molecular weight	Concentrations in influent wastewater	
			($g\ mol^{-1}$)	PAO-SBR	GAO-SBR
C-source medium[a]					
Sodium acetate[b]	127-09-3	$C_2H_3O_2Na{\cdot}3H_2O$	136.09	–	6.25 $mmol\ L^{-1}$
Sodium propionate[b]	137-40-6	$C_3H_5O_2Na$	96.06	3.57 $mmol\ L^{-1}$	–
Magnesium sulfate	7487-88-9	$MgSO_4{\cdot}7H_2O$	246.51	0.37	0.37
Calcium chloride	10043-52-4	$CaCl_2{\cdot}2H_2O$	147.02	0.10	0.10
N-source and P-source medium[a]					
Ammonium chloride	12125-02-9	NH_4Cl	53.49	1.43 $mmol\ L^{-1}$	1.43 $mmol\ L^{-1}$
Potassium dihydrogen phosphate[c]	7778-77-0	KH_2PO_4	136.09	1.61	0.07
Allyl-*N*-thiourea[d]	109-57-9	$C_4H_8N_2S$	116.18	2.00 $mg\ L^{-1}$	2.00 $mg\ L^{-1}$
Yeast extract[e]	8013-01-2	n.a	n.a	0.80	–
Peptone[e]	91079-38-8	n.a	n.a	0.80	–
Casamino acids[e]	65072-00-6	n.a	n.a	0.80	–
Trace element solution	–	–	–	0.30 $mL\ L^{-1}$	0.30 $mL\ L^{-1}$
Compound	CAS No.	Molecular formula	Molecular weight	Concentrationsin stock solution	
			($g\ mol^{-1}$)	(g per 5 L)	($mmol\ L^{-1}$)
Trace element stock solution[a, f]					
Disodium EDTA	139-33-3	$C_{10}H_{14}N_2O_8Na_2{\cdot}2H_2O$	372.25	50.00	26.86
Zinc sulfate	7733-02-0	$ZnSO_4{\cdot}7H_2O$	287.59	0.60	0.42
Manganese chloride	7773-01-5	$MnCl_2{\cdot}4H_2O$	197.92	0.60	0.61
Iron(III) chloride	7705-08-0	$FeCl_3{\cdot}6H_2O$	270.30	7.50	5.55
Sodium molybdate	7631-95-0	$Na_2MnO_4{\cdot}2H_2O$	205.92	0.30	0.29
Copper(II) sulfate	7758-98-7	$CuSO_4{\cdot}5H_2O$	249.71	0.15	0.12
Cobalt(II) chloride	7646-79-9	$CoCl_2{\cdot}6H_2O$	237.96	0.75	0.63

(continued)

Table 6.1 (continued)

Compound	CAS No.	Molecular formula	Molecular weight	Concentrationsin stock solution	
			($g\ mol^{-1}$)	(g per 5 L)	($mmol\ L^{-1}$)
Boric acid	11113-50-1	H_3BO_3	61.83	0.75	2.43
Potassium iodide	7681-11-0	KI	166.00	0.90	1.08

[a] All solutions were prepared in demineralized water. The synthetic wastewater was prepared by 1:1 mixing of 2-times concentrated C-source medium, and N-source and P-source medium

[b] The final concentrations of acetate and propionate in the influent corresponded to 12.5 $C\text{-}mmol_{Ac}\ L^{-1}$ and 10.7 $C\text{-}mmol_{Pr}\ L^{-1}$, respectively, and to 400 $mg_{COD}\ L^{-1}$. Optimal SBR start-up was operated with stepwise increase of the organic concentration. At steady state, the SBRs were fed with volumetric organic loading rates (OLR) of 200 $mg_{CODs}\ cycle^{-1}\ L_R{}^{-1}$

[c] At steady state, the COD/P ratio of the influent wastewaters amounted to 8 and 200 $g_{COD}\ g_{P\text{-}PO4}{}^{-1}$, respectively

[d] Allyl-*N*-thiourea was added to inhibit nitrification

[e] Protein complements were added in the PAO-SBR to sustain the enrichment of "*Ca.* Accumulibacter", according to the media composition used by different authors (Smolders et al. 1994a; Hesselmann et al. 1999; Hollender et al. 2002; Schuler and Jenkins 2003; Zeng et al. 2003; Lu et al. 2006; Lopez-Vazquez et al. 2009b; Marcelino et al. 2009). COD conversion factors: 1.4 $g_{COD}\ g_{yeast\ extract}{}^{-1}$, 1.4 $g_{COD}\ g_{casamino\ acids}{}^{-1}$, 1.2 $g_{COD}\ g_{peptone}{}^{-1}$

[f] The composition of the trace element solution was taken from Lopez-Vazquez et al. (2009b)

6.2.2 *Bacterial Community Compositions of PAO- and GAO-Enrichments*

Bacterial community dynamics were followed in PAO- and GAO-SBRs by terminal-restriction fragment length polymorphism (T-RFLP) analysis of eubacterial 16S rRNA encoding genes with the PCR labeled forward primer 8f[1] and unlabeled reverse primer 518r,[2] and with the *Hae*III endonuclease, according to Weissbrodt et al. (2012a). T-RFLP profiles were expressed as relative contributions of operational taxonomic units (OTU), and were presented in stacked bar plots of predominant OTUs (> 2%). Phylogenetic affiliations of OTUs were obtained by pyrosequencing-based PyroTRF-ID analysis of biomass grab samples collected on day 109 in the PAO-SBR and day 398 in the GAO-SBR, according to the PyroTRF-ID methodology (Weissbrodt et al. 2012b/Chap. 5). The pyrosequencing datasets of these particular days were analyzed with MG-RAST (Meyer et al. 2008) to decipher the bacterial microbiomes of the PAO- and GAO-sludges.

[1] 8f primer sequence: FAM-5′-AGAGTTTGATCMTGGCTCAG-3′.

[2] 518r primer sequence: 5′-ATTACCGCGGCTGCTGG-3′.

Table 6.2 Step-wise optimization of start-up conditions for the cultivation of PAO- and GAO-enrichments

Trial No.	Stepwise adaptation of operation parameters to obtain stable PAO- and GAO-enrichment cultures										Failure event
	pH (–)	T (°C)	Ac/Pr[a] (%COD)	COD (g_{COD} L^{-1})	COD/P (g_{COD} g_P^{-1})	K^+ (mg L^{-1})	Mg^{2+} (mg L^{-1})	Proteins (mg L^{-1})	SFASS[c] (mL L^{-1})	Δt_{An} (h)	
PAO-SBR											
1	6.8–7.2	18	0:100	400[c]	20	25	9	0	0	Fixed (3 h)[h]	pH oscillation (relay)[i]
2	7.1–7.5	18	100:0	400	20	25	9	0	0	Fixed (3 h)	pH oscillation (stirrer)[j]
3	7.1–7.5	15	0:100	400	20	25	9	0	0	Fixed (3 h)	Conductivity decrease[k]
4	7.1–7.9	15	0:100	400	20	25	9	0	0	Fixed (3 h)	Filamentous overgrowth[l]
5	7.1–7.5	17	Ac/Pr switches[b]	400	20	25	9	0	0	Fixed (3 h)	Conductivity decrease[k]
6	7.1–7.5	17	75:25	400	20	25	9	0	0	Fixed (3 h)	Conductivity decrease[k]
7	7.1–7.5	17	75:25	400	8[d]	63[d]	61[e]	3 × 0.8[f]	100[g]	Fixed (3 h)	Filamentous overgrowth[l]
8	7.1–8.0	17	0:100	Adapted (50 → 400)[c]	8	63	9	3 × 0.8	0	Adapted (6 → 3)[h]	(Stable)[m]
GAO-SBR											

(continued)

Table 6.2 (continued)

Trial No.	Stepwise adaptation of operation parameters to obtain stable PAO- and GAO-enrichment cultures										Failure event
	pH (–)	T (°C)	Ac/Pr[a] (%COD)	COD (g_{COD} L^{-1})	COD/P (g_{COD} g_P^{-1})	K^+ (mg L^{-1})	Mg^{2+} (mg L^{-1})	Proteins (mg L^{-1})	SFASS[c] (mL L^{-1})	Δt_{An} (h)	
1	6.8–7.2	22	100:0	400	200	9	3	0	0	Fixed (3 h)	pH oscillation (relay)[i]
2	6.3–6.7	30	100:0	400	200	9	3	0	0	Fixed (3 h)	(Stable)[m]

[a] Ratio of acetate and propionate in influent

[b] Switches between 100% acetate and 100% propionate every 10 days

[c] After initial operations with fixed concentrations of chemical oxygen demand (COD) in the influent, this variable was adapted stepwise from 50 to 400 mg_{COD} L^{-1} in 12 days in order to ensure full anaerobic VFA uptake during reactor start-up

[d] The concentration of potassium was increased together with the concentration of orthophosphate, since KH_2PO_4 was used as phosphorus source

[e] The concentration of magnesium was increased to the level of potassium, to presumably enhance the stability of the process according to Schonborn et al. (2001)

[f] The cultivation medium was supplemented with proteins, namely peptone, casamino acids, and yeast extract, since those parameter was shown to correlate with PAOs predominance (Fig. 6.3c)

[g] The cultivation medium was supplemented with sterile filtrate of activated sludge supernatant (SFASS) following Hollender et al. (2002) in order to provide presumably essential compounds that originate from the "natural" EBPR ecosystem of PAOs

[h] After initial operation with fixed anaeronic contact time, the PAO-SBR was operated in the last trial with dynamic manual control of the anaerobic contact time between 6 and 3 h according to the reactor performances recorded with conductivity

[i] Oscillations in pH regulation occurred during the first two experiments after 30 days caused by activation of pH regulation right after feeding. A timer relay was implemented to activate pH regulation only after 30 s of stirring after feeding

[j] The original stirring motor did not withstand long term SBR operation. The mechanical agitation system was modified with a RZR 2051 control stirrer (Heidolph Instruments, Germany) connected to a home-made long axis stirrer for robust continuous SBR operation on long term (1–2 years)

[k] The decrease in conductivity profiles indicated that orthophosphate-cycling activity was progressively lost

[l] Filamentous organisms proliferated in the SBR and led to deteriorated EBPR performances

[m] These operational conditions led to sustainable enrichments of PAOs and GAOs

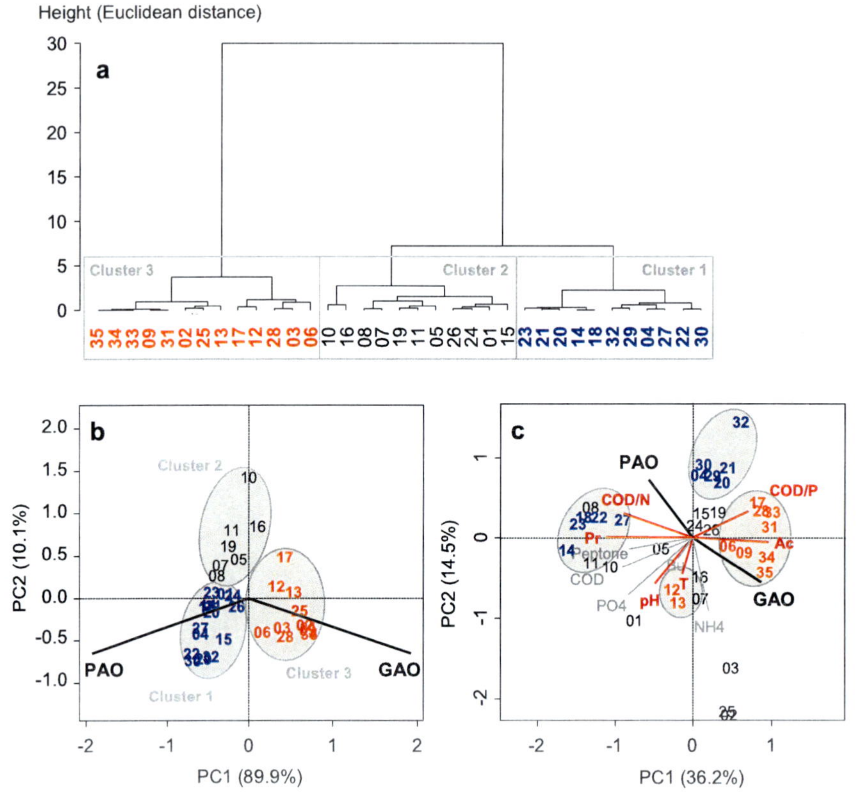

Fig. 6.3 Hierarchical clustering and principal component analyses (PCA) of cultivation conditions retrieved from literature (Additional File 6.1) selecting for either PAOs or GAOs in anaerobic-aerobic SBRs. Clusters of studies (*numbers*) were defined according to the reported relative abundances of PAOs and GAOs (**a**). According to the first PCA correlation biplot, studies affiliating with the blue cluster 1 and the orange cluster 3 reported predominance of PAOs and GAOs, respectively (**b**). The intermediate cluster 2 was related to close relative abundances of PAOs and GAOs. The second PCA correlation biplot provide information on the correlation between the cultivation conditions and the relative abundance of PAOs and GAOs obtained in the reference studies. Blue and orange colors were kept to provide information on PAOs and GAOs predominance, respectively. Vector directions and lengths indicate the extent of correlations between variables and with the reference studies

6.2.3 Conductivity-Based Anaerobic Metabolic Batch Tests

A method combining anaerobic metabolic batch tests and on-line acquisition was developed for assessing correlations between conductivity evolution and PAO fractions. Anaerobic batch tests were conducted under standardized temperature (20 °C) and pH (7.00 ± 0.05) conditions in a double-wall 200-mL PVC batch reactor with mixtures of the PAOs and GAOs sludge in volumetric ratios of 0:100, 25:75, 50:50, 75:25, and 100:0 according to Lopez-Vazquez et al. (2007a).

The synthetic medium was a 1:1 mixture of mineral and acetate solutions (Additional File 6.2). After collection of biomass of corresponding enrichments at the end of aeration, PAO- and GAO-biomass were centrifuged (1 min, 5000 rpm), washed, resuspended and mixed in 100 mL of mineral solution to achieve final concentrations of about 1 g_{VSS} L^{-1}. The suspensions were homogenized at 500 rpm, and deoxygenated by continuously sparging with dinitrogen gas at 0.5 mL min^{-1}. The pH was regulated with HCl or NaOH at 0.25 mol L^{-1}. DO, pH, temperature, and conductivity were measured on-line.

As soon as anaerobic conditions were achieved, anaerobic batches were started by addition of 100 mL of pre-deoxygenated acetate solution, and incubated during 5 h. Liquid phase samples were collected every 5 min during the first 30 min, and every 15 min afterwards.

Acetate and ionic species were analyzed by HPLC and ion chromatography. Initial volumetric rates of soluble components and conductivity were calculated based on the first 30 min of the measured profiles.

Biomass samples were collected after 30 min for enzymatic assay (see below). In the end, biomass was sampled for measurement of total (TSS), inorganic (ISS) and volatile suspended solids (VSS), and for T-RFLP analysis. Relative abundances of "*Ca.* Accumulibacter"- and "*Ca.* Competibacter"-affiliating OTUs were used to correct the theoretical composition of PAO/GAO mixtures.

An empirical relation was deduced between initial maximum biomass specific rates of conductivity evolution and fractions of "*Ca.* Accumulibacter". The method was tested with a suspension of intact and ground aerobic granular sludge collected in an EBPR-SBR.

6.2.4 *Implementation of a PAO/GAO Metabolic Model in PHREEQC*

A mathematical model was implemented in the aqueous geochemistry software PHREEQC[3] (Parkhurst 1995) for describing the anaerobic metabolic activities of PAOs and GAOs at 20 °C and pH 7.0, for computing conductivity evolution based on chemical speciation and acid-base equilibria, and for identifying the processes contributing to conductivity.

Stoichiometric and biokinetic matrices of PAO/GAO processes formulated by Smolders et al. (1995b), Manga et al. (2001), Zeng et al. (2003), Siegrist et al. (2002), and Lopez-Vazquez et al. (2009b) were adapted with the main participating ions acetate, orthophosphate, potassium, magnesium, and bicarbonate, according to Serralta et al. (2004), Seco et al. (2004), and Aguado et al. (2006) (Additional File 6.2). All concentrations, stoichiometric, and kinetic coefficients were expressed in moles as prerequisite for implementation in PHREEQC.

[3] Electronic web link: http://wwwbrr.cr.usgs.gov/projects/GWC_coupled/phreeqc/.

Two sets of simulations were run to assess the impact of the concentration of PAOs (0–250 C-mmol$_x$ L^{-1}, or 0–6 g_{VSS} L^{-1}) and of the presence of GAOs in different PAO/GAO ratios (100:0 to 0:100 m/m) on conductivity profiles.

6.2.5 Polyphosphatase Enzymatic Assay

A polyphosphatase enzymatic assay was developed based on Lindner et al. (2009) with biomass samples collected in the PAO-SBR after 20 min of anaerobic periods, when Pi release was the fastest according to conductivity.

Mixed liquor samples of 5 mL were collected in 13-mL plastic tubes, washed twice by centrifugation (1 min, 5000 rpm) and resuspension in 3 and 1 mL, respectively, of Tris-HCl buffer 50 mmol L^{-1} pH 7.5 (CAS 1185-53-1), and transferred in 2-mL cryotubes prior to centrifugation (10 s, 13'000 rpm). The biomass pellets were flash-frozen in liquid nitrogen, and stored at −80 °C. Cell extracts were prepared by thawing biomass pellets at ambient temperature in 1 mL Tris-HCl, by de-agglomerating them in a 15-mL Potter–Elvehjem Tissue Grinder (Wheaton Industries Inc., USA), by weakening cell membranes with 1 mL of fresh lysozyme solution (4 mg mL^{-1}, CAS 12650-88-3, Fluka Analytika, Switzerland), and by sonication in 10 × 5 pulses of 0.7 s and 50 W (Vibra Cell 72405, Bioblock Fisher Scientific, France) interrupted by 1-min idle on ice, and in the presence of DNAse (CAS 9003-98-9, Roche Diagnostics, Switzerland) to decrease viscosity. Enzymatic assays were conducted on crude cell extracts. Since PPK are a membrane-bound proteins (Ahn and Kornberg 1990; Akiyama et al. 1992), and PPX probably too (Akiyama et al. 1993), it was crucial to keep cell debris.

Enzymatic reactions were started by adding 10 μL of a solution of commercial polyphosphate (PP_{45}) at 10 mmol$_{PP45}$ L^{-1} (sodium phosphate glass Type 45, $P_{45}O_{136}Na_{47}$, 4'650 g mol^{-1}, S4379, Sigma Aldrich, Switzerland) to a mixture of 40 μL of cell extract and 150 μL of PPX buffer pH 7.5 (Tris-HCl 50 mmol L^{-1}, KCl 25 mmol L^{-1}, $MgCl_2 \cdot 6H_2O$ 2 mmol L^{-1}) in 1.5-mL Eppendorf tubes, and run over 20–30 min at room temperature (20 °C). PP hydrolysis was followed with off-line colorimetric measurements of the residual concentration of PP_{45}. Aliquots of 10 μL reaction mixture were collected every 2 min and diluted 5-times in ultrapure water. Aliquots of 10 μL of the diluted sample were transferred in acrylic semi-micro cuvettes (10 × 4 × 45 mm, No. 67.740, Sarstedt, Germany) containing 1 mL of Toluidine Blue O solution 6 mg L^{-1} ($C_{15}H_{16}N_3S \cdot Cl$, CAS 92-31-9, 305.83 g mol^{-1}, Sigma-Aldrich, Switzerland) in acetic acid 40 mmol L^{-1} pH ~ 1, and by measuring the metachromatic shifts from 630 nm (free toluidine) to 530 nm (toluidine-PP complex) at 20 °C in a Hitatchi spectrophotometer U-3010 (Hitachi Instruments Inc., USA). Since PP concentration is proportional to the absorbance ratio A_{530}/A_{630} in a cubic trend, measurements targeted the pseudo linear range. Maximum initial volumetric rates of PP hydrolysis were converted into biomass specific rates by normalization with the protein content of cell extracts (free of lyzozyme and DNAse) measured with the bicinchoninic BCA Protein Assay Kit (Pierce Biotechnology Inc., USA).

The enzymatic assay was used to assess correlations with the activity of PAO/GAO mixtures measured in the conductivity assay and to compare the polyphosphatase activity of cell extracts of PAOs biomass collected under anaerobic and aerobic conditions. The polyphosphatase activity was empirically related to the PAO fraction.

6.2.6 Degenerate PCR for the Screening of PPX Genes

PPX represent a very broad enzyme family in bacteria. In order to target the largest possible diversity of *ppx* genes in the full-scale EBPR sludge and in the PAO-enrichment, all bacterial PPX protein sequences present in general databases (NCBI, non-redundant) were screened by an iterative search using BlastP.[4]

All sequence entries that satisfied the criteria of sequence length (ranging from 200 to 700 amino acids) and of minimal matching score (E-value) of 1×10^{-10} were kept for further analysis. A database of 1503 PPX protein sequences was generated. Sequences were aligned with ClustalX[5] and clustered by MEGA4.[6] Two conserved amino acid stretches, named regions 1 and 2, were clearly identified in most PPX sequences (1431 and 1501 considered sequences for regions 1 and 2, respectively). Both regions were usually found separated by approximately 130 amino acids (corresponding to ~ 390 nucleotides). Sequence clusters were further aligned to produce amino acid motifs with Weblogo,[7] allowing to design degenerate primers (Table 6.3). The five forward primers (PPX-1a-F to PPX-1e-F) and two reverse primers (PPX-2a-R and PPX-2b-R) covered 97 and 92% of PPX sequences (Additional File 6.3).

DNA was extracted from biomass with the PowerSoil DNA isolation kit (Mobio, France) in two sequences of vortexing (4 s), heating (70 °C, 5 min), and beat-beating (30 s, 100 G). PCR amplifications were conducted with ten couples of degenerate primers (Table 6.3). The 10-μL PCR mixtures comprised 1 μL DNA template (45–50 ng), 0.5 μL of each primer (10 μmol L^{-1}), 2 μL Go*Taq* PCR buffer 5X, 5.65 μL nuclease free ultrapure sterile water (hereafter referred to as water, Merck-Millipore, Germany), 0.3 μL dNTPs (2.5 mmol L^{-1}), and 0.05 μL Go*Taq* DNA polymerase (0.25 U, Promega, Switzerland). The PCR program comprised initial denaturation (95 °C, 5 min) followed by 30 cycles of denaturation (95 °C, 1 min), annealing (50 °C, 1 min), and elongation (72 °C, 90 s), and by a final elongation (72 °C, 10 min). Extraction and PCR negative controls consisted of water. Amplicon qualities were assessed by Gel Red-stained (Brunschwig Chemie, Switzerland) planar agarose gel electrophoresis. Efficient primer couples were selected for enhanced PCR (50 μL) prior to cloning-sequencing. Positive amplicons were pooled and purified with the QIAquick kit (Quiagen, Switzerland), eluted in 30 μL buffer to increase their

---

[4] Electronic web link: www.ncbi.nlm.nih.gov/blast.

[5] Electronic web link: www.clustal.org.

[6] Electronic web link: www.megasoftware.net.

[7] Electronic web link: http://weblogo.berkeley.edu/logo.cgi.

Table 6.3 Degenerate primers targeting PPX encoding genes

Primer direction, name, sequence[a] and degeneracy	Protein region 1	Protein region 2	Couple no with reverse primer	
Forward primers			PPX-2a-R	PPX-2b-R
PPX-1a-F GAYTGYGGIACNAAYWS Degeneracy = 128	 DCGTN(S/T)[b]	–	I	II
PPX-1b-F GAYVTIGGIWSIAAYWS Degeneracy = 192	 D(I/L[c]/V/M)G(S/T)N(S/T)[b]	–	III	IV
PPX-1c-F GGIACNAAYAAYTGYMG Degeneracy = 64	 GTNNCR[b]	–	V	VI
PPX-1d-F ATIGAYRTNGGNWSIAAYG Degeneracy = 256	 ID(I/V)GSN(A/G)[b]	–	VII	VIII
PPX-1e-F GAYRTNGGIWSITAYWS Degeneracy = 512	 D(I/V)G(S/T)Y(S/T)[b]	–	IX	X

(continued)

Table 6.3 (continued)

Primer direction, name, sequence[a] and degeneracy	Protein region 1	Protein region 2	Couple no with reverse primer	
			PPX-2a-R	PPX-2b-R
Forward primers				
Reverse primers				
PPX-2a-R SWNSCICCNCCIARRTC Degeneracy = 512	–	 D(F/L)GG(G/A)(S/T)[b]		
PPX-2b-R SWNSCICCNCCIAYRTC Degeneracy = 512	–	 D(I/M/V)GG(G/A)(S/T)[b]		

[a] IUPAC nucleotide nomenclature: *I*: inosine; *R*: A or G; *S*: C or G; *V*: A, C or G; *W*: A or T; *Y*: C or T

[b] Only the two first bases of the last codon were included in the primer sequence

[c] Only 4 out of 6 possible leucine codons were included here. Both codons TTA and TTG were not considered here in order to reduce the degeneracy level

concentration, and their quality was assessed at 260, 230 and 280 nm (NanoDrop NG-1000, Witec, Switzerland).

Amplified gene fragments were cloned and sequenced for affiliation to putative members of the *ppx* gene cluster. The purified amplicons were ligated overnight in pGEM-T Easy Vectors (Promega, Switzerland) carrying ampicillin resistance genes, and the plasmids transformed in competent *E. coli* cells by heat shock (1 min at 42 °C, 3 min on ice), including ligation and transformation controls. Bacterial suspensions were grown in 900 μL of Luria–Bertani (LB) broth medium at 37 °C under agitation, plated on solid LB agar medium containing ampicillin (100 μg mL^{-1}), and incubated overnight at 37 °C. Microcolonies were picked and resuspended in 10 μL water, and lysed at 95 °C (5 min). Gene fragments present in the isolated plasmids were amplified in 10-μL PCR mixtures comprising 1 μL cell lysate, 0.5 μL of each T7 and SP6 primer (10 μmol L^{-1}), 1 μL GoTaq PCR buffer 5X, 6.65 μL water, 0.3 μL dNTPs (2.5 mmol L^{-1}), and 0.05 μL *Taq* DNA polymerase (PeqLab, Germany). The PCR program comprised initial denaturation (95 °C, 5 min) followed by 30 cycles of denaturation (95 °C, 45 s), annealing (50 °C, 45 s), elongation (72 °C, 1 min), and by final elongation (72 °C, 1 min). Clones containing gene fragments were selected for enhanced PCR (30 μL). Positive amplicons were purified with the MSB Spin PCRapace kit (Invitek, Stratec Molecular, Germany), eluted in 15 μL buffer, and subjected to quality assessment.

PCR for sequencing was performed in 10-μL mixtures of 200 ng purified amplicons, 1.6 μL primer T7 (1 μmol L^{-1}), 2 μL BigDye Terminator V3.1 sequencing mix (Invitrogen, Switzerland), and up to 10 μL water. The PCR program comprised initial denaturation (96 °C, 5 min), followed by 30 cycles of denaturation (96 °C, 10 s), annealing (50 °C, 5 s), and elongation (60 °C, 4 min). Amplicons were precipitated during 20 min in the dark in a mixture of 64 μL ethanol 100% and 26 μL water. Precipitation products were centrifuged (20 min, 16'000 G), and dried in 250 μL ethanol 70%, prior to a second centrifugation step (10 min, 16'000 G). After careful supernatant withdrawal, the pellets were dried in air, resuspended in 10 μL formamide, denaturated at 95 °C (2 min), and cooled on ice (3 min). The single brand analytes were sequenced in an ABI PRISM 3130xl Genetic Analyzer (Applied Biosystems, Life Technologies, USA). The pGEM-T based plasmids were sequenced with the T7 primer. The obtained insert sequences were analyzed with BlastX.[8]

[8] Electronic web link: www.ncbi.nlm.nih.gov/blast.

6.3 Results

6.3.1 *Principal Component Analysis of PAO- and GAO-Enrichment Conditions*

Hierarchical clustering and principal component analyses (PCA) of conditions reported in literature for the operation of anaerobic-aerobic SBRs (Additional File 6.1) enabled highlighting factors selecting for either PAOs or GAOs (Fig. 6.3). According to cluster positions, PCA vector directions and lengths, PAOs have best been selected with propionate, low COD/P ratios (< 20 $g_{COD}\ g_{P\text{-}PO4}^{-1}$) and high COD/N ratios (> 20 $g_{COD}\ g_{N\text{-}NH4}^{-1}$) over the reported ranges of 9–200 $g_{COD}\ g_{P\text{-}PO4}^{-1}$ and 3–39 $g_{COD}\ g_{N\text{-}NH4}^{-1}$, and at temperatures of 20 °C and below. The presence of peptone also correlated positively with predominance of PAOs. Preferential selection of GAOs has mainly been obtained with acetate, high COD/P and low COD/N ratios, and low pH. The conditions used in this study were designed based on these correlations. The PAO-SBR was operated at 17 °C, pH 7.0–8.0, in the presence of propionate, with a COD/P ratio of maximum 20 $g_{COD}\ g_{P\text{-}PO4}^{-1}$, a COD/N ratio of 20 $g_{COD}\ g_{N\text{-}NH4}^{-1}$, and amendments of peptone, casamino acids, and yeast extract. The GAO-SBR was operated at 30 °C, pH 6.3–6.7, with acetate, 185 $g_{COD}\ g_{P\text{-}PO4}^{-1}$, and 20 $g_{COD}\ g_{N\text{-}NH4}^{-1}$ (Table 6.1).

6.3.2 *Typical Profiles of Soluble Compounds Recorded in the PAO-SBR*

SBR performances were followed with on-line conductivity, and with off-line analysis of VFA, orthophosphate, nitrogen species, and other ionic compounds. Under favorable conditions, VFA were fully removed anaerobically (Fig. 6.4a). In the PAO-SBR, VFA uptake correlated with release of Pi, potassium, and magnesium ions originating from PP hydrolysis (Fig. 6.4b). The three ionic species were taken up aerobically to regenerate polyphosphate. An effective dephosphatation can be observed from the difference in influent and effluent concentrations. Conductivity evolution positively correlated with anaerobic-aerobic cycling of the three ionic species (Fig. 6.4c). Ammonium was in addition slightly removed under anaerobic (13%) and aerobic conditions (38%) most probably due to anabolic requirements of the biomass, since nitrite and nitrate were not formed in the presence of allyl-*N*-thiourea. In the GAO-SBR, acetate was fully consumed anaerobically after 20 days of reactor operation. Since GAOs do not release Pi and pH was regulated between 6.3 and 6.7, performances of the GAO-SBR could not be assessed with on-line measurements. Conductivity, pH and DO signals were still useful to detect potential major failure events.

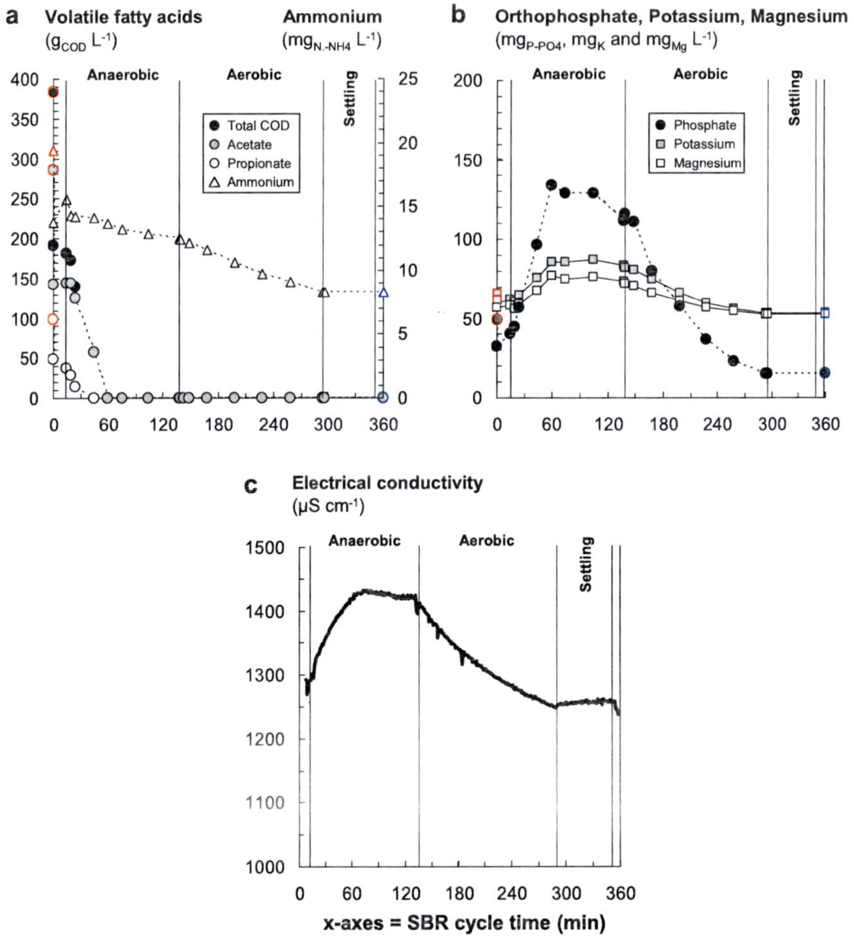

Fig. 6.4 Typical profiles of anaerobic VFA uptake and ammonium assimilation (**a**), of orthophosphate, potassium and magnesium cycling (**b**), and of electrical conductivity (**c**) recorded in the PAO-SBR for an enrichment trial with a mixture of acetate and propionate 75:25%$_{COD}$ as carbon source. ***Legend***: concentrations in the influent (*red circles*) and in the effluent (*blue circles*) of the reactor

6.3.3 *Obtaining a Stable PAO-Enrichment by Control of OLR and Anaerobic Phase Length*

Although operation conditions were optimized based on literature information, obtaining stable enrichment cultures remained a challenging task. Whereas only two trials were sufficient to obtained a stable GAO-enrichment, eight experiments were necessary to obtain a stable PAO-enrichment (Table 6.2). The two first trials had to be aborted because of technical issues related to technology transfer to the laboratory that were troubleshoot by home-made solutions. The five next experiments led to unfavorable decrease of conductivity amplitudes and filamentous proliferation after initial enrichment of "*Ca.* Accumulibacter" relatives over the first 30 days despite stepwise adaptation of pH, temperature, VFA composition, ratio of magnesium to potassium ions, and supplementation of cultivation media with sterile filtrate of activated sludge supernatant (SFASS) from the EBPR-WWTP. A successfully stable PAO-enrichment was finally obtained by progressive increase of the volumetric organic loading rate (OLR) from 25 to 200 mg_{COD} $cycle^{-1}$ $L_R{}^{-1}$ within 12 days and by dynamic manual control of the anaerobic contact time between 6 and 3 h over the whole experimental period in order to preferentially select for PAOs over OHO by ensuring full anaerobic uptake of VFA.

6.3.4 Continuous Bacterial Community Monitoring of PAO- and GAO-Enrichments

During PAO-enrichment trials 3–5, progressive failure in EBPR correlated with the dynamics of the underlying bacterial communities, as shown by the two examples in Fig. 6.5a. All phylogenetic affiliations obtained with the pyrosequencing-based PyroTRF-ID methodology are available in the Appendix. During trial 5 with VFA switches, two PAOs populations of "*Ca.* Accumulibacter" (31%) and *Acinetobacter* (21%) predominated after 10 days. However, *Zoogloea* spp. progressively increased from 7 to 12%, and *Rhizobiales* relatives (19%) became predominant after 30 days. During trial 7 with VFA mixture and supplementation with SFASS, a third PAO-related *Tetrasphaera* population that dominated the inoculation sludge remained above 30% during the first 10 days. *Xanthomonadaceae* were present in abundances of up to 15%. "*Ca.* Accumulibacter" affiliates became predominant after only 25 days (43%), but were then again outcompeted by *Zoogloea* (up to 16%) and *Rhizobiales* (up to 24%) affiliates. Optimization of start-up resulted in efficient Pi-cycling activities on long term that correlated with predominant "*Ca.* Accumulibacter" relatives (34–61%) (Fig. 6.5b). Despite steady-state operation with a fixed sludge retention time (SRT) of 8 days, oscillations were observed in accompanying *Xanthomonadaceae* (7–23%), *Herpetosiphon* (4–15%), *Sphingobacteriales* (2–10%), *Tetrasphaera* (3–10%), and *Comamonadaceae* (2–8%) relatives. *Zoogloea* (< 4%) and *Rhizobiales* (< 2%) were never abundant. An enrichment of GAOs was obtained under the conditions of the second trial (30 °C, pH 6.3–6.7, acetate) although the reactor was operated with fixed OLR of 200 mg_{CODs} $cycle^{-1}$ $L_R{}^{-1}$ and anaerobic contact time of 3 h right from start-up. *Tetrasphaera* spp. dominated during the first two weeks (25–38%), whereas "*Ca.* Competibacter" became predominant (> 35%) after 50 days (Fig. 6.5c). After 100 days, doubling the SRT from 8 to 16 days resulted in the proliferation of putative alphaproteobacterial GAOs affiliating with *Rhodospirillaceae* (30%). Different bacterial populations exhibited dynamics on the long run, e.g. "*Ca.* Competibacter" (12–47%), *Rhodospirillaceae* (3–28%), *Acidobacteriaceae* (2–20%), *Sphingobacteriales* (4–44%), *Xanthomonadaceae* (2–31%), *Rhizobiales* (2–13%), *Tetrasphaera* (3–7%), and *Armatimonadetes* (2–7%) relatives.

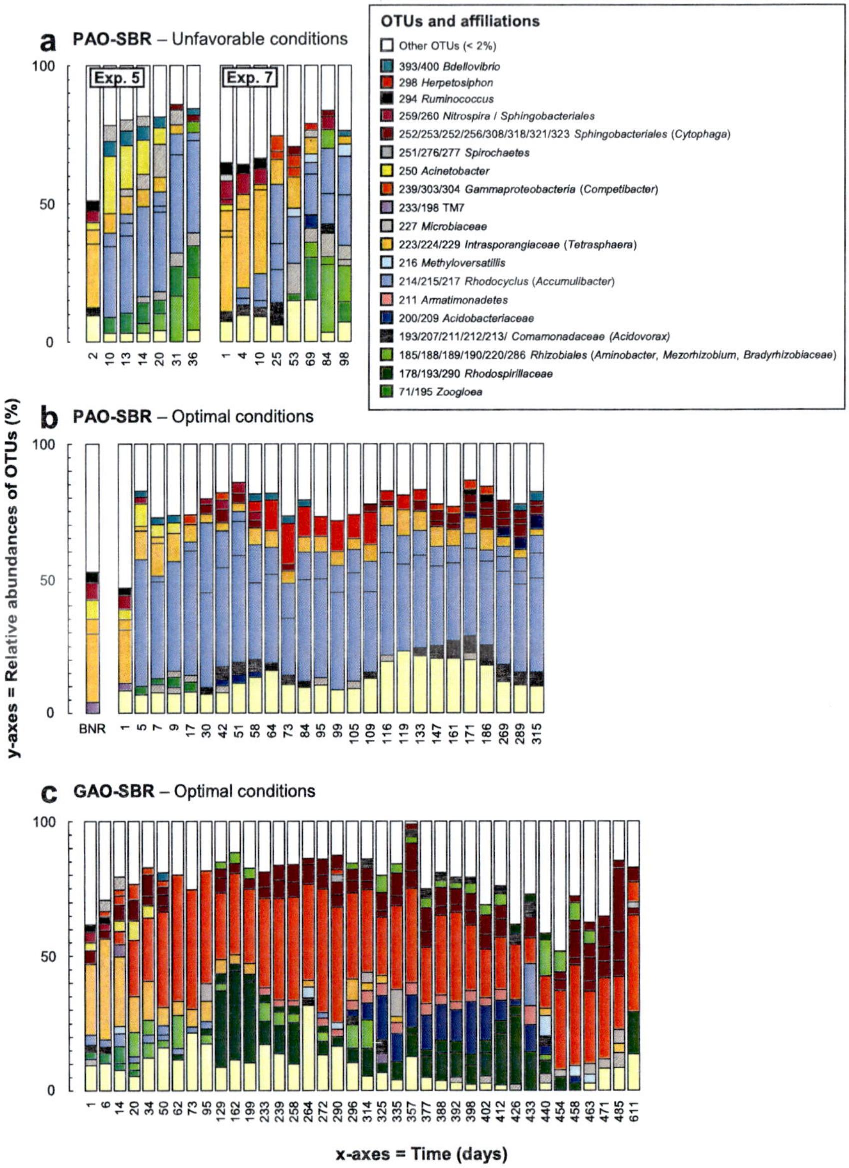

Fig. 6.5 Bacterial community dynamics during the enrichment of PAOs (**a**) and of GAOs (**b**) in activated sludge under unfavorable (*left*) and under optimal (*right*) start-up conditions

6.3.5 Composition of the Bacterial Microbiomes of the PAO- and GAO-Enrichments

The phylogenetic trees constructed on the pyrosequencing datasets of the two biomass samples collected on day 109 in the PAO-SBR and on day 398 in the GAO-SBR confirmed that both bacterial microbiomes were displaying significantly different bacterial community structures (Fig. 6.6). On day 109, the PAO-enrichment was mainly composed of *Betaproteobacteria* affiliating with *Rhodocyclales* ("*Ca.* Accumulibacter") and *Burkholderiales* (*Acidovorax*, *Ralstonia*), of *Cytophagales*, of *Flavobacteriales*, *Herpetosiphonales*, and *Xanthomonadales*. On day 398, the GAO-enrichment comprised predominant populations of *Methylococcales*, *Chromatiales* (*Nitrococcus*) and unclassed *Gammaproteobacteria* (most probably "*Ca.* Competibacter"), of alphaproteobacterial *Rhodospirillales*, *Rhodobacterales*, and *Rhizobiales* (*Methylosinus*, *Bradyrhizobium*), of *Sphingobacteriales* (*Terrimonas*, *Chitinophaga*), and *Acidobacteria* (*Acidobacterium*, "*Ca.* Koribacter").

6.3.6 Correlation Between Conductivity Profiles and PAO/GAO Ratios

Anaerobic metabolic batch tests were conducted with mixtures of sludges collected on day 315 in the PAO-SBR comprising abundant "*Ca.* Accumulibacter" relatives (50%) and on day 611 in the GAO-SBR comprising abundant GAOs (60%) affiliating with "*Ca.* Competibacter" (43%) and *Rhodospirillaceae* (17%). The measured biomass specific conductivity evolution correlated with increases in the amount of Pi released (Additional File 6.4). When only GAOs were present in the sludge, an increase in conductivity was also detected, but without concomitant Pi release. With the relatively low biomass concentrations used (0.8–1.0 g_{VSS} L^{-1}), only 100–250 mg_{COD} L^{-1} of acetate were consumed within 5 h.

Linear correlations were obtained between the maximum biomass specific rates of conductivity evolution (q_σ) and of orthophosphate release (q_{PO4}) calculated over the first 30 min, the yield of Pi-release to acetate uptake ($Y_{PO4/Ac}$) and the PAO/GAO ratio with proportionality factors of 337 $\mu S\ cm^{-1}\ h^{-1}\ g_{CODx}{}^{-1}$, 46 $mg_{P\text{-}PO4}\ h^{-1}\ g_{CODx}{}^{-1}$, and 0.51 $mg_{P\text{-}PO4}\ mg_{COD,Ac}{}^{-1}$, respectively (Fig. 6.7a and Additional File 6.4). The conductivity assay was tested with an aerobic granular sludge (AGS) collected from an EBPR-SBR. PAO fractions assessed with this method were close to those measured by T-RFLP analysis (Table 6.4). Intact and ground AGS suspensions provided identical results.

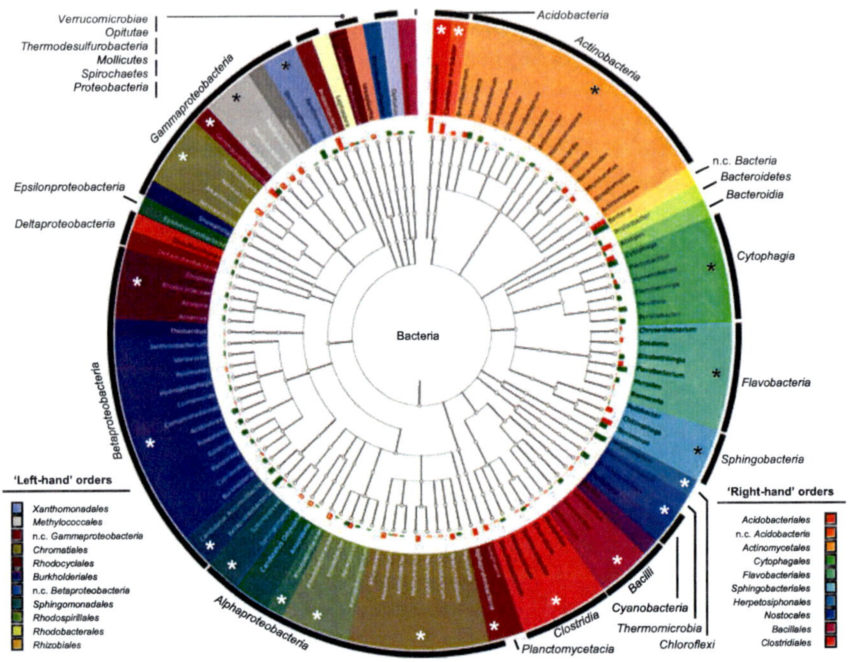

Fig. 6.6 Circular phylogenetic tree representation of the full bacterial microbiomes of PAO-SBR (*green bar plot*) and GAO-SBR (*red bar plot*) obtained after analysis with MG-RAST (Meyer et al. 2008) of the pyrosequencing datasets of biomass grab samples collected on day 109 and day 398, respectively. The RDP database (Cole et al. 2009) was used as annotation source, and a minimum identity cutoff of 97% was applied. Each bar plot is related to the number of pyrosequencing reads affiliating with the target bacterial relative. The tree is presented with classes (*outer black circle segments*), order subdivisions (*colored slices*), and bacterial genera names. Identities of target orders (*asterisks*) are given for left- and right-hand halves of the tree

6.3.7 Model-Based Evaluation of Conductivity Evolutions in Anaerobic Metabolic Batch Tests

A mathematical model was successfully implemented in PHREEQC for the description of metabolic activities and conductivity evolution in anaerobic metabolic batch tests.

With only PAOs, increases in their concentration resulted in faster volumetric conductivity evolutions, Pi-release, and acetate uptake (Fig. 6.8a and Additional File 6.5). Conductivity and Pi could be related in a linear manner with a factor 2.8 (μS cm^{-1}) (mg$_{\text{P-PO4}}$ L^{-1})$^{-1}$. Similarly to experiments, a maximum of 50 mg$_{\text{COD}}$ L^{-1} of acetate was consumed within 30 min with PAOs concentrations below 25 C-mmol$_X$ L^{-1} (1 C-mol$_X$ ~ 24.6 g$_{\text{VSS}}$). The kinetic and stoichiometric coefficients q_σ (195 μS cm^{-1} h^{-1} g$_{\text{CODx}}$$^{-1}$), $Y_{\text{PO4/Ac}}$ (0.42 mg$_{\text{P-PO4}}$ mg$_{\text{COD,Ac}}$$^{-1}$), and $Y_{\sigma/\text{Ac}}$ (1.18 μS cm^{-1} mg$_{\text{COD,Ac}}$$^{-1}$) were independent from the PAOs concentration.

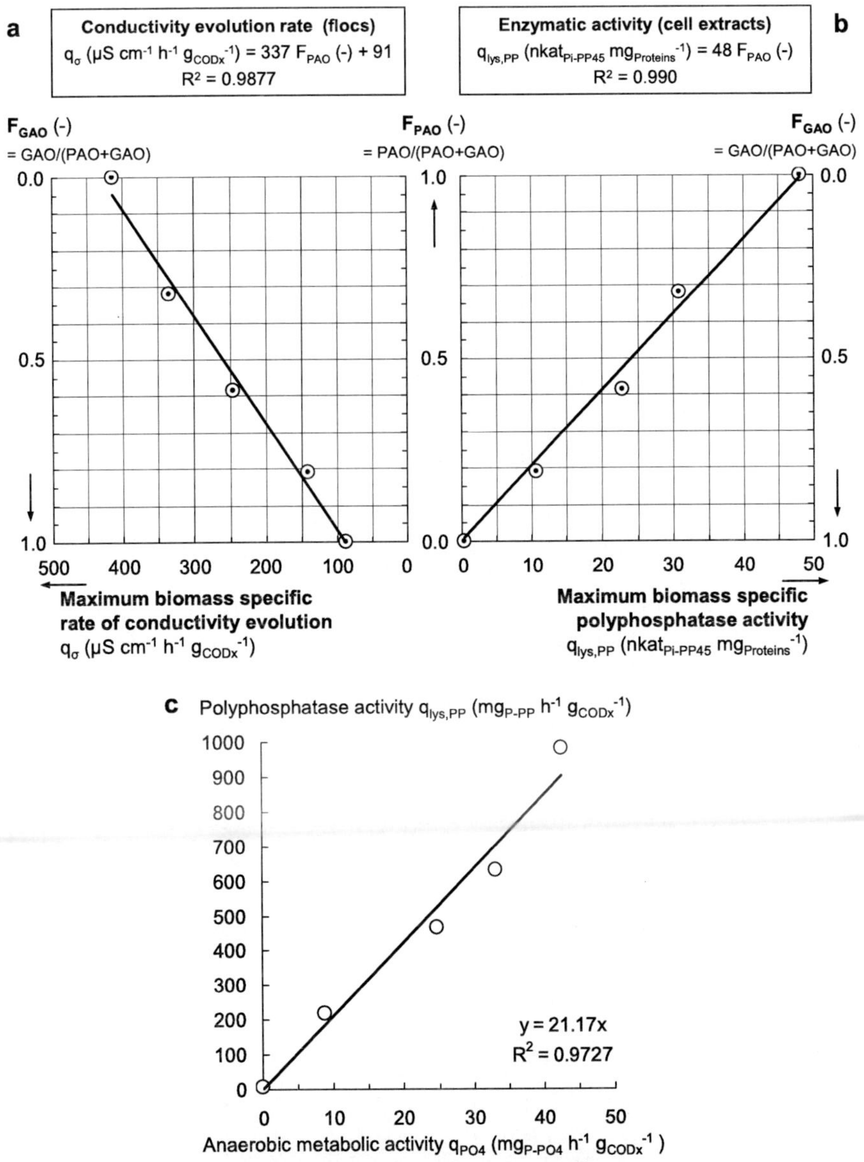

Fig. 6.7 Linear correlations obtained between the biomass specific rate of conductivity evolution (**a**), the biomass specific polyphosphatase activity (**b**), and the fraction of PAOs present in the mixtures of PAO- and GAO-enrichments. Comparison of the polyphosphate-hydrolyzing activities measured during the anaerobic batch test by monitoring orthophosphate release (*x-axis*) and by polyphosphatase enzymatic assays (*y-axis*) (**c**)

Table 6.4 Comparison of PAO fractions obtained with the conductivity assay and with T-RFLP measurements

Biomass sample	Biomass specific rate of conductivity evolution	PAO fraction obtained from conductivity assay	T-RFLP relative abundance in originating sludge		PAO fraction calculated from T-RFLP measurement
	q_σ (μS cm^{-1} h^{-1} $g_{CODx}{}^{-1}$)	$F_{PAO,Assay}$[b, c] (–)	PAOs (%)	GAOs (%)	$F_{PAO,T\text{-}RFLP}$[b, d] (–)
Intact AGS	280	0.56	24.3	20.5	0.54
Ground AGS[a]	264	0.51	24.3	20.5	0.54

[a] Conductivity assays conducted with fractions of intact and ground AGS collected from an EBPR-SBR
[b] F_{PAO} is defined as equal to the fraction of abundances of PAOs and GAOs: $X_{PAO}/(X_{PAO} + X_{GAO})$
[c] $F_{PAO,Assay}$ was obtained with the previously obtained linear relation of the biomass specific rate of conductivity evolution: q_σ (μS cm^{-1} h^{-1} $g_{CODx}{}^{-1}$) = 337 F_{PAO} (–) + 91 (Fig. 6.7)
[d] $F_{PAO,T\text{-}RFLP}$ according to the relative abundances of PAOs and GAOs measured by T-RFLP

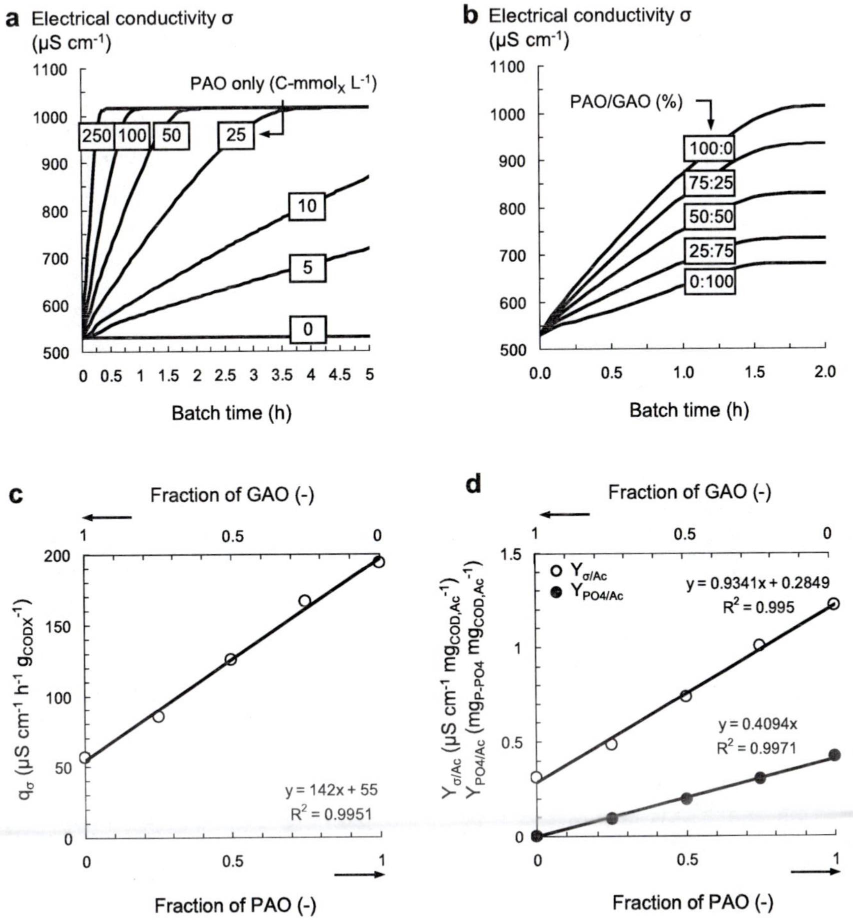

Fig. 6.8 Model-based evaluation of the correlation between the evolution of conductivity in anaerobic conditions and the relative abundance of PAOs and GAOs in the activated sludge

In the presence of GAOs in the PAO-enriched biomass, the conductivity and Pi evolution rates decreased proportionally to the PAO/GAO ratio (Fig. 6.8b and Additional File 6.5), whereas the acetate uptake rate was almost constant. Coefficients q_σ (Fig. 6.8c), $Y_{PO4/Ac}$ and $Y_{\sigma/Ac}$ (Fig. 6.8d) correlated linearly with the PAO/GAO ratio with factors of 142 $\mu S\ cm^{-1}\ h^{-1}\ g_{CODx}{}^{-1}$, 0.41 $mg_{COD,Ac}{}^{-1}$, and 0.93 $\mu S\ cm^{-1}\ mg_{COD,Ac}{}^{-1}$, respectively.

With only GAOs, the conductivity increase without concomitant Pi-release was related to bicarbonate formation during acetate uptake (Additional File 6.5). The slightly higher bicarbonate production with GAOs than with PAOs was in agreement with the stoichiometric coefficients of 0.54 and 0.33 $mol_{HCO3}\ mol_{Ac}{}^{-1}$, respectively, defined in Additional File 6.2. Bicarbonate contribution explained the y-intercepts and the slight sigmoidal shapes of the experimental and modeled lines of q_σ and $Y_{\sigma/Ac}$ (Figs. 6.7a and 6.8c, d).

6.3.8 Correlating Polyphosphatase Activity and PAO/GAO Ratios

Enzymatic assays conducted on cell extracts of the PAO/GAO mixtures and commercial PP45 revealed that the maximum biomass specific polyphosphatase activity ($q_{lys,PP}$) correlated linearly with the PAO/GAO ratio, with a factor of 48 $nmol_{Pi\text{-}PP45}\ s^{-1}\ mg_{Proteins}{}^{-1}$ (or $nkat_{Pi\text{-}PP45}\ mg_{Proteins}{}^{-1}$ in SI units) (Fig. 6.7b and Additional File 6.6). A linear correlation was thus obtained between the polyphosphatase activity and the conductivity evolution rate. An enzymatic activity of 1 $nkat_{Pi\text{-}PP45}\ mg_{Proteins}{}^{-1}$ corresponded to an increase of 7.2 $\mu S\ cm^{-1}\ h^{-1}\ g_{CODx}{}^{-1}$ (Additional File 6.6). When expressed with equivalent units by considering the protein content of the cell extracts (0.25 $g_{Proteins}\ g_{VSS}{}^{-1}$) and the standard biomass conversion factor of 1.366 $g_{CODx}\ g_{VSS}{}^{-1}$, polyphosphate hydrolysis by cell extracts was 21 times higher than by intact flocs in anaerobic batches (Fig. 6.7c). Higher polyphosphatase activity was in addition obtained with crude cell extracts containing cell debris than with only supernatant, indicating possible involvement of membrane-bound proteins in PP hydrolysis. By using the enzymatic assay, it was shown that the two samples of PAOs sludge taken in the middle of the anaerobic and the aerobic phase of an SBR cycle exhibited close polyphosphatase activities of 35 and 45 $nkat_{Pi\text{-}PP45}\ mg_{Proteins}{}^{-1}$, respectively.

6.3.9 Screening PPX Genes in Activated Sludge and PAO-Enrichment

The degenerate primers designed based on PPX protein sequences available in public databases were shown to match with PPX protein regions of the metaproteogenome provided by Wilmes et al. (2008) for an EBPR sludge dominated by "*Ca.* Accumulibacter" (Fig. 6.9a, b).

Agarose gels revealed that only couple II of the forward PPX-1a-F (DNA sequence: 5′-GAYTGYGGIACNAAYWS-3′) and reverse PPX-2b-R (5′-SWNSCICCNCCIAYRTC-3′) primers out of the ten tested (Table 6.3) resulted in amplification of gene fragments from the PAO-enrichment and full-scale EBPR sludge. Slight enhancement in PCR amplifications were obtained for couples II, IV and VI, that contained the reversed primer PPX-2b-R by optimizing the concentrations of primers and of DNA template, type of *Taq* polymerase, initial denaturation time, primer annealing temperature, and the number of PCR cycles, and by addition of the polar solvent dimethyl sulfoxide (DMSO) that inhibits formation of secondary DNA structures (Fig. 6.9c).

The optimized PCR conditions comprised, for 10-μL mixtures, 1 μL DNA template, 1 μL Peqlab high specificity buffer 10X, 1 μL of each primer (10 μmol L^{-1}), 4.65 μL water, 0.3 μL dNTPs (2.5 mmol L^{-1}), 1 μL DMSO, and 0.05 μL

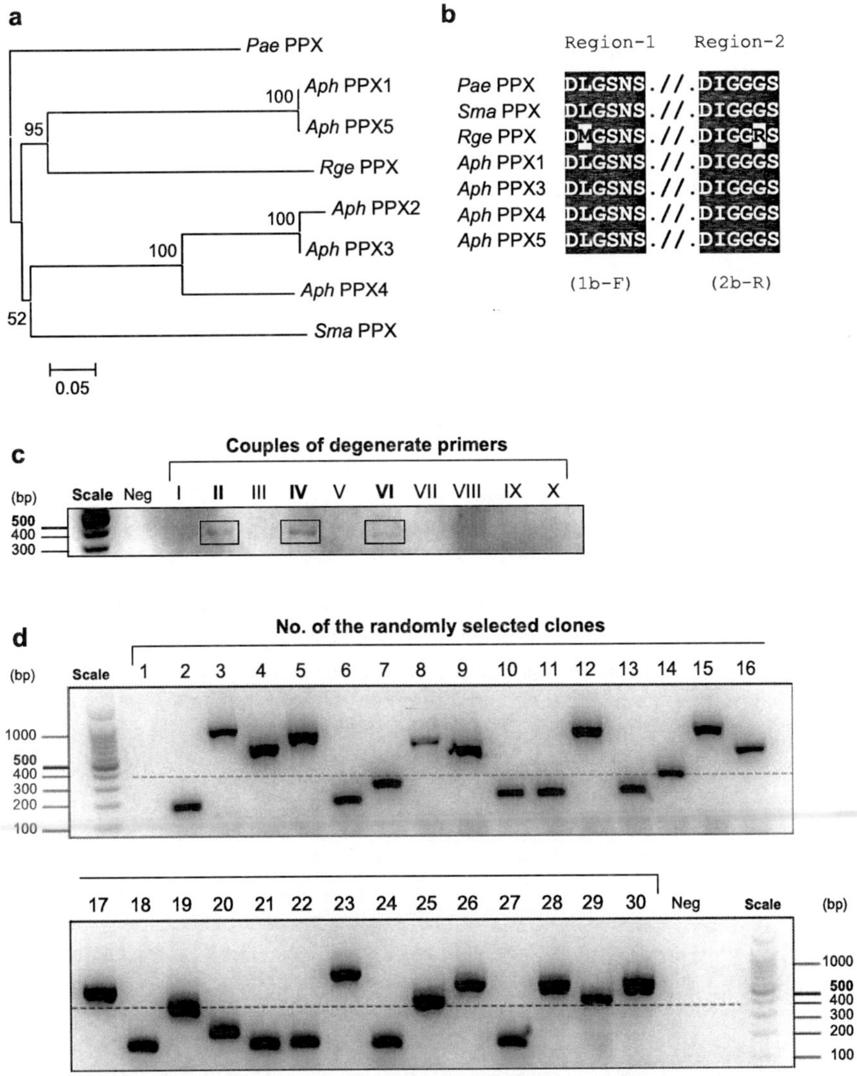

Fig. 6.9 Design and results of the degenerate PCR investigation for the detection of putative *ppx* gene fragments. Likelihood tree analysis of PPX protein sequences identified in the metagenome of "*Candidatus* Accumulibacter phosphatis" compared to three characterized long PPX sequences (**a**). Alignment of protein regions used to design degenerate primers (**b**). The primers targeting the corresponding PPX encoding regions are indicated in brackets. Legend: *Aph*: "*Ca.* Accumulibacter phosphatis"; *Pae*: *Pseudomonas aeruginosa* PAOs1 (gi: 13878636), used here to root the tree; *Rge*: *Rubrivivax gelatinosus* (gi: 7416772); *Sma*: *Serratia marcescens* (gi: 84181178). Results of the PCR amplifications of putative *ppx* gene fragments. Amplification of 500-bp fragments with the couples II, IV, and VI of forward primers with the reverse primer 2b-R after optimization of the degenerate PCR conditions (**c**). PCR conducted on isolated plasmids after cloning resulted in the amplification of fragments of various sizes of 200–1000 bp (**d**): only the clones with fragments above 400 bp were sequenced

Peqlab *Taq* DNA polymerase, with a PCR program consisting of initial denaturation (95 °C, 5 min) followed by 35 cycles of denaturation (95 °C, 1 min), annealing (52 °C, 1 min), elongation (72 °C, 90 s), and by final elongation (72 °C, 10 min). After cloning of the gene fragments amplified with primer couples II, IV, VI, and X, only half of the clones contained inserts of sizes analog to theoretical *ppx* gene fragments of 400–600 bp (Fig. 6.9d).

Sequencing of the insert of 11 plasmids did however not reveal affiliation with any putative *ppx* gene (Additional File 6.3). The detailed analysis of these sequences suggested that most amplicons were the results of a relatively low specificity of the PCR conditions as the primer PPX-2b-R was often found at both ends of the amplicons.

6.4 Discussion

6.4.1 Environmental Triggers for PAOs and GAOs Selection

The statistical analysis of cultivation conditions reported in literature was useful to identify the triggers for PAOs or GAOs selection. The type of carbon source clearly impacts on the PAO/GAO balance. Since PAOs can consume acetate and propionate at the same rate, these organisms predominate when propionate is present in the medium (Oehmen et al. 2006; Lopez-Vazquez et al. 2009b).

Compared to operation with acetate, more stable EBPR has been obtained with propionate, most probably because of energetic advantages provided by the 3-carbon compound (Thomas 2008; Gonzalez-Gil and Holliger 2011). Alphaproteobacterial GAOs affiliating with *Defluviicoccus* inside *Rhodospirillaceae* can however compete with PAOs for propionate ideally under higher mesophilic temperature and slight acidic conditions (Meyer et al. 2006; Lopez-Vazquez et al. 2009b). Operating the PAO-SBR with propionate at 17 °C and pH 7.0–8.0 preferentially selected for "*Ca.* Accumulibacter" over *Rhodospirillaceae* affiliates. In the GAO-SBR operated with acetate at 30 °C and pH 6.3–6.7, *Rhodospirillaceae* relatives were surprisingly able to compete with "*Ca.* Competibacter" as soon as the SRT was doubled from 8 to 16 days to favor the growth of GAOs according to Lopez-Vazquez et al. (2009a). This indicates that the competition between "*Ca.* Competibacter" and *Rhodospirillaceae* relatives inside the GAOs guild is not only governed by substrate dependencies, but also by growth kinetics. Additional research on the competition of gamma- and alphaproteobacterial GAOs could be useful for optimizing mixed culture processes operated for PHA production from organic matter in wastewater (Johnson et al. 2009).

Whereas the PCA confirmed that the PAO/GAO continuum depends on the COD/P ratio (Schuler and Jenkins 2003), no significant pH effect was detected, mainly because most studies have been conducted at pH 7.0–8.0. One cluster of articles revealed that GAOs were more abundant at the lowest pH values. Several authors have proposed pH as a key factor of the PAOs metabolism, since under alkaline conditions PAOs benefit from PP hydrolysis to compensate for pH gradients across cell membranes (Smolders et al. 1994a, b; Filipe et al. 2001; Schuler and Jenkins 2002; Oehmen et al. 2005a; Lopez-Vazquez et al. 2008, 2009b). The fact that PAOs could have been favored by higher COD/N remains puzzling since no effect of nitrogen on this functional group has been reported yet. Since different clades of PAOs and GAOs can denitrify (Oehmen et al. 2010b), research on the impact of nitrogen on PAO/GAO competition is required.

6.4.2 The Quest for Stable Enrichment Cultures and EBPR Processes

Although cultivation conditions analog to the ones reported in literature were applied, obtaining a stable PAO-enrichment remained challenging. The critical issue consisted in the competition between "*Ca.* Accumulibacter" and *Zoogloea-Rhizobiales* populations.

Instead of using different VFA compositions or supplementing the medium with various compounds, efficient start-up and PAO-enrichment was achieved by ensuring full anaerobic VFA uptake by dynamic control of OLR and anaerobic contact time. This is in agreement with different authors who recommended proper "anaerobic selectors" to prevent undesired proliferation of filamentous and zoogloeal organisms (Montoya et al. 2008; van Loosdrecht et al. 2008; Weissbrodt et al. 2012a). Although EBPR processes have been studied since the eighties, it is surprising that almost no study has reported on start-up conditions of EBPR reactors. Smolders et al. (1995a) reported that control of full anaerobic acetate uptake by stepwise addition of VFA can prevent the growth of OHO, but concluded that these organisms do not disturb the growth of PAOs by consuming the surplus of VFA entering into aeration phases. The microbial ecology data collected here however showed that leakage of VFA into aeration negatively impacted on the bacterial community structure and on the activity of PAO-enrichments. According to the present study, more attention should be given to start-up conditions for efficient bacterial resource management in EBPR systems, start-up procedures should be better described in methods sections of literature, and microbial ecology knowledge should be collected as a mean for troubleshooting microbial processes.

When compared to PAO- and GAO-enrichment levels of up to 70–90% reported in the literature, the abundances obtained here seem rather low (30–60%) despite full control of enrichment conditions. The explanation probably relies on the analytical technique used to measure relative abundances. Most studies have applied FISH, while T-RFLP and pyrosequencing were used here. Whereas the latter methods can be affected by DNA extraction and PCR amplification biases and do not inform on the overall abundance of active members of a functional group (Lueders and

Friedrich 2003; Rossi et al. 2009; Albertsen et al. 2012), FISH can lead to overestimations resulting from saturation of hybridized samples with exciting light (Nielsen et al. 2009). Despite T-RFLP-based relative abundances were probably underestimated, the use of the pyrosequencing-based PyroTRF-ID methodology (Weissbrodt et al. 2012b/Chap. 5) provided detailed description of the bacterial microbiomes of the PAO- and GAO-enrichments. It revealed that the two sludges displayed distant bacterial community structures at the level of predominant organisms but also at the level of accompanying populations. Bacterial communities displayed dynamic behaviors despite operation under steady-state biomass conditions. Regular monitoring of bacterial community composition is thus required to diagnose the "health" state of biological systems. Systems microbiology research is further required to understand the relationships between predominant and low abundant populations for ecological engineering of EBPR processes (Curtis and Sloan 2006; Harris et al. 2012; Nielsen et al. 2012b).

6.4.3 *Fast Assessment of PAO Fractions and EBPR Potential of Sludge*

The metabolic activities measured in the anaerobic metabolic batch tests are in agreement with the ones reported by Lopez-Vazquez et al. (2007a). In both studies, positive linear correlations were obtained between $Y_{PO4/Ac}$ and PAO/GAO fractions, with a proportionality factor of 0.51 $g_{P\text{-}PO4}\ {g_{COD,Ac}}^{-1}$.

According to Lopez-Vazquez et al. (2007a) such maximum yield is in agreement with the multiple experimental values reported in literature for EBPR processes. The fact that identical results were obtained between the two studies conducted at different locations and times indicate robustness of the anaerobic metabolic batch tests. The use of on-line conductivity monitoring is related to definite advantages. Liquid phase sampling and off-line physicochemical analysis are no more required, and the related time expenditures and experimental errors avoided. Conductivity sensors are significantly cheaper and easier to handle than standard test kits for single ions and liquid chromatography infrastructure. On-line acquisition enables high sampling frequency of data that can be reprocessed with simple real-time procedures on a standard computer. Since positive correlations were obtained between conductivity profiles, polyphosphatase activity, anaerobic metabolic activities, and PAO/GAO fractions, the developed methods can be used for fast assessment of relative abundances of active PAOs, and of the EBPR potential of activated sludge. The conductivity assay can further be optimized by reducing the working volume to conduct measurements at higher biomass concentrations, and thus to reduce the measuring time, and by avoiding pH control by designing suitable pH buffering conditions.

As stressed by Lopez-Vazquez et al. (2007a), standardization of substrate, temperature, and pH conditions are crucial for representativeness and reproducibility purposes of batch tests. Since the use of on-line conductivity has been demonstrated for real-time control of EBPR-SBRs (Maurer and Gujer 1995; Serralta et al. 2004; Aguado et al. 2006), the developed conductivity method could probably be transferred for rough on-line assessment of active PAO fractions directly in the mixed liquor during anaerobic batch phases under on-site conditions. Such an on-line method could for instance be used in heterogeneous aerobic granular sludge biofilm SBRs. The fact that identical results were obtained between conductivity measurements with intact AGS and with ground AGS samples indicated that the answer is not affected by mass transfer limitations under the tested conditions. Bassin et al. (2012) recently reported methods for assessing biological nutrient removal activities of granular sludge and biofilms with crushed aggregates. The type of system designed here for the anaerobic metabolic batches could easily be implemented for these additional objectives.

The mathematical model implemented in PHREEQC enabled detailed understanding of the processes contributing to conductivity evolution in PAO/GAO mixtures. Bicarbonate was identified as an important participating ion that predominantly contributed to conductivity evolution after orthophosphate. Whereas the linear correlations that were obtained between conductivity evolution rates, yields, and PAO/GAO fractions in metabolic batch experiments, were confirmed with the model, differences were observed in absolute values. Additional research is required to calibrate model responses to experimental measurements.

The polyphosphatase assay is interesting for determining the overall EBPR potential of activated sludge by circumventing the cell barrier. Organisms should indeed not be maintained under proper physiological conditions. Direct information is provided on the activity of enzymes produced by cells. Production of cell extract, enzymatic reactions, and spectrophotometric measurements can also relatively easily be conducted in the laboratory of a WWTP. This method is however related to higher initial investment cost for the spectrophotometer. The procedure could further be optimized with EnzChek Phosphate Assay Kits (E-6646, Molecular Probes, Life Technologies, Switzerland) that enable on-line monitoring of Pi residues produced directly in cuvettes or multi-well microplate readers.

Overall, compared to traditional microbial ecology methods that provide answers in more than a day, the methods developed here can lead to a result in less than one hour. Prior to implementation in laboratories of full-scale WWTPs, the developed methods should be validated in an analytical benchmarking survey by testing sludge sampled from different types of (non-) EBPR processes operated under various conditions.

6.4.4 The Quest for PPX Genes in Activated Sludge

At the fundamental level, contrary to meta-omics approaches, enzymatic assays provide fast and direct information on the activity of proteins expressed by organisms. Different authors have used enzymatic assays to characterize the polyphosphatase activity of proteins purified from different types of organisms (Bonting et al.

1993; Gomez-Garcia et al. 2003; Lindner et al. 2009). The genes related to the PP metabolism are widely conserved in prokaryotic and eukaryotic organisms since PP is a key metabolic compound of biological systems (Kornberg 1995; Keasling et al. 2000; Kulaev and Kulakovskaya 2000; Rao et al. 2009). The higher levels in polyphosphatase activity recorded when PAOs were present in higher abundances however revealed that the pool of enzymes related to PP hydrolysis is obviously larger in PAOs than in GAOs and other members of activated sludge systems. The identical polyphosphatase activities measured on extracts of PAOs sludge collected in anaerobic and aerobic phases indicated that the pool of enzymes present during aeration was displaying a polyphosphatase activity. This is in agreement with studies reporting hydrolysis of PP catalyzed by the PPK1 enzyme and even more by the PPK2 enzyme acting in reversed mode (Ahn and Kornberg 1990; Pramanik et al. 1999; McMahon et al. 2007; Wilmes et al. 2008; Rao et al. 2009).

He and McMahon (2011a) have detected with reverse transcriptase quantitative PCR and metatranscriptomic array analyses abundant *ppk1* gene transcripts during both anaerobic and aerobic phases, but have not been able to detect *ppx* gene transcript with the single specific probe used. In the present study, PCR and cloning-sequencing investigations with degenerate primers matching with highly conserved regions of PPX protein sequences referenced in public databases and of the metaproteogenome provided by Wilmes et al. (2008) for one EBPR sludge did not result in detection of *ppx* genes. This confirms the remaining analytical challenge mostly due to the very broad enzyme family of PPX (Kulaev and Kulakovskaya 2000). Although different authors have reported reversible involvement of PPK enzymes in Pi-cycling activities of PAOs (Pramanik et al. 1999; He and McMahon 2011b), PPX is always represented in the related conceptual models (Garcia Martin et al. 2006; Wilmes et al. 2008; He et al. 2010b). Kulaev and Kulakovskaya (2000) have indeed considered PPX as central enzymes of the PP metabolism, but have nevertheless recognized their lower involvement in prokaryotes than in eukaryotes (e.g. yeasts). When referring to Gomez-Garcia et al. (2003) who related the PPX function in cyanobacteria to the supply of Pi from PP under Pi-deprivation conditions, it might considered that PPX in PAOs would be expressed not during anaerobic conditions, but during extended aerobic starvation conditions after full uptake of Pi from the medium. In *Escherichia coli*, Akiyama et al. (1993) have detected a *ppx* gene adjacent to the *ppk* gene, indicating a possible PP operon of the two genes, and suggesting physiological relationship between PPK and PPX enzymes.

Additional research should be conducted to increase the specificity of degenerate PCR conditions for the detection of *ppx* genes. Attention should be paid on the type of sludge considered since the bacterial community composition of sludge cultivated under synthetic lab-scale conditions can significantly diverge from the one of full-scale EBPR sludge. This is illustrated by the predominance of "*Ca.* Accumulibacter" in the PAO-SBR, whereas the inoculation sludge was dominated by the glucose-fermenting and PP-accumulating *Tetrasphaera* spp. that are putatively not exhibiting the same metabolism as "*Ca.* Accumulibacter" (Nguyen et al. 2011; Nielsen et al. 2012a). Nguyen et al. (2012) have also recently shown that "*Candidatus* Halomonas phosphatis" can significantly contribute to EBPR in full-scale WWTP. Metagenomic analyses could be used for targeted identification of PPX protein regions of organisms present in the investigated sludges. Detection of *ppx* genes is a challenging work, but could provide additional knowledge on the complex metabolism of PAOs.

6.5 Conclusions

Multilevel investigations of the PAOs metabolism and of the PAO/GAO competition from bioreactor to enzymatic scales led to the following conclusions:

- Efficient start-up of PAO- and GAO-enrichment procedures requires step-wise adaptation of OLR and anaerobic contact time for ensuring full anaerobic VFA uptake and preferential selection of "*Ca.* Accumulibacter" and "*Ca.* Competibacter", respectively.
- PAO- and GAO-enrichment displayed distinct bacterial microbiomes. The former mainly comprised *Betaproteobacteria*, *Cytophagia*, and *Chloroflexi* affiliates, the latter *Gammaproteobacteria*, *Alphaproteobacteria*, *Acidobacteria*, and *Sphingobacteria* relatives. PAO-related *Tetrasphaera* spp. withstood slight acidic and higher mesophilic conditions favoring GAOs.
- Linear relations were obtained between the PAO fraction of PAO/GAO mixtures, the apparent biomass specific rate of conductivity evolution under anaerobic conditions, and the polyphosphatase activity. Such methods are dedicated for rapid assessment of relative abundance of PAOs and EBPR potential of activated and granular sludge.
- A mathematical model implemented in PHREEQC provided detailed information on the PAO/GAO metabolic activities and on the processes contributing to conductivity.
- Genes encoding for PPX were not detected by using degenerate primers designed to cover the whole diversity of *ppx* genes referenced in public databases. The design of more specific conditions targeting key PAOs populations is therefore required.

Acknowledgements This study was performed in collaboration with three outstanding master students, Jonathan May (AgroSup Dijon, France) on the design of enrichment conditions and infrastructure, Christelle Petit (Polytech' Clermont-Ferrand, France) on investigations at genetic level, and Alexandre Chabrelie (Polytech' Grenoble, France) on the development of the conductivity and enzymatic assays. Are acknowledged for their scientific collaboration: Julien Maillard on genetic and biochemical objectives (EPFL, Laboratory for Environmental Biotechnology), and Alessandro Brovelli on mathematical modeling (EPFL, Ecological Engineering Laboratory). The author is also grateful to Elsa Lacroix affiliated to these two labs for punctual advice on PHREEQC.

Appendix

The Appendix is available at the end of Chap. 5 on PyroTRF-ID.

Appendix: Phylogenetic affiliations obtained with PyroTRF-ID.

Supplementary Information

Additional File 6.1 Listing of cultivation conditions used in reference studies for the operation of anaerobic-aerobic SBR exhibiting EBPR and non-EBPR performances, and that selected for either PAOs or GAOs.

Additional File 6.2 Composition, stoichiometric and biokinetic matrices of the mathematical model implemented in PHREEQC aiming at describing PAOs and GAOs metabolic activities and electrical conductivity evolutions in anaerobic conditions.

Additional File 6.3 Design of the degenerate primers targeting for ppx genes.Additional File 6.3Design of the degenerate primers targeting for *ppx* genes.

Additional File 6.4 Detailed description of the experimental correlations obtained between the rate of conductivity evolution and the fraction of PAOs present in activated sludge.

Additional File 6.5 Detailed description of the theoretical correlations obtained after mathematical modelling in PHREEQC between conductivity profiles and relative abundances of PAOs and GAOs.

Additional File 6.6 Detailed description of the experimental correlations obtained between the polyphosphatase activity and the fraction of PAOs present in activated sludge.

References

Aguado D, Montoya T, Ferrer J, Seco A (2006) Relating ions concentration variations to conductivity variations in a sequencing batch reactor operated for enhanced biological phosphorus removal. Environ Model Softw 21(6):845–851

Ahn K, Kornberg A (1990) Polyphosphate kinase from *Escherichia coli*. Purification and demonstration of a phosphoenzyme intermediate. J Biol Chem 265(20):11734–11739

Ahn J, McIlroy S, Schroeder S, Seviour R (2009) Biomass granulation in an aerobic:anaerobic-enhanced biological phosphorus removal process in a sequencing batch reactor with varying pH. J Ind Microbiol Biotechnol 36(7):885–893

Akiyama M, Crooke E, Kornberg A (1992) The polyphosphate kinase gene of *Escherichia coli*. Isolation and sequence of the *ppk* gene and membrane location of the protein. J Biol Chem 267(31):22556–22561

Akiyama M, Crooke E, Kornberg A (1993) An exopolyphosphatase of *Escherichia coli*. The enzyme and its ppx gene in a polyphosphate operon. J Biol Chem 268(1):633–639

Albertsen M, Hansen LBS, Saunders AM, Nielsen PH, Nielsen KL (2012) A metagenome of a full-scale microbial community carrying out enhanced biological phosphorus removal. ISME J 6(6):1094–1106

Bassin JP, Kleerebezem R, Dezotti M, van Loosdrecht MCM (2012) Measuring biomass specific ammonium, nitrite and phosphate uptake rates in aerobic granular sludge. Chemosphere 89(10):1161–1168

Blackall LL, Crocetti GR, Saunders AM, Bond PL (2002) A review and update of the microbiology of enhanced biological phosphorus removal in wastewater treatment plants. Anton Leeuw Int J G 81(1–4):681–691

Bond PL, Hugenholtz P, Keller J, Blackall LL (1995) Bacterial community structures of phosphate-removing and non-phosphate-removing activated sludges from sequencing batch reactors. Appl Environ Microbiol 61(5):1910–1916

Bond PL, Erhart R, Wagner M, Keller J, Blackall LL (1999) Identification of some of the major groups of bacteria in efficient and nonefficient biological phosphorus removal activated sludge systems. Appl Environ Microbiol 65(9):4077–4084

Bonting CFC, Kortstee GJJ, Zehnder AJB (1993) Properties of polyphosphatase of Acinetobacter johnsonii 210A. Antonie Van Leeuwenhoek 64(1):75–81

Burow LC, Mabbett AN, McEwan AG, Bond PL, Blackall LL (2008) Bioenergetic models for acetate and phosphate transport in bacteria important in enhanced biological phosphorus removal. Environ Microbiol 10(1):87–98

Cole JR, Wang Q, Cardenas E, Fish J, Chai B, Farris RJ, Kulam-Syed-Mohideen AS, McGarrell DM, Marsh T, Garrity GM, Tiedje JM (2009) The Ribosomal Database Project: improved alignments and new tools for rRNA analysis. Nucleic Acids Res 37:D141–D145

Comeau Y, Rabionwitz B, Hall KJ, Oldham WK (1987) Phosphate release and uptake in enhanced biological phosphorus removal from wastewater. J Water Pollut Con F 59(7):707–715

Crocetti GR, Hugenholtz P, Bond PL, Schuler A, Keller J, Jenkins D, Blackall LL (2000) Identification of polyphosphate-accumulating organisms and design of 16S rRNA-directed probes for their detection and quantitation. Appl Environ Microbiol 66(3):1175–1182

Crocetti GR, Banfield JF, Keller J, Bond PL, Blackall LL (2002) Glycogen-accumulating organisms in laboratory-scale and full-scale wastewater treatment processes. Microbiology 148(11):3353–3364

Curtis TP, Sloan WT (2006) Towards the design of diversity: stochastic models for community assembly in wastewater treatment plants. Water Sci Technol 54(1):227–236

de Kreuk MK, Picioreanu C, Hosseini M, Xavier JB, van Loosdrecht MCM (2007) Kinetic model of a granular sludge SBR: influences on nutrient removal. Biotechnol Bioeng 97(4):801–815

Filipe CDM, Daigger GT, Grady CPL (2001) pH as a key factor in the competition between glycogen-accumulating organisms and phosphorus-accumulating organisms. Water Environ Res 73(2):223–232

Garcia Martin H, Ivanova N, Kunin V, Warnecke F, Barry KW, McHardy AC, Yeates C, He S, Salamov AA, Szeto E, Dalin E, Putnam NH, Shapiro HJ, Pangilinan JL, Rigoutsos I, Kyrpides NC, Blackall LL, McMahon KD, Hugenholtz P (2006) Metagenomic analysis of two enhanced biological phosphorus removal (EBPR) sludge communities. Nat Biotechnol 24(10):1263–1269

Gavigan J-A, Marshall LM, Dobson ADW (1999) Regulation of polyphosphate kinase gene expression in Acinetobacter baumannii 252. Microbiology 145(10):2931–2937

Gomez-Garcia MR, Losada M, Serrano A (2003) Concurrent transcriptional activation of ppa and ppx genes by phosphate deprivation in the cyanobacterium Synechocystis sp. strain PCC 6803. Biochem Biophys Res Commun 302(3):601–609

Gonzalez-Gil G, Holliger C (2011) Dynamics of microbial community structure and enhanced biological phosphorus removal of propionate- and acetate-cultivated aerobic granules. Appl Environ Microbiol 77:8041–8051

Harris JA, Baptista JDC, Curtis TP, Nelson AK, Pawlett M, Ritz K, Tyrrel SF (2012) Engineering difference: matrix design determines community composition in wastewater treatment systems. Ecol Eng 40:183–188

He S, McMahon KD (2011a) *"Candidatus* Accumulibacter" gene expression in response to dynamic EBPR conditions. ISME J 5(2):329–340

He S, McMahon KD (2011b) Microbiology of *"Candidatus* Accumulibacter" in activated sludge. Microb Biotechnol 4(5):603–619

He S, Bishop FI, McMahon KD (2010a) Bacterial community and *"Candidates* Accumulibacter" population dynamics in laboratory-scale enhanced biological phosphorus removal reactors. Appl Environ Microbiol 76(16):5479–5487

He S, Kunin V, Haynes M, Martin HG, Ivanova N, Rohwer F, Hugenholtz P, McMahon KD (2010b) Metatranscriptomic array analysis of *"Candidatus* Accumulibacter phosphatis"-enriched enhanced biological phosphorus removal sludge. Environ Microbiol 12(5):1205–1217

Hesselmann RPX, Werlen C, Hahn D, van der Meer JR, Zehnder AJB (1999) Enrichment, phylogenetic analysis and detection of a bacterium that performs enhanced biological phosphate removal in activated sludge. Syst Appl Microbiol 22(3):454–465

Hesselmann RPX, Von Rummell R, Resnick SM, Hany R, Zehnder AJB (2000) Anaerobic metabolism of bacteria performing enhanced biological phosphate removal. Water Res 34(14):3487–3494

Hollender J, Dreyer U, Kornberger L, Kämpfer P, Dott W (2002) Selective enrichment and characterization of a phosphorus-removing bacterial consortium from activated sludge. Appl Microbiol Biotechnol 58(1):106–111

Jobbagy A, Literathy B, Wong MT, Tardy G, Liu WT (2006) Proliferation of glycogen accumulating organisms induced by Fe(III) dosing in a domestic wastewater treatment plant. Water Sci Technol 54(1):101–109

Johnson K, Jiang Y, Kleerebezem R, Muyzer G, Van Loosdrecht MCM (2009) Enrichment of a mixed bacterial culture with a high polyhydroxyalkanoate storage capacity. Biomacromolecules 10(4):670–676

Keasling JD, Van Dien SJ, Trelstad P, Renninger N, McMahon K (2000) Application of polyphosphate metabolism to environmental and biotechnological problems. Biochem Mosc 65(3):324–331

Kong YH, Beer M, Rees GN, Seviour RJ (2002) Functional analysis of microbial communities in aerobic-anaerobic sequencing batch reactors fed with different phosphorus/carbon (P/C) ratios. Microbiology 148(8):2299–2307

Kornberg A (1995) Inorganic polyphosphate: toward making a forgotten polymer unforgettable. J Bacteriol 177(3):491–496

Kulaev I, Kulakovskaya T (2000) Polyphosphate and phosphate pump. Annu Rev Microbiol 54(1):709–734

Lemaire R, Yuan Z, Bernet N, Marcos M, Yilmaz G, Keller J (2008) A sequencing batch reactor system for high-level biological nitrogen and phosphorus removal from abattoir wastewater. Biodegradation 20(3):1–12

Levantesi C, Serafim LS, Crocetti GR, Lemos PC, Rossetti S, Blackall LL, Reis MAM, Tandoi V (2002) Analysis of the microbial community structure and function of a laboratory scale enhanced biological phosphorus removal reactor. Environ Microbiol 4(10):559–569

Lindner SN, Knebel S, Wesseling H, Schoberth SM, Wendisch VF (2009) Exopolyphosphatases PPX1 and PPX2 from Corynebacterium glutamicum. Appl Environ Microbiol 75(10):3161–3170

Liu WT, Nielsen AT, Wu JH, Tsai CS, Matsuo Y, Molin S (2001) In situ identification of polyphosphate- and polyhydroxyalkanoate-accumulating traits for microbial populations in a biological phosphorus removal process. Environ Microbiol 3(2):110–122

Lopez-Vazquez CM, Hooijmans CM, Brdjanovic D, Gijzen HJ, van Loosdrecht MCM (2007a) A practical method for quantification of phosphorus- and glycogen-accumulating organism populations in activated sludge systems. Water Environ Res 79(13):2487–2498

Lopez-Vazquez CM, Song YI, Hooijmans CM, Brdjanovic D, Moussa MS, Gijzen HJ, van Loosdrecht MMC (2007b) Short-term temperature effects on the anaerobic metabolism of glycogen accumulating organisms. Biotechnol Bioeng 97(3):483–495

Lopez-Vazquez CM, Hooijmans CM, Brdjanovic D, Gijzen HJ, van Loosdrecht MCM (2008) Factors affecting the microbial populations at full-scale enhanced biological phosphorus removal (EBPR) wastewater treatment plants in The Netherlands. Water Res 42(10–11):2349–2360

Lopez-Vazquez CM, Hooijmans CM, Brdjanovic D, Gijzen HJ, van Loosdrecht MCM (2009a) Temperature effects on glycogen accumulating organisms. Water Res 43(11):2852–2864

Lopez-Vazquez CM, Oehmen A, Hooijmans CM, Brdjanovic D, Gijzen HJ, Yuan Z, van Loosdrecht MCM (2009b) Modeling the PAO-GAO competition: effects of carbon source, pH and temperature. Water Res 43(2):450–462

Lu H, Oehmen A, Virdis B, Keller J, Yuan Z (2006) Obtaining highly enriched cultures of "Candidatus Accumulibacter phosphates" through alternating carbon sources. Water Res 40(20):3838–3848

Lueders T, Friedrich MW (2003) Evaluation of PCR amplification bias by terminal restriction fragment length polymorphism analysis of small-subunit rRNA and mcrA genes by using defined template mixtures of methanogenic pure cultures and soil DNA extracts. Appl Environ Microbiol 69(1):320–326

Manga J, Ferrer J, Garcia-Usach F, Seco A (2001) A modification to the Activated Sludge Model No. 2 based on the competition between phosphorus-accumulating organisms and glycogen-accumulating organisms. Water Sci Technol 43(11):161–171

Marcelino M, Guisasola A, Baeza JA (2009) Experimental assessment and modelling of the proton production linked to phosphorus release and uptake in EBPR systems. Water Res 43(9):2431–2440

Maurer M, Gujer W (1995) Monitoring of microbial phosphorus release in batch experiments using electric conductivity. Water Res 29(11):2613–2617

Maurer M, Gujer W, Hany R, Bachmann S (1997) Intracellular carbon flow in phosphorus accumulating organisms from activated sludge systems. Water Res 31(4):907–917

McMahon KD, Dojka MA, Pace NR, Jenkins D, Keasling JD (2002) Polyphosphate kinase from activated sludge performing enhanced biological phosphorus removal. Appl Environ Microbiol 68(10):4971–4978

McMahon KD, Yilmaz S, He S, Gall DL, Jenkins D, Keasling JD (2007) Polyphosphate kinase genes from full-scale activated sludge plants. Appl Microbiol Biotechnol 77(1):167–173

McMahon KD, He S, Oehmen A (2010) The microbiology of phosphorus removal. In: Seviour RJ, Nielsen PH (eds) Microbial ecology of activated sludge. IWA Publishing, London, UK, pp 281–320

Meyer RL, Saunders AM, Zeng RJ, Keller J, Blackall LL (2003) Microscale structure and function of anaerobic-aerobic granules containing glycogen accumulating organisms. FEMS Microbiol Ecol 45(3):253–261

Meyer RL, Saunders AM, Blackall LL (2006) Putative glycogen-accumulating organisms belonging to the *Alphaproteobacteria* identified through rRNA-based stable isotope probing. Microbiology 152:419–429

Meyer F, Paarmann D, D'Souza M, Olson R, Glass EM, Kubal M, Paczian T, Rodriguez A, Stevens R, Wilke A, Wilkening J, Edwards RA (2008) The metagenomics RAST server—a public resource for the automatic phylogenetic and functional analysis of metagenomes. BMC Bioinform 9:386

Mino T (2000) Microbial selection of polyphosphate-accumulating bacteria in activated sludge wastewater treatment processes for enhanced biological phosphate removal. Biochem Mosc 65(3):341–348

Mino T, Satoh H, Matsuo T (1994) Metabolisms of different bacterial populations in enhanced biological phosphate removal processes. Water Sci Technol 29(7):67–70

Mino T, van Loosdrecht MCM, Heijnen JJ (1998) Microbiology and biochemistry of the enhanced biological phosphate removal process. Water Res 32(11):3193–3207

Montoya T, Borrás L, Aguado D, Ferrer J, Seco A (2008) Detection and prevention of enhanced biological phosphorus removal deterioration caused by Zoogloea overabundance. Environ Technol 29(1):35–42

Nguyen HTT, Le VQ, Hansen AA, Nielsen JL, Nielsen PH (2011) High diversity and abundance of putative polyphosphate-accumulating Tetrasphaera-related bacteria in activated sludge systems. FEMS Microbiol Ecol 76(2):256–267

Nguyen HTT, Nielsen JL, Nielsen PH (2012) "Candidatus Halomonas phosphatis", a novel polyphosphate-accumulating organism in full-scale enhanced biological phosphorus removal plants. Environ Microbiol 14(10):2826–2837

Nielsen PH, Daims H, Lemmer H (2009) FISH handbook for biological wastewater treatment—identification and quantification of microorganisms in activated sludge and biofilms by FISH, 1st edn. IWA Publishing, London, UK

Nielsen JL, Nguyen H, Meyer RL, Nielsen PH (2012a) Identification of glucose-fermenting bacteria in a full-scale enhanced biological phosphorus removal plant by stable isotope probing. Microbiology 158(7):1818–1825

Nielsen PH, Saunders AM, Hansen AA, Larsen P, Nielsen JL (2012b) Microbial communities involved in enhanced biological phosphorus removal from wastewater—a model system in environmental biotechnology. Curr Opin Biotechnol 23(3):452–459

Oehmen A, Vives MT, Lu H, Yuan Z, Keller J (2005a) The effect of pH on the competition between polyphosphate-accumulating organisms and glycogen-accumulating organisms. Water Res 39(15):3727–3737

Oehmen A, Yuan Z, Blackall LL, Keller J (2005b) Comparison of acetate and propionate uptake by polyphosphate accumulating organisms and glycogen accumulating organisms. Biotechnol Bioeng 91(2):162–168

Oehmen A, Zeng RJ, Yuan Z, Keller J (2005c) Anaerobic metabolism of propionate by polyphosphate-accumulating organisms in enhanced biological phosphorus removal systems. Biotechnol Bioeng 91(1):43–53

Oehmen A, Saunders AM, Vives MT, Yuan Z, Keller J (2006) Competition between polyphosphate and glycogen accumulating organisms in enhanced biological phosphorus removal systems with acetate and propionate as carbon sources. J Biotechnol 123(1):22–32

Oehmen A, Lemos PC, Carvalho G, Yuan Z, Keller J, Blackall LL, Reis MAM (2007) Advances in enhanced biological phosphorus removal: from micro to macro scale. Water Res 41:2271–2300

Oehmen A, Carvalho G, Lopez-Vazquez CM, van Loosdrecht MCM, Reis MAM (2010a) Incorporating microbial ecology into the metabolic modelling of polyphosphate accumulating organisms and glycogen accumulating organisms. Water Res 44(17):4992–5004

Oehmen A, Lopez-Vazquez CM, Carvalho G, Reis MAM, van Loosdrecht MCM (2010b) Modelling the population dynamics and metabolic diversity of organisms relevant in anaerobic/anoxic/aerobic enhanced biological phosphorus removal processes. Water Res 44(15):4473–4486

Oksanen J, Kindt R, Legendre P, O'Hara B, Simpson GL, Solymos P, Stevens MHH, Wagner H (2009) Vegan: community ecology package. R package version 1.15-4. R Foundation for Statistical Computing, Vienna, Austria. http://CRAN.R-project.org/package=vegan

Parkhurst DL (1995) User's guide to PHREEQC—a computer program for speciation, reaction-path, advective-transport, and inverse geochemical calculations, vol 95-4227. US Geological Survey Water-Resources Investigations. http://wwwbrr.cr.usgs.gov/projects/GWC_coupled/phreeqc/

Pijuan M, Saunders AM, Guisasola A, Baeza JA, Casas C, Blackall LL (2004) Enhanced biological phosphorus removal in a sequencing batch reactor using propionate as the sole carbon source. Biotechnol Bioeng 85(1):56–67

Pijuan M, Guisasola A, Baeza JA, Carrera J, Casas C, Lafuente J (2006) Net P-removal deterioration in enriched PAO sludge subjected to permanent aerobic conditions. J Biotechnol 123(1):117–126

Pramanik J, Trelstad PL, Schuler AJ, Jenkins D, Keasling JD (1999) Development and validation of a flux-based stoichiometric model for enhanced biological phosphorus removal metabolism. Water Res 33(2):462–476

R-Development-Core-Team (2008) R: a language and environment for statistical computing. http://www.r-project.org/. R Foundation for Statistical Computing, Vienna

Rao NN, Gomez-Garcia MR, Kornberg A (2009) Inorganic polyphosphate: essential for growth and survival. Annu Rev Biochem 78(1):605–647

Rossi P, Gillet F, Rohrbach E, Diaby N, Holliger C (2009) Statistical assessment of variability of terminal restriction fragment length polymorphism analysis applied to complex microbial communities. Appl Environ Microbiol 75(22):7268–7270

Saito T, Brdjanovic D, van Loosdrecht MCM (2004) Effect of nitrite on phosphate uptake by phosphate accumulating organisms. Water Res 38(17):3760–3768

Satoh H, Mino T, Matsuo T (1992) Uptake of organic substrates and accumulation of polyhydroxyalkanoates linked with glycolysis of intracellular carbohydrates under anaerobic conditions in the biological excess phosphate removal processes. Water Sci Technol 26(5–6):933–942

Satoh H, Mino T, Matsuo T (1994) Deterioration of enhanced biological phosphorus removal by the domination of microorganisms without polyphosphate accumulation. Water Sci Technol 30(6 pt 6):203–211

Schonborn C, Bauer HD, Roske I (2001) Stability of enhanced biological phosphorus removal and composition of polyphosphate granules. Water Res 35:3190–3196

Schuler AJ, Jenkins D (2002) Effects of pH on enhanced biological phosphorus removal metabolisms. Water Sci Technol 46(4–5):171–178

Schuler AJ, Jenkins D (2003) Enhanced biological phosphorus removal from wastewater by biomass with different phosphorus contents, part I: experimental results and comparison with metabolic models. Water Environ Res 75(6):485–498

Seco A, Ribes J, Serralta J, Ferrer J (2004) Biological nutrient removal model no. 1 (BNRM1). Water Sci Technol 50(6):69–78

Serralta J, Borras L, Blanco C, Barat R, Seco A (2004) Monitoring pH and electric conductivity in an EBPR sequencing batch reactor. Water Sci Technol 50(10):145–152

Siegrist H, Rieger L, Koch G, Kuhni M, Gujer W (2002) The EAWAG Bio-P module for activated sludge model no. 3. Water Sci Technol 45(6):61–76

Slater FR, Johnson CR, Blackall LL, Beiko RG, Bond PL (2010) Monitoring associations between clade-level variation, overall community structure and ecosystem function in enhanced biological phosphorus removal (EBPR) systems using terminal-restriction fragment length polymorphism (T-RFLP). Water Res 44(17):4908–4923

Smolders GJF, van der Meij J, van Loosdrecht MCM, Heijnen JJ (1994a) Model of the anaerobic metabolism of the biological phosphorus removal process—stoichiometry and pH influence. Biotechnol Bioeng 43(6):461–470

Smolders GJF, van Loosdrecht MCM, Heijnen JJ (1994b) pH: keyfactor in the biological phosphorus removal process. Water Sci Technol 29(7):71–74

Smolders GJF, Bulstra DJ, Jacobs R, van Loosdrecht MCM, Heijnen JJ (1995a) A metabolic model of the biological phosphorus removal process. 2. Validation during start-up conditions. Biotechnol Bioeng 48(3):234–245

Smolders GJF, van der Meij J, van Loosdrecht MCM, Heijnen JJ (1995b) A structured metabolic model for anaerobic and aerobic stoichiometry and kinetics of the biological phosphorus removal process. Biotechnol Bioeng 47(3):277–287

Thomas MP (2008) The secret to achieving reliable biological phosphorus removal. Water Sci Technol 58(6):1231–1236

van Groenestijn JW, Bentvelsen MMA, Deinema MH, Zehnder AJB (1989) Polyphosphate-degrading enzymes in *Acinetobacter* spp. and activated sludge. Appl Environ Microbiol 55(1):219–223

van Loosdrecht MCM, Hooijmans CM, Brdjanovic D, Heijnen JJ (1997) Biological phosphate removal processes. Appl Microbiol Biotechnol 48(3):289–296

van Loosdrecht MCM, Martins AM, Ekama GA (2008) Bulking sludge. In: Henze M, van Loosdrecht MCM, Ekama GA, Brdjanovic D (eds) Biological wastewater treatment: principles, modelling and design. IWA Publishing, London, pp 291–308

Weissbrodt DG, Lochmatter S, Ebrahimi S, Rossi P, Maillard J, Holliger C (2012a) Bacterial selection during the formation of early-stage aerobic granules in wastewater treatment systems operated under wash-out dynamics. Front Microbiol 3:332

Weissbrodt DG, Shani N, Sinclair L, Lefebvre G, Rossi P, Maillard J, Rougemont J, Holliger C (2012b) PyroTRF-ID: a novel bioinformatics methodology for the affiliation of terminal-restriction fragments using 16S rRNA gene pyrosequencing. BMC Microbiol 12:306

Wentzel MC, Comeau Y, Ekama GA, van Loosdrecht MCM, Brdjanovic D (2008) Enhanced biological phosphorus removal. In: Henze M, van Loosdrecht MCM, Ekama GA, Brdjanovic D (eds) Biological wastewater treatment: principles, modelling and design. IWA Publishing, London, pp 155–220

Wilmes P, Bond PL (2004) The application of two-dimensional polyacrylamide gel electrophoresis and downstream analyses to a mixed community of prokaryotic microorganisms. Environ Microbiol 6(9):911–920

Wilmes P, Andersson AF, Lefsrud MG, Wexler M, Shah M, Zhang B, Hettich RL, Bond PL, VerBerkmoes NC, Banfield JF (2008) Community proteogenomics highlights microbial strain-variant protein expression within activated sludge performing enhanced biological phosphorus removal. ISME J 2(8):853–864

Wong MT, Liu WT (2006) Microbial succession of glycogen accumulating organisms in an anaerobic-aerobic membrane bioreactor with no phosphorus removal. Water Sci Technol 54(1):29–37

Wong MT, Tan FM, Ng WJ, Liu WT (2004) Identification and occurrence of tetrad-forming Alphaproteobacteria in anaerobic-aerobic activated sludge processes. Microbiology 150:3741–3748

Zeng RJ, van Loosdrecht MCM, Yuan ZG, Keller J (2003) Metabolic model for glycogen-accumulating organisms in anaerobic/aerobic activated sludge systems. Biotechnol Bioeng 81(1):92–105

Zhang H, Ishige K, Kornberg A (2002) A polyphosphate kinase (PPK2) widely conserved in bacteria. Proc Nat Acad Sci U S A 99(26):16678–16683

Zhou Y, Pijuan M, Zeng RJ, Lu H, Yuan Z (2008) Could polyphosphate-accumulating organisms (PAOs) be glycogen-accumulating organisms (GAOs)? Water Res 42(10–11):2361–2368

Chapter 7
Microbial Selection During Granulation of Activated Sludge Under Wash-Out Dynamics

Long start-up periods required for the development of granules from floccular sludge, and the loss of biomass in this period leading to poor nutrient removal performance are key challenges.
(Verawaty et al. 2012)

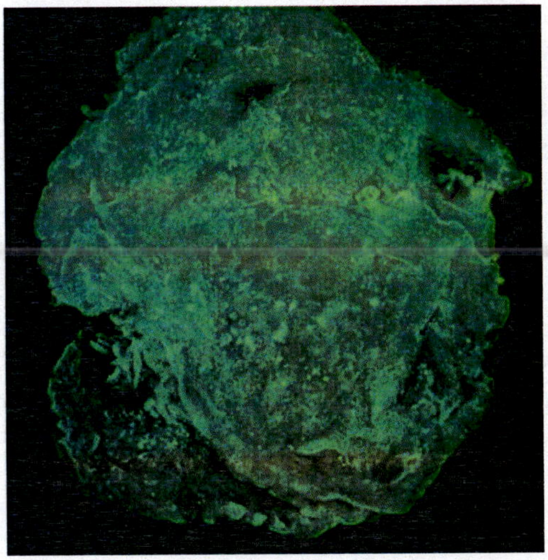

Granule nucleus

The content of this chapter was published in a modified version in: Weissbrodt DG, Lochmatter S, Ebrahimi S, Rossi P, Maillard J, Holliger C (2012) Bacterial selection during the formation of early-stage aerobic granules in wastewater treatment systems operated under wash-out dynamics. Frontiers in Microbiology 3:332. https://doi.org/10.3389/fmicb.2012.00332. Permission was granted to reuse the figure materials (© 2012 Frontiers Media S.A.).

Supplementary Information The online version contains supplementary material available at https://doi.org/10.1007/978-3-031-41009-3_7.

D. G. Weissbrodt, *Engineering Granular Microbiomes*, Springer Theses,
https://doi.org/10.1007/978-3-031-41009-3_7

7.1 Introduction

Aerobic granular sludge (AGS) wastewater treatment processes are attractive for intensive and high-rate biological nutrient removal (BNR) and secondary clarification in single sequencing batch reactors (SBR) (de Bruin et al. 2004; Giesen et al. 2012). Stable dense and fast-settling aerobic granules with tailored metabolic activities for the removal of carbon, nitrogen and phosphorus are desired for the operation of robust AGS wastewater treatment plants (WWTP). For instance, the overgrowth of filamentous organisms must be avoided in order to prevent process disturbances by the deterioration of the settling properties of aerobic granules (van Loosdrecht et al. 2008).

The formation of aerobic granules has been stimulated by reactor start-up conditions leading to the wash-out of flocculent biomass and selecting for a fast-settling biomass, namely with the combination of short settling times of 3–5 min and short hydraulic retention times (HRT) of 6 h (Beun et al. 1999). Granulation can be impacted by additional operation parameters such as the influent feeding regime, the hydrodynamic shear force, and the concentration of dissolved oxygen (DO). For a review, refer to Lee et al. (2010). Granules have been successfully cultivated with feast-famine regimes involving pulse feeding (3–5 min) followed by prolonged aeration (3–4 h) (Morgenroth et al. 1997; Beun et al. 1999; Tay et al. 2002), or anaerobic feeding (1 h) followed by aerobic starvation (2 h) (de Kreuk et al. 2005). Granulation has only been observed with up-flow superficial air velocities (SAV) above 0.010 m s^{-1}, typically between 0.025 and 0.045 m s^{-1}. High SAV induces high shear and compaction forces at the surface of granules (Zima et al. 2007), and stimulates the production of exopolymeric substances (EPS) as well as hydrophobic adhesive interactions (Liu and Tay 2002; Dulekgurgen et al. 2008).

Several studies have however reported on the deterioration of the settling properties of aerobic granules by overgrowth of filamentous microbial structures, called filamentous bulking. For a review, refer to Liu and Liu (2006). This phenomenon has been observed with volumetric organic loading rates (OLR) above 6 kg_{CODs} d^{-1} m^{-3} (Shin et al. 1992; Moy et al. 2002), with high-energy carbon sources such as carbohydrates (Morgenroth et al. 1997; Weber et al. 2007), and at higher mesophilic temperatures of 30–35 °C (Weber et al. 2007; Ebrahimi et al. 2010). Filamentous overgrowth has been limited with higher up-flow mixing or aeration velocities, and with the use of acetate as carbon source (Liu and Liu 2006). Despite the reduction of filamentous bulking with this substrate, residual filamentous structures have still been observed, and have been presumed to act as backbones for the immobilization of microbial colonies (Martins et al. 2004). In studies investigating granulation in up-flow anaerobic sludge blanket reactors, it has been observed that the same *Methanosaeta*-affiliating phylotype was constantly dominating the bacterial community during the evolution of fluffy granules to compact granules under the progressive increase in shear forces in the reactor (Grotenhuis et al. 1992; Hulshoff Pol et al.

2004). In the case of aerobic granules, analysis of bacterial compositions of fluffy and dense granules is required to assess whether different granule structures exhibit the same dominant phylotypes or not.

From the nutrient removal point of view, AGS studies initially concentrated on the formation of aerobic granules, and on the removal of organic matter (Morgenroth et al. 1997; Beun et al. 1999; Tay et al. 2002). Emphasis has then been put on achieving nitrification (Tsuneda et al. 2003), denitrification (Beun et al. 2001; Mosquera-Corral et al. 2005), dephosphatation (Lin et al. 2003), and combined BNR (de Kreuk et al. 2005; Lemaire et al. 2008; Yilmaz et al. 2008) in AGS systems. However, it has been shown that 75–100 days have been required to obtain efficient nutrient removal activities in aerobic granules after reactor start-up with flocculent activated sludge (de Kreuk et al. 2005; Xavier et al. 2007; Ebrahimi et al. 2010; Gonzalez-Gil and Holliger 2011). In these studies, only carbon has been removed during the start-up period. Some authors have achieved enhanced granulation with faster improvements in nutrient removal performances by seeding reactors with crushed pre-cultivated granules (Pijuan et al. 2011; Verawaty et al. 2012). However, the reason why phosphorus and nitrogen removal activities have been inhibited during the first 3 months of reactor start-up with flocculent inoculation sludge and wash-out conditions has not yet been further investigated. Different microbial ecology studies have mainly been conducted on mature granules (Adav et al. 2010; Gonzalez-Gil and Holliger 2011), but only little information is available on the microbial composition of early-stage AGS.

The present study aimed to investigate the bacterial community dynamics during the formation of early-stage aerobic granules (0–60 days) in bubble-column SBRs operated under conditions selecting for a fast-settling biomass. The main objective was to assess the effect of wash-out dynamics and operation conditions on the underlying bacterial selection, the shape of aerobic granules, the biomass settling properties, and the nutrient removal performances. We first focused on the differences in predominant bacterial populations between compact and fluffy granules, and on a way to avoid filamentous bulking in AGS systems. We then carried out a detailed monitoring of the start-up of one reactor to detect correlations between operation conditions, bacterial community dynamics and nutrient removal performances. The knowledge gained at the microbial ecology level enabled to determine why nutrient removal deteriorated during the start-up of the granulation process in bubble-column SBRs operated under wash-out conditions.

7.2 Material and Methods

7.2.1 *Reactor Infrastructure and Sequencing Batch Operation*

The design and the operation of the bubble-column SBRs were adapted from de Kreuk et al. (2005). The bubble-columns consisted of internal diameters of 52–62 mm, height-to-diameter ratios of 20–25, and working volumes of 2.1–3.1 L. The SBRs were operated in fixed cycles of 3 h comprising feeding of the influent wastewater through the settled sludge bed in pulse (6 min) or anaerobic regime (60 min), aeration (110 min), biomass settling (5 min or stepwise decrease from 15 to 3 min), and withdrawal of the treated effluent (remaining cycle time).

The SBRs were inoculated with 2–3 $g_{VSS}\ L^{-1}$ of flocculent activated sludge originating from full-scale WWTPs. Biomass wash-out conditions were imposed with a short HRT of 6 h in order to stimulate granulation, according to Beun et al. (1999). A volume exchange ratio of 50% was applied to this end.

The SBRs were operated under biomass dynamic conditions at undefined sludge retention time (SRT). The SRT was a function of the imposed settling time, of the height of the effluent withdrawal point, and of the intrinsic settling properties of the cultivated biomass.

The composition of the synthetic cultivation media was similar to the one used by Ebrahimi et al. (2010), and is available in Additional File 7.1 in the Supplementary Information. Acetate was supplied as sole carbon and energy source at a constant concentration between 400 and 500 $mg_{CODs}\ L^{-1}$ in the influent wastewater.

This resulted in a constant volumetric OLR of 200–250 $mg_{CODs}\ cycle^{-1}\ L_R{}^{-1}$ (or 1.6–2.0 $kg_{CODs}\ d^{-1}\ m_R{}^{-3}$ daily equivalents), and in an initial biomass specific OLR of 50–60 $mg_{CODs}\ cycle^{-1}\ g_{CODx}{}^{-1}$ (or 0.4–0.6 $kg_{CODs}\ d^{-1}\ kg_{CODs}{}^{-1}$ daily equivalents). The biomass specific OLR was a dynamic function of the residual biomass concentration evolving in the reactor.

The phosphorous and nitrogenous nutrient ratios amounted to 4.8 $g_{P\text{-}PO4}$ and 12.5 $g_{N\text{-}NH4}$ per 100 g_{CODs}, respectively.

During aeration, air was supplied at the target flow-rate with mass flow controllers (Brooks Instrument, The Netherlands), DO was not controlled and reached saturation (8–9 $mg_{O2}\ L^{-1}$), and pH was regulated at 7.0 ± 0.2 by addition of 1 M HCl or NaOH with a proportional-integral controller.

7.2.2 *Granulation Experiments*

In the first part, the granulation process was studied in 5 reactors (R1–R5) where the bacterial community compositions of early-stage AGS were analyzed in relation with the different combinations of operation parameters, as summarized in Table 7.1.

Table 7.1 Operation parameters applied during the granulation start-up experiments

Reactor[a]	Inoculation sludge[b]	Temperature (°C)	Feeding regime[d]	Up-flow SAV ($cm\ s^{-1}$)	Settling time[f] (min)
R1	OMR	23 ± 2[c]	Pulse	1.8/4.0[e]	5
R2	OMR	23 ± 2[c]	Anaerobic	1.8/4.0[e]	5
R3	BNR	20	Anaerobic	1.8	15 to 3
R4	BNR	20	Anaerobic	2.0	15 to 3
R5	BNR	30	Anaerobic	1.8	15 to 3
R6	BNR	23 ± 2[c]	Anaerobic	2.5	15 to 3

[a] R1–R5 were run in the first part of the study investigating differences in bacterial community compositions of early-stage aerobic granules. R6 was operated for detailed analysis of process conditions governing bacterial selection during granulation

[b] The reactors were inoculated with flocculent activated sludge originating either from a WWTP designed for organic matter removal (OMR) only, or from a WWTP operated for full biological nutrient removal (BNR)

[c] R1, R2 and R6 were operated at ambient temperature without temperature control

[d] The synthetic influent wastewater was supplied either in 6 min with a pulse-feeding regime, or in 60 min with an anaerobic-feeding plug-flow regime

[e] R1 and R2 were operated first with a low up-flow superficial air velocity (SAV) of 1.8 cm s^{-1}. This parameter was doubled after 4 weeks for remediating filamentous overgrowth

[f] Two different biomass settling patterns were tested with a constant settling time of 5 min, or with stepwise decrease in the settling time from 15 to 3 min

The first two reactors R1 and R2 were inoculated with a flocculent activated sludge originating from a WWTP designed for organic matter removal (OMR) only (ERM Morges, Switzerland). R1 and R2 were operated during at most 50 days at ambient temperature (23 ± 2 °C), with pulse (6 min) or anaerobic plug-flow (60 min) feeding regimes, with an initially low up-flow SAV of 1.8 cm s^{-1} during aeration, and with a constant settling time of 5 min. After having observed proliferation of fluffy granules (30 days), the up-flow SAV was increased to 4.0 cm s^{-1} in order to obtain dense granules.

The reactors R3, R4 and R5 were inoculated with a flocculent activated sludge originating for a WWTP designed for full BNR along an anaerobic-anoxic-aerobic process (ARA Thunersee, Switzerland). These reactors were operated at either low (20 °C, R3 and R4) or high (30 °C, R5) mesophilic temperature, with anaerobic plug-flow feeding (60 min), with a low up-flow SAV of 0.018–0.020 m s^{-1} during aeration, and with a stepwise decrease in the settling time from 15 to 3 min in 10–15 days. The operation of R4 and R5 has previously been described in detail by Ebrahimi et al. (2010), and lasted over 40 days. R3 was run on a shorter period of 15 days, but microbial ecology data were collected at higher frequency during the transition from flocculent to granular sludge.

In the second part, reactor R6 was operated with a combination of conditions selecting for the formation of dense fast-settling granules and a detailed monitoring of operation conditions, bacterial community compositions and nutrient removal performances was carried out. R6 was inoculated with flocculent activated sludge taken from the BNR-WWTP, and was operated during 60 days at 23 ± 2 °C and pH 7.0 ± 0.2, with anaerobic plug-flow feeding (60 min), a moderate up-flow SAV of 0.025 m s^{-1}, and a stepwise decrease in the settling time from 15 to 3 min. Temperature, pH, DO, and electrical conductivity signals were collected on-line. Concentrations of biomass present in the reactor and in the treated effluent, and microbial ecology data were collected on a daily basis. Liquid phase samples were taken every 3–5 days for physicochemical analyses of soluble compounds in the influent wastewater, in the reactor at the end of the anaerobic phase, and in the treated effluent.

7.2.3 *Characterizing Metabolic Activities of Inoculation Sludge Taken from the BNR-WWTP*

The nutrient removal capacities of the inoculation sludge taken from the BNR-WWTP were tested in anaerobic, aerobic and anoxic batch tests, and compared to the operation data of the BNR-WWTP. The tests were run at 20 °C in 2-L stirred tank reactors with a biomass concentration of 3–4 g_{VSS} L^{-1} and with similar starting nutrient concentrations as in R6.

7.2.4 *Analyses of Soluble Compounds and Biomass*

Acetate concentration was determined with a high performance liquid chromatograph equipped with an organic acids ion exclusion column ORH-801 (Transgenomics, UK) and a refraction index detector (HPLC Jasco Co-2060 Plus, Omnilab, Germany). The concentration of anions was measured with an ICS-90 ion exchange chromatograph equipped with an IonPacAS14A column and an electrical conductivity detector (Dionex, Switzerland). The concentration of cations was measured with an ICS-3000A ion exchange chromatograph equipped with an IonPacCS16 column and an electrical conductivity detector (Dionex, Switzerland). The particulate concentrations of total (TSS), volatile (VSS) and inorganic suspended solids (ISS) were measured according to Kreuk et al. (2005). Granules were observed by light microscopy.

7.2.5 *Molecular Analyses of Bacterial Community Compositions*

The compositions and dynamics of the bacterial communities were characterized by terminal-restriction fragment length polymorphism (T-RFLP) analysis targeting the v1-v3 hypervariable region of the *Eubacteria* 16S rRNA gene pool. The T-RFLP method was adapted from Rossi et al. (2009) and Ebrahimi et al. (2010), and contained the following modifications.

DNA was extracted from 100 mg of homogenized biomass samples using the Maxwell 16 Tissue DNA Purification System (Promega, Switzerland). Gene fragments of 500 bp were amplified by PCR using universal eubacterial primers: a FAM-labeled 8-F forward primer (FAM-5′-GAGTTTGATCMTGGCTCAG-3′) and an unlabeled 518-R reversed primer (5′-ATTACCGCGGCTGCTGG-3′).

The PCR program was run in a T3000 Thermocycler (Biometra GmbH, Germany) in 30 cycles comprising a longer denaturation time than used by the authors, for optimal amplification of "*Ca.* Accumulibacter"-related polyphosphate-accumulating organisms: 10 min initial denaturation (95 °C), 1 min denaturation (95 °C), 45 s primer annealing (56 °C), 2 min elongation (72 °C), 10 min final elongation (72 °C). The amplicons were purified and concentrated using Invisorb MSB Spin PCRapace purification kits (Invitek Stratec Molecular GmbH, Germany).

Amounts of 200 ng of purified PCR products were digested at 37 °C for 3 h with 0.5 units of the *Hae*III endonuclease (Promega, Switzerland). Volumes of 1 μL of digestion products were mixed with 8.5 μL of HiDi formamid and 0.5 μL of GeneScan 600-LIZ internal size standard (Applied Biosystems, USA), and were denaturated for 2 min at 95 °C. The terminal-restriction fragments (T-RFs) were separated and analyzed by capillary gel electrophoresis in an ABI Prism 3100-Avant Genetic Analyzer using a fluid POP-6 gel matrix and fluorescence laser detection (Applied Biosystems, USA). The T-RFLP profiles were aligned using the Treeflap crosstab macro (Rees et al. 2004).

The bacterial community structures were expressed as relative contributions of all operational taxonomic units (OTU) contributing to the total measured fluorescence. Predominant OTUs with relative abundances above 2% were presented in stacked bar plots for simplified visual observation. Three single biomass samples from the whole set of samples of the study were analyzed in triplicates to determine the overall relative standard deviation related to the T-RFLP method (6%). For reactor R6, biomass equivalents of target OTUs were expressed by multiplying their relative abundances by the mass of VSS present in the reactor.

7.2.6 Analysis of the Richness and Diversity of the Bacterial Community Evolving in Reactor R6

Richness and Shannon's H′ diversity indices were computed with the R software version 2.14.1 (R Development Core Team 2008) equipped with the Vegan package (Oksanen et al. 2009) based on the full T-RFLP profiles collected during experiment R6.

Mathematical geometric evolution models were fitted to the measured richness and diversity profiles with the Berkeley Madonna software (Macey et al. 2000) in order to simulate the evolution of both indices in the reactor. Standard deviation intervals on model predictions were computed from 1000 Monte Carlo simulations on underlying parameters.

7.2.7 Phylogenetic Affiliation of Operational Taxonomic Units

Predominant OTUs detected in R1–R5 were affiliated to closest bacterial relatives by using the cloning-sequencing databank developed by Ebrahimi et al. (2010) and complemented in the present study.

Two DNA extracts from biomass samples collected in R6 at day 2 (flocculent sludge) and day 59 (granular sludge) were sent to Research and Testing Laboratory (Lubbock, Texas, USA) for 454 Tag-encoded FLX amplicon pyrosequencing with a Genome Sequencing FLX System (Roche, Switzerland) using the procedure developed by Sun et al. (2011), and the same primers (8-F and 518-R) as the ones used for T-RFLP analysis.

The pyrosequencing datasets were denoised and processed with the PyroTRF-ID bioinformatics procedure developed by Weissbrodt et al. (2012), which includes sequence annotation with the Greengenes database (McDonald et al. 2012), digital T-RFLP profiling, comparison of digital and experimental T-RFLP profiles, and phylogenetic affiliation of OTUs. QIIME algorithms were used for denoising (Caporaso et al. 2010).

7.2.8 Bacterial Microbiome Analysis

The pyrosequencing datasets of the two biomass samples collected in R6 were analyzed by the metagenomics RAST server (MG-RAST) (Meyer et al. 2008) for annotation and comparative analysis. The Ribosomal Database Project (RDP) (Cole et al. 2009) was used as sequence annotation source. A minimum identity cutoff of 97% was applied in order to retain only the closest bacterial affiliations. A circular phylogenetic tree was constructed with the pyrosequencing datasets of the two samples. The tree was complemented with two bar plots representing the relative abundances of the bacterial genera in both samples. Richness and Shannon's H′ diversity indices were also computed from these datasets.

7.3 Results

7.3.1 Composition and Activity of Early-Stage Granules Cultivated from OMR-Sludge

Reactors R1 and R2 were inoculated with flocculent activated sludge taken from an aeration tank of a WWTP designed for organic matter removal only (OMR-sludge). Operation at ambient temperature (23 ± 2 °C), with acetate as carbon source, low up-flow SAV of 1.8 cm s^{-1}, and a fixed settling time of 5 min resulted in the proliferation of slow-settling fluffy early-stage granules (Fig. 7.1a). Segmented chain filamentous bacterial structures were detected by light microscopy (Fig. 7.1b). The underlying bacterial community compositions are presented in Fig. 7.2a. Closest bacterial affiliations of specific OTUs detected in these reactors are given in Additional File 7.2. The fluffy granules obtained in R1 with pulse feeding (6 min) were predominantly composed of *Burkholderiales* affiliates related to the *Sphaerotilus-Leptothrix* group (OTU-208, 23–33%), and by *Zoogloea* spp. belonging to *Rhodocyclales* (OTU-195, 12–27%). In R2 operated with anaerobic feeding (60 min), *Sphaerotilus-Leptothrix* affiliates dominated (39–50%) over *Zoogloea* spp. (2–8%) in fluffy granules.

The application of a higher up-flow SAV of 4.0 cm s^{-1} after 30 days resulted in the recovery of smooth and dense fast-settling granules. The predominant organisms shifted from *Sphaerotilus-Leptothrix* affiliates to *Zoogloea* spp. In the granules of R1, *Zoogloea* spp. (27%) and *Thaurea* spp. (OTU-217, 30%) outcompeted *Burkholderiales* (< 1%). The dense granules of R2 were highly dominated by *Zoogloea* spp. (47–57%). *Burkholderiales* decreased to 2% after day 48. OTU-185 affiliating with *Gammaproteobacteria* and with *Rhizobiales* from *Alphaproteobacteria* was present up to 17%.

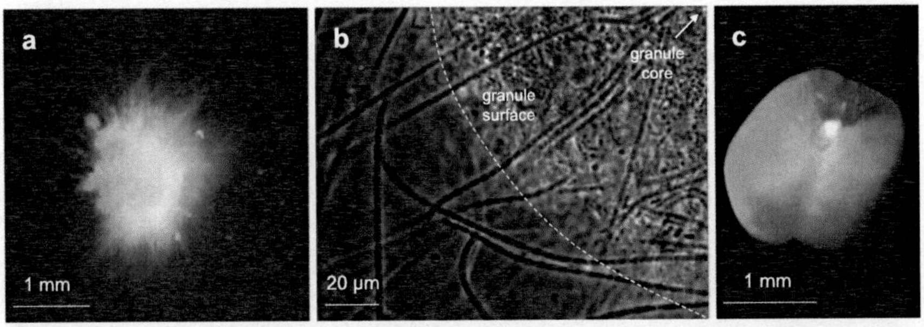

Fig. 7.1 Example of early-stage aerobic granule structures observed with light microscopy. Fluffy slow-settling granule obtained after 30 days in reactor R2 with OMR-sludge and low up-flow SAV of 1.8 cm s^{-1}, and exhibiting filamentous outer structures (**a**). Filamentous segmented chain bacterial structures interspersing across the granular biofilm observed on a sample collected on day 22 in R2 (**b**). Dense fast-settling granule present after 50 days in R6 with BNR-sludge and moderate up-flow SA of 0.025 m s^{-1}, and displaying a tulip-like folded structure around a more opaque internal core (**c**)

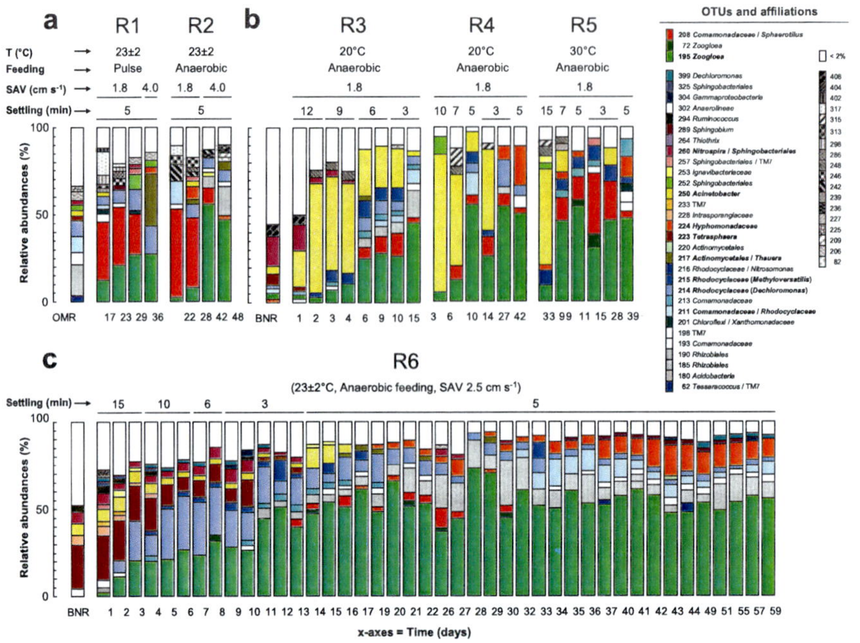

Fig. 7.2 Dynamics of predominant bacterial OTUs analyzed with T-RFLP during the six granulation experiments. Reactors R1 and R2 were inoculated with activated sludge from the OMR-WWTP (**a**). R3, R4 and R5 were inoculated with activated sludge from the BNR-WWTP (**b**). High resolution bacterial ecology data were collected from R6 to assess the effect of wash-out dynamics on bacterial selection during granulation (**c**). Main operation conditions are indicated at the top of each graph. Closest bacterial affiliations of target OTUs presented in Table 7.2 are given on the right

At physical reactor boundaries, filamentous bulking led to deteriorated sludge settling. Both reactors displayed poor nutrient removal performances. After the recovery of fast-settling granules, nutrient removal did not improve. After pulse feeding in R1, acetate was fully removed within 40 min during the aeration phase. In R2 where slow plug-flow anaerobic feeding was applied, more than 90% of the acetate leaked into the aeration phase during which it was fully consumed. Ammonium was not nitrified and biological dephosphatation did not occur. Only partial nitrogen (20%) and phosphorus removal (10%) was detected in both reactors which was probably due to anabolic requirements.

7.3.2 Composition and Activities of Early-Stage Granules Cultivated from BNR-Sludge

The reactors R3, R4 and R5 were operated with an inoculation sludge originating from a WWTP designed for full BNR, under anaerobic feeding, and with a stepwise decrease in the settling time from 15 to 3 min. The three reactors resulted in the formation of smooth and dense fast-settling granules after 9–10 days (Fig. 7.1c).

The underlying bacterial community dynamics are presented in Fig. 7.2b. The inoculation sludge from the BNR-WWTP was dominated by *Nitrospira* and *Sphingobacteriales* (OTU-260, 17%), *Tetrasphaera* spp. (OTU-223, 7%), and an unidentified OTU-408 (7%). *Zoogloea* spp. (OTU-195), *Burkholderiales* (OTU-207), OTU-210 affiliating with *Acidobacteriales* and *Firmicutes*, and OTU-214 affiliating with *Rhodocyclales*-related organisms such as *Dechloromonas* and *Methyloversatilis* spp. were present in lower abundances (2–3%).

In all three reactors, the sludge was still in the flocculent state over the first 6 days. The predominant organisms of the inoculum were replaced within 2–3 days by *Acinetobacter* spp. (OTU-250, 21–79%). Granulation correlated with the proliferation of *Zoogloea* spp. (28–55%). *Dechloromonas* (5–16%), *Methyloversatilis* (3–10%) and *Rhizobiales* (4–16%) were detected as flanking populations. *Hyphomonadaceae* affiliates were abundantly present after 42 days in R4 (23%), and after 39 days in R5 (12%).

In contrast to the operation at 20 °C in R3 and R4 where dense fast-settling granules were constantly present, the operation at 30 °C in R5 led to the proliferation of organisms affiliated to the *Sphaerotilus-Leptothrix* group (35%) and resulted in a mixture of dense and fluffy granules.

Even though a BNR-sludge was used as inoculum, nitrification and dephosphatation activities were not detected in the three AGS systems. Acetate was only consumed to a small extent during the anaerobic feeding phases (18–25%) and fully removed during the aeration phases.

7.3.3 Dynamics of Process Performance and Bacterial Populations Under Wash-Out Conditions

7.3.3.1 Process Dynamics Under Wash-Out Conditions

For reactor R6, high frequency of data collection allowed to detect correlations between operation conditions, bacterial dynamics, and BNR performances during early-stage granulation under wash-out conditions. Changes in biomass properties are presented in Fig. 7.3a, b in function of the settling time.

With initial settling times of 15 and then 10 min during the first 5 days, the activated sludge remained in the flocculent state and a biomass concentration of 2.45–2.95 g_{VSS} L^{-1} was maintained in the reactor, forming a settled sludge blanket of 15–30 cm, and the sludge retention time (SRT) amounted to 12 days.

The decrease in the settling time from 6 to 3 min at day 8 resulted in extensive biomass wash-out (Fig. 7.3c). An extremely low residual biomass concentration of 0.2 g_{VSS} L^{-1} was remaining in the system, and formed a settled sludge blanket of only 1 cm. The SRT dropped to 0.5 day, and approached the HRT of 0.25 day.

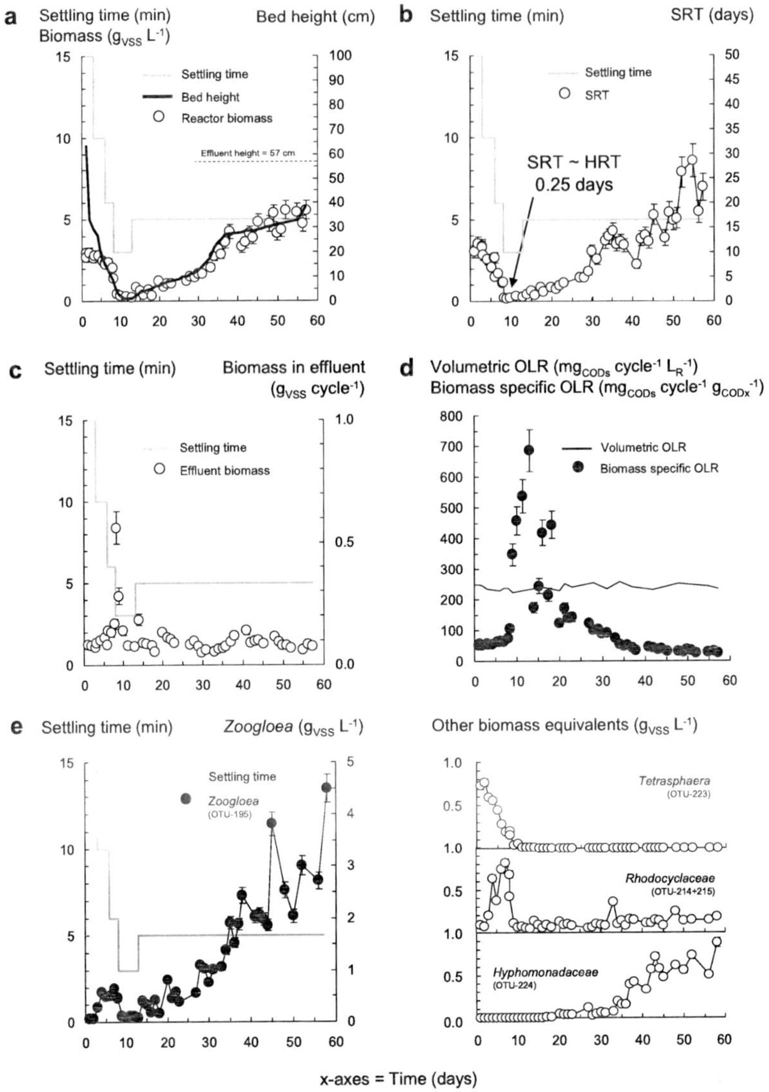

Fig. 7.3 Detailed evolution of biomass parameters during granulation in reactor R6 under wash-out dynamics, in function of the imposed settling time. The evolution of the biomass concentration and of the height of the settled sludge blanket displayed similar profiles as soon as the settling time was decreased to 3 min (**a**). The application of wash-out conditions resulted in the re-coupling of the SRT to the HRT at day 8 (**b**). A high amount of biomass was withdrawn with the effluent wastewater during wash-out from day 8 on (**c**). Under reactor operation with a constant volumetric OLR, the biomass specific OLR exhibited a drastic increase during the period of extensive wash-out between day 8 and day 20 (**d**). *Zoogloea*, *Tetrasphaera*, *Rhodocyclaceae* and *Hyphomonadaceae* affiliates displayed distinct biomass evolutions (**e**). Early-stage nuclei formed after 10 days

First granule nuclei were observed after 10 days. At day 12, the settling time was increased to 5 min as safety measure to keep the granules in the system. The granular biomass increased to 4.0 g_{VSS} L^{-1} at day 37, progressively stabilized at 5.3 ± 0.4 g_{VSS} L^{-1} after 52 days, and formed a settled sludge blanket of 32–40 cm.

The fraction of inorganic suspended solids (ISS) amounted to 38% in the inoculation sludge, and to 12% in the early-stage AGS. An example of a dense fast-settling granule present in R6 after 50 days is presented in Fig. 7.1c.

The reactor was operated with a constant volumetric OLR of 250 mg_{CODs} $cycle^{-1}$ $L_R{}^{-1}$. The biomass specific OLR however evolved with the residual biomass concentration from initially 51 mg_{CODs} $cycle^{-1}$ $g_{CODx}{}^{-1}$ to 685 mg_{CODs} $cycle^{-1}$ $g_{CODx}{}^{-1}$ between day 8 and day 20 after the intensive biomass wash-out (Fig. 7.3d). As the AGS biomass grew in the system, the biomass specific OLR progressively decreased to 20 mg_{CODs} $cycle^{-1}$ $g_{CODx}{}^{-1}$.

7.3.3.2 Bacterial Population Dynamics Under Wash-Out Conditions

Tetrasphaera spp. (OTU-223) dominated in the inoculation sludge (26%), and were progressively replaced in the flocculent sludge after 5 days by *Zoogloea* spp. (OTU-195, 21%) and by OTU-214 (29%) (Fig. 7.2c). During this initial phase, mainly *Dechloromonas* and *Comamonadaceae* relatives contributed to OTU-214, whereas "*Ca.* Accumulibacter" accounted for only 1% of this OTU. When expressed as biomass concentration equivalents, OTU-195 and OTU-214 increased during this period up to 0.6 and 0.8 g_{VSS} L^{-1}, respectively (Fig. 7.3e). The extensive biomass wash-out at day 8 resulted in the rapid decrease in the masses of all bacterial populations below 0.1 g_{VSS} L^{-1}. *Zoogloea* spp. then rapidly proliferated up to a relative abundance of 54 ± 8% in the early-stage AGS from day 15 to day 60. Other *Rhodocyclales* affiliates (OTUs 214 and 215) declined below 5% at day 26. The concentration of *Zoogloea* spp. amounted to 3.0 g_{VSS} L^{-1} after 52 days. The concentration of other *Rhodocyclales* affiliates remained low, but exhibited a slight increase from 0.06 to 0.19 g_{VSS} L^{-1} from day 10 to day 60. All bacterial affiliations obtained with the pyrosequencing-based PyroTRF-ID methodology (Weissbrodt et al. 2012/Chap. 5) are provided in Appendix.

After granulation, additional bacterial populations evolved in the AGS. The relative abundance of *Rhizobiales* (OTU-185) increased from 6% at day 11 to 26% at day 39, and stabilized subsequently at 10 ± 4% over the next 20 days. *Hyphomonadaceae* (OTU-224) were detected above 1% from day 17 on, and were present at 13 ± 3% after day 37. *Comamonadaceae* (OTU-211) increased up to 16% at day 34, and remained at 5 ± 2% until the end of the experiment. *Acinetobacter* spp. (OTU-250) were only detected during the first 17 days in relative abundances of 3–12%. OTU-260 affiliating with *Sphingobacteriales* (and *Nitrospira* probably to a less extent) was present up to 7% at day 10, but was only present at low relative abundances of < 1% to 4% in the early-stage granules. Nitrifiers were not detected above the detection limit of the T-RFLP method.

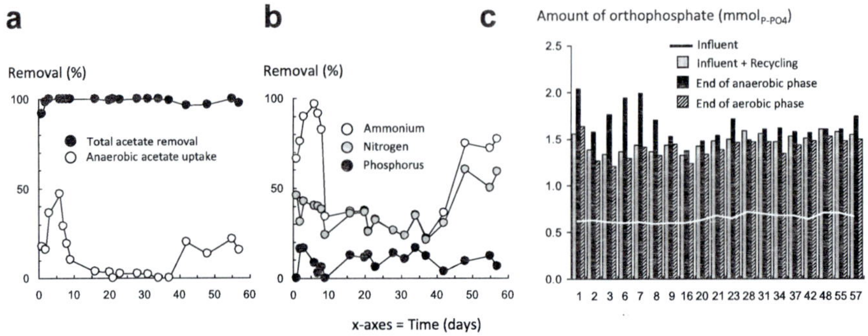

Fig. 7.4 Detailed evolution of the nutrient removal performances in reactor R6. The application of wash-out dynamics resulted in the transient loss of anaerobic acetate uptake (**a**), nitrification, nitrogen removal, and phosphorus removal performances (**b**) from day 6 to day 40. Orthophosphate cycling activities in alternating anaerobic-aerobic conditions were not detected during the same period (**c**)

WWTP operation data and metabolic batch tests indicated that the inoculation sludge was efficiently removing organic matter (95%), nitrogen (97%) and phosphorus (92%). BNR activities were detected in R6 during initial operation with a high settling time (Fig. 7.4a, b). After 6 days, 48% of the acetate load was consumed during anaerobic feeding, ammonium was efficiently nitrified to nitrate (97%), and 40% of nitrogen was removed. Two mmol of orthophosphate were cycled in alternating anaerobic-aerobic conditions (Fig. 7.4c), but only 9% of phosphorus was effectively removed.

After intensive biomass wash-out at day 8, BNR activities were lost except carbon removal. Between day 10 and day 40, less than 4% of acetate was taken up during the anaerobic feeding phase, and only 31 ± 6% of ammonium was removed, presumably by assimilation into biomass. The orthophosphate cycling activity was lost, and phosphorus removal remained at 11 ± 4% until the end of the experiment.

After day 40, ammonium and nitrogen removal recovered to 77% and 60%, respectively. However, nitrite instead of nitrate accumulated in the system. A slight increase in the anaerobic acetate uptake (up to 22%) was detected.

7.3.4 *Analysis of the Bacterial Microbiome of the Flocculent and Granular Sludges in R6*

Based on the T-RFLP data collected from R6, the bacterial community was displaying a strong decrease of 66% in richness and 52% in diversity during the start-up of the reactor (Fig. 7.5).

The bacterial community of the activated sludge taken from the full-scale BNR was associated with a richness of 53 OTUs and a diversity index of 3.3. The bacterial

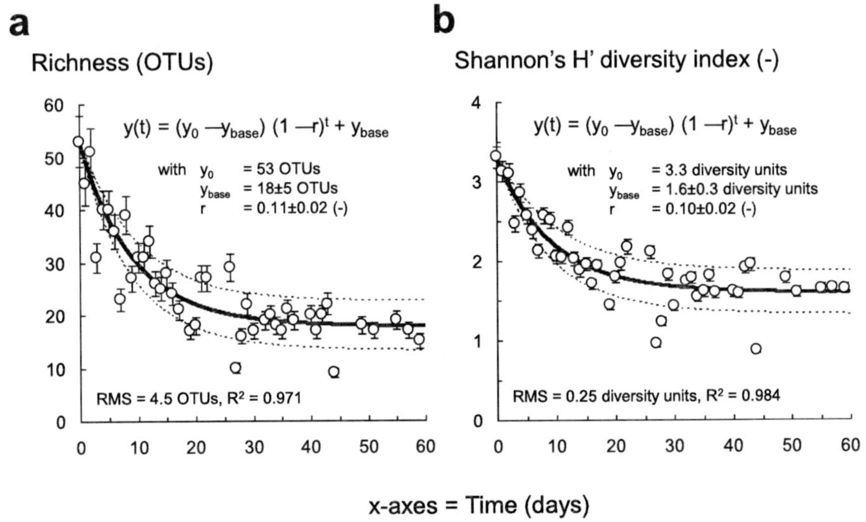

Fig. 7.5 The bacterial community present in R6 exhibited a strong decrease of about 66% in richness (**a**) and of about 52% in Shannon's H′ diversity (**b**) from inoculation with flocculent activated sludge from a full-scale BNR-WWTP fed with real wastewater to formation of early-stage granules fed with an acetate-based synthetic influent. Mathematical negative exponential growth models successfully explained the evolution of both indices during reactor start-up (R^2 = 0.97–0.98). The model trends are given with standard deviation intervals computed from 1000 Monte Carlosimulations. Legend: y_0: initial richness or diversity value, y_{base}: average final richness or diversity value after 60 days, r: negative growth rate, RMS: root mean square error

community of the early-stage granules (day 30 to day 60) was composed of 18 ± 5 OTUs, and showed a diversity index of 1.6 ± 0.3. The best fits of the mathematical geometric evolution models to the evolution of richness and diversity indices (R^2 = 0.97 and 0.98, respectively) were obtained with finite rates of decrease of 11 and 10%, respectively. According to the models, the decrease in richness and diversity before extensive biomass wash-out (day 0–8) amounted to 38% and 30%, i.e. apparent decrease rates of about 2.5 OTUs and 0.13 diversity units per day. During early-stage granulation occurring after wash-out (day 8–27), the richness and diversity decreased by another 27% and 21% (0.8 OTUs and 0.04 diversity units per day), respectively.

The pyrosequencing analyses of biomass samples collected on day 2 and day 59 confirmed that the early-stage AGS displayed a strongly reduced richness and diversity compared to the initial flocculent sludge (Table 7.2). The bacterial microbiome of the flocculent sludge on day 2 was composed of 50 orders and 170 genera that were evenly distributed (3.9 diversity units). The bacterial microbiome of the early-stage AGS was composed of only 20 orders and 57 genera that were unevenly distributed (1.1 diversity units). This analysis also showed that the T-RFLP method was covering at least 85% of the diversity obtained by pyrosequencing. Within the *Rhodocyclales* order, *Zoogloea* affiliates became very predominant in the early-stage AGS, and *Dechloromonas*-related organisms that were abundant in the flocculent sludge at day

2 were replaced by "*Ca.* Accumulibacter" and *Azoarcus* relatives in the early-stage AGS at day 59. Representations of full bacterial microbiomes are provided in the Additional Files 7.3 and 7.4.

Table 7.2 Summary of the main bacterial orders and genera identified by pyrosequencing analysis of the flocculent sludge and early-stage AGS samples taken from R6

Main bacterial orders	%[a]	Main bacterial genera and relative abundances
Flocculent sludge (day 2)		
Rhodocyclales	24	*Zoogloea* (10%), *Dechloromonas* (9.7%), *Methyloversatilis* (1.4%), *Azoarcus* (1.3%), *Thauera* (1.2%)
Actinomycetales	10	*Tetrasphaera* (3.4%), *Terrabacter* (3.2%), *Nocardia* (1.0%), *Micrococcus* (0.4%), *Streptomyces* (0.3%)
Burkholderiales	9	*Acidovorax* (3.7%), *Diaphorobacter* (1.4%), *Burkholderia* (1.2%), *Alcaligenes* (0.7%), *Hydrogenophaga* (0.4%),
Pseudomonadales	9	*Acinetobacter* (8.0%), *Pseudomonas* (0.7%)
Bacillales	6	*Brevibacillus* (5.3%), *Trichococcus* (0.3%)
Chromatiales	5	*Thiolamprovum* (1.7%), *Allochromatium* (1.7%), *Halochromatium* (0.7%)
Rhodobacterales	5	*Azospirillum* (2.0%)
Rhodospirillales	3	*Rhodobacter* (3.7%), *Rhodobaca* (0.7%)
Rhizobiales	2	*Methylosinus* (1.1%), *Rhodopseudomonas* (0.5%), *Methylocystis* (0.3%), *Bradyrhizobiaceae* (0.1%)
Sphingobacteriales	2	*Terrimonas* (1.3%), *Chitinophaga* (0.9%)
Sphingomonadales	2	*Sphingomonas* (1.8%)
Bacteroidales	2	*Butyricimonas* (2.0%)
38 residual orders (< 2%)	21	137 residual genera
Early-stage AGS (day 59)		
Rhodocyclales	84	*Zoogloea* (80%), "*Ca.* Accumulibacter" (3.6%), *Azoarcus* (2.8%), *Thauera* (0.6%)
Burkholderiales	4	*Massilia* (2.5%), *Comamonas* (0.5%), *Acidovorax* (0.5%)
Flavobacteriales	3	*Flavobacterium* (2.8%)
Xanthomonadales	2	*Stenotrophomonas* (1.0%), *Pseudoxanthomonas* (0.3%), *Dyella* (0.2%)
Neisseriales	2	*Aquitalea* (1.4%)
15 residual orders (< 2%)	5	45 residual genera

[a] Relative abundances of bacterial orders obtained after mapping in MG-RAST (Meyer et al. 2008). Phylogenetic tree of the full bacterial microbiome and sector graph representations are available in the Additional Files 7.3 and 7.4

At the level of the nitrifiers, the pyrosequencing analysis enabled detection of ammonium- (AOB) and nitrite-oxidizing bacteria (NOB). AOB were only detected at relative abundances below 0.5% in the flocculent sludge, namely *Nitrosococcus* (0.24%), *Nitrosomonas* (0.12%), and *Nitrosovibrio* spp. (0.06%), and represented a biomass concentration of 0.012 g_{VSS} L^{-1}. The NOB-related *Nitrospira* spp. were detected in higher abundance (1.02%) than *Nitro-bacter* spp. (0.06%). The two genera together accounted for a biomass concentration of 0.032 g_{VSS} L^{-1}. The AOB and NOB present in the flocculent sludge were not detected in the early-stage granules. Only the AOB *Nitrosospira* spp. were detected at 0.03%, and accounted for 0.002 g_{VSS} L^{-1} on this particular day.

7.4 Discussion

7.4.1 Fluffy and Dense Fast-Settling Granules Harbored Different Predominant Phylotypes

Unfavorable filamentous bulking occurring during early-stage granulation was related to the application of an insufficient SAV (1.8 cm s^{-1}) in the case of an inoculum taken from OMR-WWTP, or when operation was conducted at higher mesophilic temperature (30 °C).

The bacterial community of slow-settling fluffy granules was dominated by filamentous *Sphaerotilus* and *Leptothrix* bacterial genera. These organisms are known to cause severe filamentous bulking in conventional WWTPs (Richard et al. 1985). During the formation of compact flocs and granular biofilms, the proliferation of filamentous organisms towards the outside of microbial aggregates is enhanced by substrate gradients generated by diffusion limitations across the biofilm matrices (Martins et al. 2004; Liu and Liu 2006). The ecology data showed that filamentous bulking can also occur with acetate as carbon source, and not only with carbohydrates that have been proposed as main bulking vectors (Liu and Liu 2006).

The application of a more intensive SAV (4.0 cm s^{-1}) was successful for the recovery of smooth and dense fast-settling granules. In the study of McSwain et al. (2004), filamentous overgrowth was counteracted by high shear forces. In analogy to chlorine addition in conventional WWTPs, high shear forces helped to break the superficial filamentous structures. Specific remedial actions that suppress the cause of filamentous proliferation are however preferred for sustainable reactor operation (van Loosdrecht et al. 2008).

The inoculation sludge taken from the BNR-WWTP was beneficial for the production of compact granules at 20 °C with a low SAV. At full-scale level, this corresponds to definite energetic advantages. With the BNR-sludge, fluffy granules were only observed at 30 °C. The growth kinetics of filamentous bacteria are enhanced at such temperature (Richard et al. 1985). In BNR-WWTPs, the successive anaerobic, anoxic and aerobic zones are clearly separated. The readily biodegradable substrates are fully removed by polyphosphate-accumulating organisms (PAOs) under anaerobic conditions, and are not available in the aerobic zone for fast-growing heterotrophs such as filamentous bacteria (van Loosdrecht et al. 2008). Thus, BNR-sludge exhibits a lower filamentous bulking potential than OMR-sludge, and can be advantageous for the granulation process. Ensuring full anaerobic acetate uptake in AGS-SBRs might also favorably suppress filamentous overgrowth.

Dense fast-settling early-stage aerobic granules were dominated by *Zoogloea* relatives. In contrast to fluffy and dense anaerobic granules that have been both dominated by *Methanosaeta* spp. (Grotenhuis et al. 1992; Hulshoff Pol et al. 2004), fluffy and dense aerobic granules were composed of different predominant phylotypes. *Zoogloea* spp. have also previously been detected in other granulation studies involving wash-out conditions (Etterer 2006; Li et al. 2008; Ebrahimi et al. 2010; Gonzalez-Gil and Holliger 2011).

7.4.2 The Possible Role of Rhodocyclales-Related Organisms in Granulation

The T-RFLP and metagenomics analyses revealed that *Rhodocyclales*-affiliated *Zoogloea*, *Dechloromonas*, *Thauera* and *Rhodocyclus* spp. were abundant in the communities of fast-settling early-stage granules.

Acinetobacter spp. were present during the transition from flocs to granules with anaerobic feeding. The *Rhodocyclales*-affiliated organisms share some physiological properties in BNR-WWTPs (Hesselsoe et al. 2009). They produce EPS and store poly-β-hydroxyalkanoates (PHAs) when high organic loads are present under aerobic conditions, hold an arsenal of surface adhesins, and form flocs and biofilms (Sich and Van Rijn 1997; Allen et al. 2004; Dugan et al. 2006; Oshiki et al. 2008; Nielsen et al. 2010; Seviour et al. 2012). Acetate was abundantly present under aerobic conditions due to pulse feeding, or to incomplete anaerobic uptake. Feast-famine regimes and high shear stress also stimulate EPS production during granulation (Liu and Tay 2002; Dulekgurgen et al. 2008; Seviour et al. 2010).

In contrast to flocculent sludge settling that can suffer from *Zoogloea*-mediated viscous bulking (Norberg and Enfors 1982; van Niekerk et al. 1987; Lajoie et al. 2000), AGS settling was not hampered by the proliferation of *Zoogloea* relatives. High shear stress and compaction forces generated by up-flow aeration (Zima et al. 2007) were likely to counteract viscous bulking. In addition, storage compounds such as PHAs confer higher density and settling velocity to bacterial cells (Mas et al. 1985; Schuler et al. 2001). PHA storage was confirmed by confocal laser scanning microscopy analysis with Nile Red staining of cross-sectioned granules dominated by *Zoogloea* spp. (data not shown).

Hence, the physiology of *Zoogloea*-like and other *Rhodocyclales*-affiliated organisms might be relevant for the cohesion of granular biofilms. However, microbial aggregation is probably not restricted to single organisms, and specific process conditions could select for other organisms with similar functions (Bossier and Verstraete 1996; Beun et al. 1999; Wang et al. 2009).

7.4.3 *Wash-Out Conditions as Drastic Bacterial Selection Pressure During Aerobic Granulation*

Even though inoculation with BNR-sludge and anaerobic feeding were combined, active PAOs and nitrifiers were outcompeted by *Zoogloea* spp. during start-up. Two tentative explanations of this specific bacterial selection were formulated from the results of R3–R5 based on wash-out dynamics.

First, the wash-out dynamics resulted in an insufficient SRT that did not enable bacterial populations with lower growth rates such as PAOs and nitrifiers to maintain themselves in the system.

Second, during anaerobic feeding, the influent wastewater was not long enough in contact with the low residual biomass after wash-out. With a constant volumetric OLR and a fixed anaerobic plug-flow feeding phase, a large acetate fraction was present during aeration and selected for fast-growing *Zoogloea* spp. over PAOs. The data collected with R6 were used to confirm these explanations, and are discussed hereafter.

Tetrasphaera spp. and other *Rhodocyclales*-affiliated organisms such as *Dechloromonas*, *Methyloversatilis* spp., and *Rhodocyclus* spp. to a lower extent, were able to compete with *Zoogloea* spp. for the carbon source when 2.45 $g_{VSS}\ L^{-1}$ and 30 cm of settled flocculent biomass was initially present in the system. By considering a bed porosity of 0.5 and an influent flow rate of 21 mL min^{-1}, each volume fraction of the influent wastewater was in contact with the settled biomass during 15 min on average. With this contact time, 50% of acetate was removed under anaerobic conditions with concomitant release of orthophosphate showing that PAO activity was still present. "*Ca.* Accumulibacter" was only present in low abundance in the flocculent sludge (0.1–0.5%). However, additional organisms could have contributed to the detected PAO activity.

Tetrasphaera spp. have been described as putative PAOs in full-scale BNR-WWTP, but their underlying dephosphatating metabolism has not yet been deciphered (Nielsen et al. 2012). *Dechloromonas* spp. have been described as an accompanying guild of "*Ca.* Accumulibacter", and have been proposed as putative PAOs as well (Kong et al. 2007; Oehmen et al. 2010). The genus *Methyloversatilis* that affiliates to *Rhodocyclales* has only recently been discovered, and has been shown to metabolize nitrogen (Kalyuzhnaya et al. 2006; Baytshtok et al. 2008; Kittichotirat et al. 2011). However, more research is required on its metabolism under alternating anaerobic-aerobic conditions.

The combination of a low settling time (3 min) and a low HRT (6 h) resulted in intensive biomass wash-out. The SRT dropped to a value close to the HRT, and the reactor system was governed by the hydraulic properties. Aerobic heterotrophic organisms such as *Zoogloea*, *Dechloromonas*, *Acinetobacter* and filamentous *Burkholderiales* affiliates that are related to maximum growth rates of 0.229–0.690 h^{-1} (Lau et al. 1984; van Niekerk et al. 1987; Logan et al. 2001; Kim and Pagilla 2003) that are above 1/HRT, were able to proliferate over slower-growing PAOs [0.042 h^{-1}, Henze et al. (1999)] and nitrifiers [0.017–0.046 h^{-1}, Xavier et al. (2007)].

During reactor start-up, strong decreases in richness and diversity were observed. The apparent decrease rates were about 3 times higher before than after wash-out, indicating that the use of a synthetic wastewater with acetate as sole carbon source significantly contributed to the change in the bacterial community structure before wash-out. Gonzalez-Gil and Holliger (2011) have also reported that early-stage AGS cultivated with acetate or propionate as sole carbon and energy sources displayed half of the richness of the inoculation sludge. Winkler et al. (2011) have reported that, although distant, denaturing gradient gel electrophoresis profiles of bacterial communities of a conventional WWTP and of a pilot AGS reactor fed with the same urban wastewater exhibited similar richness and eveness.

After wash-out, only 0.3 g_{VSS} L^{-1} of biomass was remaining in the system and the biomass specific OLR increased by a factor of 13 from 51 to 685 mg_{CODs} $cycle^{-1}$ $g_{CODx}{}^{-1}$. Granulation started with a biomass specific OLR above 2.7 kg_{CODs} d^{-1} $kg_{CODx}{}^{-1}$ equivalents, which is in agreement with the bottom value of 1.3 kg_{CODs} d^{-1} $kg_{CODx}{}^{-1}$ considered by Morgenroth et al. (1997) to enable sludge granulation. However, with a settled biomass height of only 1 cm, the contact time with the influent wastewater was extremely short (30 s).

The fixed anaerobic plug-flow feeding phase thus resulted in the leakage of more than 90% of the acetate load into the aeration phase where it was available for fast-growing aerobic heterotrophs. This also explains why *Zoogloea* spp. outcompeted "*Ca.* Accumulibacter", and why phosphorus was not removed. Deteriorated dephosphatation has also been correlated in flocculent sludge SBRs with *Zoogloea* proliferation over PAOs caused by the concomitant presence of acetate as electron donor and oxygen or nitrate as terminal electron acceptors (Fang et al. 2002; Montoya et al. 2008). Proper anaerobic selector operation has been recommended to suppress this zoogloeal overgrowth (van Loosdrecht et al. 2008).

7.5 Conclusions

The detailed microbial ecology investigation involving T-RFLP, pyrosequencing and PyroTRF-ID analyses conducted in this study in combination with a bioprocess engineering approach showed that:

- Slow-settling fluffy granules and dense fast-settling early-stage granules cultivated under wash-out dynamics were displaying distinct predominant phylotypes, namely filamentous *Burkholderiales* affiliates and *Zoogloea* relatives, respectively.
- Filamentous bulking could be remediated by the application of intensive up-flow aeration, or by the use of an inoculation sludge taken from a BNR-WWTP.
- A combination of insufficient SRT and of leakage of acetate into the aeration phase was the cause for the proliferation of *Zoogloea* spp. in dense fast-settling granules, and for the deterioration of BNR performances which has been commonly observed by different authors during granulation start-ups. It is however not certain that *Zoogloea*-like organisms are essential in granule formation.
- Additional research is needed to determine if they are required to stimulate early-stage granulation in BNR systems, or if granules can be cultivated without their involvement. Furthermore, optimal operation conditions should be elucidated for maintaining a balance between organisms with granulation propensity and nutrient removing organisms in order to form granules with BNR activities in short start-up periods.

Acknowledgements The author obtained a grant from the EPFL FEE Foundation for part of this study. ERM Morges and ARA Thunersee for providing the inoculation sludge. Corinne Weis, Yoan Rappaz, Sébastien Gabus, and Emmanuelle Rohrbach for excellent assistance in reactor operation and molecular ecology analyses. Scot E. Dowd, Yan Sun and Lars Koenig from the Research and Testing Laboratory (Lubbock, Texas, USA) for pyrosequencing analyses and advice.

Appendix

The Appendix is available at the end of Chap. 5 on PyroTRF-ID.

Appendix: Phylogenetic affiliations obtained with PyroTRF-ID.

Supplementary Information

Additional File 7.1 Composition of the cultivation media.

Additional File 7.2 Cloning-sequencing databank constructed for OTUs detected in R1–R5.

Additional File 7.3 Phylogenetic tree of the full bacterial microbiome constructed with the two pyrosequencing datasets.

Additional File 7.4 Sector graph representation of the bacterial composition of the two bacterial microbiomes of flocculent sludge and granular sludge.

References

Adav SS, Lee DJ, Lai JY (2010) Microbial community of acetate utilizing denitrifiers in aerobic granules. Appl Microbiol Biotechnol 85(3):753–762

Allen MS, Welch KT, Prebyl BS, Baker DC, Meyers AJ, Sayler GS (2004) Analysis and glycosyl composition of the exopolysaccharide isolated from the floc-forming wastewater bacterium Thauera sp. MZ1T. Environ Microbiol 6(8):780–790

Baytshtok V, Kim S, Yu R, Park H, Chandran K (2008) Molecular and biokinetic characterization of methylotrophic denitrification using nitrate and nitrite as terminal electron acceptors. Water Sci Technol 58(2):359–365

Benson DA, Karsch-Mizrachi I, Lipman DJ, Ostell J, Sayers EW (2011) GenBank. Nucleic Acids Res 39(suppl 1):D32–D37

Beun JJ, Hendriks A, van Loosdrecht MCM, Morgenroth E, Wilderer PA, Heijnen JJ (1999) Aerobic granulation in a sequencing batch reactor. Water Res 33(10):2283–2290

Beun JJ, Heijnen JJ, van Loosdrecht MCM (2001) N-removal in a granular sludge sequencing batch airlift reactor. Biotechnol Bioeng 75(1):82–92

Bossier P, Verstraete W (1996) Triggers for microbial aggregation in activated sludge? Appl Microbiol Biotechnol 45(1–2):1–6

Caporaso JG, Kuczynski J, Stombaugh J, Bittinger K, Bushman FD, Costello EK, Fierer N, Pena AG, Goodrich JK, Gordon JI, Huttley GA, Kelley ST, Knights D, Koenig JE, Ley RE, Lozupone CA, McDonald D, Muegge BD, Pirrung M, Reeder J, Sevinsky JR, Turnbaugh PJ, Walters WA, Widmann J, Yatsunenko T, Zaneveld J, Knight R (2010) QIIME allows analysis of high-throughput community sequencing data. Nat Methods 7(5):335–336

Cole JR, Wang Q, Cardenas E, Fish J, Chai B, Farris RJ, Kulam-Syed-Mohideen AS, McGarrell DM, Marsh T, Garrity GM, Tiedje JM (2009) The Ribosomal Database Project: improved alignments and new tools for rRNA analysis. Nucleic Acids Res 37:D141–D145

de Kreuk MK, Heijnen JJ, van Loosdrecht MCM (2005) Simultaneous COD, nitrogen, and phosphate removal by aerobic granular sludge. Biotechnol Bioeng 90(6):761–769

de Bruin LMM, de Kreuk MK, van der Roest HFR, Uijterlinde C, van Loosdrecht MCM (2004) Aerobic granular sludge technology: An alternative to activated sludge? Water Sci Technol 49(11–12):1–7

Dugan P, Stoner D, Pickrum H (2006) The genus *Zoogloea*. In: Dworkin M, Falkow S, Rosenberg E, Schleifer KH, Stackebrandt E (eds) The prokaryotes, vol 7. Springer, New York, pp 960–970

Dulekgurgen E, Artan N, Orhon D, Wilderer PA (2008) How does shear affect aggregation in granular sludge sequencing batch reactors? Relations between shear, hydrophobicity, and extracellular polymeric substances. Water Sci Technol 58(2):267–276

Ebrahimi S, Gabus S, Rohrbach-Brandt E, Hosseini M, Rossi P, Maillard J, Holliger C (2010) Performance and microbial community composition dynamics of aerobic granular sludge from sequencing batch bubble column reactors operated at 20°C, 30°C, and 35°C. Appl Microbiol Biotechnol 87:1555–1568

Etterer TJ (2006) Formation, structure and function of aerobic granular sludge. Ph.D. thesis, Technische Universität München

Fang HHP, Zhang T, Liu Y (2002) Characterization of an acetate-degrading sludge without intracellular accumulation of polyphosphate and glycogen. Water Res 36(13):3211–3218

Giesen A, Niermans R, van Loosdrecht MCM (2012) Aerobic granular biomass: the new standard for domestic and industrial wastewater treatment? Water 21 4:28–30

Gonzalez-Gil G, Holliger C (2011) Dynamics of microbial community structure and enhanced biological phosphorus removal of propionate- and acetate-cultivated aerobic granules. Appl Environ Microbiol 77:8041–8051

Grotenhuis JTC, Plugge CM, Stams AJM, Zehnder AJB (1992) Hydrophobicities and electrophoretic mobilities of anaerobic bacterial isolates from methanogenic granular sludge. Appl Environ Microbiol 58(3):1054–1056

Henze M, Gujer W, Mino T, Matsuo T, Wentzel MC, Marais GVR, Van Loosdrecht MCM (1999) Activated sludge model No.2d, ASM2d. Water Sci Technol 39(1):165–182

Hesselsoe M, Fureder S, Schloter M, Bodrossy L, Iversen N, Roslev P, Nielsen PH, Wagner M, Loy A (2009) Isotope array analysis of *Rhodocyclales* uncovers functional redundancy and versatility in an activated sludge. ISME J 3(12):1349–1364

Hulshoff Pol LW, De Castro Lopes SI, Lettinga G, Lens PNL (2004) Anaerobic sludge granulation. Water Res 38(6):1376–1389

Kalyuzhnaya MG, De Marco P, Bowerman S, Pacheco CC, Lara JC, Lidstrom ME, Chistoserdova L (2006) *Methyloversatilis universalis* gen. nov., sp. nov., a novel taxon within the *Betaproteobacteria* represented by three methylotrophic isolates. Int J Syst Evol Microbiol 56(11):2517–2522

Kim H, Pagilla KR (2003) Competitive growth of Gordonia and Acinetobacter in continuous flow aerobic and anaerobic/aerobic reactors. J Biosci Bioeng 95(6):577–582

Kittichotirat W, Good NM, Hall R, Bringel F, Lajus A, Médigue C, Smalley NE, Beck D, Bumgarner R, Vuilleumier S, Kalyuzhnaya MG (2011) Genome sequence of *Methyloversatilis universalis* FAM5 T, a methylotrophic representative of the order *Rhodocyclales*. J Bacteriol 193(17):4541–4542

Kong Y, Xia Y, Nielsen JL, Nielsen PH (2007) Structure and function of the microbial community in a full-scale enhanced biological phosphorus removal plant. Microbiology 153(12):4061–4073

Lajoie CA, Layton AC, Gregory IR, Sayler GS, Don ET, Meyers AJ (2000) Zoogleal clusters and sludge dewatering potential in an industrial activated-sludge wastewater treatment plant. Water Environ Res 72(1):56–64

Lau AO, Strom PF, Jenkins D (1984) Growth kinetics of Sphaerotilus natans and a floc former in pure and dual continuous culture. J Water Pollut Con F 56(1):41–51

Lee D-J, Chen Y-Y, Show K-Y, Whiteley CG, Tay J-H (2010) Advances in aerobic granule formation and granule stability in the course of storage and reactor operation. Biotechnol Adv 28(6):919–934

Lemaire R, Yuan Z, Blackall LL, Crocetti GR (2008) Microbial distribution of Accumulibacter spp. and Competibacter spp. in aerobic granules from a lab-scale biological nutrient removal system. Environ Microbiol 10(2):354–363

Li AJ, Yang SF, Li XY, Gu JD (2008) Microbial population dynamics during aerobic sludge granulation at different organic loading rates. Water Res 42(13):3552–3560

Lin YM, Liu Y, Tay JH (2003) Development and characteristics of phosphorus-accumulating microbial granules in sequencing batch reactors. Appl Microbiol Biotechnol 62(4):430–435

Liu Y, Liu Q-S (2006) Causes and control of filamentous growth in aerobic granular sludge sequencing batch reactors. Biotechnol Adv 24(1):115–127

Liu Y, Tay JH (2002) The essential role of hydrodynamic shear force in the formation of biofilm and granular sludge. Water Res 36:1653–1665

Logan BE, Zhang H, Mulvaney P, Milner MG, Head IM, Unz RF (2001) Kinetics of perchlorate- and chlorate-respiring bacteria. Appl Environ Microbiol 67(6):2499–2506

Macey R, Oster G, Zahnley T (2000) Berkeley Madonna user's guide. University of California, Berkeley, CA, USA

Martins AMP, Picioreanu C, Heijnen JJ, van Loosdrecht MCM (2004) Three-dimensional dual-morphotype species modeling of activated sludge flocs. Environ Sci Technol 38(21):5632–5641

Mas J, Pedros-Alio C, Guerrero R (1985) Mathematical model for determining the effects of intracytoplasmic inclusions on volume and density of microorganisms. J Bacteriol 164(2):749–756

McDonald D, Price MN, Goodrich J, Nawrocki EP, DeSantis TZ, Probst A, Andersen GL, Knight R, Hugenholtz P (2012) An improved Greengenes taxonomy with explicit ranks for ecological and evolutionary analyses of bacteria and archaea. ISME J 6(3):610–618

McSwain BS, Irvine RL, Wilderer PA (2004) The influence of settling time on the formation of aerobic granules. Water Sci Technol 50(10):195–202

Meyer F, Paarmann D, D'Souza M, Olson R, Glass E, Kubal M, Paczian T, Rodriguez A, Stevens R, Wilke A, Wilkening J, Edwards R (2008) The metagenomics RAST server—a public resource for the automatic phylogenetic and functional analysis of metagenomes. BMC Bioinform 9(1):386
Montoya T, Borrás L, Aguado D, Ferrer J, Seco A (2008) Detection and prevention of enhanced biological phosphorus removal deterioration caused by Zoogloea overabundance. Environ Technol 29(1):35–42
Morgenroth E, Sherden T, van Loosdrecht MCM, Heijnen JJ, Wilderer PA (1997) Aerobic granular sludge in a sequencing batch reactor. Water Res 31(12):3191–3194
Mosquera-Corral A, de Kreuk MK, Heijnen JJ, van Loosdrecht MCM (2005) Effects of oxygen concentration on N-removal in an aerobic granular sludge reactor. Water Res 39(12):2676–2686
Moy BYP, Tay JH, Toh SK, Liu Y, Tay STL (2002) High organic loading influences the physical characteristics of aerobic sludge granules. Lett Appl Microbiol 34(6):407–412
Nielsen PH, Mielczarek AT, Kragelund C, Nielsen JL, Saunders AM, Kong Y, Hansen AA, Vollertsen J (2010) A conceptual ecosystem model of microbial communities in enhanced biological phosphorus removal plants. Water Res 44(17):5070–5088
Nielsen PH, Saunders AM, Hansen AA, Larsen P, Nielsen JL (2012) Microbial communities involved in enhanced biological phosphorus removal from wastewater—a model system in environmental biotechnology. Curr Opin Biotechnol 23(3):452–459
Norberg AB, Enfors SO (1982) Production of extracellular polysaccharide by Zoogloea ramigera. Appl Environ Microbiol 44(5):1231–1237
Oehmen A, Carvalho G, Lopez-Vazquez CM, van Loosdrecht MCM, Reis MAM (2010) Incorporating microbial ecology into the metabolic modelling of polyphosphate accumulating organisms and glycogen accumulating organisms. Water Res 44(17):4992–5004
Oksanen J, Kindt R, Legendre P, O'Hara B, Simpson GL, Solymos P, Stevens MHH, Wagner H (2009) Vegan: community ecology package. R package version 1.15-4. R Foundation for Statistical Computing, Vienna, Austria. http://CRAN.R-project.org/package=vegan
Oshiki M, Onuki M, Satoh H, Mino T (2008) PHA-accumulating microorganisms in full-scale wastewater treatment plants. Water Sci Technol 58(1):13–20
Pijuan M, Werner U, Yuan Z (2011) Reducing the startup time of aerobic granular sludge reactors through seeding floccular sludge with crushed aerobic granules. Water Res 45(16):5075–5083
R Development Core Team (2008) R: a language and environment for statistical computing. R Foundation for Statistical Computing, Vienna, Austria. http://cran.r-project.org/
Rees G, Baldwin D, Watson G, Perryman S, Nielsen D (2004) Ordination and significance testing of microbial community composition derived from terminal restriction fragment length polymorphisms: application of multivariate statistics. Antonie Van Leeuwenhoek 86(4):339–347
Richard M, Hao O, Jenkins D (1985) Growth kinetics of Sphaerotilus species and their significance in activated sludge bulking. J Water Pollut Con F 57(1):68–81
Rossi P, Gillet F, Rohrbach E, Diaby N, Holliger C (2009) Statistical assessment of variability of terminal restriction fragment length polymorphism analysis applied to complex microbial communities. Appl Environ Microbiol 75(22):7268–7270
Schuler AJ, Jenkins D, Ronen P (2001) Microbial storage products, biomass density, and setting properties of enhanced biological phosphorus removal activated sludge. Water Sci Technol 43(1):173–180
Seviour T, Donose BC, Pijuan M, Yuan Z (2010) Purification and conformational analysis of a key exopolysaccharide component of mixed culture aerobic sludge granules. Environ Sci Technol 44(12):4729–4734
Seviour T, Yuan Z, van Loosdrecht MCM, Lin Y (2012) Aerobic sludge granulation: a tale of two polysaccharides? Water Res 46(15):4803–4813
Shin HS, Lim KH, Park HS (1992) Effect of shear stress on granulation in oxygen aerobic upflow sludge bed reactors. Water Sci Technol 26(3–4):601–605
Sich H, Van Rijn J (1997) Scanning electron microscopy of biofilm formation in denitrifying, fluidised bed reactors. Water Res 31(4):733–742

Sun Y, Wolcott RD, Dowd SE (2011) Tag-encoded FLX amplicon pyrosequencing for the elucidation of microbial and functional gene diversity in any environment. Methods Mol Biol 733:129–141
Tay JH, Liu QS, Liu Y (2002) Aerobic granulation in sequential sludge blanket reactor. Water Sci Technol 46(4–5):13–18
Tsuneda S, Nagano T, Hoshino T, Ejiri Y, Noda N, Hirata A (2003) Characterization of nitrifying granules produced in an aerobic upflow fluidized bed reactor. Water Res 37(20):4965–4973
van Loosdrecht MCM, Martins AM, Ekama GA (2008) Bulking sludge. In: Henze M, van Loosdrecht MCM, Ekama GA, Brdjanovic D (eds) Biological wastewater treatment: principles, modelling and design. IWA Publishing, London, pp 291–308
van Niekerk AM, Jenkins D, Richard MG (1987) The competitive growth of Zoogloea ramigera and type 021N in activated sludge and pure culture—a model for low F: M bulking. J Water Pollut Con F 59(5):262–273
Verawaty M, Pijuan M, Yuan Z, Bond PL (2012) Determining the mechanisms for aerobic granulation from mixed seed of floccular and crushed granules in activated sludge wastewater treatment. Water Res 46(3):761–771
Wang J, Zhang Z, Wu W (2009) Research advances in aerobic granular sludge. Acta Scien Circum 29(3):449–473
Weber SD, Ludwig W, Schleifer KH, Fried J (2007) Microbial composition and structure of aerobic granular sewage biofilms. Appl Environ Microbiol 73(19):6233–6240
Weissbrodt DG, Shani N, Sinclair L, Lefebvre G, Rossi P, Maillard J, Rougemont J, Holliger C (2012) PyroTRF-ID: a novel bioinformatics methodology for the affiliation of terminal-restriction fragments using 16S rRNA gene pyrosequencing. BMC Microbiol 12:306
Winkler MKH, Kleerebezem R, de Bruin LMM, Habermacher J, Abbas B, van Loosdrecht MCM (2011) Microbial diversity differences in aerobic granular sludge in comparison to conventional treatment plant. In: Qi Z (ed) IWA biofilm specialist conference 2011 processes in biofilms, Tongji University, Shanghai, China
Xavier JB, de Kreuk MK, Picioreanu C, van Loosdrecht MCM (2007) Multi-scale individual-based model of microbial and byconversion dynamics in aerobic granular sludge. Environ Sci Technol 41(18):6410–6417
Yilmaz G, Lemaire R, Keller J, Yuan Z (2008) Simultaneous nitrification, denitrification, and phosphorus removal from nutrient-rich industrial wastewater using granular sludge. Biotechnol Bioeng 100(3):529–541
Zima BE, Diez L, Kowalczyk W, Delgado A (2007) Sequencing batch reactor (SBR) as optimal method for production of granular activated sludge (GAS)—fluid dynamic investigations. Water Sci Technol 55(8–9):151–158

Chapter 8
Bacterial and Structural Dynamics During the Bioaggregation of Aerobic Granular Biofilms

(Attached) microbial growth originates in a mixture of slime and zoogloeal bacteria, i.e. microorganisms that form gelatinous aggregates.
(Characklis 1973)

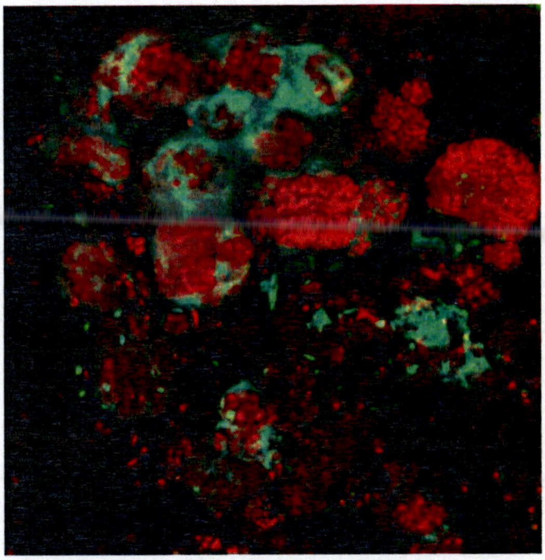

Microbial colonies in an EPS matrix

The content of this chapter was published in a modified version in: Weissbrodt DG, Neu TR, Kuhlicke U, Rappaz Y, Holliger C (2013) Assessment of bacterial and structural dynamics in aerobic granular biofilms. Frontiers in Microbiology 4:175 (2013). https://doi.org/10.3389/fmicb.2013.00175. Permission was granted to reuse the figure materials (© 2013 Frontiers Media S.A.).

Supplementary Information The online version contains supplementary material available at https://doi.org/10.1007/978-3-031-41009-3_8.

D. G. Weissbrodt, *Engineering Granular Microbiomes*, Springer Theses,
https://doi.org/10.1007/978-3-031-41009-3_8

8.1 Introduction

Aerobic granular sludge (AGS) used for intensified biological nutrient removal (BNR) from wastewater (Giesen et al. 2012) comprises suspended biofilm particles, called aerobic granules, formed by microbial self-aggregation (Morgenroth et al. 1997). Although some full-scale plants are installed world-wide, the granulation mechanism at the microbial community level is not yet fully understood and improved knowledge of this phenomenon may enable further process optimization.

Nutrient removal deficiency and other process instabilities during granulation have been observed in several studies (Morgenroth et al. 1997; de Kreuk et al. 2005; Liu and Liu 2006; Gonzalez-Gil and Holliger 2011). Wash-out conditions that have been recommended for the selection of fast-settling biomass (Beun et al. 1999) have been shown to result in the deterioration of the settling properties and of the BNR performances caused by bacterial community imbalances with overgrowth of filamentous or zoogloeal populations, respectively (Weissbrodt et al. 2012a). Nevertheless, granule formation has been positively correlated with proliferation of Zoogloea spp. under wash-out conditions (Etterer 2006; Adav et al. 2009; Ebrahimi et al. 2010; Weissbrodt et al. 2012a). Although Zoogloea can produce cohesive extracellular polymeric substances (EPS) (Seviour et al. 2012), it remains unclear whether these organisms are required to initiate granulation. Shifts in predominant populations in AGS systems have further been related to specific operation parameters. For instance, nutrient composition, temperature, carbon source, and selective excess sludge removal can impact the competition of polyphosphate- (PAOs) and glycogen-accumulating organisms (GAOs) related to "Candidati Accumulibacter and Competibacter phosphates", respectively (de Kreuk and van Loosdrecht 2004; Ebrahimi et al. 2010; Gonzalez-Gil and Holliger 2011; Winkler et al. 2011a; Bassin et al. 2012). Since some clades have been reported to denitrify, PAOs and GAOs can in addition be impacted by the type of terminal electron acceptor present in the medium (Kuba et al. 1997; Oehmen et al. 2010).

Granule structures comprise stratification of microbial niches oriented along radial substrate and microhabitat gradients (de Kreuk et al. 2005). Whereas bacterial community dynamics at reactor scale (Liu et al. 1998) can be assessed by terminal-restriction fragment length polymorphism (T-RFLP) analysis, biofilm architecture and microbial arrangements can be examined by confocal laser scanning microscopy (CLSM) combined with fluorescence in situ hybridization (FISH) (Wagner et al. 1994; Wuertz et al. 2004; Nielsen et al. 2009; Neu et al. 2010). Localization of bacterial populations along dissolved oxygen (DO) gradients in granules has been investigated with the latter methods (Tsuneda et al. 2003; Kishida et al. 2006; Lemaire et al. 2008; Yilmaz et al. 2008; Gao et al. 2010; Filali et al. 2012). Whereas ammonium-oxidizing organisms (AOOs) have formed clusters near the surface, no consensus has been found on the position of PAOs and GAOs within the granules and their involvement in denitrification. Similarly to biofilms (Lawrence et al. 1996; Okabe et al. 1997; Alpkvist and Klapper 2007), granules are likely to exhibit complex spatial architecture depending on specific process conditions. Since granular biofilms have been composed of specific matrices of exopolymeric substances such as 'granulan' (Seviour et al. 2011) or alginate (Lin et al. 2010), the combination of fluorescence lectin-binding analysis (FLBA) and CLSM (Neu and Lawrence 1999; Staudt et al. 2003) could be relevant for mapping glycoconjugates in the granular biofilm matrix. Such analysis has previously been conducted for the staining of global exopolysaccharide matrices in different types of granules (Tay et al. 2003; McSwain et al. 2005; Adav et al. 2008). Since lectins display specific binding properties, a screening of different lectins can provide additional information on the type of polysaccharide residues present in granular biofilms.

The present research was conducted to elucidate the dynamics of the bacterial communities and of the structures of bioaggregates during transitions from activated sludge flocs to early-stage nuclei and to mature granular biofilms. Granulation was studied in one bubble-column (BC-SBR) and two stirred-tank (PAO-SBR, GAO-SBR) anaerobic-aerobic sequencing batch reactors operated for fast AGS cultivation and for enriching "*Ca.* Accumulibacter" and "*Ca.* Competibacter" in activated sludge, respectively. T-RFLP, pyrosequencing, CLSM, lectin-binding, and FISH methods were combined to investigate the mechanisms of bacterial selection, granule formation, and maturation in relation with the evolution of process variables. A conceptual ecological model was developed, and used to identify key operation factors for cultivating AGS with BNR activities. Since granulation in the GAO-SBR occurred unexpectedly after only more than 450 days, the detailed datasets collected from the BC-SBR and PAO-SBR were used to this end.

8.2 Material and Methods

8.2.1 Bubble-Column SBR Operation Under Wash-Out Dynamics

The BC-SBR was operated at 23 ± 2 °C under wash-out conditions as reported in Weissbrodt et al. (2012a). This 2.5-L single-wall PVC reactor (height-to-diameter H/D ratio of 28) was inoculated with 3 g_{VSS} L^{-1} of activated sludge from a BNR-WWTP (Thunersee, Switzerland).

The fixed 3-h SBR cycles comprised anaerobic feeding (60 min), aeration (110 min), settling (stepwise decrease from 15 to 3 min), and withdrawal (remaining cycle time; volume exchange ratio of 50%). Wash-out was generated by short settling and hydraulic retention times (HRT, 6 h). The sludge retention time (SRT) was not controlled. The synthetic wastewater composition comprised 4.8 $g_{P\text{-}PO4}$ and 12.5 $g_{N\text{-}NH4}$ per 100 g_{CODs} of acetate (Additional File 8.1 in Supplementary Information). The system was fed over 220 days with a constant volumetric organic loading rate (OLR, 250 mg_{CODs} $cycle^{-1}$ L_R^{-1}). The biomass specific OLR depended on the remaining biomass (initially 50 mg_{CODs} $cycle^{-1}$ g_{CODx}^{-1}). Aeration phases were run with up-flow superficial gas velocities of 0.025 m s^{-1}, free DO evolution up to saturation, and pH 7.0 ± 0.2. Temperature, DO, pH, and electrical conductivity were recorded on-line.

8.2.2 *Stirred-Tank PAO-SBR and GAO-SBR Operation Under Steady State*

The PAO- and GAO-SBRs were run to cultivate activated sludge enrichments over 1–2 years. The 2.5-L double-wall glass reactors (Applikon Biotechnology, The Netherlands, H/D = 1.3) were inoculated with 3 g_{VSS} L^{-1} of BNR activated sludge, and operated according to Lopez-Vazquez et al. (2009b). SBR cycles comprised N_2-flush (7 min), pulse feeding (7.3 min), N_2-flush (5 min), anaerobic, aerobic and settling phases (for timing see below), and withdrawal (5 min; 50%). During the anaerobic and aerobic (3.5 ± 0.5 mg_{O2} L^{-1}) phases, the reactor content was stirred at 300 rpm. Nitrification was inhibited in both reactors by addition of allyl-N-thiourea (Additional File 8.1). SRTs were controlled by purging excess sludge after aeration.

The PAO-SBR was operated at 17 °C and pH 7.0–8.0, with 12-h HRT and 8-days SRT at steady state, and with propionate, as well as with 9 g_{CODs} $g_{P\text{-}PO4}{}^{-1}$ in the influent wastewater following Schuler and Jenkins (2003). Enhanced anaerobic propionate uptake and ortho-phosphate cycling activities were ensured by stepwise adaptation of the volumetric OLR from 15 to 200 mg_{CODs} $cycle^{-1}$ $L_R{}^{-1}$ in 12 days, and by proper control of the anaerobic and aerobic contact times (3–5 h) based on on-line conductivity profiles (Maurer and Gujer 1995; Aguado et al. 2006). Since fast-settling biomass formed after 30 days, the settling time was decreased from 60 to 10 min to save cycle time, and to prevent prolonged endogenous respiration.

The GAO-SBR was operated at 30 °C and pH 6.5 ± 0.2, with 12-h HRT and longer 16-days SRT at steady state (Lopez-Vazquez et al. 2009a), acetate, and 200 g_{COD} $g_{P\text{-}PO4}{}^{-1}$. The volumetric OLR, anaerobic phase length, and settling time were fixed at 200 mg_{CODs} $cycle^{-1}$ $L_R{}^{-1}$, 3 h, and 60 min since start-up, respectively.

8.2.3 *Analyses of Soluble Compounds and Biomass*

Concentrations of volatile fatty acids (VFA) and inorganic ions were measured by ion exclusion high performance liquid chromatography and ion exchange chromatography, respectively. Sludge compositions were characterized as fractions of total (TSS), volatile (VSS), and inorganic suspended solids (ISS). For details refer to Weissbrodt et al. (2012a).

8.2.4 *Molecular Analyses of Bacterial Community Compositions*

Bacterial community dynamics were investigated by terminal-restriction fragment length polymorphism (T-RFLP) analysis (Weissbrodt et al. 2012a). Biomass samples were homoge-nized by grinding, aliquoted in 1.5-mL Eppendorf tubes, and stored at − 20 °C. Eubacterial 16S rRNA gene pools were targeted and amplified by PCR with the labeled 8f (FAM-5′-AGAGTTTGATCMTGGCTCAG-3′) and unlabeled 518r (5′-ATTACCGCGGCTGCTGG-3′) primers.

Amplicons were digested with *Hae*III. Operational taxonomic units (OTU) were affiliated to phylotypes with the PyroTRF-ID methodology (Weissbrodt et al. 2012b) applied to biomass grab samples collected on days 2 and 59 (BC-SBR), 109 (PAO-SBR), and 398 (GAO-SBR). Greengenes (DeSantis et al. 2006) was used as mapping database in the PyroTRF-ID pipeline. The pyrosequencing dataset of a mature AGS sample (BC-II) originating from a precedent reactor operated under similar conditions was used to affiliate OTUs at later stage in the BC-SBR. T-RFLP profiles were generated with predominant OTUs (> 2%).

Biomass equivalents of OTUs were estimated by multiplying relative abundances with the VSS concentration. Richness and Shannon's H' diversity indices were computed in R (R Development Core Team 2008).

8.2.5 *Confocal Laser Scanning Microscopy Analyses of Flocs and Granules*

Structural transitions from flocs to granules were examined with CLSM. Collected biomass samples were washed twice in phosphate buffer saline (PBS) pH 7.4, and stored at 0–5 °C in paraformaldehyde 4% (m/v in PBS). Fluorescent dyes were screened for mapping bioaggregates (Additional File 8.2). Rhodamine 6G was optimal for time series. Glycoconjugates were detected in selected AGS samples of the BC-SBR by FLBA, according to Staudt et al. (2003) and Zippel and Neu (2011).

Spatial bacterial dynamics were followed by FISH-CLSM using rRNA oligonucleotide probes selected from probeBase (Loy et al. 2003) and targeting *Zoogloea*, "*Ca.* Accumulibacter" and "*Ca.* Competibacter" (BC-SBR), "*Ca.* Accumulibacter" and *Zoogloea* (PAO-SBR), as well as "*Ca.* Competibacter" (GAO-SBR) (Additional File 8.3). Samples were hybridized according to Nielsen et al. (2009). Granules smaller than 2 mm were cross-sectioned with a scalpel at ambient temperature in 0.5–1.0 mm deep CoverWell chambers mounted on microscopy slides (Life Technologies, Switzerland). Bigger granules were cryosectioned (80 μm) in a cryotome CM3050S (Leica, Germany) after freezing at − 26 °C in Tissue-Tek OCT compound (Sakura, The Netherlands).

The CLSM used was a TCS SP5X (Leica, Germany) equipped with upright microscope, an acusto optical beam splitter, and a supercontinuum light source. The system was controlled by the LAS AF software version 2.6.1. Samples were examined by the objective lenses 10×0.3 NA, 20×0.5 NA (overview), and 63×1.2 NA (high resolution). Excitation and emission wavelengths of fluorochromes are given in Additional Files 8.2 and 8.3. The CLSM reflection signal was collected as structural reference, and in order to distinguish voids from unstained regions. Multi-channel datasets were recorded in sequential mode to avoid cross-talk (Zippel and Neu 2011). Images were collected by optical sample sectioning over thicknesses and stepsizes of 5–215 and 1–8 μm, respectively.

Digital Leica Image Files (.lif) were processed in the Imaris software. Digital images were mainly represented as maximum intensity projections (MIP) with the Easy 3-D mode. Specific biofilm structures were examined in three dimensions either with the XYZ projection mode or the volume mode. The color intensity levels of the tagged image files generated in Imaris were slightly adjusted in the Photoshop software for improved resolution and contrast in the digital image datasets.

8.3 Results

8.3.1 Process and Bacterial Community Dynamics in the Bubble-Column SBR

The formation and maturation of granules in the BC-SBR operated under wash-out conditions were linked to dynamics in bacterial community compositions (Fig. 8.1A and Appendix A) and in process variables (Fig. 8.2) over 220 days. Start-up over the first 60 days was described in details in Weissbrodt et al. (2012a/Chap. 7).

During start-up, the biomass was washed out extensively and the sludge retention time (SRT) dropped after having decreased the settling time below 6 min (Fig. 8.2A–C). Operation with constant volumetric organic loading rate (OLR) of 247 ± 14 mg_{CODs} $cycle^{-1}$ $L_R{}^{-1}$ led to an extreme peak in the biomass-specific OLR of up to 685 mg_{CODs} $cycle^{-1}$ $g_{CODx}{}^{-1}$ between day 10 and day 30 (Fig. 8.2D). Initial nucleation occurred after 10 days. Wash-out prevailed up to day 30 when AGS began to accumulate. Whereas *Tetrasphaera* (OTU-223) declined from 26 (day 1) to < 2% (day 15), *Zoogloea* (OTU-195) dominated early-stage granules up to day 90 (37–79%, 6.7 g_{VSS} L^{-1}) (Figs. 8.1A and 8.2E). AGS accumulated stepwise up to 11.7 g_{TSS} L^{-1} with 90% of volatile suspended solids (VSS) and to a 50-cm bed height, together with progressive increase in the SRT (Fig. 8.2B), while the biomass specific OLR stabilized below 20 mg_{CODs} $cycle^{-1}$ $g_{CODx}{}^{-1}$. During this period, *Zoogloea* predominated over "*Ca.* Accumulibacter"-related *Rhodocyclaceae* affiliates (OTUs 214–215) and "*Ca.* Competibacter"-related *Gammaproteobacteria* (OTU-239) (Fig. 8.2E, F). This correlated positively with low levels of anaerobic acetate uptake (0–30%), anaerobic orthophosphate release to acetate uptake ($Y_{P/C,An} = 0$–0.10 P-mol $C\text{-}mol_{Ac}{}^{-1}$),

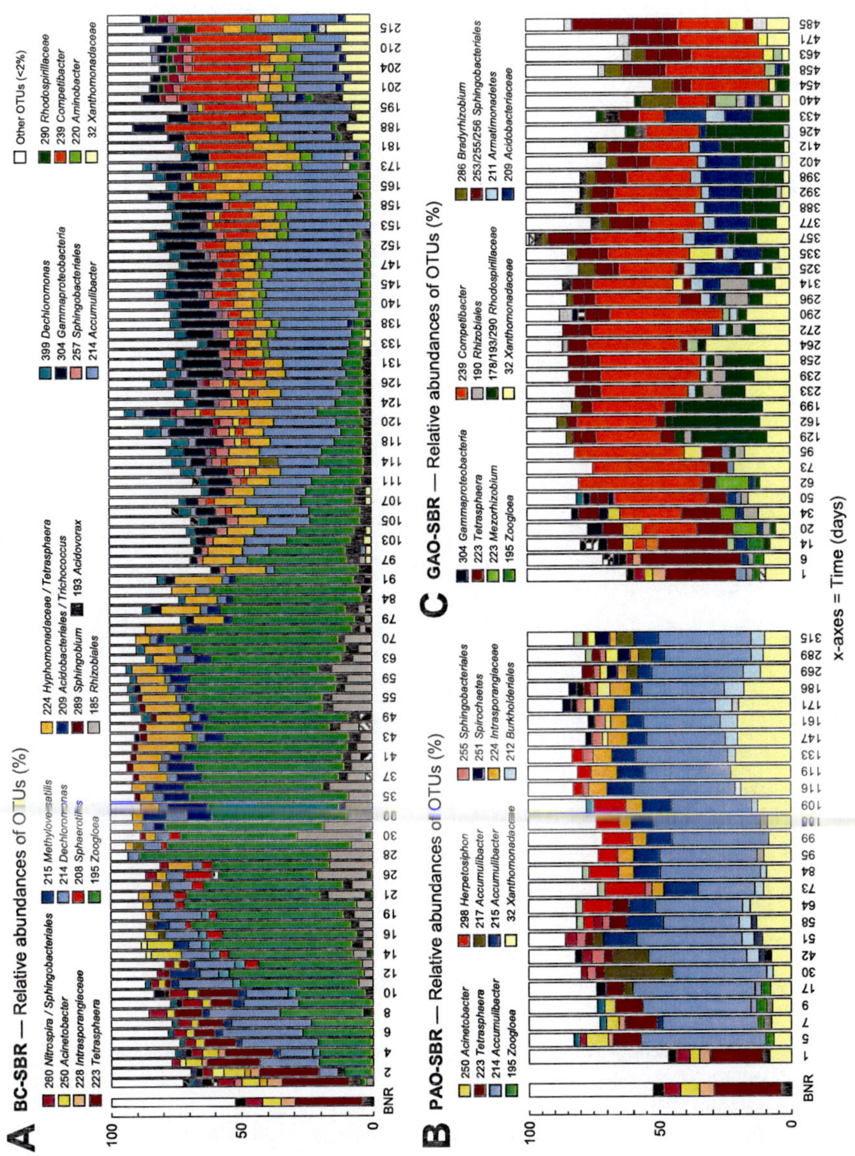

Fig. 8.1 Bacterial community dynamics in BC-SBR (**A**), PAO-SBR (**B**), and GAO-SBR (**C**). The data of the two latter reactors were recalled from Weissbrodt et al. (2014/Chap. 6). Phylotypes were identified with PyroTRF-ID

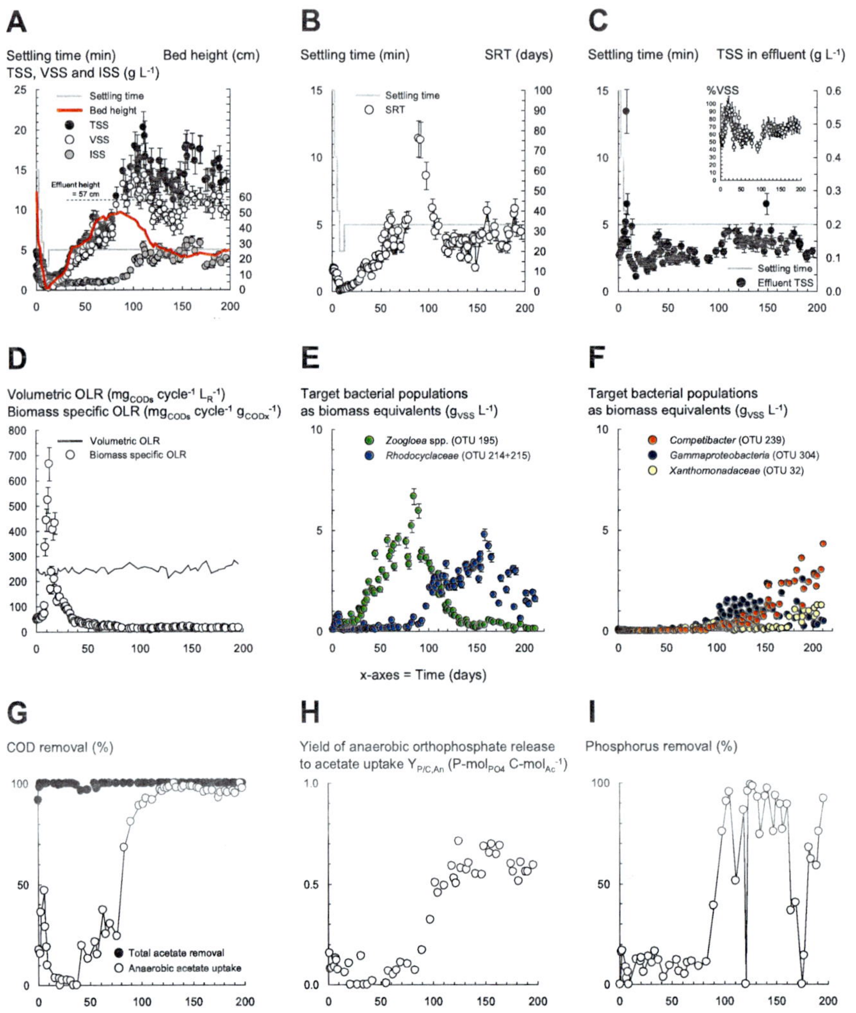

Fig. 8.2 Dynamics in process variables during the operation of the BC-SBR under wash-out conditions. Evolution of the concentration of total (TSS), volatile (VSS, i.e. biomass), and inorganic suspended solids (ISS) during granule formation and maturation (**A**). Evolution of the sludge retention time (SRT) under dynamic conditions (**B**). Concentration and composition of the TSS in the effluent wastewater (**C**). Variations of the biomass specific organic loading rate (OLR) during operation under wash-out conditions with constant volumetric OLR (**D**). Shifts in predominant bacterial populations during granule formation and maturation (**E, F**). Dynamics in dephosphatation (**G–I**), nitrification and nitrogen removal (**J–K**) activities measured from inoculation to maturation. Rough estimation of the evolution of the sum of OTUs possibly affiliating with denitrifiers, according to the physiological properties reported in literature for the underlying phylotypes detected with PyroTRF-ID (**l**)

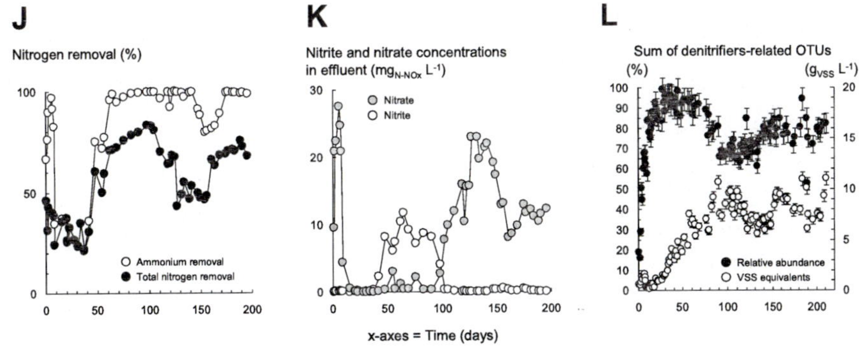

Fig. 8.2 (continued)

phosphorus removal (10 ± 3%), and inorganic suspended solids (ISS, 0.9 ± 0.1 g_{ISS} L^{-1}, 15 ± 3%) (Fig. 8.2G–I).

Dephosphatation improved after day 90 (Fig. 8.2I) together with a shift in predominance from *Zoogloea* (< 5%, 0.5 g_{VSS} L^{-1}) to "*Ca.* Accumulibacter" (33 ± 5%, 3.2 ± 0.6 g_{VSS} L^{-1}) (Fig. 8.2E, F). Full anaerobic acetate uptake (Fig. 8.2G), high apparent $Y_{P/C,An}$ (0.60 ± 0.07 P-mol C-$mol_{Ac}{}^{-1}$) (Fig. 8.2H), and an increase in ISS (4.8 ± 0.8 g_{ISS} L^{-1}, 32 ± 1%) (Fig. 8.2A) were detected at the same time. Dephosphatation reached almost 100% at day 100. Without regular purge of excess sludge, accumulation of AGS correlated with transiently high SRT of up to 75 days (Fig. 8.2B). Despite further accumulation up to 20 g_{TSS} L^{-1} after day 100, the bed remained at 40–50 cm (Fig. 8.2A). After day 110, the bed dropped below 30 cm together with TSS decrease in the reactor (< 15 g_{TSS} L^{-1}) and increase in the effluent (0.155 ± 0.028 g_{TSS} $L_{Eff}{}^{-1}$), and spontaneous SRT stabilization at 24 ± 5 days. After day 150, dephosphatation became instable (82 ± 25%) despite efficient anaerobic acetate uptake (96 ± 2%). $Y_{P/C,An}$ slightly decreased to 0.50 P-mol C-$mol_{Ac}{}^{-1}$. "*Ca.* Competibacter" (OTU-239, 20–30%, 3.6 g_{VSS} L^{-1}) were more abundant than "*Ca.* Accumulibacter" (10–15%, 0.6–1.9 g_{VSS} L^{-1}) from day 173 on.

Nitrification and nitrogen removal were inactive during wash-out, but started to recover since day 40 on by reaching conversions of up to 90 and 70%, respectively, together with the accumulation of AGS and with the lowering of the biomass specific OLR (Fig. 8.2J). Nitrite and then nitrate accumulated between days 50–150 (Fig. 8.2K). Only one nitrifier-related OTU was present in mature granules above the minimum detection limit of the T-RFLP method. OTU-260 affiliating with *Sphingobacteriales* and nitrite-oxidizing *Nitrospira* increased from 0.5 ± 0.4% (days 30–60) to 1.6 ± 0.7% (days 110–200). Pyrosequencing analyses were sensitive enough to reveal shifts in nitrifying populations. The flocculent sludge sample collected on day 2 comprised AOB related to *Nitrosomonas* (0.12%), *Nitrosococcus* (0.24%), and *Nitrosovibrio* (0.06%), and NOB related to *Nitrobacter* (0.06%) and *Nitrospira* (1.02%). Only the ammonium-oxidizing *Nitrosospira* (0.03%) was detected in the early-stage AGS on day 59. The mature AGS sample BC-II was composed of *Nitrosomonas* (0.1%), *Nitrospira* (0.2%), and *Nitrococcus*-related *Gammaproteobacteria* detected in a significantly high abundance of 6.2%. Efficient nitrogen removal between days 50–100 (71–83%) and 150–200 (63–75%) correlated with high totals of denitrifiers-related OTUs of up to 10 g_{VSS} L^{-1} (Fig. 8.2L).

8.3.2 Process and Bacterial Dynamics in the Stirred-Tank PAO-SBR and GAO-SBR

Operation of the PAO-SBR under control of the anaerobic contact time (5–3 h), biomass concentration (> 1.5 g_{VSS} L^{-1}), SRT (stabilization from 5 to 8 days in 15 days), and volumetric OLR from 15 to 200 mg_{CODs} $cycle^{-1}$ L_R^{-1} (Fig. 8.3A–D) ensured full anaerobic propionate uptake, and resulted in the rapid enrichment of "*Ca.* Accumulibacter" in the activated sludge (48% on day 5) (Figs. 8.1B and 8.3E).

Enhanced orthophosphate-cycling activities were obtained as displayed by the high conductivity amplitudes of up to 350 μS cm^{-1} (Fig. 8.3F) together with high $Y_{P/C,An}$ of 0.56–0.64 P-mol $C\text{-}mol_{Pr}^{-1}$. Steady state was reached after 15 days, with gradual SRT stabilization by purging excess sludge. Interestingly, fast-settling biomass nuclei (< 500 μm) formed after 20 days. By decreasing the settling time from 60 to 10 min to save cycle time, nuclei evolved towards 1–2 mm large granules over the next 40 days. *Tetrasphaera* declined from 22 to 3% within 30 days, but remained between 2 and 10% in the system (Fig. 8.1B). The enrichment displayed constant predominance of "*Ca.* Accumulibacter" (41 ± 6%), and significant level of *Xanthomonadaceae* (OTU-32, 7–20%) right from start-up. *Herpetosiphon* (OTU-298) amounted to 7–17% on days 60–120. *Zoogloea* was only detected up to day 20 with relative abundances below 4%.

In the GAO-SBR, biomass remained flocculent over more than 450 days. With constant volumetric OLR of 200 mg_{CODs} $cycle^{-1}$ L_R^{-1}, full anaerobic acetate uptake was obtained from day 20 onward (data not shown). *Tetrasphaera* dominated over the first 2 weeks (31–47%), and remained at 5% abundance up to day 200 (Fig. 8.1C). "*Ca.* Competibacter" prevailed in the enrichment (22–59%). Despite operation at steady-state, accompanying guilds displayed quite high dynamics. For example abundances of *Alphaproteobacteria* related to *Rhodospirillaceae* (OTUs 178, 193 and 290, 2–40%), *Rhizobiales* (OTU-190, 2–12%), *Bradyrhizobium* (OTU-285, 2–8%), and *Sphingomonas* (OTU-287, 2–20%), as well as *Acidobacteriaceae* (OTU-209, 7–23%) and *Thiobacillus* relatives (OTU-216, 21% on day 433) fluctuated considerably during rector operation. After more than 450 days, fast-settling nuclei (< 500 μm) were observed in the system, and evolved towards 1–2 mm granules from day 480 onward. Granulation correlated with transient over-aeration caused by biofilm growth on DO sensors. An increase in *Sphingobacteriales* relatives up to more than 40% (OTUs 253–256) was observed during granule formation and *Zoogloea*, "*Ca.* Accumulibacter", and *Rhodocyclaceae* relatives were almost absent (< 5%).

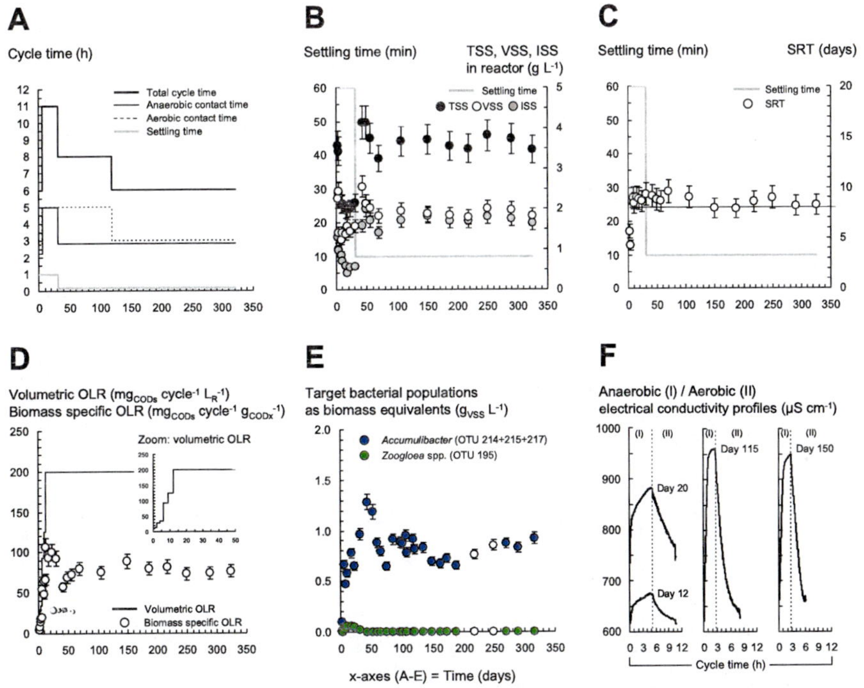

Fig. 8.3 Process and bacterial dynamics in the PAO-SBR. The reactor was operated with proper control of anaerobic and aerobic phases (**A**), with low variations in total (TSS), volatile (VSS) and inorganic suspended solids (ISS) (**B**), and with stable sludge retention time (SRT) (**C**). The volumetric organic loading rate (OLR) was increased stepwise in 15 days (**D**) for rapid and preferential selection of "*Ca.* Accumulibacter" (**E**) with enhanced anaerobic-aerobic orthophosphate-cycling activities as displayed by electrical conductivity profiles (**F**). On **E** the two points on days 217 and 249 were estimated based on the average relative abundances of "*Ca.* Accumulibacter"—(43 ± 5%) and *Zoogloea*—(0%) related OTUs at steady-state

8.3.3 Structural and Bacterial Transitions from Flocs to Granules in the Bubble-Column SBR

CLSM examinations with Rhodamine 6G staining revealed that amorphous flocs (150–200 µm) initially present in the BC-SBR (Fig. 8.4, day 1) underwent transformation by swelling of microbial colonies around flocs (day 9), followed by granulation of nuclei with dense rounded structures of 450–750 µm (day 23).

Early-stage granules (850–1500 µm) displayed smooth and folded biofilm structures (day 30). Between days 50–140, round and compact microcolonies (10–100 µm) followed by larger ones (120–300 µm) proliferated from the inner core of granules outwards. Granular biofilm detachment was detected on day 60, and increased during maturation (day 112). Mature granules comprised internal voids, large biofilm clusters, and slimy interfacial matrices (days 209 and 218).

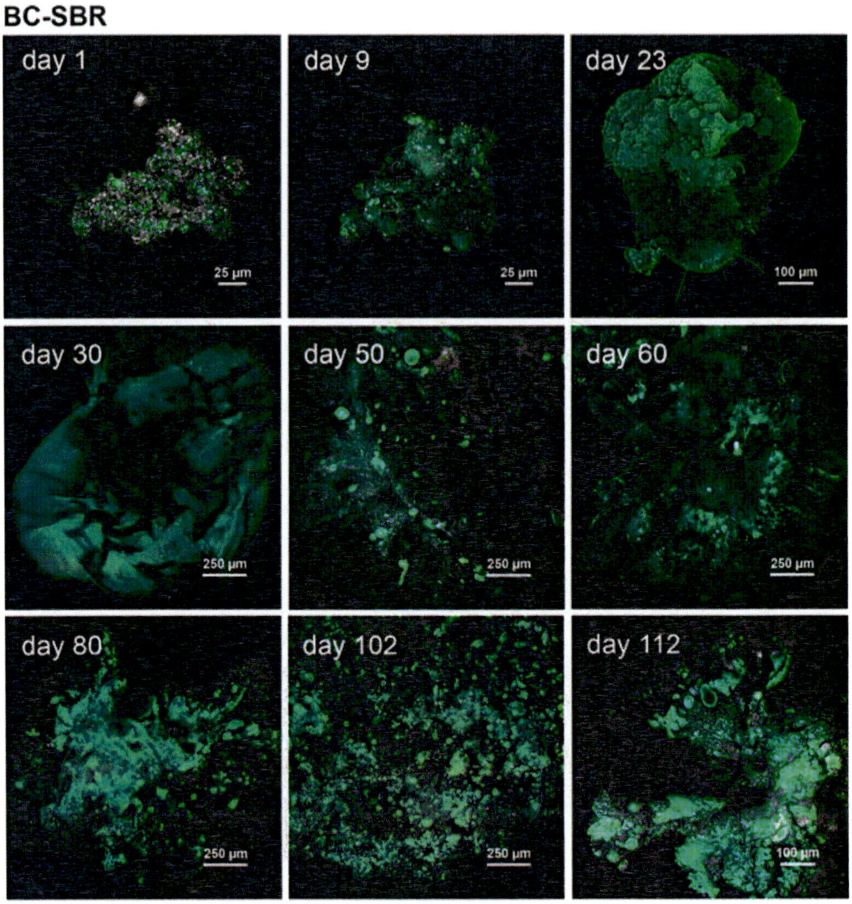

Fig. 8.4 Temporal evolution of the architecture of bioaggregates from activated sludge flocs (day 1) to early-stage (day 23) and mature granular biofilms (day 102) in the BC-SBR operated under wash-out conditions. CLSM datasets were recorded on full bioaggregates from samples taken from day 1 to day 23, as well as on day 218. Granules from samples taken between day 30 and 140 were analyzed on cross-sections. The sample taken on day 209 was analyzed as 80-µm cryosection. The green fluorescent dye Rhodamine 6G was used to map cells and biofilm matrices. In 8 bit data sets, 256 green levels were allocated to this dye. The reflection signal was used as reference with 256 grey/white color allocation. On day 9, swelling of microbial colonies around the floc structure can be observed. Early-stage granule nuclei on day 23 were 4–5 times bigger than flocs, and displayed compact biofilm aggregation. On day 30, early-stage granules were composed of a continuous biofilm displaying homogenous cell distribution and folded structures. From day 30 to day 102, the internal architecture of granules evolved with the proliferation of dense microcolonies from the granule core outwards. Larger microbial clusters appeared in the structure of granules between day 112 and 140. Detachment phenomena contributed to the heterogeneous structure of mature granules (days 112 and 209). After more than 200 days, mature granules exhibited aggregation of dense biofilm clusters, internal voids, and eroded surface slimy structures

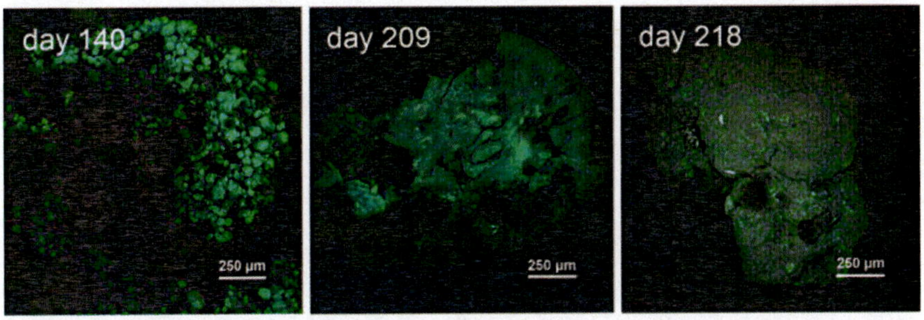

Fig. 8.4 (continued)

Different glycoconjugates were detected by fluorescent lectin-binding analysis (FLBA) of cross-sections of granules collected on days 85 and 105 (Fig. 8.5A and Additional File 8.2). Lectin characteristics are provided in Additional File 8.2. FLBA showed (i) embedded continuous matrices revealed by STA, PHA-L, and IAA lectins, (ii) matrices surrounding larger microbial clusters (HAA, LEA, SBA), (iii) matrices of microcolony interfaces (WGA, LcH), and (iv) direct binding to cell surfaces (ConA). HAA also revealed filamentous sheaths, and ConA the outwards palisade-like orientation of the biofilm continuum that surrounded dense colony clusters. Further interesting structures were detected in the architecture of mature granules with the use of other fluorescent probes (Fig. 8.5B), e.g. spherical dense colony clusters of about 140 μm stained with SYPRO Red, aggregation of cells containing bright reflecting intracellular storage compounds after staining membranes with FM4-64, and extracellular DNA stained with DDAO.

FISH-CLSM confirmed T-RFLP analyses and provided information on spatial dynamics of active bacteria inside the structure of bioaggregates (Fig. 8.6). *Zoogloea* proliferated during granulation as microcolonies of 20–45 μm swelling around flocs (day 6), and formed the early-stage granular biofilm continuum (day 50). It can be observed on the latter picture that early-stage granules displayed loose core and dense surface aggregation. *Zoogloea* disappeared from the biofilm architecture of mature granules, as displayed on day 170. PAOs and GAOs were present as microcolonies (< 10 μm) in the flocculent sludge. After 100 days, PAOs and GAOs established over *Zoogloea* from the granule cores outwards by forming large and dense clusters (> 300 μm) exhibiting bright and low reflection, respectively. After initial presence in flocs as compact microcolonies of up to 30 μm (day 6), AOOs proliferated as dense microcolonies (20–70 μm) across granules after 105 days and into wider biofilm matrices near the granule surface after 170 days.

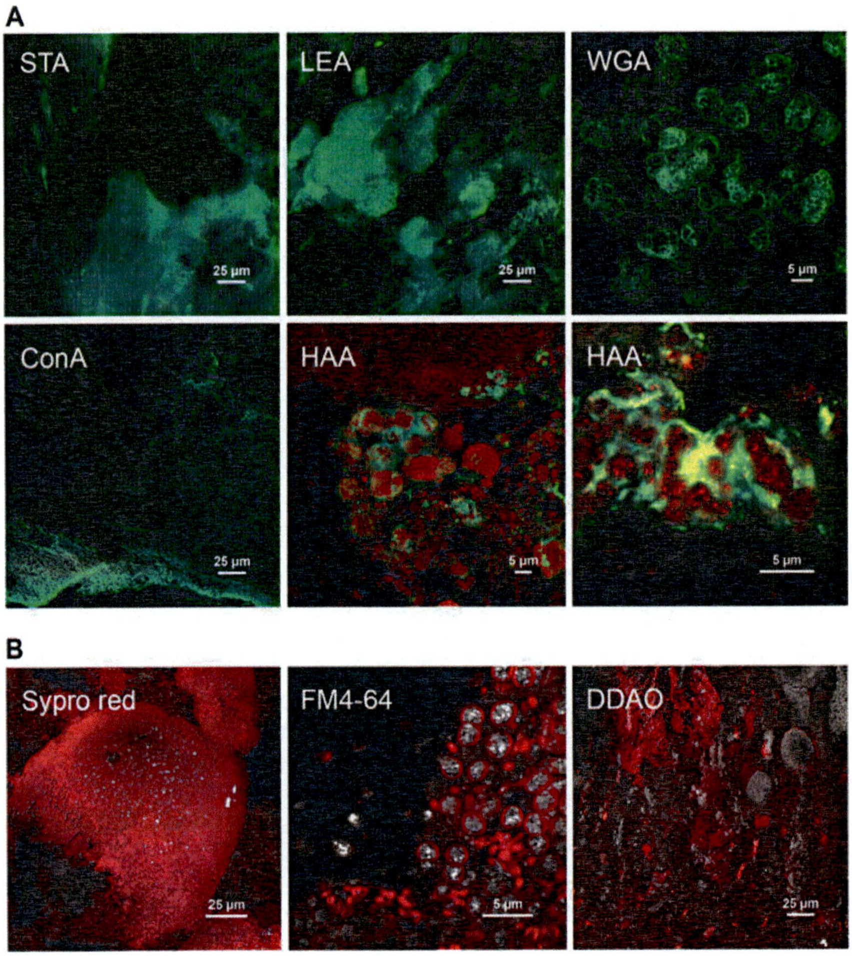

Fig. 8.5 Examples of cellular and extracellular features detected in cross-sectioned granular biofilms collected after 85 days in the BC-SBR (**A**). Selected glycoconjugate signals by means of FLBA. The STA lectin revealed the presence of a wide glycoconjugate matrix across the granule sections. Biofilm growth directions from the inner core to the outer sphere can be observed on the left part of the image with the growth lines displayed with the lectin staining. LEA showed dense microcolonies surrounded by a glycoconjugate matrix. WGA clearly showed glycoconjugate matrices surrounding specific types of microcolonies. ConA stained cell surface glycoconjugates also indicating biofilm growth lines and directions. HAA was used in combination with SYTO 60 and FM4-64 fluorescent probes. Color allocations: lectins binding to glycoconjugates (green), cell staining (red). Further structures detected with additional fluorescent probes in the architecture of mature granules collected after 111 days in the BC-SBR (**B**). Dense spherical microbial clusters stained with SYPRO Red. Detection of microbial cells comprising bright reflecting intracellular storage compounds (cell membranes were stained with FM4-64). Presence of extracellular DNA stained with DDAO in the biofilm matrix and around cell clusters. Color allocations: fluorescent probes (red), reflection signal (grey)

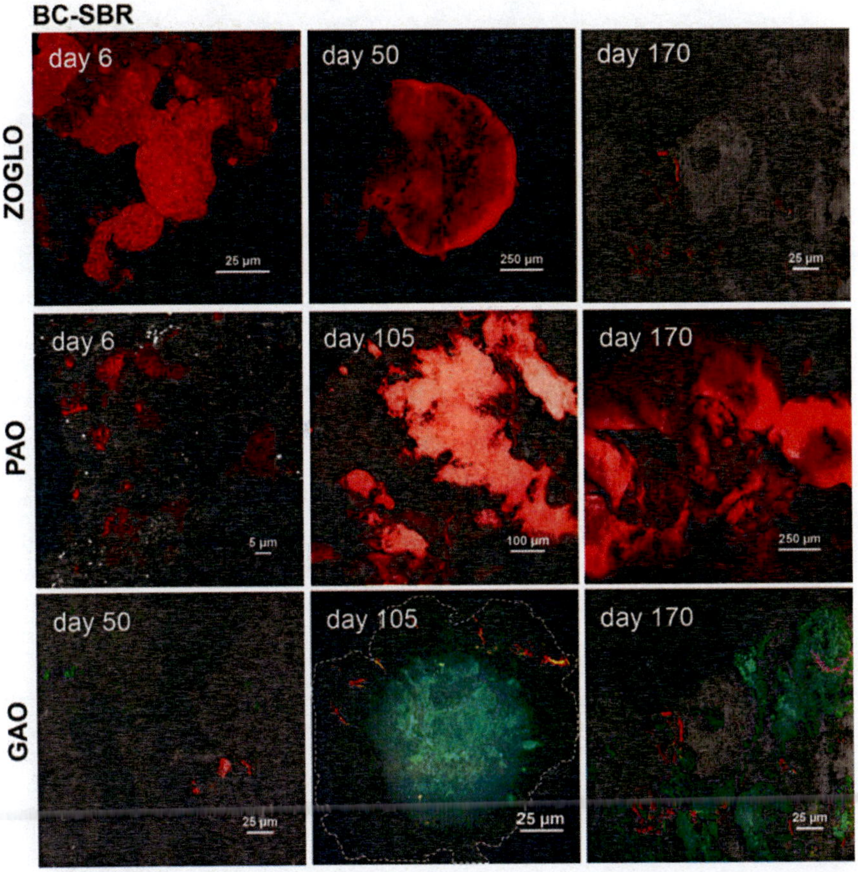

Fig. 8.6 FISH-CLSM analysis of spatial dynamics of *Zoogloea* spp. (ZOGLO, red), "*Ca.* Accumulibacter" affiliates (PAO, red), "*Ca.* Competibacter" relatives (GAO, green), and ammonium-oxidizing organisms (AOO, green) during granule formation and maturation in the BC-SBR. On day 6, colonies of *Zoogloea* spp. swell around the floc structure. Microcolonies of PAOs and AOOs were also detected inside flocs on the same day. On day 50, early-stage granules were composed of a smooth biofilm continuum dominated by *Zoogloea* spp. PAOs and GAOs were only low abundant on this day. The green-fluorescent GAO gene probe was used in combination with the red fluorescent PAO (day 50) and ZOGLO (days 105 and 170) gene probes. Biofilm growth lines from granule core outwards can also be observed on the *Zoogloea*-related picture. Denser cell aggregation was detected at the edge of the granule as well. After 105 days, PAOs and GAOs proliferated inside mature granules from inner core outwards. AOOs proliferated across granules as dense microcolony clusters. After 170 days, granules were predominated by PAOs and GAOs exhibiting bright and low reflection, respectively. AOOs were also present as wider population matrices near granule edges. *Zoogloea* spp. were only detected in low abundances in interstices between the different other microbial clusters

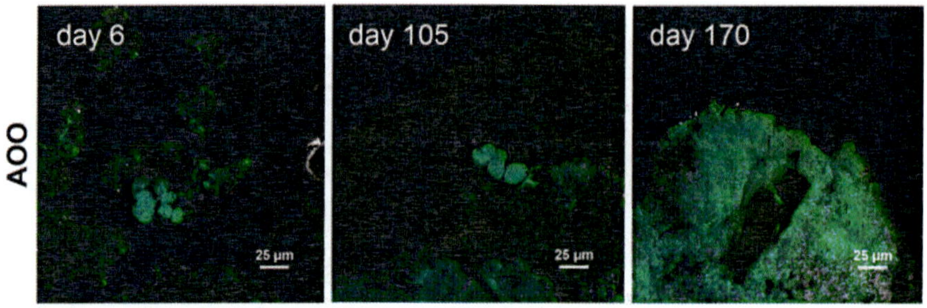

Fig. 8.6 (continued)

8.3.4 Granulation in the Stirred-Tank PAO-SBR and GAO-SBR

According to Fig. 8.7, granulation occurred in the PAO-SBR successively (i) by floc smothering (day 30), (ii) by proliferation of round and dense clusters of 90–180 μm in flocs (day 49), by formation of smooth and dense nuclei that evolved up to 1.3-mm early-stage granules (day 62) and 1.5–2.0 mm mature granules (day 205). Mature granules displayed folded biofilm structures that contained aggregation of cells in dense biofilm clusters (day 205 and 215).

FISH-CLSM measurements confirmed that *Zoogloea* were almost absent during granulation in the PAO-SBR, by forming only low abundant colonies (10–20 μm) in flocs (Fig. 8.8, day 5) and patches at the surface of granule nuclei (day 31). *Zoogloea* were only present in biofilm interstices after 62 days. PAOs dominated during granulation by forming small dense clusters (10–120 μm) in smooth globular flocs (day 31), and larger dense clusters (<500 μm) in early-stage granules (day 62). PAOs occupied the entire cross-sections of mature granules after 108 days. PAO cells (1.5 μm) had a high content of bright-reflecting intracellular storage compounds, and were more densely aggregated over the first 200–300 μm from the granule surface.

In the GAO-SBR, flocs turned into compact nuclei of 200–600 μm after 484 days that evolved up to larger 1.3–1.5 mm granules (Fig. 8.7, day 496). Granules present in the GAO-SBR were heterogeneous conglomerates of large and dense clusters of 50–350 μm comprising 2.0–2.5 μm sized cells (day 510). FISH-CLSM analyses revealed that granulation occurred in this reactor under predominance of GAOs across granules (Fig. 8.8).

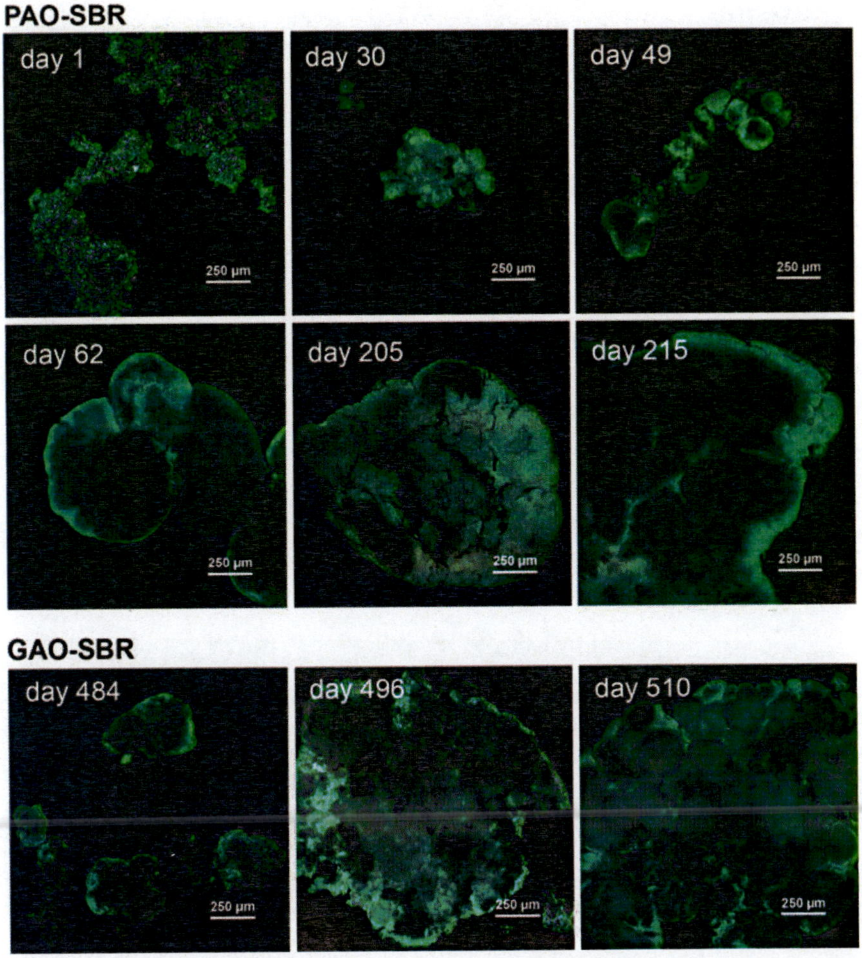

Fig. 8.7 Temporal evolution of the architecture of bioaggregates from flocs to early-stage granules up to mature granules in the stirred-tank PAO-SBR and GAO-SBR operated under steady-state conditions. Granule formation occurred in the PAO-SBR by smothering of flocs (day 30) and proliferation of dense microbial clusters around the floc structure (day 49). Dense granule nuclei obtained after 62 days evolved towards mature granules exhibiting folded biofilm structures (day 205) and aggregation of big, dense and homogeneous clusters (day 215). In the GAO-SBR dense nuclei (day 484) evolved in 2 weeks toward 4–5 times larger granular biofilms (day 496). On day 510, granules were characterized as heterogeneous conglomerates of dense microbial clusters

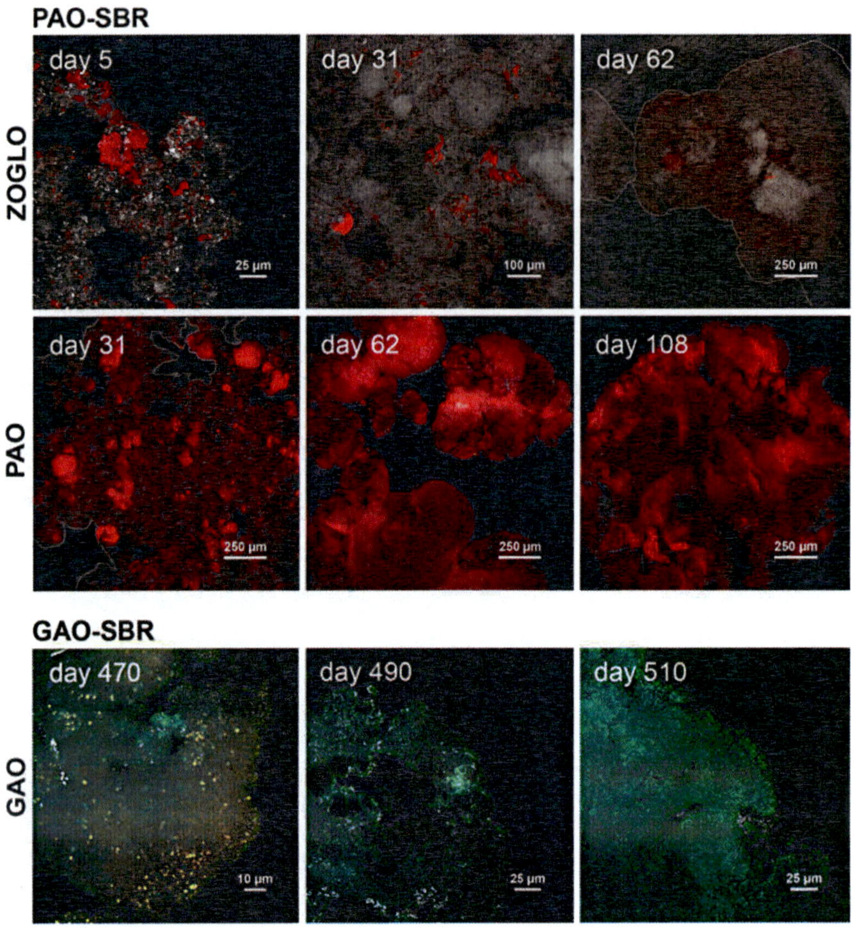

Fig. 8.8 FISH-CLSM analyses of spatial dynamics targeting bacterial populations in the PAO-SBR and in the GAO-SBR. *Zoogloea* spp. (ZOGLO, red) were never abundant during granulation in the PAO-SBR. Active "*Ca.* Accumulibacter" affiliates (PAO, red) predominated during granule formation in the PAO-SBR, and formed the dense bacterial clusters that proliferated in the floc structure (day 31). Later it led to the formation of heterogeneous dense granules composed of bright reflecting clusters (day 62). The image of day 108 exhibits "*Ca.* Accumulibacter" coverage over the whole granule cross-section. "*Ca.* Competibacter" (GAO, green) was predominant during granulation in the GAO-enrichment. On day 470, the microbial cell membranes were stained with the fluorescent probe FM4-64 (red/yellow) after hybridization with the GAO gene probe. On day 490, the granules were composed of different GAO populations present in a continuous matrice and in dense microbial clusters, respectively. On day 510, granules displayed predominance of GAO populations with round cells

8.3.5 *Correlation Between Granule Structures and Predominant Populations*

Based on the different structural time series described above for the three SBRs, the structure of granules correlated with the predominant organisms involved. A compilation of main structural differences is provided in Fig. 8.9.

Over the first 60 days in the BC-SBR, *Zoogloea*-dominated granules displayed smooth external and internal granular biofilm architectures (A, B). In this case, granules displayed homogeneous embedment of cells in a biofilm matrix growing outwards like petals (C). Granules dominated by slower-growing PAOs and GAOs displayed heterogeneous aggregation of population clusters (D–F). In the GAO-SBR, cauliflower-like structures were observed from the surface of granules (E). Different cell distributions were observed in cross-sections of early-stage granules collected on day 50 (G) and of mature granules collected on day 111 (H) in the BC-SBR. Mature granules dominated by PAOs in the BC-SBR displayed heterogeneous internal architectures closer to the ones of granules cultivated in the PAO-SBR (I). Additional peculiar structures presented in Fig. 8.9 are the initial microcolonies that swelled in flocs after 6 days in the BC-SBR (J), the filamentous populations that were detected after 11 days in the flocs of the PAO-SBR (K), and the slimy matrices present at the surface of mature granules after 209 days in the BC-SBR (L).

8.3.6 *Three-Dimensional Analyses of Granule Structures*

Specific biofilm and microbial structures examined by CLSM were finally analyzed in three dimensions (3-D) either as XYZ projections or volumetric presentations (Fig. 8.10).

The early-stage granule obtained after 20 days in the BC-SBR and previously presented in Fig. 8.9A displayed surface roughness with biofilm protuberances and valleys (A). During the transition from early-stage to mature granules after 80 days, the internal architecture of granules was composed of various types of microcolonies such as dense spherical microbial clusters proliferating from the granule core and populations aggregated in looser amorphous structures (B). After 111 days, a zoom on a microcolony revealed a spherical shape and denser cell aggregation at the edge of the microcolony (C). On the same day, dual staining with SYTOX Green (nucleic acids label) and Nile Red (label for hydrophobic components such as poly-β-hydroxyalkanoates) enabled to detect different types of cells based on their membrane and intracellular properties (D). The volumetric presentation of the granular aggregates obtained in the GAO-SBR on day 484 exhibit the cauliflower-like or mushroom-like cluster evolving from the granule core outwards (E). Other volumetric presentations display the 3-D orientation of the glycoconjugate matrices detected with the

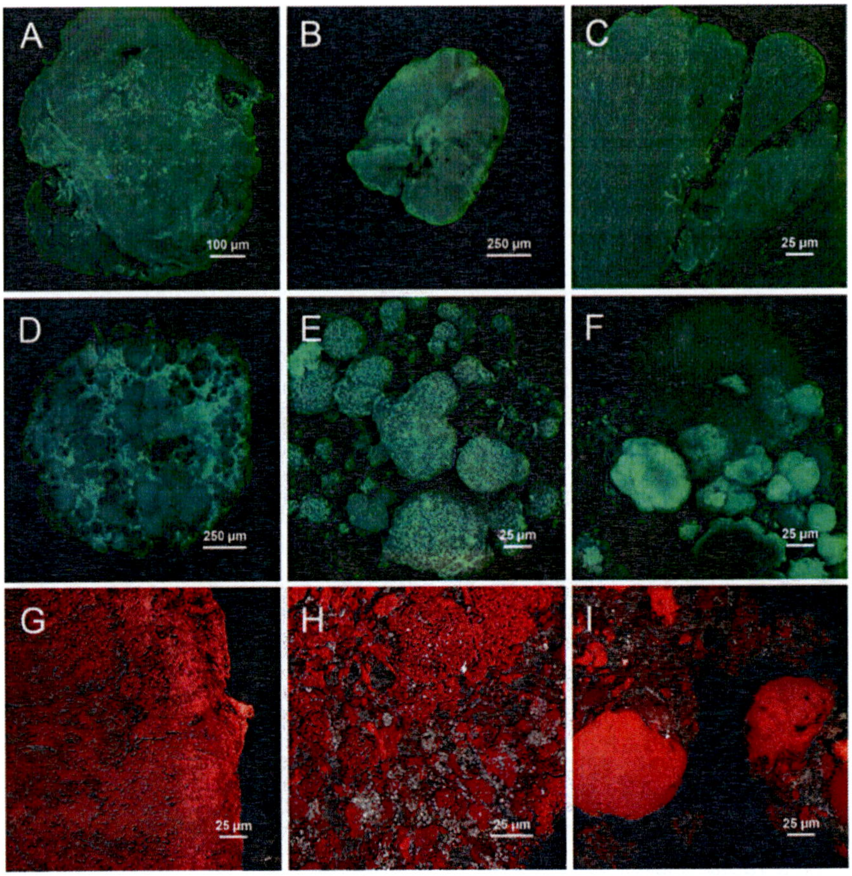

Fig. 8.9 Main differences in the architectures of granules depending on the predominant bacterial populations involved. Under predominance of *Zoogloea* spp., smooth granular biofilm continua were obtained in early-stage granules collected after 20 days (**A**) and 60 days (**B**) in the BC-SBR. The formation and the outgrowth direction of biofilm petals can be observed in (**C**). The biofilm was composed of a homogeneous cell distribution in a gel matrix. Under conditions favoring PAO and GAO growth heterogeneous granular biofilm conglomerates were obtained as exemplified by day 503 (**D**) and day 484 (**E**) in the GAO-SBR, and by day 140 in the BC-SBR (**F**). Color allocation A-F: green = Rhodamine 6G. The cross-section of an early-stage granule collected on day 50 in the BC-SBR and hybridized with *Zoogloea*-targeted gene probes (**G**) revealed a more homogeneous cell composition than the highly-heterogeneous cross-section of a mature granule collected on day 111 in the same reactor and stained with SYPRO Red (**H**). **G** Also provides information on the palisade-like biofilm outgrowth lines, and on the denser cell aggregation on the first 75 µm from the granule edge. The heterogeneous structure of mature granules obtained in the BC-SBR, that is actually dominated by PAOs on day 111, is similar to the one of granules cultivated in the PAO-SBR after 31 days (**I**, color allocation: red = PAO gene probe). Further structures were detected by staining with Rhodamine 6G, such as the globular microcolonies swelling around flocs after 6 days in the BC-SBR (**J**), filaments detected in the bioaggregates of the PAO-SBR after day 11 (**K**), and slimy structures at the surface of mature granules collected after 209 days in the BC-SBR (**I**)

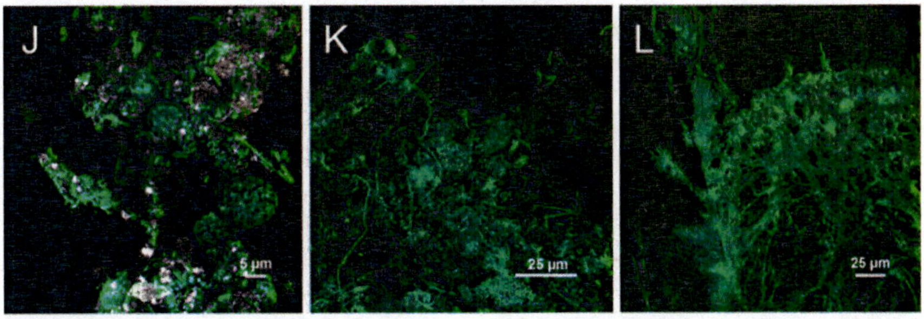

Fig. 8.9 (continued)

WGA lectin around microcolony interfaces (F) and the finger-like population structures detected in early-stage nuclei obtained after 15 days in the BC-SBR that was dominated by *Zoogloea* (G).

8.4 Discussion

8.4.1 Granulation Can Occur Under Wash-Out and Steady-State Conditions

Granulation of activated sludge occurred under wash-out and steady-state conditions with different SBR designs. In this study, wash-out conditions in the BC-SBR resulted in fast granulation after 2 weeks, which agrees with the granule formation periods reported by Beun et al. (1999). Mosquera-Corral et al. (2011) have succeeded to cultivate granules in a fully aerated stirred-tank SBR by imposing wash-out with short settling time (1 min) and HRT (6 h) right from start-up.

In the present study, spontaneous granulation occurred in the stirred-tank SBRs run at steady-state to cultivate PAO- and GAO-enrichments. This phenomenon has also been observed in various stirred-tank SBRs operated for BNR (Dangcong et al. 1999; de Villiers and Pretorius 2001; Dulekgurgen et al. 2003; You et al. 2008; Barr et al. 2010). Granulation in the GAO-SBR correlated with over-aeration that could have induced a physiological stress by increased endogenous respiration processes. Although 3-times slower than with wash-out, spontaneous granulation occurred in the PAO-SBR operated at steady-state with controlled SRT (8–10 days), mild settling and 12-h HRT for selection of active "*Ca.* Accumulibacter". Fast-settling nuclei formed despite initially long 60-min settling periods. Such settling enhancement can be explained by the fact that dephosphatating sludge is more dense than conventional activated sludge because of the presence of organic and inorganic intracellular polymers inside PAO cells (Schuler et al. 2001). The decrease of the settling time from 60 to 10 min probably further selected for faster settling aggregates and full granulation.

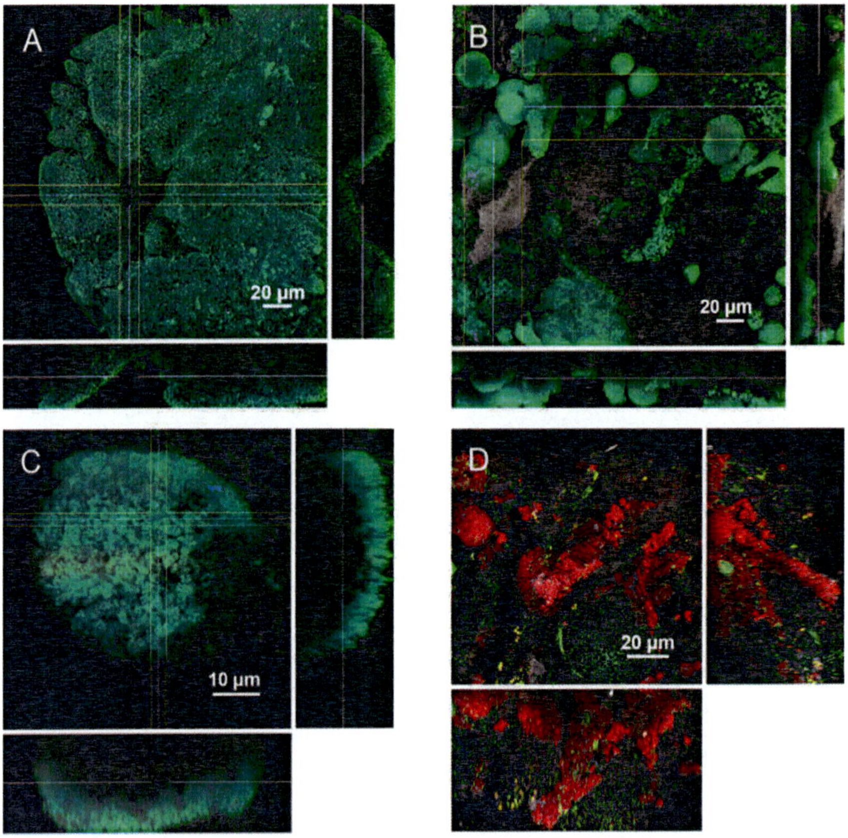

Fig. 8.10 Three-dimensional examinations of specific structures of granular biofilms with XYZ projections (**A**–**D**) and 3D volume projection (**E**–**G**). The smooth early-stage granular biofilms on day 20 in the BC-SBR exhibited a surface roughness with biofilm protuberances and valleys (**A**). During the transition on day 80 from early-stage to mature granules in the BC-SBR, the granular biofilm was composed of heterogeneous microbial clusters proliferating from the granule core outward (**B**). The spherical shapes of the dense microcolonies can be observed. A zoom on a spherical microcolony in a mature granule on day 111 in the BC-SBR revealed denser cell aggregation at the edge of the colony (**C**). Dual staining with SYTOX Green (nucleic acids) and Nile Red (hydrophobic cellular components and poly-β-hydroxyalkanoates) revealed different cell types in the heterogeneous structure of mature granular biofilms on day 111 in the BC-SBR (**D**). The cauliflower-like structure of heterogeneous granules present in the GAO-SBR is comparable to the mushroom-like structures of biofilm growth on a solid substratum under substrate-limiting conditions (**E**, grid size of 20 μm). 3D distribution of glycoconjugate matrices highlighted by the WGA lectin surrounding interfaces of microcolony clusters (**F**, grid size of 5 μm). Finger-like zoogloeal population structures were detected in the architecture of aggregates collected on day 15 in the BC-SBR (**G**, grid size of 5 μm). Color allocation in **A**–**C** and **E**–**G**: Rhodamine 6G (green) and reflection (gray/white)

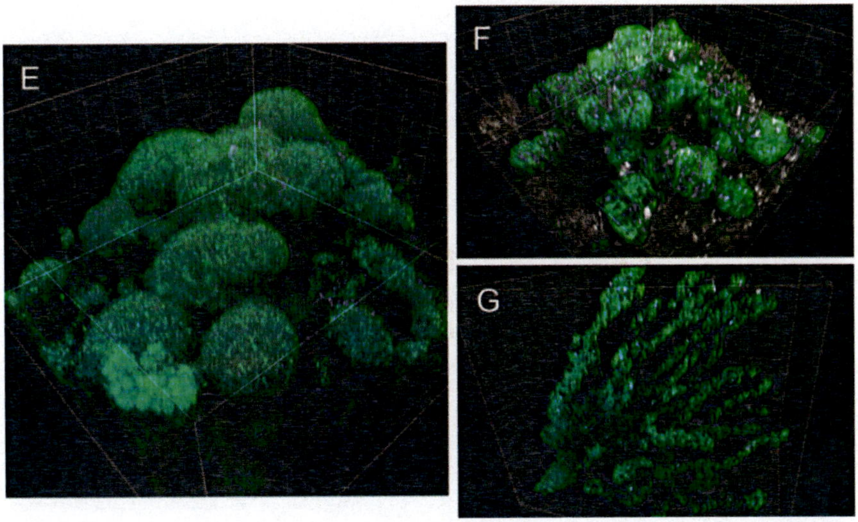

Fig. 8.10 (continued)

Hence, although wash-out conditions are efficient for fast granulation, granulation can also occur spontaneously. In addition, wash-out conditions can lead to prolonged deteriorated BNR since proper conditions to select for active PAOs and nitrifiers are often not fulfilled (Weissbrodt et al. 2012a). Intermediate conditions between wash-out and steady-state should therefore be investigated for rapid formation of granules with good BNR activities.

8.4.2 Bacterial Selection Mechanisms During Granulation

The complex and operation-specific bacterial community dynamics that were highlighted during granulation in the BC-SBR and PAO-SBR were compiled in a conceptual ecological model. In the BC-SBR (Fig. 8.11A), shifts in predominant bacterial populations correlated with the evolutions of process variables. As long as wash-out prevailed in the BC-SBR, important leakage of VFA into aeration phases was detected. This resulted from the application of fixed volumetric OLR and anaerobic plug-flow feeding phase lengths. The presence of COD equivalents under aerobic conditions selected for *Zoogloea*, what was already shown in Weissbrodt et al. (2012a).

When AGS accumulated and enabled low biomass specific OLR and longer anaerobic contact time, "*Ca.* Accumulibacter" proliferated and efficient dephosphatation was observed. Later deterioration of dephosphatation and proliferation of "*Ca.* Competibacter" can probably be related to saturation of "*Ca.* Accumulibacter" cells with polyphosphate under too high non-controlled SRT (Gonzalez-Gil and Holliger

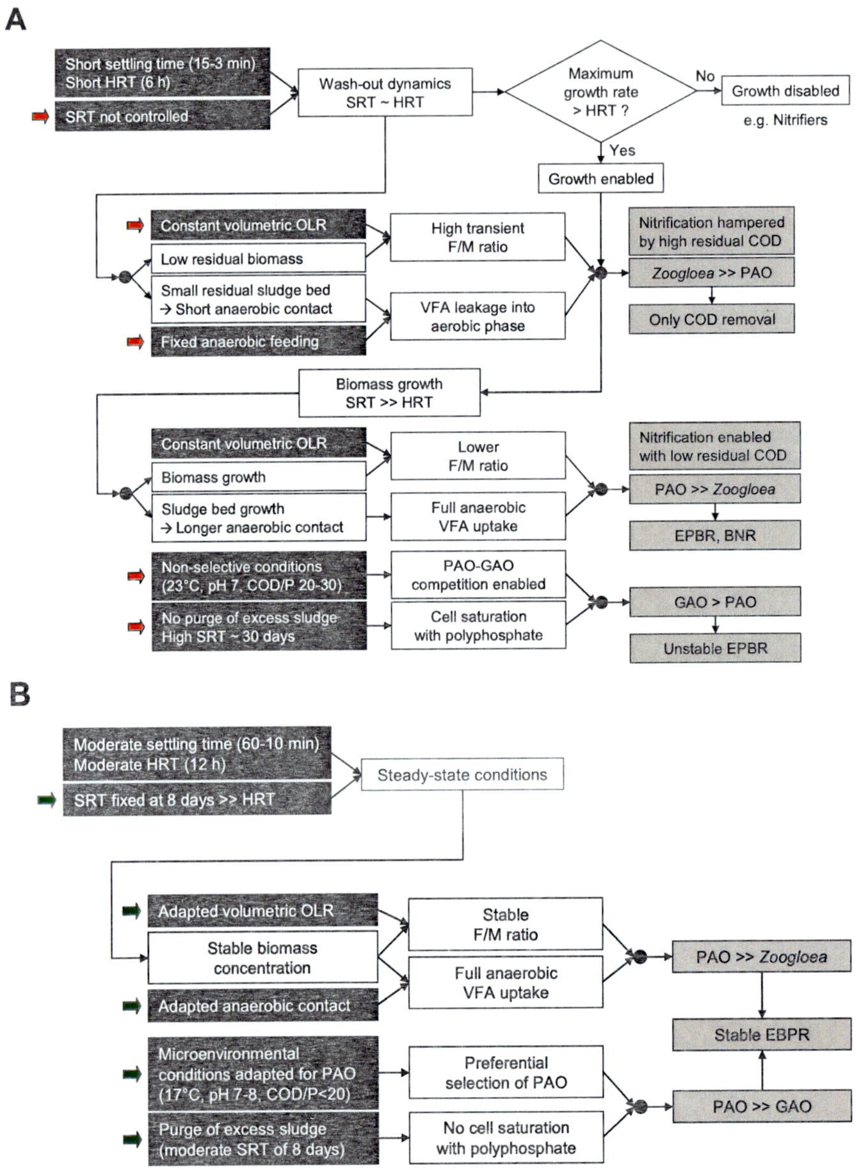

Fig. 8.11 Conceptual ecological model of bacterial selection during the formation and the maturation of AGS in the anaerobic-aerobic BC-SBR operated under wash-out conditions (**A**), and in the anaerobic-aerobic stirred-tank PAO-SBR operated under steady-state conditions (**B**). Red arrows display the weaknesses of the operation under wash-out dynamics. Green arrows display the advantages of the operation under steady-state conditions

2011). Whereas GAO can establish under non-selective microenvironmental conditions (23 °C, pH 6.8–7.2) (Lopez-Vazquez et al. 2009b), operation at mature stage without purge of excess sludge is detrimental to PAOs.

Since active granules also formed in the specific PAO-SBR, lessons about granulation can be learned from this reactor system too (Fig. 8.11B). Moderate SRT (8–10 days) and full control of anaerobic VFA uptake by adapting the volumetric OLR and the anaerobic contact time, and moderate SRT (8–10 days) successfully selected for "*Ca.* Accumulibacter". Such conditions prevented the proliferation of *Zoogloea*. This agrees with anaerobic selection strategies used to prevent viscous bulking phenomena in BNR activated sludge systems (van Loosdrecht et al. 2008). Specific micro-environmental conditions (17 °C, pH 7–8, propionate) also prevented the growth of "*Ca.* Competibacter". Lopez-Vazquez et al. (2007) have investigated the correlation between the yield of orthophosphate release to the uptake of acetate under anaerobic conditions ($Y_{P/C,An}$) at 20 °C and pH 7.0 and the fraction of active PAOs and GAOs present in activated sludge. The high level of this parameter (0.5–0.6 P-mol C-mol_{Pr}^{-1}) detected in the present study confirmed the presence of an abundant active PAO population in the reactor.

Further optimization of granulation processes should therefore target full control of system behavior, by taking advantage of the flexibility of the SBR technology (Wilderer and McSwain 2004). Uncontrolled evolution of process variables was the main weakness of the BC-SBR. Since a proper anaerobic "selector" was key for selecting for "*Ca.* Accumulibacter" in granules, the anaerobic contact time can, as safety measure, be extended by adding an anaerobic mixed batch phase after plug-flow feeding. On-line electrical conductivity profiles can be used for real-time control according to what was done for the PAO-SBR in analogy to Maurer and Gujer (1995) and Aguado et al. (2006). Since aerobic nitrifiers only proliferate at low COD concentrations, the nitrifying activity would also profit from strict anaerobic VFA uptake. Recovery of nitrifying activity was indeed observed in the uncontrolled BC-SBR together with AGS accumulation and recovery of biological dephosphatation activity.

8.4.3 Bacterial Ecology Considerations

Since *Tetrasphaera* have been reported as important phosphorus-removing and glucose-fermenting organisms of full scale BNR-WWTP (Nielsen et al. 2012a, 2012b), their regular presence in AGS-SBRs can be explained by the alternation of anaerobic-aerobic conditions and the high content of exopolysaccharides present in granules.

The longest persistence of *Tetrasphaera* in the GAO-SBR at high abundances revealed that this population can cope with operation at higher mesophilic temperature and slight acidic pH, what could be relevant for dephosphatation of warm and acidic wastewaters. The GAO-enrichment culture revealed abundant *Rhodospirillaceae*-related *Alphaproteobacteria. Defluviicoccus vanus* which belongs to *Rhodospirillaceae* has been reported as a putative alphaproteobacterial GAO (Meyer et al. 2006), that is selected for with propionate as substrate (Lopez-Vazquez et al. 2009b). Here, *Rhodospirillaceae* relatives were abundant despite the presence of only acetate as VFA substrate.

Nitrogen removal in the BC-SBR correlated with dynamics of denitrifiers. Although clades of "*Ca.* Accumulibacter" (and "*Ca.* Competibacter") can denitrify and are desired for denitrifying dephosphatation (Yilmaz et al. 2008; Oehmen et al. 2010), denitrification in anaerobic-aerobic AGS-SBRs is not restricted to only PAOs and GAOs. Other denitrifiers that presumably do not take up VFA anaerobically were present in the full anaerobic-aerobic AGS ecosystems. Denitrifying metabolic activities and utilization of exopolysaccharides as electron donors in AGS systems should be investigated further based on previous knowledge gained from activated sludge systems (Finkmann et al. 2000; Thomsen et al. 2007; Ni et al. 2009; Nielsen et al. 2012b).

The richness and diversity patterns, which are commonly used to characterize biological and wastewater systems (Liu et al. 1997; Borcard et al. 2011; Gonzalez-Gil and Holliger 2011; Winkler et al. 2011b), were used to compare the evolution of the overall bacterial community structure under wash-out and steady state conditions (Additional File 8.4). In addition to the drop in richness and diversity during early-stage granulation reported in Weissbrodt et al. (2012a), wash-out conditions more intensively impacted on these indices. Under steady-state conditions in the PAO-SBR, granulation occurred with stable and relatively high community indices, despite first decrease due to synthetic laboratory conditions in the very beginning.

8.4.4 Granulation Mechanisms Depend on Process Conditions and Predominant Organisms

Granulation occurred under predominance of *Zoogloea* (BC-SBR), "*Ca.* Accumulibacter" (PAO-SBR), as well as "*Ca.* Competibacter" and *Sphingobacteriales* relatives (GAO-SBR). Thus, *Zoogloea* seem not to be essential for granule formation, answering an open question from earlier work where *Zoogloea* was predominant in dense fast-settling granules in different start-up experiments of AGS BC-SBRs (Weissbrodt et al. 2012a). According to Hesselsoe et al. (2009) and Nielsen et al. (2010), the predominant organisms observed in granules share physiological functions required for biofilm formation, namely production of exopolysaccharides and of surface adhesins.

Whereas *Zoogloea*, *Xanthomonadaceae*, *Sphingomonadales*, and *Rhizobiales* relatives can contribute to the production of e.g. zooglan, xanthan, and sphingan exopolysaccharides (Lee et al. 1997; Pollock et al. 1998; Denner et al. 2001; Dow et al. 2003), *Sphingobacteriales* affiliates consume exopolysaccharides (Matsuyama et al. 2008). Granules can therefore be considered as exopolysaccharide-based ecosystems of bacterial producers and consumers. Seviour et al. (2011; 2012) have isolated a specific exopolysaccharide (granulan) from GAO-dominated granules, whereas Lin et al. (2008; 2010) have highlighted alginate as key granular exopolymer. The here applied in situ assessment of specific glycoconjugates by FLBA can inform here on the type of abundant sugar residues (Zippel and Neu 2011) present in the granules examined. N-acetyl-glucosamine (GlcNAc) multimers were observed in the biofilm continuum (STA lectin) and monomers in matrices surrounding microcolonies (LEA, WGA), N-acetyl-galactosamine (GalNAc) around bacterial clusters (HAA, SBA), and α-glucose or α-mannose at (*Zoogloea*-) cell surfaces dispersed in the continuum (ConA). Whereas the GAO-related granulan heteropolysaccharide isolated by Seviour et al. (2012) comprises GalNAc residues, the exopolysaccharides produced by *Zoogloea ramigera* contain glucose and galactose. The diversity of polymer matrices detected here indicated however that granules are composed of a complex mixture of glycoconjugates and that there is probably not only one key exopolymer present in all aerobic granules. Since exopolysaccharide presence depends on predominant microorganisms involved and on specific growth and operation conditions (Nielsen et al. 2004), functional screening should operate on a systems microbiology approach. Glycoconjugates and bacterial populations could be co-localized by combined lectin-binding and FISH analyses. The microbial ecology data collected in this study indicated that only a low number of microorganisms would have to be screened in BNR AGS systems.

Different granulation mechanisms were observed depending on process conditions and predominant organisms involved. Predominance of fast-growing *Zoogloea* under non-limiting substrate conditions resulted in homogenous granular biofilm matrices of cells dispersed in a gel that was formed by microcolony swelling around flocs that embedded further proliferation of dense clusters of slower-growing nitrifiers, PAOs, and GAOs. This picture is similar to experimental and model-based descriptions of multispecies biofilms and underlying cooperative and non-cooperative interactions (Picioreanu et al. 2000, 2004; Alpkvist et al. 2006; Alpkvist and Klapper 2007; Xavier and Foster 2007). Fast-growing heterotrophic competitors apparently formed the embedding biofilm continuum by significant production of exopolysaccharides and rapid proliferation outwards. Biofilm growth against substrate gradients explains the palisade-like orientation of cell lines, which has also been shown in methanogenic granules (Batstone et al. 2006). This converges to the first hypothesis of Barr et al. (2010) on the granulation mechanism by microcolony outgrowth, and on the early statement of Characklis (1973) that attached '*microbial growth originates in a mixture of slime and zoogloeal bacteria (microorganisms that form gelatinous aggregates)*'.

Proliferation of nitrifiers, PAOs, and GAOs as dense clusters transported by the zoogloeal matrix relies on slower growth rates (de Kreuk and van Loosdrecht 2004; Okabe et al. 2004). Accumulation of "*Ca.* Accumulibacter" resulted in denser granular biofilms, something which can explain, in addition to higher polyphosphate contents, the slight decrease in bed height in the BC-SBR despite increase of the biomass concentration between days 90–110. At mature stage in all reactors, the "*Ca.*

Accumulibacter"- and "*Ca.* Competibacter"-dominated granules displayed heterogeneous aggregation of dense bacterial clusters in a cauliflower-like structure. Formation of heterogeneous biofilms has also been related to growth rate considerations under substrate limitations (Picioreanu et al. 2000; Alpkvist et al. 2006) that occurred as soon as full anaerobic feast and aerobic growth under starvation were achieved. Whereas heterogeneous architectures of mature granules can be explained by microcolony re-aggregation after detachment (Barr et al. 2010). Differences in bacterial physiologies can also lead to formation of different granular shapes by proliferation in single granules of either fast-growing organisms in a smooth continuum or slower-growing ones in heterogeneous dense clusters.

Biofilm growth is limited by detachment (Morgenroth 2008). Detachment not only occurred at granule surfaces, but on entire biofilm whorls starting from the core of granules. Such detachment can result in dual access of substrates by the surface and the core of granules, what is comparable to biofilms growing on porous membranes with substrate penetrating from both sides (Downing and Nerenberg 2008), and should be considered in mass transport phenomena across granules. Similarly to planar biofilms and to what has also been observed by different authors (Lemaire et al. 2008; Lee et al. 2009; Barr et al. 2010), granules exhibit more complex structures than stratified architectures considered in mathematical models. Granules heterogeneities can also explain the differences in the spatial organization of target organisms in the granular biofilm ecosystems depending on the process operation.

8.5 Conclusions

The knowledge gained on the complex bacterial and structural dynamics during granule formation and maturation leads to the following findings:

- *Zoogloea*, "*Ca.* Accumulibacter", and "*Ca.* Competibacter" genera can form granules and therefore *Zoogloea* are not essential for granulation.
- Granulation mechanisms depend on operation and predominant organisms. *Zoogloea* form homogenous biofilms embedding the development of nitrifiers, PAO, and GAO colonies. "*Ca.* Accumulibacter" and "*Ca.* Competibacter" form heterogeneous aggregates of dense clusters.
- Mature granules display complex internal architectures with interspersing channels and detachment that can favor growth of bacterial clusters from granule core outwards.
- *Zoogloea* proliferate under wash-out when fixed volumetric OLR and anaerobic feeding phases results in incomplete anaerobic uptake of volatile fatty acids. "*Ca.* Accumulibacter" proliferate when accumulation of AGS enables low biomass specific OLR and sufficient anaerobic plug-flow contact time.

- Granulation of active "*Ca.* Accumulibacter" populations was possible under operation at steady-state with full control of PAO-selective conditions, although 3-times slower than under wash-out conditions. Purge of excess sludge enables stabilizing the sludge age, bacterial community, and dephosphatation.
- Further studies targeting rapid formation of actively dephosphatating granules from flocculent sludge should find a compromise between wash-out and steady-state conditions.

Acknowledgements Thomas R. Neu and Ute Kuhlicke for motivating research collaboration on CLSM investigations of granular biofilms at UFZ Magdeburg, Germany. 1st -ever PhD Mobility Award of the EFPL Doctoral Program in Civil and Environmental Engineering for collaboration with Helmholtz-Centre for Environmental Research, UFZ Magdeburg, Germany, 2011. Best Poster Presentation Award, IWA Biofilm Conference, Tongji University, Shanghai, China, 2011.

Appendix

The Appendix is available at the end of Chap. 5 on PyroTRF-ID.

Appendix A Phylogenetic affiliations obtained with PyroTRF-ID.

Supplementary Information

Additional File 8.1 Composition of the synthetic influent wastewaters fed in the BC-SBR, and in the stirred-tank PAO-SBR and GAO-SBR.

Additional File 8.2 Efficiency of fluorescent dyes and specificity of glycoconjugate-binding lectins for structural analysis of flocs and granules with CLSM.

Additional File 8.3 FISH probes used to follow with CLSM the temporal and spatial evolutions of *Zoogloea*, "*Ca.* Accumulibacter", "*Ca.* Competibacter", and ammonium-oxidizing organisms (AOOs) during granule formation.

Additional File 8.4 Evolutions of the richness and Shannon's H' diversity indices of the bacterial communities present in the BC-SBR, PAO-SBR, and GAO-SBR computed based on the measured T-RFLP profiles.

References

Adav SS, Lee DJ, Tay JH (2008) Extracellular polymeric substances and structural stability of aerobic granule. Water Res 42(6–7):1644–1650

Adav SS, Lee DJ, Lai JY (2009) Functional consortium from aerobic granules under high organic loading rates. Bioresour Technol 100(14):3465–3470

Aguado D, Montoya T, Ferrer J, Seco A (2006) Relating ions concentration variations to conductivity variations in a sequencing batch reactor operated for enhanced biological phosphorus removal. Environ Model Softw 21(6):845–851

Alpkvist E, Klapper I (2007) A multidimensional multispecies continuum model for heterogeneous biofilm development. Bull Math Biol 69(2):765–789

Alpkvist E, Picioreanu C, van Loosdrecht MCM, Heyden A (2006) Three-dimensional biofilm model with individual cells and continuum EPS matrix. Biotechnol Bioeng 94(5):961–979

Barr JJ, Cook AE, Bond PL (2010) Granule formation mechanisms within an aerobic wastewater system for phosphorus removal. Appl Environ Microbiol 76(22):7588–7597

Bassin JP, Winkler MKH, Kleerebezem R, Dezotti M, van Loosdrecht MCM (2012) Improved phosphate removal by selective sludge discharge in aerobic granular sludge reactors. Biotechnol Bioeng 109(8):1919–1928

Batstone DJ, Picioreanu C, van Loosdrecht MCM (2006) Multidimensional modelling to investigate interspecies hydrogen transfer in anaerobic biofilms. Water Res 40(16):3099–3108

Beun JJ, Hendriks A, van Loosdrecht MCM, Morgenroth E, Wilderer PA, Heijnen JJ (1999) Aerobic granulation in a sequencing batch reactor. Water Res 33(10):2283–2290

Borcard D, Gillet F, Legendre P (2011) Numerical ecology with R, 1st edn. Springer-Verlag GmbH, Heidelberg

Characklis WG (1973) Attached microbial growths: I. Attachment and growth. Water Res 7(8):1113–1127

Crocetti GR, Hugenholtz P, Bond PL, Schuler A, Keller J, Jenkins D, Blackall LL (2000) Identification of polyphosphate-accumulating organisms and design of 16S rRNA-directed probes for their detection and quantitation. Appl Environ Microbiol 66(3):1175–1182

Crocetti GR, Banfield JF, Keller J, Bond PL, Blackall LL (2002) Glycogen-accumulating organisms in laboratory-scale and full-scale wastewater treatment processes. Microbiology 148(11):3353–3364

Dangcong P, Bernet N, Delgenes JP, Moletta R (1999) Aerobic granular sludge-a case report. Water Res 33:890–893

de Kreuk MK, van Loosdrecht MCM (2004) Selection of slow growing organisms as a means for improving aerobic granular sludge stability. Water Sci Technol 49:9–17

de Kreuk M, Heijnen JJ, van Loosdrecht MCM (2005) Simultaneous COD, nitrogen, and phosphate removal by aerobic granular sludge. Biotechnol Bioeng 90(6):761–769

de Villiers GH, Pretorius WA (2001) Abattoir effluent treatment and protein production. Water Sci Technol 43(11):243–250

Denner EBM, Paukner S, Kampfer P, Moore ERB, Abraham WR, Busse HJ, Wanner G, Lubitz W (2001) *Sphingomonas pituitosa* sp. nov., an exopolysaccharide-producing bacterium that secretes an unusual type of sphingan. Int J Syst Evol Microbiol 51(3):827–841

DeSantis TZ, Hugenholtz P, Larsen N, Rojas M, Brodie EL, Keller K, Huber T, Dalevi D, Hu P, Andersen GL (2006) Greengenes, a chimera-checked 16S rRNA gene database and workbench compatible with ARB. Appl Environ Microbiol 72(7):5069–5072

Dow JM, Crossman L, Findlay K, He YQ, Feng JX, Tang JL (2003) Biofilm dispersal in *Xanthomonas campestris* is controlled by cell-cell signaling and is required for full virulence to plants. Proc Nat Acad Sci USA 100(19):10995–11000

Downing LS, Nerenberg R (2008) Effect of oxygen gradients on the activity and microbial community structure of a nitrifying, membrane-aerated biofilm. Biotechnol Bioeng 101(6):1193–1204

Dulekgurgen E, Ovez S, Artan N, Orhon D (2003) Enhanced biological phosphate removal by granular sludge in a sequencing batch reactor. Biotechnol Lett 25(9):687–693

Ebrahimi S, Gabus S, Rohrbach-Brandt E, Hosseini M, Rossi P, Maillard J, Holliger C (2010) Performance and microbial community composition dynamics of aerobic granular sludge from sequencing batch bubble column reactors operated at 20, 30, and 35°C. Appl Microbiol Biotechnol 87:1555–1568

Etterer TJ (2006) Formation, structure and function of aerobic granular sludge. PhD thesis, Technische Universität München

Filali A, Bessiere Y, Sperandio M (2012) Effects of oxygen concentration on the nitrifying activity of an aerobic hybrid granular sludge reactor. Water Sci Technol 65(2):289–295

Finkmann W, Altendorf K, Stackebrandt E, Lipski A (2000) Characterization of N_2O-producing Xanthomonas-like isolates from biofilters as *Stenotrophomonas nitritireducens* sp. nov., *Luteimonas mephitis* gen. nov., sp. nov. and *Pseudoxanthomonas broegbernensis* gen. nov., sp. nov. Int J Syst Evol Microbiol 50(1):273–282

Gao JF, Chen RN, Su K, Zhang Q, Peng YZ (2010) Formation and reaction mechanism of simultaneous nitrogen and phosphorus removal by aerobic granular sludge. Huanjing Kexue/environ Sci 31(4):1021–1029

Giesen A, Niermans R, van Loosdrecht MCM (2012) Aerobic granular biomass: the new standard for domestic and industrial wastewater treatment? Water 21(4):28–30

Gonzalez-Gil G, Holliger C (2011) Dynamics of microbial community structure and enhanced biological phosphorus removal of propionate- and acetate-cultivated aerobic granules. Appl Environ Microbiol 77:8041–8051

Hesselmann RPX, Werlen C, Hahn D, van der Meer JR, Zehnder AJB (1999) Enrichment, phylogenetic analysis and detection of a bacterium that performs enhanced biological phosphate removal in activated sludge. Syst Appl Microbiol 22(3):454–465

Hesselsoe M, Fureder S, Schloter M, Bodrossy L, Iversen N, Roslev P, Nielsen PH, Wagner M, Loy A (2009) Isotope array analysis of *Rhodocyclales* uncovers functional redundancy and versatility in an activated sludge. ISME J 3(12):1349–1364

Hollender J, Dreyer U, Kornberger L, Kämpfer P, Dott W (2002) Selective enrichment and characterization of a phosphorus-removing bacterial consortium from activated sludge. Appl Microbiol Biotechnol 58(1):106–111

Kishida N, Kim J, Tsuneda S, Sudo R (2006) Anaerobic/oxic/anoxic granular sludge process as an effective nutrient removal process utilizing denitrifying polyphosphate-accumulating organisms. Water Res 40(12):2303–2310

Kong YH, Ong SL, Ng WJ, Liu WT (2002) Diversity and distribution of a deeply branched novel proteobacterial group found in anaerobic-aerobic activated sludge processes. Environ Microbiol 4(11):753–757

Kuba T, van Loosdrecht MCM, Heijnen JJ (1997) Biological dephosphatation by activated sludge under denitrifying conditions: pH influence and occurrence of denitrifying dephosphatation in a full-scale waste water treatment plant. Water Sci Technol 36(12):75–82

Lawrence JR, Korber DR, Wolfaardt GM (1996) Heterogeneity of natural biofilm communities. Cells Mater 6(1–3):175–191

Lee JW, Yeomans WG, Allen AL, Gross RA, Kaplan DL (1997) Production of zoogloea gum by *Zoogloea ramigera* with glucose analogs. Biotechnol Lett 19(8):799–802

Lee CC, Lee DJ, Lai JY (2009) Labeling enzymes and extracellular polymeric substances in aerobic granules. J Taiwan Inst Chem E 40(5):505–510

Lemaire R, Webb RI, Yuan Z (2008) Micro-scale observations of the structure of aerobic microbial granules used for the treatment of nutrient-rich industrial wastewater. ISME J 2(5):528–541

Lin YM, Wang L, Chi ZM, Liu XY (2008) Bacterial alginate role in aerobic granular bio-particles formation and settleability improvement. Sep Sci Technol 43:1642–1652

Lin Y, de Kreuk M, van Loosdrecht MCM, Adin A (2010) Characterization of alginate-like exopolysaccharides isolated from aerobic granular sludge in pilot-plant. Water Res 44(11):3355–3364

Liu Y, Liu Q-S (2006) Causes and control of filamentous growth in aerobic granular sludge sequencing batch reactors. Biotechnol Adv 24(1):115–127

Liu WT, Marsh TL, Cheng H, Forney LJ (1997) Characterization of microbial diversity by determining terminal restriction fragment length polymorphisms of genes encoding 16S rRNA. Appl Environ Microbiol 63(11):4516–4522

Liu WT, Marsh TL, Forney LJ (1998) Determination of the microbial diversity of anaerobic-aerobic activated sludge by a novel molecular biological technique. Water Sci Technol 37(4–5):417–422

Lopez-Vazquez CM, Hooijmans CM, Brdjanovic D, Gijzen HJ, van Loosdrecht MCM (2007) A practical method for quantification of phosphorus- and glycogen-accumulating organism populations in activated sludge systems. Water Environ Res 79(13):2487–2498

Lopez-Vazquez CM, Hooijmans CM, Brdjanovic D, Gijzen HJ, van Loosdrecht MCM (2009a) Temperature effects on glycogen accumulating organisms. Water Res 43(11):2852–2864

Lopez-Vazquez CM, Oehmen A, Hooijmans CM, Brdjanovic D, Gijzen HJ, Yuan Z, van Loosdrecht MCM (2009b) Modeling the PAO-GAO competition: effects of carbon source, pH and temperature. Water Res 43(2):450–462

Loy A, Horn M, Wagner M (2003) ProbeBase: an online resource for rRNA-targeted oligonucleotide probes. Nucl Acids Res 31(1):514–516

Loy A, Schulz C, Lucker S, Schopfer-Wendels A, Stoecker K, Baranyi C, Lehner A, Wagner M (2005) 16S rRNA gene-based oligonucleotide microarray for environmental monitoring of the betaproteobacterial order *Rhodocyclales*. Appl Environ Microbiol 71(3):1373–1386

Lu H, Oehmen A, Virdis B, Keller J, Yuan Z (2006) Obtaining highly enriched cultures of "Candidatus Accumulibacter phosphates" through alternating carbon sources. Water Res 40(20):3838–3848

Marcelino M, Guisasola A, Baeza JA (2009) Experimental assessment and modelling of the proton production linked to phosphorus release and uptake in EBPR systems. Water Res 43(9):2431–2440

Matsuyama H, Katoh H, Ohkushi T, Satoh A, Kawahara K, Yumoto I (2008) *Sphingobacterium kitahiroshimense* sp. nov., isolated from soil. Int J Syst Evol Microbiol 58(7):1576–1579

Maurer M, Gujer W (1995) Monitoring of microbial phosphorus release in batch experiments using electric conductivity. Water Res 29(11):2613–2617

McSwain BS, Irvine RL, Hausner M, Wilderer PA (2005) Composition and distribution of extracellular polymeric substances in aerobic flocs and granular sludge. Appl Environ Microbiol 71(2):1051–1057

Meyer RL, Saunders AM, Blackall LL (2006) Putative glycogen-accumulating organisms belonging to the *Alphaproteobacteria* identified through rRNA-based stable isotope probing. Microbiology 152:419–429

Mobarry BK, Wagner M, Urbain V, Rittmann BE, Stahl DA (1996) Phylogenetic probes for analyzing abundance and spatial organization of nitrifying bacteria. Appl Environ Microbiol 62(6):2156–2162

Morgenroth E (2008) Modelling biofilms. In: Henze M, van Loosdrecht MCM, Ekama GA, Brdjanovic D (eds) Biological wastewater treatment: principles, modelling and design. IWA Publishing, London, pp 457–492

Morgenroth E, Sherden T, van Loosdrecht MCM, Heijnen JJ, Wilderer PA (1997) Aerobic granular sludge in a sequencing batch reactor. Water Res 31(12):3191–3194

Mosquera-Corral A, Arrojo B, Figueroa M, Campos JL, Mendez R (2011) Aerobic granulation in a mechanical stirred SBR: treatment of low organic loads. Water Sci Technol 64(1):155–161

Neu TR, Lawrence JR (1999) Lectin-binding analysis in biofilm systems. Methods Enzymol 310:145–152

Neu TR, Manz B, Volke F, Dynes JJ, Hitchcock AP, Lawrence JR (2010) Advanced imaging techniques for assessment of structure, composition and function in biofilm systems. FEMS Microbiol Ecol 72(1):1–21

Ni BJ, Fang F, Rittmann BE, Yu HQ (2009) Modeling microbial products in activated sludge under feast-famine conditions. Environ Sci Technol 43(7):2489–2497

Nielsen PH, Thomsen TR, Nielsen JL (2004) Bacterial composition of activated sludge: importance for floc and sludge properties. Water Sci Technol 49(10):51–58

Nielsen PH, Daims H, Lemmer H (2009) FISH handbook for biological wastewater treatment: identification and quantification of microorganisms in activated sludge and biofilms by FISH, 1st edn. IWA Publishing, London, UK

Nielsen PH, Mielczarek AT, Kragelund C, Nielsen JL, Saunders AM, Kong Y, Hansen AA, Vollertsen J (2010) A conceptual ecosystem model of microbial communities in enhanced biological phosphorus removal plants. Water Res 44(17):5070–5088

Nielsen JL, Nguyen H, Meyer RL, Nielsen PH (2012a) Identification of glucose-fermenting bacteria in a full-scale enhanced biological phosphorus removal plant by stable isotope probing. Microbiology 158(7):1818–1825

Nielsen PH, Saunders AM, Hansen AA, Larsen P, Nielsen JL (2012b) Microbial communities involved in enhanced biological phosphorus removal from wastewater: a model system in environmental biotechnology. Curr Opin Biotechnol 23(3):452–459

Oehmen A, Carvalho G, Lopez-Vazquez CM, van Loosdrecht MCM, Reis MAM (2010) Incorporating microbial ecology into the metabolic modelling of polyphosphate accumulating organisms and glycogen accumulating organisms. Water Res 44(17):4992–5004

Okabe S, Yasuda T, Watanabe Y (1997) Uptake and release of inert fluorescence particles by mixed population biofilms. Biotechnol Bioeng 53(5):459–469

Okabe S, Kindaichi T, Ito T, Satoh H (2004) Analysis of size distribution and areal cell density of ammonia-oxidizing bacterial microcolonies in relation to substrate microprofiles in biofilms. Biotechnol Bioeng 85(1):86–95

Picioreanu C, van Loosdrecht MCM, Heijnen JJ (2000) A theoretical study on the effect of surface roughness on mass transport and transformation in biofilms. Biotechnol Bioeng 68(4):355–369

Picioreanu C, Kreft JU, van Loosdrecht MCM (2004) Particle-based multidimensional multispecies biofilm model. Appl Environ Microbiol 70(5):3024–3040

Pollock TJ, Van Workum WAT, Thorne L, Mikolajczak MJ, Yamazaki M, Kijne JW, Armentrout RW (1998) Assignment of biochemical functions to glycosyl transferase genes which are essential for biosynthesis of exopolysaccharides in Sphingomonas strain S88 and *Rhizobium leguminosarum*. J Bacteriol 180(3):586–593

R Development Core Team (2008) R: A language and environment for statistical computing. R Foundation for Statistical Computing, Vienna. http://cran.r-project.org/

Rossello Mora RA, Wagner M, Amann R, Schleifer KH (1995) The abundance of *Zoogloea ramigera* in sewage treatment plants. Appl Environ Microbiol 61(2):702–707

Schuler AJ, Jenkins D (2003) Enhanced biological phosphorus removal from wastewater by biomass with different phosphorus contents, part I: experimental results and comparison with metabolic models. Water Environ Res 75(6):485–498

Schuler AJ, Jenkins D, Ronen P (2001) Microbial storage products, biomass density, and setting properties of enhanced biological phosphorus removal activated sludge. Water Sci Technol 43(1):173–180

Seviour TW, Lambert LK, Pijuan M, Yuan Z (2011) Selectively inducing the synthesis of a key structural exopolysaccharide in aerobic granules by enriching for "Candidatus Competibacter phosphatis." Appl Microbiol Biotechnol 92(6):1297–1305

Seviour T, Yuan Z, van Loosdrecht MCM, Lin Y (2012) Aerobic sludge granulation: a tale of two polysaccharides? Water Res 46(15):4803–4813

Smolders GJF, van der Meij J, van Loosdrecht MCM, Heijnen JJ (1994) Model of the anaerobic metabolism of the biological phosphorus removal process: stoichiometry and pH influence. Biotechnol Bioeng 43(6):461–470

Staudt C, Horn H, Hempel DC, Neu TR (2003) Screening of lectins for staining lectin-specific glycoconjugates in the EPS of biofilms. In: Lens P, Moran AP, Mahony T, Stoodley P, O'Flaherty V (eds) Biofilms in medicine, industry and environmental biotechnology. Integrated environmental technology series. IWA Publishing, London, UK, pp 308–326

Tay JH, Tay STL, Ivanov V, Pan S, Jiang HL, Liu QS (2003) Biomass and porosity profiles in microbial granules used for aerobic wastewater treatment. Lett Appl Microbiol 36:297–301

Thomsen TR, Kong Y, Nielsen PH (2007) Ecophysiology of abundant denitrifying bacteria in activated sludge. FEMS Microbiol Ecol 60(3):370–382

Tsuneda S, Nagano T, Hoshino T, Ejiri Y, Noda N, Hirata A (2003) Characterization of nitrifying granules produced in an aerobic upflow fluidized bed reactor. Water Res 37(20):4965–4973

van Loosdrecht MCM, Martins AM, Ekama GA (2008) Bulking sludge. In: Henze M, van Loosdrecht MCM, Ekama GA, Brdjanovic D (eds) Biological wastewater treatment: principles, modelling and design. IWA Publishing, London, pp 291–308

Wagner M, Amann R, Lemmer H, Manz W, Schleifer KH (1994) Probing activated sludge with fluorescently labeled rRNA targeted oligonucleotides. Water Sci Technol 29(7):15–23

Weissbrodt DG, Lochmatter S, Ebrahimi S, Rossi P, Maillard J, Holliger C (2012a) Bacterial selection during the formation of early-stage aerobic granules in wastewater treatment systems operated under wash-out dynamics. Front Microbiol 3:332

Weissbrodt DG, Shani N, Sinclair L, Lefebvre G, Rossi P, Maillard J, Rougemont J, Holliger C (2012b) PyroTRF-ID: a novel bioinformatics methodology for the affiliation of terminal-restriction fragments using 16S rRNA gene pyrosequencing. BMC Microbiol 12:306

Weissbrodt DG, Maillard J, Brovelli A, Chabrelie A, May J, Holliger C (2014) Multilevel correlations in the biological phosphorus removal process: from bacterial enrichment to conductivity-based metabolic batch tests and polyphosphatase assays. Biotechnol Bioeng 111(12):2421–2435

Wilderer PA, McSwain BS (2004) The SBR and its biofilm application potentials. Water Sci Technol 50(10):1–10

Winkler MKH, Bassin JP, Kleerebezem R, de Bruin LMM, van den Brand TPH, van Loosdrecht MCM (2011a) Selective sludge removal in a segregated aerobic granular biomass system as a strategy to control PAO-GAO competition at high temperatures. Water Res 45(11):3291–3299

Winkler MKH, Kleerebezem R, de Bruin LMM, Habermacher J, Abbas B, van Loosdrecht MCM (2011b) Microbial diversity differences in aerobic granular sludge in comparison to conventional treatment plant. In: Qi Z (ed) IWA biofilm specialist conference 2011b processes in biofilms. Tongji University, Shanghai.

Wuertz S, Okabe S, Hausner M (2004) Microbial communities and their interactions in biofilm systems: an overview. Water Sci Technol 49(11–12):327–336

Xavier JB, Foster KR (2007) Cooperation and conflict in microbial biofilms. Proc Nat Acad Sci USA 104(3):876–881

Yilmaz G, Lemaire R, Keller J, Yuan Z (2008) Simultaneous nitrification, denitrification, and phosphorus removal from nutrient-rich industrial wastewater using granular sludge. Biotechnol Bioeng 100(3):529–541

You Y, Peng Y, Yuan ZG, Li XY, Peng YZ (2008) Cultivation and characteristic of aerobic granular sludge enriched by phosphorus accumulating organisms. Huanjing Kexue/environ Sci 29(8):2242–2248

Zeng RJ, Van Loosdrecht MCM, Yuan Z, Keller J (2003) Metabolic model for glycogen-accumulating organisms in anaerobic/aerobic activated sludge systems. Biotechnol Bioeng 81(1):92–105

Zippel B, Neu TR (2011) Characterization of glycoconjugates of extracellular polymeric substances in tufa-associated biofilms by using fluorescence lectin-binding analysis. Appl Environ Microbiol 77(2):505–516

Chapter 9
Linking Bacterial Populations and Nutrient Removal in the Granular Sludge Ecosystem

"The concept of a "stable" microbial community should be replaced by that of a cooperative community continuum." *(Verstraete et al. 2007)*

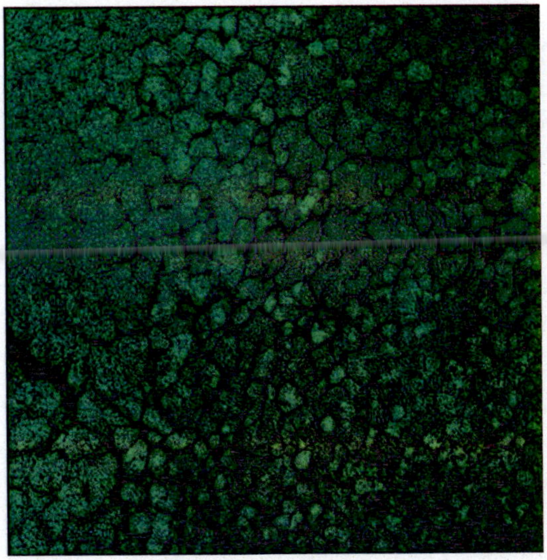

Microbial ecosystem

The content of this chapter was published in a modified version in: Weissbrodt DG, Shani N, Holliger C (2014) Linking bacterial population dynamics and nutrient removal in the granular sludge biofilm ecosystem engineered for wastewater treatment. FEMS Microbiol Ecol 88(3):579–95. https://doi.org/10.1111/1574-6941.12326. Permission was granted to reuse the figure materials (© 2014 Oxford University Press).

Supplementary Information The online version contains supplementary material available at https://doi.org/10.1007/978-3-031-41009-3_9.

D. G. Weissbrodt, *Engineering Granular Microbiomes*, Springer Theses,
https://doi.org/10.1007/978-3-031-41009-3_9

9.1 Introduction

Efficient management of bacterial resource in aerobic granular sludge (AGS) sequencing batch reactors (SBR) for high-rate biological nutrient removal (BNR) from wastewater requires understanding of the impact of process conditions on bacterial community dynamics.

Similarly to conventional activated sludge systems (Lopez-Vazquez et al. 2009b; Oehmen et al. 2010b), AGS reactors have been subjected to the competition of polyphosphate—(PAOs) and glycogen-accumulating organisms (GAOs) related to "*Candidati* Accumulibacter and Competibacter phosphates", respectively. Both PAOs and GAOs have established in mature AGS under alternating anaerobic plug-flow feeding and aerobic starvation conditions (de Kreuk and van Loosdrecht 2004). PAOs have been preferentially selected with temperatures below 20 °C, non-limiting phosphorus conditions in the influent wastewater, and selective purge of the upper fractions of the settled AGS bed (de Kreuk and van Loosdrecht 2004; Ebrahimi et al. 2010; Winkler et al. 2011). GAOs have mainly proliferated under phosphorus-limiting conditions, higher mesophilic temperatures, and when PAOs have been saturated with polyphosphate under operation without control of the sludge retention time (SRT) what has been shown in Weissbrodt et al. (2013/Chap. 8) and by Gonzalez-Gil and Holliger (2011).

Certain clades of "*Ca.* Accumulibacter" and "*Ca.* Competibacter" are able to denitrify (Kuba et al. 1996; Oehmen et al. 2010a, 2010b) and denitrifying PAOs are advantageous for simultaneous phosphorus and nitrogen removal due to savings in chemical oxygen demand (COD) and lower sludge production. Simultaneous BNR has been implemented in AGS-SBRs by selecting for different bacterial niches along gradients of dissolved nutrients and oxygen (DO) inside granular biofilms (Kishida et al. 2006; de Kreuk et al. 2007; Lemaire et al. 2008; Yilmaz et al. 2008). These studies have however not converged towards a uniform conclusion on the involvement of either PAOs or GAOs in denitrification. At the actual state of knowledge, an adequate balance of these organisms should be maintained in AGS for efficient BNR. Lemaire et al. (2008) have reported high performances with 48 ± 18% of PAOs and 26 ± 8% of GAOs forming a PAO/GAO ratio of 1.9 ± 0.5. According to Weissbrodt et al. (2013/Chap. 8), the denitrifying bacterial community of AGS systems is not restricted to only PAOs and GAOs. By using the pyrosequencing-based PyroTRF-ID method (Weissbrodt et al. 2012b/Chap. 5), denitrifying populations of *Xanthomonadaceae*, *Rhodocyclaceae*, *Comamonadaceae*, *Rhizobiales*, *Sphingomonadales*, and *Sphingobacteriales* have been detected in significant abundances despite full anaerobic-aerobic conditions. This method has also offered higher resolution for the identification of nitrifiers. The various AGS studies cited above have indeed reported full nitrification in BNR AGS-SBRs with ammonium-oxidizing organisms (AOOs) and nitrite-oxidizing organisms (NOOs) present in abundances close to minimum detection limits of traditional microbial ecology methods.

Based on the diversity of bacterial populations present in AGS, this study provides detailed understanding of their dynamics in mature anaerobic-aerobic AGS-SBRs operated for BNR under fluctuations in operation variables. A multivariate statistical approach was used to investigate multilevel correlations between operation variables, BNR performances, and bacterial community compositions, and to identify predominant populations and accompanying populations involved. Based on these correlations, optimized operation conditions are proposed for efficient bacterial resource management and BNR in AGS systems.

9.2 Material and Methods

9.2.1 Operation of Anaerobic–Aerobic AGS-SBRs at 20 and 25 °C

Two double-wall glass bubble-column SBRs were operated over 5 months for BNR at 20 (SBR-20) and 25 °C (SBR-25) under alternating anaerobic-aerobic conditions and under fluctuations in acetate concentration and nutrient ratios in the influent wastewater, and in DO setpoint (Table 9.1).

The fixed 3-h SBR cycles adapted from de Kreuk et al. (2005) comprised plug-flow anaerobic feeding (60 min), up-flow aeration (110 min), settling (3 min), withdrawal (4 min, volume exchange ratio of 50%), and idle (3 min), and were related to 6-h hydraulic retention times (HRT). The composition of the cultivation medium has been described previously (Weissbrodt et al. 2012a). During aeration, gas headspaces were recirculated with membrane vacuum pumps (KNF Neuberger AG, Switzerland)

Table 9.1 Ranges of fluctuations in operation variables in SBR-20 and SBR-25

Operation variables	Units	SBR-20	SBR-25
Acetate concentration	$mg_{CODs}\ L^{-1}$	379–577	329–595
Volumetric OLR[1]	$mg_{CODs}\ L_r^{-1}\ cycle^{-1}$	190–289	164–298
Daily volumetric OLR[1]	$kg_{CODs}\ m_r^{-3}\ d^{-1}$	1.5–2.3	1.3–2.4
F/M ratio[1]	$mg_{CODs}\ g_{CODx}^{-1}\ cycle^{-1}$	12–18	10–24
Daily sludge loading rate[1]	$kg_{CODs}\ m^{-3}_{granules}\ d^{-1}$	10–15	11–20
COD/P ratio[2]	$g_{CODs}\ g_P^{-1}$	19–41	20–35
COD/N ratio[2]	$g_{CODs}\ g_N^{-1}$	8–14	8–12
N/P ratio[2]	$g_N\ g_P^{-1}$	2.1–4.0	2.1–3.5
Dissolved oxygen[3]	$\%O_2$ saturation	60 or 80	80 or 100

[1]The acetate concentration in the influent wastewater was expressed as volumetric organic loading rate (OLR), food-to-microorganism ratio (F/M), and daily sludge loading rate for comparison with ranges reported in literature, e.g. de Kreuk et al. (2007). These conversions were obtained by considering the 50% volume exchange ratio, the eight SBR cycles per day, the average biomass concentration of 12 $g_{VSS}\ L_r^{-1}$ (≡ 16 $g_{CODx}\ L_r^{-1}$), and the estimated volume of granules of 0.157 and 0.123 $m^3_{granules}\ m_r^{-3}$ present in SBR-20 and SBR-25, respectively (Table 9.2)

Table 9.2 Characteristics and system analysis of SBR-20 and SBR-25

Parameter	Symbol and formula	Units	SBR-20	SBR-25
Operation temperature	T	°C	20	25
Reactor characteristics				
Total reactor height	H	m	1.4	1.4
Internal diameter	D_i	m	$55 \cdot 10^{-3}$	$59 \cdot 10^{-3}$
Working volume	V_r	m^3	$2.5 \cdot 10^{-3}$	$3.5 \cdot 10^{-3}$
Height of the liquid phase	H_L	m	1.050	1.282
H/D ratio	$H/D = H \cdot D_i^{-1}$	–	26	24
H_L/D ratio	$H_L/D = H_L \cdot D_i^{-1}$	–	19	22
Reactor cross-section	$A_{cross} = \pi \cdot D_i^2/4$	m^2	$2.38 \cdot 10^{-3}$	$2.73 \cdot 10^{-3}$
Aeration characteristics				
Dissolved oxygen saturation in water	$S_{O2}(T)$*	$mg_{O2}\ L^{-1}$	9.1	8.3
Measured gas flowrate	Q_{gas}	$L\ min^{-1}$	2.2	2.5
Superficial gas velocity	$SGV = Q_{gas} \cdot A_{cross}^{-1}$	$m\ s^{-1}$	0.0154	0.0153
Theoretical k_La formulas[1]	$k_La(20\ °C, < 0.050\ m\ s^{-1}) = 2 \cdot SGV$	s^{-1}	0.031	0.031
	$k_La(20\ °C, > 0.050\ m\ s^{-1}) = 0.32 \cdot SGV^{0.7}$	s^{-1}	0.017	0.017
	$k_La(T) = k_La(20\ °C) \cdot \theta^{(T - 20\ °C)}$	s^{-1}	0.031	0.035
	with $\theta = 1.020 - 1.024$	–		
Measured k_La value[2]	$k_La(T)$	s^{-1}	0.014	0.007
AGS biofilm characteristics				
Measured height of settled AGS bed	H_{bed}	m	0.30	0.35
Settled AGS bed volume	$V_{bed} = H_{bed} \cdot A_{cross}$	m^3	$0.714 \cdot 10^{-3}$	$0.956 \cdot 10^{-3}$
Measured average Ferret diameter of granules[3]	d_p	mm	1.5 ± 0.9 (N = 77)	2.9 ± 1.5 (N = 40)
Average granule volume[3]	$V_p = 1/6 \cdot \pi \cdot d_p^3$	$m^3\ particle^{-1}$	$0.18 \cdot 10^{-8}$	$1.28 \cdot 10^{-8}$
Number of granules in reactor[4]	$N_p = V_{bed} \cdot (1 - \varepsilon) \cdot V_p^{-1}$	particles	$2.222 \cdot 10^5$	$0.411 \cdot 10^5$
Total volume of AGS biomass	$V_{biomass} = V_{bed} \cdot \varepsilon$	m_p^3	$0.393 \cdot 10^{-3}$	$0.430 \cdot 10^{-3}$
AGS biomass to reactor volume ratio	$V_{biomass}/V_r$	$m_p^3\ m_r^{-3}$	0.157	0.123

(continued)

Table 9.2 (continued)

Parameter	Symbol and formula	Units	SBR-20	SBR-25
Average surface area per granule[3]	$A_p = \pi \cdot d_p{}^2$	m^2 $particle^{-1}$	$0.71 \cdot 10^{-5}$	$2.64 \cdot 10^{-5}$
Total biofilm surface area	$A_{biofilm} = A_p \cdot N_p$	m^2	1.57	1.20
Biofilm area to biomass volume ratio	$A_{biofilm}/V_{biomass}$	m^2 $m_p{}^{-3}$	3995	2791
Biofilm area to reactor volume ratio	$A_{biofilm}/V_r$	m^2 $m_r{}^{-3}$	628	343

[1]The theoretical volumetric oxygen mass transfer coefficients (k_La) were calculated according to the formula developed by Heijnen and Van't Riet (1984) for ranges of superficial gas velocities (SGV) of 0–0.050 and 0.050–0.300 $m \cdot s^{-1}$, respectively
[2]The k_La-measurements were conducted with the dynamic method according to Mena et al. (2005) at operation temperature in the three-phase AGS-SBRs
[3]Granules were considered as non-porous spherical particles for simplified computations
[4]The average porosity of the AGS bed ($\varepsilon = 0.45$) measured in Weissbrodt et al. (2017/Chap. 11) was considered here

with superficial gas velocities (SGV) of 0.015 m s^{-1}. DO was regulated by feedback proportional-integral control with addition air or dinitrogen gas with mass flow controllers (Brooks Instrument, USA), and pH was maintained at 7.0 ± 0.3 by addition of HCl or NaOH 1 mol L^{-1}.

The reactors were inoculated with pre-cultivated mature AGS composed of 28% of PAOs and 25% of GAOs, and that exhibited efficient BNR activities. Prior to starting investigations on the dynamics of BNR performances and of bacterial communities, the AGS were grown during a preliminary phase of 65 days up to optimal biomass concentrations of 18 ± 2 g_{TSS} L^{-1} and bed heights of 30–35 cm enabling full anaerobic uptake of acetate.

Steady-state conditions were then achieved by fixing the sludge retention time (SRT) to 20 ± 3 days by purge of excess sludge, and this corresponded to the starting experimental day. During the first 115 days, excess sludge was wasted manually every 6 days at the bottom of the settled AGS beds. From this day onwards, a fraction of mixed liquors was purged in an automated way at the end of each aeration phase. Changes in operation variables lasted over at least one sludge age. Characteristics of both SBRs are compiled in Table 9.2.

9.2.2 *Analyses of Soluble Compounds and Particulate Biomass*

Methods for the analysis of concentrations of acetate, anionic and cationic solutes, and of total (TSS), volatile (VSS) and inorganic suspended solids (ISS) were described in Weissbrodt et al. (2012a). Granule size distributions (Feret's diameter) were measured after 100 days by digital image analysis in ImageJ 1.45 s (NIH 2012) according to Beun et al. (1999).

9.2.3 Molecular Analyses of Bacterial Community Compositions

Bacterial community composition of AGS was investigated by T-RFLP analysis of eubacterial 16S rRNA encoding genes with the labeled 8f and the unlabeled 518r PCR primers (FAM-5′-AGAGTTTGATCMTGGCTCAG-3′ and 5′-ATTACCGCGGCTGCTGG-3′), and with the *Hae*III endonuclease, according to Weissbrodt et al. (2012a). The T-RFLP profiles were expressed as relative contributions of operational taxomic units (OTU), and were presented in stacked bar plots of predominant OTUs (> 2%) for facilitated visual observation.

Closest phylogenetic affiliations of OTUs were obtained by 454 pyrosequencing-based PyroTRF-ID analysis of selected biomass samples collected on days 66, 84, and 114 in SBR-20, and on day 125 in SBR-25, as well as on day 6 from excess AGS purged from SBR-20, according to Weissbrodt et al. (2012b/Chap. 2).

9.2.4 Clustering and Multivariate Statistical Analyses of Operation, BNR, and Datasets

Hierarchical clustering and multivariate statistical analyses of operation, BNR, and T-RFLP datasets were computed in the R software (R-Development-Core-Team 2008) equipped with the additional packages Vegan (Oksanen et al. 2009), Heatplus (Ploner 2011), and heatmap.plus (Day 2007), according to the methodology detailed by Borcard et al. (2011).

Hierarchical clustering analyses were computed with the Ward's minimum variance method in order to define major groups of daily reactor states. Principal component (PCA) and multiple factor analyses (MFA) were used to assess the relationships between the datasets and to represent their common evolution in a graphical way during reactor operation. As results of MFA, the interconnections of datasets were represented in group representations, the evolutions of the reactors states were graphed in individual factor maps, and the extent of correlations between the three datasets were displayed in correlation circles with proportional vector directions and lengths.

Spearman's rank correlation coefficients, RV-coefficients (Robert and Escoufier 1976), and *p*-values with 10,000 permutations were computed as measures of similarities to quantify and to assess the statistical significance of the observed correlations. Pair-wise correlations between each OTU and each operation and performance parameter were graphed in heatmaps with proportional color intensities in order to detect conditions that select for clusters of OTUs.

9.3 Results

9.3.1 Nutrient Removal Performances at 20 and 25 °C

The BNR dynamics in SBR-20 and SBR-25 were delineated in six periods according to the main changes in operation variables during 5 months of reactor operation (Fig. 9.1). For each period, the average levels of operation and BNR parameters are compiled in Table 9.3.

After 20 days of adaptation of reactors to fixed SRT conditions (*limit 1* on Fig. 9.1), changes in the DO set point and fluctuations in the composition of the influent wastewater of SBR-20 impacted on BNR. Operation between days 20–45 with 60% DO, 442 ± 28 g_{COD} L_{Inf}^{-1}, 33 ± 3 g_{COD} g_P^{-1}, and 11 ± 1 g_{COD} g_N^{-1} (Fig. 9.1a) resulted in efficient nitrogen ($69 \pm 3\%$) and phosphorus removal ($96 \pm 3\%$) (Fig. 9.1b). Nitrogen removal decreased to 44% after an increase in DO from 60 to 80% on day 45 (*limit 2*), but recovered up to 70% after DO set-back to 60% on day 75 (*limit 3*). Decreases in nutrient ratios to 20 ± 1 g_{COD} g_P^{-1} and 9 ± 1 g_{COD} g_N^{-1} between days 80–100 (*limits 4–5*) resulted in successive deficiencies in nitrogen (47%) and phosphorus removal (69%) between days 100–130. After day 130 (*limit 6*), nitrogen (50%) and phosphorus removal (90%) recovered with increases in acetate concentration and in nutrient ratios up to 540 g_{COD} L_{Inf}^{-1}, 25 g_{COD} g_P^{-1}, and 10 g_{COD} g_N^{-1}, respectively. Actetate was fully taken up ($99 \pm 1\%$) and ammonium nitrified ($99 \pm 1\%$) during the whole experimental period (Fig. 9.1b).

In SBR-25, changes in DO impacted on both nitrification and dephosphatation. Fluctuations in nutrient ratios affected BNR, as well. Nitrification ($98 \pm 1\%$) and nitrogen removal ($83 \pm 6\%$) were efficient up to day 45 with 100% DO, 555 ± 20 g_{COD} L_{Inf}^{-1}, 27 ± 3 g_{COD} g_P^{-1}, and 9 ± 1 g_{COD} g_N^{-1} (Fig. 9.1c, d). Nitrification (51%) and dephosphatation (43%) declined after a decrease in DO to 80% on day 45 (*limit 2*), but recovered after DO set-back to 100% on day 75 (*limit 3*). Although nitrification recovered, nitrogen removal continued to decline. Switches from 540 ± 23 to 379 ± 9 mg_{CODs} L_{Inf}^{-1} and from 27 ± 3 to 21 ± 1 g_{COD} g_P^{-1} between days 90–100 (*limits 4–5*) resulted in poor nitrogen (< 30%) and phosphorus removal (< 70%). Whereas nitrification decreased to 90% after day 130 (*limit 6*), nitrogen and phosphorus removal improved to more than 50 and 90%, respectively. This correlated with slight increases in acetate concentration (413 mg_{COD} L_{Inf}^{-1}) and COD/P ratio (24 g_{COD} g_P^{-1}). Full anaerobic acetate uptake ($99 \pm 1\%$) was obtained during the whole experimental period.

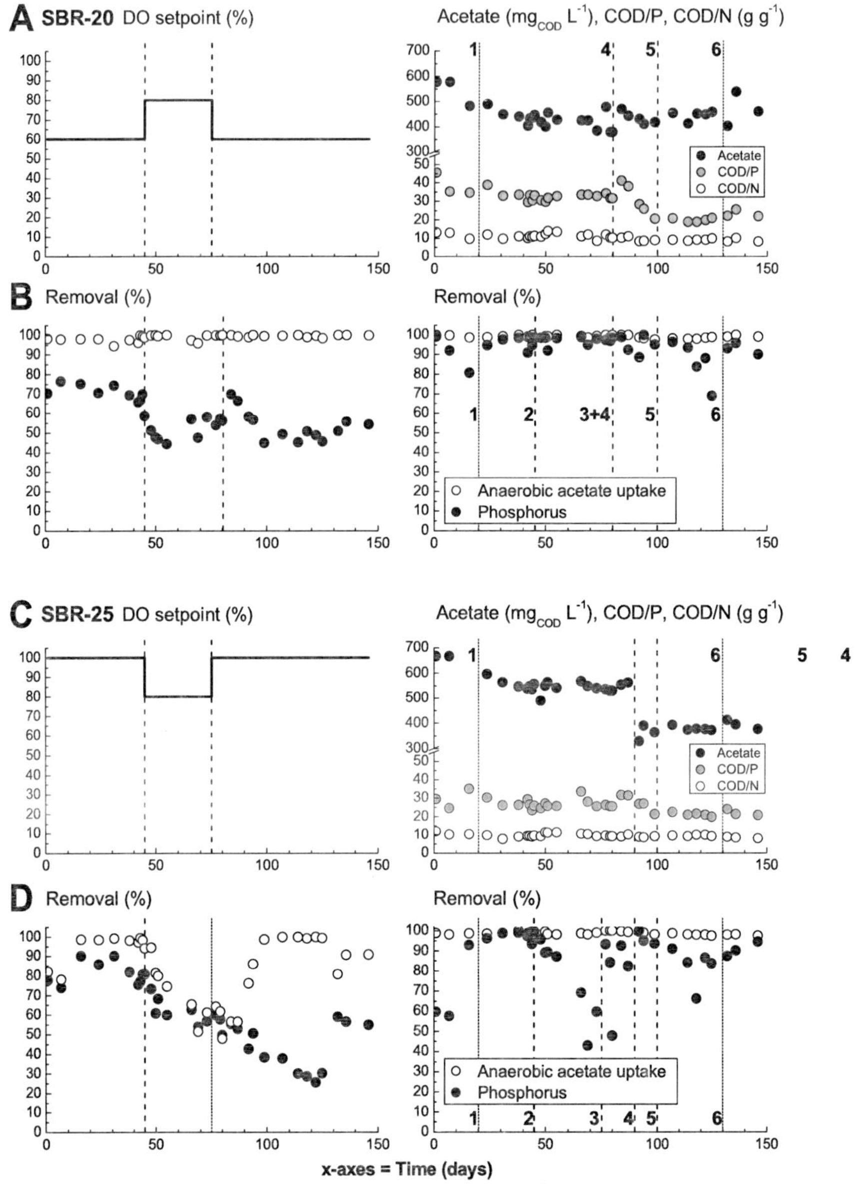

Fig. 9.1 Impact of switches in dissolved oxygen (DO) set-points and of fluctuations in the influent wastewater compositions on biological nutrient removal in SBR-20 (**a**, **b**) and SBR-25 (**c**, **d**). The main changes in operation variables are indicated by numbered vertical straight lines

Table 9.3 Average BNR performances during different periods of reactor operation according to changes in DO set-points, fluctuations in nutrient concentrations, and ratios of nutrients in the influent wastewater

Part 1

Time period (day–day)	Average levels of operation variables[1]					
	DO (%)	Acetate (mg_{COD} L^{-1})	P-PO4 (mg L^{-1})	N-NH4 (mg L^{-1})	COD/P (g g^{-1})	COD/N (g g^{-1})
SBR-20						
(1) 0–20	60	545 ± 55	14 ± 2	47 ± 4	39 ± 6	12 ± 2
(2) 20–45	60	442 ± 28	13 ± 1	42 ± 3	33 ± 3	11 ± 1
(3) 45–80	80	424 ± 30	13 ± 1	37 ± 4	32 ± 1	12 ± 2
(4) 80–100	60	426 ± 31	↗[4]	45 ± 6	↘[4]	10 ± 1
(5) 100–125	60	446 ± 19	23 ± 1	49 ± 1	20 ± 1	9 ± 1
(6) 125–150	60	467 ± 69	20 ± 2	54 ± 3	23 ± 2	9 ± 1
SBR-25						
(1) 0–20	100	685 ± 32	24 ± 3	64 ± 7	30 ± 5	11 ± 1
(2) 20–45	100	555 ± 20	21 ± 2	61 ± 6	27 ± 2	9 ± 1
(3) 45–80	80	541 ± 23	20 ± 1	53 ± 4	27 ± 3	10 ± 1
(4) 80–100	100	↘[4]	↘[4]	↘[4]	↘[4]	9 ± 1
(5) 100–125	100	379 ± 9	18 ± 1	40 ± 2	21 ± 1	10 ± 1
(6) 125–150	100	395 ± 18	18 ± 1	46 ± 1	22 ± 2	9 ± 1

Part 2

Time period (day–day)	Average levels of BNR performances[2]				Average	Average
	Anaerobic acetate uptake (%)	P-removal (%)	Nitrification (%)	N-removal (%)	PAO/GAO ratio[3] (–)	Abundance OTU-32[4] (%)
SBR-20						
(1) 0–20	99 ± 1	91 ± 10	98 ± 1	74 ± 3	2.7 ± 0.5	24 ± 4
(2) 20–45	99 ± 1	96 ± 3	98 ± 2	69 ± 3	1.6 ± 0.1	22 ± 4
(3) 45–80	99 ± 1	97 ± 2	99 ± 1	52 ± 5	1.6 ± 0.7	24 ± 6
(4) 80–100	99 ± 1	96 ± 4	100 ± 1	59 ± 8	2.0 ± 0.5	19 ± 2
(5) 100–125	98 ± 1	86 ± 11	99 ± 1	48 ± 3	1.5 ± 0.3	23 ± 2
(6) 125–150	99 ± 1	93 ± 3	100 ± 0	54 ± 3	1.5 ± 0.2	20 ± 4
SBR-25						
(1) 0–20	99 ± 1	70 ± 20	86 ± 11	81 ± 9	1.6 ± 0.8	33 ± 6
(2) 20–45	99 ± 1	98 ± 2	98 ± 2	82 ± 5	1.6 ± 0.8	32 ± 3
(3) 45–80	99 ± 1	79 ± 18	70 ± 13	61 ± 6	1.6 ± 0.4	32 ± 5
(4) 80–100	99 ± 1	85 ± 19	70 ± 20	48 ± 7	2.0 ± 1.0	30 ± 10

(continued)

Table 9.3 (continued)

Part 2

Time period (day–day)	Average levels of BNR performances[2]				Average	Average
	Anaerobic acetate uptake (%)	P-removal (%)	Nitrification (%)	N-removal (%)	PAO/GAO ratio[3] (–)	Abundance OTU-32[4] (%)
(5) 100–125	98 ± 1	82 ± 9	100 ± 1	30 ± 5	1.2 ± 0.2	28 ± 5
(6) 125–150	98 ± 1	91 ± 4	87 ± 6	57 ± 2	2.6 ± 1.1	33 ± 5

[1]Average levels of operation variables that fluctuated during reactor operation, namely dissolved oxygen setpoint (DO) during aeration phases, concentrations of chemical oxygen demand (COD), orthophosphate phosphorus (P-PO4), and ammonium nitrogen (N-NH4), and nutrient ratios in the influent wastewater (COD/P and COD/N)

[2]Average biological nutrient removal (BNR) performances for each period of change in operation variables

[3]Average ratio of "*Ca.* Accumulibacter" and *Comeptibacter* populations

[4]Average abundance of *Xanthomonadaceae* relatives (OTU-32)

[5]Progressive increase (↗) or decrease (↘) in the level of operation variables over the time period

9.3.2 *Overall Bacterial Community Compositions at 20 and 25 °C*

The T-RFLP profiles in Fig. 9.2 exhibit the dynamics of bacterial communiies of SBR-20 and SBR-25. The six periods of fluctuations in operation variables were superimposed. Phylogenetic affiliations of OTUs obtained with PyroTRF-ID are given in the Appendix A.

The AGS bacterial communities were composed of the same predominant populations over five months. *Xanthomonadaceae* (OTU-32), "*Ca.* Accumulibacter" (OTU-214) and "*Ca.* Competibacter" (OTU-239) affiliates dominated with average relative abundances of 16–28, 19–35, and 10–26% in SBR-20 (Fig. 9.2a), and 19–40, 18–39, and 8–24% in SBR-25 (Fig. 9.2b), respectively. The average PAO/GAO ratio was almost identical in SBR-20 (1.8 ± 0.6) and in SBR-25 (1.7 ± 0.8). Main differences between the two reactors were that *Xanthomonadaceae* affiliates were more abundant at 25 °C and that *Nitrospira* (OTU-260, 4–8%) and *Sphingobacteriales* affiliates (many OTUs between 252 and 364 bp; 5–10%) were more abundant at 20 °C. After the preliminary period of 65 days of AGS cultivation under dynamic conditions, *Xanthomonadaceae* (33%) and "*Ca.* Competibacter" (18%) were already more abundant in SBR-25 than in SBR-20 where "*Ca.* Accumulibacter" dominated (31%). Other populations of *Intrasporangiaceae* (OTU-223, 2–9%) and *Aminobacter* (OTU-220, 1–5%) exhibited comparable abundances at 20 and 25 °C. In SBR-20, "*Ca.* Accumulibacter" evolved proportionally to the acetate concentration and to the nutrients ratios, and that "*Ca.* Competibacter" displayed an inverse behavior.

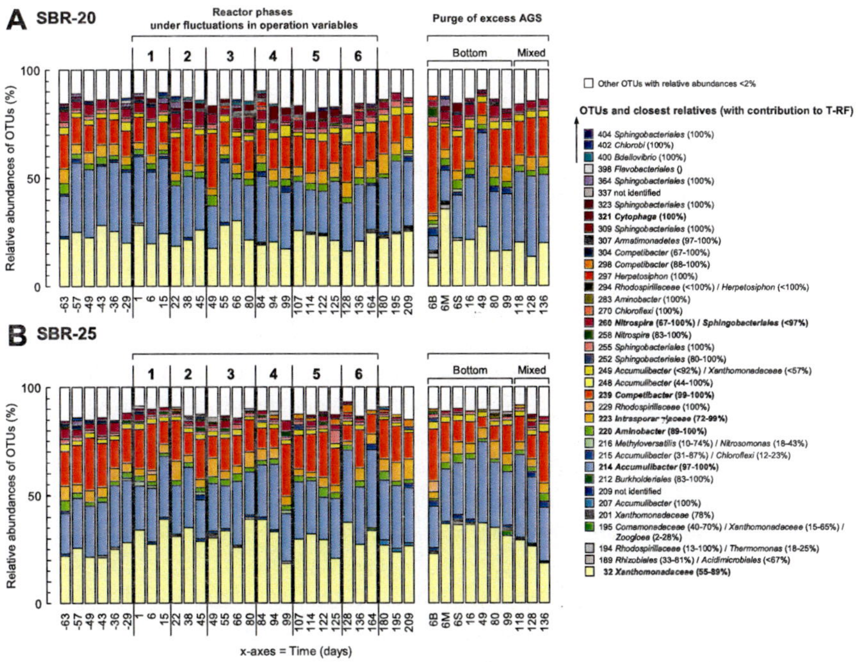

Fig. 9.2 Evolution of bacterial community compositions of AGS present in the reactors and in the purge streams during operation of SBR-20 (**a**) and SBR-25 (**b**). The same mature AGS was seeded in both reactors and grown during a preliminary period of 65 days (samples − 63 to − 29) under uncontrolled SRT to form sufficient bed heights. Experiments at fixed SRT started on day 1 and lasted over 150 days. Excess AGS was removed manually from the bottom of the reactors during the first 115 days, and then in an automated way from the aerated mixed liquors. The AGS were kept in the reactors over 50 additional days under the final conditions. On day 6, T-RFLP profiles were also measured on purged granules sorted by size, namely small granules below 1 mm (*S*), medium granules between 1 and 3 mm (*M*), and big granules above 3 mm (*B*). Closest bacterial affiliations obtained with PyroTRF-ID (General Appendix GA.IV) are given together with their relative contribution to the corresponding OTUs

In SBR-25, *Xanthomonadaceae* relatives were slightly more abundant with higher acetate load, nutrient ratios, and nitrogen removal.

Comparison with the bacterial community compositions of the excess AGS purged manually between days 15–99 at the bottom of the settled beds revealed that "*Ca.* Accumulibacter" were more abundant than "*Ca.* Competibacter" in the bottom fractions (PAO/GAO ratio = 3.8 ± 2.0) than in the mixed liquors of SBR-25 (1.8 ± 0.8) (Fig. 9.2). In SBR-20, no evident differences were found between the two sludge fractions with the exception of one high PAO/GAO ratio of 7.7 detected in the bottom sludge on day 49. A similar observation was made on this particular day in SBR-25. Granule size sorting on day 6 in both reactors revealed that "*Ca.* Competibacter" were more abundant in granules bigger than 2 mm (PAO/GAO ratios = 0.3–1.0), than in smaller granules (PAO/GAO ratios = 1.3–2.5). Aerated mixed AGS purged from the middle of the reactors phases after day 115 displayed close PAO/GAO ratios (1.3–3.1) to the ones of the AGS present in the reactors (1.4–3.1).

9.3.3 Hierarchical Clustering and PCA Revealed Distant Behaviors of SBR-20 and SBR-25

According to hierarchical clustering of operation and BNR datasets displayed in Additional File 9.1, SBR-20 (*Y symbols*) and SBR-25 (*O symbols*) exhibited distant behaviors up to day 90, but shared similar ones from then onwards (*blue clusters*). On the contrary, similar bacterial community structures were obtained during the pre-cultivation of AGS beds, but distant ones during operation at fixed SRT.

A principal component analysis (PCA; Fig. 9.3) showed that the distinct initial behaviors of SBR-20 (cluster Y1) and SBR-25 (clusters O1-O3) were explained by opposite correlations. Cluster Y1 positively correlated with operation at higher nutrient ratios, and with higher phosphorus removal and nitrification, whereas clusters O1 and O2 inversely correlated with these performances. Cluster O2 positively correlated with higher nitrogen removal as a consequence of higher acetate concentrations in the influent wastewater. Later stages of SBR-20 (days 92–146) and SBR-25 (days 99–125) corresponding to periods 5–6 affiliated with the same cluster (Y2 = O3) that correlated with high nitrate concentrations in the effluent, i.e. efficient nitrification but poor nitrogen removal. The nitrification vector explained the vertical separation in Fig. 9.3 between clusters Y1–Y2 (positive correlation) and clusters O1–O2 (inverse correlation). Contrary to SBR-25, ammonium was fully and constantly nitrified in SBR-20. Since the nitrification and phosphorus removal vectors were opposed to the temperature and DO vectors, these two operation variables were suspected to have impacted on the performances of aerobic processes in SBR-25, together with the higher ammonium and orthophosphate concentrations in the influent of this reactor. The nitrogen removal vector explained the horizontal separation between clusters Y1–O2 and cluster Y2 = O3, and correlated with higher acetate concentration, COD/P, and COD/N.

9.3.4 Correlations Between Operation, BNR, and Bacterial Community Datasets

The extent and significance of correlations between operation and BNR parameters were confirmed by pair-wise Spearman's rank correlation coefficients and p-values (Additional file 9.2).

The two full datasets correlated significantly for both SBRs. The BNR performances were mainly explained by the influent orthophosphate concentration and the

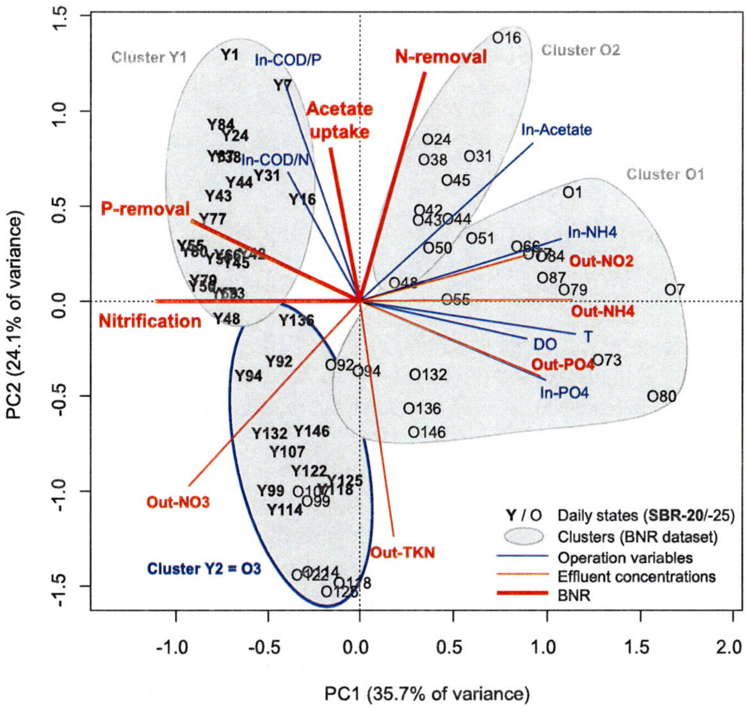

Fig. 9.3 Principal component analysis (PCA) of the concatenated datasets of operation variables and BNR performances of SBR-20 (*bold Y symbols*) and SBR-25 (*light O symbols*). The gray ellipses represent the five main clusters (Y1–Y2, O1–O4) of daily reactor states obtained after hierarchical clustering analysis of the BNR dataset in Additional File 9.1. The cluster "Y2 = O3" indicates common behavior of both SBRs during later operation. The direction and the length of the red vectors indicate the extent of positive and inverse correlations between the operation variables and the BNR performances (nutrient removal and residual concentrations in effluent) for each cluster of daily states

nutrient ratios in SBR-20, and by the influent acetate and ammonium concentrations, and COD/P in SBR-25.

According to pair-wise correlations, dephosphatation was efficient in SBR-20 during periods of higher DO, lower orthophosphate concentration, and higher nutrient ratios. Nitrogen was efficiently removed with higher acetate concentration and nutrient ratios. Since full anaerobic acetate uptake and nitrification were always obtained, the correlations displayed in this table only explained minor fluctuations in these performances.

In SBR-25, nitrification was efficient during periods of higher DO, lower concentrations of acetate and ammonium, and lower COD/P. Nitrogen was efficiently removed under the opposite conditions. Only low correlations were obtained for dephosphatation. Since COD/N was almost constant, the related correlation coefficient should not be considered.

Significant correlations between bacterial community structures and almost all operation and BNR parameters, except DO and anaerobic acetate uptake, were only obtained for SBR-20. Highest correlations were obtained with orthophosphate concentration, nutrient ratios, and nitrogen removal.

9.3.5 Identification of Bacterial Relatives Sharing Similar Behavior

Since significant correlations were obtained for SBR-20, a multiple factor analysis (MFA) was conducted on the three datasets in order to detect correlations between the dynamics of bacterial relatives and the fluctuations in operation variables and BNR performances (Additional File 9.3).

Anaerobic acetate uptake and nitrification were not considered in the MFA since full conversions with only minor variations were constantly obtained. The interpretation mainly focused on phosphorus and nitrogen removal. The groups representation confirmed that the three datasets were closely connected (Additional File 9.3a). The same type of analysis revealed that only the operation and BNR datasets were connected in SBR-25 (data not shown).

The evolution of each dataset in SBR-20 were represented in an individual factor map (Additional File 9.3b) where the daily states of each dataset were localized in a statistical space (*colored points*) and the overall daily reactor states were averaged over the three datasets (*centroid black points*).

The superimposition of the clusters of daily bacterial community compositions obtained by hierarchical clustering (Additional File 9.3c) to the individual factor maps displayed structuring of the bacterial community dynamics in three successive periods.

Since individual factor maps and correlation circles (Additional File 9.3d) were computed in related spaces, multilevel correlations were provided by positions of clusters, and direction and length of vectors of operation variables, BNR performances, and OTUs.

The heatmap in Fig. 9.4 provides a compact view of the clusters of bacterial relatives, comprising dominant populations and accompanying guilds that shared similar pair-wise correlations with single operation variables and BNR performances. The obtained clusters of bacterial relatives were superimposed in Additional File 9.3d to the correlation circle of the bacterial community dataset.

The MFA revealed that operation during the first 40 days (cluster Y1) with higher acetate concentration, COD/P, and COD/N resulted in efficient nitrogen removal, and was related to the bacterial cluster I (Additional File 9.3d). Cluster I was composed of predominant "*Ca.* Accumulibacter" (OTU-214) and *Nitrospira* (OTU-260) relatives accompanied by populations of *Bdellovibrio*, *Thermomonas*, *Methyloversatilis*, *Nitrosomonas,* and others (Fig. 9.4). The next period up to day 100 (cluster Y2) with higher DO resulted in efficient phosphorus removal but less efficient

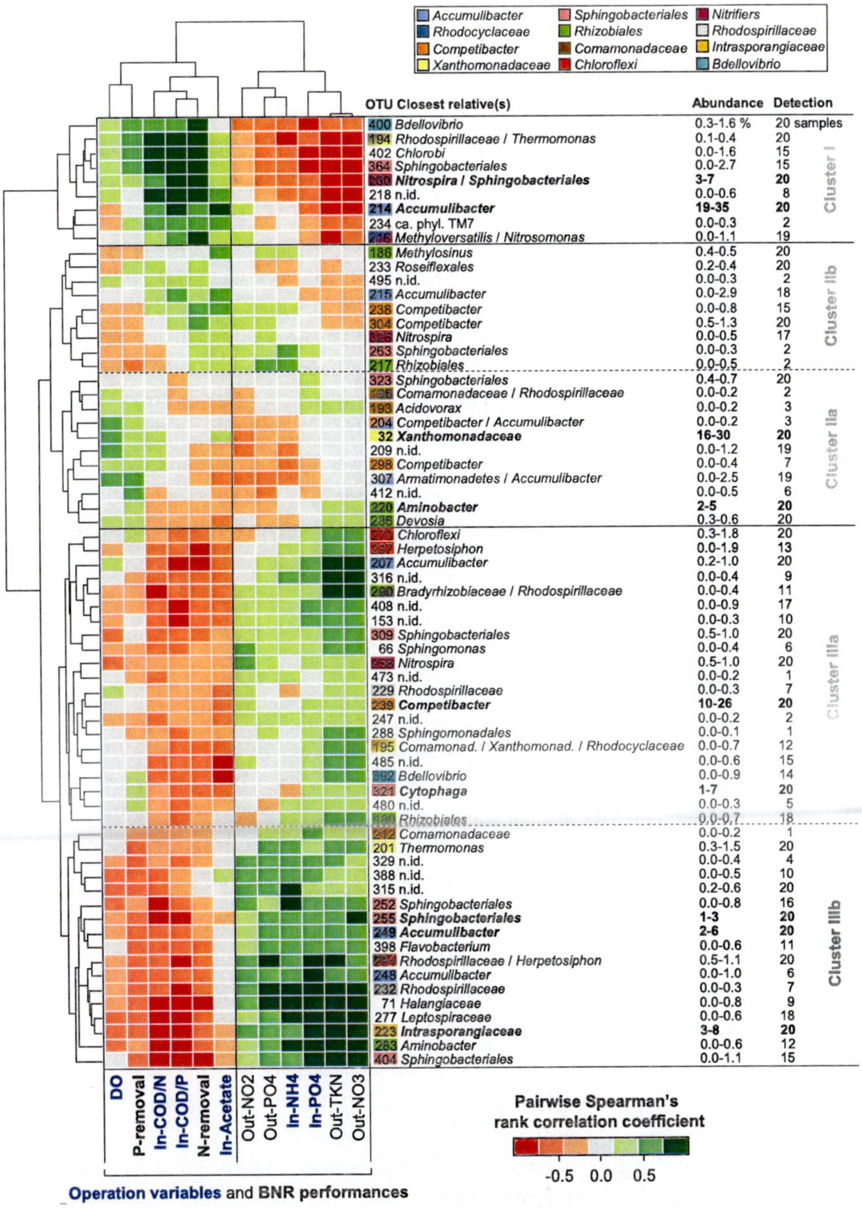

Fig. 9.4 Heat map representation of the pair-wise Spearman's rank correlations between OTUs and operation and BNR parameters in SBR-20. The top dendrogram defines clusters of operation and BNR parameters correlating with each other. The left dendrogram defines clusters of OTUs sharing similar correlations with operation and BNR parameters. Each OTU is given with the closest bacterial affiliation(s), with the minimum–maximum range of relative abundance, and the number over a maximum of 20 samples in which the OTU was detected

nitrogen removal, and in higher relative abundances of predominant *Xanthomonadaceae* (OTU-32) and *Aminobacter* (OTU-220) relatives inside sub-cluster IIa, and in lower abundances of non-dominant populations of "*Ca.* Accumulibacter", "*Ca.* Competibacter", *Nitrospira*, *Sphingobacteriales*, *Armatimonadetes*, and *Rhizobiales* inside sub-cluster IIb. Operation over the last 50 days with lower acetate concentration and nutrient ratios was related to deficient nitrogen and phosphorus removal, and higher abundances in bacterial relatives belonging to cluster III. Sub-cluster IIIa was composed of the predominant "*Ca.* Competibacter" (OTU-239) and *Sphingobacteriales*-related *Cytophaga* (OTU-321) affiliates, as well as accompanying populations of *Chloroflexi*, "*Ca.* Accumulibacter", *Rhodospirillaceae*, *Nitrospira*, and *Rhizobiales* relatives, and was directly opposed to the vector of the predominant "*Ca.* Accumulibacter" affiliate (OTU-214). Sub-cluster IIIb comprised *Intrasporangiaceae* (OTU-223) relatives and secondary "*Ca.* Accumulibacter" (OTU-249) and *Sphingobacteriales* (OTU-255) affiliates, as well as accompanying populations of *Thermomonas*, *Rhodospirillaceae*, and *Aminobacter*. Overall, the heatmap of Fig. 9.4 indicates that enhanced BNR performances were obtained under conditions selecting for bacterial relatives of clusters I and II, and that BNR deteriorated when conditions where more favorable for the members of cluster III.

9.4 Discussion

Since wastewater treatment systems are operated on long term with variations in influent wastewater composition and environmental conditions, the interest of this study relies on the knowledge gained on the impact of operation variables and their fluctuations on BNR performances and bacterial community structures of anaerobic-aerobic AGS-SBRs.

9.4.1 Importance of Biomass-Related Steady-State Conditions in AGS Systems

AGS-SBRs have mainly been operated under conditions of low settling time and low HRT to cultivate and maintain a fast-settling granular biomass in the system by washing out slow-settling suspended flocs (Beun et al. 1999). Wash-out conditions not only exert a pressure on sludge for the selection of fast-settling granules but also on bacterial communities that can lead to inefficient BNR until a sufficient amount of AGS has accumulated (Weissbrodt et al. 2012a).

According to Weissbrodt et al. (2012a; 2013/Chaps. 7 and 8), the granulation process has been subjected to dynamics in the SRT, with a SRT as low as the HRT during the transition from flocs to granules, and with a SRT decoupling from the HRT during AGS accumulation and reaching values higher than 30 days at mature stage. Further operation of acetate-fed AGS-SBRs without purge of excess sludge has led to instable dephosphatation that was linked under too high SRT to saturation of PAOs by intracellular polyphosphate and to the establishment of GAOs (de Kreuk et al. 2007; Gonzalez-Gil and Holliger 2011).

In the present study, the operation of the two acetate-fed SBR-20 and SBR-25 with fixed SRTs of 20 days by regular purge of excess AGS was not only leading to effective phosphorus removal, but also to stable overall bacterial community compositions. Over a period of five months, the bacterial communities were indeed composed of the same predominant phylotypes, and were characterized by average PAO/GAO ratios of 1.8 ± 0.6 and 1.7 ± 0.8, respectively, that are likely to favor efficient dephosphatation when compared to the ratios reported by Lemaire et al. (2008).

Decreasing gradients in granule size and in PAO/GAO ratio have been observed along AGS beds, and selective purge of the upper part of the settled AGS bed has led to improved dephosphatation even at higher mesophilic temperatures (Winkler et al. 2011; Bassin et al. 2012). The operation of SBR-20 and SBR-25 with manual purge of bottom bed fractions did however not affect the PAO/GAO balance and dephosphatation. It was only confirmed in SBR-25 that "*Ca.* Accumulibacter" were more abundant in bottom fractions, and no clear gradient in granule size were observed along AGS beds. Gradients in granule size and in PAO/GAO ratios along AGS beds are thus system specific.

Although a selective purge has been efficient for decoupling the SRT of specific bacterial competitors in bubble-column SBRs, its application in full-scale systems designed with H/D ratios near unity remains puzzling. The purge of excess AGS from aerated mixed liquors tested during later SBR operations may be more easily achievable, and suitable for maintaining a precise sludge age. The different approaches are nevertheless in agreement with Morgenroth and Wilderer (1999) who concluded that controlled biomass removal is the key parameter to achieve biological dephosphatation in biofilm systems.

9.4.2 *Impact of AGS Bed Volume and Granule Size on Oxygenation of Biomass*

An explanation for the distinct nitrification performances displayed by SBR-20 and SBR-25 during the first 90 days is provided hereafter. According to the physiological properties of nitrifiers in activated sludge (Wild et al. 1971; Ekama and Wentzel 2008; Gujer 2010), nitrification should have been more efficient in SBR-25 operated at higher mesophilic temperature and higher DO. However, full nitrification was obtained in SBR-20 with 60% DO whereas nitrification deteriorated in SBR-25 already with 80% DO. Other authors have previously shown the impact of floc and biofilm properties on the mass transfer of solutes (e.g. DO) from the bulk liquid phase to bacterial cells (Stenstrom and Song 1991; Manser et al. 2005; Mosquera-Corral et al. 2005; Liu and Tay 2007; Gujer 2008; Morgenroth 2008; Filali et al. 2012). An analysis of the three-phase systems of SBR-20 and SBR-25 was conducted here to determine the efficiency of the transfert of oxygen from the air gas bubbles to the bulk liquid phase (gas-liquid interface, G/L) and from the bulk liquid phase to the biofilm (liquid-film interface, L/F) (Table 9.2).

At the gas-liquid interface (G/L), theoretical coefficients of volumetric mass transfer (k_La) of oxygen of 0.031 and 0.035 s^{-1} were calculated at 20 and 25 °C, respectively, under operation with identical up-flow superficial gas velocities (SGV) of 0.015 m s^{-1}, based on Heijnen and Van't Riet (1984). Twice and 5-times lower k_La were however measured, respectively, with the dynamic method (Mena et al. 2005). Contrary to DO saturation that decreases with temperature, k_La should have been more efficient at higher temperature (Heijnen and Van't Riet 1984; Vogelaar et al. 2000). The presence of solid particles has further been described to affect the gas-liquid mass transfer of oxygen in shallow bubble columns by increasing the coalescence of gas bubbles (Mena et al. 2005; Lau et al. 2010). This effect is however inversely proportional to the size of granules. According to the authors, although bigger granules were present in SBR-25, the lower k_La measured in this reactor can be related to the slightly higher liquid height to reactor diameter ratio, and to the higher total volume of granular biomass present in the system.

At the liquid-biofilm interface (L/F), the total theoretical biofilm surface area available for the DO transfer was almost twice lower in SBR-25, as a consequence of bigger granules despite higher total volume of AGS. Since Moy et al. (2002) have found a positive correlation between granule size and organic loading rate, bigger granules in SBR-25 were possibly explained by initial operation of this reactor with more than 650 mg_{COD} L^{-1} of acetate. Temperature has further not been described to affect the granule size (de Kreuk et al. 2007; Song et al. 2009; Ebrahimi et al. 2010), and both reactors were operated with identical SGV.

Overall, aerobic processes of nitrification and dephosphatation were affected in SBR-25 by limited mass transfer of oxygen at both gas-liquid interface caused by the presence of a higher amount of AGS particles and liquid-biofilm interface caused by the presence of bigger granules, i.e. lower interfacial biofilm surface area. Although the size and shape of granules are hardly controllable (de Kreuk et al. 2007), these parameters should therefore be measured periodically for feed-back adaptation of operation variables to ensure optimal BNR.

9.4.3 Impact of Fluctuations in Operation Variables in AGS Systems

The use of a numerical ecology approach including hierarchical clustering and multivariate statistical analyses, such as the ones applied for the investigation of natural systems (Borcard et al. 2011), was efficient for structuring the operation, BNR, and bacterial community datasets, for detecting multilevel correlations, for highlighting clusters of predominant bacterial relatives and accompanying populations sharing common behaviors in the AGS-SBRs, and for identifying the conditions that select for target populations. Rees et al. (2004) and Wang et al. (2012) have previously related complex T-RFLP and pyrosequencing patterns of microbial communities to operation conditions via statistical computations.

Relative to the fluctuation ranges of wastewater composition encountered during this study, phosphorus and nitrogen were efficiently removed in the AGS-SBRs operated with a biomass concentration of 12 g_{VSS} L^{-1} and a SRT of 20 days with at least 500 mg_{COD} L^{-1} (i.e. sludge loading rate above 14 kg_{CODs} d^{-1} $m^{-3}{}_{granules}$), 30 g_{COD} $g_P{}^{-1}$, and 10 g_{COD} $g_N{}^{-1}$. With lower levels, these heterotrophic processes were limited in carbon. In a modeling study of de Kreuk et al. (2007) investigating impacts of SRT, sludge loading rate, DO and particle size on BNR in AGS-SBRs, a sludge loading rate of 13 kg_{CODs} d^{-1} $m^{-3}{}_{granules}$ has been optimal for nitrogen removal, whereas a decrease to 9 kg_{CODs} d^{-1} $m^{-3}{}_{granules}$ has led to lower accumulation of poly-β-hydroxyalkanoates by PAOs, to an increase of SRT from 20 to 38 days, and to deteriorated dephosphatation. The stoichiometric analysis of Kuba et al. (1996) including biomass growth has shown that full anoxic dephosphatation by denitrifying PAOs in activated sludge SBRs was obtained with 3.4–3.8 g_{COD} $g_N{}^{-1}$ for fixed 27 g_{COD} $g_P{}^{-1}$ (i.e. 400 mg_{CODs} L^{-1} for 4–7 g_N $g_P{}^{-1}$). In the case of separated dephosphatation and denitrification by different organisms, 600–774 mg_{CODs} L^{-1} were however required for 4–7 g_N $g_P{}^{-1}$. Since the influents of SBR-20 and SBR-25 were composed of 2–4 g_N $g_P{}^{-1}$, an acetate supply below 500 mg_{CODs} L^{-1} would have been sufficient for anoxic dephosphatation, but limiting for separate processes. According to the limited performances measured at lower concentrations, phosphorus and nitrogen removals were thus not conducted by only PAOs.

The effect of the COD/P ratio has been investigated on the PAO/GAO balance in activated sludge and AGS, and 20 and 200 g_{COD} $g_P{}^{-1}$ have typically been used to enrich for "*Ca.* Accumulibacter" and "*Ca.* Competibacter", respectively (Schuler and Jenkins 2003; de Kreuk and van Loosdrecht 2004; Oehmen et al. 2007). It was observed here that "*Ca.* Competibacter" was able to compete with "*Ca.* Accumulibacter" even at COD/P ratios below 30 g_{COD} $g_P{}^{-1}$. Schuler and Jenkins (2003) have considered the PAO/GAO competition on a continuum of the two metabolisms along the COD/P axis. Ratios below 8.3 g_{COD} $g_P{}^{-1}$ have been efficient for non-phosphorus-limited selection of active PAOs, but have led to incomplete dephosphatation because of phosphorus in excess in the influent. Increasing COD/P ratios have resulted in phosphorus-limited PAO activity and in a potential for GAOs to establish in the sludge, but in effective dephosphatation because of non-excessive phosphorus in the influent. Ratios above 25 g_{COD} $g_P{}^{-1}$ have been recommended for safe operation of dephosphatation systems. The results presented here were thus in agreement with this threshold.

The better efficiency of nitrogen removal in SBR-20 at lower DO can be explained by a decreased penetration of DO in the granules that can result in bigger anoxic biofilm volumes available for denitrification (Manser et al. 2005; Mosquera-Corral et al. 2005; Volcke et al. 2012). De Kreuk et al. (2007) have modeled that 1.2–1.4 mm granules and specific biofilm surface areas of 600–700 m^2 $m_r{}^{-3}$ were optimal for balanced aerobic and anoxic processes with a DO of 2 mg_{O2} L^{-1} in the medium.

In the case of SBR-20, the granules characteristics fitted with these ranges (1.5 ± 0.9 mm and 628 m^2 m_r^{-3}, Table 9.2), and thus oxygen could have been saved by operation with a DO below 60% (5.5 mg_{O2} L^{-1}). Oxygenation strategies should nevertheless integrate impacts on dephosphatation, since a positive correlation was detected between this process and DO.

9.4.4 A Structured Bacterial Community Continuum

According to multilevel correlations, SBR-20 displayed a bacterial community structure correlating with the operation and BNR parameters. The mass transfer limitations detected in SBR-25 could explain the low correlations between bacterial community structures and BNR efficiencies found for this reactor. Under mass transfer limitations, populations present in granules would not have been able to sense fluctuations in environmental variables as fast as in SBR-20.

In the latter reactor, the conditions favoring BNR were selecting for the predominant relatives within populations of "*Ca.* Accumulibacter" (PAOs), *Nitrosomonas* (AOOs), *Nitrospira* (NOOs), *Xanthomonadaceae*, *Aminobacter*. Whereas *Xanthomonadaceae* belongs to the denitrifying community of activated sludge and AGS systems (Thierry et al. 2004; Adav et al. 2010), *Aminobacter* inside *Rhizobiales* can accumulate poly-β-hydroxyalkanoates and reduce nitrate to nitrite (Urakami et al. 1992; Kondo et al. 2009), and are also able to degrade xenobiotic substances such as estradiols and ethylenediaminetetraacetic acid (EDTA) (Bucheli-Witschel and Egli 2001; Yu et al. 2007; Yuan and VanBriesen 2008). EDTA was indeed present in relatively high amounts in the cultivation media adapted from de Kreuk et al. (2005), and degradation was observed under aerobic conditions in both SBRs. Conditions leading to deteriorated BNR were more favorable for "*Ca.* Competibacter" (GAOs), *Sphingobacteriales* (*Cytophaga*), and *Intrasporangiaceae* relatives. *Sphingobacteriales* affiliates are widely present in activated sludge plants, and metabolize macromolecules such as polysaccharides and proteins (Kirchman 2002). The abundant extracellular polymeric substances present in granular biofilms have been suspected to favour their establishment in AGS (Weissbrodt et al. 2013/Chap. 8). The *Intrasporangiaceae* affiliates, such as the glucose-fermenting and putative polyphosphate-accumulating *Tetrasphaera* spp., have been reported to proliferate under similar conditions as "*Ca.* Competibacter". Higher abundances of the gammaproteobacterial "*Ca.* Competibacter" and *Xanthomonadaceae* relatives (e.g. *Thermomonas*, *Pseudoxanthomonas*) at 25 °C also met with physiological descriptions provided in literature (Thierry et al. 2004; Lopez-Vazquez et al. 2009a).

The low-abundant accompanying populations were heterogeneously distributed among the different clusters. Other "*Ca.* Accumulibacter", "*Ca.* Competibacter", *Sphingobacteriales*, and *Aminobacter* relatives were found in the antagonistic clusters. Some flanking populations were mainly affiliated to one particular cluster. *Chlorobi*, *Methyloversatilis*, nitrogen-fixing *Rhizobiales* (*Methylosinus*,

Devosia), *Roseiflexales*, denitrifying *Comamonadaceae* (*Acidovorax*), *Armatimonadetes*, and TM7 relatives affiliated with the "*Ca.* Accumulibacter" cluster. *Chloroflexi* (*Herpetosiphon, Anaerolineae*), *Rhodospirillaceae, Flavobacterium, Halangiaceae*, and *Leptospiraceae* populations affiliated with the "*Ca.* Competibacter" cluster. *Defluviicoccus* spp. inside *Rhodospirillaceae* has for instance been reported as a putative alphaproteobacterial GAOs (Meyer et al. 2006; Nielsen et al. 2010). The fact that "*Ca.* Accumulibacter" and "*Ca.* Competibacter" can be composed of different clades with different aerobic and anoxic metabolisms (Oehmen et al. 2010b) can explain the presence of multiple OTUs related to these organisms. Additional research on fine-scale dynamics of different clades would require the use of molecular methods targeting functional genes (Kim et al. 2010; Slater et al. 2010; Gonzalez-Gil and Holliger 2011).

The different organisms detected here have also been detected in other BNR AGS systems (Adav et al. 2010), and belong to the core microbiome of full-scale BNR activated sludge systems (Nielsen et al. 2012; Wang et al. 2012). According to Nielsen et al. (2012), the dynamics resulting from alternating anaerobic-aerobic conditions select for versatile populations active during the various reactor phases. Proper management of the bacterial resource in wastewater treatment systems has been recommended by considering a bacterial community as a cooperative continuum (Verstraete et al. 2007). A conceptual model of bacterial interactions within the core microbiome of activated sludge has been proposed by Nielsen et al. (2010). Although AGS-SBR operation under defined synthetic conditions induces a decrease in the bacterial diversity (Weissbrodt et al. 2013/Chap. 8), the number of low-abundant flanking populations detected here is in agreement with these approaches.

9.5 Conclusions

The detailed investigations conducted on the multilevel correlations between operation variables, BNR performances, and bacterial community dynamics in mature anaerobic-aerobic AGS-SBRs led to the following conclusions:

- Operation under stead-state biomass conditions at a moderate SRT of 20 days by purging excess AGS was crucial for effective phosphorus removal and for stabilizing the overall composition of bacterial communities.
- Limitation of oxygen mass transfer was an issue in the granular biofilm reactors, but was system specific. Adaptation of DO setpoints to the amount and characteristics of granules is required for optimized aerobic and anoxic processes.
- Efficient BNR was obtained with concentrations of acetate above 500 $mg_{COD}\ L^{-1}$, COD/P ratios above 30 $g_{COD}\ g_P{}^{-1}$, and COD/N ratios above 10 $g_{COD}\ g_N{}^{-1}$. Thus, denitrification was apparently not restricted to only PAOs in the anaerobic-aerobic AGS-SBRs, but relied on a broad denitrifying community.

- The bacterial community structure was identified according to correlations between the dynamics of clusters of bacterial relatives and the fluctuations in operation and BNR parameters. The predominant and flanking populations present in AGS belonged to the core bacterial microbiome of BNR activated sludge processes.
- The bacterial community continuum was composed of two major opponent clusters comprising populations of (i) "*Ca.* Accumulibacter", *Nitrospira*, *Xanthomonadaceae*, and *Aminobacter*, and of (ii) "*Ca.* Competibacter", *Sphingobacteriales*, *Cytophaga*, and *Intrasporangiaceae*, respectively. The first cluster correlated to conditions leading to efficient BNR, whereas the second was linked to periods of inefficient BNR.
- Surprisingly, *Xanthomonadaceae* affiliates were predominant in these BNR granules. Further physiological studies should determine how these denitrifiers can thrive in the granular ecosystem under anaerobic-aerobic conditions.

Acknowledgements Noam Shani (EPFL-LBE and Agroscope Liebefeld-Posieux, Switzerland) for collaboration on multivariate statistical analyses. François Gillet (Université de Franche-Comté, France) and Daniel Borcard (Université de Montréal, Canada) for advice on numerical ecology. Scot E. Dowd and the Research and Testing Laboratory (Lubbock, USA) for pyrosequencing analyses.

Appendix

The Appendix is available at the end of Chap. 5 on PyroTRF-ID.

Appendix A Phylogenetic affiliations obtained with PyroTRF-ID.

Supplementary Information

Additional File 9.1 Hierarchical clustering analyses of operation, BNR, and bacterial community datasets of SBR-20 and SBR-25.

Additional File 9.2 Correlation coefficients used to quantify the relations between the three datasets.

Additional File 9.3 Multiple factor analysis (MFA) providing a picture of the relationships among operation, BNR, and bacterial community datasets of SBR-20.

References

Adav SS, Lee DJ, Lai JY (2010) Microbial community of acetate utilizing denitrifiers in aerobic granules. Appl Microbiol Biotechnol 85(3):753–762

Bassin JP, Winkler MKH, Kleerebezem R, Dezotti M, van Loosdrecht MCM (2012) Improved phosphate removal by selective sludge discharge in aerobic granular sludge reactors. Biotechnol Bioeng 109(8):1919–1928

Beun JJ, Hendriks A, van Loosdrecht MCM, Morgenroth E, Wilderer PA, Heijnen JJ (1999) Aerobic granulation in a sequencing batch reactor. Water Res 33(10):2283–2290

Borcard D, Gillet F, Legendre P (2011) Numerical ecology with R, 1st edn. Springer-Verlag GmbH, Heidelberg

Bucheli-Witschel M, Egli T (2001) Environmental fate and microbial degradation of aminopoly-carboxylic acids. FEMS Microbiol Rev 25(1):69–106

Day A (2007) heatmap.plus: Heatmaps with more sensible behavior. R package version 1.3. R Foundation for Statistical Computing, Vienna. http://cran.r-project.org/web/packages/heatmap.plus/index.html

de Kreuk MK, van Loosdrecht MCM (2004) Selection of slow growing organisms as a means for improving aerobic granular sludge stability. Water Sci Technol 49:9–17

de Kreuk MK, Heijnen JJ, van Loosdrecht MCM (2005) Simultaneous COD, nitrogen, and phosphate removal by aerobic granular sludge. Biotechnol Bioeng 90(6):761–769

de Kreuk MK, Picioreanu C, Hosseini M, Xavier JB, van Loosdrecht MCM (2007) Kinetic model of a granular sludge SBR: influences on nutrient removal. Biotechnol Bioeng 97(4):801–815

Ebrahimi S, Gabus S, Rohrbach-Brandt E, Hosseini M, Rossi P, Maillard J, Holliger C (2010) Performance and microbial community composition dynamics of aerobic granular sludge from sequencing batch bubble column reactors operated at 20, 30, and 35°C. Appl Microbiol Biotechnol 87:1555–1568

Ekama GA, Wentzel MC (2008) Nitrogen removal. In: Henze M, van Loosdrecht MCM, Ekama GA, Brdjanovic D (eds) Biological wastewater treatment: principles, modelling and design. IWA Publishing, London, pp 87–138

Filali A, Bessiere Y, Sperandio M (2012) Effects of oxygen concentration on the nitrifying activity of an aerobic hybrid granular sludge reactor. Water Sci Technol 65(2):289–295

Gonzalez-Gil G, Holliger C (2011) Dynamics of microbial community structure and enhanced biological phosphorus removal of propionate- and acetate-cultivated aerobic granules. Appl Environ Microbiol 77:8041–8051

Gujer W (2008) Systems analysis for water technology. Springer Verlag, Berlin, Heidelberg

Gujer W (2010) Nitrification and me: a subjective review. Water Res 44(1):1–19

Heijnen JJ, Van't Riet K (1984) Mass transfer, mixing and heat transfer phenomena in low viscosity bubble column reactors. Chem Eng J 28(2):B21–B42

Kim JM, Lee HJ, Kim SY, Song JJ, Park W, Jeon CO (2010) Analysis of the fine-scale population structure of "*Candidatus* accumulibacter phosphatis" in enhanced biological phosphorus removal sludge, using fluorescence in situ hybridization and flow cytometric sorting. Appl Environ Microbiol 76(12):3825–3835

Kirchman DL (2002) The ecology of cytophaga–flavobacteria in aquatic environments. FEMS Microbiol Ecol 39(2):91–100

Kishida N, Kim J, Tsuneda S, Sudo R (2006) Anaerobic/oxic/anoxic granular sludge process as an effective nutrient removal process utilizing denitrifying polyphosphate-accumulating organisms. Water Res 40(12):2303–2310

Kondo T, Tsuneda S, Ebie Y, Inamori Y, Xu K (2009) Characterization of the microbial community in the anaerobic/oxic/anoxic process combined with sludge ozonation and phosphorus adsorption. JWET 7(3):155–162

Kuba T, van Loosdrecht MCM, Heijnen JJ (1996) Phosphorus and nitrogen removal with minimal COD requirement by integration of denitrifying dephosphatation and nitrification in a two-sludge system. Water Res 30(7):1702–1710

Lau R, Mo R, Beverly Sim WS (2010) Bubble characteristics in shallow bubble column reactors. Chem Eng Res Des 88(2):197–203

Legendre P, Gallagher E (2001) Ecologically meaningful transformations for ordination of species data. Oecologia 129(2):271–280

Lemaire R, Yuan Z, Blackall LL, Crocetti GR (2008) Microbial distribution of *Accumulibacter* spp. and *Competibacter* spp. in aerobic granules from a lab-scale biological nutrient removal system. Environ Microbiol 10(2):354–363

Liu Y-Q, Tay J-H (2007) Cultivation of aerobic granules in a bubble column and an airlift reactor with divided draft tubes at low aeration rate. Biochem Eng J 34(1):1–7

Lopez-Vazquez CM, Hooijmans CM, Brdjanovic D, Gijzen HJ, van Loosdrecht MCM (2009a) Temperature effects on glycogen accumulating organisms. Water Res 43(11):2852–2864

Lopez-Vazquez CM, Oehmen A, Hooijmans CM, Brdjanovic D, Gijzen HJ, Yuan Z, van Loosdrecht MCM (2009b) Modeling the PAO-GAO competition: effects of carbon source, pH and temperature. Water Res 43(2):450–462

Manser R, Gujer W, Siegrist H (2005) Consequences of mass transfer effects on the kinetics of nitrifiers. Water Res 39(19):4633–4642

Mena PC, Pons MN, Teixeira JA, Rocha FA (2005) Using image analysis in the study of multiphase gas absorption. Chem Eng Sci 60(18):5144–5150

Meyer RL, Saunders AM, Blackall LL (2006) Putative glycogen-accumulating organisms belonging to the *Alphaproteobacteria* identified through rRNA-based stable isotope probing. Microbiology 152:419–429

Morgenroth E (2008) Modelling biofilms. In: Henze M, van Loosdrecht MCM, Ekama GA, Brdjanovic D (eds) Biological wastewater treatment: principles, modelling and design. IWA Publishing, London, pp 457–492

Morgenroth E, Wilderer PA (1999) Controlled biomass removal: the key parameter to achieve enhanced biological phosphorus removal in biofilm systems. Water Sci Technol 39(7):33–40

Mosquera-Corral A, de Kreuk MK, Heijnen JJ, van Loosdrecht MCM (2005) Effects of oxygen concentration on N-removal in an aerobic granular sludge reactor. Water Res 39(12):2676–2686

Moy BYP, Tay JH, Toh SK, Liu Y, Tay STL (2002) High organic loading influences the physical characteristics of aerobic sludge granules. Lett Appl Microbiol 34(6):407–412

Nielsen PH, Mielczarek AT, Kragelund C, Nielsen JL, Saunders AM, Kong Y, Hansen AA, Vollertsen J (2010) A conceptual ecosystem model of microbial communities in enhanced biological phosphorus removal plants. Water Res 44(17):5070–5088

Nielsen PH, Saunders AM, Hansen AA, Larsen P, Nielsen JL (2012) Microbial communities involved in enhanced biological phosphorus removal from wastewater: a model system in environmental biotechnology. Curr Opin Biotechnol 23(3):452–459

NIH (2012) ImageJ, image processing and analysis in Java. PUblisher. http://imagej.nih.gov/ij/

Oehmen A, Lemos PC, Carvalho G, Yuan Z, Keller J, Blackall LL, Reis MAM (2007) Advances in enhanced biological phosphorus removal: from micro to macro scale. Water Res 41:2271–2300

Oehmen A, Carvalho G, Lopez-Vazquez CM, van Loosdrecht MCM, Reis MAM (2010a) Incorporating microbial ecology into the metabolic modelling of polyphosphate accumulating organisms and glycogen accumulating organisms. Water Res 44(17):4992–5004

Oehmen A, Lopez-Vazquez CM, Carvalho G, Reis MAM, van Loosdrecht MCM (2010b) Modelling the population dynamics and metabolic diversity of organisms relevant in anaerobic/anoxic/aerobic enhanced biological phosphorus removal processes. Water Res 44(15):4473–4486

Oksanen J, Kindt R, Legendre P, O'Hara B, Simpson GL, Solymos P, Stevens MHH, Wagner H (2009) Vegan: community ecology package. R package version 1.15-4. R Foundation for Statistical Computing, Vienna. http://CRAN.R-project.org/package=vegan

Ploner A (2011) Heatplus: heatmaps with row and/or column covariates and colored clusters. R package version 2.1.0. R Foundation for Statistical Computing, Vienna. http://www.bioconductor.org/packages/2.10/bioc/html/Heatplus.html

R-Development-Core-Team (2008) R: a language and environment for statistical computing. R Foundation for Statistical Computing, Vienna. http://www.r-project.org/

Rees G, Baldwin D, Watson G, Perryman S, Nielsen D (2004) Ordination and significance testing of microbial community composition derived from terminal restriction fragment length polymorphisms: application of multivariate statistics. Antonie Van Leeuwenhoek 86(4):339–347

Robert P, Escoufier Y (1976) A unifying tool for linear multivariate statistical methods: the RV-coefficient. J Royal Stat Soc Ser C Appl Stat 25(3):257–265

Schuler AJ, Jenkins D (2003) Enhanced biological phosphorus removal from wastewater by biomass with different phosphorus contents, part I: experimental results and comparison with metabolic models. Water Environ Res 75(6):485–498

Slater FR, Johnson CR, Blackall LL, Beiko RG, Bond PL (2010) Monitoring associations between clade-level variation, overall community structure and ecosystem function in enhanced biological phosphorus removal (EBPR) systems using terminal-restriction fragment length polymorphism (T-RFLP). Water Res 44(17):4908–4923

Song Z, Ren N, Zhang K, Tong L (2009) Influence of temperature on the characteristics of aerobic granulation in sequencing batch airlift reactors. J Environ Sci 21(3):273–278

Stenstrom MK, Song SS (1991) Effects of oxygen transport limitation on nitrification in the activated sludge process. Res J Water Pollut Control Federat 63(3):208–219

Thierry S, Macarie H, Iizuka T, Geissdorfer W, Assih EA, Spanevello M, Verhe F, Thomas P, Fudou R, Monroy O, Labat M, Ouattara AS (2004) *Pseudoxanthomonas mexicana* sp. nov. and *Pseudoxanthomonas japonensis* sp. nov., isolated from diverse environments, and emended descriptions of the genus *Pseudoxanthomonas* Finkmann et al. 2000 and of its type species. Int J Syst Evol Microbiol 54(6):2245–2255

Urakami T, Araki H, Oyanagi H, Suzuki K-I, Komagata K (1992) Transfer of *Pseudomonas aminovorans* (den Dooren de Jong 1926) to *Aminobacter* gen. nov. as *Aminobacter aminovorans* comb. nov. and description of *Aminobacter aganoensis* sp. nov. and *Aminobacter niigataensis* sp. nov. Int J Syst Bacteriol 42(1):84–92

Verstraete W, Wittebolle L, Heylen K, Vanparys B, de Vos P, van de Wiele T, Boon N (2007) Microbial resource management: the road to go for environmental biotechnology. Eng Life Sci 7(2):117–126

Vogelaar JCT, Klapwijk A, van Lier JB, Rulkens WH (2000) Temperature effects on the oxygen transfer rate between 20 and 55°C. Water Res 34(3):1037–1041

Volcke EIP, Picioreanu C, De Baets B, van Loosdrecht MCM (2012) The granule size distribution in an anammox-based granular sludge reactor affects the conversion - Implications for modeling. Biotechnol Bioeng 109(7):1629–1636

Wang X, Hu M, Xia Y, Wen X, Ding K (2012) Pyrosequencing analysis of bacterial diversity in 14 wastewater treatment systems in China. Appl Environ Microbiol 78(19):7042–7047

Weissbrodt DG, Lochmatter S, Ebrahimi S, Rossi P, Maillard J, Holliger C (2012a) Bacterial selection during the formation of early-stage aerobic granules in wastewater treatment systems operated under wash-out dynamics. Front Microbiol 3:332

Weissbrodt DG, Shani N, Sinclair L, Lefebvre G, Rossi P, Maillard J, Rougemont J, Holliger C (2012b) PyroTRF-ID: a novel bioinformatics methodology for the affiliation of terminal-restriction fragments using 16S rRNA gene pyrosequencing. BMC Microbiol 12:306

Weissbrodt DG, Neu TR, Kuhlicke U, Rappaz Y, Holliger C (2013) Assessment of bacterial and structural dynamics in aerobic granular biofilms. Front Microbiol 4:175

Weissbrodt DG, Holliger C, Morgenroth E (2017) Modeling hydraulic transport and anaerobic uptake by PAOs and GAOs during wastewater feeding in EBPR granular sludge reactors. Biotechnol Bioeng 114(8):1688–1702

Wild HE, Sawyer CN, McMahon TC (1971) Factors affecting nitrification kinetics. J Water Pollut Con F 43(9):1845–1854

Winkler MKH, Bassin JP, Kleerebezem R, de Bruin LMM, van den Brand TPH, van Loosdrecht MCM (2011) Selective sludge removal in a segregated aerobic granular biomass system as a strategy to control PAO-GAO competition at high temperatures. Water Res 45(11):3291–3299

Yilmaz G, Lemaire R, Keller J, Yuan Z (2008) Simultaneous nitrification, denitrification, and phosphorus removal from nutrient-rich industrial wastewater using granular sludge. Biotechnol Bioeng 100(3):529–541
Yu CP, Roh H, Chu KH (2007) 17beta-estradiol-degrading bacteria isolated from activated sludge. Environ Sci Technol 41(2):486–492
Yuan Z, VanBriesen J (2008) Bacterial growth yields on EDTA, NTA, and their biodegradation intermediates. Biodegradation 19(1):41–52

Chapter 10
Factors Selecting for Polyphosphate- and Glycogen-Accumulating Organisms in Granular Sludge Sequencing Batch Reactors

"Bacterial competition: Surviving and thriving in the microbial jungle."
(Hibbing et al. 2010)

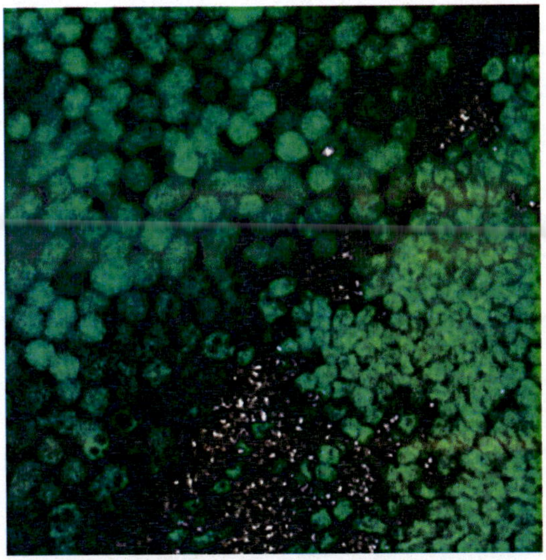

Bacterial competition

The content of this chapter was published in a modified version in: Weissbrodt DG, Schneiter GS, Fürbringer JM, Holliger C (2013) Identification of trigger factors selecting for polyphosphate- and glycogen-accumulating organisms in aerobic granular sludge sequencing batch reactors. Water Res 47(19):7006–18. https://doi.org/10.1016/j.watres.2013.08.043. Permission was granted to reuse the figure materials (© 2013 Elsevier Ltd.).

Supplementary Information The online version contains supplementary material available at https://doi.org/10.1007/978-3-031-41009-3_10.

D. G. Weissbrodt, *Engineering Granular Microbiomes*, Springer Theses,
https://doi.org/10.1007/978-3-031-41009-3_10

10.1 Introduction

Aerobic granular sludge (AGS) wastewater treatment processes are attractive for high-rate biological nutrient removal (BNR) (de Kreuk et al. 2005; Yilmaz et al. 2008). One key element of BNR processes is the preferential selection of polyphosphate-accumulating organisms (PAOs) over glycogen-accumulating competitors (GAOs) (Oehmen et al. 2010a). Whereas the betaproteobacterial "*Candidatus* Accumulibacter phosphatis" has been identified as one key PAOs in lab-scale systems, other organisms such as *Tetrasphaera* spp. within *Actinobacteria* and "*Candidatus* Halomonas phosphatis" within *Gammaproteobacteria* have been proposed as possibly major PAOs of full-scale BNR plants (Maszenan et al. 2000; Kong et al. 2005; Nguyen et al. 2011, 2012). The GAOs guild comprises "*Candidatus* Competibacter phosphatis" inside *Gammaproteobacteria* and the alphaproteobacterial genus *Defluviicoccus* belonging to *Rhodospirillaceae* (Crocetti et al. 2002; Meyer et al. 2006).

Based on the extensive knowledge that has been gained on the metabolism of "*Ca.* Accumulibacter" and "*Ca.* Competibacter" from activated sludge enrichments, Lopez-Vazquez et al. (2009b) and Oehmen et al. (2010a) have integrated in metabolic models the combined effects of pH, temperature, composition of volatile fatty acids (VFAs), and terminal electron acceptors such as dissolved oxygen (DO), nitrate, and nitrite on the PAO/GAO competition. The authors have shown for instance that VFA mixtures of acetate and propionate (Ac/Pr) of 50:50%$_{COD}$ and 75:25%$_{COD}$ preferentially select for PAOs over gamma- and alphaproteobacterial GAOs quite independently from pH (6.5–7.5) and temperature (15–30 °C). "*Ca.* Competibacter" and *Defluviicoccus* relatives have only been able to surpass "*Ca.* Accumulibacter" under combined higher mesophilic (30 °C) and acidic (pH 6.5–7.0) conditions with either only acetate or only propionate in the influent wastewater. Different clades of PAOs and GAOs have in addition displayed selective metabolic activities depending on the nature of terminal electron acceptors (Oehmen et al. 2010a).

Despite that the AGS technology has already been scaled up for full-scale applications (Giesen et al. 2012), only little information is available on the PAO/GAO competition in AGS sequencing batch reactors (SBR). GAOs have been able to outcompete PAOs in AGS systems under phosphorus-limiting conditions in the influent wastewater (de Kreuk and van Loosdrecht 2004) or at 30 °C (Ebrahimi et al. 2010). Gonzalez-Gil and Holliger (2011) have reported preferential selection of a clade of "*Ca.* Accumulibacter" (Type II) and stable dephospha-tation in an AGS-SBR operated without purge of excess sludge and fed with propionate as sole carbon source. Winkler et al. (2011) have recently observed in a bubble-column AGS-SBR that PAOs have mostly been present in the fast-settling granules present in the bottom fractions of the settled AGS bed, while GAOs have been predominantly present in the upper fractions. The authors have been able to select for PAOs and stable dephosphatation in this type of system even at 30 °C by selective purge of upper bed fractions (Bassin et al. 2012).

The present study further aimed to identify the predominant effects of operation parameters and micro-environmental conditions on the PAO/GAO competition within the bacterial community of bubble-columns AGS-SBRs operated for BNR. First investigations were conducted to assess the impact of the VFA composition on the shifts in PAO and GAO populations under dynamic conditions in a reactor operated without purge of excess sludge under non-selective pH (7.0) and room temperature conditions (23 ± 2 °C). A Plackett–Burman multifactorial experimental design (Plackett and Burman 1946; Fisher 1971; Box et al. 1978) was then adopted in order to screen for the main factors within a set of six parameters chosen based on information available in literature about the PAO/GAO competition and on previous knowledge gained on these guilds in the present research project. Eight reactor experiments were conducted under steady-state conditions during three sludge ages. Based on the identified factors, strategies are proposed for optimizing the operation of AGS systems for an efficient selection of PAOs and BNR.

10.2 Material and Methods

10.2.1 Experimental Set-Up Under Dynamic Conditions

A 2.5-L bubble-column AGS-SBR (Fig. 10.1a) was operated at room temperature (23 ± 2 °C) with pre-cultivated mature AGS under alternation of anaerobic plug-flow feeding phase of a VFA-based synthetic influent wastewater (60 min), and of aerobic starvation phase (110 min). The 3-h SBR cycles were complemented with a settling phase (5 min), and a withdrawal phase (5 min; volume exchange ratio of 50%). The hydraulic retention time (HRT) amounted to 6 h. The reactor consisted of a single-wall PVC column with a height-to-diameter ratio (H/D) of 22.

The reactor was operated under dynamic conditions under uncontrolled sludge retention time (SRT) that varied between 20 and 70 days depending on the operation conditions. The starvation phase was fully aerated without DO control, and pH was regulated at 7.0 ± 0.2 by dosing HCl or NaOH at 1 mol L^{-1}.

The initial composition of the synthetic influent wastewater was identical to the one used in (Weissbrodt et al. 2012a/Chap. 7). The PAO/GAO dynamics were followed along successive changes in the VFA composition with three Ac/Pr mixtures (100:0, 0:100 and 50:50% COD equivalents) while maintaining the COD concentration constant at 500 $mg_{COD}\ L_{Inf}^{-1}$. Each VFA mixture was run during ~ 50 days. A fourth experiment was conducted over 25 days with an increased COD concentration of 750 $mg_{COD}\ L_{Inf}^{-1}$ with a 50:50% Ac/Pr mixture.

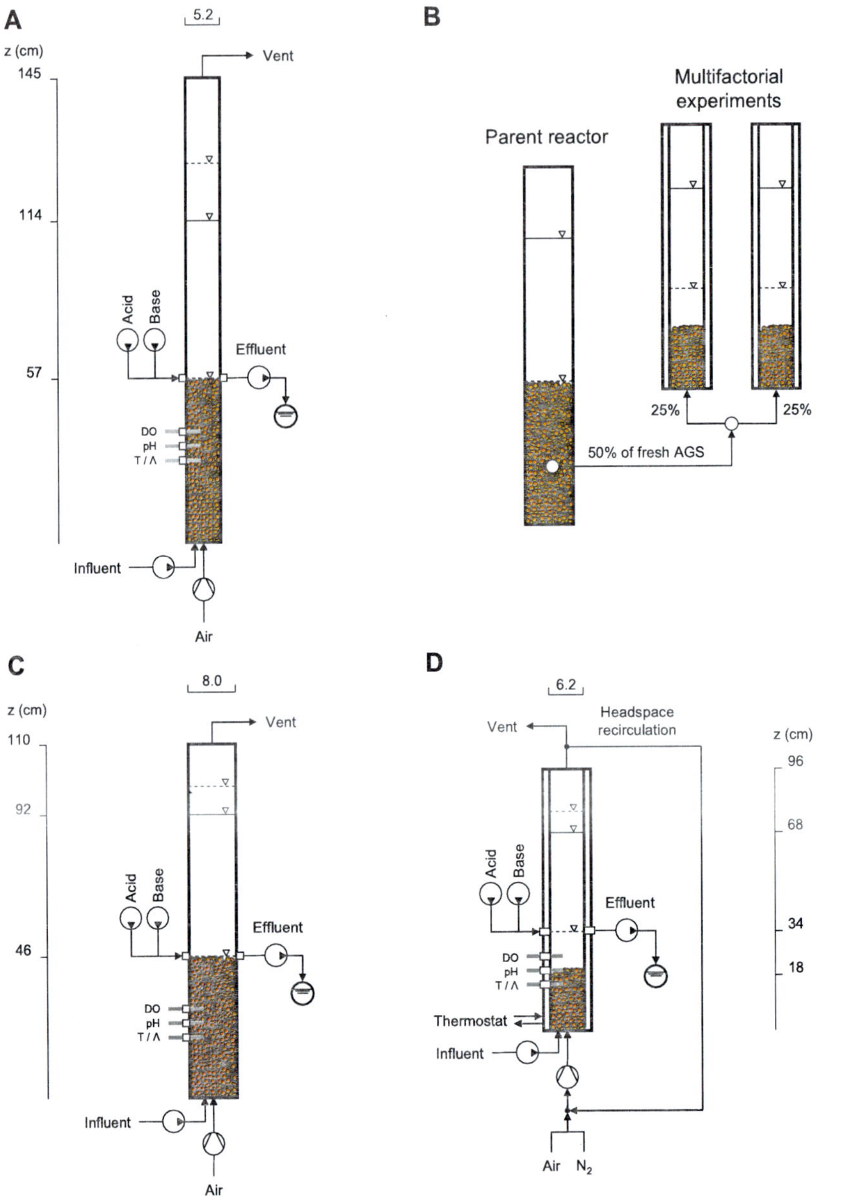

Fig. 10.1 Scheme of the bubble-column SBR infrastructure used for the first part of the study investigating the effect of COD composition and load on bacterial dynamics (**a**), and for the second part of the study focusing on the multifactorial analysis of the PAO/GAO competition in AGS (**b**–**d**). In the latter case, 50% of mature AGS cultivated in the parent SBR was used to inoculate the daughter SBRs in equal fractions (**b**). Design of the parent bubble-column SBR operated to maintain the mature AGS used as inoculum for the multifactorial experiments (**c**). Design of the daughter bubble-column SBRs operated to conduct the multifactorial experiments (**d**)

10.2.2 *Multifactorial Experiments Under Steady-State Conditions*

Six parameters were selected from previous information available for multifactorial experiments: COD concentration, Ac/Pr ratio, COD/P ratio, pH, temperature, as well as redox conditions during the starvation phase (either fully aerobic conditions, or alternating aerobic/anoxic conditions).

According to the Plackett–Burman experimental design, N + 1 runs are conducted for N parameters, and runs are performed at the borders of the chosen experimental ranges. Mathematical Plackett–Burman generators are available for numbers of runs that are a multiple of 4. Eight runs were thus required for the assessment of the effect of six parameters. Since only six parameters were selected, the additional empty column of the experimental matrix was used to check for interactions between parameters and to compute analyses of variance (ANOVA).

The selected minimum–maximum working ranges of the tested parameters are summarized in Table 10.1, with justifications on their choice. The construction of the Plackett–Burman experimental design is given in Additional File 10.1 in the Supporting Information, and the final design in Table 10.2. Experiments were performed in a random order. A 'center experiment' was finally conducted to check for system responses with intermediate parameter values, i.e. with 900 $mg_{CODs}\ L_{Inf}{}^{-1}$, Ac/Pr 75:25%, 95 $g_{CODs}\ g_P{}^{-1}$, pH 7.0, 22 °C, and 40% DO during the starvation phase.

Each experiment was run over 50 days (3 sludge ages) in order to reach steady-state. A total of 7 months were required to gain information on the selected factors. In a full factorial design, $2^6 = 64$ experiments and 4.5 years would have been required.

Two double-wall glass bubble-column AGS-SBRs (Fig. 10.1b–d) were used for running the experiments, and consisted of the same reactor design: working volume of 1.75 L, internal diameter of 62 mm, H/D ratio of 11.

The composition of the synthetic influent wastewater was initially based on the one used in the first part of the study. Adaptations in the concentrations of acetate, propionate and orthophosphate were done according to the experimental matrix (Additional File 10.1). The ammonium concentration was kept constant at 50 $mg_{N\text{-}NH4}$ $L_{Inf}{}^{-1}$. The COD/N and the N/P ratios in the feed therefore depended on the COD and orthophosphate concentrations applied. Each experiment was started with a inoculum at 15 $g_{VSS}\ L^{-1}$ by sampling approximately 450 mL of fresh mature AGS maintained in a parent SBR over the 200 days of the whole experimental period (see below).

The 4-h SBR cycles consisted of the following phases: N_2-flush (5 min), pulse feeding (5 min), N_2-flush (5 min), anaerobic feast phase (95 min), starvation phase (120 min), settling (5 min), and withdrawal (5 min; 50% volume exchange ratio). The HRT amounted to 8 h. The reactors were fully mixed during both anaerobic feast phase and starvation phase by up-flow gas sparging in order to control pH by dosing HCl or NaOH at 0.5 mol L^{-1}.

The anaerobic phases were achieved by recirculating the reactor gas headspace at a flow-rate of 3 L min^{-1}, and with addition of N_2.

Table 10.1 Minimum–maximum experimental ranges of the parameters selected for the multifactorial analysis of the PAO/GAO competition in AGS systems

Parameters (X_i)		Type[1]	Units	Range[2]		Published range[3]
				Min-value (− 1)	Max-value (+ 1)	
1	COD[4]	Continuous	$mg_{CODs}\ L_{Inf}^{-1}$	600	1200	100–5000
2	Ac/Pr ratio	Continuous	%COD	50:50	100:0	0:100–100:0
3	COD/P ratio[4]	Continuous	$g_{CODs}\ g_P^{-1}$	35	100	8–250
4	pH[5]	Continuous	(–)	6.5	7.5	3.0–9.0
5	Temperature[6]	Continuous	°C	15	28	8–55
6	Redox[7]	Binomial	(–)	Ox/Ax	Ox	0–100% O_2-sat

[1]Type of parameter range: continuous (range of real values) or binomial (one or the other state)
[2]Minimum (− 1) and maximum values (+ 1) are based on ranges tested in literature for PAO/GAO competition. Safety margins were considered to avoid drastic conditions
[3]Parameter ranges reported in literature
[4]The minimum COD concentration and COD/P ratio were chosen based on Weissbrodt et al. (2014b/Chap. 9) to avoid COD-limited dephosphatation and denitrification processes. The minimum value is twice higher than in pre-settled urban wastewater. An acetate concentration of 425 $mg_{CODs}\ L_R^{-1}$ has been reported to hamper GAOs (Lopez-Vazquez et al. 2007), corresponding to 850 $mg_{CODs}\ L_{Inf}^{-1}$. The maximum value considers the effect of high-loaded wastewater from e.g. food industries
[5]pH was defined to avoid limited nitrification (at pH < 6.5), and precipitation of orthophosphates (at pH > 7.5)
[6]The maximum temperature was selected to avoid detrimental effects on both PAOs and GAOs (at T > 30 °C)
[7]Redox conditions during the starvation phase: full aeration (Ox) or aerobic/anoxic shifts (Ox/Ax)

Table 10.2 Multifactorial experiments: operational conditions and main system responses after 50 days of experimental runs

Parameters	Units	Runs								
		1	2	3	4	5	6	7	8	Cente
Factors (X_i)										
COD	(mg_{CODs} L^{-1})	1200	600[1]	600	1200	600	1200	1200	600	900
Ac/Pr ratio	(% COD)	100:0	100:0	50:50	50:50	100:0	50:50	100:0	50:50	75:2
COD/P ratio	(g_{CODs} g_P^{-1})	150	150	150	35	35	150	35	35	93
pH	(-)	6.5	7.5	7.5	7.5	6.5	6.5	7.5	6.5	7.0
T	(°C)	28	15	28	28	28	15	15	15	22
Redox[3]	(-)	Ox/Ax	Ox	Ox/Ax	Ox	Ox	Ox	Ox/Ax	Ox/Ax	40%
Responses after 50 days (Y_j)										
Biomass										
Bed height	(cm)	22	14	20	23	9	23	19	19	19
Biomass	(g_{CODx} L^{-1})	16	6	15	18	8	17	23	19	21
Granule diameter	(mm)	-0.66	-0.22	-0.64	-0.23	-1.38	+0.43	-0.09	-0.12	n.a.
COD removal										
F/M ratio[5]	(mg_{CODs} g_{CODx}^{-1})	39	56	20	40	33	36	27	16	21
$q_{VFA,An}$* [6]	(mg_{CODs} g_{CODx}^{-1})	39	52	20	40	33	10	27	16	21
COD (anaerobic)	(%)	100	93	100	100	100	28	100	100	100
Ac (anaerobic)	(%)	100	93	100	100	100	8	100	100	100
Pr (anaerobic)	(%)	-	-	100	100	-	49	-	100	100
P-removal										
P/M ratio[5]	(mg_P g_{CODx}^{-1})	0.26	0.36	0.14	1.11	0.99	0.21	1.03	0.50	0.25
$Y_{P/C,An}$	(mol_P C-mol_{VFA}^{-1})	0.07	0.12	0.21	0.54	0.01	0.18	0.64	0.15	0.07
q_{P-PO4}* [6]	(mg_P g_{CODx}^{-1})	0.00	0.36	0.05	0.88	0.11	0.09	0.60	0	0
P-removal	(%)	0	100	38	80	11	44	58	0	47
N-removal										
N/M ratio[5]	(mg_N g_{CODx}^{-1})	1.82	3.68	1.82	1.66	2.45	1.11	1.22	1.24	1.08
q_{N-NH4}* [6]	(mg_N g_{CODx}^{-1})	1.51	1.97	1.61	1.64	0.54	0.84	0.97	0.59	0.50
$q_{N,Tot}$* [6]	(mg_N g_{CODx}^{-1})	1.51	1.84	1.53	1.62	0.54	0.83	0.97	0.58	0.50
Nitrification	(%)	83	54	88	99	22	76	80	48	47
N-removal	(%)	83	50	84	97	22	75	80	47	46
Populations										
Accumulibacter	OTU-214 (%) [6]	13	45	25	45	16	4	30	10	17
Accumulibacter	OTU-249 (%)	5	6	7	8	4	3	5	6	10
Tetrasphaera	OTU-224 (%)	11	5	5	2	14	4	8	8	5
Competibacter	OTU-239 (%)	27	5	34	12	32	1	11	22	20
Xanthomonadaceae	OTU-32 (%)	12	18	15	25	18	27	25	22	10
Rhodospirillaceae	OTU-229 (%)	<1	1	2	<1	2	2	5	10	8
Aminobacter	OTU-221 (%)	<1	7	3	2	2	27	2	2	5
Zoogloea	OTU-195 (%)	8	<1	<1	<1	<1	10	<1	6	<1

[1]Minimum (−1, *black*) and maximum (+1, *white*) values of experimental ranges (Table 10.1)
[2]Additional experiment conducted at the center values of the experimental ranges of the factors
[3]Redox conditions during starvation phase: full aerobic (Ox) or aerobic/anoxic shifts (Ox/Ax)
[4]A DO setpoint at 40% O_2-saturation was set during starvation phases of the center experiment
[5]Carbonaceous (F/M), phosphorus (P/M) and nitrogenous (N/M) food-to-microorganism ratios
[6]Apparent biomass specific conversion rates (q_i*): anaerobic VFA uptake rate ($q_{VFA,An}$*), orthophosphate uptake rate (q_{P-PO4}*), ammonium removal rate (q_{N-NH4}*), and nitrogen removal rate (q_{N-NH4}*)
[7]Relative abundances of OTUs affiliating with phylotypes of interest, as analyzed by T-RFLP

Depending on the experimental matrix, the starvation phase was conducted either with full aerobic conditions (Ox) by adding air in the headspace recirculation flow, or with 10-min aerobic and anoxic switches (Ox/Ax) by adding either air or N_2 in the headspace recirculation flow in analogy to a strategy proposed by Lochmatter et al. (2012) to promote nitrogen removal. In the latter case, the starvation phase was terminated with 20 min of full aeration to enhance dephosphatation.

When experimental conditions enabled AGS growth, the biomass concentration and the sludge age were maintained around $15 \pm 2\ g_{VSS}\ L^{-1}$ and 15–20 days, respectively, by automated purge of excess sludge from the aerated mixed liquors according to (Weissbrodt et al. 2014b/Chap. 9).

Temperature was controlled with a thermostat, and pH ($\pm$ 0.2 units) was regulated by dosing HCl or NaOH at 0.5 mol L^{-1} during all mixed phases. Temperature, pH, DO and electrical conductivity were recorded on-line.

10.2.3 Operation of the Parent AGS-SBR

The parent AGS-SBR (Fig. 10.1b–d) was operated under similar conditions as the reactor used in the first part of the study.

The 3-h SBR cycles comprised anaerobic feeding (60 min), full aerobic starvation (110 min), settling (5 min), and withdrawal (5 min). Propionate was supplied as sole carbon source at an influent concentration of 600 $mg_{CODs}\ L^{-1}$. The reactor consisted of a single-wall PVC bubble-column. A working volume of 4.725 L was used in order to cultivate sufficient mature AGS for the inoculation of the two daughter SBRs.

After the inoculation of each series of multifactorial experiments (performed two by two), one half of the sludge bed of the parent SBR was remaining for recultivating fresh mature AGS for the next series. It was crucial to maintain a sufficient amount of sludge in the parent SBR for preventing large variations in the bacterial community composition between series of experiments.

The parent SBR was operated at room temperature (23 $\pm$ 2 °C) and pH 7.0 $\pm$ 0.2 by dosing HCl or NaOH at 1 mol L^{-1}. Temperature, pH, DO and electrical conductivity were recorded on-line, in order to check for the stability of the system on the long term.

10.2.4 Chemical and Molecular Analyses

The reactors used to conduct experiments were sampled twice a week, and the parent SBR once a week, for liquid phase, biomass and molecular biology analyses.

Evolutions of VFAs and inorganic ions were measured by chromatography. Biomass was sampled for measurements of total (TSS), inorganic (ISS) and volatile suspended solids (VSS).

The size distribution of granules (Ferret's diameter) was followed over time by digital processing of pictures of granules using ImageJ (NIH 2012).

Bacterial community compositions and dynamics were followed by terminal-restriction fragment length polymorphism (T-RFLP) analysis of eubacterial 16S rRNA gene fragments amplified by PCR with the fluorescently labeled forward primer 8f (FAM-5′-AGAGTTTGATCMTGGCTCAG-3′) and the reverse primer 518r (5′-ATTACCGCGGCTGCTGG-3′), and digested with the *Hae*III restriction enzyme.

Operational taxonomic units (OTU) were affiliated to closest bacterial relatives based on the extended phylogenetic information collected during this research project with the pyrosequencing-based PyroTRF-ID methodology described in Weissbrodt et al. (2012b/Chap. 5). Details on analytical methods are available in Weissbrodt et al. (2012a).

10.2.5 Data Analyses

The effects of operation factors on the AGS systems were assessed based on the evolutions of the biomass concentration, settled AGS bed height, the granule diameter, BNR performances, yield of orthophosphate-release to VFA-uptake under anaerobic conditions ($Y_{P/C,An}$), and relative abundances of predominant operational taxonomic units (OTU).

For the investigations under dynamic conditions, effects of VFA composition and load on the PAO/GAO dynamics were reported in a descriptive analysis.

For the multifactorial investigations, after initial descriptive analysis, data were analyzed in details with the 'half-effect' approach used in multifactorial screening designs (Plackett and Burman 1946; Fisher 1971; Box et al. 1978). The relative half-effect of each factor of the experimental matrix was computed by multiplication of the transposed information matrix (mean vector plus experimental matrix) with each response vector of interest. System responses were analyzed at the biomass level, at the nutrient removal level, and at the bacterial community level.

A simplified linear model $Y = a_0 \pm \Sigma\ (a_i\ X_i)$ was considered in a first assumption for the description of the dependency of the system responses (Y) to the factors (X_i), where a_0 is the mean response, and a_i the half-effect of each factor.

A half-normal plot of the absolute half-effects was computed for each system response in order to detect significant factors that were standing out of the normal experimental variability (or stochastic noise) (Taylor 1994; Taam 2008). In a half-normal plot, effects that are aligned on a straight line cannot be distinguished from the noise, whereas divergent effects are significant and can be taken into account.

An ANOVA was performed in order to quantify the significance of each effect on the system responses (Fisher 1971) with the following definition: significant if p-value < 5%, highly significant if p-value < 1%. The significant factors were conserved in the linear model structures. The quality of the linear models was assessed by a second ANOVA on model estimates. The seventh column of the experimental matrix was intentionally left empty to keep one degree of freedom for the calculation of the residuals between the model estimates and the measured system responses, and to compute ANOVAs.

The Plackett–Burman design assumes that no interaction exist between parameters. The degree of freedom of the empty column was thus also used to check for interactions.

10.3 Results

10.3.1 *Single Effects of the COD Composition and Load on the PAO/GAO Competition*

The dynamics of the bacterial communities and of the nutrient removing activities are presented in Fig. 10.2 in relation to the Ac/Pr ratio and the COD concentration in the influent. The successive experimental periods are separated by vertical lines. The initial 140 days before the first period correspond to reactor start-up to cultivate a mature AGS as presented in Weissbrodt et al. (2012a; 2013/Chaps. 7 and 8). The corresponding data were again presented in Fig. 10.2 for global view.

10.3.2 *Effect of the VFA Composition*

The bed height and biomass concentration (Fig. 10.2a) were low with only acetate (days 140–200: 23 ± 1 cm and 14 ± 4 g_{VSS} L^{-1}, respectively), and increased linearly after the propionate switch (days 200–245: 23–41 cm and 12–25 g_{VSS} L^{-1}). With an Ac/Pr mixture 50:50% (days 245–295), the bed height grew up to 50 cm but with a 5-time lower rate (0.2 cm d^{-1}), whereas the biomass concentration remained constant at 25 ± 2 g_{VSS} L^{-1}. The fraction of inorganic suspended solids (ISS) increased from 32 ± 1% with acetate to 37 ± 2% with propionate, and remained above 35% with the Ac/Pr mixture.

Acetate was efficiently taken up during anaerobic feeding (97 ± 1%), despite a transient decrease to 93% on day 170, (Fig. 10.2a). Propionate was fully removed anaerobically (99 ± 1%). The Ac/Pr mixture was efficiently consumed anaerobically (99 ± 1%), but propionate (99 ± 1%) more efficiently than acetate (97–100%). The yield of orthophosphate release to VFA uptake under anaerobic conditions ($Y_{P/C,An}$) was as high as 0.64 ± 0.06 $mol_{P\text{-}PO4}$ $C\text{-}mol_{VFA}{}^{-1}$ with acetate, but twice lower with propionate (0.35 ± 0.05 $mol_{P\text{-}PO4}$ $C\text{-}mol_{VFA}{}^{-1}$), indicating in this case that less energy was required from polyphosphate hydrolysis for the uptake of propionate. This yield increased to 0.53 ± 0.13 $mol_{P\text{-}PO4}$ $C\text{-}mol_{VFA}{}^{-1}$ after switching to the Ac/Pr mixture. Dephosphatation was high with acetate (80–97%) but unstable between days 164–177, was efficient with propionate (91 ± 6%), and was unstable with the Ac/Pr mixture (59–99%) with local negative peaks. Ammonium was fully nitrified over the whole experimental period, except around day 160 where it decreased transiently to 81%. Nitrogen removal was initially low with acetate (49 ± 3%) but increased up to 72% on later stage. It decreased to 47 ± 6% with propionate, and 35% afterwards with the Ac/Pr mixture.

At the bacterial community level (Fig. 10.2b), "*Ca.* Accumulibacter" (OTU-214) initially amounted to 34 ± 6% with acetate, but did not remain stable (final value at 16%). This population increased to 30% with propionate, and dominated (32 ± 6%)

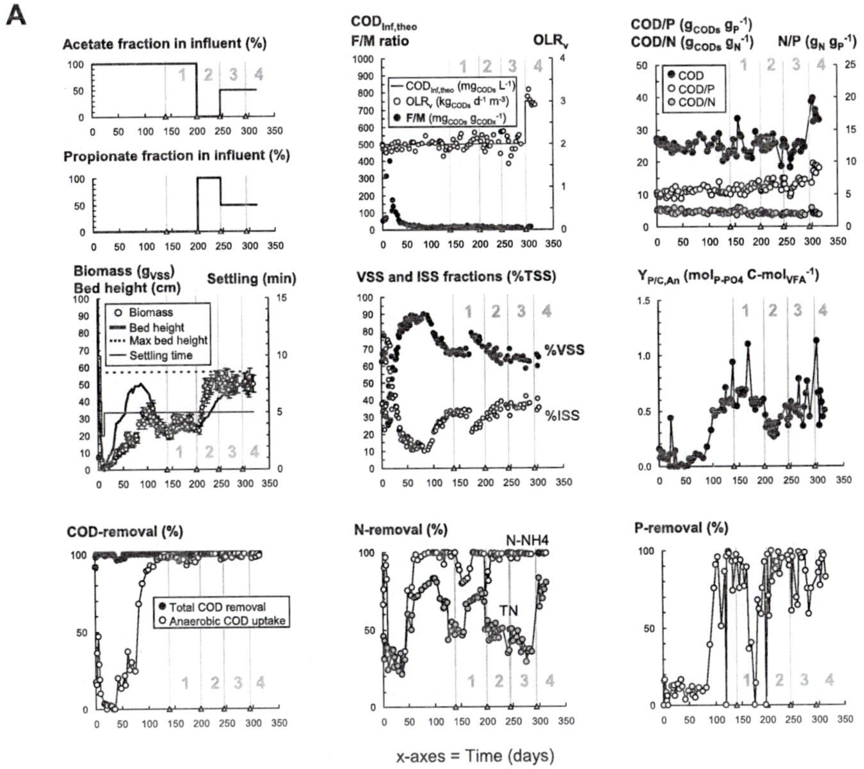

Fig. 10.2 Single effects of the COD composition and load on the biomass dynamics and on the nutrient removal performances (**a**, previous page), on the dynamics of predominant (**b**) and subdominant bacterial populations (**c**) in AGS. The first 120 days comprised the start-up phase for granulation of activated sludge flocs into early-stage granules and into mature granules (recall from Weissbrodt et al. (2012a; 2013/Chaps. 7 and 8). After reaching the first pseudo steady-state (140 days after reactor start-up), the four experiment series investigated during the current study at different Ac/Pr ratios and COD loads are indicated with grey vertical bars. *Phase 1*: Ac/Pr = 100:0%, 500 mg_{CODs} L^{-1}; *Phase 2*: Ac/Pr = 0:100%, 500 mg_{CODs} L^{-1}; *Phase 3*: Ac/Pr = 50:50%, 500 mg_{CODs} L^{-1}; *Phase 4*: Ac/Pr = 50:50%, 750 mg_{CODs} L^{-1}. Note: The T-RFLP method did not enable to distinguish between OTUs 223 and 224. Pyrosequencing samples taken on day 2 revealed that a *Tetrasphaera*-related OTU-223 was abundant. After 60 days a *Hyphomonadaceae*-related OTU-224 was more important. This OTU-224 however affiliated with *Tetrasphaera* in mature AGS. For the present study between days 150–300, *Tetrasphaera* was thus considered

with the Ac/Pr mixture. A second "*Ca.* Accumulibacter"-related OTU-249 proliferated up to 11% under the latter conditions. The third PAOs population affiliating to *Tetrasphaera* spp. (OTU-224) evolved between 5 and 16% with acetate, remained stable at 6 ± 2% with propionate, and decreased to 3% with the Ac/Pr mixture. "*Ca.* Competibacter" (OTU-239) interestingly followed a similar trend with increasing abundances between 6 and 25% with acetate, decreasing to 15% with propionate and further to 6% with the Ac/Pr mixture. *Xanthomonadaceae* relatives (OTU-32)

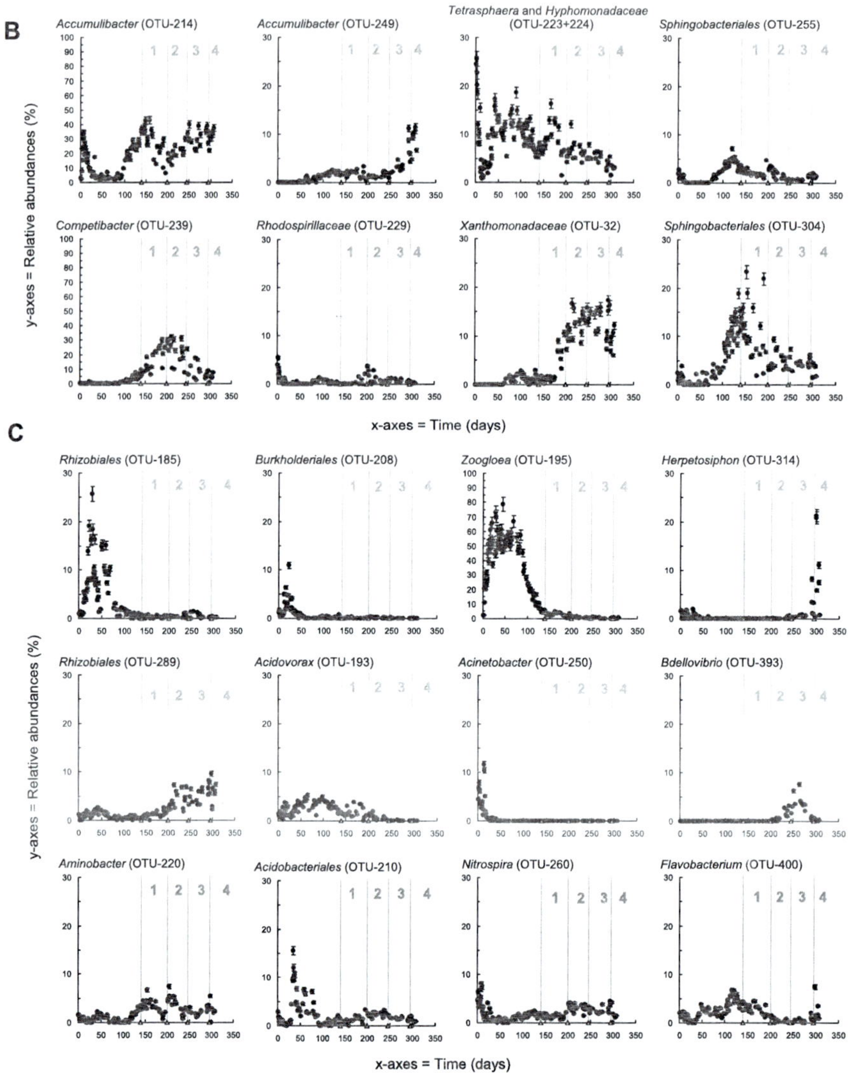

Fig. 10.2 (continued)

proliferated to 10% after day 170 with acetate, and remained at 12 ± 3% in the presence of propionate.

At the level of accompanying populations (Fig. 10.2c), *Acidovorax* spp. (OTU-193, 4%) were more abundant with acetate. Increase in *Zoogloea* spp. (OTU-195, 5%) was detected under periods of incomplete COD removal under anaerobic conditions. Propionate slightly favored *Acidobacteriales* (OTU-210, 3%), *Rhodospirillaceae*

(OTU-229, 4%), nitrite-oxidizing *Nitrospira* (OTU-260, 5%), *Bdellovibrio* (OTU-393, 3%), and *Rhizobiales* (OTU-289) relatives. Another *Rhizobiales* population (OTU-185, 2%) was favored by the Ac/Pr mixture.

10.3.3 Effect of the COD Load

After increasing the COD concentration of the Ac/Pr mixture from 500 to 750 mg_{CODs} L_{Inf}^{-1} on day 295, a higher bed (54 cm) near the upper limit (effluent discharging point) and a higher biomass concentration (27 g_{VSS} L^{-1}) were obtained. The ISS fraction decreased from 38 to 29%. COD was efficiently consumed anaerobically, with full propionate uptake but still with slight variations in acetate uptake (97–100%). Dephosphatation was high but unstable (80–98%). After initial adaptation to higher load, $Y_{P/C,An}$ was similar to the previous phase. Ammonium was fully nitrified, and up to 75 ± 3% of nitrogen was removed.

10.3.4 Performances of the Parent SBR Operated to Maintain a Fresh Mature AGS Inoculum

The BNR performances and bacterial community compositions of the parent SBR reactor over 250 days are presented in Additional File 10.2. After five sludge samplings to inoculate multifactorial runs every 50 days the AGS bed recovered at ever faster rates of 0.5–1.4 cm day^{-1}. Propionate was efficiently taken up under anaerobic conditions (98 ± 2%). $Y_{P/C,An}$ amounted on average to 0.29 ± 0.16 $mol_{P\text{-}PO4}$ C-mol_{VFA}^{-1}, but was affected by sludge sampling. Dephosphatation displayed similar trends between 67 and 100%. Instabilities were mainly observed during periods after the first two sludge samplings. The apparent biomass specific rate of dephosphatation amounted on average to 0.49 ± 0.16 $mg_{P\text{-}PO4}$ g_{CODx}^{-1} per cycle. Nitrification was affected by sludge sampling, but recovered to 100% (1.07 ± 0.30 mg_N g_{CODx}^{-1}) at the end of each period. Again, less variation was measured after the first two periods. Nitrogen removal fluctuated between 40 and 90% (0.79 ± 0.23 mg_N g_{CODx}^{-1}). Highest removal correlated with transiently higher COD concentration (700 mg_{CODs} L_{Inf}^{-1}).

"*Ca.* Accumulibacter" (OTU-214) dominated (31 ± 11%) over almost the 250 days, but with two decreases to 10% during periods 3 and 4. The second "*Ca.* Accumulibacter" affiliate (OTU-249) was initially low (7 ± 2%), but increased to 12% whilst OTU-214 decreased. *Tetrasphaera* relatives (OTU-224) were not abundant (3 ± 1%). "*Ca.* Competibacter" (OTU-239) was initially low (5 ± 3%) but increased to 21% during period 4. *Xanthomonadaceae* (OTU-32) was abundant during the 250 days (20 ± 5%). *Rhizobiales* (OTU-289) and *Bdellovibrio* (OTU-393) relatives progressively declined after period 1. *Aminobacter* relatives (OTU-220) were not abundant (4 ± 1%). Nitrifiers were below the lower detection limit. Only *Nitrospira* (OTU-260) amounted between 1 and 5% with highest initial abundances. *Herpetosiphon* relatives within *Chloroflexi* (OTUs 296 and 298) increased at later stage.

10.3.5 Multifactorial Assessment of Bacterial Competition and BNR Performances

The main responses observed in terms of BNR performance and bacterial population compositions were compiled in Table 10.2. The BNR performances and population dynamics measured of the multifactorial runs are displayed in Fig. 10.3 and Additional File 10.3. For comparison between runs, influent conditions were expressed as carbonaceous (F/M), phosphorous (P/M), and nitrogenous (N/M) food-to-microorganism ratios, and BNR performances as apparent biomass specific rates of acetate ($q_{Ac}{}^{*}$), propionate ($q_{Pr}{}^{*}$), orthophosphate ($q_{P}{}^{*}$), and nitrogen ($q_{N}{}^{*}$) removal in addition to percentages.

A rapid observation of these data indicated that the conditions of run 4 with combination of high COD concentration, Ac/Pr mixture, low COD/P ratio, alkaline pH, higher mesophilic temperature, and full aeration during starvation resulted in efficient biomass growth, BNR performances, and selection for "*Ca.* Accumulibacter". This population was actually present in abundances above 20% in all runs conducted at higher pH, and below 20% under acidic conditions. "*Ca.* Competibacter" were mainly present in abundances above 20% in runs operated at 28 °C. For all runs, nitrification and nitrogen removal exhibited identical performances, indicating that nitrification was the limiting step. An additional observation revealed that when COD was transiently not fully taken up under anaerobic conditions, propionate was always more efficiently consumed than acetate.

The datasets were analyzed in a statistical way according to the experimental design. The relative half-effects computed for each operation parameter on the measured responses are presented in Fig. 10.4. Half-normal plots used to identify factors with effects out of the experimental variability are shown in Fig. 10.5. ANOVA p-values are provided in brackets in this figure as a measure of the significance of the contribution of main factors to responses (significant and highly significant if p-value < 5 and 1%). Linear model relations between responses and main factors are described in Additional File 10.4. A second ANOVA provided information on the adequacy of the linear models, based on the ratio of the sum of squares of model estimates to measured responses (R^2) and on p-values computed on model estimates.

At the biomass level (Figs. 10.4a and 10.5a), higher growth was obtained with higher COD, aerobic/anoxic shifts during starvation, Ac/Pr mixture, and low COD/P ratio. Temperature, COD and Ac/Pr ratio highly impacted on the granule diameter,

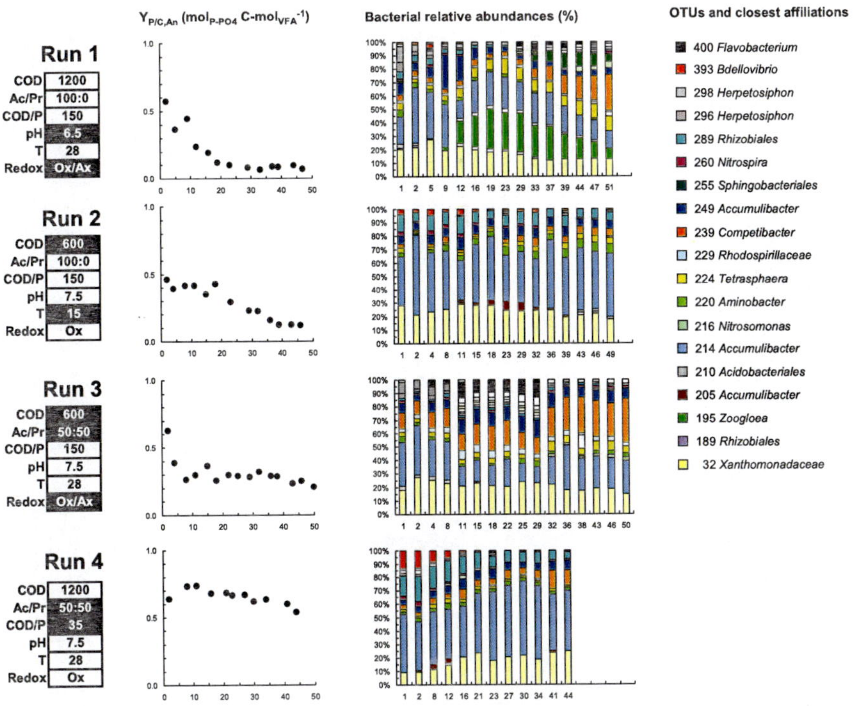

Fig. 10.3 Multifactorial effects on the yield of orthophosphate release to VFA uptake under anaerobic conditions ($Y_{P/C,\ An}$) and on the bacterial dynamics in AGS and. More details are available in Additional File 10.3

obtaining smaller ones with higher mesophilic temperature, 600 $mg_{CODs}\ L_{Inf}^{-1}$, and Ac/Pr 100:0%.

At the BNR level (Figs. 10.4c, d and 10.5c), higher nitrification and nitrogen removal were obtained under higher mesophilic and alkaline conditions. Nitrogen removal was limited by nitrification. High dephosphatation was obtained with alkaline pH, full aerobic starvation, high COD, low temperature. Higher orthophosphate-release activities were obtained with alkaline pH, high COD, and low COD/P ratio.

At the bacterial community level (Figs. 10.4b and 10.5b), "*Ca.* Accumulibacter" (OTU-214) was preferentially selected under alkaline conditions. Even though other parameter effects were close or belonged to the normal distribution, full aerobic starvation conditions was positively influencing "*Ca.* Accumulibacter". One third effect (a_7) was potentially impacting on "*Ca.* Accumulibacter", but was related to the seventh empty column of the experimental matrix (see below). *Tetrasphaera* spp. were favored with only acetate, and under acidic conditions. "*Ca.* Competibacter" was selected by higher mesophilic temperature, lower COD concentration, and aerobic/anoxic shifts. Highest ratios of "*Ca.* Accumulibacter" to "*Ca.*

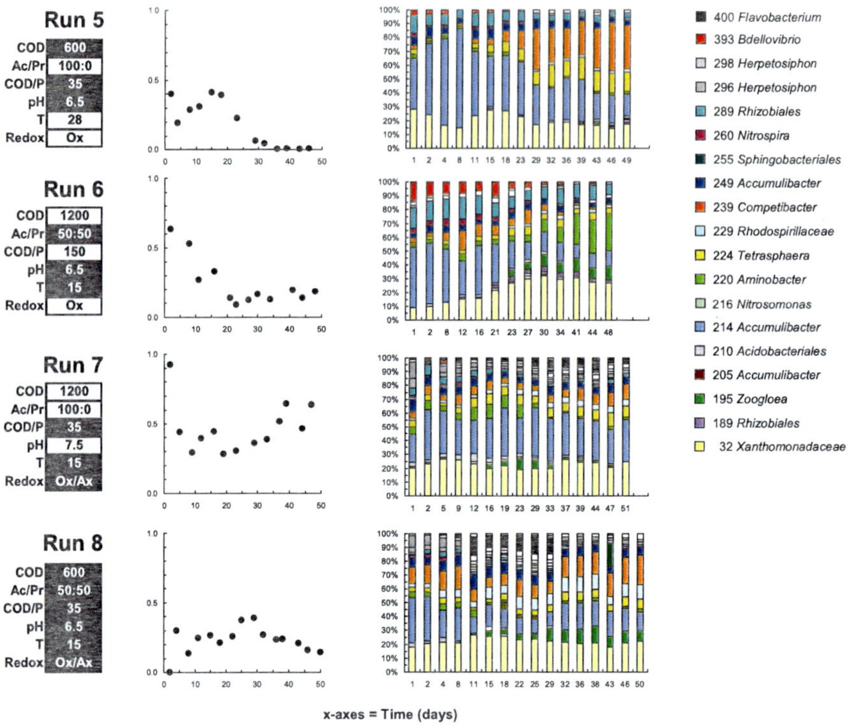

Fig. 10.3 (continued)

Competibacter" populations were obtained with alkaline pH, low temperature, and full aerobic starvation conditions.

Zoogloea was mainly favored under acidic pH, high COD concentration, and high COD/P ratio, but when carbon was leaking from the anaerobic to the aerobic phase. No significant correlation was however obtained for *Xanthomonadaceae* relatives and accompanying populations.

Significant linear models dependencies were obtained for the system responses related to biomass (R^2 = 0.958–0.996, *p*-values = 0.000–0.007%), BNR (0.974–0.996, 0.000–0.002%), and predominant organisms (0.941–0.990, 0.000–0.005%%; Additional File 10.4). The 'center experiment' was conducted with all operation parameters set at the center values of the minimum–maximum experimental ranges. Responses measured after 50 days in this experiment were close to the theoretical values provided by the models for the biomass, bed height, and bacterial populations. Poorer BNR performances were however measured, indicating a possible negative impact of the fixed 40% DO setpoint.

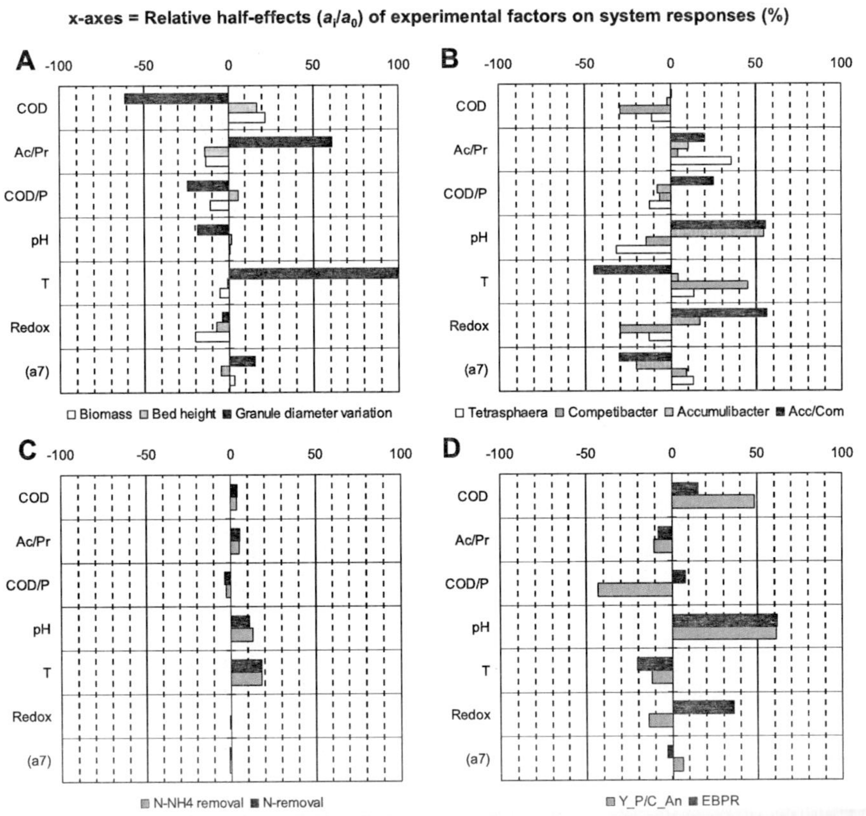

Fig. 10.4 Relative half effects of each individual experimental factor on the system responses: granule diameter (**a**), PAO and GAO bacterial populations (**b**), ammonium and nitrogen removal (**c**), EBPR and $Y_{P/C,An}$ (**d**)

10.3.6 Potential Parameter Interaction Impacting "Ca. Accumulibacter"

A hidden factor a_7 related to the seventh empty column of the experimental matrix was identified with a potential impact on "*Ca.* Accumulibacter". A matrix of aliases was computed for the adopted Plackett–Burman experimental design in order to check for possible first order interactions within the set of six tested parameters (Additional File 10.5).

According to this analysis, the seventh factor can be associated by three possible underlying first order interactions between COD and temperature (X_{15}), Ac/Pr and COD/P ratios (X_{23}), and pH and redox conditions (X_{46}). The Plackett–Burman design can however not provide information on which interaction effectively occurred. However, since pH was the most dominant factor impacting on "*Ca.* Accumulibacter" and since redox conditions were also identified as one other factor by ANOVA, the

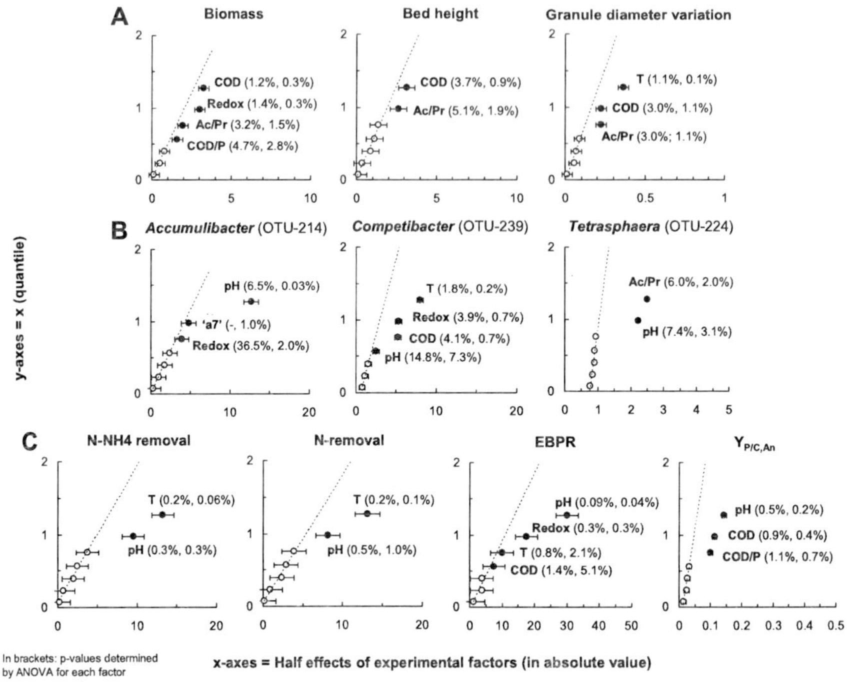

Fig. 10.5 Half normal-plots of effects related to biomass parameters (**a**), target PAOs and GAOs (**b**), and BNR (**c**) computed in order to identify significant factors outside the stochastic noise. The effects that are aligned on a straight line are normally distributed, and cannot be differentiated from the normal experimental variation. The related factors are discarded from the set of parameters, and can be used to determine the experimental error. Only effects that deviate from the straight line are out of noise, and the related parameters are taken into consideration as significant factors impacting on the target system response. Two p-values computed by ANOVA are given in brackets for the significant factors. The first p-value was obtained after computing an ANOVA by considering the six factors and by keeping the a7 factor empty. The second p-value was obtained after computing an ANOVA by considering only the significant factors (including the a7 factor) that were out of the normal distribution. A *p*-value expresses the probability that an effect is obtained by chance. *P*-values < 0.05 (or 5%) and < 0.01 (or 1%) are related to significant and highly significant effects, respectively. The combination of both half-normal plot and ANOVA approaches were powerful to identify the significant factors that acted on the system responses

interaction X_{46} between these two parameters was taken into account in the linear model (Additional File 10.4). Including this first order interaction in the linear models of both "*Ca.* Accumulibacter"-related OTUs 214 and 249 improved the quality of the model responses ($R^2 = 0.995$, *p*-value = 0.001%). This would mean that pH could have different effects on these populations depending on the redox conditions imposed during starvation.

10.4 Discussion

10.4.1 *Trigger Factors of "*Ca.* Accumulibacter" Selection and EBPR in Anaerobic–Aerobic AGS-SBRs*

Experimental and mathematical modeling investigations conducted during this extended study on the impact of operation parameters on bacterial competition and performances in anaerobic-aerobic AGS-SBRs revealed that pH was the main trigger factor of the PAO/GAO competition and of BNR.

Different authors have previously reported that pH has an important effect on the metabolisms of PAOs and GAOs and on enhanced biological phosphorus removal (EBPR) in activated sludge systems (Smolders et al. 1994; Filipe et al. 2001; Oehmen et al. 2005a; Lopez-Vazquez et al. 2009b). Preferential selection of "*Ca.* Accumulibacter" and efficient EBPR were obtained here at alkaline pH. A high PAO/GAO ratio in AGS-SBRs was positively correlated with operation at pH 7.5 and 15 °C, almost independently from the Ac/Pr ratio. Since the biomass specific acetate uptake rate of PAOs was independent from pH in the model, preferential selection of "*Ca.* Accumulibacter" was explained by the fact that GAOs display slower acetate uptake kinetics and higher maintenance requirements under alkaline conditions. According to multifactorial experiments and modelling, "*Ca.* Accumulibacter" was able to compete with "*Ca.* Competibacter" even at higher mesophilic temperature after application of alkaline conditions. Increasing the alkalinity of the influent wastewater would therefore be recommended for promoting acetate uptake by PAOs at any process temperature. Further modeling research could inform on the pH gradients establishing in the settled AGS bed during anaerobic plug-flow feeding phases, and their impact on anaerobic conversions (Weissbrodt et al. 2017).

Multifactorial experiments also revealed that full aerobic conditions were required with fixed 2-h starvation phase in order to obtain efficient EBPR in addition to "*Ca.* Accumulibacter" selection. In the case of strictly aerobic clades of "*Ca.* Accumulibacter", such observation would indicate that a longer starvation phase should compensate for the time used for anoxic shifts. On the other hand, even though some clades can denitrify, anoxic heterotrophic microbial processes are around 40% less energy efficient than aerobic ones (Henze et al. 2000; Oehmen et al. 2010b). The periodical application of anoxic conditions during starvation was therefore likely to lead to lower volumetric conversion rates and to incomplete orthophosphate uptake.

Under full aerobic starvation, additional conditions of run 4 with alkaline pH, higher mesophilic temperature, non-limiting phosphorus conditions, and an Ac/Pr mixture was optimal for sustaining biomass growth, proliferation of "*Ca.* Accumulibacter", and efficient removal of relatively high nutrient loads (40 mg_{CODs}, 1.11 $mg_{P\text{-}PO4}$ and 1.66 $mg_{N\text{-}NH4}$ $g_{CODx}{}^{-1}$) in AGS-SBRs. These pH, COD/P and Ac/Pr levels have also been shown to favor PAOs in flocculent sludge (Schuler and Jenkins 2003; Oehmen et al. 2005a; Lopez-Vazquez et al. 2009b). PAO-enrichments have also been favorably cultivated in Weissbrodt et al. (2014a/Chap. 6) with alkaline pH, in the presence of propionate, and with low COD/P ratio. In AGS, de Kreuk and van

Loosdrecht (2004) have previously reported that PAOs were present in high abundances at pH 7.0 and 20 °C when phosphorus was not limiting. It has been shown in Weissbrodt et al. (2014a/Chap. 6) that "*Ca.* Accumulibacter" and stable EBPR were obtained with the values above 30 $g_{COD}\ g_{P}^{-1}$ in the lower COD/P range.

The presence of propionate as single VFA source or in mixtures with acetate was shown in the first AGS-SBR operated under dynamic conditions to favor "*Ca.* Accumulibacter" and to lead to more stable EBPR. According to the metabolic properties of PAOs and GAOs described by Lopez-Vazquez et al. (2009b), the presence of propionate only indirectly favors "*Ca.* Accumulibacter" by disadvantaging "*Ca.* Competibacter" which cannot take up propionate as fast as acetate. Experiments revealed in addition that propionate was always more efficiently consumed than acetate in the AGS-SBRs. The lower $Y_{P/C,An}$ measured with propionate in the dynamic SBR indicated that less polyphosphate was hydrolyzed and therefore less energy used to take up and condense propionate as poly-β-hydroxyalkanoates (PHAs). Without periodical purge of biomass to maintain a fixed SRT, substantial biomass growth and more stable EBPR were obtained as soon as the carbon source was switched from acetate to propionate, leading to a bed height close to the maximum capacity. With identical COD in the influent, this means that a larger fraction of PHAs stored from propionate was available as carbon and energy source for the synthesis of biomass. If an AGS-SBR is operated without regular purge of excess sludge, this also means that the type of carbon source will impact on the SRT. Such energy consideration has also been used by Gonzalez-Gil and Holliger (2011) to explain more efficient selection for "*Ca.* Accumulibacter" and EBPR in an AGS-SBR operated with propionate than in a second one operated with acetate.

Both studies are in agreement with authors reporting more energy-efficient metabolism of communities enriched in "*Ca.* Accumulibacter" with the 3-carbons propionate than with the 2-carbons acetate (Oehmen et al. 2005b; Burow et al. 2008; Lopez-Vazquez et al. 2009b; Wexler et al. 2009; Zhou et al. 2009, 2010; He and McMahon 2011).

10.4.2 *Tetrasphaera*-Related PAOs Can Withstand GAOs-Selective Conditions

According to the multifactorial results, *Tetrasphaera* spp., that possibly belong to major PAOs of full-scale BNR-WWTPs, might withstand slight acidic conditions that typically select for "*Ca.* Competibacter" over "*Ca.* Accumulibacter".

This confirmed previous observation made in Weissbrodt et al. (2014a/Chap. 6) that *Tetrasphaera* spp. were able to maintain more longer under the conditions of the GAO-enrichment than of the PAO-enrichment. *Tetrasphaera* spp. belong to polysaccharide-hydrolyzing and glucose-fermenting bacteria (Nielsen et al. 2010, 2012). The process of glucose fermentation into VFAs in open mixed cultures has been more efficient at acidic pH (Temudo et al. 2007; Temudo 2008). Although glucose was not supplied into the AGS-SBRs, *Tetrasphaera* spp. might consume exopolysaccharides present in the granular biofilm matrices such as described in Weissbrodt et al. (2014b/Chap. 9). These organisms have also been describe to use the VFA fermentation products that they produce (Nielsen et al. 2010), what is efficient on an energetic point of view. Additional research is required to unravel the physiological properties of *Tetrasphaera* spp.

10.4.3 Trigger Factors of (Unfavorable) GAOs Selection

Conditions selecting for "*Ca.* Competibacter" in the AGS-SBR were shown to be detrimental to "*Ca.* Accumulibacter" and EBPR. Results of the dynamic AGS-SBR first indicated that "*Ca.* Competibacter" was able to compete with "*Ca.* Accumulibacter" under non-selective conditions with pH 7.0, 23 °C, and acetate conditions, and even more under too long SRTs resulting from the absence of purge of excess sludge.

A similar observation has also been made by Gonzalez-Gil and Holliger (2011). According to the multifactorial experiments, "*Ca.* Competibacter" proliferation was mainly favoured by higher mesophilic conditions when only acetate was supplied, and quite independently from the COD/P ratio. Ebrahimi et al. (2010) have previously shown that higher mesophilic temperatures favored GAOs over PAOs in an AGS-SBR operated at neutral pH. These results are in agreement with previous knowledge gained from activated sludge systems (Filipe et al. 2001; Zeng et al. 2003; Oehmen et al. 2005a; Lopez-Vazquez et al. 2009a, 2009b), and with the model.

Higher abundance of "*Ca.* Competibacter" was also linked to aerobic/anoxic shifts during starvation, and to lower COD in the influent. A concentration of acetate above 13.3 C-mmol$_{Ac}$ L_R^{-1} in the bulk liquid phase has been pointed out by Lopez-Vazquez et al. (2007) to induce a temporary inhibition of the anaerobic metabolism of GAOs. The total VFA concentration used in the multifactorial runs amounted to 8.7 and 18.75 C-mmol$_{VFA}$ L_R^{-1} at the beginning of each anaerobic mixed batch phase. The upper concentration could have disadvantaged "*Ca.* Competibacter". Additional research is required to unravel the effect of VFA load on the metabolism of "*Ca.* Competibacter". In contrast to the model predictions of Lopez-Vazquez et al. (2009b), "*Ca.* Competibacter" was still able to compete with "*Ca.* Accumulibacter" when an Ac/Pr mixture was supplied, as exemplified by runs 3 and 8. These runs were in addition operated with aerobic/anoxic shifts during starvation and with lower COD concentration. Such conditions were identified in the statistical approach as GAOs-selective.

The putatively GAOs-related *Rhodospirillaceae* affiliates slightly proliferated at pH 7.0 when only propionate was fed in the dynamic AGS-SBR. This is also in agreement with the metabolic model formulations of Lopez-Vazquez et al. (2009b). During multifactorial runs with either only acetate or Ac/Pr mixture, this population was never predominant. The cited study has reported proliferation of these alphaproteobacterial GAOs with Ac/Pr 50:50% only under combined acidic and higher mesophilic conditions, which were not tested here.

10.4.4 The Intriguing Presence of *Xanthomonadaceae* in Anaerobic–Aerobic AGS-SBRs

Other *Gammaproteobacteria* affiliating with *Xanthomonadaceae* were present in abundances of 10–30% in the different AGS-SBRs.

Underlying *Pseudoxanthomonas* and *Thermomonas* relatives identified with PyroTRF-ID have been described as denitrifiers present in EBPR and BNR activated sludge plants (Jorgensen and Pauli 1995; Merzouki et al. 1999; Mergaert et al. 2003; Thierry et al. 2004; Yu et al. 2009), and have also been detected in the denitrifying community of AGS by Adav et al. (2010).

Whereas Adav et al. operated the AGS-SBR with pulse feeding and aerobic conditions, it is quite intriguing that *Xanthomonadaceae* relatives were present in important amount in the anaerobic-aerobic AGS-SBRs of the research project.

According to Weissbrodt et al. (2014b/Chap. 9), within the bacterial continuum of AGS, these organisms belonged to the "*Ca.* Accumulibacter"-related cluster which was correlated with conditions of efficient BNR. In Weissbrodt et al. (2013/Chap. 8), *Xanthomonadaceae* proliferated in the AGS-SBR as soon as the biomass concentration was sufficiently high to ensure full anaerobic acetate uptake. However, no reference study has reported anaerobic uptake of VFAs by these organisms. Additional research is required to determine whether *Xanthomonadaceae* relatives can display a GAOs metabolism. It therefore remains unclear how *Xanthomonadaceae* relatives can thrive in anaerobic-aerobic SBRs. *Xanthomonadaceae* relatives are known to further involve lytic enzymes to harvest saccharide residues (Aparna et al. 2009). It could be therefore hypothesized that *Xanthomonadaceae* could proliferate on lysis products originating from endogenous respiration processes during starvation.

The genus *Stenotrophomonas* has also been described to degrade EDTA aerobically (Kaparullina et al. 2009). Since an equivalent of 95 $mg_{CODs}\ L^{-1}$ of EDTA was present in the influent define based on conditions reported in literature, i.e. one fifth of the initial VFA concentration, EDTA might be considered as nutritive source. EDTA degradation was indeed observed (data not shown). Finally, *Xanthomonadaceae* affiliates have also exhibited enhanced adhesion and co-aggregation propensity (Malik et al. 2003). These potentially virulent organisms have actually been described to comprise an arsenal of adhesins, to produce extracellular polymeric substances, such as xanthan, and to form biofilms (Sutherland 2001; Timke et al. 2004; Mhedbi-Hajri et al. 2011). Their presence could contribute to granular biofilm cohesion.

10.4.5 *Toward Efficient BNR in Anaerobic–Aerobic AGS-SBRs*

In addition to insight on mechanisms of bacterial selection, this study also enabled to identify operation factors that significantly impacted on BNR performances. Alkaline conditions were shown to favor not only dephosphatation but also nitrification and nitrogen removal.

During the multifactorial runs, full nitrification was only obtained with conditions of run 4, i.e. alkaline and higher mesophilic conditions. This is in agreement with the activated sludge knowledge (Henze et al. 2000). In the dynamic SBR, higher nitrogen removal was obtained after the increase from 500 to 750 $mg_{CODs}\ L_{Inf}^{-1}$. By referring to the stoichiometric study of Kuba et al. (1996), this confirmed the observation made in Weissbrodt et al. (2014b/Chap. 9) that heterotrophic dephosphatation and denitrification processes were limited in COD at lower concentration, and that denitrification is probably not restricted to PAOs or GAOs but to a broader denitrifying community.

The multifactorial experiments revealed in addition that the length of the starvation phase should be extended when alternating aerobic and anoxic conditions are applied, what is in agreement with different authors (Kishida et al. 2008; Yilmaz et al. 2008; Chen et al. 2011). Therefore, as mentioned in Weissbrodt et al. (2014b/Chap. 9), AGS processes should benefit from the flexibility of the SBR technology (Wilderer and McSwain 2004).

Despite the fact that propionate enabled higher AGS growth (up to the effluent discharging point), purge of excess sludge was crucial in the operation of AGS-SBRs. This is in agreement with Morgenroth and Wilderer (1999), Winkler et al. (2011) and to previous results presented in Weissbrodt et al. (2014b/Chap. 9).

In the same study, it was shown that the granule diameter can affect the oxygenation efficiency and the aerobic processes. The morphology of granules was shown here to be impacted by operation parameters. The multifactorial runs confirmed a direct positive correlation between the granule diameter, the COD load, and temperature.

10.5 Conclusions

The combination of detailed experimental, multifactorial, and modeling investigations on the effect of operation parameters on bacterial selection and BNR performances in mature AGS-SBRs led to the following findings:

- At the reactor system boundary, the predominant factors that govern the competition of "*Ca.* Accumulibacter" and "*Ca.* Competibacter" in AGS are similar to the one that impact on flocculent activated sludge EBPR systems.

- The pH was identified as the main trigger factor of the PAO/GAO competition and of BNR performances. Optimal selection of "*Ca.* Accumulibacter" was obtained under alkaline conditions independently from the chosen temperature conditions.
- Under anaerobic plug-flow feeding conditions, maintaining a relatively high level of alkalinity in the influent would therefore be key.
- Propionate was more energetically efficient than acetate. The presence of propionate either as single VFA source or in mixtures with acetate sustained biomass growth, "*Ca.* Accumulibacter" selection, and EBPR.
- The putatively major PAOs-related *Tetrasphaera* spp, can withstand GAOs-selective conditions, namely acidic pH and higher mesophilic temperature.
- The size of aerobic granules was positively correlated with the COD concentration and the temperature of the influent wastewater.
- With fixed 4-h SBR cycles including a 2-h starvation phase, EBPR was only efficient under full aerobic starvation conditions. If alternating aerobic/anoxic starvation conditions are imposed, the length of the starvation phase should be extended for full BNR.

Acknowledgements Jean-Marie Fuerbringer from Mechanical Engineering Institute at EPFL for advice on multifactorial design of experiments.

Supplementary Information

Additional File 10.1 Construction of the Plackett–Burman experimental design.

Additional File 10.2 Overall data collected for the parent reactor from reactor to biomass and bacterial community scales.

Additional File 10.3 Overall data collected for each multifactorial experiment from reactor to biomass and bacterial community scales.

Additional File 10.4 Multifactorial assessment of the dynamics of biomass, bacterial community and nutrient removal in AGS systems: comparison of model estimates to measured system responses.

Additional File 10.5 Detection of potential aliases inside the Placket–Burman experimental design adopted in the current study.

References

Adav SS, Lee DJ, Lai JY (2010) Microbial community of acetate utilizing denitrifiers in aerobic granules. Appl Microbiol Biotechnol 85(3):753–762

Aparna G, Chatterjee A, Sonti RV, Sankaranarayanan R (2009) A cell wall-degrading esterase of *Xanthomonas oryzae* requires a unique substrate recognition module for pathogenesis on rice. Plant Cell Online 21(6):1860–1873

Bassin JP, Winkler MKH, Kleerebezem R, Dezotti M, van Loosdrecht MCM (2012) Improved phosphate removal by selective sludge discharge in aerobic granular sludge reactors. Biotechnol Bioeng 109(8):1919–1928

Box GEP, Hunter WJ, Hunter J (1978) Statistics for experimenters: an introduction to design, data analysis and model building. John Wyley and Sons, New York

Burow LC, Mabbett AN, Blackall LL (2008) Anaerobic glyoxylate cycle activity during simultaneous utilization of glycogen and acetate in uncultured accumulibacter enriched in enhanced biological phosphorus removal communities. ISME J 2(10):1040–1051

Chen F-Y, Liu Y-Q, Tay J-H, Ning P (2011) Operational strategies for nitrogen removal in granular sequencing batch reactor. J Hazard Mater 189(1–2):342–348

Crocetti GR, Banfield JF, Keller J, Bond PL, Blackall LL (2002) Glycogen-accumulating organisms in laboratory-scale and full-scale wastewater treatment processes. Microbiology 148(11):3353–3364

de Kreuk MK, van Loosdrecht MCM (2004) Selection of slow growing organisms as a means for improving aerobic granular sludge stability. Water Sci Technol 49:9–17

de Kreuk MK, Heijnen JJ, van Loosdrecht MCM (2005) Simultaneous COD, nitrogen, and phosphate removal by aerobic granular sludge. Biotechnol Bioeng 90(6):761–769

Ebrahimi S, Gabus S, Rohrbach-Brandt E, Hosseini M, Rossi P, Maillard J, Holliger C (2010) Performance and microbial community composition dynamics of aerobic granular sludge from sequencing batch bubble column reactors operated at 20, 30, and 35°C. Appl Microbiol Biotechnol 87:1555–1568

Filipe CDM, Daigger GT, Grady CPL (2001) pH as a key factor in the competition between glycogen-accumulating organisms and phosphorus-accumulating organisms. Water Environ Res 73(2):223–232

Fisher RA (1971) The design of experiments, 9th edn. Macmillian, London

Giesen A, Niermans R, van Loosdrecht MCM (2012) Aerobic granular biomass: the new standard for domestic and industrial wastewater treatment? Water 21(4):28–30

Gonzalez-Gil G, Holliger C (2011) Dynamics of microbial community structure and enhanced biological phosphorus removal of propionate- and acetate-cultivated aerobic granules. Appl Environ Microbiol 77:8041–8051

He S, McMahon KD (2011) Microbiology of "*Candidatus* Accumulibacter" in activated sludge. Microbial Biotechnol 4(5):603–619

Henze M, Gujer W, Mino T, van Loosdrecht MCM (2000) Activated sludge models ASM1, ASM2, ASM2d and ASM3. IWA Publishing, London

Hibbing ME, Fuqua C, Parsek MR, Peterson SB (2010) Bacterial competition: surviving and thriving in the microbial jungle. Nat Rev Microbiol 8(1):15–25

Jorgensen KS, Pauli ASL (1995) Polyphosphate accumulation among denitrifying bacteria in activated sludge. Anaerobe 1(3):161–168

Kaparullina E, Doronina N, Chistyakova T, Trotsenko Y (2009) *Stenotrophomonas chelatiphaga* sp. nov., a new aerobic EDTA-degrading bacterium. Syst Appl Microbiol 32(3):157–162

Kishida N, Tsuneda S, Sakakibara Y, Kim JH, Sudo R (2008) Real-time control strategy for simultaneous nitrogen and phosphorus removal using aerobic granular sludge. Water Sci Technol 58(2):445–450

Kong Y, Nielsen JL, Nielsen PH (2005) Identity and ecophysiology of uncultured actinobacterial polyphosphate-accumulating organisms in full-scale enhanced biological phosphorus removal plants. Appl Environ Microbiol 71(7):4076–4085

Kuba T, van Loosdrecht MCM, Heijnen JJ (1996) Phosphorus and nitrogen removal with minimal COD requirement by integration of denitrifying dephosphatation and nitrification in a two-sludge system. Water Res 30(7):1702–1710

Lochmatter S, Weissbrodt DG, Holliger C (2012) Aeration strategies for enhanced N and P-removal with aerobic granular biofilm. In: Ren N (ed) IWA nutrient removal and recovery conference 2011 Trends in NRR, Harbin, China

Lopez-Vazquez CM, Hooijmans CM, Brdjanovic D, Gijzen HJ, van Loosdrecht MCM (2007) A practical method for quantification of phosphorus- and glycogen-accumulating organism populations in activated sludge systems. Water Environ Res 79(13):2487–2498

Lopez-Vazquez CM, Hooijmans CM, Brdjanovic D, Gijzen HJ, van Loosdrecht MCM (2009a) Temperature effects on glycogen accumulating organisms. Water Res 43(11):2852–2864

Lopez-Vazquez CM, Oehmen A, Hooijmans CM, Brdjanovic D, Gijzen HJ, Yuan Z, van Loosdrecht MCM (2009b) Modeling the PAOs-GAOs competition: effects of carbon source, pH and temperature. Water Res 43(2):450–462

Malik A, Sakamoto M, Hanazaki S, Osawa M, Suzuki T, Tochigi M, Kakii K (2003) Coaggregation among non-flocculating bacteria isolated from activated sludge. Appl Environ Microbiol 69(10):6056–6063

Maszenan AM, Seviour RJ, Patel BKC, Schumann P, Burghardt J, Tokiwa Y, Stratton HM (2000) Three isolates of novel polyphosphate-accumulating Gram-positive cocci, obtained from activated sludge, belong to a new genus, *Tetrasphaera* gen. nov., and description of two new species, *Tetrasphaera japonica* sp. nov. and *Tetrasphaera australiensis* sp. nov. Int J Syst Evol Microbiol 50(2):593–603

Mergaert J, Cnockaert MC, Swings J (2003) *Thermomonas fusca* sp. nov. and *Thermomonas brevis* sp. nov., two mesophilic species isolated from a denitrification reactor with poly(E-caprolactone) plastic granules as fixed bed, and emended description of the genus Thermomonas. Int J Syst Evol Microbiol 53(6):1961–1966

Merzouki M, Delgenes JP, Bernet N, Moletta R, Benlemlih M (1999) Polyphosphate-accumulating and denitrifying bacteria isolated from anaerobic-anoxic and anaerobic-aerobic sequencing batch reactors. Curr Microbiol 38(1):9–17

Meyer RL, Saunders AM, Blackall LL (2006) Putative glycogen-accumulating organisms belonging to the *Alphaproteobacteria* identified through rRNA-based stable isotope probing. Microbiology 152:419–429

Mhedbi-Hajri N, Jacques M-A, Koebnik R (2011) Adhesion mechanisms of plant-pathogenic *Xanthomonadaceae*. In: Linke D, Goldman A (eds) Bacterial adhesion: advances in experimental medicine and biology, vol 715. Springer, New York, pp 71–89

Morgenroth E, Wilderer PA (1999) Controlled biomass removal: the key parameter to achieve enhanced biological phosphorus removal in biofilm systems. Water Sci Technol 39(7):33–40

Nguyen HTT, Le VQ, Hansen AA, Nielsen JL, Nielsen PH (2011) High diversity and abundance of putative polyphosphate-accumulating Tetrasphaera-related bacteria in activated sludge systems. FEMS Microbiol Ecol 76(2):256–267

Nguyen HTT, Nielsen JL, Nielsen PH (2012) Candidatus Halomonas phosphatis", a novel polyphosphate-accumulating organism in full-scale enhanced biological phosphorus removal plants. Environ Microbiol 14(10):2826–2837

Nielsen PH, Mielczarek AT, Kragelund C, Nielsen JL, Saunders AM, Kong Y, Hansen AA, Vollertsen J (2010) A conceptual ecosystem model of microbial communities in enhanced biological phosphorus removal plants. Water Res 44(17):5070–5088

Nielsen JL, Nguyen H, Meyer RL, Nielsen PH (2012) Identification of glucose-fermenting bacteria in a full-scale enhanced biological phosphorus removal plant by stable isotope probing. Microbiology 158(7):1818–1825

NIH (2012) ImageJ, image processing and analysis in Java. PUblisher. http://imagej.nih.gov/ij/

Oehmen A, Vives MT, Lu H, Yuan Z, Keller J (2005a) The effect of pH on the competition between polyphosphate-accumulating organisms and glycogen-accumulating organisms. Water Res 39(15):3727–3737

Oehmen A, Zeng RJ, Yuan Z, Keller J (2005b) Anaerobic metabolism of propionate by polyphosphate-accumulating organisms in enhanced biological phosphorus removal systems. Biotechnol Bioeng 91(1):43–53

Oehmen A, Carvalho G, Lopez-Vazquez CM, van Loosdrecht MCM, Reis MAM (2010a) Incorporating microbial ecology into the metabolic modelling of polyphosphate accumulating organisms and glycogen accumulating organisms. Water Res 44(17):4992–5004

Oehmen A, Lopez-Vazquez CM, Carvalho G, Reis MAM, van Loosdrecht MCM (2010b) Modelling the population dynamics and metabolic diversity of organisms relevant in anaerobic/anoxic/aerobic enhanced biological phosphorus removal processes. Water Res 44(15):4473–4486

Plackett RL, Burman JP (1946) The design of optimum multifactorial experiments. Biometrika 33(4):305–325

Schuler AJ, Jenkins D (2003) Enhanced biological phosphorus removal from wastewater by biomass with different phosphorus contents, part I: experimental results and comparison with metabolic models. Water Environ Res 75(6):485–498

Smolders GJF, van Loosdrecht MCM, Heijnen JJ (1994) pH: Keyfactor in the biological phosphorus removal process. Water Sci Technol 29(7):71–74

Sutherland IW (2001) Biofilm exopolysaccharides: a strong and sticky framework. Microbiology 147(1):3–9

Taam W (2008) Half-normal plot encyclopedia of statistics in quality and reliability. John Wiley & Sons Ltd., Amsterdam

Taylor GAR (1994) Analysis of experiments by using half-normal plots. J R Stat Soc Ser D 43(4):529–536

Temudo MF, Kleerebezem R, Van Loosdrecht M (2007) Influence of the pH on (Open) mixed culture fermentation of glucose: a chemostat study. Biotechnol Bioeng 98(1):69–79

Temudo MF (2008) Directing product formation by mixed culture fermentation. PhD thesis, Delft University of Technology

Thierry S, Macarie H, Iizuka T, Geissdorfer W, Assih EA, Spanevello M, Verhe F, Thomas P, Fudou R, Monroy O, Labat M, Ouattara AS (2004) *Pseudoxanthomonas mexicana* sp. nov. and *Pseudoxanthomonas japonensis* sp. nov., isolated from diverse environments, and emended descriptions of the genus Pseudoxanthomonas Finkmann et al. 2000 and of its type species. Int J Syst Evol Microbiol 54(6):2245–2255

Timke M, Wolking D, Wang-Lieu NQ, Altendorf K, Lipski A (2004) Microbial composition of biofilms in a brewery investigated by fatty acid analysis, fluorescence in situ hybridisation and isolation techniques. Appl Microbiol Biotechnol 66(1):100–107

Weissbrodt DG, Lochmatter S, Ebrahimi S, Rossi P, Maillard J, Holliger C (2012a) Bacterial selection during the formation of early-stage aerobic granules in wastewater treatment systems operated under wash-out dynamics. Front Microbiol 3:332

Weissbrodt DG, Shani N, Sinclair L, Lefebvre G, Rossi P, Maillard J, Rougemont J, Holliger C (2012b) PyroTRF-ID: a novel bioinformatics methodology for the affiliation of terminal-restriction fragments using 16S rRNA gene pyrosequencing. BMC Microbiol 12:306

Weissbrodt DG, Neu TR, Kuhlicke U, Rappaz Y, Holliger C (2013) Assessment of bacterial and structural dynamics in aerobic granular biofilms. Front Microbiol 4:175

Weissbrodt DG, Maillard J, Brovelli A, Chabrelie A, May J, Holliger C (2014a) Multilevel correlations in the biological phosphorus removal process: from bacterial enrichment to conductivity-based metabolic batch tests and polyphosphatase assays. Biotechnol Bioeng 111(12):2421–2435

Weissbrodt DG, Shani N, Holliger C (2014b) Linking bacterial population dynamics and nutrient removal in the granular sludge biofilm ecosystem engineered for wastewater treatment. FEMS Microbiol Ecol 88(3):579–595

Weissbrodt DG, Holliger C, Morgenroth E (2017) Modeling hydraulic transport and anaerobic uptake by PAOs and GAOs during wastewater feeding in EBPR granular sludge reactors. Biotechnol Bioeng 114(8):1688–1702

Wexler M, Richardson DJ, Bond PL (2009) Radiolabelled proteomics to determine differential functioning of Accumulibacter during the anaerobic and aerobic phases of a bioreactor operating for enhanced biological phosphorus removal. Environ Microbiol 11(12):3029–3044

Wilderer PA, McSwain BS (2004) The SBR and its biofilm application potentials. Water Sci Technol 50(10):1–10

Winkler MKH, Bassin JP, Kleerebezem R, de Bruin LMM, van den Brand TPH, van Loosdrecht MCM (2011) Selective sludge removal in a segregated aerobic granular biomass system as a strategy to control PAOs-GAOs competition at high temperatures. Water Res 45(11):3291–3299

Yilmaz G, Lemaire R, Keller J, Yuan Z (2008) Simultaneous nitrification, denitrification, and phosphorus removal from nutrient-rich industrial wastewater using granular sludge. Biotechnol Bioeng 100(3):529–541

Yu L, Liu Y, Wang G (2009) Identification of novel denitrifying bacteria *Stenotrophomonas* sp. ZZ15 and *Oceanimonas* sp. YC13 and application for removal of nitrate from industrial wastewater. Biodegradation 20(3):391–400

Zeng RJ, Van Loosdrecht MCM, Yuan Z, Keller J (2003) Metabolic model for glycogen-accumulating organisms in anaerobic/aerobic activated sludge systems. Biotechnol Bioeng 81(1):92–105

Zhou Y, Pijuan M, Zeng RJ, Yuan Z (2009) Involvement of the TCA cycle in the anaerobic metabolism of polyphosphate accumulating organisms (PAOss). Water Res 43(5):1330–1340

Zhou Y, Pijuan M, Oehmen A, Yuan Z (2010) The source of reducing power in the anaerobic metabolism of polyphosphate accumulating organisms (PAOss): a mini-review. Water Sci Technol 61(7):1653–1662

Chapter 11
Modeling the Hydraulic Transport of Wastewater and Anaerobic Uptake of Organics by PAOs and GAOs During the Feeding of a Granular Sludge Reactor

Is modelling of biological wastewater treatment a mature technology?
(Gujer 2011)

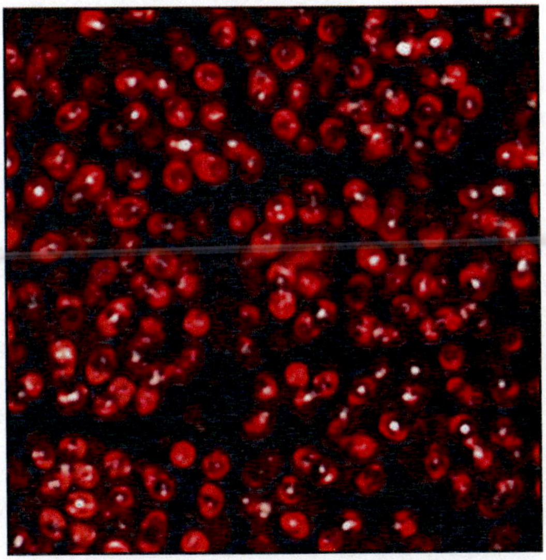

Polyphosphate-accumulating organisms

Weissbrodt DG, Holliger C, Morgenroth E (2017) Modelling hydraulic transport and anaerobic uptake by PAOs and GAOs during wastewater feeding in EBPR granular sludge reactors. Biotechnol Bioeng 114:1688–1702. https://doi.org/10.1002/bit.26295. Permission was granted to reuse the figure materials (© 2017 Wiley Periodicals, Inc.).

Supplementary Information The online version contains supplementary material available at https://doi.org/10.1007/978-3-031-41009-3_11.

D. G. Weissbrodt, *Engineering Granular Microbiomes*, Springer Theses,
https://doi.org/10.1007/978-3-031-41009-3_11

11.1 Introduction

New-generation environmental bioprocesses using granular sludge emerged as technological breakthrough toward high-rate biological nutrient removal (BNR) from wastewater and secondary clarification in single sequencing batch reactors (SBR) (Giesen et al. 2012). Similar to activated sludge systems, key bacterial populations should be selected and activated in the underlying bacterial microbiome for an efficient elimination of carbon, nitrogen and phosphorus loads (Nielsen et al. 2010; Oehmen et al. 2010a; Weissbrodt et al. 2014c). By analogy to any open mixed-culture environmental biotechnology system (Rittmann et al. 2006; Verstraete et al. 2007), the performance of granular sludge SBRs relies on an efficient management of the underlying bacterial resource mediated by dedicated engineering of operational conditions (Weissbrodt 2012; Weissbrodt et al. 2014a).

Preferential selection of polyphosphate-accumulating organisms (PAOs) is required for enhanced biological phosphorus removal (EBPR). PAOs are selected over ordinary heterotrophic organisms (OHOs) by uncoupling the supplies of organic electron donors such as volatile fatty acids (VFAs) and terminal electron acceptors such as dissolved oxygen, nitrite, or nitrate in tanks or phases in series (Comeau et al. 1986; Wentzel et al. 2008). Glycogen-accumulating organisms (GAOs) compete for VFAs under anaerobic conditions (Cech and Hartman 1993; Mino et al. 1995). "*Candidati* Accumulibacter and Competibacter phosphates" (referred to as *Accumulibacter* and *Competibacter*) are model PAOs and GAOs, respectively (Hesselmann et al. 1999; Crocetti et al. 2002).

Mathematical models have been developed to investigate EBPR in activated sludge (Rieger et al. 2001; Guisasola et al. 2004; Houweling et al. 2010; Hauduc et al. 2013), and mechanisms of selections of PAOs and GAOs integrated in metabolic models (Yagci et al. 2003; Whang et al. 2007; Lopez-Vazquez et al. 2009; Oehmen et al. 2010a, 2010b; Lanham et al. 2014). Factors such as carbon sources, pH, temperature, and redox conditions direct the selection for either PAOs or GAOs.

Mechanisms of PAO/GAO selection are analogous in activated sludge and granular sludge systems (Weissbrodt et al. 2014b). Granular sludge SBRs engineered for EBPR are operated by alternating phases of slow up-flow feeding across the granule blanket under anaerobic conditions, i.e. commonly called anaerobic feeding, and aeration (de Kreuk and van Loosdrecht 2004; Rocktäschel et al. 2013). Full control of anaerobic phase length is crucial for maintaining active populations of PAOs and preventing proliferation of unfavorable fast-growing aerobic OHOs such as filamentous and zoogloeal ones (Weissbrodt et al. 2012, 2013a). Operation with dedicated control of loading and anaerobic conditions is efficient to this end (Weissbrodt et al. 2013a; Lochmatter and Holliger 2014). Temperature and pH are key factors triggering selection and BNR in granular sludge (Weissbrodt et al. 2013b). Alkaline conditions select for *Accumulibacter* and EBPR, acidic and mesophilic ones for *Competibacter* and deteriorated performances.

Bacterial selection is primarily governed by an efficient uptake of the carbon source under anaerobic conditions. Granular sludge systems require targeted design of the anaerobic feeding selector together with control of environmental conditions in the bulk (Weissbrodt et al. 2014a). System analysis and mathematical modeling investigations were conducted here in order to specifically investigate (i) the hydraulic transport patterns of wastewater under rapid and slow feeding regimes across a bed of granular sludge and (ii) the metabolic processes of PAOs and GAOs under anaerobic slow up-flow feeding conditions. Residence time distributions (RTD) were recorded in tracer experiments and used to calibrate hydraulic transport models. Metabolic formulations were coupled to assess impacts of feeding time, pH, and temperature on overall conversion of VFA and individual contributions of PAOs and GAOs. The hydraulic-metabolic model can sustain the design of bed geometries and feeding conditions for directing selection and EBPR in granular sludge. It provides an interesting base toward more complex modeling investigations under full-scale conditions.

11.2 Material and Methods

11.2.1 Experimental Set-Up

The lab-scale reactor design was described in details earlier (Weissbrodt et al. 2013b). It consisted of a segmented double-wall glass column with an internal diameter of 59 mm, working height of 57 cm, and working volume of 1.6 L (Fig. 11.1a). The granular sludge bed comprised 543 g (Additional File 11.1 in the Supplementary Information) of wet mature EBPR granules with an average density of 1120 kg m^{-3} of biofilm. Granules size distribution (Tijhuis et al. 1994; de Kreuk et al. 2007; Lochmatter et al. 2013) was characterized by a median equivalent diameter of 1.1 mm (first-third quartiles = 0.3–2.2 mm) homogenously distributed across the height of 33 cm. Temperature was maintained at 20 °C with a thermostat.

11.2.2 Hydraulic Residence Time Distribution Experiments

Tracer experiments were conducted under continuous up-flows of tap water supplied under rapid feeding regime at a flow-rate of 200 mL min^{-1} (superficial and interstitial velocities of 4.4 and 9.4 m h^{-1}, respectively) and under slow feeding regime at a flow-rate of 20 mL min^{-1} (0.44 and 0.94 m h^{-1}). The effluent was weighted continuously for accurate measurement of flow-rates at steady state (197 and 20.6 mL min^{-1}, respectively). The theoretical hydraulic retention times (HRT) inside the bed and column amounted to 2.2 and 5.6 min under rapid feeding, and to 22 and 56 min

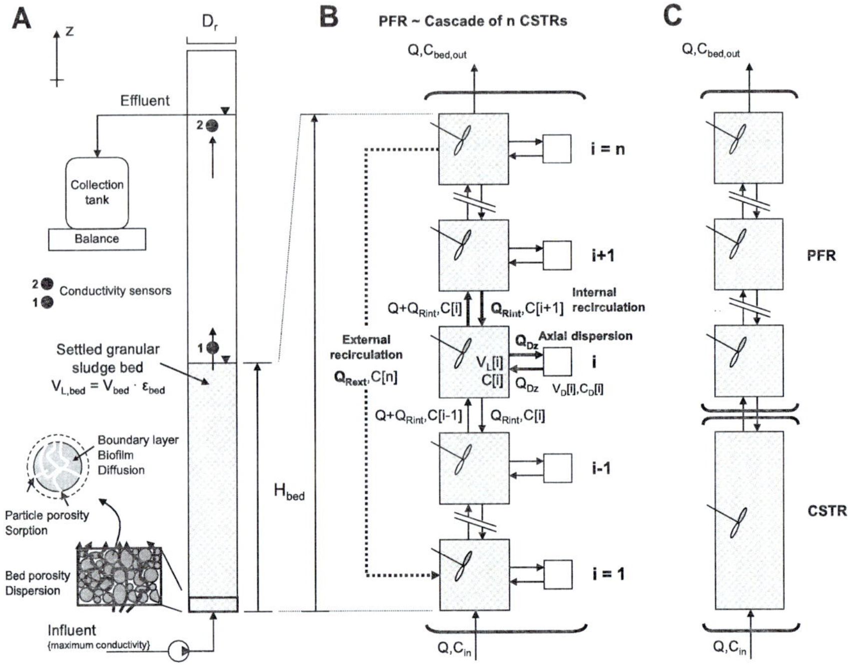

Fig. 11.1 Experimental setup and conceptual model for RTD investigations. After a step-change in NaBr influent concentration, two RTDs were recorded with electrical conductivity sensors positioned at the bed outlet (34 cm) and at the reactor outlet (57 cm) (**a**). Conceptual model of the hydraulic transport across the porous bed medium (**b**). The plug-flow behavior was approached by a cascade of n CSTRs. Axial dispersion was modelled either with side mixing process or with internal recirculation process. External recirculation was used to assess possible periodicity in the exit profile. Scheme of a large CSTR compartment connected to a PFR compartment used as second model structure B (Table 11.1) to simulate the RTD under rapid feeding conditions (**c**). *Legend* granular sludge bed height (H_{bed}), internal reactor diameter (D_r), total bed volume (V_{bed}), bed porosity (ε_{bed}), volume of liquid phase in the bed ($V_{L,bed}$), influent flow-rate (Q), amount of tracer rapidd (E_T), solute concentration (C), internal (Q_{Rint}) and external (Q_{Rext}) recirculation flow-rates, axial dispersion flow-rate (Q_{Dz}), volume of stagnant fluid (V_D)

under slow feeding, respectively. Stabilization of the flow-through system was monitored over three HRTs prior to each tracer experiment using two electrical conductivity sensors (Jumo, Type 202,922, 10/cm cell constant, max 200 mS/cm, Germany) positioned at the bed (34 cm) and reactor (57 cm) outlets.

RTDs were recorded under the two feeding regimes by step-change of the inlet concentration of an inert tracer of NaBr, following Gujer (2008). Contrary to Dirac tracer rapid, the step-change method requires a moderately concentrated salt solution with a density (1002 kg m^{-3}) close to water, averts density-related disturbances, and prevents toxicity and inhibition of the biomass (Anderson et al. 1991; Gujer 2008). After stabilization of the conductivity signal of the tap water flow (286 μS cm^{-1}), the influent source was switched to a solution of NaBr ~ 5 g L^{-1} at t_o. The accurate concentration of 4.45 g L^{-1} was obtained from conductivity ($\sigma = 4450\ \mu S\ cm^{-1}$) with C_{NaBr} (g L^{-1}) = 0.001·σ (μS cm^{-1}) over 0–5 g L^{-1} ($R^2 = 0.9991$). The breakthrough curves of tracer were recorded at bed and reactor outlets as cumulative frequency functions of residence time F(τ) until new plateaus were reached in the conductivity profiles. The frequency of residence time f(τ) – or probability density function – was obtained by computing the first derivative of F(τ) (van Genuchten 1980; Levenspiel 1999; Martin 2000).

Data interpretation concentrated on F(τ) whereas f(τ) provided indicative peak position and shape. Numeric differentiation of F(τ) is indeed often inaccurate and leads to noisy signals.

11.2.3 Hydraulic Transport Model

One-dimensional mathematical models were implemented in Berkeley Madonna (Macey et al. 2000) and calibrated to RTDs under the two feeding regimes (Fig. 11.1b). Plug-flow transport across the porous medium of granular sludge was formulated as interconnected network of perfectly mixed regions representing liquid interstices of the bed. Model structures were formulated for computational saving as cascades of continuous-flow stirred tank reactors (CSTR) (Deans and Lapidus 1960; Hinduja et al. 1980; Gujer 2008) inflected with relations from up-flow anaerobic sludge blankets, packed beds, and porous media (Smith et al. 1996; Albuquerque and Santana 2004; Zeng et al. 2005; Quaff and Guha 2011). Flow transport across porous media is affected by dispersion (Hoogendoorn and Lips 1965; Hinduja et al. 1980). Axial dispersion parallel to the z-direction of flow was modeled either by side mixing or by internal recirculation. In the first case, fluid is exchanged between stagnant (dead volume) and flowing (active volume) parts of the void bed region. The second case was modeled with an internal recirculation flow-rate (Q_{Rint}). An external recirculation term (Q_{Rext}) was included to assess periodicities in exit tracer functions.

The theoretical mass balances in Eq. (11.1) based on Gujer (2008) and Smith et al. (1996) describe the evolution of the tracer concentration (C_A) with time (t) during transport in a cascade of n CSTRs in function of the inlet tracer concentration ($C_{A,Inf}$), total bed volume (V_{bed}) and porosity (ε), influent (Q), internal recirculation (Q_{Rint}), external recirculation (Q_{Rext}) and axial dispersion (Q_{Dz}) flow-rates, and volume (F_{DV}) and flow-rate (F_{QV}) fractions of axial dispersion. The models were fitted to experimental RTDs by step-wise tuning of parameters followed by fine-scale automated curve fit. Dispersion coefficients were estimated according to Gujer's procedure (2008). Root mean square errors (RMSE) were computed between model predictions and measured responses. The modeled distributions were characterized with descriptive statistics by computation of 1st order moment (mean residence time, τ_m), and 2nd (variance, σ^2), 3rd (skewness, g_1) and 4th (kurtosis, g_2) order central moments (Table 11.1). The mode (τ_{max}) resulted from peak maximum in f(τ). The lower quartile (25%), median (50%), and upper quartile (75%) were obtained from

F(τ). Dimensionless probability functions were computed by normalization with the mean residence time τ_m for comparison between feeding conditions and ideal characteristic curves of CSTRs and PFRs.

$$\frac{dC_{A,1}}{dt} = \frac{Q}{V_{L,i}} \cdot C_{A,Inf} + \frac{Q_{Rint}}{V_{L,i}} \cdot C_{A,2} + \frac{Q_{Rext}}{V_{L,i}} \cdot C_{A,n} - \frac{Q + Q_{Rint} + Q_{Rext}}{V_{L,i}} \cdot C_{A,1} + \frac{Q_{Dz}}{V_{L,i}} \cdot \left(C_{A,D,1} - C_{A,1}\right)$$

$$\left.\frac{dC_{A,i}}{dt}\right|_{i=2...(n-1)} = \frac{Q + Q_{Rint} + Q_{Rext}}{V_{L,i}} \cdot \left(C_{A,i-1} - C_{A,i}\right) + \frac{Q_{Rint}}{V_{L,i}} \cdot \left(C_{A,i+1} - C_{A,i}\right) + \frac{Q_{Dz}}{V_{L,i}} \cdot \left(C_{A,D,i} - C_{A,i}\right)$$

$$\frac{dC_{A,n}}{dt} = \frac{Q + Q_{Rint} + Q_{Rext}}{V_{L,i}} \cdot \left(C_{A,n-1} - C_{A,n}\right) + \frac{Q_{Dz}}{V_{L,i}} \cdot \left(C_{A,D,n} - C_{A,n}\right)$$

$$\frac{dC_{A,d,i}}{dt} = \frac{Q_{Dz}}{V_{D,i}} \cdot \left(C_{A,i} - C_{A,d,i}\right)$$

$$\text{with } C_{A,Inf} = \begin{cases} pulse(area, f, r) & \text{if t} = \text{t}_0 \\ 0 & \text{if t} > \text{t}_0 \end{cases} \quad \text{for Dirac rapid experiments}$$

$$area = \frac{E_A}{Q}, \quad f = \text{t}_0 = 0, \quad r = 10'000 \text{ min}$$

$$V_{L,i} = \frac{V_{L,bed} - V_D}{n} = \frac{V_{bed} \cdot \varepsilon_{bed} \cdot (1 - F_{DV})}{n}$$

$$Q_{Dz} = Q \cdot F_{DQ} \tag{11.1}$$

11.2.4 Metabolic Model

The calibrated hydraulic transport model was coupled to a structured biomass model in order to describe the metabolic activities of PAOs and GAOs during anaerobic slow feeding conditions in function of pH and temperature. Stoichiometry and biokinetics were translated from Lopez-Vazquez et al. (2009) and Oehmen et al. (2010a) into mass concentrations and COD-based units (Additional File 11.2). These complex metabolisms depend on the dynamics of intracellular storage polymers of inorganic polyphosphate, glycogen, and poly-β-hydroxyalkanoates (PHAs) that are not transported with the flow of wastewater. The simulation time, which accounts for the feeding time, is thus a key parameter since conversions cannot reach a steady-state. Simulations at hydrodynamic steady-state would only result in acetate transport across the bed without consumption.

The combined hydraulic-metabolic model accounts for the following system simplifications toward a compromise between research questions and computation workload. The granular sludge was considered as homogenous biomass of active cells that were all in contact with substrates. The properties of granular biofilms systems with mass transfer limitations and segregation of biomass at micro- and macroscales (Morgenroth 2008b) were not formulated. Flow resistances and diffusion phenomena were nevertheless taken into account by RTDs. Fractions of PAOs and GAOs were homogeneously distributed across the bed, without consideration of potential heterogeneity in microbial distribution within granules and along beds (Lemaire et al. 2008; Winkler et al. 2011). *Accumulibacter* and *Competibacter* were considered as model populations of the PAO and GAO guilds, respectively. Other affiliates were only low abundant in the VFA-fed granular sludge used (Weissbrodt et al. 2013b). Simulations were run at constant pH values, whereas pH gradient might occur in function of microbial conversions and buffering capacity of wastewater. Speciation and acid-base equilibria could be integrated in PHREEQC (Parkhurst 1995; Weissbrodt et al. 2014b).

Reference simulation conditions were based on experimental ones (Weissbrodt et al. 2013b), namely 20 °C, pH 7.0, full anaerobiosis, acetate concentration of 1000 $mg_{COD}\ L_{Inf}^{-1}$, COD/P ratio in influent of 20 $g_{COD}\ g_P^{-1}$, bed height of 50 cm, identical PAO and GAO fractions of 25%, average amount of total suspended solids (TSS) of 1.75 $g_{TSS}\ cm_{bed}^{-1}$, fraction of volatile suspended solids (VSS) of 67%, fraction of active cells of 50% in the VSS, and feeding time of 60 min. TSS, VSS, and relative abundances of PAOs and GAOs were confirmed for the granular sludge used. The simulated bed height was set sufficiently long to display the different biokinetic states across the bed without significant impact on the computation time. The concentration profiles of soluble and particulate components and the volumetric rates of acetate uptake were represented along the bed in function of n CSTRs in series with a conversion factor of 0.66 cm $CSTR^{-1}$ (50 CSTRs set for a bed of 33 cm). Simulations were run to assess individual and combined impacts of feeding time (15–150 min), temperature (10–30 °C) and pH (6.0–8.0) on bed loading, overall conversions, and PAO/GAO competition.

11.3 Results and Discussion

11.3.1 *Wastewater Adopts a Plug-Flow Regime with Dispersion When Flowing Across Granular Sludge*

The RTDs recorded by step-change in tracer were described with high confidence with model structures of plug-flow with dispersion identified for the two rapid and slow up-flow feeding conditions across the granular sludge bed (Table 11.1, RMSE < 0.010 and $R^2 = 0.999$).

Rapid feeding at a flow-rate of 200 mL min^{-1} (superficial and interstitial velocities of 4.4 and 9.4 m h^{-1}) was related to a first model structure MR1 of a cascade of 50 CSTRs with axial dispersion of 33% in volume and 2% in flow-rate (Figs. 11.1b and 11.2a, Table 11.1-MR1). Similar fit was obtained with a hybrid structure MR2 comprising one large mixed zone contributing to dispersion and representing 47% of bed pore volume, connected to one plug-flow zone representing 53% of bed pore volume modeled as cascade of 200 CSTRs (Fig. 11.1c and Table 11.1-MR2).

Slow feeding at 20 mL min^{-1} (0.44 and 0.94 m h^{-1}) was related to a first model structure MS1 of a cascade of 50 CSTRs with axial dispersion of 36% in volume and 7% in flow-rate (Fig. 11.2b, Table 11.1-MS1). Similar fit was obtained with a hybrid structure MS2 including dispersion as internal recirculation flow-rate of 38 mL min^{-1} (Table 11.1-MS2).

11.3.2 *Analysis of Axial Dispersion Explains Differences Between Rapid and Slow Feeding Regimes*

Dimensionless probability functions enabled comparison between feeding conditions (Fig. 11.3a) and ideal regimes (Fig. 11.3b, c). Following the identified model structures of plug-flow with dispersion, the normalized RTDs laid between characteristic curves of CSTRs and PFRs for the two feeding regimes. The profiles displayed differences in peak skewness, kurtosis, and dimensionless variance ($\sigma_\theta{}^2 = \sigma^2/\tau_m{}^2$) (Table 11.1). The distributions displayed strong tailing with positive peak skewness and mode positioned at the lower quartile under rapid feeding, whereas more symmetrical and slightly closer to a PFR profile under slow feeding.

These differences were explained by distinct axial dispersion patterns inside the bed according to dimensionless numbers (van der Laan 1958; Albuquerque and Santana 2004; Gujer 2008) (Table 11.2). The dimensionless axial dispersion number (N_{Dz}) estimated from the dimensionless variance by iteration with the Newton–Raphson method was almost 30 times higher under rapid (0.138) than slow (0.047) feeding regime. Axial dispersion was nonetheless significant under both cases

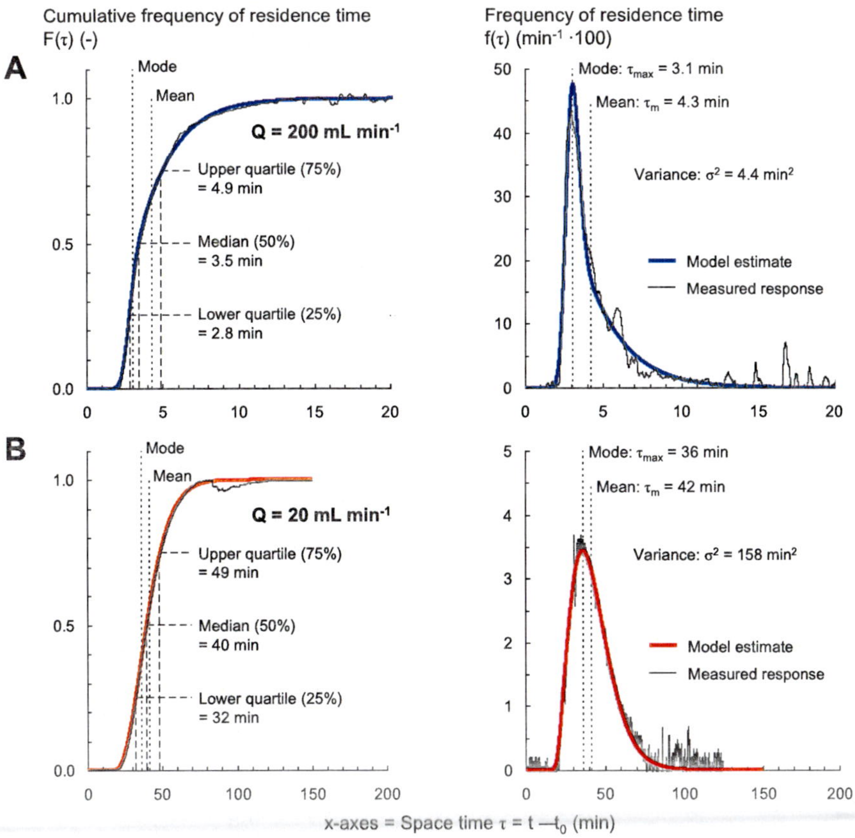

Fig. 11.2 Hydraulic RTDs inside the granular sludge bed obtained with the step change method under rapid feeding conditions with an influent flow-rate of 200 mL min^{-1} (**a**) and under slow anaerobic slow feeding conditions with a flow-rate of 20 mL min^{-1} (**b**). Cumulative frequency functions F(τ) of the residence time τ are given on the left side. The frequency functions f(τ) on the right side were obtained by numerical differentiation of F(τ), and therefore contain background noise. The calibrated hydraulic transport models provided in Table 11.1 for rapid (MR1, *blue*) and slow (MS1, *red*) plug-flow feeding regimes with dispersion (Fig. 11.1b) are superimposed to the experimental data (*black lines*)

according to typical N_{Dz} summarized for small (0.000–0.002), intermediate (0.002–0.025), and large amounts of dispersion (0.025–0.200) within the range delineated by ideal plug (0) and completely mixed (∞) flows. The N_{Dz} characteristic for the slow feeding matched with values of 0.040–0.050 reported for a packed bed bioreactor with similar dimensions and flow-rate (Albuquerque and Santana 2004).

Absolute dispersion coefficients of 10 and 0.35 $m^2\ d^{-1}$ were retrieved from N_{Dz} for rapid and slow feeding, respectively. Reynolds numbers in the bed (Re_{bed}) revealed lower transient flow regime under rapid feeding (1.6) and laminar flow regime under slow feeding (0.17). Higher back-mixing of the fluid phase occurred in the bed under

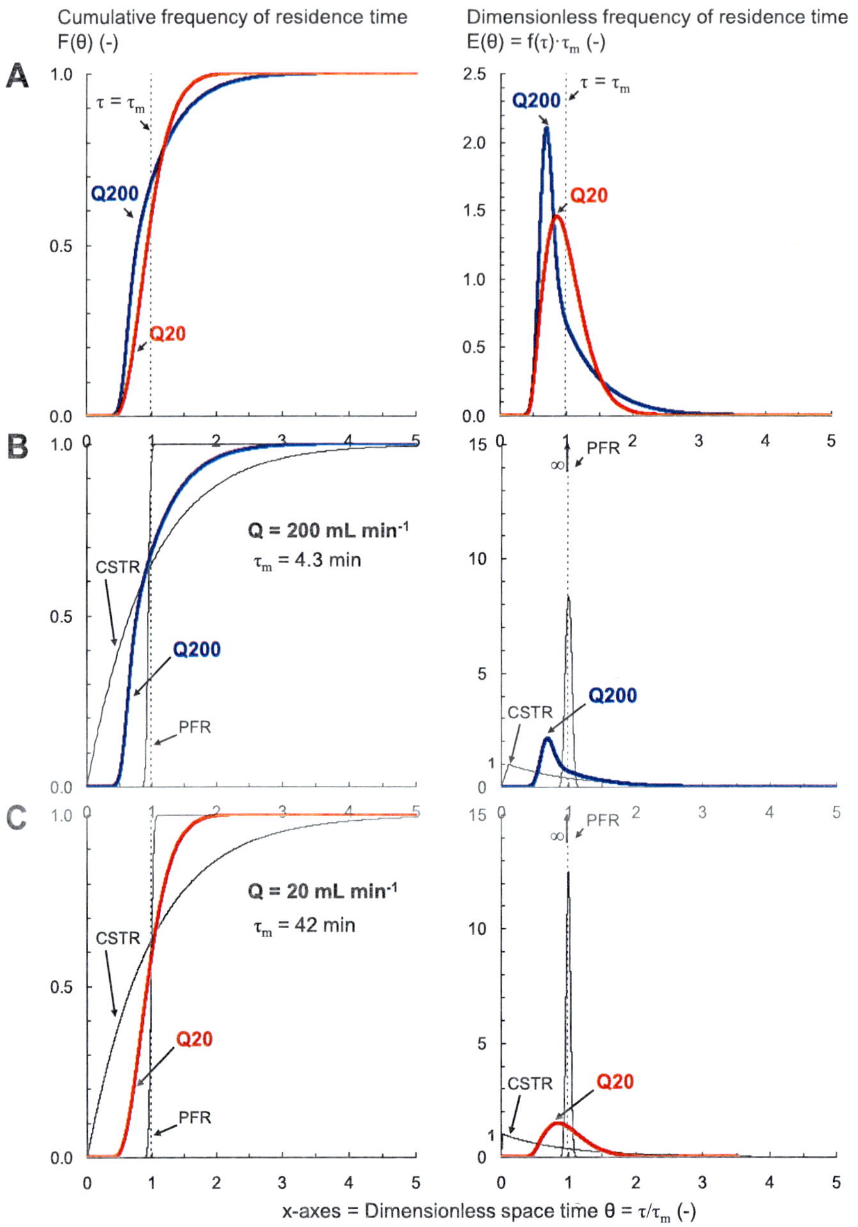

Fig. 11.3 Dimensionless analysis of the modelled RTDs obtained with 200 mL min^{-1} (Q200) and 20 mL min^{-1} (Q20) (**a**). The space time and the frequency of residence time functions were normalized with the mean residence time of the models (τ_m*). The dimensionless functions of the granular sludge system were compared to the ones obtained for ideal CSTR and PFR (**b** and **C**). After peak maximum, the distribution obtained with 200 mL min^{-1} exhibited a tailing dispersion close to the CSTR effect. A more symmetrical PFR-like distribution was obtained with 20 mL min^{-1} but with important axial dispersion

rapid feeding (internal recirculation factor $F_{R,int}*$ of 6.4) and explained the tailing observed, than under slow feeding (1.9).

11.3.3 Hidden Transport Phenomena Might Be Explained by Radial Dispersion and Permeability

Model structures used to describe hydrodynamics inside tubular packed bed reactors mostly consider dispersion in the z-direction of flow. However, dispersion also comprises a radial component transverse to the z-direction (Hinduja et al. 1980; Gunn 1987). The contribution of radial dispersion under rapid and slow feeding regimes was assessed with dimensionless axial and radial Peclet (Pe_z and Pe_t) and dispersion (N_{Dz} and N_{Dt}) numbers computed from the dimensionless variances ($\sigma_\theta^2 = \sigma^2/\tau_m^2$) (Tables 11.1 and 11.2). The values confirmed that both flow regimes were related to high dispersion (Dias et al. 2011) and that the radial contributions ($N_{Dt} = 0.219$ and 0.299) were 1.6 and 6.4 times higher than the axial ones, respectively. Radial dispersion can thus not be neglected in granular sludge beds, and more generally in systems characterized by low Peclet numbers (Rexwinkel et al. 1997).

The analysis of dispersion partly explains the difference between the bed porosity used to fit the model ($\varepsilon* = 0.90$–0.93) and the measured porosity ($\varepsilon = 0.46$) (Table 11.1). The permeability and packing of particles can further impact flow dynamics in beds related to Re_b below 10 (Gunn 1987), and could contribute to the porosity alias. Granular biofilms comprise about 95% of water, are permeable to salts, and need 2–5 min to get as dense as the surrounding solution (Morgenroth 2008b; Winkler et al. 2012). Permeation might be important under slow feeding conditions. A packing of granules could be sensitive to flow conditions since granules display specific gravities relatively close to the densities of fresh and salt water. A median superficial velocity of 6 (first-third quartiles = 0.5–20) m h^{-1} corresponding to a flow-rate of 273 (21–904) mL min^{-1} would be required to fluidize the granular sludge of specific density of 1.12 and median diameter of 1.1 (0.3–2.2) mm according to Ergun's equation (Hinduja et al. 1980) (Additional File 11.3). In contrast to slow feeding conditions (20 mL min^{-1}), rapid feeding conditions (200 mL min^{-1}) can thus impact the bed compactness. The advection-dispersion model could also integrate retardation linked to sorption of solutes to biomass (Albuquerque and Santana 2004).

Overall, granular sludge beds are heterogeneous and complex cases of packed bed reactors. Additional research would be required for a comprehensive description of transport phenomena. A more complex structure could be implemented with a network of interconnected compartments along bed height and diameter, but more computational power would be required. The simplified model structure was nonetheless considered here for metabolic coupling, since the modeled flow pattern was calibrated with tracer experiment.

Table 11.1 Description of the column system and model parameters used to describe the hydraulic RTDs inside the settled granular sludge bed under rapid feeding and anaerobic slow feeding regime

Parameter	Formula	Units	Value
System parameters[1]			
Reactor diameter	D_r	mm	59
Total cross section area[2]	$A = ¼ \cdot \pi \cdot D_r^2$	m^2	$2.7 \cdot 10^{-3}$
Settled bed height at rest	H_{bed}	cm	33
Position of conductivity sensors	H_{sensor}	cm	34, 57
Measured total bed volume	V_{bed}	mL	923
Mass of granules present in bed	$m_{P,Tot}$	g	543
Average mass of individual granules	m_P	mg	17
Average number of granules in bed	$N_P = m_{P,Tot} \cdot m_P^{-1}$	(–)	$3.2 \cdot 10^4$
Granule diameter distribution (median; first-third quartiles)	d_P	mm	1.1 (0.3–2.2)
Measured granular sludge bed pore volume at rest	$V_{L,bed}$	mL	433
Measured granular sludge bed porosity at rest	$\varepsilon = V_{L,bed} \cdot V_{bed}^{-1}$	–	0.46
Average packing angle	α		82
Interstitial cross section area	$A_i = A \cdot \varepsilon$	m^2	$1.3 \cdot 10^{-3}$
Influent concentration of NaBr tracer	C_{Inf}	$g\ L^{-1}$	4.45

Feeding regime	Formula	Units	Rapid	Slow
Measured influent flow-rate	Q	$mL\ min^{-1}$	*197*	*20.6*
Superficial velocity	$v_S = Q \cdot A^{-1}$	$m\ s^{-1}$	$1.2 \cdot 10^{-3}$	$1.2 \cdot 10^{-4}$
Interstitial velocity	$v_i = Q \cdot A_i^{-1} = v_S \cdot \varepsilon^{-1}$	$m\ s^{-1}$	$2.6 \cdot 10^{-3}$	$2.6 \cdot 10^{-4}$
Reynolds number in the bed	Re_{bed}	(–)	1.6 (transient)	0.17 (laminar)
Linear pressure loss due to bed	$\Delta P_{bed}/H_{bed}$	$(Pa\ m^{-1})$	552	54

Parameters[3] of model structures[4]	Formula	Units	*MR1*	*MR2*	*MS1*	*MS2*
Root mean square error[5]	RMSE	(–)	0.007	0.008	0.005	0.007
R-square[5]	R^2	(–)	0.999	0.999	0.999	0.999
Volume fraction of the PFR section	F_{PFR}	(–)	1.00	0.53	1.00	1.00
Granular sludge bed porosity	ε^*	(–)	0.90	0.89	0.93	0.93
Granular sludge bed pore volume	$V_{bed,L}^* = V_{bed} \cdot \varepsilon^*$	mL	831	821	858	858
Number of CSTRs in series in PFR	n	(–)	50	200	50	50

(continued)

Table 11.1 (continued)

Parameters[3] of model structures[4]	Formula	Units	*MR1*	*MR2*	*MS1*	*MS2*
As fraction of influent flow-rate	$F_{R,int} = R_{int} \cdot Q^{-1}$	(–)	–	–	–	1.9
Fraction of axial dispersion volume	F_{DV}	(–)	0.33	–	0.36	–
Fraction of axial dispersion flow-rate	F_{DQ}	(–)	0.02	–	0.07	–
Internal recirculation flow-rate	Q_{Rint}	mL min^{-1}	–	–	–	38
External recirculation flow-rate	Q_{Rext}	mL min^{-1}	–	–	–	–
RTD characteristics[6]						
1st order moment (mean residence time)	M_1 or τ_m	min	4.3	4.4	42	42
Granular sludge bed pore volume	$V_{bed,L}{}^* = Q \cdot \tau_m$	mL	848	867	869	869
2nd order central moment (variance)	M_2 or σ^2	min^2	4.4	3.9	158	155
3rd order central moment	M_3	min^3	18	15	$1.76 \cdot 10^3$	$1.69 \cdot 10^3$
4th order central moment	M_4	min^4	169	134	$1.02 \cdot 10^5$	$1.04 \cdot 10^5$
Skewness	$g_1 = M_3 \cdot {M_2}^{3/2}$	(–)	2.0	2.0	0.9	0.9
Kurtosis	$g_2 = M_4 \cdot {M_2}^2 - 3$	(–)	5.7	5.9	1.1	1.3
Dimensionless variance (modelled RTD)	${\sigma_\theta}^2 = \sigma^2/{\tau_m}^2$	(–)	0.238	0.202	0.090	0.088

[1]Fixed parameters related to the column system used to conduct the RTD experiments
[2]In italic: calculated from previous parameter
Packing angle, Reynolds number in bed, and head loss obtained with the Idelchik method (Fried and Idelchik 1989; LMNOEng. 2001)
$\varepsilon = 1 - \pi/6 \cdot (1 - \cos\alpha)^{-1} \cdot (1 + 2 \cdot \cos\alpha)^{-1/2}$ (valid for $60° < \alpha < 90°$)
$Re_{bed} = 0.45 \cdot v_S \cdot d_P \cdot \rho_W \cdot {\mu_W}^{-1} \cdot (1 - \varepsilon)^{-1} \cdot \varepsilon^{-1/2}$ (valid for any flow regime; laminar $Re_{bed} < 1$, transient $1 < Re_{bed} < 10'000$, turbulent $10'000 < Re_{bed}$)
$\Delta P_{bed} = 1.53 \cdot L \cdot \varepsilon^{-4.2} \cdot d_P \cdot (30/Re_{bed} + 3/{Re_{bed}}^{0.7} + 0.3) \cdot \rho_W \cdot {v_S}^2/2$ (valid for any flow regime)
[3]Values of model parameters obtained for the best fit of the simulated RTDs to the experimental datasets
[4]Different model structures led to optimal description of the experimental datasets: one PFR with axial dispersion (**A**), one CSTR and one PFR in series (**B**), one PFR with internal recirculation (**C**). PFR were approached by cascades of n CSTRs according to Gujer (2008)
[5]The root mean square error (RMSE) and R-square (R^2) values were computed in order to assess the goodness of fit of the models to the experimental datasets. An optimal fit is related to RSME value close to 0, and to a R^2
[6]Statistical characteristics obtained by computation of moments of the simulated probability distribution functions (RTDs)

Table 11.2 Analysis of dispersion of the fluid in the granular sludge bed under rapid and anaerobic slow feeding conditions, based on the dimensionless variance as input parameter

Estimated parameters			Feeding regime	
			Rapid	Slow
Dimensionless variance (modelled RTD)	$\sigma_\theta{}^2 = \sigma^2/\tau_m{}^2$	(–)	0.238	0.090
Axial dispersion				
Total dimensionless axial dispersion number[1]	N_{Dz}	(–)	0.138	0.047
Total axial dispersion coefficient[1]	$D_z = N_{Dz} \cdot v_i \cdot H_{bed}$	$m^2\ d^{-1}$	10.4	0.36
Dimensionless axial Péclet number[2]	$Pe_z = 1/N_{Dz}$	(–)	7.2	21.3
Volume fraction of stagnant fluid[3]	S	(–)	0.35	0.24
Dimensionless rate of exchange of fluid[3]	F	(–)	0.88	1.20
Stagnant volume per unit total bed volume	$\varepsilon_b = S \cdot \varepsilon$	(–)	0.16	0.11
Flowing (active) volume per unit total bed volume	$\varepsilon_a = (1 - S) \cdot \varepsilon = \varepsilon - \varepsilon_b$	(–)	0.30	0.35
Contribution of model elements to axial dispersion[4]				
Axial dispersion number for n CSTRs in series	$N_{Dz,C} = f(n)$	(–)	0.010	0.010
Axial dispersion number for internal recirculation	$N_{Dz,R} = N_{Dz} - N_{Dz,C}$	(–)	0.128	0.037
Internal recirculation flow-rate	$Q_{Rint}{}^* = N_{Dz,R} \cdot n \cdot Q$	$mL\ min^{-1}$	$1.3 \cdot 10^3$	38
As fraction of influent flow-rate	$F_{R,int}{}^* = Q_{Rint}{}^* \cdot Q^{-1}$	(–)	6.4	1.9
Axal dispersion coefficient for internal recirculation	$D_{z,int} = Q_{Rint}{}^* \cdot H_{bed} \cdot A^{-1} \cdot n^{-1}$	$m^2\ d^{-1}$	5.1	0.15
Radial dispersion				
Dimensionless radial Péclet number[2,3]	Pe_t	(–)	4.6	3.4
Total dimensionless radial dispersion number	$N_{Dt} = 1/Pe_t$	(–)	0.219	0.299
Total radial dispersion coefficient	$D_t = N_{Dt} \cdot v_i \cdot H_{bed}$	$m^2\ d^{-1}$	16.5	2.3

The RTDs simulated with model structures A and C of Table 11.1 were taken into consideration for the two flow regimes
[1]The total dimensionless axial dispersion number was obtained by iteration according to the formula provided by van der Laan (1958) and Gujer (2008): $\sigma^2/\tau_m{}^2 = 2 \cdot N_{Dz} - 2 \cdot N_{Dz}{}^2 \cdot (1 - \exp(-1/N_{Dz}))$
[2]Dimensionless Peclet numbers represent the ratio of the rate of advection to the rate of diffusion
[3]Dimensionless groups computed according to mathematical relations developed by Hinduja et al. (1980) for axial and radial dispersion:

$$S = \varepsilon_b / \varepsilon$$

$$F = 1.75(\varepsilon_a/\varepsilon)^2 \cdot (1-\varepsilon)/\varepsilon = 1.75 \cdot (1-S)^2 \cdot (1-\varepsilon)/\varepsilon\} \rightarrow S/(1-S)$$

$$= (1.75 \cdot 1/Pe_z \cdot (1-\varepsilon)/\varepsilon)^{1/2}$$

$$F = S^2 \cdot Pe_z$$

$Pe_t = 2 \cdot (F \cdot L)^{-1}$ with $L = 0.5$ for a packing of spherical particles
[4]Assessment of the contribution of model elements to axial dispersion according to the methodology developed by Gujer (2008). The author provides values for $N_{Dz,C}$ for a range of numbers of CSTRs in series

11.3.4 The Raw Influent Wastewater Does Not Mix with the Supernatant Phase

On-line conductivity measurements revealed identical RTD profiles at the bed (34 cm) and column (57 cm) outlets, this under both rapid and slow feeding regimes (Additional File 11.4). Plug-flow hydraulic transport can be considered in the supernatant section above the bed, indicating no mixing with the raw influent wastewater. Lag times of 3.3 and 32 min were measured between the two sensors under rapid and slow regimes, respectively. On the application point of view, a safe fill-and-draw phase (Arnz et al. 2000) could thus be safely implemented in granular sludge SBRs. Such phase could last 1.25 times the HRT inside the bed without release of inlet wastewater in the effluent of the present column system (geometrical ratios: $H_L/D = 10$, $H_{bed}/D = 6$, $H_{bed}/H_L = 0.6$).

11.3.5 The Feeding Time Impacts on Acetate Uptake by PAOs and GAOs in Granular Sludge Beds

Impacts of feeding time on acetate uptake patterns by PAOs and GAOs are provided hereafter. The conversion states after an anaerobic slow feeding phase length of 60 min traditionally implemented in lab-scale granular sludge SBRs operated for EBPR (de Kreuk and van Loosdrecht 2004) are presented in Additional File 11.5. Reference simulations conditions comprised a bed of 50 cm, 20 °C, pH 7.0, 1.75 g_{TSS} cm^{-1}, 67% VSS, 50% of active cells, and 25% of PAOs and 25% of GAOs homogenously distributed along the bed. Under non selective conditions of temperature and pH, both organisms consumed acetate at equal rates. Simulations conducted over a range of feeding times from 15 to 150 min under constant slow regime (20 mL min^{-1}) revealed significant impact on the individual contributions of PAOs (Fig. 11.4a) and GAOs (Fig. 11.4b) and on the overall consumption of acetate along the bed (Fig. 11.4c). PAOs were able to sustain a maximum duration of acetate feeding of 120 min, whereas GAOs 60 min. This was related to differences in the depletion dynamics of glycogen and polyphosphate in the two organisms. PAOs have the advantage to involve both polyphosphate and glycogen in their anaerobic metabolism as source of energy and reducing equivalents for the transport and condensation of VFAs as PHAs. GAO conversions switched off in the first bed quarter as soon as glycogen got depleted and led to limited acetate uptake across the bed, while PAO conversions were still active over the lower bed half.

The overall acetate breakthrough profiles (Fig. 11.4c) revealed a maximum feeding time limit of 120 min toward constant bed loading. The first 20 cm of bed got inactive after 150 min because of full depletion of glycogen and polyphosphate overall. In the present system, a bed of 33 cm combined with a feeding time of 60 min was a good tradeoff for full bed loading, non-limited conversions across the bed, and full acetate uptake under anaerobic feeding. Similar conditions have led to efficient

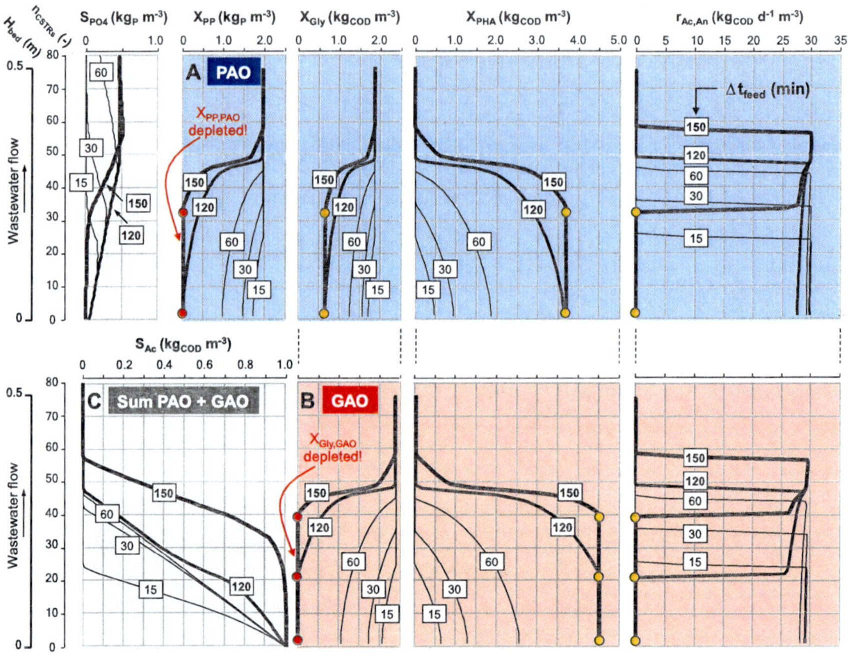

Fig. 11.4 Impact of the feeding time on the anaerobic metabolism of PAO (**a**—*blue shade*) and GAO (**b**—*red shade*) across the granular sludge bed, at 20 °C and pH 7.0. Breakthrough curve of acetate in function of the feeding time (**c**—*grey shade*). Prolonged supply of acetate during e.g. 120 or 150 min resulted in the depletion of glycogen in GAO and polyphosphate in PAO in the lower half of the bed (delineated by red points), that switched off the biochemical conversions in this bed section (delineated by yellow points). With a bed of 50 cm comprising 1.75 gTSS cm^{-1}, 67% of VSS, 25% of PAOs, and 25% of GAOs, acetate was still fully taken up in the upper half. However, the acetate present in the fluid phase of the lower half was not degraded. In such a case, an additional anaerobic idle or mixed phase should be implemented after the feeding period in order to fully remove acetate prior to switching to aeration. An anaerobic slow feeding period of 60 min is a good tradeoff, and a bed of 33 cm would be sufficient. *Legend* volumetric anaerobic acetate uptake rate ($r_{Ac,An}$) by PAOs and by GAOs; concentration of intracellular PHAs (X_{PHA}), glycogen (X_{Gly}), and polyphosphate (X_{PP}) in PAOs and in GAOs; concentration of acetate (S_{Ac}) and orthophosphate (S_{PO4}) in the flowing fluid phase

PAO-GAO selection and EBPR in earlier research (Weissbrodt et al. 2014c). An analogous model can be used in practice for optimizing the bed design and feeding phase length in granular sludge SBRs operated for EBPR toward efficient anaerobic consumption of organic substrates prior to aeration (de Kreuk and van Loosdrecht 2004; Weissbrodt et al. 2013a, 2014a; Lochmatter and Holliger 2014).

Competition between the model PAO genus *Accumulibacter* and GAO genus *Competibacter* for the uptake of acetate during anaerobic slow feeding was investigated by coupling the calibrated hydraulic transport model to the metabolic model proposed by Lopez-Vazquez et al. (2009). In contrast to activated sludge systems, granular biomass does not move with the wastewater under the investigated plug-flow feeding regime. Intracellular storage products of inorganic polyphosphate, glycogen, and poly-β-hydroxyalkanoates (PHAs) are not transported in the water flow, contrary to dissolved compounds such as acetate and orthophosphate. The conversions in the sludge bed will therefore always be dynamic: increasing the feeding phase length will never lead to steady-state conversions. The only steady-state that can be reached is when intracellular stocks of polyphosphate and glycogen are fully depleted meaning that acetate is no more taken up. The feeding time is therefore key parameter to consider in plug-flow biological systems such as EBPR ones involving cycling of intracellular polymers.

11.3.6 *Temperature and pH Impact on Acetate Uptake by PAOs and GAOs in Granular Sludge Beds*

Previous research highlighted temperature and pH as key factors triggering bacterial selection in granular sludge EBPR SBRs (Weissbrodt et al. 2013b). The combined hydraulic-metabolic model was used to investigate the underlying effects on acetate uptake by PAOs and GAOs during anaerobic slow feeding (Fig. 11.5). Temperature and pH dependencies of the maximum biomass specific rates of acetate uptake are presented in Fig. 11.5a based on kinetics developed by Lopez-Vazquez et al. (2009). *Accumulibacter* preferentially take up acetate under slightly alkaline (pH 7.0–8.0) and psychrophilic conditions (10–20 °C). *Competibacter* are mainly active under acidic (pH 6.0–7.0) and higher mesophilic (20–30 °C) conditions. The rate of acetate uptake is proportional to temperature for both populations from 10 to 20 °C. Temperature also impacts on maintenance rates under anaerobic conditions. Both affiliates spend 2.4 ATP-mmol h^{-1} C-mol_X^{-1} at 20 °C. *Accumulibacter* requires twice less maintenance (0.9 ATP-mmol h^{-1} C-mol_X^{-1}) than *Competibacter* at 10–20 °C, whereas the situation is reversed at 20–30 °C with *Accumulibacter* spending 5.9 ATP-mmol h^{-1} C-mol_X^{-1}.

During anaerobic slow feeding under mesophilic temperatures of 20–30 °C, full uptake of acetate was obtained with a minimum bed height of 30 cm and with GAOs being more efficient than PAOs under non-selective pH conditions. Significantly lower volumetric rates of acetate uptake were observed for both populations along decreasing temperatures from 20 to 10 °C. PAOs (decrease factor of 1.7–2.5 fold) had certain competitive advantage over GAOs (2.3–sevenfold) (Fig. 11.5b). Almost twice higher beds of granular sludge should be planned for full anaerobic conversions under psychrophilic conditions compared to the situation at 20 °C. This has practical implication for the operation of granular sludge SBRs in winter.

PAOs consumed acetate at maximum volumetric rates of 30 kg_{COD} d^{-1} m^{-3} independently of pH within the bed (Fig. 11.5c). In contrast, the rate of acetate uptake by GAOs was directly impacted by pH with selective advantage and maximum uptake rates of 32 kg_{COD} d^{-1} m^{-3} at pH 6.0–6.5. Alkaline conditions were detrimental to this

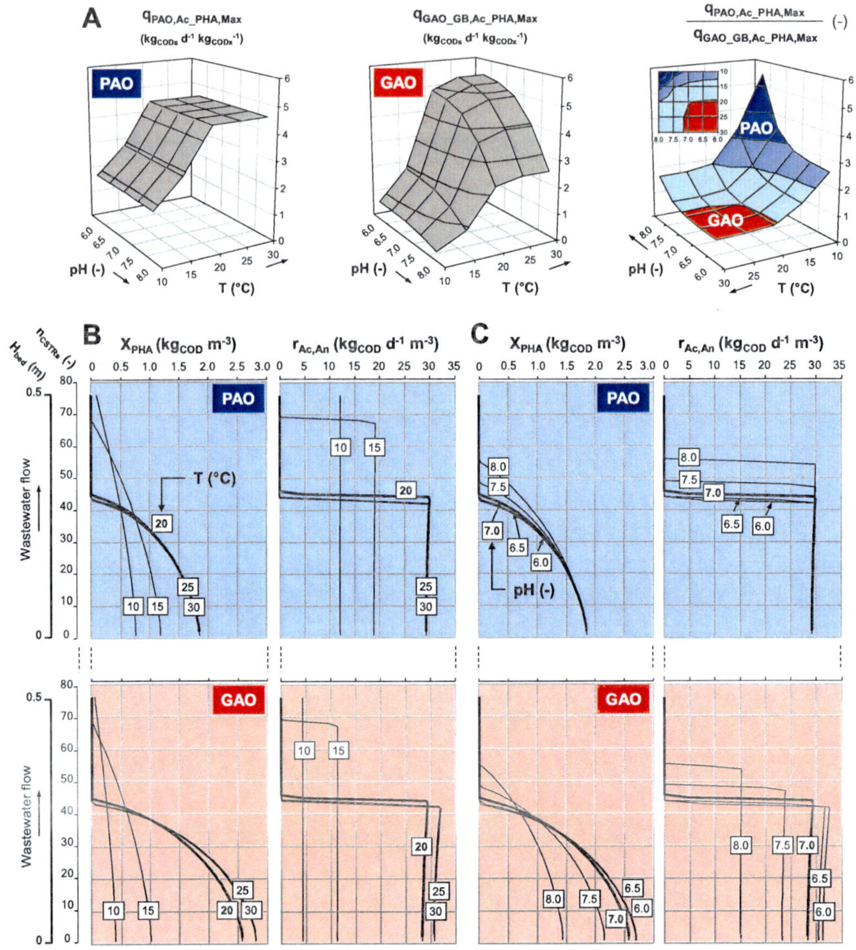

Fig. 11.5 Impact of temperature and pH on the anaerobic metabolisms of PAOs and GAOs. The theoretical combined impact of temperature and pH was computed as contour plot to highlight the dependency of the maximum biomass specific rate of acetate uptake by PAOs and GAOs under anaerobic conditions (**a**). For PAOs, this rate is only dependent on temperature, whereas for GAOs both temperature and pH affect this rate. pH was assumed fixed in these computations, whereas this parameter may evolve along the bed depending on the local water chemistry. The ratio of the PAO- and GAO-related rates provides information on optimal combination of temperature and pH conditions for preferential selection of either PAOs (*blue shade*) or GAOs (*red shade*). Impact of temperature (**b**) and pH (**c**) on the anaerobic conversions across the granular sludge bed during feeding. *Legend* volumetric acetate uptake rate under anaerobic conditions ($r_{Ac,An}$), concentration of intracellular poly-β-hydroxyalkanoates (X_{PHA}), number of CSTRs in series used to model the plug-flow transport (n_{CSTRs}), height of the settled granular sludge bed (H_{bed})

population. A bed height of 30 cm was required for full acetate uptake by the consortium at pH 6.0–7.0. Only slightly higher beds up to 36 cm were required for full uptake under alkaline conditions. This difference in height originates from the amount of energy as ATP required for the transport of acetate across the cell membranes (Smolders et al. 1994). The stoichiometric yields of hydrolysis of glycogen (by GAOs) and polyphosphate (by PAOs) account for this pH dependency (Additional File 11.2). Higher hydrolysis is required to compensate pH gradients across the cell membrane under alkaline conditions for the same amount of acetate taken up (Additional File 11.6).

Simulations under combined conditions of temperature and pH displayed preferential uptake of acetate by *Accumulibacter* in the bed independently of temperature (Fig. 11.6). Operation at 10–15 °C was detrimental to *Competibacter*, even at pH 6.0. Combined acidic and higher mesophilic conditions selected for acetate uptake by *Competibacter*. The difference between the volumetric rates of acetate uptake by GAOs and PAOs ($\Delta r_{Ac,An(GAO\text{-}PAO)} = 6\ kg_{COD}\ d^{-1}\ m^{-3}$) was however only half of the difference observed at pH 8.0 that favored PAOs ($\Delta r_{Ac,An(PAO\text{-}GAO)} = 12\text{–}13\ kg_{COD}\ d^{-1}\ m^{-3}$). A short bed of 28 cm was required for full uptake at 30 °C and pH 6.0. Decreasing the temperature significantly impacted on bed height requirement. Increasing pH resulted in only small increases in required bed height. The modeling results met with previous knowledge in the sense that pH is a key factor triggering the PAO/GAO competition in both activated sludge and granular sludge systems (Lopez-Vazquez et al. 2009; Oehmen et al. 2010a; Weissbrodt et al. 2013b). More generally, the same organisms compose the microbiome of the two types of sludges (Nielsen et al. 2010; Weissbrodt et al. 2014c) and thus analogous mechanisms of selection apply (Weissbrodt et al. 2014a). Addition of lime in the influent might significantly help to select for PAOs and EBPR in granular sludge.

11.3.7 Integration in the Granular Sludge Research and Practice

This hydraulic-metabolic model can serve as a base for further research aiming to investigate the impact wastewater compositions and environmental conditions on anaerobic metabolisms across granular sludge beds. This modeling framework has practical implications for the design of bed geometries and feeding conditions notably in relation to local temperature and pH levels toward efficient bacterial selection and EBPR performances. The hydraulic-metabolic model might further be made more complex depending on specific objectives.

Different metabolic behaviors have been reported for the PAO and GAO guilds in function of the VFA composition (Lopez-Vazquez et al. 2009). The impact of carbon sources in granular sludge (Gonzalez-Gil and Holliger 2011; Weissbrodt et al. 2013b) can be tested in silico. More specifically, an anaerobic selector is required to remove the full load of organic substrates prior to switching to the aeration phase in

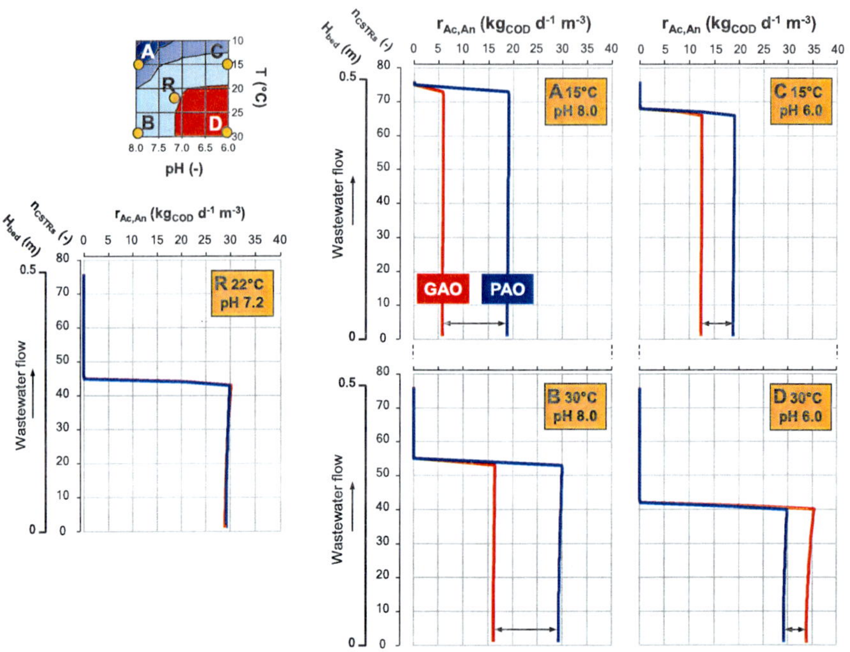

Fig. 11.6 Simulated combined impact of temperature and pH during the anaerobic slow feeding phase. A pH-value of 8.0 was efficient for preferential anaerobic acetate uptake by PAOs independently from temperature (**a, b**). Higher volumetric rate was nevertheless obtained at 30 °C than at 15 °C. Combined conditions at 15 °C and pH 6.0 were still favorable for PAOs (**c**). The combination of pH 6.0 and 30 °C resulted in predominant GAO activity (**d**). The difference between the GAO- and PAO-related rates at this optimum is however less than the differences obtained with predominant PAO activity. PAOs and GAOs displayed identical acetate uptake rates with operation at 22 °C and pH 7.2 (**e**)

order to efficiently select for active PAOs at the expense of unfavorable OHOs that induce filamentous bulking or deteriorated BNR (Ebrahimi et al. 2010; Weissbrodt et al. 2012, 2013a, 2013b). The present approach is decisive toward optimizing the bed geometry and feeding conditions in order to prevent the release of organic loads in the supernatant. An anaerobic mixed batch phase can also be implemented after the anaerobic slow feeding phase as safety measure in order to polish the anaerobic treatment before aeration (Rocktäschel et al. 2013; Weissbrodt et al. 2013a; Lochmatter and Holliger 2014). Translation into full-scale conditions would need consideration of slowly biodegradable particulate substrates which can make up to 40–60% of the total COD of the influents originating from areas with short and steep sewers (Koch et al. 2000; Vollertsen and Hvitved-Jacobsen 2005; Weissbrodt et al. 2014a). The impact of real wastewater conditions on granulation is an important field of investigation (de Kreuk et al. 2010; Rocktäschel et al. 2013; Wagner and Da Costa 2013; Sturm and Faraj 2014) (Derlon et al. in prep., Wagner et al. in prep.). Anaerobic microbial processes of hydrolysis into soluble substrates and subsequent fermentation into VFAs prior to uptake should be investigated by analogy to activated sludge (Carucci et al. 1994; Rieger et al. 2001; Weissbrodt et al. 2014a) and could be included in the model.

Gradients of PAOs and GAOs can occur inside granules (Lemaire et al. 2008) and along granular sludge beds (Winkler et al. 2011; Bassin et al. 2012) depending on specific operational conditions. The model structure could be adapted accordingly in order to investigate effects of biomass segregation at micro- and macroscale on conversions (Morgenroth 2008a, 2008b). Gradients of pH might establish across granular sludge beds in function of active metabolisms. Analogous model structure could be implemented in software including acid-base equilibria and speciation such as PHREEQC (Parkhurst 1995; Weissbrodt et al. 2014b). The PAO/GAO competition is also governed by environmental and operational factors under aerobic and anoxic conditions (Oehmen et al. 2010a). Nitrite and nitrate electron acceptors originating from nitrification and incomplete denitrification in previous SBR cycles can be present in the bulk during feeding. Model adaptations could account for anoxic processes. Definitely, the identified model structures of the feeding phase can be coupled to formulations of models of granular sludge SBRs (de Kreuk et al. 2007; Ni and Yu 2010; Su et al. 2013; Kagawa et al. 2014) toward full scale. Advances in molecular methods have finally suggested complex bacterial compositions of EBPR ecosystems with additional PAOs and GAOs to be considered (Meyer et al. 2006; Forbes et al. 2009; Nielsen et al. 2010, 2012; Nguyen et al. 2012; Albertsen et al. 2013; Kristiansen et al. 2013; Lanham et al. 2013; Weissbrodt et al. 2014c).

11.4 Conclusions

The hydraulic-metabolic model developed in the present study enabled to specifically investigate the hydraulic transport patterns and mechanism of selection of PAOs and GAOs using different feeding approaches of a granular sludge SBR. This led to the following conclusions:

- The plug-flow model with dispersion explained with high confidence the hydraulic transport of wastewater recorded as RTDs across the bed of granular sludge in the column-type reactor system for both rapid and slow feeding conditions.
- The combined hydraulic-metabolic model enabled to highlight that PAOs can be selected over GAOs independently of temperature by maintaining slightly alkaline conditions during anaerobic slow feeding, e.g., by dosing lime in the influent.
- In contrast to pH, temperature induced a much stronger impact on the bed height requirements toward full uptake of organic substrate under anaerobic conditions.
- Overall, this modeling framework can help designing bed geometries and feeding conditions for an efficient management of the bacterial resource in granular sludge. It can enable optimal bed loading and efficient organic consumption during anaerobic slow feeding.

Acknowledgements This study was financed by the Swiss National Science Foundation (Grants no. 205321-120536 and 200020-138148) and internal funds at ETH Zürich and Eawag.

Supplementary Information

Additional File 11.1 Determination of the exact mass of granules in the bed

Additional File 11.2 PAO/GAO metabolic model

Additional File 11.3 Minimum fluidization of granular sludge

Additional File 11.4 Residence time distributions at the bed and column outlets

Additional File 11.5 Conversions of intracellular storage products along the bed

Additional File 11.6 Impact of pH on polyphosphate hydrolysis requirements.

References

Albertsen M, Saunders AM, Nielsen KL, Nielsen PH (2013) Metagenomes obtained by 'deep sequencing': what do they tell about the enhanced biological phosphorus removal communities? Water Sci Technol 68(9):1959–1968

Albuquerque A, Santana F (2004) Hydrodynamic behaviour of a biological packed bed under different hydraulic and organic loading. In: Reis AH, Miguel AF (eds) International conference on applications of porous media, Evora, Portugal, 2004. CGE-UE, pp 319–327

Anderson GK, Campos CMM, Chernicharo CAL, Smith LC (1991) Evaluation of the inhibitory effects of lithium when used as a tracer for anaerobic digesters. Water Res 25(7):755–760

Arnz P, Esterl S, Nerger C, Delgado A, Wilderer PA (2000) Simultaneous loading and draining as a means to enhance efficacy of sequencing biofilm batch reactors. Water Res 34(5):1763–1766

Bassin JP, Winkler MKH, Kleerebezem R, Dezotti M, van Loosdrecht MCM (2012) Improved phosphate removal by selective sludge discharge in aerobic granular sludge reactors. Biotechnol Bioeng 109(8):1919–1928

Carucci A, Majone M, Ramadori R, Rossetti S (1994) Dynamics of phosphorus and organic substrates in anaerobic and aerobic phases of a sequencing batch reactor. Water Sci Technol 30(6 pt 6):237–246

Cech JS, Hartman P (1993) Competition between polyphosphate and polysaccharide accumulating bacteria in enhanced biological phosphate removal systems. Water Res 27(7):1219–1225

Comeau Y, Hall KJ, Hancock REW, Oldham WK (1986) Biochemical model for enhanced biological phosphorus removal. Water Res 20(12):1511–1521

Corominas L, Rieger L, Takacs I, Ekama G, Hauduc H, Vanrolleghem PA, Oehmen A, Gernaey KV, van Loosdrecht MCM, Comeau Y (2010) New framework for standardized notation in wastewater treatment modelling. Water Sci Technol 61:841–857

Crocetti GR, Banfield JF, Keller J, Bond PL, Blackall LL (2002) Glycogen-accumulating organisms in laboratory-scale and full-scale wastewater treatment processes. Microbiology 148(11):3353–3364

de Kreuk MK, van Loosdrecht MCM (2004) Selection of slow growing organisms as a means for improving aerobic granular sludge stability. Water Sci Technol 49(11–12):9–17 de Kreuk MK, Picioreanu C, Hosseini M, Xavier JB, van Loosdrecht MCM (2007) Kinetic model of a granular sludge SBR: Influences on nutrient removal. Biotechnol Bioeng 97(4):801–815

de Kreuk MK, Kishida N, Tsuneda S, van Loosdrecht MCM (2010) Behavior of polymeric substrates in an aerobic granular sludge system. Water Res 44(20):5929–5938

Deans HA, Lapidus L (1960) A computational model for predicting and correlating the behavior of fixed-bed reactors: I. Derivation of model for nonreactive systems. AICHE J 6(4):656–663

Dias PM, Sukasam K, Jin JH (2011) Radiotracer applications in wastewater treatment plants, vol 49. In: International atomic energy agency (IAEA), Vienna, Austria

Ebrahimi S, Gabus S, Rohrbach-Brandt E, Hosseini M, Rossi P, Maillard J, Holliger C (2010) Performance and microbial community composition dynamics of aerobic granular sludge from sequencing batch bubble column reactors operated at 20, 30, and 35°C. Appl Microbiol Biotechnol 87(4):1555–1568

Forbes CM, O'Leary ND, Dobson AD, Marchesi JR (2009) The contribution of 'omic'-based approaches to the study of enhanced biological phosphorus removal microbiology: minireview. FEMS Microbiol Ecol 69(1):1–15

Fried E, Idelchik IE (1989) Flow resistance: a design guide for engineers. Hemisphere Pub. Co., New York

Giesen A, Niermans R, van Loosdrecht MCM (2012) Aerobic granular biomass: the new standard for domestic and industrial wastewater treatment? Water 21(4):28–30

Gonzalez-Gil G, Holliger C (2011) Dynamics of microbial community structure of and enhanced biological phosphorus removal by aerobic granules cultivated on propionate or acetate. Appl Environ Microbiol 77(22):8041–8051

Guisasola A, Pijuan M, Baeza JA, Carrera J, Casas C, Lafuente J (2004) Aerobic phosphorus release linked to acetate uptake in bio-P sludge: process modeling using oxygen uptake rate. Biotechnol Bioeng 85(7):722–733

Gujer W (2008) Systems analysis for water technology. Springer, Berlin, Heidelberg

Gujer W (2011) Is modeling of biological wastewater treatment a mature technology? Water Sci Technol 63(8):1739–1743

Gunn DJ (1987) Axial and radial dispersion in fixed beds. Chem Eng Sci 42(2):363–373

Hauduc H, Rieger L, Oehmen A, van Loosdrecht MCM, Comeau Y, Héduit A, Vanrolleghem PA, Gillot S (2013) Critical review of activated sludge modeling: state of process knowledge, modeling concepts, and limitations. Biotechnol Bioeng 110(1):24–46

Hesselmann RPX, Werlen C, Hahn D, van der Meer JR, Zehnder AJB (1999) Enrichment, phylogenetic analysis and detection of a bacterium that performs enhanced biological phosphate removal in activated sludge. Syst Appl Microbiol 22(3):454–465

Hinduja MJ, Sundaresan S, Jackson R (1980) Crossflow model of dispersion in packed bed reactors. AICHE J 26(2):274–281

Hoogendoorn CJ, Lips J (1965) Axial mixing of liquid in gas-liquid flow through packed beds. Can J Chem Eng 43(3):125–131

Houweling D, Comeau Y, Takacs I, Dold P (2010) Uncertainty and variability in enhanced biological phosphorus removal (EBPR) stoichiometry: consequences for process modelling and optimization. Water Sci Technol 61(7):1793–1800

Kagawa Y, Tahata J, Kishida N, Matsumoto S, Picioreanu C, van Loosdrecht MCM, Tsuneda S (2014) Modeling the nutrient removal process in aerobic granular sludge system by coupling the reactor- and granule-scale models. Biotechnol Bioeng 4:217–219

Koch G, Kuhni M, Gujer W, Siegrist H (2000) Calibration and validation of activated sludge model no. 3 for Swiss municipal wastewater. Water Res 34(14):3580–3590

Kristiansen R, Nguyen HTT, Saunders AM, Nielsen JL, Wimmer R, Le VQ, McIlroy SJ, Petrovski S, Seviour RJ, Calteau A, Nielsen KL, Nielsen PH (2013) A metabolic model for members of the genus Tetrasphaera involved in enhanced biological phosphorus removal. ISME J 7(3):543–554

Lanham AB, Oehmen A, Saunders AM, Carvalho G, Nielsen PH, Reis MAM (2013) Metabolic versatility in full-scale wastewater treatment plants performing enhanced biological phosphorus removal. Water Res 47(19):7032–7041

Lanham AB, Oehmen A, Saunders AM, Carvalho G, Nielsen PH, Reis MAM (2014) Metabolic modelling of full-scale enhanced biological phosphorus removal sludge. Water Res 66:283–295

Lemaire R, Yuan Z, Blackall LL, Crocetti GR (2008) Microbial distribution of *Accumulibacter* spp. and *Competibacter* spp. in aerobic granules from a lab-scale biological nutrient removal system. Environ Microbiol 10(2):354–363

Levenspiel O (1999) Chemical reaction engineering, 3rd edn. John Wiley & Sons, Hoboken, NJ, USA

LMNOEng (2001) Packed bed calculator. PUblisher. http://www.lmnoeng.com/Groundwater/PackedBed.htm

Lochmatter S, Holliger C (2014) Optimization of operation conditions for the startup of aerobic granular sludge reactors biologically removing carbon, nitrogen, and phosphorous. Water Res 59:58–70

Lochmatter S, Gonzalez-Gil G, Holliger C (2013) Optimized aeration strategies for nitrogen and phosphorus removal with aerobic granular sludge. Water Res 47(16):6187–6197

Lopez-Vazquez CM, Oehmen A, Hooijmans CM, Brdjanovic D, Gijzen HJ, Yuan Z, van Loosdrecht MCM (2009) Modeling the PAO-GAO competition: Effects of carbon source, pH and temperature. Water Res 43(2):450–462

Macey R, Oster G, Zahnley T (2000) Berkeley Madonna user's guide. University of California, Berkeley, CA, USA

Martin AD (2000) Interpretation of residence time distribution data. Chem Eng Sci 55(23):5907–5917

Meyer RL, Saunders AM, Blackall LL (2006) Putative glycogen-accumulating organisms belonging to the Alphaproteobacteria identified through rRNA-based stable isotope probing. Microbiology 152(2):419–429

Mino T, Liu WT, Kurisu F, Matsuo T (1995) Modelling glycogen storage and denitrification capability of microorganisms in enhanced biological phosphate removal processes. Water Sci Technol 31(2):25–34

Morgenroth E (2008a) Biofilm reactors. In: Henze M, van Loosdrecht MCM, Ekama GA, Brdjanovic D (eds) Biological wastewater treatment: principles, modelling and design. IWA Publishing, London, pp 493–511

Morgenroth E (2008b) Modelling biofilms. In: Henze M, van Loosdrecht MCM, Ekama GA, Brdjanovic D (eds) Biological wastewater treatment: principles, modelling and design. IWA Publishing, London, pp 457–492

Nguyen HTT, Nielsen JL, Nielsen PH (2012) Candidatus Halomonas phosphatis", a novel polyphosphate-accumulating organism in full-scale enhanced biological phosphorus removal plants. Environ Microbiol 14(10):2826–2837

Ni BJ, Yu HQ (2010) Mathematical modeling of aerobic granular sludge: a review. Biotechnol Adv 28(6):895–909

Nielsen PH, Mielczarek AT, Kragelund C, Nielsen JL, Saunders AM, Kong Y, Hansen AA, Vollertsen J (2010) A conceptual ecosystem model of microbial communities in enhanced biological phosphorus removal plants. Water Res 44(17):5070–5088

Nielsen PH, Saunders AM, Hansen AA, Larsen P, Nielsen JL (2012) Microbial communities involved in enhanced biological phosphorus removal from wastewater: a model system in environmental biotechnology. Curr Opin Biotechnol 23(3):452–459

Oehmen A, Carvalho G, Lopez-Vazquez CM, van Loosdrecht MCM, Reis MAM (2010a) Incorporating microbial ecology into the metabolic modelling of polyphosphate accumulating organisms and glycogen accumulating organisms. Water Res 44(17):4992–5004

Oehmen A, Lopez-Vazquez CM, Carvalho G, Reis MAM, van Loosdrecht MCM (2010b) Modelling the population dynamics and metabolic diversity of organisms relevant in anaerobic/anoxic/aerobic enhanced biological phosphorus removal processes. Water Res 44(15):4473–4486

Parkhurst DL (1995) User's guide to PHREEQC: a computer program for speciation, reaction-path, advective-transport, and inverse geochemical calculations US geological survey water-resources investigations, vol 95–4227. http://wwwbrr.cr.usgs.gov/projects/GWC_coupled/phreeqc/

Quaff A, Guha S (2011) Evaluation of mixing and performance of lab-scale upflow anaerobic sludge blanket reactors treating domestic wastewater. J Environ Eng 137(5):322–331

Rexwinkel G, Heesink ABM, van Swaaij WPM (1997) Mass transfer in packed beds at low Peclet numbers: wrong experiments or wrong interpretations? Chem Eng Sci 52(21–22):3995–4003

Rieger L, Koch G, Kuhni M, Gujer W, Siegrist H (2001) The EAWAG Bio-P module for activated sludge model No. 3. Water Res 35(16):3887–3903

Rittmann BE, Hausner M, Loffler F, Love NG, Muyzer G, Okabe S, Oerther DB, Peccia J, Raskin L, Wagner M (2006) A vista for microbial ecology and environmental biotechnology. Environ Sci Technol 40(4):1096–1103

Rocktäschel T, Klarmann C, Helmreich B, Ochoa J, Boisson P, Soerensen KH, Horn H (2013) Comparison of two different anaerobic feeding strategies to establish a stable aerobic granulated sludge bed. Water Res 47(17):6423–6431

Smith LC, Elliot DJ, James A (1996) Mixing in upflow anaerobic filters and its influence on performance and scale-up. Water Res 30(12):3061–3073

Smolders GJF, van der Meij J, van Loosdrecht MCM, Heijnen JJ (1994) Model of the anaerobic metabolism of the biological phosphorus removal process: stoichiometry and pH influence. Biotechnol Bioeng 43(6):461–470

Sturm B, Faraj R (2014) The effect of slowly biodegradable cabon on the performance of aerobic granular sludge. Paper presented at the IWA specialist conference on global challenges: sustainable wastewater treatment and resource recovery, Kathmandu, Nepal

Su KZ, Ni BJ, Yu HQ (2013) Modeling and optimization of granulation process of activated sludge in sequencing batch reactors. Biotechnol Bioeng 110(5):1312–1322

Tijhuis L, Van Benthum WAJ, Van Loosdrecht MCM, Heijnen JJ (1994) Solids retention time in spherical biofilms in a biofilm airlift suspension reactor. Biotechnol Bioeng 44(8):867–879

van der Laan ET (1958) Notes on the diffusion-type model for the longitudinal mixing in flow. Chem Eng Sci 7:187–191

van Genuchten MT (1980) Determining transport parameters from solute displacement experiments. USDA-SEA-AR and University of California, Riverside, CA, USA, p 27

Verstraete W, Wittebolle L, Heylen K, Vanparys B, de Vos P, van de Wiele T, Boon N (2007) Microbial resource management: the road to go for environmental biotechnology. Eng Life Sci 7(2):117–126

Vollertsen J, Hvitved-Jacobsen T (2005) Biodegradability of wastewater: a method for COD-fractionation. Water Sci Technol 45(3):25–34

Wagner J, Da Costa RHR (2013) Aerobic granulation in a sequencing batch reactor using real domestic wastewater. J Environ Eng 139(11):1391–1396

Weissbrodt DG, Lochmatter S, Ebrahimi S, Rossi P, Maillard J, Holliger C (2012) Bacterial selection during the formation of early-stage aerobic granules in wastewater treatment systems operated under wash-out dynamics. Front Microbiol 3:332

Weissbrodt DG, Neu TR, Kuhlicke U, Rappaz Y, Holliger C (2013a) Assessment of bacterial and structural dynamics in aerobic granular biofilms. Front Microbiol 4:175

Weissbrodt DG, Schneiter GS, Furbringer JM, Holliger C (2013b) Identification of trigger factors selecting for polyphosphate- and glycogen-accumulating organisms in aerobic granular sludge sequencing batch reactors. Water Res 47(19):7006–7018

Weissbrodt DG, Derlon N, Wagner J, Holliger C, Morgenroth E (2014a) A consolidated approach of flocculent and granular sludge systems under the perspective of bacterial resource management. Paper presented at the water environment federation's annual technical exhibition and conference (WEFTEC), New Orleans, USA

Weissbrodt DG, Maillard J, Brovelli A, Chabrelie A, May J, Holliger C (2014b) Multilevel correlations in the biological phosphorus removal process: from bacterial enrichment to conductivity-based metabolic batch tests and polyphosphatase assays. Biotechnol Bioeng 111(12):2421–2435

Weissbrodt DG, Shani N, Holliger C (2014c) Linking bacterial population dynamics and nutrient removal in the granular sludge biofilm ecosystem engineered for wastewater treatment. FEMS Microbiol Ecol 88(3):579–595

Weissbrodt DG (2012) Bacterial resource management for nutrient removal in aerobic granular sludge wastewater treatment systems. PhD thesis No. 5641, Ecole Polytechnique Fédérale de Lausanne, Switzerland

Wentzel MC, Comeau Y, Ekama GA, van Loosdrecht MCM, Brdjanovic D (2008) Enhanced biological phosphorus removal. In: Henze M, van Loosdrecht MCM, Ekama GA, Brdjanovic D (eds) Biological wastewater treatment: principles, modelling and design. IWA Publishing, London, pp 155–220

Whang LM, Filipe CDM, Park JK (2007) Model-based evaluation of competition between polyphosphate- and glycogen-accumulating organisms. Water Res 41(6):1312–1324

Winkler MKH, Bassin JP, Kleerebezem R, de Bruin LMM, van den Brand TPH, van Loosdrecht MCM (2011) Selective sludge removal in a segregated aerobic granular biomass system as a strategy to control PAO-GAO competition at high temperatures. Water Res 45(11):3291–3299

Winkler MKH, Bassin JP, Kleerebezem R, van der Lans RGJM, van Loosdrecht MCM (2012) Temperature and salt effects on settling velocity in granular sludge technology. Water Res 46(12):3897–3902

Yagci N, Artan N, Cokgor EU, Randall CW, Orhon D (2003) Metabolic model for acetate uptake by a mixed culture of phosphate- and glycogen-accumulating organisms under anaerobic conditions. Biotechnol Bioeng 84(3):359–373

Zeng RJ, Yuan Z, Keller J (2003) Model-based analysis of anaerobic acetate uptake by a mixed culture of polyphosphate-accumulating and glycogen-accumulating organisms. Biotechnol Bioeng 83(3):293–302

Zeng Y, Mu SJ, Lou SJ, Tartakovsky B, Guiot SR, Wu P (2005) Hydraulic modeling and axial dispersion analysis of UASB reactor. Biochem Eng J 25(2):113–123

Chapter 12
Concluding Remarks and Outlook

Principles for an Ecological Engineering of Nutrient Removal in Granular Sludge Processes

Bacterial resource management: the seed for granular sludge.
(This study)

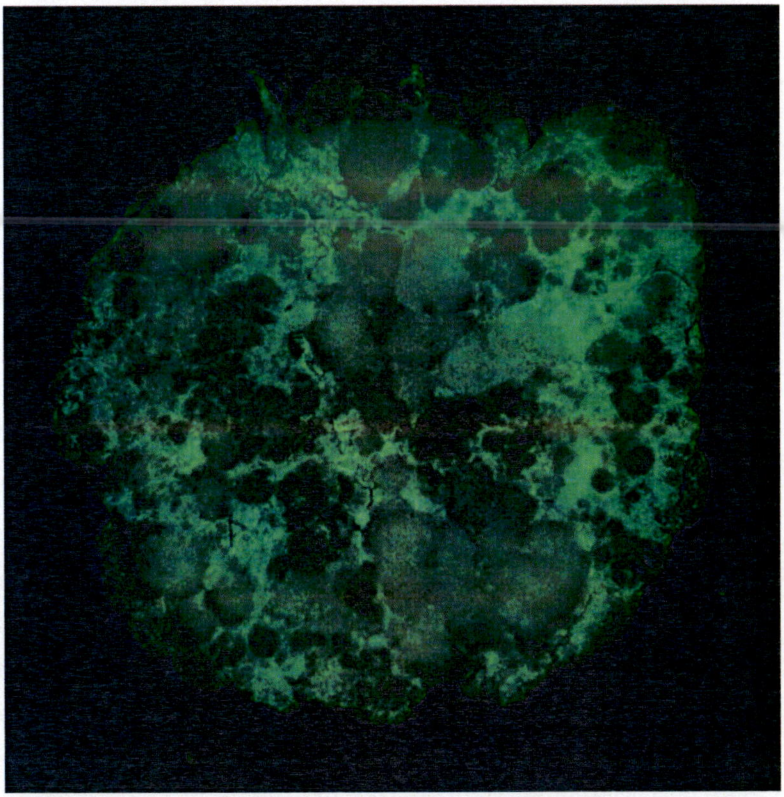

Engineering granular microbiomes

D. G. Weissbrodt, *Engineering Granular Microbiomes*, Springer Theses,
https://doi.org/10.1007/978-3-031-41009-3_12

12.1 General Conclusions

The aerobic granular sludge (AGS) wastewater treatment technology is becoming a new standard for high-rate biological nutrient removal (BNR) and secondary clarification in single sequencing batch reactors (SBR) (Giesen et al. 2012; Pronk et al. 2015b; van der Roest and van Loosdrecht 2012; Winkler et al. 2018). BNR granular sludge stimulates not only the water chain sustainability by intensifying WWTPs, but also the circular economy by recovering resources and biomaterials from wastewater (van der Roest et al. 2015; van Loosdrecht and Brdjanovic 2014; Winkler and van Loosdrecht 2022).

Fundamental research was conducted here in an integrative system approach to investigate the underlying mechanisms of microbial selection, bioaggregation, and nutrient conversions in granular sludge. The main findings open interesting perspectives for examinations of metabolic functionalities in the BNR granular sludge microbiome. They sustain the development of strategies for an optimal management of the microbial resource fostering efficient process performances.

The outcomes have been importantly discussed with the scientific and engineering communities toward knowledge utilization in the microbial ecology, environmental biotechnology, and (waste)water engineering sector at international and regional levels (Guimarães et al. 2016, 2017b; Regmi et al. 2017; Weissbrodt 2018, 2019; Weissbrodt et al. 2014a, 2020a; Weissbrodt and Holliger 2014). The scientific findings were translated into an applied methodology for an ecological engineering of BNR granular sludge processes. This approach has notably been successfully tested at pilot level to sustain granulation and BNR with real wastewater under coastal warm climate conditions in Brazil (Guimarães et al. 2018).

The main conclusions of the thesis are further discussed in relation to the advances made on microbial community engineering principles of BNR granular sludge over the last 10 years (2012–2022).

12.1.1 The BNR Granular Sludge Technology Is Economically Attractive Through Process Intensification and Integration

The AGS technology has been related with definite savings in land area, construction costs, and operation costs (de Bruin et al. 2004). This process using the propensity of microorganisms to spontaneously aggregate and to connect metabolic niches for an almost simultaneous nutrient removal provides keys for intensifying the treatment of urban and industrial wastewaters (Caluwé et al. 2017; Winkler and van Loosdrecht 2022). Its economic value has been recognized by technology analysts (Radauer et al. 2012). According to the economic analysis performed in introduction to this book, the implementation of the AGS technology in a Swiss wastewater treatment plant (WWTP) of 200,000 person-equivalents operated for full BNR can relate to theoretical savings of 0.45 CHF m^{-3} and up to 6.3 mioCHF y^{-1} (1 Swiss franc CHF $\approx$ 1.05 USD) (Weissbrodt and Holliger 2013, 2014), requiring validation on site.

The AGS technology has been efficiently developed using sequencing batch reactors (SBRs) with currently more than 90 Nereda® full-scale plants installed in 20 countries world-wide.[1] Granulation is further studied for the retrofitting of existing WWTPs (Daigger et al. 2018; Downing et al. 2022; Wei et al. 2020), with interesting advances in continuous-flow mode using different types of selectors (Avila et al. 2021; Miyake et al. 2023; Regmi et al. 2022; Santorio et al. 2021; Strubbe et al. 2022; Sun et al. 2021; Xu et al. 2022; Yan et al. 2021).

Overall, BNR granular sludge offers interesting perspectives for intensifying and integrating processes for the protection of the aquatic environment. It results in savings in both energy and space, enabling the upgrade of WWTPs with additional process extensions, e.g., for the treatment of micropollutants such as fostered by the latest revision of the Swiss Waters Protection Ordinance[2] (Eggen et al. 2014; Weissbrodt and Holliger 2013).

12.1.2 The Engineering of BNR Granular Sludge Systems Requires an Advanced Management of Microbiomes

According to the current annual publication growth rate of 7.2% (0.03 log per year) computed in the state of the art section of this book based on the scientometrics approach of Price (1963) and Larsen and von Ins (2010) and the Scopus® literature database, the AGS research will progressively reach over the next 10–20 years a mature steady state in scientific productivity comparable to the one on the up-flow anaerobic sludge blanket (UASB) technology (Weissbrodt and Holliger 2013, 2014). The extended literature review of this book highlighted that intensive research should be conducted at the interface between the granular biofilm and bacterial community level as a base for tailoring conditions for efficient management of the bacterial resource for BNR in AGS systems.

The interconnection between the physical, chemical, and biological sciences is an essential component for stimulating the development of robust environmental biotechnology processes like granular sludge (Winkler et al. 2018), alongside a fundamental mechanistic understanding of phenomena. Bringing systems microbiology at wastewater treatment lines (Cerruti et al. 2021) will deliver new opportunities for better understanding, designing, and managing processes. The implementation of a design-build-test-learn (DBTL) cycle will drive the engineering of robust microbiomes for environmental services (Lawson et al. 2019).

---

[1] Royal HaskoningDHV (status 2022) https://nereda.royalhaskoningdhv.com/.

[2] Swiss Confederation, 814.201 Waters Protection Ordinance 1998 (status 1st January 2021). https://www.fedlex.admin.ch/eli/cc/1998/2863_2863_2863/en.

12.1.3 Combined Wet-Lab and Dry-Lab Molecular Workflows Enable an In-Depth Analysis of Microbiomes and Selection Mechanisms

The combined wet-lab and dry-lab procedures for the profiling of the bacterial community of granular sludge were importantly used to identify microorganisms involved in granulation and BNR processes. In the transition period from classical molecular fingerprinting to modern amplicon sequencing (both based on the profiling of the 16S rRNA gene pool of the community), the PyroTRF-ID methodology (Weissbrodt et al. 2012b) was designed to combine terminal-restriction fragment length polymorphism (T-RFLP) and pyrosequencing data. This workflow provided a high resolution in community fingerprints, but also enabled to track specific microorganisms along the process timeseries in relationship to process operation and performances (Weissbrodt et al. 2014c), in a rational amount of time and analytical research costs. Based on the initial sequencing of a biomass sample collected from the system of interest, PyroTRF-ID can be used to design experimental T-RFLP strategies by optimization of community fingerprints after screening different restriction enzymes. Since different organisms can contribute to the same T-RF, it is also recommended to sequence samples on a regular basis and to check for closest relatives in PyroTRF-ID. The combination of wet-lab and dry-lab biomolecular workflows allows for investigating microbial communities in engineered systems with high phylogenetic and temporal resolution. It helps detecting and tracking the predominant phylotypes, prior to further examining their genetic potential and expressed metabolisms using functional analyses.

Over the last 10 years, 16S rRNA gene amplicon sequencing has established as an efficient workflow to profile bacterial communities (Cerruti et al. 2021; Karst et al. 2016; Weissbrodt et al. 2020a), obtaining simultaneously the relative abundances and phylogenetic information of the taxa composing the microbiome. This was based on developments of sequencing workflows (e.g., Illumina MiSeq and Oxford Nanopore MinION), field guides and databases of microorganisms from wastewater environments (e.g., MiDAS), and bioinformatics workflows and R packages for data visualization (Albertsen et al. 2015; Andersen et al. 2018; Dueholm et al. 2022; McIlroy et al. 2015), facilitating the access to and understanding of community datasets in relation to operational and environmental metadata (Weissbrodt et al. 2014c). While amplicon sequencing provides high throughput, the fast throughput of T-RFLP still needs to be challenged by the new generation of molecular workflows. Amplicon sequencing can be either done in house if a sequencer is available (but sequencing runs are economically launched when enough samples are available) or through outsourced service by a sequencing facility (but data are obtained typically after 4–6 weeks on average). Microbial ecology methods comprise a range of ecophysiological techniques to examine and validate metabolisms (Nielsen et al. 2016; Nielsen and McMahon 2014; Weissbrodt et al. 2020a). In addition, genome-centric metagenomics (through the obtaining of metagenome-assembled genomes,

MAGs) (Adler et al. 2022; Albertsen et al. 2013a; Singleton et al. 2021) and metaproteomics (Cydzik-Kwiatkowska et al. 2022; Kleikamp et al. 2021) provide high resolution on microorganisms, their genetic potential, and their expressed metabolic functions within microbiomes in single sequencing and mass spectrometry runs.

The mass of information delivered by these big data provides unprecedented access to microbial understanding, while the processing of the data necessitates time, computational infrastructure, and new expertise in bioinformatics that needs to be integrated in study programs of biotechnologists and environmental engineers. Making microbial and functional fingerprinting fast-throughput and at process line remains an important objective toward a microbially-informed process control (Cerruti et al. 2021; Stokholm-Bjerregaard et al. 2021; Weissbrodt et al. 2015). DNA-based analyses are progressively entering the environmental engineering sector, such as recently used to unravel bulking phenomena in a WWTP and develop operational solutions (Fumasoli et al. 2022). Molecular analyses were used to detect the progressive selection of filamentous bacteria, before their effective over-proliferation as filamentous phenotypes.

The analysis of other informational molecules from DNA to RNA and proteins (besides lipids and sugars) extracted from biomasses will provide higher resolution on microorganisms and their expressed functionalities. Together with the implement of analytical methods, bioinformatics and biostatistics workflows should be democratized to facilitate the processing and visualization of data for extracting the essential information on microorganisms and metabolisms to support process management. In addition to stoichiometry and kinetics used for process design and operation, molecular analyses provide an in-depth understanding of the process ecology.

12.1.4 Complex Microbial Networks Can Be Rationalized into Conceptual Ecosystem Models as Basis for Functional Analyses

The investigation of microbial communities at high resolution using new-generation molecular and bioinformatics approaches leads to the generation of substantial amounts of sequencing data (Weissbrodt et al. 2012b). Rationalizing the high resolution obtained is required to be meaningful in the engineering context. Complex bacterial community compositions can for instance be efficiently rationalized as population networks in conceptual ecosystem models (Nielsen et al. 2010; Weissbrodt et al. 2014c). The integration of systems microbiology and environmental biotechnology in the discipline of microbiome science and engineering should target the investigation of single lineages of interest and their interactions within the complex network of the microbial community (Cerruti et al. 2021; Lawson et al. 2019).

The microbial ecosystem model of BNR granular sludge developed in this work (Weissbrodt et al. 2014c) and further refined elsewhere (Weissbrodt et al. 2020a; Winkler et al. 2018) serves as an excellent basis for directing analyses of functional genetic traits and validations of metabolic properties (Fig. 12.1). It supports the design of microbial community engineering strategies (Weissbrodt et al. 2014a; Weissbrodt and Holliger 2014). This ecosystem model displays interesting microbial niches and metabolic features to examine.

It can help investigate the diversity and selection of other PAOs alongside "*Candidatus* Accumulibacter" and their effective contribution to EBPR, using volatile fatty acids (VFAs) or other sources of carbon and electrons (Adler and Holliger 2020; Elahinik et al. 2022; Layer et al. 2019; Wagner et al. 2015), and using either dissolved oxygen or nitrogen oxides as terminal electron acceptors (Rubio-Rincon et al. 2019).

Nitrification has recently been revisited with the contribution of complete ammonium oxidizers (comammox) affiliating with the genus *Nitrospira* (Costa et al. 2006; van Kessel et al. 2015). This genus, traditionally described as a nitrite-oxidizing bacterium (NOB), was detected in the granules in larger abundances than the ammonium-oxidizing bacterium (AOB) *Nitrosomonas*. Investigating the involvement of comammox organisms in the nitrification processes in granular sludge is an interesting objective.

A diversity of denitrifiers was detected in the granules. However, these chemoorganoheterotrophic organisms require an organic source of electrons and carbon for growth. In the AGS reactors, the organic substrate (here VFAs) is supplied under anaerobic conditions to stimulate their preferential storage by PAOs and GAOs, outcompeting other heterotrophic organisms. The actual capability of the identified denitrifiers (else than the known DPAOs and DGAOs) to store VFAs into PHAs (Ding et al. 2023; Oshiki et al. 2008) under anaerobic conditions and sustain the subsequent reduction of nitrogen oxides during the aeration phase in the anoxic biovolume of granules requires further examination. Across the denitrification network, the study of generalist and specialist microbial niches involved in the production and consumption of nitrous oxide (N_2O) will become an important task for managing greenhouse N_2O emissions from activated sludge and granular sludge WWTPs (Conthe et al. 2018; Dobbeleers et al. 2018; Finkmann et al. 2000; Gruber et al. 2021; Kampschreur et al. 2009; Lochmatter et al. 2014; van Dijk et al. 2021; Wunderlin et al. 2012; Yoon et al. 2019).

The ecosystem model can further help delineate research questions on microbial populations involved in the production and consumption of extracellular polymeric substances (EPS) within the biofilm matrix of granules (Lin et al. 2010; Weissbrodt et al. 2013a). One focus is the actual contribution of the slow-growing PAOs to EPS formation in the granules (Guimarães et al. 2016; Weissbrodt 2019), and the detailed study of their EPS biosynthesis and biodegradation metabolisms (Tomás-Martínez et al. 2022a, c).

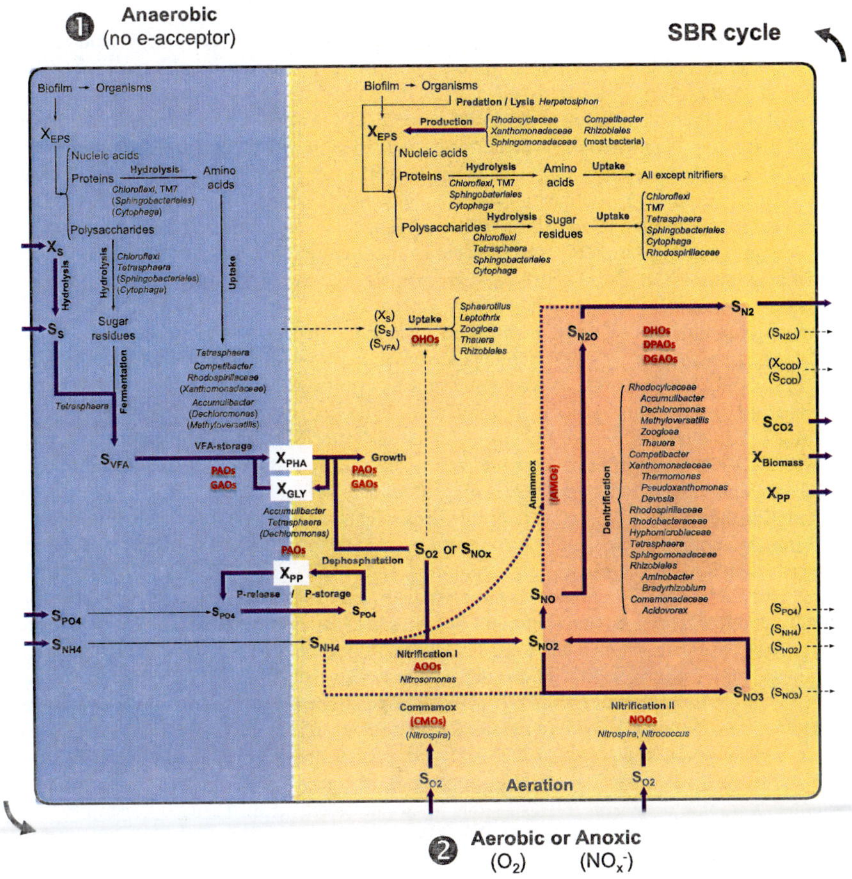

Anaerobic
(no e-acceptor)
SBR cycle
Biofilm → Organisms
X_{EPS}
Nucleic acids
Proteins
Hydrolysis
Amino acids
Chloroflexi, TM7
(Sphingobacteriales)
(Cytophaga)
Polysaccharides
X_S
Hydrolysis
Chloroflexi
Tetrasphaera
(Sphingobacteriales)
(Cytophaga)
Uptake
S_S
Sugar residues
Fermentation
Tetrasphaera
Tetrasphaera
Competibacter
Rhodospirillaceae
(Xanthomonadaceae)
Accumulibacter
(Dechloromonas)
(Methyloversatilis)
S_{VFA}
VFA-storage
PAOs
GAOs
X_{PHA}
X_{GLY}
Growth
PAOs
GAOs
Accumulibacter
Tetrasphaera
(Dechloromonas)
PAOs
Dephosphatation
X_{PP}
P-release / P-storage
S_{PO4}
S_{PO4}
S_{PO4}
S_{NH4}
S_{NH4}
Biofilm → Organisms
Predation / Lysis Herpetosiphon
Production
Rhodocyclaceae
Xanthomonadaceae
Sphingomonadaceae
Competibacter
Rhizobiales
(most bacteria)
X_{EPS}
Nucleic acids
Proteins
Hydrolysis
Chloroflexi, TM7
Sphingobacteriales
Cytophaga
Amino acids
Uptake
All except nitrifiers
Polysaccharides
Hydrolysis
Chloroflexi
Tetrasphaera
Sphingobacteriales
Cytophaga
Sugar residues
Uptake
Chloroflexi
TM7
Tetrasphaera
Sphingobacteriales
Cytophaga
Rhodospirillaceae
(X_S)
(S_S)
(S_{VFA})
Uptake
OHOs
Sphaerotilus
Leptothrix
Zoogloea
Thauera
Rhizobiales
S_{O2} or S_{NOx}
S_{N2}
S_{N2O}
DHOs
DPAOs
DGAOs
Rhodocyclaceae
Accumulibacter
Dechloromonas
Methyloversatilis
Zoogloea
Thauera
Competibacter
Xanthomonadaceae
Thermomonas
Pseudoxanthomonas
Devosia
Rhodospirillaceae
Rhodobacteraceae
Hyphomicrobiaceae
Tetrasphaera
Sphingomonadaceae
Rhizobiales
Aminobacter
Bradyrhizobium
Comamonadaceae
Acidovorax
Denitrification
Anammox
(AMOs)
S_{NO}
S_{NO2}
S_{NO3}
Nitrification I
AOOs
Nitrosomonas
Commamox
(CMOs)
(Nitrospira)
Nitrification II
NOOs
Nitrospira, Nitrococcus
S_{O2}
Aeration
S_{O2}
(S_{N2O})
(X_{COD})
(S_{COD})
S_{CO2}
$X_{Biomass}$
X_{PP}
(S_{PO4})
(S_{NH4})
(S_{NO2})
(S_{NO3})
Aerobic or Anoxic
(O_2) (NO_x^-)

◂**Fig. 12.1** Conceptual model of the microbial ecosystem of BNR granular sludge (modified from Weissbrodt et al. 2014c; Winkler et al. 2018). An evolved version is available elsewhere (Weissbrodt et al. 2020a). It provides an excellent basis for functional analyses, besides rationalizing its microbial and metabolic network. AGS-SBRs are optimally operated under temporal alternation of anaerobic feeding (*blue section*) and aeration phases (*yellow section*). Anoxic biovolumes for denitrification or anammox (*red section*) can be established in granules in function of the penetration of dissolved oxygen (O_2). The main microbial guilds metabolizing nutrients are displayed in *red*: polyphosphate-accumulating organisms (PAOs), glycogen-accumulating organisms (GAOs), ordinary heterotrophic organisms (OHOs), aerobic ammonium-oxidizing organisms (AOOs), nitrite-oxidizing organisms (NOOs), denitrifying heterotrophic organisms (DHOs), denitrifying PAOs and GAOs (DPAOs, DGAOs). Bacterial populations detected in BNR granular sludge are given in italics. The network of biochemical conversions of organic matter, phosphorus, and nitrogen is displayed with *thick purple arrows*. Other metabolic processes are given with *thin black arrows*. Uptake of P and N for anabolic requirements are not represented. A prolonged anaerobic contact time between the pre-settled influent and the biomass allows for the hydrolysis of slowly biodegradable particulate substrates (X_s) into readily biodegradable soluble substrates (S_S), their fermentation into volatile fatty acids (VFAs), and prior to their preferential uptake by PAOs (e.g., "*Ca.* Accumulibacter" and *Tetrasphaera*) and GAOs (e.g., "*Ca.* Competibacter"). Orthophosphate (PO_4^{3-}) is released by PAOs during the uptake of VFAs into poly-β-hydroxyalkanoates (PHAs). Leakage of biodegradable COD into the aeration phase should be prevented to avoid the proliferation of organisms that lead to filamentous bulking (e.g., *Sphaerotilus*, *Leptothrix*) and of OHOs (e.g., *Zoogloea*, *Thauera*) which outcompete PAOs and GAOs. Aeration provides dissolved oxygen to stimulate the aerobic growth of PAOs (and GAOs) which remove orthophosphate, and to activate the oxidation of ammonium into nitrite (NO_2^-) and nitrate (NO_3^-) by nitrifiers. A disproportion between ammonium- (AOOs) and nitrite-oxidizing organisms (NOOs) was detected in BNR granular sludge. The higher abundance of NOOs might be explained by some *Nitrospira* populations conducting a complete ammonium oxidation (comammox, CMOs) into nitrate. Clades of denitrifying PAOs and GAOs are presumed to denitrify in granules, but sequencing displayed a wider diversity of possible denitrifiers. Full nitrogen removal can arise from the metabolic contributions of generalist and specialist denitrifying heterotrophic organisms (DHOs) reducing nitrate to nitrite, nitrous (NO) and nitric (N_2O) oxides, and dinitrogen (N_2). The ecophysiology of these DHO populations needs to get examined, notably for their ability to store VFAs into PHAs under anaerobic conditions as electron donors and carbon sources for the follow-up NO_x reductions. Anaerobic ammonium-oxidizing organisms (anammox, AMOs) can further contribute to N-removal. Besides BNR metabolisms, microorganisms produce extracellular polymeric substances (EPS) that bind them in the biofilm matrix of granules. The microbial ecosystem of granules can be composed of EPS producers and consumers. EPS metabolisms can be active under the different redox conditions. EPS biodegradation can also relate to the general biomass decay processes. The metabolic functions of microorganisms displayed in parentheses should be confirmed by ecophysiological investigations

In the one-water one-health context (Miłobedzka et al. 2021; Prata et al. 2022), additional microbial genotypes and phenotypes involved in the biodegradation of xenobiotic compounds (Amorim et al. 2014; Falås et al. 2016; Margot et al. 2016; Winkler et al. 2018) and in the development of antimicrobial resistance (Calderón-Franco et al. 2022; Li et al. 2020; Pallares-Vega et al. 2021; Sabri et al. 2020) and other horizontal gene transfer mechanisms (Wuertz et al. 2004), as well as pathogens and viruses (Barrios-Hernández et al. 2021) are becoming important aspects to study within the granular sludge microbiome.

12.1.5 Granulation Mechanisms and Granular Biofilm Architectures Rely on the Main Microorganisms and Physiologies Involved

Structural investigations with confocal laser scanning microscopy (CLSM) revealed different granulation mechanisms in function of the dominant populations involved under wash-out conditions in the bubble column and under steady-state conditions in the stirred-tank SBRs (Weissbrodt et al. 2012a, 2013b).

The predominance of fast-growing *Zoogloea* spp. resulted in the formation of smooth and homogeneous granular biofilms by swelling and outgrowth of *Zoogloea* colonies around flocs (Weissbrodt et al. 2012a). Progressive formation of heterogeneous biofilm structures developing from granule cores outwards was detected after a particle diameter of more than 500 μm was reached, indicating possible limitations in mass transfer of substrate (Picioreanu et al. 2000).

Dense clusters of slower-growing populations of nitrifiers, "*Ca.* Accumulibacter", and "*Ca.* Competibacter" further proliferated inside the zoogloeal biofilm continuum from granule cores outwards (Weissbrodt et al. 2013a) similarly to the observations made on multispecies planar biofilms (Alpkvist and Klapper 2007). The predominance of "*Ca.* Accumulibacter" in the PAO-SBR and "*Ca.* Competibacter" in the GAO-SBR resulted in the formation of heterogeneous granules by development of dense clusters around flocs. The selection for slow-growing organisms like PAOs and GAOs, but also nitrifiers or anammox populations, is key for a stable granular aggregation (de Kreuk and van Loosdrecht 2004; Guimarães et al. 2018). CLSM measurements revealed the compactness of microcolonies formed by "*Ca.* Accumulibacter" and "*Ca.* Competibacter" (Weissbrodt et al. 2013a). Such dense aggregation can lead to optimal characteristics of granular aggregates for intensified and efficient granular sludge processes.

Within the architectures of granules (Gonzalez-Gil and Holliger 2014; Kirkland et al. 2020), the different types and densities of microcolonies formed by the different microbial niches across granule and biofilm architectures can further impact on the diffusional properties of substrates (Pérez et al. 2020; Poot et al. 2016; van den Berg et al. 2020, 2021; van Dijk et al. 2022), and can impact kinetics—e.g., observed K_s values (Baeten et al. 2018; Shaw et al. 2015). A careful consideration of metric scales (molecular, micro, meso, macro) and temporal scales (Cerruti et al. 2021; Morgenroth and Milferstedt 2009; Young et al. 2016) is crucial for process understanding and control. Microbial ecology can be directly combined with bioprocess engineering in a "Process-2-Process" integration (i.e., microbial processes vs. engineered process).

12.1.6 Aminosugars Are Key Components Among the Complex Chemical Composition of EPS Matrices of Microbial Biofilms and Granules

A diversity of glycoconjugate matrices were detected in the architecture of granules by fluorescence lectin-binding analysis (FLBA) and CLSM (Weissbrodt et al. 2013a). These EPS were observed in the biofilm continuum, in matrices surrounding microcolonies, around bacterial cluster, and at cell surfaces dispersed in the biofilm continuum. The exopolysaccharide composition of granules is more complex than a "tale" of the only two granulan and alginate polymers (Seviour et al. 2012). Besides sugar residues like α-glucose and α-mannose, aminosugars (like N-acetylglucosamine, N-acetylgalactosamine, and N-acetylneuraminic acids also known as sialic acids) were highlighted here by FLBA-CLSM as important components of the EPS binding microbial cells in granules and biofilms.

The recovery of biomaterials out of wastewater using granular sludge (Lin et al. 2015; Tavares Ferreira et al. 2021; van Loosdrecht and Brdjanovic 2014) is an important part of novel incentives for the transformation of WWTPs into water resource recovery facilities (WRRFs) (Daigger 2011, 2017; Guest et al. 2009; Kehrein et al. 2020a, b; Weissbrodt et al. 2020b), also called water resource factories or sewage treatment and resource recovery plants—StaRRE (derived from the French semantic *Station de Récupération des Ressources de l'Eau—StaRRE*) (Lalumière 2016; Vanrolleghem 2015; Weissbrodt 2018). Alginate-like exopolymers derived from wastewater are biomaterials that behave like alginate, providing interesting applications for hydrophobic coatings, flame retardants or even artificial leather[3] in the civil engineering, paper and textile industries, among others (Amorim de Carvalho et al. 2021; Kim et al. 2020; Lin et al. 2015; Sepúlveda-Mardones et al. 2019). The first process for the recovery of Kaumera Nereda® Gum[4] from surplus granular sludge has been installed in Zutphen in the Netherlands (Dutch Water Sector 2019), sustaining interesting market prospects.

EPS are functional exopolymers produced by microorganisms in the community, and harbor compositions as complex as microbiomes. New methods have been developed to extract and "sequence" the monomeric compositions of the EPS (Felz et al. 2016, 2019; Seviour et al. 2019), across the continuum of sugars, aminosugars, glycoproteins and proteins. This involves, e.g., high-performance anion-exchange chromatography with pulsed amperometric detection (HPAEC-PAD) that has been used for characterizing oligosaccharides distributions in alginates (Ballance et al. 2005).

---

[3] Delft student team WaterSkins won the Bio-based Innovation Student Challenge Europe with artificial leather made out of Kaumera® https://www.tudelft.nl/en/2021/tnw/making-artificial-leather-while-processing-wastewater.

[4] Kaumera Nereda® Gum, a biomaterial extracted from Nereda® granules https://kaumera.com/.

Further investigations should address relationships with microbial populations involved. Combining chemical ecology and microbial ecology, correlations have been drawn between EPS production and predominant microorganisms like "*Ca.* Accumulibacter" selected in the granular sludge process (Guimarães et al. 2016, 2017a; Schambeck et al. 2020). The association of "EPS sequencing" and glycomics with DNA/RNA sequencing (metagenomics, metatranscriptomics) and protein mass spectrometry (metaproteomics) in a multi-omics approach will provide access to EPS metabolisms in microbial communities.

Different strategies can be used by analyzing either the EPS diversity (but facing its chemical complexity as main challenge) or specific components of the EPS for a more targeted mechanistic understanding and management of their metabolic pathways and turnover toward a possible directed production (Tomás-Martínez et al. 2021, 2022c). For instance, sialic acids and nonulosonic acids (NulOs) have been lately targeted as interesting substances from the EPS of granules (Boleij et al. 2020; de Graaff et al. 2019; Kleikamp et al. 2020), based on their properties as recognition molecules on microbial surfaces, for their involvement in pathogenicity and biofilm formation, and even toward medical applications, e.g., as sepsis treatment drugs neutralizing histones (Tomás-Martínez et al. 2022a, b).

12.1.7 *Wash-Out Conditions Propel Granulation but Affect the Microbial Community Balance, Leading to Process Failures*

The formation of aerobic granules has been stimulated under wash-out conditions selecting for a fast-settling biomass (Beun et al. 1999). However, BNR performances often deteriorated during start-ups (Ebrahimi et al. 2010; Gonzalez-Gil and Holliger 2011; Weissbrodt et al. 2012a), requiring optimization of start-up conditions (Lochmatter and Holliger 2014).

Wash-out conditions were shown to exert a selection pressure not only on physical properties of granules, but also on the underlying bacterial community compositions (Weissbrodt et al. 2012a). Improper control of organic loading rates leads to the unfavorable proliferation of fast-growing aerobic heterotrophs, and to hindered settling resulting from filamentous bulking and to BNR process failures because of the outcompetition of populations harboring nutrient-removing metabolic capacities (Aqeel et al. 2016; Guimarães et al. 2018; He et al. 2019; Wagner et al. 2015; Weissbrodt et al. 2012a, 2013b).

Filamentous bacterial populations affiliating with *Burkholderiales* proliferated when insufficient superficial gas velocity (SGV, 1.8 cm s^{-1}) was applied, or at higher mesophilic temperature (30 °C). Their overgrowth led to unfavorable filamentous bulking phenomena, namely the formation of fluffy granules exhibiting deteriorated settling behavior (Weissbrodt et al. 2012a).

Zoogloea spp. proliferated during the formation of fast-settling dense granules obtained with higher SGV (4.0 cm s^{-1}), or after inoculation with activated sludge sludge taken from a BNR-WWTP (Weissbrodt et al. 2012a). *Zoogloea*-like organisms producing extracellular polymeric substances (EPS) were presumed as important organisms in the granulation process. However, *Zoogloea* proliferation over PAOs and nitrifiers was identified as the main cause of BNR deficiency during start-up (Weissbrodt et al. 2013a). The presence of high residual concentrations of VFA in aeration phases preferentially selected for *Zoogloea* spp.

"*Ca.* Accumulibacter" proliferated as soon as sufficient biomass was present in the system for full uptake of acetate during the anaerobic feeding phase (Weissbrodt et al. 2013a). Operation at mature stage without excess sludge removal involved "*Ca.* Competibacter" in competition. It is recommended to start up AGS-SBRs with a dynamic control of the organic loading rate (OLR), anaerobic contact time, and sludge retention time (SRT) in order to select for active PAOs in granules, and to balance the microbiome for full BNR (Lochmatter and Holliger 2014; Weissbrodt et al. 2013a, 2014b).

Interestingly, granulation has further been used as a method to improve activated sludge settling and to overcome typical bulking problems in WWTPs treating food-industry effluents (Avila et al. 2021; Caluwé et al. 2022; Daigger et al. 2018; Henriet et al. 2017; Meunier et al. 2016). Capturing the drivers of bioaggregation on the continuum of flocs, granules and biofilms is an important fundamental task toward the design of robust environmental biotechnologies (Aqeel et al. 2019; Weissbrodt et al. 2014a; Wilén et al. 2018). Managing granulation requires a thorough understanding of physical, chemical, and microbial phenomena (Winkler et al. 2018).

12.1.8 PAOs Proliferate with an Anaerobic Selector Designed to Fully Remove Organics Prior to Aeration, and Aggregate Densely

Granules spontaneously formed in stirred-tank SBRs operated for enriching for PAOs and GAOs in activated sludge under steady-state conditions (Weissbrodt et al. 2013a, 2014b). "*Ca.* Accumulibacter" and "*Ca.* Competibacter" exhibited a "spontaneous" granulation potential. Wash-out conditions and *Zoogloea* spp. were shown not to be crucial for AGS formation. Microbial aggregation is not restricted to single organisms (Bossier and Verstraete 1996).

Granules dominated by "*Ca.* Accumulibacter" were obtained under operation with controlled OLR, anaerobic phase length, and SRT of 8 days (Weissbrodt et al. 2013a, 2014b). A fast-settling biomass formed spontaneously after 30 days of operation with a settling time as long as 60 min. Nuclei evolved towards granules after decreasing the settling time to 10 min in order to save on SBR cycle time. Optimal EBPR activities were obtained right from start-up. Granules formed after more than 450 days of operation in the flocculent sludge enrichment of "*Ca.* Competibacter" (Weissbrodt et al. 2013a, 2014b). This resulted from a period of uncontrolled over-aeration caused by biofilm proliferation on the dissolved oxygen (DO) sensor in the GAOs-SBR.

Granules were therefore obtained in stirred tanks SBRs with height to diameter (H/D) ratios close to 1, similar to earlier report (Mosquera-Corral et al. 2011). Granulation should further be investigated with this design, which is more favorable for practice than column-based systems. The granulation process should be optimized in a compromise between wash-out and steady-state conditions to select for a bacterial community with BNR and biofilm-forming activities (Lochmatter and Holliger 2014; Weissbrodt et al. 2013a). Beyond the shear forces (de Graaff et al. 2020a), the selection for slow-growing organisms like PAOs, GAOs, nitrifiers or anammox bacteria, among others, are beneficial for stabilizing granules (Aqeel et al. 2019; de Kreuk and van Loosdrecht 2004; Guimarães et al. 2018; Henriet et al. 2016; Weissbrodt et al. 2013a; Winkler et al. 2018).

The design of an efficient anaerobic selector is crucial for an optimal granulation and needs to be carefully considered in function of the complexity of the influent wastewater (de Kreuk et al. 2010; Ekholm et al. 2022; Guimarães et al. 2018; Layer et al. 2022; Toja Ortega et al. 2022; Wagner et al. 2015). Such as shown by the conceptual ecosystem of granular sludge (Weissbrodt et al. 2014c, 2020a; Winkler et al. 2018), particulate substrates need first to get hydrolyzed into soluble substrates and fermented into VFAs prior to their uptake by PAOs and GAOs under anaerobic conditions. The extent of the substrate diffusion inside the granules also governs the stability of the bioaggregates (Pronk et al. 2015a). The anaerobic contact time between the wastewater and the biomass needs to be carefully designed to enable a full anaerobic consumption of the biodegradable COD prior to aeration, and an efficient of PAOs and GAOs which stabilize granules on top of contributing to BNR.

Design can be sustained by mathematical modelling (Baeten et al. 2019; van Dijk et al. 2020, 2022), such as exemplified by the hydraulic-biokinetic model developed in this work and explained hereafter (Weissbrodt et al. 2017). If P-removal is targeted, selection for PAOs should be driven. If P-removal is not a treatment objective (depending on local regulations on emissions), selection for GAOs can be a target since also contributing to COD-removal, and N-removal in the case of denitrifying GAOs.

12.1.9 pH Triggers the Competitive Selection of PAOs and GAOs, and EBPR in Granular Sludge Systems

A Plackett–Burman experimental design helped screen for the main environmental and operational factors that trigger the competition between PAOs and GAOs in AGS-SBRs (Weissbrodt et al. 2013a, b). pH exerted the most significant impact on bacterial selection in AGS systems.

"*Ca.* Accumulibacter" and EBPR were mainly favored with slight alkaline pH, but also at temperature below 20 °C and with a VFA mixture of acetate and propionate. "*Ca.* Competibacter" proliferated at slightly acidic pH, at higher mesophilic temperature close to 30 °C, and with acetate (Weissbrodt et al. 2013b). Optimal conditions for *Tetrasphaera* spp., a major PAOs of full-scale EBPR-WWTPs (Nguyen et al. 2011), were similar to the ones of GAOs (Weissbrodt et al. 2013b). Similar physiological observations have been made for "*Ca.* Accumulibacter", "*Ca.* Competibacter", and *Tetrasphaera* relatives in the PAO- and GAO-enrichments (Weissbrodt et al. 2013a, 2014b). Physiological properties of PAOs and GAOs in AGS were in agreement with information available on activated sludge (Oehmen et al. 2010a).

Efficient nitrification-denitrification was obtained under slightly alkaline and higher mesophilic conditions (Weissbrodt et al. 2013b). Full BNR was obtained under non-limiting COD conditions with food-to-microorganism (F/M) ratios above 25 $mg_{CODs}\ g_{CODx}{}^{-1}$. The length of the starvation phase was finally shown to impact on the BNR performances depending on the redox conditions applied (Weissbrodt et al. 2013b). With fixed 2 h starvation, dephosphatation was only complete with full aeration. The application of alternating aerobic-anoxic starvation conditions under fixed starvation phase length led to partial BNR, since nitrate and nitrite are less efficient electron acceptors than oxygen (Henze et al. 2000). As an alternative to simultaneous nitrification and denitrification (SND) (de Kreuk et al. 2005) which is often limited (Layer et al. 2020), the aeration strategy by alternating nitrification and denitrification (AND) can lead to efficient nitrogen and phosphorus removal in granular sludge systems (Lochmatter et al. 2013) as soon as the starvation phase length is controlled based on residual concentrations of nitrogenous and orthophosphate dissolved compounds.

The online monitoring of off-gases like the O_2 consumed and CO_2 released by the microbial metabolisms, but also the greenhouse gases N_2O and CH_4, will become an important task for the control of granular sludge SBR processes (Baeten et al. 2021). Within the denitrification network, monitoring of N_2O emissions needs a close attention (Gruber et al. 2022; Lochmatter et al. 2014; Mannina et al. 2016; van Dijk et al. 2021; Wunderlin et al. 2013). Nitrogen balances in lab-scale reactors are mainly only made based on the measurements of ammonium, nitrite and nitrate, assuming that all the nitrogen oxides are converted to N_2 during the denitrification process. More analytical consideration of NO and N_2O intermediates via off-gas analyses should provide a better overview of the effective denitrification performance of the process. In Switzerland, WWTPs have recently been shown to contribute to 1–2% of the countrywide emissions of N_2O (i.e., 410–1690 tons of N_2O–N per year) (Gruber et al. 2021). Besides the agriculture and the biotechnology industry, among others, WWTPs need to carefully monitor N_2O emissions.

12.1.10 Active PAO Fractions and EBPR Potential Are Rapidly Measured by Conductivity-Based Metabolic Batch Test and Polyphosphatase Assay

Enhanced biological phosphorus removal (EBPR) relies on the preferential selection of active PAOs (Lopez-Vazquez et al. 2009; Oehmen et al. 2007). The combination of PAOs-selective environmental conditions (17 °C, pH 7.0–8.0, propionate as carbon source, and a nutrient COD:P:N composition of 100:11:5 m/m) with dynamic control of the volumetric organic loading rate (OLR) and the anaerobic contact time during start-up was optimal to cultivate a stable PAO-enrichment (Weissbrodt et al. 2013a, 2014b). Efficient management of the PAO resource is also based on regular analysis of the abundances of these organisms and their GAO competitors (Weissbrodt et al. 2014b). However, molecular methods generally provide data in more than a day up to several weeks depending on the availability of instruments and expertise, what could limit their implementation in laboratories of WWTPs or at process line. Besides high-throughput, the development of fast-throughput methods is needed toward process surveillance and control based on microbial detection.

Linear correlations were obtained between the PAO/GAO ratio and the EBPR potential of sludge, and the volumetric rate of electrical conductivity evolution in anaerobic metabolic batch tests and polyphosphatase enzymatic activities of protein extracts from biomass (Weissbrodt et al. 2014b). Standardized conductivity-based metabolic batch tests and polyphosphatase assays are therefore proposed for a rapid and low-cost assessment of active PAOs fractions together with the dephosphatation potential of activated and granular sludges.

Other assays have recently been efficient to accurately quantify the polyphosphate pool of an EBPR biomass, by uncoupling uptake of acetic acid and the anaerobic release of phosphorus with two stress-inducing methods using acetate addition at low pH and the presence of EDTA in solution (Feng et al. 2020). The recent development of molecular imprinted polymer based sensors for the measurement of phosphate from wastewater (Warwick et al. 2014) can provide a nice alternative to electrical conductivity (Aguado et al. 2006; Maurer and Gujer 1995) in order to specifically track orthophosphate evolutions out of the electrical noise related to the pool of ions present in wastewater.

12.1.11 Modelling Hydraulic Transport and Biokinetics During Up-Flow Feeding Helps Design Selection Pressures in Granular Sludge SBRs

According to hydraulic transport modelling calibrated with residence time distributions (RTD) (Weissbrodt et al. 2017), the influent wastewater adopts a plug-flow regime with axial dispersion under both pulse feeding (9 m h^{-1}) and slow anaerobic up-flow feeding (0.9 m h^{-1}) across a settled bed of AGS in the bubble-column SBR.

This hydraulic transport model was efficiently combined with a biokinetic model for characterizing the competition of PAOs and GAOs for VFAs under the anaerobic phase (Lopez-Vazquez et al. 2009; Weissbrodt et al. 2017). This anaerobic phase is crucial to govern the selection of PAOs and GAOs and to maintain the stability of granules and BNR, and the robustness of the process (Guimarães et al. 2018; Lochmatter and Holliger 2014; Weissbrodt et al. 2013b).

Simulation outputs revealed that a bed height to diameter ratio (H_{bed}/D) of 6 was optimal for full bed loading and efficient acetate uptake under reference simulation conditions (20 °C, pH 7.0, 1.75 g_{TSS} cm^{-1} of bed, 65% VSS, 50% EPS, 25% PAOs, 25% GAOs). Simulation at lower temperature resulted in a doubled bed height requirement, whereas pH did not have a significant effect on this geometric need. A full uptake of VFAs under anaerobic conditions is required to select for PAOs in BNR systems. The modeling strategy can sustain the design of efficient selection pressures.

Recently, numerical models have been developed to better understand the mechanisms of granulations and sustain the design and control of BNR granular sludge processes at full scale, integrating phenomena like hydrodynamic regimes, selective feeding (i.e., race for substrate between granules of different settling properties), substrate types and penetration depth, microbial selection, granule breakage, and selective sludge wasting (Derlon et al. 2022; van Dijk et al. 2020, 2022). Selection pressures therefore apply to both the microorganisms and the bioaggregates.

Integrating the knowledge gained over a century of activated sludge (Jenkins and Wanner 2014) and almost half a century of granular sludge processes across redox conditions is important for developing robust environmental biotechnologies (Weissbrodt et al. 2014a). At the "IWA Biofilms: Granular Sludge Conference 2018",[5] we built for the first time the redox bridge across anaerobic, anoxic, and aerobic granular sludges. Knowledge can further be integrated using mathematical models (Baeten et al. 2019).

12.1.12 Microbial Composition and BNR Are Functions of Granule Metrics and Operational Fluctuations, to Manage with Control Strategies

Systems analyses and multivariate numerical investigations revealed the impacts of granule metrics and operational variations at macro, meso and micro scales I BNR granular sludge SBRs (Weissbrodt et al. 2014c).

Following the analysis of mass transfer phenomena in bubble columns according to Heijnen and Van't Riet (1984), the amount and size of granules obtained at these conditions in the granular sludge impacted the oxygen mass transfer and the aerobic

[5] de Kreuk and Weissbrodt (conference chairs), International Water Association, Biofilms Specialist Group, "IWA Biofilms: Granular Sludge Conference 2018", Delft, The Netherlands. www.granularsludgeconference.org.

processes of nitrification and dephosphatation (Weissbrodt et al. 2014c). Relationships between granule size and BNR efficiencies have been mainly discussed in the context of the establishment of aerobic and anoxic biovolumes in granules (de Kreuk et al. 2005; Layer et al. 2020; Mosquera-Corral et al. 2005; Nguyen Quoc et al. 2021), but hardly in terms of the effect of granule size on mass transfer efficiencies in the reactor. Feed-back modulation of aeration should be achieved based on regular measurements of particle size distributions and physical properties of granules.

Fluctuations in organic and nutrient loadings affected BNR. An efficient BNR was obtained with acetate concentrations above 500 $mg_{COD}\ L^{-1}$ and nutrient ratios above 30 $g_{COD}\ g_P^{-1}$ and 10 $g_{COD}\ g_N^{-1}$ (Weissbrodt et al. 2014c). Applying stoichiometric calculations following Kuba et al. (1996), it was deduced that denitrification was not carried out by only PAOs in the AGS-SBRs. A diversity of denitrifiers was identified by amplicon sequencing (Weissbrodt et al. 2012b, 2014c). A closer look at their effective denitrification functionality in the granular sludge should be given using, e.g., genome-centric metagenomics (Pessi et al. 2022; Vanwonterghem et al. 2016; Wang et al. 2022; Zhang et al. 2020), ecophysiological analyses with substrate labelling (e.g., DNA/RNA/protein-SIP) (Bellini et al. 2017; Bradford et al. 2018; Jehmlich et al. 2008; Kleiner et al. 2021), and metaproteomics (Hawley et al. 2014; Kleikamp et al. 2022; Rodríguez-Ramos et al. 2022; Wilmes et al. 2015).

The bacterial community continuum of the BNR granular sludge was composed of two main opponent clusters of populations (Weissbrodt et al. 2014c). The first cluster was predominantly composed of "*Ca.* Accumulibacter", *Nitrospira*, *Xanthomonadaceae*, and *Aminobacter*, whose abundance was stimulated by operational conditions that resulted in an efficient BNR. The second cluster mainly comprised "*Ca.* Competibacter", *Sphingobacteriales*, *Cytophaga*, and *Intrasporangiaceae* (genus *Tetrasphaera*) affiliates, and their abundance correlated with periods of poor BNR.

Developing not-only feed-back but also feed-forward control strategies based on analytical information collected across scales on biochemical conversion, granule characteristics, and microbial compositions, is important for maintaining robust granular sludge processes. The combination of these multi-scale data with mathematical models (van Dijk et al. 2020, 2022) can become powerful for process management.

12.1.13 Microbial Resource Management in Granular Sludge Relies on an Ecological Engineering of the Process, Using the SBR Flexibility

The robust design and operation of BNR granular sludge systems relies on the strong consideration of microbial selection, besides physical-chemical and hydraulic phenomena (Val del Rio et al. 2016; Weissbrodt et al. 2014a; Weissbrodt and Holliger 2014; Winkler et al. 2018). The predominant microorganisms selected in the process were shown to impact both the architecture of bioaggregates and the BNR performances. Multi-scale interactions (Cerruti et al. 2021; Morgenroth and Milferstedt

2009; Oehmen et al. 2007; Weissbrodt et al. 2014b; Young et al. 2016) from process to granules, microbiome, and metabolisms underlie the design of strategies of microbial community engineering.

All operational conditions should be designed and controlled toward the selection and activation of functional guilds and populations of interest inside the microbial community network of granular sludge. One should take advantage of the flexibility of the SBR technology (Morgenroth and Wilderer 1998; Wilderer and McSwain 2004) in order to combine an optimal set of selection pressures to manage the microbial resource (Marzorati et al. 2011; Verstraete et al. 2007; Vlaeminck et al. 2012) and the BNR conversions. Microbial ecology principles must be integrated in the engineering framework (Cerruti et al. 2021; Kleerebezem and van Loosdrecht 2007; Lawson et al. 2019).

With the advent of multi-omics applied to (waste)water treatment (Cerruti et al. 2021; McDaniel et al. 2021; Mino and Satoh 2006; Narayanasamy et al. 2015; Weissbrodt et al. 2020a), the always more in-depth data that can be collected from the microbiomes of environmental biotechnologies should be translated into models and concepts for engineering. By analogy to sewer- and plant-wide mathematical modelling (Elawwad et al. 2019; Grau et al. 2007; Solon et al. 2017), here comes the concept of a plant-wide systems microbiology for the wastewater industry (Cerruti et al. 2021).

Engineering granular sludge microbiomes is an integrative science that requires a close connection between microbial ecology and process engineering.

12.2 Microbial Diversity of Aerobic Granular Sludge: Is It Different from Activated Sludge?

Within an international fusion talk given at the *Aerobic Granular Sludge Workshop* of the *IWA Biofilm Conference 2011* (Weissbrodt et al. 2011), we were asked to assess whether granular sludge and activated sludge present significant differences at microbial ecology level. The elucidation of the BNR granular sludge microbiome enabled to address this question.

12.2.1 Activated Sludge and Granular Sludge Harbor Similar Microbiomes

The bacterial communities of AGS systems are composed of predominant and flanking populations present in the activated sludge microbiome of BNR WWTPs (Adav et al. 2010; Ebrahimi et al. 2010; Nielsen et al. 2012b; Wang et al. 2012; Weissbrodt et al. 2014c, 2020a; Winkler et al. 2018). Using the activated sludge ecosystem model as reference (Nielsen et al. 2010, 2012b), the conceptual model of the BNR granular sludge ecosystem (Weissbrodt et al. 2014c, 2020a; Winkler et al. 2018) was developed to draw ecological relationships between microorganisms and to examine their functional metabolic properties in the microbial community network (Fig. 12.1).

12.2.2 Microbiome Compositions from Synthetic to Real Wastewater

When AGS-SBRs are run in the lab with low-complexity synthetic wastewater and controlled pH and temperature, the microbial community compositions of the granular sludge are obviously distant from the ones of the real activated sludges used as inoculum. The extent of these differences is a function of the operation conditions applied.

The use of single VFAs leads to an initial drop in the richness and diversity of the microbial community. The application of wash-out conditions to foster a fast granulation exacerbates this microbial imbalance, further leading to impaired BNR is not enough biomass is maintained in the system and/or if the process operation (e.g., SBR phase length) is not adapted to such low residual biomass concentration. In contrast, granulation can occur under steady-state conditions without a significant impact on the community structure, such as shown in the PAO-enrichment SBR where "*Ca.* Accumulibacter" spontaneously formed dense aggregates and granules (Weissbrodt et al. 2014b).

Bacterial communities of a pilot AGS-SBR and a full-scale activated sludge system that were both operated for BNR from the same municipal wastewater have exhibited similar richness but dissimilar composition, indicating possible functional redundancy (Winkler et al. 2012c). Since other factors than the composition and temperature of the wastewater were shown to impact on the microbiome (Weissbrodt et al. 2013b, 2014c), further investigations should unravel the microbiology of BNR activated sludge and granular sludge processes operated under identical conditions with real wastewater. Pilot systems (de Kreuk and van Loosdrecht 2006; Derlon et al. 2016; Guimarães et al. 2018; Layer et al. 2019; Luiz de Sousa Rollemberg et al. 2020) and full-scale Nereda® WWTPs (Ali et al. 2019; Calderón-Franco et al. 2022; Ekholm et al. 2022; Pallares-Vega et al. 2021; Pronk et al. 2015b, 2020; Sabri et al. 2020) installed world-wide can help uncover the microbiomes of granular sludges and delineate their specificities across regions, in comparison with activated sludges. The world catalogue of wastewater microorganisms (Dueholm et al. 2022) now offers a strong opportunity to disentangle WWTP microbiomes.

From the full-scale perspective, an often overlooked aspect that needs to get further examined is the impact of microbial immigration from the influent into the sludge tank on the BNR functionalities of activated sludge and granular sludge processes (Ali et al. 2019; Frigon and Wells 2019; Guo et al. 2019). Both deterministic and probabilistic phenomena shape the performance of biological wastewater treatment installations (Cerruti et al. 2021; Ofiteru et al. 2010).

12.2.3 The Saga of Polyphosphate-Accumulating Organisms

The type of predominant PAOs involved was one major difference between the compositions of the activated sludge from the full-scale BNR WWTP used as inoculum (*Tetrasphaera* spp.) and of the AGS and PAO-enrichment cultures grown in the lab on VFA-based synthetic wastewater ("*Ca.* Accumulibacter") (Weissbrodt et al. 2013a, b, 2014b, c). With a VFA-based artificial feed, "*Ca.* Accumulibacter" was the predominant in both the PAO-SBR and AGS-SBRs, similarly to most studies having investigated EBPR enrichment cultures (McMahon et al. 2010; Oehmen et al. 2010b; Weissbrodt et al. 2014b).

Tetrasphaera has been often detected in EBPR and BNR sludges treating real municipal streams, and postulated as PAO; it has been described as a glucose-fermenting organism with an ability to store polyphosphate (Hanada et al. 2002; Liu et al. 2001; Maszenan et al. 2000; Nguyen et al. 2011, 2015; Nielsen et al. 2012a; Onda and Takii 2002). It has been recently reported as abundantly present in granules from full-scale BNR granular sludge installations (Cydzik-Kwiatkowska et al. 2022; Ekholm et al. 2022).

According to quantitative proteomics and NMR-metabolomics (Herbst et al. 2019), *Tetrasphaera* can efficiently adapt to redox dynamics in BNR WWTPs. *Tetrasphaera* differs from "*Ca.* Accumulibacter" by its way to store polyphosphate by combining glucose fermentation and polyphosphate storage, but not involving PHA cycling. "*Ca.* Accumulibacter" typically stores VFAs as PHAs under anaerobic conditions hydrolyzing polyphosphate as energy source and glycogen as reducing equivalents; it grows under aerobic conditions on the PHA pool as electron donor and carbon source (Comeau et al. 1986; Páez-Watson et al. 2023; Seviour et al. 2003; van Loosdrecht et al. 1997). According to a genome-based metabolic model (Kristiansen et al. 2013), *Tetrasphaera* takes up glucose and partly stores it as glycogen using the energy generated from polyphosphate hydrolysis and substrate fermentation under anaerobic conditions. Under aerobic conditions (and presumably anoxic ones as well since able to denitrify), glycogen is catabolized to deliver cellular energy for growth and polyphosphate storage. *Tetrasphaera* can metabolize more complex resources than the traditionally VFAs used to enrich for "*Ca.* Accumulibacter", like sugars but also proteins or amino acids even storing them intracellularly (Marques et al. 2017; Nguyen et al. 2015). "*Ca.* Accumulibacter" may store glucose as glycogen under anaerobic conditions (Goel et al. 1998; Ziliani et al. 2023).

The storage compounds involved by *Tetrasphaera* depend on the phylodiversity of clades, some also involving PHAs (Close et al. 2021). Recently, using a large collection of metagenome-assembled genomes (MAGs) binned from full-scale BNR WWTPs in Denmark (Singleton et al. 2021), a fine-scale examination of clades resulted in the reclassification of the *Tetrasphaera* affiliates that have been abundantly present in these installations as the candidate genus "*Ca.* Phosphoribacter" (Singleton et al. 2022).

From the investigations of factors selecting for PAOs and GAOs in granular sludge (Weissbrodt et al. 2013b), *Tetrasphaera* was interestingly selected under conditions favorable to GAOs like "*Ca.* Competibacter", namely slightly acidic and mesophilic conditions. *Tetrasphaera* has been enriched in activated sludge and granular sludge by replacing or amending the VFA feed with the supply of monomeric and polymeric fermentable compounds like glucose and casamino acids and even starch and peptone in the influent of anaerobic-aerobic EBPR bioreactors (Adler and Holliger 2020; Close et al. 2021). According to the BNR granular sludge ecosystem model (Fig. 12.1), since a fermenter, *Tetrasphaera* can be efficiently associated with "*Ca.* Accumulibacter" in the BNR microbiome. Their co-existence delivers robustness to the EBPR performance on complex wastewater (Adler and Holliger 2020).

Different methods have been used to examine the polyphosphate storage traits of microorganisms from activated sludge (Fernando et al. 2019; Gunther et al. 2009; Hupfer et al. 2008; Liu et al. 2001; Nguyen et al. 2015). DAPI staining and tetracycline labeling can reveal polyphosphate pools in cells detected by fluorescence microscopy, CLSM or flow cytometry. NMR, microautoradiography, or Raman-FISH provide more resolution for detecting and (semi-)quantifying the polyphosphate pools. Within the microbial diversity of WWTPs, other PAOs like *Dechloromonas* (which is often co-present with "*Ca.* Accumulibacter"), "*Ca.* Halomonas phosphatis", and several other taxa have been proposed based by combining phylogenetic and functional methods (Nguyen et al. 2012; Stokholm-Bjerregaard et al. 2017).

The "saga of PAOs" continues, evolving together with the advent of novel analytical techniques (Cerruti et al. 2021; Weissbrodt 2022). Since inorganic polyphosphate is a key metabolic compound of biological systems (Kornberg 1995), the PAOs guild is probably more complex than generally described. The initial focus on *Acinetobacter* during the eighties to nineties (van Groenestijn et al. 1989) was followed by the era of "*Ca.* Accumulibacter" since the development of FISH. Sequencing, mass spectrometry, and ecophysiology techniques are progressively providing a better overview of microbiomes of EBPR systems and their main contributing PAOs. The importance now is to quantify the exact contribution of each PAO population to phosphorus removal, in yield and rate terms. The future will tell whether "*Ca.* Accumulibacter" remains the defender or whether other PAOs guilds are becoming important challengers for the phosphate resource.

The design and operation of activated sludge and granular sludge processes for EBPR and BNR will have to account for the predominant PAOs involved (Barnard et al. 2017; Weissbrodt 2019), since their selection mechanisms and metabolisms differ from taxon to taxon. These fascinating microbial ecology insights on PAOs sustains new insights for wastewater engineering and innovation.

12.2.4 Impacts of Lineage Differentiation Within Nitrifiers

An important difference between activated sludge and granular sludge has been detected at the level of nitrifiers. Different compositions of nitrifying guilds were detected in this work by amplicon sequencing in flocculent inoculation sludge, early-stage granules, and mature granules. Nitrite-oxidizing organisms (NOOs) were present in higher abundances than aerobic ammonium oxidizing organisms (AOOs) in granules. *Nitrospira* was the predominant NOO genus in the VFA-fed mature AGS. Disproportions in the NOO/AOO ratio have been previously reported in granular sludge (Winkler et al. 2012a). Higher abundances of *Nitrobacter* spp. Have been related to either to mixotrophic growth by utilization of acetate (Bock et al. 1990; Freitag et al. 1987; Zare et al. 2012) or to additional supply of nitrite by denitrification.

The results obtained on the impact of granule size on oxygen mass transfer (Weissbrodt et al. 2014c) combined with the general knowledge on nitrification (Gujer 2010) lead to the conclusion that nitrification behaves differently in cell suspensions than when cells are buried in biofilms. During nitrification in a hybrid AGS reactor operated with granules and flocs, an equal distribution of AOOs in granules and flocs has been observed, but the slower-growing NOOs were preferentially located in granules (Filali 2011; Filali et al. 2012). The simultaneous presence of granules and flocs resulted in more stable volumetric nitrification rates and lower aeration requirements. Further developments on AOO and NOO competition have been made from granular to moving-bed biofilm reactors (MBBRs), notably in the context of partial nitritation and anammox (Hoekstra et al. 2019; Lackner et al. 2010, 2014; Laureni et al. 2019; Poot et al. 2016; Reino et al. 2016; Soler-Jofra et al. 2019; Vlaeminck et al. 2012). Process control strategies have been designed based on the microbial segregation of AOOs and NOOs, as well as anammox organisms (AMOs) between floc and granule/biofilm fractions and in function of concentration profiles of dissolved oxygen and nitrogen compounds in the systems.

Differences in selection for either *Nitrospira* or *Nitrobacter* inside the NOO guild can relate to system and operational specificities. *Nitrospira* has outcompeted *Nitrobacter* in a pilot biofilm reactor operated with real urban wastewater (Daims et al. 2000). *Nitrospira* benefit from a mixotrophic lifestyle with metabolization of organic compounds and storage of glycogen (Lucker et al. 2010; Nielsen et al. 2012b). Competition between *Nitrospira* and *Nitrobacter* is also related to the affinity for nitrite (Nogueira and Melo 2006): *Nitrobacter* dominates when nitrite is abundant, whereas *Nitrospira* is selected limiting nitrite conditions.

The recent elucidation of comammox organisms (CMOs) affiliating with the metabolically versatile genus *Nitrospira* with an ability to completely oxidize ammonium into nitrate (Daims et al. 2015; Kleerebezem and Lücker 2021; Koch et al. 2015; Lawson and Lücker 2018; Lucker et al. 2010; Palomo et al. 2018; van Kessel et al. 2015) provides an interesting new avenue for the explanation of its overabundance versus AOOs like *Nitrosomonas*. Like for EBPR, managing nitrification processes requires a good metabolic understanding of the main nitrifiers selected in the process. On the metabolic labour and process perspectives (Costa et al. 2006), one question relates to whether the management of aeration should be different when the nitrification work is divided in two guilds (AOOs + NOOs) or completed by one guild (CMOs). Aeration management is an important aspect of partial nitritation in one-stage systems (De Clippeleir et al. 2014; Dobbeleers et al. 2017; Joss et al. 2011; Lochmatter et al. 2014). Since outcompeting NOOs and preventing nitrate formation can be challenging (Perez et al. 2009, 2014), some strategies are currently targeting a partial denitrification and anammox (Gao et al. 2022; Le et al. 2019).

Finally, ammonium-oxidizing archaea (AOA) can contribute besides AOB to the oxidation of ammonium. AOA substantially contribute to the nitrogen cycle in aquatic environments but are not frequently targeted by molecular analyses in nutrient loaded municipal WWTPs. AOA can establish their niche, e.g., under low ammonium and saline conditions (Erguder et al. 2008; Fitzgerald et al. 2015; Holmes et al. 2019; Landreau et al. 2020; Lee et al. 2014; Limpiyakorn et al. 2011; Muñoz-Palazon et al. 2018; Oishi et al. 2012; Sauder et al. 2012, 2017; Ye and Zhang 2011; Zhang et al. 2016).

The nitrogen cycle offers (Jetten 2008; Kuypers et al. 2018) plenty of opportunities for scientific elucidations and engineering developments, such as within granular sludge. Microbial ecology is an essential component for managing nitrogen conversions in these microbial assemblies and processes.

12.2.5 Granulation to Overcome Bulking Sludge

Filamentous bulking and zoogloeal viscous bulking are two examples of proliferation of undesired organisms leading to deteriorated settling performances of activated sludge (Eikelboom and Geurkink 2002; Martins et al. 2004b; van Loosdrecht et al. 2020; van Niekerk et al. 1987; Wilén et al. 2008).

Proliferation of filamentous bacteria penetrating outside flocs is favored by readily biodegradable substrates present in residual concentrations during aeration, by higher mesophilic temperatures, and by substrate gradients generated by diffusion limitations across compact flocs or biofilms (Martins et al. 2004b). In AGS systems, filamentous bulking resulted from the overgrowth of filamentous bacteria like *Sphaerotilus* or *Lepthotrix* into slow-settling fluffy granules after transitions from flocs to granules (Liu and Liu 2006; Weissbrodt et al. 2012a). Similar to chlorine dosed in activated sludge tanks, higher SGV enabled to transiently remediate filamentous bulking in AGS systems, but at the expense of higher energy requirements. Chlorination or higher aeration velocity only transiently remediate the problem. Specific remedial actions are needed to suppress the growth of the unwanted organisms (van Loosdrecht et al. 2020). The development of aerobic granulation made substantial advancement from studies to overcome bulking sludge (Martins et al. 2004a; Tijhuis et al. 1994). Granulation and the use of the aforementioned anaerobic selector has further been used as a tool to remediate bulking, e.g., by *Thiothrix* and to maintain an active sludge, notably in industrial WWTPs (Caluwé et al. 2022; De Vleeschauwer et al. 2019; Henriet et al. 2017). Granulation has also been successfully achieved with a sludge dominated by *Thiothrix* in a sulfide-rich seawater (de Graaff et al. 2020b). The key for maintaining compact granules is to achieve a substrate consumption rate lower than its transport rate (Pronk et al. 2015a).

Zoogloeal bulking in activated sludge has been characterized by entrapment of water by the exopolysaccharides produced in excess by *Zoogloea*, *Thauera*, or *Acinetobacter* relatives that are selected by high substrate concentrations under aerobic conditions (Lajoie et al. 2000). When readily biodegradable COD leaks into the aeration tank (in a flow-through system) or aeration phase (in a SBR), these *Zoogloea*-like fast-growing heterotrophs scavenge the substrate and disturb the EBPR performance by outcompeting PAOs (Montoya et al. 2008; Weissbrodt et al. 2012a). In granular sludge, proliferation of *Zoogloea* however related to the formation of fast-settling, dense and smooth granules, and did not lead to deteriorated settling (Weissbrodt et al. 2012a). High shear and compaction forces generated by upflow aeration (Zima et al. 2007) were possibly counteracting water entrapment by exopolysaccharides produced by *Zoogloea*. Since granule formation is related to EPS production and hydrophobic interactions (Dulekgurgen et al. 2008; Liu and Tay 2002; Seviour et al. 2012), *Zoogloea* was thus initially proposed as a model organism for granulation. However, their overabundance is not desired in anaerobic-aerobic granular sludge SBRs since hampering BNR, notably during start-up (Lochmatter and Holliger 2014; Weissbrodt et al. 2012a).

While uncoupling the rates of substrate transport and consumption, the enforcement of a full uptake of the readily biodegradable COD under anaerobic conditions by the operation of a well-functioning anaerobic selector prevents the proliferation of both filamentous and *Zoogloea*-like organisms, selects for PAOs (and GAOs), maintains a stable granule cohesion and settleability, and enables BNR. This can be achieved using different strategies: (i) in an anaerobic plug-flow regime, where the wastewater is fed across the bed of granules and the contaminant load can be adapted to the biomass concentration; (ii) in an anaerobic batch phase in a stirred tank or by recirculation of the off-gas, whose length is controlled by electrical conductivity or online measurements of residual phosphate; (iii) in a combination of both (de Kreuk and van Loosdrecht 2004; Guimarães et al. 2018; Haaksman et al. 2020; Lochmatter and Holliger 2014; Pronk et al. 2015a, b; Weissbrodt et al. 2014b, 2017).

12.2.6 Linking Organisms to Metabolic Functions

Since microbial aggregation is not restricted to single organisms (Bossier and Verstraete 1996), knowledge on functional triggers is required to unravel the granulation phenomena. The formation of bacterial community assemblies are driven by protein functions present in the cells, rather than the nature of the bacteria (Burke et al. 2011). Functional niche partitioning has been studied with respect to genes but not to phylogeny in algal-associated bacterial communities. Out of metatranscriptomics and metaproteomics datasets collected from granular sludge, about 600 messenger RNAs (mRNAs) and 200 proteins were statistically different between activated sludge and AGS (Barr 2010; Barr et al. 2011).

Like biofilm formation (Beloin and Ghigo 2005; Niba et al. 2007; Sauer 2003), granulation should involve genes and proteins expressed to regulate signal transduction, cell motility, EPS production, and cell membrane biogenesis, among others. Further research should determine whether a universal pool of functional genes and proteins can be associated to granule formation independently from phylotypes present in the sludge. Different granulation mechanisms were however detected here by CLSM in function of the predominant organisms involved. Since organisms lie behind functions and bacterial community composition can exert an effect on floc and sludge properties (Nielsen et al. 2004; Wagner et al. 2006), relating functions to organisms seems worthwhile for unraveling granulation mechanisms. Multi meta-omics investigations via genome-centric metagenomics, metatranscriptomics, and metaproteomics (Albertsen et al. 2013a, b; Bagchi et al. 2016; Barr et al. 2015; Dueholm et al. 2023; Kleikamp et al. 2021, 2022; Narayanasamy et al. 2015; Rodríguez et al. 2015) can be combined to this end.

This also leads to interesting avenues for unravelling the EPS metabolism of microorganisms selected in granular sludge, like "*Ca.* Accumulibacter" and others, beyond correlations between population enrichment levels and EPS mass fractions (Guimarães et al. 2017a; Kleikamp et al. 2020; Tomás-Martínez et al. 2022b, c).

Microbial ecology methods have opened doors to visit not the metabolic repertoire of microorganisms, beyond their phylogenies, in natural and engineered microbiomes. More than 30 methods are available to address various investigation questions and increase the analytical resolution on microorganisms and their metabolisms in microbial communities, such as reviewed in different works (Agrawal et al. 2018; Cerruti et al. 2021; Christensen et al. 1999; Gilbride et al. 2006; Goel et al. 2011; Gu et al. 2010; Hofman-Bang et al. 2003; Karst et al. 2016; Mahendra et al. 2012; McMahon et al. 2009, 2010; Muyzer and Ramsing 1995; Onuki et al. 2000; Weissbrodt et al. 2020a). Harnessing their selection principles and metabolic power by an ecological engineering for controllable outputs (Bernstein 2019; Lawson et al. 2019; Lindemann et al. 2016) is an essential part of the development of new-generation environmental biotechnologies.

12.2.7 Impact of Protozoans and Phages on the Bacterial Community

BNR conversions in activated sludge, granular sludge, and biofilms are mainly conducted by the bacterial populations that metabolize carbon, nitrogen, and phosphorus. Here, the factors of community assembly were investigated by assessing the direct link between operational parameters, granular structures, and bacterial compositions. Microbial communities are not only composed of bacteria. Impacts of other biological factors such as predation by bacterivorous protists (Hahn and Höfle 2001) and infection by bacteriophages (Withey et al. 2005) should be considered on shaping bacterial community structures.

Colonization of aerobic granules by protozoans has been reported, but mainly when complex wastewaters with particulate substrates have been used (de Kreuk et al. 2010; Schwarzenbeck et al. 2004; Weber et al. 2007). Protozoans can be active in the hydrolysis of particulate substrates. The grazing activity of these eukaryotic microorganisms also impacts microbial compositions and physiologies in planktonic and sessile communities (Adiba et al. 2010; Derlon et al. 2012; Huws et al. 2005; Koton-Czarnecka and Chrost 2002; Matz and Kjelleberg 2005; Pernthaler et al. 1997; Posch et al. 1999; Ronn et al. 2002; Wey et al. 2008). Protozoans may even act as a biological matrix for the development of granules (Weber et al. 2007).

Bacteriophages also shape microbial communities (Khan et al. 2006; Shapiro and Kushmaro 2011; Shapiro et al. 2010). Following the "killing the winner" theory, bacteriophage activity selectively affects "*Ca.* Accumulibacter" in enrichment cultures and impairs EBPR activity (Barr et al. 2010). Phage therapy has been used to remediate bulking in activated sludge, to control biofilm formation, and to suppress biofouling in membrane bioreactors treating wastewater (Bhattacharjee et al. 2015; Choi et al. 2011; Kotay et al. 2011; Mathieu et al. 2019; Motlagh et al. 2016). Phages further carry and transfer antimicrobial resistance among wastewater microbiomes (Calero-Cáceres and Muniesa 2016; Gunathilaka et al. 2017; Lekunberri et al. 2017; Subirats et al. 2016).

Activated sludge, granular sludge, and biofilm systems are highly heterogeneous, beyond the diversity of bioaggregates that they can harbor, when considering the different types of microorganisms and other biological agents that interplay in their microbiomes. Unravelling inter-kingdom relationships is an interesting prospect for better managing and using relationships between bacteria, eukaryotes, and viruses in robust mixed-culture processes.

12.3 Recommendations for Managing the Microbial Resource in Granular Sludge for Nutrient Removal from Wastewater

Scientific findings were collectively translated into an applied methodology for managing the microbial resource in BNR granular sludge systems (Fig. 12.2).

As a starting basis, the knowledge gained from a century of activated sludge and half a century of granular sludge across redox conditions should not be forgotten when aiming at designing and managing BNR granular sludge processes.

It addresses in eight milestones a full process control for an optimal microbial selection, bioaggregation, and nutrient conversions, by taking advantage of the flexibility of the SBR technology. Concepts can certainly be translated to flow-through

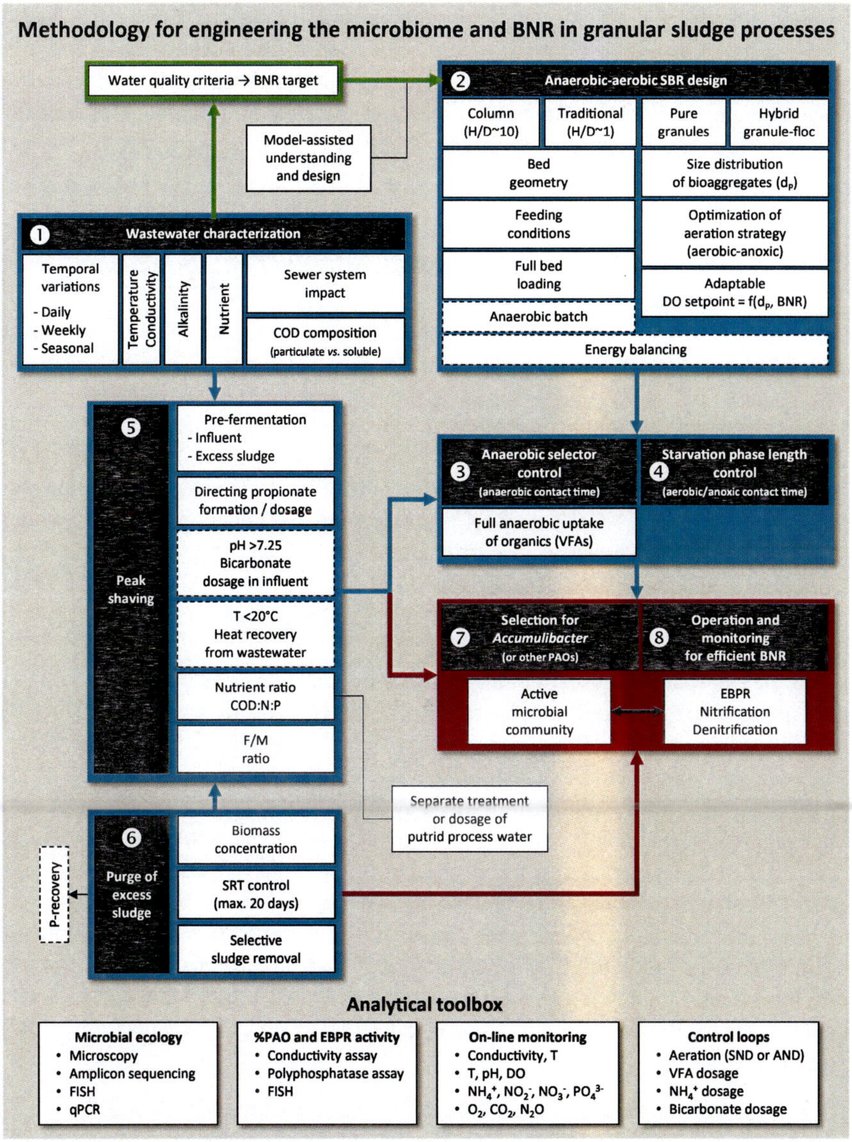

Fig. 12.2 Methodology in 8 milestones for an optimal management of the microbial resource for BNR in granular sludge wastewater treatment systems. Legend: biological nutrient removal (BNR), chemical oxygen demand (COD), ratios of COD to phosphorus (COD/P) and to ammonium nitrogen (COD/N) in the influent wastewater, reactor height to diameter ratio (H/D), granule diameter (d_P), dissolved oxygen (DO), simultaneous or alternating nitrification and denitrification (SND, AND), food to microorganism ratio (F/M), sludge retention time (SRT), volatile fatty acids (VFA), polyphosphate-accumulating organisms (PAOs)

granular sludge systems as well. The methodology can be transposed from laboratory scale to testing at pilot scale, and validation at full scale. It has been used to promote granulation in a pilot SBR treating domestic wastewater under costal warm climate conditions in Brazil (i.e., mesophilic temperatures and diluted streams through frequent rain events) (Guimarães et al. 2018).

12.3.1 Milestones for an Ecological Engineering of BNR Granular Sludge

1. **Characterize the influent wastewater**

Measurements of the composition of organic matter inform on the fraction of particulate and soluble CODs. The composition of the influent COD depends on regions, the geometry of sewer systems, and the conditions prevailing in the sewer (Huisman et al. 2003; Nielsen et al. 1992). The short and steep gravity sewers present in Switzerland result in a large fraction of particulate organic matter (40–60% of COD) in the influent. Long and flat anaerobic sewers of, e.g., Denmark, The Netherlands or Australia can lead to higher fermentation in the sewer and a higher concentration of fermentation products like VFAs entering the WWTP.

Further measurements should target the C–P–N nutrient composition, alkalinity, and temperature of the wastewater. Electrical conductivity measurements can provide information on the salinity of the wastewater, since impacting granular sludge settleability (Winkler et al. 2012b) and PAO-GAO metabolisms (Welles et al. 2014, 2015). Measurement of sulfide transported by seawater that infiltrate in sewer systems of coastal regions can help predict impairments in PAO metabolism (Rubio-Rincon et al. 2017) and counteract the proliferation of filamentous bacteria like *Thiothrix* by good granulation practices (de Graaff et al. 2020b).

Measuring the daily, weekly, and seasonal variations of wastewater parameters is important for predicting microbial dysbiosis and process disturbances, and for providing opportunities for feed-forward control. Considering both the wastewater data and the local effluent quality criteria is needed to set the BNR performances to be achieved.

2. **Design the granular sludge SBRs assisted by mathematical modelling**

Different reactor designs can be used to process wastewater with granular sludge. SBRs can be operated in a column with reactor height to diameter (H/D) ratio of about 10 or in a more traditional tank with a H/D of 1. Granular towers have been used for anaerobic digestion like UASB or EGSB reactors (van Lier et al. 2016), and for the treatment of industrial wastewater with granular biofilm carriers (Heijnen et al. 1992). While most lab-scale AGS investigations have been performed in bubble columns, the implementation of the Nereda® process at full scale has been made in tanks of a H/D ~ 1 (Pronk et al. 2015b; van der Roest and van Loosdrecht 2012).

Mathematical modelling can sustain the understanding of microbial selection, granulation, and BNR phenomena, as well as the design of the granular sludge bed geometries and flow conditions (Derlon et al. 2022; van Dijk et al. 2020, 2022; Weissbrodt et al. 2017). From this thesis, targeting a full bed loading and a full VFA uptake during the plug-flow anaerobic feeding phase or in a (subsequent) anaerobic batch phase is important to select for PAOs (or GAOs if P-removal is not an objective), stabilize the granules, and maintain an active microbial community for an efficient BNR.

Another part of the SBR design consists in the consideration of the type of bioaggregates, acknowledging that at pilot and full scale mainly a hybrid granules-flocs sludge is produced (Derlon et al. 2016; Filali et al. 2012; Wei et al. 2020). Periodical measurements of the bioaggregates size distribution can be useful for optimizing the aeration strategy. In a fill/draw full-scale SBR system, granules can be maintained and flocs out-selected by a good design of the up-flow hydraulic flow regime in the bed (van Dijk et al. 2020). In a continuous-flow granular sludge system, the plug-flow regime and intermittent mixing can help select for granules.

3. **Control the anaerobic selector for a full uptake of organics**

Anaerobic-aerobic operation should preferentially select for PAOs (or GAOs). A good design and operation of the anaerobic selector is important to foster the conversion of the biodegradable COD, involving hydrolysis of particulate organic matter into soluble substrates and their fermentation into VFAs prior to full uptake by PAOs (or GAOs). An additional anaerobic batch phase homogenized by mechanical mixing (or off-gas recirculation) can be implemented after the feeding phase for polishing the VFA uptake prior to switching to the aeration phase.

A good design of the hydraulic flow regime is needed to select granules and remove flocs, to avoid the mixing of the influent with the effluent, and to enable a sufficient contact time between the wastewater and the biomass (Derlon et al. 2016; van Dijk et al. 2020; Weissbrodt et al. 2017). If a tank design is selected, a mixed anaerobic batch phase can be directly implemented after (fast) feeding the wastewater; its length can be controlled based on online measurements of electrical conductivity or phosphate.

If the influent is rich in more complex organics, the process can be extended with a hydrolysis and fermentation step. This can be achieved either in a separate buffer tank or directly in the granular sludge reactor by setting a longer anaerobic contact time between the wastewater and the biomass (either with slower pumping or by integrating an anaerobic batch phase after feeding).

4. **Control the aeration phase length for full BNR**

On-line measurements can be used for a dynamic control of the lengths of both the anaerobic phase and the aeration phase. Besides traditional simultaneous nitrification and denitrification (SND) (de Kreuk et al. 2005; Layer et al. 2020; Yilmaz et al. 2008), nitrogen removal can be efficiently conducted by alternating nitrification and denitrification (AND), i.e., by alternating aerobic and anoxic conditions by on/off aeration control during the starvation phase (Lochmatter et al. 2013; Weissbrodt et al. 2014b).

This can lead to aeration savings, overcome COD limitations for denitrification, and lead to improved BNR performances.

In hybrid systems, an in situ microfiltration operation could be implemented for avoiding release of flocculent biomass into the treated effluent (Sperandio et al. 2005), but with energy needed for the membrane filtration. The choice of the different process options should be based on an energy balance analysis integrating pumping, mixing, and aeration aspects over the process.

5. **Shave variations in the influent**

Similar to conventional activated sludge systems (Siegrist et al. 2000), fluctuations in influent parameters can be "shaved" by dosing specific compounds in order to maintain a balance, e.g., in COD to nutrient ratios and in microbial activities.

For instance, propionate can be dosed to maintain sufficiently high food-to-microorganism (F/M), COD/P, and COD/N ratios. The daily ammonium load can be averaged by supplying ammonium from putrid process water in periods of low ammonium (e.g., during night) (Siegrist et al. 2000).

Slightly alkaline conditions (pH ~ 7.5) can be maintained by supplying bicarbonate in the influent to sustain the proliferation of "*Ca.* Accumulibacter".

In the case of a warm wastewater, heat exchangers can be implemented upstream on the sewer network in order to recover heat (Durrenmatt and Wanner 2008; Wanner et al. 2005), but the effect of a possible temperature decrease on nitrification should be assessed.

6. **Purge excess granular sludge and control the sludge age**

Purge of excess AGS is crucial for an effective P-removal, and for obtaining a stable and cooperative bacterial community on long term.

A controlled biomass removal is the key parameter for EBPR in biofilm reactors (Morgenroth and Wilderer 1999). Purge of excess granular sludge can be easily achieved from the aerated mixed liquor (Weissbrodt et al. 2014c).

In a column reactor, a selective purge of excess sludge from the upper fraction of settled AGS beds can be tested to remove GAO-dominated granules and enhance the selection for PAOs (Bassin et al. 2012; Winkler et al. 2011). GAOs can also be out selected using a side stream system (Stokholm-Bjerregaard 2016).

Excess AGS should also be purged to maintain a stable biomass concentration and F/M ratio. An optimal EBPR can be achieved in the granular sludge system operated with a SRT of maximum 20 days. At higher SRT, microbial cells can get saturated with polyphosphate, leading to a possible EBPR impairment.

7. **Select for PAOs in the BNR microbiome**

Selecting for PAOs is important for granule cohesion and EBPR. The predominant PAOs proliferating in the system are different with a VFA-based wastewater or with a more complex wastewater rich in amino acid or proteins, (poly)sugars, and particulate organic matter,

"*Ca.* Accumulibacter" are efficiently selected by a VFA-based feed, by temperatures below 20 °C, a slightly alkaline pH (~ 7.5), and a relatively low COD/P ratio

(~ 20–50 g_{COD} g^{-1} P), on top of the anaerobic-aerobic alternation. Other PAOs like the fermenting *Tetrasphaera* (or "*Ca.* Phosphoribacter") are selected by more complex organic matter (sugars, amino acids, proteins), and by mesophilic (> 25 °C) and slightly acidic (~ 6.5) conditions. Process operation should be adapted to the predominant PAO involved and its metabolic traits.

To select for "*Ca.* Accumulibacter", VFAs should ideally be provided to the system. In the case of a Swiss wastewater, the particulate COD can be solubilized by acidified in a pre-fermentation step to produce VFAs. The influent wastewater can also be amended with an external source of VFA originating from fermentation of excess sludge or molasses (Moser-Engeler et al. 1998; Zeng et al. 2006). The presence of propionate in the influent is energetically favorable to stimulate the growth of the granular biomass, the selection for "*Ca.* Accumulibacter", and a stable EBPR activity. Dosage of propionate in the influent can be achieved by directing product formation in the fermenter (Rombouts et al. 2019; Temudo 2008; Temudo et al. 2008; Thomas 2008). GAO overgrowth can be controlled in a side stream as well (Stokholm-Bjerregaard 2016).

Fluctuations in operation variables such as the influent composition impact the microbial community composition, on the relative abundance of PAOs, and on the BNR performances. Shaving the variation peaks can help maintain a stable process. Controlling the sludge age is an important tool to maintain an active EBPR biomass by removing the PAOs filled with polyphosphate with the waste sludge.

8. **Manage a stable BNR process**

Collectively, the proposed methodology can lead to a stable, cooperative, and active microbial community required for a long-lasting, efficient BNR in the granular sludge SBRs. The "health" state of the biological systems should be controlled periodically using an analytical toolbox of (molecular) microbial ecology methods (Weissbrodt et al. 2020a). Depending on the level of instrumentation and expertise available on site, biological samples can be collected and sent to an external facility for analysis (e.g., amplicon sequencing, meta-omics).

Amplicon sequencing (e.g., twice a month under pseudo steady-state conditions) can provide interesting insights on the microbiome balance in relation to operation and BNR performances (Cerruti et al. 2021; Fumasoli et al. 2022; Nielsen et al. 2012b; Seviour et al. 2003; Stokholm-Bjerregaard et al. 2021; Weissbrodt et al. 2014c). qPCR can provide a higher sensitivity on specific microbial populations of interest, such as to anticipate the progressive proliferation of unfavorable filamentous organisms (on top of microscopy).

More frequent data can be collected on the fraction of active PAOs and on the EBPR potential of the granular sludge with rapid methods like electrical conductivity-based anaerobic metabolic batch tests and polyphosphatase enzymatic assays (Weissbrodt et al. 2014b). On-line monitoring of electrical conductivity, temperature, pH, DO, and the concentrations of ammonium, nitrite and nitrate is important for implementing control loops. Additional analyses of off-gases (e.g., O_2, CO_2, N_2O) (Baeten et al. 2021; Lochmatter et al. 2014; van Dijk et al. 2021) provide additional insights on metabolic processes and stoichiometric balances.

Relatively high concentrations of suspended solids of up to 100 mg_{TSS} L^{-1} have been regularly measured in effluents of granular sludge systems (notably during start-ups). Lamella settlers can be considered to polish the treated effluents.

12.4 Research and Engineering Perspectives

Significant advances in the understanding of the mechanisms of microbial selection, bioaggregation, and BNR have been made to design robust granular sludge technologies. The integration of microbial ecology principles has been one central point for the ecological engineering of granular sludge processes. Engineering microbiomes requires a strong association of fundamental science and engineering practice. BNR granular sludge offers a diversity of possibilities for scientific and applied investigations, where the association of systems microbiology, environmental biotechnology, and mathematical modelling can make a difference. Some avenues are given hereafter.

The ecosystem model of BNR granular sludge highlighted several interesting targets among the microbial diversity and metabolic networks. The PAO and GAO guilds require a deeper examination of microbial populations able to exhibit a PAO/GAO metabolism; this can be supported by the binning and annotation of MAGs from granular sludge populations followed by functional analyses by metatranscriptomics and/or metaproteomics, and validation by ecophysiology analyses. The role and control of comammox in nitrification in granular sludge is of interest. The ability of the diverse set of denitrifiers to store VFAs under anaerobic conditions require a closer look, using substrate labelling analyses.

Granulation mechanisms can get further examined under the multi meta-omics focus to address the contributions of specific microbial populations to the formation of the biofilm architecture. Unravelling the EPS metabolisms of granular sludge populations is of high scientific interest for the fundamental understanding and regulation of metabolic pathways and the selective production of EPS compounds and biomaterials to reinject in the circular economy. Phototrophic mixed-culture biotechnologies can further benefit from granulation principles to intensify (in)organic carbon and nutrient capture from used streams.

Translation of the granular sludge technology from SBR to continuous flow-through systems will support the upgrade and intensification of existing WWTP infrastructures. Granulation is applicable not only to municipal but also to industrial streams, eventually leading to an efficient sludge management. Good granulation practices are needed to sustain the processes.

The granular sludge technology was implemented at full scale in several locations world-wide over the last 10 years. These installations show promising performances. Understanding better the triggers of a successful operation at full scale under the regional constrains is an important aspect. Important considerations can relate to, e.g., impact of sewer, conversions of particulate organic substrates, diluted wastewater under frequent rain events, warm and cold climates, and salinity among many other targets to be delineated under local constrains.

WWTPs face increasing challenges to remove not only nutrients, but also the cocktail of chemical and biological contaminants emitted in wastewater catchment areas. Process integration should rely on a strong understanding and management of the microbial resource and its metabolic power. An efficient combination of physical, chemical, and biological principles is an important endpoint for a sustainable water chain. The ecological engineering principles developed from activated sludge to granular sludge and biofilm systems can be translated to any water engineering system. Engineering microbiomes is an essential component of sustainability designs to protect the environment, safeguard water resources, and promote health and the circular economy.

References

Adav SS, Lee DJ, Lai JY (2010) Microbial community of acetate utilizing denitrifiers in aerobic granules. Appl Microbiol Biotechnol 85(3):753–762

Adiba S, Nizak C, van Baalen M, Denamur E, Depaulis F (2010) From grazing resistance to pathogenesis: the coincidental evolution of virulence factors. PLoS ONE 5(8):e11882

Adler A, Holliger C (2020) Multistability and reversibility of aerobic granular sludge microbial communities upon changes from simple to complex synthetic wastewater and back. Front Microbiol 11:574361

Adler A, Poirier S, Pagni M, Maillard J, Holliger C (2022) Disentangle genus microdiversity within a complex microbial community by using a multi-distance long-read binning method: example of *Candidatus* Accumulibacter. Environ Microbiol 24(4):2136–2156

Agrawal S, Seuntjens D, Cocker P, Lackner S, Vlaeminck SE (2018) Success of mainstream partial nitritation/anammox demands integration of engineering, microbiome and modeling insights. Curr Opin Biotechnol 50:214–221

Aguado D, Montoya T, Ferrer J, Seco A (2006) Relating ions concentration variations to conductivity variations in a sequencing batch reactor operated for enhanced biological phosphorus removal. Environ Model Softw 21(6):845–851

Albertsen M, Hugenholtz P, Skarshewski A, Nielsen KL, Tyson GW, Nielsen PH (2013a) Genome sequences of rare, uncultured bacteria obtained by differential coverage binning of multiple metagenomes. Nat Biotechnol 31(6):533–538

Albertsen M, Saunders AM, Nielsen KL, Nielsen PH (2013b) Metagenomes obtained by 'deep sequencing'—what do they tell about the enhanced biological phosphorus removal communities? Water Sci Technol 68(9):1959–1968

Albertsen M, Karst SM, Ziegler AS, Kirkegaard RH, Nielsen PH (2015) Back to basics—the influence of DNA extraction and primer choice on phylogenetic analysis of activated sludge communities. PLoS ONE 10(7):e0132783

Ali M, Wang Z, Salam KW, Hari AR, Pronk M, van Loosdrecht MCM, Saikaly PE (2019) Importance of species sorting and immigration on the bacterial assembly of different-sized aggregates in a full-scale aerobic granular sludge plant. Environ Sci Technol 53(14):8291–8301

Alpkvist E, Klapper I (2007) A multidimensional multispecies continuum model for heterogeneous biofilm development. Bull Math Biol 69(2):765–789

Amorim CL, Maia AS, Mesquita RBR, Rangel AOSS, van Loosdrecht MCM, Tiritan ME, Castro PML (2014) Performance of aerobic granular sludge in a sequencing batch bioreactor exposed to ofloxacin, norfloxacin and ciprofloxacin. Water Res 50:101–113

Amorim de Carvalho C, Ferreira dos Santos A, Tavares Ferreira TJ, Sousa Aguiar Lira VN, Mendes Barros AR, Bezerra dos Santos A (2021) Resource recovery in aerobic granular sludge systems: is it feasible or still a long way to go? Chemosphere 274:129881

Andersen KS, Kirkegaard RH, Karst SM, Albertsen M (2018) ampvis2: an R package to analyse and visualise 16S rRNA amplicon data. bioRxiv 299537

Aqeel H, Basuvaraj M, Hall M, Neufeld JD, Liss SN (2016) Microbial dynamics and properties of aerobic granules developed in a laboratory-scale sequencing batch reactor with an intermediate filamentous bulking stage. Appl Microbiol Biotechnol 100(1):447–460

Aqeel H, Weissbrodt DG, Cerruti M, Wolfaardt GM, Wilén B-M, Liss SN (2019) Drivers of bioaggregation from flocs to biofilms and granular sludge. Environ Sci Water Res Technol 5:2072–2089

Avila I, Freedman D, Johnston J, Wisdom B, McQuarrie J (2021) Inducing granulation within a full-scale activated sludge system to improve settling. Water Sci Technol 84(2):302–313

Baeten JE, van Loosdrecht MCM, Volcke EIP (2018) Modelling aerobic granular sludge reactors through apparent half-saturation coefficients. Water Res 146:134–145

Baeten JE, Batstone DJ, Schraa OJ, van Loosdrecht MCM, Volcke EIP (2019) Modelling anaerobic, aerobic and partial nitritation-anammox granular sludge reactors—a review. Water Res 149:322–341

Baeten JE, van Dijk EJH, Pronk M, van Loosdrecht MCM, Volcke EIP (2021) Potential of off-gas analyses for sequentially operated reactors demonstrated on full-scale aerobic granular sludge technology. Sci Total Environ 787:147651

Bagchi S, Lamendella R, Strutt S, van Loosdrecht MCM, Saikaly PE (2016) Metatranscriptomics reveals the molecular mechanism of large granule formation in granular anammox reactor. Sci Rep 6:28327

Ballance S, Holtan S, Aarstad OA, Sikorski P, Skjåk-Braek G, Christensen BE (2005) Application of high-performance anion-exchange chromatography with pulsed amperometric detection and statistical analysis to study oligosaccharide distributions—a complementary method to investigate the structure and some properties of alginates. J Chromatogr A 1093(1–2):59–68

Barnard JL, Dunlap P, Steichen M (2017) Rethinking the mechanisms of biological phosphorus removal. Water Environ Res 89(11):2043–2054

Barr JJ (2010) The microbial ecology of granular sludge in enhanced biological phosphorus removal. PhD thesis, The University of Queensland

Barr JJ, Slater FR, Fukushima T, Bond PL (2010) Evidence for bacteriophage activity causing community and performance changes in a phosphorus-removal activated sludge. FEMS Microbiol Ecol 74(3):631–642

Barr JJ, Hastie ML, Fukushima T, Plan MR, Gorman JJ, Tyson GW, Bond PL (2011) Metaproteomic analysis of laboratory scale phophorus removal reactors reveals functional insights of aerobic granular sludge. In: Qi Z (ed) IWA biofilm specialist conference 2011: processes in biofilms. Tongji University, Shanghai, China

Barr JJ, Dutilh BE, Skennerton CT, Fukushima T, Hastie ML, Gorman JJ, Tyson GW, Bond PL (2015) Metagenomic and metaproteomic analyses of *Accumulibacter phosphatis*-enriched floccular and granular biofilm. Environ Microbiol 18(1):273–287

Barrios-Hernández ML, Bettinelli C, Mora-Cabrera K, Vanegas-Camero M-C, Garcia H, van de Vossenberg J, Prats D, Brdjanovic D, van Loosdrecht MCM, Hooijmans CM (2021) Unravelling the removal mechanisms of bacterial and viral surrogates in aerobic granular sludge systems. Water Res 195:116992

Bassin JP, Winkler MKH, Kleerebezem R, Dezotti M, van Loosdrecht MCM (2012) Improved phosphate removal by selective sludge discharge in aerobic granular sludge reactors. Biotechnol Bioeng 109(8):1919–1928

Bellini MI, Kumaresan D, Tarlera S, Murrell JC, Fernández-Scavino A (2017) Identification of active denitrifiers by DNA-stable isotope probing and amplicon sequencing reveals *Betaproteobacteria* as responsible for attenuation of nitrate contamination in a low impacted aquifer. FEMS Microbiol Ecol 94(2):fix181

Beloin C, Ghigo J-M (2005) Finding gene-expression patterns in bacterial biofilms. Trends Microbiol 13(1):16–19

Bernstein HC (2019) Reconciling ecological and engineering design principles for building microbiomes. mSystems 4(3):e00106–19

Beun JJ, Hendriks A, van Loosdrecht MCM, Morgenroth E, Wilderer PA, Heijnen JJ (1999) Aerobic granulation in a sequencing batch reactor. Water Res 33(10):2283–2290

Bhattacharjee AS, Choi J, Motlagh AM, Mukherji ST, Goel R (2015) Bacteriophage therapy for membrane biofouling in membrane bioreactors and antibiotic-resistant bacterial biofilms. Biotechnol Bioeng 112(8):1644–1654

Bock E, Koops H-P, Muller U, Rudert M (1990) A new facultatively nitrite oxidizing bacterium, *Nitrobacter vulgaris* sp. nov. Arch Microbiol 153(2):105–110

Boleij M, Kleikamp H, Pabst M, Neu TR, van Loosdrecht MCM, Lin Y (2020) Decorating the anammox house: sialic acids and sulfated glycosaminoglycans in the extracellular polymeric substances of anammox granular sludge. Environ Sci Technol 54(8):5218–5226

Bossier P, Verstraete W (1996) Triggers for microbial aggregation in activated sludge? Appl Microbiol Biotechnol 45(1–2):1–6

Bradford LM, Vestergaard G, Táncsics A, Zhu B, Schloter M, Lueders T (2018) Transcriptome-stable isotope probing provides targeted functional and taxonomic insights into microaerobic pollutant-degrading aquifer microbiota. Front Microbiol 9:2696

Burke C, Steinberg P, Rusch D, Kjelleberg S, Thomas T (2011) Bacterial community assembly based on functional genes rather than species. Proc Nat Acad Sci USA 108(34):14288–14293

Calderón-Franco D, Sarelse R, Christou S, Pronk M, van Loosdrecht MCM, Abeel T, Weissbrodt DG (2022) Metagenomic profiling and transfer dynamics of antibiotic resistance determinants in a full-scale granular sludge wastewater treatment plant. Water Res 219:118571

Calero-Cáceres W, Muniesa M (2016) Persistence of naturally occurring antibiotic resistance genes in the bacteria and bacteriophage fractions of wastewater. Water Res 95:11–18

Caluwé M, Dobbeleers T, D'Aes J, Miele S, Akkermans V, Daens D, Geuens L, Kiekens F, Blust R, Dries J (2017) Formation of aerobic granular sludge during the treatment of petrochemical wastewater. Bioresour Technol 238:559–567

Caluwé M, Goossens K, Seguel Suazo K, Tsertou E, Dries J (2022) Granulation strategies applied to industrial wastewater treatment: from lab to full-scale. Water Sci Technol 85(9):2761–2771

Cerruti M, Guo B, Delatolla R, de Jonge N, Hommes-de Vos van Steenwijk A, Kadota P, Lawson CE, Mao T, Oosterkamp MJ, Sabba F, Stokholm-Bjerregaard M, Watson I, Frigon D, Weissbrodt DG (2021) Plant-wide systems microbiology for the wastewater industry. Environ Sci Water Res Technol 7(10):1687–1706

Choi J, Kotay SM, Goel R (2011) Bacteriophage-based biocontrol of biological sludge bulking in wastewater. Bioeng Bugs 2(4):214–217

Christensen BB, Sternberg C, Andersen JB, Palmer RJ Jr, Nielsen AT, Givskov M, Molin S (1999) Molecular tools for study of biofilm physiology. Methods Enzymol 310:20–42

Close K, Marques R, Carvalho VCF, Freitas EB, Reis MAM, Carvalho G, Oehmen A (2021) The storage compounds associated with Tetrasphaera PAO metabolism and the relationship between diversity and P removal. Water Res 204:117621

Comeau Y, Hall KJ, Hancock REW, Oldham WK (1986) Biochemical model for enhanced biological phosphorus removal. Water Res 20(12):1511–1521

Conthe M, Kuenen JG, Kleerebezem R, van Loosdrecht MCM (2018) Exploring microbial N_2O reduction: a continuous enrichment in nitrogen free medium. Environ Microbiol Rep 10(1):102–107

Costa E, Pérez J, Kreft J-U (2006) Why is metabolic labour divided in nitrification? Trends Microbiol 14(5):213–219

Cydzik-Kwiatkowska A, de Jonge N, Poulsen JS, Nielsen JL (2022) Unravelling gradient layers of microbial communities, proteins, and chemical structure in aerobic granules. Sci Total Environ 829:154253

Daigger GT (2011) Changing paradigms: from wastewater treatment to resource recovery. Proc Water Environ Fed 2011(6):942–957

Daigger GT (2017) Flexibility and adaptability: essential elements of the WRRF of the future. Water Pract Technol 12(1):156–165

Daigger GT, Redmond E, Downing L (2018) Enhanced settling in activated sludge: design and operation considerations. Water Sci Technol 78(2):247–258

Daims H, Nielsen PH, Nielsen JL, Juretschko S, Wagner M (2000) Novel *Nitrospira*-like bacteria as dominant nitrite-oxidizers in biofilms from wastewater treatment plants: diversity and in situ physiology. Water Sci Technol 41(4–5):85–90

Daims H, Lebedeva EV, Pjevac P, Han P, Herbold C, Albertsen M, Jehmlich N, Palatinszky M, Vierheilig J, Bulaev A, Kirkegaard RH, Bergen MV, Rattei T, Bendinger B, Nielsen PH, Wagner M (2015) Complete nitrification by *Nitrospira* bacteria. Nature 528(7583):504–509

de Bruin LMM, de Kreuk MK, van der Roest HFR, Uijterlinde C, van Loosdrecht MCM (2004) Aerobic granular sludge technology: an alternative to activated sludge? Water Sci Technol 49(11–12):1–7

De Clippeleir H, Han M, Al-Omari A, Bott C, Wett B, Murthy S (2014) From nitrite shunt to mainstream deammonification strategy: pilot-scale demonstration. In: WEF (ed) Water environment federation's annual technical exhibition and conference (WEFTEC), New Orleans, LA

de Graaff DR, Felz S, Neu TR, Pronk M, van Loosdrecht MCM, Lin Y (2019) Sialic acids in the extracellular polymeric substances of seawater-adapted aerobic granular sludge. Water Res 155:343–351

de Graaff DR, van Dijk EJH, van Loosdrecht MCM, Pronk M (2020a) Strength characterization of full-scale aerobic granular sludge. Environ Technol 41(13):1637–1647

de Graaff DR, van Loosdrecht MCM, Pronk M (2020b) Stable granulation of seawater-adapted aerobic granular sludge with filamentous *Thiothrix* bacteria. Water Res 175:115683

de Kreuk MK, van Loosdrecht MCM (2004) Selection of slow growing organisms as a means for improving aerobic granular sludge stability. Water Sci Technol 49(11–12):9–17

de Kreuk MK, van Loosdrecht MCM (2006) Formation of aerobic granules with domestic sewage. J Environ Eng ASCE 132(6):694–697

de Kreuk MK, Heijnen JJ, van Loosdrecht MCM (2005) Simultaneous COD, nitrogen, and phosphate removal by aerobic granular sludge. Biotechnol Bioeng 90(6):761–769

de Kreuk MK, Kishida N, Tsuneda S, van Loosdrecht MCM (2010) Behavior of polymeric substrates in an aerobic granular sludge system. Water Res 44(20):5929–5938

De Vleeschauwer F, Caluwé M, Dobbeleers T, Stes H, Dockx L, Kiekens F, D'Aes J, Copot C, Dries J (2019) Performance and stability of a dynamically controlled EBPR anaerobic/aerobic granular sludge reactor. Bioresour Technol 280:151–157

Derlon N, Peter-Varbanets M, Scheidegger A, Pronk W, Morgenroth E (2012) Predation influences the structure of biofilm developed on ultrafiltration membranes. Water Res 46(10):3323–3333

Derlon N, Wagner J, da Costa RHR, Morgenroth E (2016) Formation of aerobic granules for the treatment of real and low-strength municipal wastewater using a sequencing batch reactor operated at constant volume. Water Res 105:341–350

Derlon N, Garcia Villodres M, Kovács R, Brison A, Layer M, Takács I, Morgenroth E (2022) Modelling of aerobic granular sludge reactors: the importance of hydrodynamic regimes, selective sludge removal and gradients. Water Sci Technol 86(3):410–431

Ding K, Xu L, Chen Y, Li W, Chai X, Dai X, Wu B (2023) Mechanistic insights into polyhydroxyalkanoate-enhanced denitrification capacity of microbial community: evolution of community structure and intracellular electron transfer of nitrogen metabolism. Sci Total Environ 856(Pt 1):159147

Dobbeleers T, D'aes J, Miele S, Caluwé M, Akkermans V, Daens D, Geuens L, Dries J (2017) Aeration control strategies to stimulate simultaneous nitrification-denitrification via nitrite during the formation of aerobic granular sludge. Appl Microbiol Biotechnol 101(17):6829–6839

Dobbeleers T, Caluwé M, Daens D, Geuens L, Dries J (2018) Evaluation of two start-up strategies to obtain nitrogen removal via nitrite and examination of the nitrous oxide emissions for different nitritation levels during the treatment of slaughterhouse wastewater. J Chem Technol Biotechnol 93(2):569–576

Downing L, Redmond E, Avila I (2022) When density is desirable. Water Online, September 9, 2022

Dueholm MKD, Nierychlo M, Andersen KS, Rudkjøbing V, Knutsson S, Arriaga S, Bakke R, Boon N, Bux F, Christensson M, Chua ASM, Curtis TP, Cytryn E, Erijman L, Etchebehere C, Fatta-Kassinos D, Frigon D, Garcia-Chaves MC, Gu AZ, Horn H, Jenkins D, Kreuzinger N, Kumari S, Lanham A, Law Y, Leiknes TO, Morgenroth E, Muszyński A, Petrovski S, Pijuan M, Pillai SB, Reis MAM, Rong Q, Rossetti S, Seviour R, Tooker N, Vainio P, van Loosdrecht M, Vikraman R, Wanner J, Weissbrodt D, Wen X, Zhang T, Nielsen PH, Albertsen M, Nielsen PH, MiDAS GC (2022) MiDAS 4: a global catalogue of full-length 16S rRNA gene sequences and taxonomy for studies of bacterial communities in wastewater treatment plants. Nat Commun 13(1):1908

Dueholm MKD, Besteman M, Zeuner EJ, Riisgaard-Jensen M, Nielsen ME, Vestergaard SZ, Heidelbach S, Bekker NS, Nielsen PH (2023) Genetic potential for exopolysaccharide synthesis in activated sludge bacteria uncovered by genome-resolved metagenomics. Water Res 229:119485

Dulekgurgen E, Artan N, Orhon D, Wilderer PA (2008) How does shear affect aggregation in granular sludge sequencing batch reactors? Relations between shear, hydrophobicity, and extracellular polymeric substances. Water Sci Technol 58(2):267–276

Durrenmatt DJ, Wanner O (2008) Simulation of the wastewater temperature in sewers with TEMPEST. Water Sci Technol 57(11):1809–1815

Dutch Water Sector (2019) World's first waste water treatment plant to produce biopolymer. Kaumera Water & Technology, Netherlands

Ebrahimi S, Gabus S, Rohrbach-Brandt E, Hosseini M, Rossi P, Maillard J, Holliger C (2010) Performance and microbial community composition dynamics of aerobic granular sludge from sequencing batch bubble column reactors operated at 20 °C, 30 °C, and 35 °C. Appl Microbiol Biotechnol 87(4):1555–1568

Eggen RI, Hollender J, Joss A, Schärer M, Stamm C (2014) Reducing the discharge of micropollutants in the aquatic environment: the benefits of upgrading wastewater treatment plants. Environ Sci Technol 48(14):7683–7689

Eikelboom DH, Geurkink B (2002) Filamentous micro-organisms observed in industrial activated sludge plants. Water Sci Technol 46(1–2):535–542

Ekholm J, Persson F, de Blois M, Modin O, Pronk M, van Loosdrecht MCM, Suarez C, Gustavsson DJI, Wilén B-M (2022) Full-scale aerobic granular sludge for municipal wastewater treatment—granule formation, microbial succession, and process performance. Environ Sci Water Res Technol 8(12):3138–3154

Elahinik A, Haarsma M, Abbas B, Pabst M, Xevgenos D, van Loosdrecht MCM, Pronk M (2022) Glycerol conversion by aerobic granular sludge. Water Res 227:119340

Elawwad A, Matta M, Abo-Zaid M, Abdel-Halim H (2019) Plant-wide modeling and optimization of a large-scale WWTP using BioWin's ASDM model. J Water Process Eng 31:100819

Erguder TH, Boon N, Vlaeminck SE, Verstraete W (2008) Partial nitrification achieved by pulse sulfide doses in a sequential batch reactor. Environ Sci Technol 42(23):8715–8720

Falås P, Wick A, Castronovo S, Habermacher J, Ternes TA, Joss A (2016) Tracing the limits of organic micropollutant removal in biological wastewater treatment. Water Res 95:240–249

Felz S, Al-Zuhairy S, Aarstad OA, van Loosdrecht MCM, Lin YM (2016) Extraction of structural extracellular polymeric substances from aerobic granular sludge. J vis Exp 115:e54534

Felz S, Vermeulen P, van Loosdrecht MCM, Lin YM (2019) Chemical characterization methods for the analysis of structural extracellular polymeric substances (EPS). Water Res 157:201–208

Feng C, Welles L, Zhang X, Pronk M, de Graaff D, van Loosdrecht M (2020) Stress-induced assays for polyphosphate quantification by uncoupling acetic acid uptake and anaerobic phosphorus release. Water Res 169:115228

Fernando EY, McIlroy SJ, Nierychlo M, Herbst F-A, Petriglieri F, Schmid MC, Wagner M, Nielsen JL, Nielsen PH (2019) Resolving the individual contribution of key microbial populations to enhanced biological phosphorus removal with Raman–FISH. ISME J 13(8):1933–1946

Filali A (2011) Analyse et modélisation du traitement de l'azote dans un procédé de granulation aérobie hybride. PhD thesis, INSA

Filali A, Bessiere Y, Sperandio M (2012) Effects of oxygen concentration on the nitrifying activity of an aerobic hybrid granular sludge reactor. Water Sci Technol 65(2):289–295

Finkmann W, Altendorf K, Stackebrandt E, Lipski A (2000) Characterization of N_2O-producing Xanthomonas-like isolates from biofilters as *Stenotrophomonas nitritireducens* sp. nov., *Luteimonas mephitis* gen. nov., sp. nov. and *Pseudoxanthomonas broegbernensis* gen. nov., sp. nov. Int J Syst Evol Microbiol 50(1):273–282

Fitzgerald CM, Camejo P, Oshlag JZ, Noguera DR (2015) Ammonia-oxidizing microbial communities in reactors with efficient nitrification at low-dissolved oxygen. Water Res 70:38–51

Freitag A, Rudert M, Bock E (1987) Growth of *Nitrobacter* by dissimilatoric nitrate reduction. FEMS Microbiol Lett 48(1–2):105–109

Frigon D, Wells G (2019) Microbial immigration in wastewater treatment systems: analytical considerations and process implications. Curr Opin Biotechnol 57:151–159

Fumasoli A, Keller A, Gubser N, Weissbrodt DG (2022) DNA-analyse im belebtschlamm. Aqua Gas 1:2–8

Gao R, Peng Y, Li J, Liu Y, Deng L, Li W, Kao C (2022) Mainstream partial denitrification-anammox (PD/A) for municipal sewage treatment from moderate to low temperature: reactor performance and bacterial structure. Sci Total Environ 806:150267

Giesen A, Niermans R, van Loosdrecht MCM (2012) Aerobic granular biomass: the new standard for domestic and industrial wastewater treatment? Water 21(4):28–30

Gilbride KA, Lee DY, Beaudette LA (2006) Molecular techniques in wastewater: understanding microbial communities, detecting pathogens, and real-time process control. J Microbiol Methods 66(1):1–20

Goel R, Mino T, Satoh H, Matsuo T (1998) Intracellular storage compounds, oxygen uptake rates and biomass yield with readily and slowly degradable substrates. Water Sci Technol 38(8–9):85–93

Goel R, Kotay SM, Butler CS, Torres CI, Mahendra S (2011) Molecular biological methods in environmental engineering. Water Environ Res 83(10):927–955

Gonzalez-Gil G, Holliger C (2011) Dynamics of microbial community structure and enhanced biological phosphorus removal of propionate- and acetate-cultivated aerobic granules. Appl Environ Microbiol 77:8041–8051

Gonzalez-Gil G, Holliger C (2014) Aerobic granules: microbial landscape and architecture, stages, and practical implications. Appl Environ Microbiol 80(11):3433–3441

Grau P, de Gracia M, Vanrolleghem PA, Ayesa E (2007) A new plant-wide modelling methodology for WWTPs. Water Res 41(19):4357–4372

Gruber W, von Känel L, Vogt L, Luck M, Biolley L, Feller K, Moosmann A, Krähenbühl N, Kipf M, Loosli R, Vogel M, Morgenroth E, Braun D, Joss A (2021) Estimation of countrywide N_2O emissions from wastewater treatment in Switzerland using long-term monitoring data. Water Res X 13:100122

Gruber W, Magyar PM, Mitrovic I, Zeyer K, Vogel M, von Känel L, Biolley L, Werner RA, Morgenroth E, Lehmann MF, Braun D, Joss A, Mohn J (2022) Tracing N_2O formation in full-scale wastewater treatment with natural abundance isotopes indicates control by organic substrate and process settings. Water Res X 15:100130

Gu AZ, Nerenberg R, Sturm BM, Chul P, Goel R (2010) Molecular methods in biological systems. Water Environ Res 82(10):908–930

Guest JS, Skerlos SJ, Barnard JL, Beck MB, Daigger GT, Hilger H, Jackson SJ, Karvazy K, Kelly L, Macpherson L, Mihelcic JR, Pramanik A, Raskin L, van Loosdrecht MCM, Yeh D, Love NG (2009) A new planning and design paradigm to achieve sustainable resource recovery from wastewater. Environ Sci Technol 43(16):6126–6130

Guimarães LB, Gubser NR, Lin Y, Pronk M, Welles L, Albertsen M, Daudt GC, Geleijnse MAA, da Costa RHR, Nielsen PH, van Loosdrecht MCM, Weissbrodt DG (2016) Exopolysaccharides biorefining from used water: an enterprise in the microbiome of granular sludge. In: van Loosdrecht MCM et al (eds) 13th IWA leading edge conference on water and wastewater technologies: evaluating impacts of innovation. International Water Association, Jerez de la Fontera, Spain

Guimarães LB, Gubser NR, Lin Y, Zlopasa J, Albertsen M, Akaboci TRV, da Costa RHR, Nielsen PH, van Loosdrecht MCM, Weissbrodt DG (2017a) Alginate-like exopolymer recovery from granular sludge biofilm reactor under conditions selecting for slow-growing PAOs. In: Casey et al (eds) 10th international conference on biofilm reactors. International Water Association, Dublin, Ireland

Guimarães LB, Wagner J, Akaboci TRV, Daudt GC, Nielsen PH, van Loosdrecht MCM, Weissbrodt DG, da Costa RHR (2017b) From failures to successful granular sludge process: hints for real wastewater treatment under coastal warm climate. In: Alvarez et al (eds) 14th IWA leading edge conference on water and wastewater technologies: evaluating impacts of innovation. International Water Association, Florianopolis, Brazil

Guimarães LB, Wagner J, Akaboci TRV, Daudt GC, Nielsen PH, van Loosdrecht MCM, Weissbrodt DG, da Costa RHR (2018) Elucidating performance failures in use of granular sludge for nutrient removal from domestic wastewater in a warm coastal climate region. Environ Technol 41(15):1896–1911

Gujer W (2010) Nitrification and me—a subjective review. Water Res 44(1):1–19

Gunathilaka GU, Tahlan V, Mafiz AI, Polur M, Zhang Y (2017) Phages in urban wastewater have the potential to disseminate antibiotic resistance. Int J Antimicrob Agents 50(5):678–683

Gunther S, Trutnau M, Kleinsteuber S, Hause G, Bley T, Roske I, Harms H, Muller S (2009) Dynamics of polyphosphate-accumulating bacteria in wastewater treatment plant microbial communities detected via DAPI (4′,6′-diamidino-2-phenylindole) and tetracycline labeling. Appl Environ Microbiol 75(7):2111–2121

Guo B, Liu C, Gibson C, Frigon D (2019) Wastewater microbial community structure and functional traits change over short timescales. Sci Total Environ 662:779–785

Haaksman VA, Mirghorayshi M, van Loosdrecht MCM, Pronk M (2020) Impact of aerobic availability of readily biodegradable COD on morphological stability of aerobic granular sludge. Water Res 187:116402

Hahn MW, Höfle MG (2001) Grazing of protozoa and its effect on populations of aquatic bacteria. FEMS Microbiol Ecol 35(2):113–121

Hanada S, Liu WT, Shintani T, Kamagata Y, Nakamura K (2002) *Tetrasphaera elongata* sp. nov., a polyphosphate-accumulating bacterium isolated from activated sludge. Int J Syst Evol Microbiol 52(Pt 3):883–887

Hawley AK, Brewer HM, Norbeck AD, Paša-Tolić L, Hallam SJ (2014) Metaproteomics reveals differential modes of metabolic coupling among ubiquitous oxygen minimum zone microbes. Proc Natl Acad Sci USA 111(31):11395–11400

He Q, Zhang J, Gao S, Chen L, Lyu W, Zhang W, Song J, Hu X, Chen R, Wang H, Yu J (2019) A comprehensive comparison between non-bulking and bulking aerobic granular sludge in microbial communities. Bioresour Technol 294:122151

Heijnen JJ, Van't Riet K (1984) Mass transfer, mixing and heat transfer phenomena in low viscosity bubble column reactors. Chem Eng J 28(2):B21–B42

Heijnen JJ, van Loosdrecht MCM, Mulder A, Tijhuis L (1992) Formation of biofilms in a biofilm air-lift suspension reactor. Water Sci Technol 26(3–4):647–654

Henriet O, Meunier C, Henry P, Mahillon J (2016) Improving phosphorus removal in aerobic granular sludge processes through selective microbial management. Bioresour Technol 211:298–306

Henriet O, Meunier C, Henry P, Mahillon J (2017) Filamentous bulking caused by *Thiothrix* species is efficiently controlled in full-scale wastewater treatment plants by implementing a sludge densification strategy. Sci Rep 7(1):1430

Henze M, Gujer W, Mino T, van Loosdrecht MCM (2000) Activated sludge models ASM1, ASM2, ASM2d and ASM3. IWA Publishing, London

Herbst FA, Dueholm MS, Wimmer R, Nielsen PH (2019) The proteome of *Tetrasphaera elongata* is adapted to changing conditions in wastewater treatment plants. Proteomes 7(2):16

Hoekstra M, Geilvoet SP, Hendrickx TLG, van Erp Taalman Kip CS, Kleerebezem R, van Loosdrecht MCM (2019) Towards mainstream anammox: lessons learned from pilot-scale research at WWTP Dokhaven. Environ Technol 40(13):1721–1733

Hofman-Bang J, Zheng D, Westermann P, Ahring BK, Raskin L (2003) Molecular ecology of anaerobic reactor systems. Adv Biochem Eng Biotechnol 81:151–203

Holmes DE, Dang Y, Smith JA (2019) Nitrogen cycling during wastewater treatment. Adv Appl Microbiol 106:113–192

Huisman JL, Krebs P, Gujer W (2003) Integral and unified model for the sewer and wastewater treatment plant focusing on transformations. Water Sci Technol 47(12):65–71

Hupfer M, Glöss S, Schmieder P, Grossart HP (2008) Methods for detection and quantification of polyphosphate and polyphosphate accumulating microorganisms in aquatic sediments. Int Rev Hydrobiol 93(1):1–30

Huws SA, McBain AJ, Gilbert P (2005) Protozoan grazing and its impact upon population dynamics in biofilm communities. J Appl Microbiol 98(1):238–244

Jehmlich N, Schmidt F, von Bergen M, Richnow H-H, Vogt C (2008) Protein-based stable isotope probing (protein-SIP) reveals active species within anoxic mixed cultures. ISME J 2(11):1122–1133

Jenkins D, Wanner J (eds) (2014) Activated sludge—100 years and counting. London

Jetten MSM (2008) The microbial nitrogen cycle. Environ Microbiol 10(11):2903–2909

Joss A, Derlon N, Cyprien C, Burger S, Szivak I, Traber J, Siegrist H, Morgenroth E (2011) Combined nitritation-anammox: advances in understanding process stability. Environ Sci Technol 45(22):9735–9742

Kampschreur MJ, Temmink H, Kleerebezem R, Jetten MS, van Loosdrecht MC (2009) Nitrous oxide emission during wastewater treatment. Water Res 43(17):4093–4103

Karst SM, Albertsen M, Kirkegaard RH, Dueholm MS, Nielsen PH (2016) Molecular methods. In: van Loosdrecht MCM, Nielsen PH, Lopez-Vazquez CM, Brdjanovic D (eds) Experimental methods in wastewater treatment. IWA Publishing, London, pp 285–323

Kehrein P, van Loosdrecht M, Osseweijer P, Garfí M, Dewulf J, Posada J (2020a) A critical review of resource recovery from municipal wastewater treatment plants—market supply potentials, technologies and bottlenecks. Environ Sci Water Res Technol 6(4):877–910

Kehrein P, van Loosdrecht M, Osseweijer P, Posada J, Dewulf J (2020b) The SPPD-WRF framework: a novel and holistic methodology for strategical planning and process design of water resource factories. Sustainability 12(10):4168

Khan MA, Satoh H, Mino T (2006) Occurrence and significance of bacteriophages in biological wastewater treatment processes. In: Reclaiming the desert: towards a sustainable environment in arid lands. Proceedings of the 3rd UAE-Japan symposium on sustainable GCC environment and water resources, EWR 2006, pp 295–303

Kim NK, Mao N, Lin R, Bhattacharyya D, van Loosdrecht MCM, Lin Y (2020) Flame retardant property of flax fabrics coated by extracellular polymeric substances recovered from both activated sludge and aerobic granular sludge. Water Res 170:115344

Kirkland CM, Krug JR, Vergeldt FJ, van den Berg L, Velders AH, Seymour JD, Codd SL, van As H, de Kreuk MK (2020) Characterizing the structure of aerobic granular sludge using ultra-high field magnetic resonance. Water Sci Technol 82(4):627–639

Kleerebezem R, Lücker S (2021) Cyclic conversions in the nitrogen cycle. Front Microbiol 12:622504

Kleerebezem R, van Loosdrecht MC (2007) Mixed culture biotechnology for bioenergy production. Curr Opin Biotechnol 18(3):207–212

Kleikamp HBC, Lin Y, McMillan DGG, Geelhoed JS, Naus-Wiezer SNH, van Baarlen P, Saha C, Louwen R, Sorokin DY, van Loosdrecht MCM, Pabst M (2020) Tackling the chemical diversity of microbial nonulosonic acids—a universal large-scale survey approach. Chem Sci 11:3074

Kleikamp HBC, Pronk M, Tugui C, Guedes da Silva L, Abbas B, Lin YM, van Loosdrecht MCM, Pabst M (2021) Database-independent *de novo* metaproteomics of complex microbial communities. Cell Syst 12(5):375–383.e5

Kleikamp HBC, Grouzdev D, Schaasberg P, van Valderen R, van der Zwaan R, van de Wijgaart R, Lin Y, Abbas B, Pronk M, van Loosdrecht MCM, Pabst M (2022) Comparative metaproteomics demonstrates different views on the complex granular sludge microbiome. bioRxiv 2022.03.07.483319

Kleiner M, Kouris A, Jensen M, Liu Y, McCalder J, Strous M (2021) Ultra-sensitive protein-SIP to quantify activity and substrate uptake in microbiomes with stable isotopes. Microbiome 11:24

Koch H, Lucker S, Albertsen M, Kitzinger K, Herbold C, Spieck E, Nielsen PH, Wagner M, Daims H (2015) Expanded metabolic versatility of ubiquitous nitrite-oxidizing bacteria from the genus *Nitrospira.* Proc Natl Acad Sci USA 112(36):11371–11376

Kornberg A (1995) Inorganic polyphosphate: toward making a forgotten polymer unforgettable. J Bacteriol 177(3):491–496

Kotay SM, Datta T, Choi J, Goel R (2011) Biocontrol of biomass bulking caused by *Haliscomenobacter hydrossis* using a newly isolated lytic bacteriophage. Water Res 45(2):694–704

Koton-Czarnecka M, Chrost RJ (2002) Measurement of protozoan grazing on bacteria by means of [3H-thymidine]—labeled natural assemblages of lake bacteria. Pol J Environ Stud 11(4):385–393

Kristiansen R, Nguyen HTT, Saunders AM, Nielsen JL, Wimmer R, Le VQ, McIlroy SJ, Petrovski S, Seviour RJ, Calteau A, Nielsen KL, Nielsen PH (2013) A metabolic model for members of the genus *Tetrasphaera* involved in enhanced biological phosphorus removal. ISME J 7(3):543–554

Kuba T, van Loosdrecht MCM, Heijnen JJ (1996) Phosphorus and nitrogen removal with minimal COD requirement by integration of denitrifying dephosphatation and nitrification in a two-sludge system. Water Res 30(7):1702–1710

Kuypers MMM, Marchant HK, Kartal B (2018) The microbial nitrogen-cycling network. Nat Rev Microbiol 16(5):263–276

Lackner S, Terada A, Horn H, Henze M, Smets BF (2010) Nitritation performance in membrane-aerated biofilm reactors differs from conventional biofilm systems. Water Res 44(20):6073–6084

Lackner S, Gilbert EM, Vlaeminck SE, Joss A, Horn H, van Loosdrecht MCM (2014) Full-scale partial nitritation/anammox experiences—an application survey. Water Res 55:292–303

Lajoie CA, Layton AC, Gregory IR, Sayler GS, Don ET, Meyers AJ (2000) Zoogleal clusters and sludge dewatering potential in an industrial activated-sludge wastewater treatment plant. Water Environ Res 72(1):56–64

Lalumière A (2016) Le PEX StaRRE: un nouveau programme destiné aux ouvrages d'assainissement des eaux usées. Vecteur Environ Mars 58–59

Landreau M, Byson SJ, You H, Stahl DA, Winkler MKH (2020) Effective nitrogen removal from ammonium-depleted wastewater by partial nitritation and anammox immobilized in granular and thin layer gel carriers. Water Res 183:116078

Larsen P, von Ins M (2010) The rate of growth in scientific publication and the decline in coverage provided by Science Citation Index. Scientometrics 84(3):575–603

Laureni M, Weissbrodt DG, Villez K, Robin O, de Jonge N, Rosenthal A, Wells G, Nielsen JL, Morgenroth E, Joss A (2019) Biomass segregation between biofilm and flocs improves the control of nitrite-oxidizing bacteria in mainstream partial nitritation and anammox processes. Water Res 154:104–116

Lawson CE, Lücker S (2018) Complete ammonia oxidation: an important control on nitrification in engineered ecosystems? Curr Opin Biotechnol 50:158–165

Lawson CE, Harcombe WR, Hatzenpichler R, Lindemann SR, Löffler FE, O'Malley MA, García Martín H, Pfleger BF, Raskin L, Venturelli OS, Weissbrodt DG, Noguera DR, McMahon KD (2019) Common principles and best practices for engineering microbiomes. Nat Rev Microbiol 17:725–741

Layer M, Adler A, Reynaert E, Hernandez A, Pagni M, Morgenroth E, Holliger C, Derlon N (2019) Organic substrate diffusibility governs microbial community composition, nutrient removal performance and kinetics of granulation of aerobic granular sludge. Water Res X 4:100033

Layer M, Villodres MG, Hernandez A, Reynaert E, Morgenroth E, Derlon N (2020) Limited simultaneous nitrification-denitrification (SND) in aerobic granular sludge systems treating municipal wastewater: mechanisms and practical implications. Water Res X 7:100048

Layer M, Brison A, Villodres MG, Stähle M, Házi F, Takács I, Morgenroth E, Derlon N (2022) Microbial conversion pathways of particulate organic substrate conversion in aerobic granular sludge systems: limited anaerobic conversion and the essential role of flocs. Environ Sci Water Res Technol 8(6):1236–1251

Le T, Peng B, Su C, Massoudieh A, Torrents A, Al-Omari A, Murthy S, Wett B, Chandran K, deBarbadillo C, Bott C, De Clippeleir H (2019) Nitrate residual as a key parameter to efficiently control partial denitrification coupling with anammox. Water Environ Res 91(11):1455–1465

Lee CO, Boe-Hansen R, Musovic S, Smets B, Albrechtsen HJ, Binning P (2014) Effects of dynamic operating conditions on nitrification in biological rapid sand filters for drinking water treatment. Water Res 64:226–236

Lekunberri I, Villagrasa M, Balcázar JL, Borrego CM (2017) Contribution of bacteriophage and plasmid DNA to the mobilization of antibiotic resistance genes in a river receiving treated wastewater discharges. Sci Total Environ 601–602:206–209

Li DC, Gao JF, Zhang SJ, Gao YQ, Sun LX (2020) Emergence and spread patterns of antibiotic resistance genes during two different aerobic granular sludge cultivation processes. Environ Int 137:105540

Limpiyakorn T, Sonthiphand P, Rongsayamanont C, Polprasert C (2011) Abundance of amoA genes of ammonia-oxidizing archaea and bacteria in activated sludge of full-scale wastewater treatment plants. Bioresour Technol 102(4):3694–3701

Lin Y, de Kreuk MK, van Loosdrecht MCM, Adin A (2010) Characterization of alginate-like exopolysaccharides isolated from aerobic granular sludge in pilot-plant. Water Res 44(11):3355–3364

Lin YM, Nierop KGJ, Girbal-Neuhauser E, Adriaanse M, van Loosdrecht MCM (2015) Sustainable polysaccharide-based biomaterial recovered from waste aerobic granular sludge as a surface coating material. Sustain Mater Technol 4:24–29

Lindemann SR, Bernstein HC, Song H-S, Fredrickson JK, Fields MW, Shou W, Johnson DR, Beliaev AS (2016) Engineering microbial consortia for controllable outputs. ISME J 10(9):2077–2084

Liu Y, Liu Q-S (2006) Causes and control of filamentous growth in aerobic granular sludge sequencing batch reactors. Biotechnol Adv 24(1):115–127

Liu Y, Tay J-H (2002) The essential role of hydrodynamic shear force in the formation of biofilm and granular sludge. Water Res 36(7):1653–1665

Liu WT, Nielsen AT, Wu JH, Tsai CS, Matsuo Y, Molin S (2001) In situ identification of polyphosphate- and polyhydroxyalkanoate-accumulating traits for microbial populations in a biological phosphorus removal process. Environ Microbiol 3(2):110–122

Lochmatter S, Holliger C (2014) Optimization of operation conditions for the startup of aerobic granular sludge reactors biologically removing carbon, nitrogen, and phosphorous. Water Res 59:58–70

Lochmatter S, Gonzalez-Gil G, Holliger C (2013) Optimized aeration strategies for nitrogen and phosphorus removal with aerobic granular sludge. Water Res 47(16):6187–6197

Lochmatter S, Maillard J, Holliger C (2014) Nitrogen removal over nitrite by aeration control in aerobic granular sludge sequencing batch reactors. Int J Environ Res Public Health 11(7):6955

Lopez-Vazquez CM, Oehmen A, Hooijmans CM, Brdjanovic D, Gijzen HJ, Yuan Z, van Loosdrecht MCM (2009) Modeling the PAO-GAO competition: effects of carbon source, pH and temperature. Water Res 43(2):450–462

Lucker S, Wagner M, Maixner F, Pelletier E, Koch H, Vacherie B, Rattei T, Damste JSS, Spieck E, Le Paslier D, Daims H (2010) A *Nitrospira* metagenome illuminates the physiology and evolution of globally important nitrite-oxidizing bacteria. Proc Natl Acad Sci USA 107(30):13479–13484

Luiz de Sousa Rollemberg S, Queiroz de Oliveira L, Nascimento de Barros A, Igor Milen Firmino P, Bezerra dos Santos A (2020) Pilot-scale aerobic granular sludge in the treatment of municipal wastewater: optimizations in the start-up, methodology of sludge discharge, and evaluation of resource recovery. Bioresour Technol 311:123467

Mahendra S, Gedalanga P, Kotay SM, Torres CI, Butler CS, Goel R (2012) Advancements in molecular techniques and applications in environmental engineering. Water Environ Res 84(10):814–844

Mannina G, Ekama G, Caniani D, Cosenza A, Esposito G, Gori R, Garrido-Baserba M, Rosso D, Olsson G (2016) Greenhouse gases from wastewater treatment—a review of modelling tools. Sci Total Environ 551–552:254–270

Margot J, Lochmatter S, Barry DA, Holliger C (2016) Role of ammonia-oxidizing bacteria in micropollutant removal from wastewater with aerobic granular sludge. Water Sci Technol 73(3):564–575

Marques R, Santos J, Nguyen H, Carvalho G, Noronha JP, Nielsen PH, Reis MAM, Oehmen A (2017) Metabolism and ecological niche of *Tetrasphaera* and *Ca.* Accumulibacter in enhanced biological phosphorus removal. Water Res 122:159–171

Martins AMP, Heijnen JJ, van Loosdrecht MCM (2004a) Bulking sludge in biological nutrient removal systems. Biotechnol Bioeng 86(2):125–135

Martins AMP, Pagilla K, Heijnen JJ, van Loosdrecht MCM (2004b) Filamentous bulking sludge—a critical review. Water Res 38(4):793–817

Marzorati M, Read S, Arends JB, Verstraete W, Boon N (2011) Microbial resource management (MRM): theory and practical tools to exploit bacterial capabilities. Commun Agric Appl Biol Sci 76(2):27–29

Maszenan AM, Seviour RJ, Patel BKC, Schumann P, Burghardt J, Tokiwa Y, Stratton HM (2000) Three isolates of novel polyphosphate accumulating Gram-positive cocci, obtained from activated sludge, belong to a new genus, *Tetrasphaera* gen. nov., and description of two new species, *Tetrasphaera japonica* sp. nov. and *Tetrasphaera australiensis* sp. nov. Int J Syst Evol Microbiol 50(2):593–603

Mathieu J, Yu P, Zuo P, Da Silva MLB, Alvarez PJJ (2019) Going viral: emerging opportunities for phage-based bacterial control in water treatment and reuse. Acc Chem Res 52(4):849–857

Matz C, Kjelleberg S (2005) Off the hook—how bacteria survive protozoan grazing. Trends Microbiol 13(7):302–307

Maurer M, Gujer W (1995) Monitoring of microbial phosphorus release in batch experiments using electric conductivity. Water Res 29(11):2613–2617

McDaniel EA, Wahl SA, Si I, Pinto A, Ziels R, Nielsen PH, McMahon KD, Williams RBH (2021) Prospects for multi-omics in the microbial ecology of water engineering. Water Res 205:117608

McIlroy SJ, Saunders AM, Albertsen M, Nierychlo M, McIlroy B, Hansen AA, Karst SM, Nielsen JL, Nielsen PH (2015) MiDAS: the field guide to the microbes of activated sludge. Database 2015:bav062

McMahon KD, Gu AZ, Nerenberg R, Sturm BM (2009) Molecular methods in biological systems. Water Environ Res 81(10):986–1002

McMahon KD, He S, Oehmen A (2010) The microbiology of phosphorus removal. In: Seviour RJ, Nielsen PH (eds) Microbial ecology of activated sludge. IWA Publishing, London, pp 281–320

Meunier C, Henriet O, Schroonbroodt B, Boeur JM, Mahillon J, Henry P (2016) Influence of feeding pattern and hydraulic selection pressure to control filamentous bulking in biological treatment of dairy wastewaters. Bioresour Technol 221:300–309

Miłobedzka A, Ferreira C, Vaz-Moreira I, Calderón-Franco D, Gorecki A, Purkrtova S, Jan B, Dziewit L, Singleton CM, Nielsen PH, Weissbrodt DG, Manaia CM (2021) Monitoring antibiotic resistance genes in wastewater environments: the challenges of filling a gap in the one-health cycle. J Hazard Mater 424:127407

Mino T, Satoh H (2006) Wastewater genomics. Nat Biotechnol 24(10):1229–1230

Miyake M, Hasebe Y, Furusawa K, Shiomi H, Inoue D, Ike M (2023) Pilot-scale demonstration of aerobic granular sludge augmentation applied to continuous-flow activated sludge process for the treatment of low-strength municipal wastewater. J Water Process Eng 51(1):103392

Montoya T, Borrás L, Aguado D, Ferrer J, Seco A (2008) Detection and prevention of enhanced biological phosphorus removal deterioration caused by Zoogloea overabundance. Environ Technol 29(1):35–42

Morgenroth E, Milferstedt K (2009) Biofilm engineering: linking biofilm development at different length and time scales. Rev Environ Sci Biotechnol 8(3):203–208

Morgenroth E, Wilderer PA (1998) Sequencing batch reactor technology: concepts, design and experiences. J Chart Inst Water Environ Manag 12(5):314–321

Morgenroth E, Wilderer PA (1999) Controlled biomass removal—the key parameter to achieve enhanced biological phosphorus removal in biofilm systems. Water Sci Technol 39(7):33–40

Moser-Engeler R, Udert KM, Wild D, Siegrist H (1998) Products from primary sludge fermentation and their suitability for nutrient removal. Water Sci Technol 38(1):265–273

Mosquera-Corral A, de Kreuk MK, Heijnen JJ, van Loosdrecht MCM (2005) Effects of oxygen concentration on N-removal in an aerobic granular sludge reactor. Water Res 39(12):2676–2686

Mosquera-Corral A, Arrojo B, Figueroa M, Campos JL, Mendez R (2011) Aerobic granulation in a mechanical stirred SBR: treatment of low organic loads. Water Sci Technol 64(1):155–161

Motlagh AM, Bhattacharjee AS, Goel R (2016) Biofilm control with natural and genetically-modified phages. World J Microbiol Biotechnol 32(4):1–10

Muñoz-Palazon B, Rodriguez-Sanchez A, Castellano-Hinojosa A, Gonzalez-Lopez J, van Loosdrecth MCM, Vahala R, Gonzalez-Martinez A (2018) Quantitative and qualitative studies of microorganisms involved in full-scale autotrophic nitrogen removal performance. AIChE J 64(2):457–467

Muyzer G, Ramsing NB (1995) Molecular methods to study the organization of microbial communities. Water Sci Technol 32:1–9

Narayanasamy S, Muller EE, Sheik AR, Wilmes P (2015) Integrated omics for the identification of key functionalities in biological wastewater treatment microbial communities. Microb Biotechnol 8(3):363–368

Nguyen HTT, Le VQ, Hansen AA, Nielsen JL, Nielsen PH (2011) High diversity and abundance of putative polyphosphate-accumulating *Tetrasphaera*-related bacteria in activated sludge systems. FEMS Microbiol Ecol 76(2):256–267

Nguyen HTT, Nielsen JL, Nielsen PH (2012) "*Candidatus* Halomonas phosphatis", a novel polyphosphate-accumulating organism in full-scale enhanced biological phosphorus removal plants. Environ Microbiol 14(10):2826–2837

Nguyen HT, Kristiansen R, Vestergaard M, Wimmer R, Nielsen PH (2015) Intracellular accumulation of glycine in polyphosphate-accumulating organisms in activated sludge, a novel storage mechanism under dynamic anaerobic-aerobic conditions. Appl Environ Microbiol 81(14):4809–4818

Nguyen Quoc B, Wei S, Armenta M, Bucher R, Sukapanpotharam P, Stahl DA, Stensel HD, Winkler MKH (2021) Aerobic granular sludge: impact of size distribution on nitrification capacity. Water Res 188:116445

Niba ETE, Naka Y, Nagase M, Mori H, Kitakawa M (2007) A genome-wide approach to identify the genes involved in biofilm formation in *E. coli*. DNA Res 14(6):237–246

Nielsen PH, McMahon KD (2014) Microbiology and microbial ecology of the activated sludge process. In: Jenkins D, Wanner J (eds) Activated sludge—100 years and counting. IWA Publishing, London, p 464

Nielsen PH, Raunkjaer K, Norsker NH, Jensen NA, Hvitved-Jacobsen T (1992) Transformation of wastewater in sewer systems—a review. Water Sci Technol 25(6):17–31

Nielsen PH, Thomsen TR, Nielsen JL (2004) Bacterial composition of activated sludge—importance for floc and sludge properties. Water Sci Technol 49(10):51–58

Nielsen PH, Mielczarek AT, Kragelund C, Nielsen JL, Saunders AM, Kong Y, Hansen AA, Vollertsen J (2010) A conceptual ecosystem model of microbial communities in enhanced biological phosphorus removal plants. Water Res 44(17):5070–5088

Nielsen JL, Nguyen H, Meyer RL, Nielsen PH (2012a) Identification of glucose-fermenting bacteria in a full-scale enhanced biological phosphorus removal plant by stable isotope probing. Microbiology 158(7):1818–1825

Nielsen PH, Saunders AM, Hansen AA, Larsen P, Nielsen JL (2012b) Microbial communities involved in enhanced biological phosphorus removal from wastewater—a model system in environmental biotechnology. Curr Opin Biotechnol 23(3):452–459

Nielsen JL, Seviour RJ, Nielsen PH (2016) Microscopy. In: van Loosdrecht MCM, Nielsen PH, Lopez-Vazquez CM, Brdjanovic D (eds) Experimental methods in wastewater treatment. IWA Publishing, London, pp 263–284

Nogueira R, Melo LF (2006) Competition between *Nitrospira* spp. and *Nitrobacter* spp. in nitrite-oxidizing bioreactors. Biotechnol Bioeng 95(1):169–175

Oehmen A, Lemos PC, Carvalho G, Yuan Z, Keller J, Blackall LL, Reis MAM (2007) Advances in enhanced biological phosphorus removal: from micro to macro scale. Water Res 41:2271–2300

Oehmen A, Carvalho G, Lopez-Vazquez CM, van Loosdrecht MCM, Reis MAM (2010a) Incorporating microbial ecology into the metabolic modelling of polyphosphate accumulating organisms and glycogen accumulating organisms. Water Res 44(17):4992–5004

Oehmen A, Lopez-Vazquez CM, Carvalho G, Reis MAM, van Loosdrecht MCM (2010b) Modelling the population dynamics and metabolic diversity of organisms relevant in anaerobic/anoxic/aerobic enhanced biological phosphorus removal processes. Water Res 44(15):4473–4486

Ofiteru ID, Lunn M, Curtis TP, Wells GF, Criddle CS, Francis CA, Sloan WT (2010) Combined niche and neutral effects in a microbial wastewater treatment community. Proc Natl Acad Sci USA 107(35):15345–15350

Oishi R, Hirooka K, Otawa K, Tada C, Nakai Y (2012) Ammonia-oxidizing Archaea in laboratory-scale activated sludge systems for wastewater of low- or high-ammonium concentration. Anim Sci J 83(7):571–576

Onda S, Takii S (2002) Isolation and characterization of a Gram-positive polyphosphate-accumulating bacterium. J Gen Appl Microbiol 48(3):125–133

Onuki M, Satoh H, Mino T, Matsuo T (2000) Application of molecular methods to microbial community analysis of activated sludge. Water Sci Technol 42:17–22

Oshiki M, Onuki M, Satoh H, Mino T (2008) PHA-accumulating microorganisms in full-scale wastewater treatment plants. Water Sci Technol 58(1):13–20

Páez-Watson T, van Loosdrecht MCM, Wahl SA (2023) Predicting the impact of temperature on metabolic fluxes using resource allocation modelling: application to polyphosphate accumulating organisms. Water Res 228(Pt A):119365

Pallares-Vega R, Hernandez Leal L, Fletcher BN, Vias-Torres E, van Loosdrecht MCM, Weissbrodt DG, Schmitt H (2021) Annual dynamics of antimicrobials and resistance determinants in flocculent and aerobic granular sludge treatment systems. Water Res 190:116752

Palomo A, Pedersen AG, Fowler SJ, Dechesne A, Sicheritz-Pontén T, Smets BF (2018) Comparative genomics sheds light on niche differentiation and the evolutionary history of comammox *Nitrospira*. ISME J 12(7):1779–1793

Perez J, Costa E, Kreft JU (2009) Conditions for partial nitrification in biofilm reactors and a kinetic explanation. Biotechnol Bioeng 103(2):282–295

Perez J, Lotti T, Kleerebezem R, Picioreanu C, van Loosdrecht MC (2014) Outcompeting nitrite-oxidizing bacteria in single-stage nitrogen removal in sewage treatment plants: a model-based study. Water Res 66:208–218

Pérez J, Laureni M, van Loosdrecht MCM, Persson F, Gustavsson DJI (2020) The role of the external mass transfer resistance in nitrite oxidizing bacteria repression in biofilm-based partial nitritation/anammox reactors. Water Res 186:116348

Pernthaler J, Posch T, Simek K, Vrba J, Amann R, Psenner R (1997) Contrasting bacterial strategies to coexist with a flagellate predator in an experimental microbial assemblage. Appl Environ Microbiol 63(2):596–601

Pessi IS, Viitamäki S, Virkkala A-M, Eronen-Rasimus E, Delmont TO, Marushchak ME, Luoto M, Hultman J (2022) In-depth characterization of denitrifier communities across different soil ecosystems in the tundra. Environ Microbiome 17(1):30

Picioreanu C, van Loosdrecht MCM, Heijnen JJ (2000) A theoretical study on the effect of surface roughness on mass transport and transformation in biofilms. Biotechnol Bioeng 68(4):355–369

Poot V, Hoekstra M, Geleijnse MAA, van Loosdrecht MCM, Pérez J (2016) Effects of the residual ammonium concentration on NOB repression during partial nitritation with granular sludge. Water Res 106:518–530

Posch T, Simek K, Vrba J, Pernthaler J, Nedoma J, Sattler B, Sonntag B, Psenner R (1999) Predator-induced changes of bacterial size-structure and productivity studied on an experimental microbial community. Aquat Microb Ecol 18(3):235–246

Prata JC, Ribeiro AI, Rocha-Santos T (2022) Chapter 1—An introduction to the concept of One Health. In: Prata JC, Ribeiro AI, Rocha-Santos T (eds) One Health. Academic Press, pp 1–31

Price DJdS (1963) Little science, big science. The impact of the explosive growth of science on the wolrd we live in. Columbia University Press, New York

Pronk M, Abbas B, Al-Zuhairy SH, Kraan R, Kleerebezem R, van Loosdrecht MC (2015a) Effect and behaviour of different substrates in relation to the formation of aerobic granular sludge. Appl Microbiol Biotechnol 99(12):5257–5268

Pronk M, de Kreuk MK, de Bruin B, Kamminga P, Kleerebezem R, van Loosdrecht MCM (2015b) Full scale performance of the aerobic granular sludge process for sewage treatment. Water Res 84:207–217

Pronk M, van Dijk EJH, van Loosdrecht MCM (2020) Aerobic granular sludge. In: Chen GH, van Loosdrecht MCM, Ekama GA, Brdjanovic D (eds) Biological wastewater treatment: principles, modelling and design, 2nd edn. IWA Publishing, London

Radauer A, Rivoire L, Eparvier P, Schwanbeck H (2012) Presenting the (economic) value of patents nominated for the European Inventor Award 2012—inventor file Mark van Loosdrecht. Technopolis Group Austria, Vienna, pp 1–10

Regmi P, DeBarbadillo C, Weissbrodt DG (2017) Biofilm reactor technology and design. In: Krause TL et al (eds) Design of water resource recovery facilities—MOP 8. WEF manual of practice No. 8, ASCE manuals and reports on engineering practice No. 76, 6th edn. Water Environment Federation, American Society of Civil Engineers and Environmental and Water Resources Institute, McGraw-Hill Education, Alexandria VA, Reston VA, New York

Regmi P, Sturm B, Hiripitiyage D, Keller N, Murthy S, Jimenez J (2022) Combining continuous flow aerobic granulation using an external selector and carbon-efficient nutrient removal with AvN control in a full-scale simultaneous nitrification-denitrification process. Water Res 210:117991

Reino C, Suárez-Ojeda ME, Pérez J, Carrera J (2016) Kinetic and microbiological characterization of aerobic granules performing partial nitritation of a low-strength wastewater at 10 °C. Water Res 101:147–156

Rodríguez E, García-Encina PA, Stams AJM, Maphosa F, Sousa DZ (2015) Meta-omics approaches to understand and improve wastewater treatment systems. Rev Environ Sci Biotechnol 14(3):385–406

Rodríguez-Ramos JA, Borton MA, McGivern BB, Smith GJ, Solden LM, Shaffer M, Daly RA, Purvine SO, Nicora CD, Eder EK, Lipton M, Hoyt DW, Stegen JC, Wrighton KC (2022) Genome-resolved metaproteomics decodes the microbial and viral contributions to coupled carbon and nitrogen cycling in river sediments. mSystems 7(4):e00516–e00522

Rombouts JL, Mos G, Weissbrodt DG, Kleerebezem R, van Loosdrecht MCM (2019) Diversity and metabolism of xylose and glucose fermenting microbial communities in sequencing batch or continuous culturing. FEMS Microbiol Ecol 95(2):fiz233

Ronn R, McCaig AE, Griffiths BS, Prosser JI (2002) Impact of protozoan grazing on bacterial community structure in soil microcosms. Appl Environ Microbiol 68(12):6094–6105

Rubio-Rincon FJ, Lopez-Vazquez CM, Welles L, van Loosdrecht MC, Brdjanovic D (2017) Sulphide effects on the physiology of *Candidatus* Accumulibacter phosphatis type I. Appl Microbiol Biotechnol 101(4):1661–1672

Rubio-Rincon FJ, Weissbrodt DG, Lopez-Vazquez CM, Welles L, Abbas B, Albertsen M, Nielsen PH, van Loosdrecht MCM, Brdjanovic D (2019) "*Candidatus* Accumulibacter delftensis": a clade IC novel polyphosphate-accumulating organism without denitrifying activity on nitrate. Water Res 161:136–151

Sabri NA, van Holst S, Schmitt H, van der Zaan BM, Gerritsen HW, Rijnaarts HHM, Langenhoff AAM (2020) Fate of antibiotics and antibiotic resistance genes during conventional and additional treatment technologies in wastewater treatment plants. Sci Total Environ 741:140199

Santorio S, Couto AT, Amorim CL, Val del Rio A, Arregui L, Mosquera-Corral A, Castro PML (2021) Sequencing versus continuous granular sludge reactor for the treatment of freshwater aquaculture effluents. Water Res 201:117293

Sauder LA, Peterse F, Schouten S, Neufeld JD (2012) Low-ammonia niche of ammonia-oxidizing archaea in rotating biological contactors of a municipal wastewater treatment plant. Environ Microbiol 14(9):2589–2600

Sauder LA, Albertsen M, Engel K, Schwarz J, Nielsen PH, Wagner M, Neufeld JD (2017) Cultivation and characterization of Candidatus *Nitrosocosmicus exaquare*, an ammonia-oxidizing archaeon from a municipal wastewater treatment system. ISME J 11(5):1142–1157

Sauer K (2003) The genomics and proteomics of biofilm formation. Genome Biol 4(6):219

Schambeck CM, Magnus BS, de Souza LCR, Leite WRM, Derlon N, Guimaraes LB, da Costa RHR (2020) Biopolymers recovery: dynamics and characterization of alginate-like exopolymers in an aerobic granular sludge system treating municipal wastewater without sludge inoculum. J Environ Manage 263:110394

Schwarzenbeck N, Erley R, Wilderer PA (2004) Aerobic granular sludge in an SBR-system treating wastewater rich in particulate matter. Water Sci Technol 49(11–12):41–46

Sepúlveda-Mardones M, Campos JL, Magrí A, Vidal G (2019) Moving forward in the use of aerobic granular sludge for municipal wastewater treatment: an overview. Rev Environ Sci Biotechnol 18(4):741–769

Seviour RJ, Mino T, Onuki M (2003) The microbiology of biological phosphorus removal in activated sludge systems. FEMS Microbiol Rev 27(1):99–127

Seviour T, Yuan Z, van Loosdrecht MCM, Lin Y (2012) Aerobic sludge granulation: a tale of two polysaccharides? Water Res 46(15):4803–4813

Seviour T, Derlon N, Dueholm MS, Flemming H-C, Girbal-Neuhauser E, Horn H, Kjelleberg S, van Loosdrecht MCM, Lotti T, Malpei MF, Nerenberg R, Neu TR, Paul E, Yu H, Lin Y (2019) Extracellular polymeric substances of biofilms: suffering from an identity crisis. Water Res 151:1–7

Shapiro OH, Kushmaro A (2011) Bacteriophage ecology in environmental biotechnology processes. Curr Opin Biotechnol 22(3):449–455

Shapiro OH, Kushmaro A, Brenner A (2010) Bacteriophage predation regulates microbial abundance and diversity in a full-scale bioreactor treating industrial wastewater. ISME J 4(3):327–336

Shaw A, Takacs I, Pagilla K, Riffat R, DeClippeleir H, Wilson C, Murthy S (2015) Toward universal half-saturation coefficients: describing extant K_s as a function of diffusion. Water Environ Res 87(5):387–391

Siegrist H, Brack T, Koch G, Nussbaumer A, Gujer W (2000) Optimization of nutrient removal in the WWTP Zurich-Werholzli. Water Sci Technol 41(9):63–71

Singleton CM, Petriglieri F, Kristensen JM, Kirkegaard RH, Michaelsen TY, Andersen MH, Kondrotaite Z, Karst SM, Dueholm MS, Nielsen PH, Albertsen M (2021) Connecting structure to function with the recovery of over 1000 high-quality metagenome-assembled genomes from activated sludge using long-read sequencing. Nat Commun 12(1):2009

Singleton CM, Petriglieri F, Wasmund K, Nierychlo M, Kondrotaite Z, Petersen JF, Peces M, Dueholm MS, Wagner M, Nielsen PH (2022) The novel genus, '*Candidatus* Phosphoribacter', previously identified as *Tetrasphaera*, is the dominant polyphosphate accumulating lineage in EBPR wastewater treatment plants worldwide. ISME J 16(6):1605–1616

Soler-Jofra A, Wang R, Kleerebezem R, van Loosdrecht MCM, Pérez J (2019) Stratification of nitrifier guilds in granular sludge in relation to nitritation. Water Res 148:479–491

Solon K, Flores-Alsina X, Kazadi Mbamba C, Ikumi D, Volcke EIP, Vaneeckhaute C, Ekama G, Vanrolleghem PA, Batstone DJ, Gernaey KV, Jeppsson U (2017) Plant-wide modelling of phosphorus transformations in wastewater treatment systems: impacts of control and operational strategies. Water Res 113:97–110

Sperandio M, Masse A, Espinosa-Bouchot MC, Cabassud C (2005) Characterization of sludge structure and activity in submerged membrane bioreactor. Water Sci Technol 52(10–11):401–408

Stokholm-Bjerregaard M (2016) Control of GAOs in wastewater treatment plants with enhanced biological phosphorus removal. PhD thesis, Aalborg University

Stokholm-Bjerregaard M, McIlroy SJ, Nierychlo M, Karst SM, Albertsen M, Nielsen PH (2017) A critical assessment of the microorganisms proposed to be important to enhanced biological phosphorus removal in full-scale wastewater treatment systems. Front Microbiol 8:718

Stokholm-Bjerregaard MA, Hansen AA, Strandbæk H, Vølund I, Thornberg D, Hughes L, Nielsen PH (2021) Intelligent and dynamic control of optimal WWTP operation from microbial sequencing. In: IWA (ed) IWA world water congress & exhibition, Copenhagen, Denmark. IWA Publishing

Strubbe L, Pennewaerde M, Baeten JE, Volcke EIP (2022) Continuous aerobic granular sludge plants: better settling versus diffusion limitation. Chem Eng J 428(4):131427

Subirats J, Sànchez-Melsió A, Borrego CM, Balcázar JL, Simonet P (2016) Metagenomic analysis reveals that bacteriophages are reservoirs of antibiotic resistance genes. Int J Antimicrob Agents 48(2):163–167

Sun Y, Angelotti B, Brooks M, Wang ZW (2021) Feast/famine ratio determined continuous flow aerobic granulation. Sci Total Environ 750:141467

Tavares Ferreira TJ, Luiz de Sousa Rollemberg S, Nascimento de Barros A, Machado de Lima JP, Bezerra dos Santos A (2021) Integrated review of resource recovery on aerobic granular sludge systems: possibilities and challenges for the application of the biorefinery concept. J Environ Manage 291:112718

Temudo MF (2008) Directing product formation by mixed culture fermentation. PhD thesis, Delft University of Technology

Temudo MF, Muyzer G, Kleerebezem R, van Loosdrecht MCM (2008) Diversity of microbial communities in open mixed culture fermentations: impact of the pH and carbon source. Appl Microbiol Biotechnol 80(6):1121–1130

Thomas MP (2008) The secret to achieving reliable biological phosphorus removal. Water Sci Technol 58(6):1231–1236

Tijhuis L, Rekswinkel E, van Loosdrecht MCM, Heijnen JJ (1994) Dynamics of population and biofilm structure in the biofilm airlift suspension reactor for carbon and nitrogen removal. Water Sci Technol 29(10–11):377–384

Toja Ortega S, van den Berg L, Pronk M, de Kreuk MK (2022) Hydrolysis capacity of different sized granules in a full-scale aerobic granular sludge (AGS) reactor. Water Res X 16:100151

Tomás-Martínez S, Kleikamp HBC, Neu TR, Pabst M, Weissbrodt DG, van Loosdrecht MCM, Lin Y (2021) Production of nonulosonic acids in the extracellular polymeric substances of "*Candidatus* Accumulibacter phosphatis". Appl Microbiol Biotechnol 105(8):3327–3338

Tomás-Martínez S, Chen LM, Neu TR, Weissbrodt DG, van Loosdrecht MCM, Lin Y (2022a) Catabolism of sialic acids in an environmental microbial community. FEMS Microbiol Ecol 98(5):fiac047

Tomás-Martínez S, Chen LM, Pabst M, Weissbrodt DG, van Loosdrecht MCM, Lin Y (2022b) Enrichment and application of extracellular nonulosonic acids containing polymers of *Accumulibacter*. Appl Microbiol Biotechnol 107(2-3):931–941

Tomás-Martínez S, Zwolsman EJ, Merlier F, Pabst M, Lin Y, van Loosdrecht MCM, Weissbrodt DG (2022c) Turnover of the extracellular polymeric matrix in an EBPR microbial community. Appl Microbiol Biotechnol 107(5-6):1997–2009

Val del Rio A, Winkler MKH, Volcke EI, Meunier C, Baeten J, Henriet O, Del Moro G, Suárez-Ojeda ME, De Sanctis M, Dries J, Castro P, Weissbrodt DG, Di Iaconi C (2016) Aerobic granular sludge reactors: from fundamentals to full application In: Lema et al (eds) Water 2020. IWA Publishing, London

van den Berg L, Kirkland CM, Seymour JD, Codd SL, van Loosdrecht MCM, de Kreuk MK (2020) Heterogeneous diffusion in aerobic granular sludge. Biotechnol Bioeng 117(12):3809–3819

van den Berg L, van Loosdrecht MCM, de Kreuk MK (2021) How to measure diffusion coefficients in biofilms: a critical analysis. Biotechnol Bioeng 118(3):1273–1285

van der Roest HF, van Loosdrecht MCM (2012) Water purification, the new standard: purely based on character Delft outlook. Magazine of Delft University of Technology. TU Delft, Delft, pp 6–11

van der Roest H, van Loosdrecht M, Langkamp EJ, Uijterlinde C (2015) Recovery and reuse of alginate from granular Nereda sludge. Water 21:48

van Dijk EJH, Pronk M, van Loosdrecht MCM (2020) A settling model for full-scale aerobic granular sludge. Water Res 186:116135

van Dijk EJH, van Loosdrecht MCM, Pronk M (2021) Nitrous oxide emission from full-scale municipal aerobic granular sludge. Water Res 198:117159

van Dijk EJH, Haaksman VA, van Loosdrecht MCM, Pronk M (2022) On the mechanisms for aerobic granulation—model based evaluation. Water Res 216:118365

van Groenestijn J, Zuidema M, van de Worp JJM, Deinema MH, Zehnder AJB (1989) Influence of environmental parameters on polyphosphate accumulation in Acinetobacter sp. Anton Van Leeuw Int J Gen Mol Microbiol 55(1):67–82

van Kessel MA, Speth DR, Albertsen M, Nielsen PH, Op den Camp HJ, Kartal B, Jetten MS, Lucker S (2015) Complete nitrification by a single microorganism. Nature 528(7583):555–559

van Lier JB, van der Zee FP, Frijters CT, Ersahin ME (2016) Development of anaerobic high-rate reactors, focusing on sludge bed technology. Adv Biochem Eng Biotechnol 156:363–395

van Loosdrecht MCM, Brdjanovic D (2014) Anticipating the next century of wastewater treatment. Science 344(6191):1452–1453

van Loosdrecht MCM, Hooijmans CM, Brdjanovic D, Heijnen JJ (1997) Biological phosphate removal processes. Appl Microbiol Biotechnol 48(3):289–296

van Loosdrecht MCM, Martins AM, Ekama GA (2020) Bulking sludge. In: Henze M, van Loosdrecht MCM, Ekama GA, Brdjanovic D (eds) Biological wastewater treatment: principles, modelling and design, 2nd edn. IWA Publishing, London, pp 475–496

van Niekerk AM, Jenkins D, Richard MG (1987) The competitive growth of *Zoogloea ramigera* and type 021N in activated sludge and pure culture—a model for low F:M bulking. J Water Pollut Control Fed 59(5):262–273

Vanrolleghem P (2015) Les StaRRE de type conventionnel: une bonne option pour éliminer beaucoup d'azote. Vecteur Environ Janvier 56

Vanwonterghem I, Jensen PD, Rabaey K, Tyson GW (2016) Genome-centric resolution of microbial diversity, metabolism and interactions in anaerobic digestion. Environ Microbiol 18(9):3144–3158

Verstraete W, Wittebolle L, Heylen K, Vanparys B, de Vos P, van de Wiele T, Boon N (2007) Microbial resource management: the road to go for environmental biotechnology. Eng Life Sci 7(2):117–126

Vlaeminck SE, De Clippeleir H, Verstraete W (2012) Microbial resource management of one-stage partial nitritation/anammox. Microb Biotechnol 5(3):433–448
Wagner M, Nielsen PH, Loy A, Nielsen JL, Daims H (2006) Linking microbial community structure with function: fluorescence in situ hybridization-microautoradiography and isotope arrays. Curr Opin Biotechnol 17(1):83–91
Wagner J, Weissbrodt DG, Manguin V, Ribeiro da Costa RH, Morgenroth E, Derlon N (2015) Effect of particulate organic substrate on aerobic granulation and operating conditions of sequencing batch reactors. Water Res 85:158–166
Wang X, Hu M, Xia Y, Wen X, Ding K (2012) Pyrosequencing analysis of bacterial diversity in 14 wastewater treatment systems in China. Appl Environ Microbiol 78(19):7042–7047
Wang D, Meng Y, Meng F (2022) Genome-centric metagenomics insights into functional divergence and horizontal gene transfer of denitrifying bacteria in anammox consortia. Water Res 224:119062
Wanner O, Panagiotidis V, Clavadetscher P, Siegrist H (2005) Effect of heat recovery from raw wastewater on nitrification and nitrogen removal in activated sludge plants. Water Res 39(19):4725–4734
Warwick C, Guerreiro A, Wood E, Kitson J, Robinson J, Soares A (2014) A molecular imprinted polymer based sensor for measuring phosphate in wastewater samples. Water Sci Technol 69(1):48–54
Weber SD, Ludwig W, Schleifer KH, Fried J (2007) Microbial composition and structure of aerobic granular sewage biofilms. Appl Environ Microbiol 73(19):6233–6240
Wei SP, Stensel HD, Nguyen Quoc B, Stahl DA, Huang X, Lee PH, Winkler MKH (2020) Flocs in disguise? High granule abundance found in continuous-flow activated sludge treatment plants. Water Res 179:115865
Weissbrodt DG (2018) StaRRE—stations de récupération des ressources de l'eau. Aqua Gas 1:20–24
Weissbrodt DG (2019) Engineering the microbiome of EBPR granular sludge for nutrient removal and biorefinery. In: WEF (ed) WEF forum 2019: James Barnard research conference on emerging themes in biological phosphorus removal and recovery. Water Environment Federation, Austin, TX
Weissbrodt DG (2022) Microbial ecology principles for urban circular economies—polyphosphate-accumulating organisms: nutrient cycling and beyond. In: ISME (ed) 18th international symposium on microbial ecology (ISME18), Lausanne, Switzerland
Weissbrodt DG, Holliger C (2013) Intensification du traitement biologique des eaux usées: technologie à boues granulaires et Gestion des ressources bactériennes. Bull l'ARPEA J Romand L'environ 256:12–22
Weissbrodt DG, Holliger C (2014) Towards management of the bacterial resource for nutrient removal in granular sludge biofilm systems. In: Fatone et al (eds) Keynote lecture at the 2nd IWA specialized conference on ecotechnologies for wastewater treatment. University of Verona, Italy
Weissbrodt DG, Winkler MKH, Barr JJ (2011) Microbial diversity of aerobic granules: is it different from activated sludge? In: Qi Z (ed) IWA biofilm conference 2011: processes in biofilms. Tongji University, Shanghai
Weissbrodt DG, Lochmatter S, Ebrahimi S, Rossi P, Maillard J, Holliger C (2012a) Bacterial selection during the formation of early-stage aerobic granules in wastewater treatment systems operated under wash-out
Weissbrodt DG, Shani N, Sinclair L, Lefebvre G, Rossi P, Maillard J, Rougemont J, Holliger C (2012b) PyroTRF-ID: a novel bioinformatics methodology for the affiliation of terminal-restriction fragments using 16S rRNA gene pyrosequencing data. BMC Microbiol 12:306
Weissbrodt DG, Neu TR, Kuhlicke U, Rappaz Y, Holliger C (2013a) Assessment of bacterial and structural dynamics in aerobic granular biofilms. Front Microbiol 4:175

Weissbrodt DG, Schneiter GS, Fürbringer JM, Holliger C (2013b) Identification of trigger factors selecting for polyphosphate- and glycogen-accumulating organisms in aerobic granular sludge sequencing batch reactors. Water Res 47(19):7006–7018

Weissbrodt DG, Derlon N, Morgenroth E, Holliger C (2014a) A consolidated approach of flocculent and granular sludge systems under the perspective of bacterial resource management. Proc Water Environ Fed 2014(19):5008–5009

Weissbrodt DG, Maillard J, Brovelli A, Chabrelie A, May J, Holliger C (2014b) Multilevel correlations in the biological phosphorus removal process: from bacterial enrichment to conductivity-based metabolic batch tests and polyphosphatase assays. Biotechnol Bioeng 111(12):2421–2435

Weissbrodt DG, Shani N, Holliger C (2014c) Linking bacterial population dynamics and nutrient removal in the granular sludge biofilm ecosystem engineered for wastewater treatment. FEMS Microbiol Ecol 88(3):579–595

Weissbrodt DG, Morgenroth E, Wells GF, Goel RK, Nielsen PH, van Loosdrecht MCM (2015) How far can genetic signatures be used to anticipate and trigger the behavior of environmental biotechnology systems in the water engineering domain? Proc Water Environ Fed 2015(10):6170–6170

Weissbrodt DG, Holliger C, Morgenroth E (2017) Modeling hydraulic transport and anaerobic uptake by PAOs and GAOs during wastewater feeding in EBPR granular sludge reactors. Biotechnol Bioeng 114(8):1688–1702

Weissbrodt DG, Laureni M, van Loosdrecht MCM, Comeau Y (2020a) Basic microbiology and metabolism. In: Chen GH, van Loosdrecht MCM, Ekama GA, Brdjanovic D (eds) Biological wastewater treatment: principles, modelling and design, 2nd edn. IWA Publishing, London

Weissbrodt DG, Winkler MKH, Wells GF (2020b) Responsible science, engineering and education for water resource recovery and circularity. Environ Sci Water Res Technol 6(8):1952–1966

Welles L, Lopez-Vazquez CM, Hooijmans CM, van Loosdrecht MC, Brdjanovic D (2014) Impact of salinity on the anaerobic metabolism of phosphate-accumulating organisms (PAO) and glycogen-accumulating organisms (GAO). Appl Microbiol Biotechnol 98(17):7609–7622

Welles L, Lopez-Vazquez CM, Hooijmans CM, van Loosdrecht MC, Brdjanovic D (2015) Impact of salinity on the aerobic metabolism of phosphate-accumulating organisms. Appl Microbiol Biotechnol 99(8):3659–3672

Wey JK, Scherwass A, Norf H, Arndt H, Weitere M (2008) Effects of protozoan grazing within river biofilms under semi-natural conditions. Aquat Microb Ecol 52(3):283–296

Wilderer PA, McSwain BS (2004) The SBR and its biofilm application potentials. Water Sci Technol 50(10):1–10

Wilén BM, Lumley D, Mattsson A, Mino T (2008) Relationship between floc composition and flocculation and settling properties studied at a full scale activated sludge plant. Water Res 42(16):4404–4418

Wilén B-M, Liébana R, Persson F, Modin O, Hermansson M (2018) The mechanisms of granulation of activated sludge in wastewater treatment, its optimization, and impact on effluent quality. Appl Microbiol Biotechnol 102(12):5005–5020

Wilmes P, Heintz-Buschart A, Bond PL (2015) A decade of metaproteomics: where we stand and what the future holds. Proteomics 15(20):3409–3417

Winkler MKH, Bassin JP, Kleerebezem R, de Bruin LMM, van den Brand TPH, van Loosdrecht MCM (2011) Selective sludge removal in a segregated aerobic granular biomass system as a strategy to control PAO-GAO competition at high temperatures. Water Res 45(11):3291–3299

Winkler MKH, Bassin JP, Kleerebezem R, Sorokin DY, van Loosdrecht MCM (2012a) Unravelling the reasons for disproportion in the ratio of AOB and NOB in aerobic granular sludge. Appl Microbiol Biotechnol 94(6):1657–1666

Winkler MKH, Bassin JP, Kleerebezem R, van der Lans RGJM, van Loosdrecht MCM (2012b) Temperature and salt effect on settling velocity in granular sludge technology. Water Res 46(12):3897–3902

Winkler MKH, Kleerebezem R, de Bruin LMM, Verheijen PJT, Abbas B, Habermacher J, van Loosdrecht MCM (2012c) Microbial diversity differences within aerobic granular sludge and activated sludge flocs. Appl Microbiol Biotechnol 97(16):7447–7458

Winkler MKH, Meunier C, Henriet O, Mahillon J, Suarez-Ojeda ME, Del Moro G, De Sanctis M, Di Iaconi C, Weissbrodt DG (2018) An integrative review of granular sludge for the biological removal of nutrients and of recalcitrant organic matter from wastewater. Chem Eng J 336:489–502

Winkler M-KH, van Loosdrecht MCM (2022) Intensifying existing urban wastewater. Science 375(6579):377–378

Withey S, Cartmell E, Avery LM, Stephenson T (2005) Bacteriophages—potential for application in wastewater treatment processes. Sci Total Environ 339(1–3):1–18

Wuertz S, Okabe S, Hausner M (2004) Microbial communities and their interactions in biofilm systems: an overview. Water Sci Technol 49(11–12):327–336

Wunderlin P, Mohn J, Joss A, Emmenegger L, Siegrist H (2012) Mechanisms of N_2O production in biological wastewater treatment under nitrifying and denitrifying conditions. Water Res 46(4):1027–1037

Wunderlin P, Siegrist H, Joss A (2013) Online N_2O measurement: the next standard for controlling biological ammonia oxidation? Environ Sci Technol 47(17):9567–9568

Xu D, Li J, Liu J, Qu X, Ma H (2022) Advances in continuous flow aerobic granular sludge: a review. Process Saf Environ Prot 163:27–35

Yan JL, Cui YW, Huang JL (2021) Continuous flow reactors for cultivating aerobic granular sludge: configuration innovation, principle and research prospect. J Chem Technol Biotechnol 96(10):2721–2734

Ye L, Zhang T (2011) Ammonia-oxidizing bacteria dominates over ammonia-oxidizing archaea in a saline nitrification reactor under low DO and high nitrogen loading. Biotechnol Bioeng 108(11):2544–2552

Yilmaz G, Lemaire R, Keller J, Yuan Z (2008) Simultaneous nitrification, denitrification, and phosphorus removal from nutrient-rich industrial wastewater using granular sludge. Biotechnol Bioeng 100(3):529–541

Yoon S, Song B, Phillips RL, Chang J, Song MJ (2019) Ecological and physiological implications of nitrogen oxide reduction pathways on greenhouse gas emissions in agroecosystems. FEMS Microbiol Ecol 95(6):fiz066

Young B, Banihashemi B, Forrest D, Kennedy K, Stintzi A, Delatolla R (2016) Meso and micro-scale response of post carbon removal nitrifying MBBR biofilm across carrier type and loading. Water Res 91:235–243

Zare M, Heidari MH, Pouresmaeili F, Niyyati M, Moradi M (2012) Introducing a novel facultative nitrifying bacterium, Nitrobacter hamadaniensis. Afr J Microbiol Res 6(24):5126–5133

Zeng RJ, Yuan Z, Keller J (2006) Effects of solids concentration, pH and carbon addition on the production rate and composition of volatile fatty acids in prefermenters using primary sewage sludge. Water Sci Technol 53(8):263–269

Zhang D, Vahala R, Wang Y, Smets BF (2016) Microbes in biological processes for municipal landfill leachate treatment: community, function and interaction. Int Biodeterior Biodegrad 113:88–96

Zhang S, Zhang Z, Xia S, Ding N, Long X, Wang J, Chen M, Ye C, Chen S (2020) Combined genome-centric metagenomics and stable isotope probing unveils the microbial pathways of aerobic methane oxidation coupled to denitrification process under hypoxic conditions. Bioresour Technol 318:124043

Ziliani A, Bovio-Winkler P, Cabezas A, Etchebehere C, Garcia HA, López-Vázquez CM, Brdjanovic D, van Loosdrecht MCM, Rubio-Rincón FJ (2023) Putative metabolism of *Ca.* Accumulibacter via the utilization of glucose. Water Res 229:119446

Zima BE, Diez L, Kowalczyk W, Delgado A (2007) Sequencing batch reactor (SBR) as optimal method for production of granular activated sludge (GAS)—fluid dynamic investigations. Water Sci Technol 55(8–9):151–158

Author Biography

Water and Environmental Life Science & Engineering—WAT-ELSE?

David Weissbrodt receiving the Luce Grivat Award presented by Professor Marilyne Andersen, Dean of the School of Architecture, Civil and Environmental Engineering during the ENAC Research Day 2014 at Ecole Polytechnique Fédérale de Lausanne. Photo © Alain Herzog/EPFL, Switzerland

David Gregory Weissbrodt is a full professor of Microbial Ecology at the Department of Biotechnology and Food Science of the Norwegian University of Science and Technology. Weissbrodt is driven by his passion for envirobiochemical science, engineering, and education dedicated to preserving water resources and safeguarding the environmental health. He develops responsible innovations for better sanitation, (waste)water treatment, and resource recovery contributing to the establishment of a circular economy and fostering an ecologically balanced and healthy society.

His vision aligns with global initiatives aimed at "changing paradigms" to transform wastewater treatment plants into water resource factories, while addressing the challenge of preventing the spread of emerging contaminants like xenobiotic

D. G. Weissbrodt, *Engineering Granular Microbiomes*, Springer Theses,
https://doi.org/10.1007/978-3-031-41009-3

substances (chemicals, pharmaceuticals) and xenogenetic elements (antibiotic resistance, genetically modified materials) across waterways. He explores the connections of biological, chemical, and physical processes within water and health microbiomes, seeking to inform the development of sustainable engineering solutions and policies. Through the integration of environmental systems biology and biotechnology, he harnesses the metabolic power of natural microbiomes into engineered mixed-culture processes. His research outputs are translated into bio-based principles for sustaining the environmental, industrial, and urban ecology, and public health services, and utilized in international co-operation programs across regions, like Europe, Brazil, and India.

Weissbrodt has been recognized for his teaching commitment and didactical enthusiasm to train the next scientific and engineering leaders in the field. By employing problem-based and peer learning methodologies, he encourages young professionals to creatively reflect on scientific fundamentals and technological solutions. This empowers them to contribute innovatively to solving grand challenges and advancing sustainable development goals at the water nexus.

David Weissbrodt was trained in the Swiss Alps as an envirobiochemical engineer and microbiologist. He received his Dipl.-Ing. degree in Chemical Engineering and Life Technologies (2003) from HES-SO Valais/Wallis (Urban Frei, Frans Zonnevijlle) in Switzerland, complemented by a research stay in medical bioscience at the Hochschule Mannheim (Jürgen Backhaus) and Ruprecht-Karls-University of Heidelberg (Jost Jonas) in Germany. Following professional experience in the chemical industry, drinking water sector, and research laboratories for Electrochemical Engineering (Christos Comninellis) and Air and Soil Pollution at EPFL (Hubert van den Bergh), he pursued his M.Sc. in Environmental Sciences and Engineering (2006) at EPFL. As a research assistant in Environmental Chemistry (Christa McArdell) and Process Engineering (Hansruedi Siegrist) at Eawag—the Swiss Federal Institute of Aquatic Science and Technology, he studied the emission and elimination of pharmaceutical substances in hospital wastewater. Weissbrodt earned his Ph.D. in Environment (2013) from EPFL, conducted on bacterial resource management for nutrient removal in granular sludge systems at the Laboratory for Environmental Biotechnology (Christof Holliger). During his postdoctoral research on "process ecogenomics" at ETH Zürich and the Process Engineering Department at Eawag (Eberhard Morgenroth), he investigated N_2O emissions from partial nitritation and anammox systems using meta-omics. As a Swiss NSF fellow, he joined the Environmental Biotechnology Section of Delft University of Technology in the Netherlands (Mark van Loosdrecht) and the Center for Microbial Communities at Aalborg University in Denmark (Per Halkjær Nielsen). His focus was on elucidating the systems microbiology of exopolymer biorefinery from wastewater. Weissbrodt completed his tenure track as an assistant professor of Environmental Life Science Engineering at the TU Delft Department of Biotechnology. He developed a research line on microbiome science and engineering related to exopolymer biosynthesis in granular sludge, nutrient capture by purple phototrophic bacteria, and antimicrobial resistance development in wastewater environments. In 2022, he joined the Research

Division "Analysis and Control of Microbial Systems" as a full professor of Microbial Ecology at the Department of Biotechnology and Food Science of the Norwegian University of Science and Technology, where he furthers his science on microbial and chemical processes involved in nutrient removal, resource capture, chemical impacts, and antibiotic resistance development in water systems. Weissbrodt serves in the elected board of the Professional Association of Water and Air Protection of Western Switzerland (ARPEA) and in the management committees of the specialist groups Microbial Ecology & Water Engineering (MEWE), Biofilms, and Environmental Engineering Education (E3) of the International Water Association (IWA). He organised and chaired several international events like the 1st Symposium on Microbiological Methods for Waste & Water Resource Recovery 2017 for the 175th Anniversary of the TU Delft—Technology for Life (with Marjet Oosterkamp), the IWA Biofilms: Granular Sludge Conference 2018 (with Merle de Kreuk), and the 9th IWA Microbial Ecology & Water Engineering Specialist Conference 2021 (with Mark van Loosdrecht). He has been honored with several distinctions in education and research, including accolades by the TU Delft student association '*Practische Studie*' for his teaching in environmental biotechnology and microbiology and by his peers with the inaugural 2019 IWA MEWE Early Career Researcher Award received at the University of Hiroshima in Japan.

Current address and correspondence

Prof. David Weissbrodt

Research Division "Analysis and Control of Microbial Systems"
Department of Biotechnology and Food Science, Faculty of Natural Sciences
Norwegian University of Science and Technology (NTNU)
IBT, Kjemi 3, Gløshaugen, Sem Sælands vei 8, 7034 Trondheim, Norway
T: +47 7355 9098 | E: david.weissbrodt@ntnu.no, weissbrodt.lab@gmail.com
URL: https://www.ntnu.edu/employees/david.weissbrodt
Twitter: @WeissbrodtLab
GS: https://scholar.google.com/citations?user=yGqI2rMAAAAJ&hl
Output: http://publicationslist.org/david.g.weissbrodt

Index

D. G. Weissbrodt, *Engineering Granular Microbiomes*, Springer Theses,
https://doi.org/10.1007/978-3-031-41009-3

B

D

N

O

P

R

S

Y

Z